Introduction to Computational Chemistry

Introduction to Computational Chemistry

Third Edition

Frank Jensen
Department of Chemistry, Aarhus University, Denmark

Registered Office: John Wiley & Sons, Ltd, The Atrium, Southern Gate, Chichester, West Sussex, PO19 8SQ, UK

Editorial Offices: 9600 Garsington Road, Oxford, OX4 2DQ, UK
The Atrium, Southern Gate, Chichester, West Sussex, PO19 8SQ, UK
111 River Street, Hoboken, NJ 07030-5774, USA

For details of our global editorial offices, for customer services and for information about how to apply for permission to reuse the copyright material in this book please see our website at www.wiley.com/wiley-blackwell.

Library of Congress Cataloging-in-Publication Data

Names: Jensen, Frank, author.
Title: Introduction to computational chemistry / Frank Jensen.
Description: Third edition. | Chichester, UK ; Hoboken, NJ : John Wiley & Sons, 2017. | Includes index.
Identifiers: LCCN 2016039772 (print) | LCCN 2016052630 (ebook) | ISBN 9781118825990 (pbk.) |
ISBN 9781118825983 (pdf) | ISBN 9781118825952 (epub)
Subjects: LCSH: Chemistry, Physical and theoretical–Data processing. | Chemistry, Physical and theoretical–Mathematics.
Classification: LCC QD455.3.E4 J46 2017 (print) | LCC QD455.3.E4 (ebook) | DDC 541.0285–dc23
LC record available at https://lccn.loc.gov/2016039772

A catalogue record for this book is available from the British Library.

ISBN: 9781118825990

Wiley also publishes its books in a variety of electronic formats. Some content that appears in print may not be available in electronic books.

Set in 10/12pt WarnockPro by Aptara Inc., New Delhi, India

10 9 8 7 6 5 4 3 2 1

Contents

Preface to the First Edition

Computational chemistry is rapidly emerging as a subfield of theoretical chemistry, where the primary focus is on solving chemically related problems by calculations. For the newcomer to the field, there are three main problems:

(1) Deciphering the code. The language of computational chemistry is littered with acronyms, what do these abbreviations stand for in terms of underlying assumptions and approximations?
(2) Technical problems. How does one actually run the program and what to look for in the output?
(3) Quality assessment. How good is the number that has been calculated?

Point (1) is part of every new field: there is not much to do about it. If you want to live in another country, you have to learn the language. If you want to use computational chemistry methods, you need to learn the acronyms. I have tried in the present book to include a good fraction of the most commonly used abbreviations and standard procedures.

Point (2) is both hardware and software specific. It is not well suited for a textbook, as the information rapidly becomes out of date. The average lifetime of computer hardware is a few years, the time between new versions of software is even less. Problems of type (2) need to be solved "on location". I have made one exception, however, and have included a short discussion of how to make *Z*-matrices. A *Z*-matrix is a convenient way of specifying a molecular geometry in terms of internal coordinates, and it is used by many electronic structure programs. Furthermore, geometry optimizations are often performed in *Z*-matrix variables, and since optimizations in a good set of internal coordinates are significantly faster than in Cartesian coordinates, it is important to have a reasonable understanding of *Z*-matrix construction.

As computer programs evolve they become easier to use. Modern programs often communicate with the user in terms of a graphical interface, and many methods have become essential "black box" procedures: if you can draw the molecule, you can also do the calculation. This effectively means that you no longer have to be a highly trained theoretician to run even quite sophisticated calculations.

The ease with which calculations can be performed means that point (3) has become the central theme in computational chemistry. It is quite easy to run a series of calculations that produce results that are absolutely meaningless. The program will not tell you whether the chosen method is valid for the problem you are studying. Quality assessment is thus an absolute requirement. This, however, requires much more experience and insight than just running the program. A basic understanding of the theory behind the method is needed, and a knowledge of the performance of the method for other systems. If you are breaking new ground, where there is no previous experience, you need a way of calibrating the results.

The lack of quality assessment is probably one of the reasons why computational chemistry has (had) a somewhat bleak reputation. "If five different computational methods give five widely different results, what has computational chemistry contributed? You just pick the number closest to experiments and claim that you can reproduce experimental data accurately." One commonly sees statements of the type "The theoretical results for property X are in disagreement. Calculation at the CCSD(T)/6-31G(d,p) level predicts that…, while the MINDO/3 method gives opposing results. There is thus no clear consent from theory." This is clearly a lack of understanding of the quality of the calculations. If the results disagree, there is a very high probability that the CCSD(T) results are basically correct, and the MINDO/3 results are wrong. If you want to make predictions, and not merely reproduce known results, you need to be able to judge the quality of your results. This is by far the most difficult task in computational chemistry. I hope the present book will give some idea of the limitations of different methods.

Computers don't solve problems, people do. Computers just generate numbers. Although computational chemistry has evolved to the stage where it often can be competitive with experimental methods for generating a value for a given property of a given molecule, the number of possible molecules (there are an estimated 10^{200} molecules with a molecular weight less than 850) and their associated properties is so huge that only a very tiny fraction will ever be amenable to calculations (or experiments). Furthermore, with the constant increase in computational power, a calculation that barely can be done today will be possible on medium-sized machines in 5–10 years. Prediction of properties with methods that do not provide converged results (with respect to theoretical level) will typically only have a lifetime of a few years before being surpassed by more accurate calculations.

The real strength of computational chemistry is the ability to generate data (e.g. by analyzing the wave function) from which a human may gain *insight*, and thereby rationalize the behavior of a large class of molecules. Such insights and rationalizations are much more likely to be useful over a longer period of time than the raw results themselves. A good example is the concept used by organic chemists with molecules composed of functional groups, and representing reactions by "pushing electrons". This may not be particularly accurate from a quantum mechanical point of view, but it is very effective in rationalizing a large body of experimental results, and has good predictive power.

Just as computers do not solve problems, mathematics by itself does not provide insight. It merely provides formulas, a framework for organizing thoughts. It is in this spirit that I have tried to write this book. Only the necessary (obviously a subjective criterion) mathematical background has been provided, the aim being that the reader should be able to understand the premises and limitations of different methods, and follow the main steps in running a calculation. This means that in many cases I have omitted to tell the reader of some of the finer details, which may annoy the purists. However, I believe the large overview is necessary before embarking on a more stringent and detailed derivation of the mathematics. The goal of this book is to provide an overview of commonly used methods, giving enough theoretical background to understand why, for example, the AMBER force field is used for modeling proteins but MM2 is used for small organic molecules, or why coupled cluster inherently is an iterative method, while perturbation theory and configuration interaction inherently are non-iterative methods, although the CI problem in practice is solved by iterative techniques.

The prime focus of this book is on calculating molecular structures and (relative) energies, and less on molecular properties or dynamical aspects. In my experience, predicting structures and energetics are the main uses of computational chemistry today, although this may well change in the coming years. I have tried to include most methods that are already extensively used, together with some that I expect to become generally available in the near future. How detailed the methods are described depends partly on how practical and commonly used the methods are (both in terms of computational resources and software), and partly reflects my own limitations in terms of understanding. Although simulations (e.g. molecular dynamics) are becoming increasingly powerful tools, only a very rudimentary introduction is provided in Chapter 16. The area is outside my expertise, and several excellent textbooks are already available.

Computational chemistry contains a strong practical element. Theoretical methods must be translated into working computer programs in order to produce results. Different algorithms, however, may have different behaviors in practice, and it becomes necessary to be able to evaluate whether a certain type of calculation can be carried out with the available computers. The book thus contains some guidelines for evaluating what type of resources are necessary for carrying out a given calculation.

The present book grew out of a series of lecture notes that I have used for teaching a course in computational chemistry at Odense University, and the style of the book reflects its origin. It is difficult to master all disciplines in the vast field of computational chemistry. A special thanks to H. J. Aa. Jensen, K. V. Mikkelsen, T. Saue, S. P. A. Sauer, M. Schmidt, P. M. W. Gill, P.-O. Norrby, D. L. Cooper, T. U. Helgaker and H. G. Petersen for having read various parts of the book and providing input. Remaining errors are of course my sole responsibility. A good part of the final transformation from a set of lecture notes to the present book was done during a sabbatical leave spent with Prof. L. Radom at the Research School of Chemistry, Australia National University, Canberra, Australia. A special thanks to him for his hospitality during the stay.

A few comments on the layout of the book. Definitions, acronyms or common phrases are marked in *italic*; these can be found in the index. Underline is used for emphasizing important points. Operators, vectors and matrices are denoted in **bold**, scalars in normal text. Although I have tried to keep the notation as consistent as possible, different branches in computational chemistry often use different symbols for the same quantity. In order to comply with common usage, I have elected sometimes to switch notation between chapters. The second derivative of the energy, for example, is called the force constant k in force field theory; the corresponding matrix is denoted $\mathbf{F}$ when discussing vibrations, and called the Hessian $\mathbf{H}$ for optimization purposes.

I have assumed that the reader has no prior knowledge of concepts specific to computational chemistry, but has a working understanding of introductory quantum mechanics and elementary mathematics, especially linear algebra, vector, differential and integral calculus. The following features specific to chemistry are used in the present book without further introduction. Adequate descriptions may be found in a number of quantum chemistry textbooks (J. P. Lowe, *Quantum Chemistry*, Academic Press, 1993; I. N. Levine, *Quantum Chemistry*, Prentice Hall, 1992; P. W. Atkins, *Molecular Quantum Mechanics*, Oxford University Press, 1983).

(1) The Schrödinger equation, with the consequences of quantized solutions and quantum numbers.
(2) The interpretation of the square of the wave function as a probability distribution, the Heisenberg uncertainty principle and the possibility of tunneling.
(3) The solutions for the hydrogen atom, atomic orbitals.
(4) The solutions for the harmonic oscillator and rigid rotor.
(5) The molecular orbitals for the H_2 molecule generated as a linear combination of two s-functions, one on each nuclear centre.
(6) Point group symmetry, notation and representations, and the group theoretical condition for when an integral is zero.

I have elected to include a discussion of the variational principle and perturbational methods, although these are often covered in courses in elementary quantum mechanics. The properties of angular momentum coupling are used at the level of knowing the difference between a singlet and triplet state. I do not believe that it is necessary to understand the details of vector coupling to understand the implications.

Although I have tried to keep each chapter as self-contained as possible, there are unavoidable dependencies. The part in Chapter 3 describing HF methods is a prerequisite for understanding Chapter 4. Both these chapters use terms and concepts for basis sets which are treated in Chapter 5. Chapter 5, in turn, relies on concepts in Chapters 3 and 4, that is these three chapters form the core for understanding modern electronic structure calculations. Many of the concepts in Chapters 3 and 4 are also used in Chapters 6, 7, 9, 11 and 15 without further introduction, although these five chapters probably can be read with some benefits without a detailed understanding of Chapters 3 and 4. Chapter 8, and to a certain extent also Chapter 10, are fairly advanced for an introductory textbook, such as the present, and can be skipped. They do, however, represent areas that are probably going to be more and more important in the coming years. Function optimization, which is described separately in Chapter 14, is part of many areas, but a detailed understanding is not required for following the arguments in the other chapters. Chapters 12 and 13 are fairly self-contained, and form some of the background for the methods in the other chapters. In my own course I normally take Chapters 12, 13 and 14 fairly early in the course, as they provide background for Chapters 3, 4 and 5.

If you would like to make comments, advise me of possible errors, make clarifications, add references, etc., or view the current list of misprints and corrections, please visit the author's website (URL: http://bogense.chem.ou.dk/~icc).

Preface to the Second Edition

The changes relative to the first edition are as follows:

- Numerous misprints and inaccuracies in the first edition have been corrected. Most likely some new ones have been introduced in the process; please check the book website for the most recent correction list and feel free to report possible problems. Since web addresses have a tendency to change regularly, please use your favourite search engine to locate the current URL.
- The methodologies and references in each chapter have been updated with new developments published between 1998 and 2005.
- More extensive referencing. Complete referencing is impossible, given the large breadth of subjects. I have tried to include references that preferably are recent, have a broad scope and include key references. From these the reader can get an entry into the field.
- Many figures and illustrations have been redone. The use of color illustrations has been deferred in favor of keeping the price of the book down.
- Each chapter or section now starts with a short overview of the methods, described without mathematics. This may be useful for getting a feel for the methods, without embarking on all the mathematical details. The overview is followed by a more detailed mathematical description of the method, including some key references that may be consulted for more details. At the end of the chapter or section, some of the pitfalls and the directions of current research are outlined.
- Energy units have been converted from kcal/mol to kJ/mol, based on the general opinion that the scientific world should move towards SI units.
- Furthermore, some chapters have undergone major restructuring:
  - Chapter 16 (Chapter 13 in the first edition) has been greatly expanded to include a summary of the most important mathematical techniques used in the book. The goal is to make the book more self-contained, that is relevant mathematical techniques used in the book are at least rudimentarily discussed in Chapter 16.
  - All the statistical mechanics formalism has been collected in Chapter 13.
  - Chapter 14 has been expanded to cover more of the methodologies used in molecular dynamics.
  - Chapter 12 on optimization techniques has been restructured.
  - Chapter 6 on density functional methods has been rewritten.
  - A new Chapter 1 has been introduced to illustrate the similarities and differences between classical and quantum mechanics, and to provide some fundamental background.
  - A rudimentary treatment of periodic systems has been incorporated in Chapters 3 and 14.
  - A new Chapter 17 has been introduced to describe statistics and QSAR methods.

- I have tried to make the book more modular, that is each chapter is more self-contained. This makes it possible to use only selected chapters, for example for a course, but has the drawback of repeating the same things in several chapters, rather than simply cross-referencing.

Although the modularity has been improved, there are unavoidable interdependencies. Chapters 3, 4 and 5 contain the essentials of electronic structure theory, and most would include Chapter 6 describing density functional methods. Chapter 2 contains a description of empirical force field methods, and this is tightly coupled to the simulation methods in Chapter 14, which of course leans on the statistical mechanics in Chapter 13. Chapter 1 on fundamental issues is of a more philosophical nature, and can be skipped. Chapter 16 on mathematical techniques is mainly for those not already familiar with this, and Chapter 17 on statistical methods may be skipped as well.

Definitions, acronyms and common phrases are marked in *italic*. In a change from the first edition, where underlining was used, *italic* text has also been used for emphasizing important points.

A number of people have offered valuable help and criticisms during the updating process. I would especially like to thank S. P. A. Sauer, H. J. Aa. Jensen, E. J. Baerends and P. L. A. Popelier for having read various parts of the book and provided input. Remaining errors are of course my sole responsibility.

Specific Comments on the Preface to the First Edition

Bohacek *et al.*[1] have estimated the number of possible compounds composed of H, C, N, O and S atoms with 30 non-hydrogen atoms or fewer to be 10^{60}. Although this number is so large that only a very tiny fraction will ever be amenable to investigation, the concept of functional groups means that one does not need to evaluate all compounds in a given class to determine their properties. The number of alkanes meeting the above criteria is $\sim10^{10}$: clearly these will all have very similar and well-understood properties, and there is no need to investigate all 10^{10} compounds.

Reference

1 R. S. Bohacek, C. McMartin and W. C. Guida, *Medicinal Research Reviews* **16** (1), 3–50 (1996).

Preface to the Third Edition

The changes relative to the second edition are as follows:

Numerous misprints and inaccuracies in the second edition have been corrected. Most likely some new ones have been introduced in the process, please check the book website for the most recent correction list and feel free to report possible problems.

http://www.wiley.com/go/jensen/computationalchemistry3

- Methodologies and references in each chapter have been updated with new developments published between 2005 and 2015.
- Semi-empirical methods have been moved from Chapter 3 to a separate Chapter 7.
- Some specific new topics that have been included:

  1. Polarizable force fields
  2. Tight-binding DFT
  3. More extensive DFT functionals, including range-separated and dispersion corrected functionals
  4. More extensive covering of excited states
  5. More extensive time-dependent molecular properties
  6. Accelerated molecular dynamics methods
  7. Tensor decomposition methods
  8. Cluster analysis
  9. Reduced scaling and reduced prefactor methods.

A reoccuring request over the years for a third edition has been: "It would be very useful to have recommendations on which method to use for a given type of problem." I agree that this would be useful, but I have refrained from it for two main reasons:

1. Problems range from very narrow ones for a small set of systems, to very broad ones for a wide set of systems, and covering these and all intermediate cases even rudementary is virtually impossible.
2. Making recommendations like "*do not use method XXX because it gives poor results*" will immediately invoke harsh responses from the developers of method XXX, showing that it gives good results for a selected subset of problems and systems.

A vivid example of the above is the pletora of density functional methods where a particular functional often gives good results for a selected subset of systems and properties, but may fail for other subsets of systems and properties, and no current functional provides good results for all systems and properties. I have limited the recommendations to point out well-known deficiencies.

A similar problem is present when selecting references. I have selected references based on three overriding principles:

1. References to work containing reference data, such as experimental structural results, or ground-breaking work, such as the Hohenberg–Koch theorem, are to the original work.
2. Early in each chapter or subsection, I have included review-type papers, where these are available.
3. Lacking review-type papers, I have selected one or a few papers that preferably are recent, but must at the same time also be written in a scholarly style, and should contain a good selection of references.

The process of literature searching has improved tremendously over the years, and having a few entry points usually allows searching both backwards and forwards to find other references within the selected topic.

In relation to the quoted number of compounds possible for a given number of atoms, Ruddigkeit *et al.* have estimated the number of plausible compounds composed of H, C, N, O, S and a halogen with up to 17 non-hydrogen atoms to be 166×10^9.[1]

Reference

1 L. Ruddigkeit, R. van Deursen, L. C. Blum and J.-L. Reymond, *Journal of Chemical Information and Modeling* **52** (11), 2864–2875 (2012).

1

Introduction

Chemistry is the science dealing with construction, transformation and properties of molecules. Theoretical chemistry is the subfield where mathematical methods are combined with fundamental laws of physics to study processes of chemical relevance.[1–7]

Molecules are traditionally considered as "composed" of atoms or, in a more general sense, as a collection of charged particles, positive nuclei and negative electrons. The only important physical force for chemical phenomena is the Coulomb interaction between these charged particles. Molecules differ because they contain different nuclei and numbers of electrons, or because the nuclear centers are at different geometrical positions. The latter may be "chemically different" molecules such as ethanol and dimethyl ether or different "conformations" of, for example, butane.

Given a set of nuclei and electrons, theoretical chemistry can attempt to calculate things such as:

- Which geometrical arrangements of the nuclei correspond to stable molecules?
- What are their relative energies?
- What are their properties (dipole moment, polarizability, NMR coupling constants, etc.)?
- What is the rate at which one stable molecule can transform into another?
- What is the time dependence of molecular structures and properties?
- How do different molecules interact?

The only systems that can be solved exactly are those composed of only one or two particles, where the latter can be separated into two pseudo one-particle problems by introducing a "center of mass" coordinate system. Numerical solutions to a given accuracy (which may be so high that the solutions are essentially "exact") can be generated for many-body systems, by performing a very large number of mathematical operations. Prior to the advent of electronic computers (i.e. before 1950), the number of systems that could be treated with a high accuracy was thus very limited. During the 1960s and 1970s, electronic computers evolved from a few very expensive, difficult to use, machines to become generally available for researchers all over the world. The performance for a given price has been steadily increasing since and the use of computers is now widespread in many branches of science. This has spawned a new field in chemistry, *computational chemistry*, where the computer is used as an "experimental" tool, much like, for example, an NMR (nuclear magnetic resonance) spectrometer.

Computational chemistry is focused on obtaining results relevant to chemical problems, not directly at developing new theoretical methods. There is of course a strong interplay between traditional theoretical chemistry and computational chemistry. Developing new theoretical models may

Introduction to Computational Chemistry, Third Edition. Frank Jensen.

Companion Website: http://www.wiley.com/go/jensen/computationalchemistry3

enable new problems to be studied, and results from calculations may reveal limitations and suggest improvements in the underlying theory. Depending on the accuracy wanted, and the nature of the system at hand, one can today obtain useful information for systems containing up to several thousand particles. One of the main problems in computational chemistry is selecting a suitable level of theory for a given problem and to be able to evaluate the quality of the obtained results. The present book will try to put the variety of modern computational methods into perspective, hopefully giving the reader a chance of estimating which types of problems can benefit from calculations.

1.1 Fundamental Issues

Before embarking on a detailed description of the theoretical methods in computational chemistry, it may be useful to take a wider look at the background for the theoretical models and how they relate to methods in other parts of science, such as physics and astronomy.

A very large fraction of the computational resources in chemistry and physics is used in solving the so-called *many-body problem*. The essence of the problem is that two-particle systems can in many cases be solved exactly by mathematical methods, producing solutions in terms of analytical functions. Systems composed of more than two particles cannot be solved by analytical methods. Computational methods can, however, produce approximate solutions, which in principle may be refined to any desired degree of accuracy.

Computers are not smart – at the core level they are in fact very primitive. Smart programmers, however, can make sophisticated computer programs, which may make the computer appear smart, or even intelligent. However, the basics of any computer program consist of doing a few simple tasks such as:

- Performing a mathematical operation (adding, multiplying, square root, cosine, etc.) on one or two numbers.
- Determining the relationship (equal to, greater than, less than or equal to, etc.) between two numbers.
- Branching depending on a decision (add two numbers if $N > 10$, else subtract one number from the other).
- Looping (performing the same operation a number of times, perhaps on a set of data).
- Reading and writing data from and to external files.

These tasks are the essence of any programming language, although the syntax, data handling and efficiency depend on the language. The main reason why computers are so useful is the sheer speed with which they can perform these operations. Even a cheap off-the-shelf personal computer can perform billions (10^9) of operations per second.

Within the scientific world, computers are used for two main tasks: performing numerically intensive calculations and analyzing large amounts of data. The latter can, for example, be pictures generated by astronomical telescopes or gene sequences in the bioinformatics area that need to be compared. The numerically intensive tasks are typically related to simulating the behavior of the real world, by a more or less sophisticated computational model. The main problem in simulations is the multiscale nature of real-world problems, often spanning from subnanometers to millimeters (10^{-10}–10^{-3}) in spatial dimensions and from femtoseconds to milliseconds (10^{-15}–10^{-3}) in the time domain.

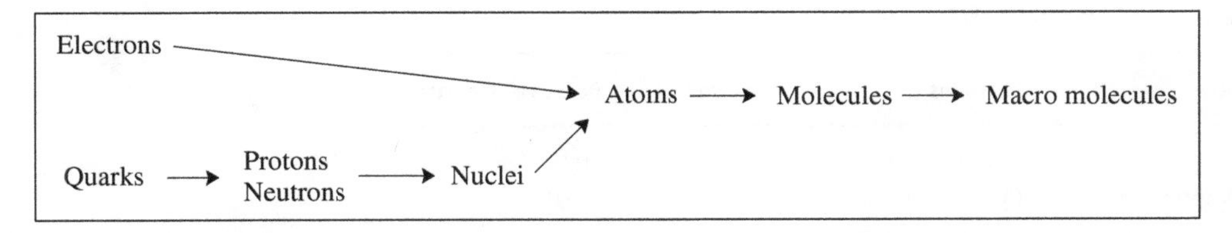

Figure 1.1 Hierarchy of building blocks for describing a chemical system.

1.2 Describing the System

In order to describe a system we need four fundamental features:

- System description. What are the fundamental units or "particles" and how many are there?
- Starting condition. Where are the particles and what are their velocities?
- Interaction. What is the mathematical form for the forces acting between the particles?
- Dynamical equation. What is the mathematical form for evolving the system in time?

The choice of "particles" puts limitations on what we are ultimately able to describe. If we choose atomic nuclei and electrons as our building blocks, we can describe atoms and molecules, but not the internal structure of the atomic nucleus. If we choose atoms as the building blocks, we can describe molecular structures, but not the details of the electron distribution. If we choose amino acids as the building blocks, we may be able to describe the overall structure of a protein, but not the details of atomic movements (see Figure 1.1).

The choice of starting conditions effectively determines what we are trying to describe. The complete phase space (i.e. all possible values of positions and velocities for all particles) is huge and we will only be able to describe a small part of it. Our choice of starting conditions determines which part of the phase space we sample, for example which (structural or conformational) isomer or chemical reaction we can describe. There are many structural isomers with the molecular formula C_6H_6, but if we want to study benzene, we should place the nuclei in a hexagonal pattern and start them with relatively low velocities.

The interaction between particles in combination with the dynamical equation determines how the system evolves in time. At the fundamental level, the only important force at the atomic level is the electromagnetic interaction. Depending on the choice of system description (particles), however, this may result in different effective forces. In force field methods, for example, the interactions are parameterized as stretch, bend, torsional, van der Waals, etc., interactions.

The dynamical equation describes the time evolution of the system. It is given as a differential equation involving both time and space derivatives, with the exact form depending on the particle masses and velocities. By solving the dynamical equation the particles' position and velocity can be predicted at later (or earlier) times relative to the starting conditions, that is how the system evolves in the phase space.

1.3 Fundamental Forces

The interaction between particles can be described in terms of either a force (**F**) or a potential (**V**). These are equivalent, as the force is the derivative of the potential with respect to the position **r**:

$$\mathbf{F}(\mathbf{r}) = -\frac{\partial \mathbf{V}}{\partial \mathbf{r}} \tag{1.1}$$

Table 1.1 Fundamental interactions.

Name	Particles	Range (m)	Relative strength
Strong interaction	Quarks	$<10^{-15}$	100
Weak interaction	Quarks, leptons	$<10^{-15}$	0.001
Electromagnetic	Charged particles	∞	1
Gravitational	Mass particles	∞	10^{-40}

Current knowledge indicates that there are four fundamental interactions, at least under normal conditions, as listed in Table 1.1.

Quarks are the building blocks of protons and neutrons, and lepton is a common name for a group of particles including the electron and the neutrino. The strong interaction is the force holding the atomic nucleus together, despite the repulsion between the positively charged protons. The weak interaction is responsible for radioactive decay of nuclei by conversion of neutrons to protons (β decay). The strong and weak interactions are short-ranged and are only important within the atomic nucleus.

Both the electromagnetic and gravitational interactions depend on the inverse distance between the particles and are therefore of infinite range. The electromagnetic interaction occurs between all charged particles, while the gravitational interaction occurs between all particles with a mass, and they have the same overall functional form:

$$\mathbf{V}_{\text{elec}}(\mathbf{r}_{ij}) = C_{\text{elec}}\frac{q_i q_j}{r_{ij}}$$
$$\mathbf{V}_{\text{grav}}(\mathbf{r}_{ij}) = -C_{\text{grav}}\frac{m_i m_j}{r_{ij}} \tag{1.2}$$

In SI units $C_{\text{elec}} = 9.0 \times 10^9$ N m^2/C^2 and $C_{\text{grav}} = 6.7 \times 10^{-11}$ N m^2/kg^2, while in atomic units $C_{\text{elec}} = 1$ and $C_{\text{grav}} = 2.4 \times 10^{-43}$. On an atomic scale, the gravitational interaction is completely negligible compared with the electromagnetic interaction. For the interaction between a proton and an electron, for example, the ratio between $\mathbf{V}_{\text{elec}}$ and $\mathbf{V}_{\text{grav}}$ is 10^{39}. On a large macroscopic scale, such as planets, the situation is reversed. Here the gravitational interaction completely dominates and the electromagnetic interaction is absent.

On a more fundamental level, it is believed that the four forces are really just different manifestations of a single common interaction, because of the relatively low energy regime we are living in. It has been shown that the weak and electromagnetic forces can be combined into a single unified theory, called *quantum electrodynamics* (QED). Similarly, the strong interaction can be coupled with QED into what is known as the *standard model*. Much effort is being devoted to also include the gravitational interaction into a grand unified theory, and *string theory* is currently believed to hold the greatest promise for such a unification.

Only the electromagnetic interaction is important at the atomic and molecular level, and in the large majority of cases, the simple Coulomb form (in atomic units) is sufficient:

$$\mathbf{V}_{\text{Coulomb}}(\mathbf{r}_{ij}) = \frac{q_i q_j}{r_{ij}} \tag{1.3}$$

Within QED, the Coulomb interaction is only the zeroth-order term and the complete interaction can be written as an expansion in terms of the (inverse) velocity of light, *c*. For systems where relativistic effects are important (i.e. containing elements from the lower part of the periodic table) or when high accuracy is required, the first-order correction (corresponding to an expansion up to $1/c^2$) for the electron–electron interaction may be included:

$$\mathbf{V}_{\text{elec}}(\mathbf{r}_{12}) = \frac{1}{r_{12}}\left[1 - \frac{1}{2}\left(\boldsymbol{\alpha}_1 \cdot \boldsymbol{\alpha}_2 + \frac{(\boldsymbol{\alpha}_1 \cdot \mathbf{r}_{12})(\boldsymbol{\alpha}_2 \cdot \mathbf{r}_{12})}{r_{12}^2}\right)\right] \tag{1.4}$$

The first-order correction is known as the *Breit* term, and $\boldsymbol{\alpha}_1$ and $\boldsymbol{\alpha}_2$ represent velocity operators. The first term in Equation (1.4) can be considered as a magnetic interaction between two electrons, but the whole Breit correction describes a "retardation" effect, since the interaction between distant particles is "delayed" relative to interactions between close particles, owing to the finite value of c (in atomic units, $c \sim 137$).

1.4 The Dynamical Equation

The mathematical form for the dynamical equation depends on the mass and velocity of the particles and can be divided into four regimes (see Figure 1.2).

Newtonian mechanics, exemplified by Newton's second law ($\mathbf{F} = m\mathbf{a}$), applies for "heavy", "slow-moving" particles. Relativistic effects become important when the velocity is comparable to the speed of light, causing an increase in the particle mass m relative to the rest mass m_0. A pragmatic borderline between Newtonian and relativistic (Einstein) mechanics is $\sim{}^1/_3c$, corresponding to a relativistic correction of a few percent.

Light particles display both wave and particle characteristics and must be described by quantum mechanics, with the borderline being approximately the mass of a proton. Electrons are much lighter and can only be described by quantum mechanics, while atoms and molecules, with a few exceptions, behave essentially as classical particles. Hydrogen (protons), being the lightest nucleus, represents a borderline case, which means that quantum corrections in some cases are essential. A prime example is the tunnelling of hydrogen through barriers, allowing reactions involving hydrogen to occur faster than expected from transition state theory.

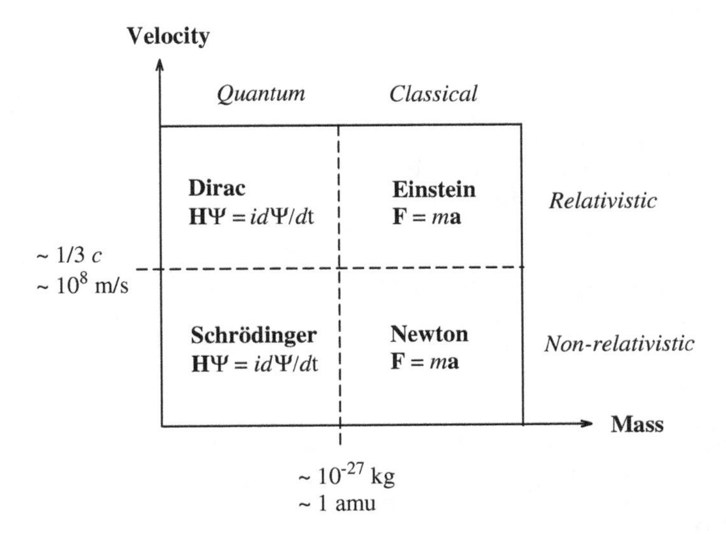

Figure 1.2 Domains of dynamical equations.

A major difference between quantum and classical mechanics is that classical mechanics is *deterministic* while quantum mechanics is *probabilistic* (more correctly, quantum mechanics is also deterministic, but the interpretation is probabilistic). Deterministic means that Newton's equation can be integrated over time (forward or backward) and can predict where the particles are at a certain time. This, for example, allows prediction of where and when solar eclipses will occur many thousands of years in advance, with an accuracy of meters and seconds. Quantum mechanics, on the other hand, only allows calculation of the *probability* of a particle being at a certain place at a certain time. The probability function is given as the square of a wave function, $P(\mathbf{r},t) = \Psi^2(\mathbf{r},t)$, where the wave function Ψ is obtained by solving either the Schrödinger (non-relativistic) or Dirac (relativistic) equation. Although they appear to be the same in Figure 1.2, they differ considerably in the form of the operator $\mathbf{H}$.

For classical mechanics at low velocities compared with the speed of light, Newton's second law applies:

$$\mathbf{F} = \frac{\mathrm{d}\mathbf{p}}{\mathrm{d}t} \tag{1.5}$$

If the particle mass is constant, the derivative of the momentum $\mathbf{p}$ is the mass times the acceleration:

$$\begin{aligned} \mathbf{p} &= m\mathbf{v} \\ \mathbf{F} &= \frac{\mathrm{d}\mathbf{p}}{\mathrm{d}t} = m\frac{\mathrm{d}\mathbf{v}}{\mathrm{d}t} = m\mathbf{a} \end{aligned} \tag{1.6}$$

Since the force is the derivative of the potential (Equation (1.1)) and the acceleration is the second derivative of the position $\mathbf{r}$ with respect to time, it may also be written in a differential form:

$$-\frac{\partial \mathbf{V}}{\partial \mathbf{r}} = m\frac{\partial^2 \mathbf{r}}{\partial t^2} \tag{1.7}$$

Solving this equation gives the position of each particle as a function of time, that is $\mathbf{r}(t)$.

At velocities comparable to the speed of light, Newton's equation is formally unchanged, but the particle mass becomes a function of the velocity, and the force is therefore not simply a constant (mass) times the acceleration:

$$m = \frac{m_0}{\sqrt{1 - v^2/c^2}} \tag{1.8}$$

For particles with small masses, primarily electrons, quantum mechanics must be employed. At low velocities, the relevant equation is the time-dependent Schrödinger equation:

$$\mathbf{H}\Psi = i\frac{\partial \Psi}{\partial t} \tag{1.9}$$

The Hamiltonian operator is given as a sum of kinetic and potential energy operators:

$$\begin{aligned} \mathbf{H}_{\text{Schrödinger}} &= \mathbf{T} + \mathbf{V} \\ \mathbf{T} &= \frac{\mathbf{p}^2}{2m} = -\frac{1}{2m}\nabla^2 \end{aligned} \tag{1.10}$$

Solving the Schrödinger equation gives the wave function as a function of time, and the probability of observing a particle at a position $\mathbf{r}$ and time t is given as the square of the wave function:

$$P(\mathbf{r}, t) = \Psi^2(\mathbf{r}, t) \tag{1.11}$$

For light particles moving at a significant fraction of the speed of light, the Schrödinger equation is replaced by the Dirac equation:

$$\mathbf{H}\Psi = i\frac{\partial \Psi}{\partial t} \tag{1.12}$$

Although it is formally identical to the Schrödinger equation, the Hamiltonian operator is significantly more complicated:

$$\mathbf{H}_{\text{Dirac}} = (c\boldsymbol{\alpha} \cdot \mathbf{p} + \boldsymbol{\beta} mc^2) + \mathbf{V} \tag{1.13}$$

The $\boldsymbol{\alpha}$ and $\boldsymbol{\beta}$ are 4×4 matrices and the relativistic wave function consequently has four components. Traditionally, these are labelled the *large* and *small* components, each having an α and β spin function (note the difference between the $\boldsymbol{\alpha}$ and $\boldsymbol{\beta}$ matrices and α and β spin functions). The large component describes the electronic part of the wave function, while the small component describes the positronic (electron antiparticle) part of the wave function, and the $\boldsymbol{\alpha}$ and $\boldsymbol{\beta}$ matrices couple these components. In the limit of $c \to \infty$, the Dirac equation reduces to the Schrödinger equation, and the two large components of the wave function reduce to the α and β spin-orbitals in the Schrödinger picture.

1.5 Solving the Dynamical Equation

Both the Newton/Einstein and Schrödinger/Dirac dynamical equations are differential equations involving the derivative of either the position vector or wave function with respect to time. For two-particle systems with simple interaction potentials $\mathbf{V}$, these can be solved analytically, giving $\mathbf{r}(t)$ or $\Psi(\mathbf{r},t)$ in terms of mathematical functions. For systems with more than two particles, the differential equation must be solved by numerical techniques involving a sequence of small finite time steps.

Consider a set of particles described by a position vector $\mathbf{r}_i$ at a given time t_i. A small time step Δt later, the positions can be calculated from the velocities, acceleration, hyperaccelerations, etc., corresponding to a Taylor expansion with time as the variable

$$\mathbf{r}_{i+1} = \mathbf{r}_i + \mathbf{v}_i(\Delta t) + \tfrac{1}{2}\mathbf{a}_i(\Delta t)^2 + \tfrac{1}{6}\mathbf{b}_i(\Delta t)^3 + \cdots \tag{1.14}$$

The positions a small time step Δt earlier were (replacing Δt with $-\Delta t$)

$$\mathbf{r}_{i-1} = \mathbf{r}_i - \mathbf{v}_i(\Delta t) + \tfrac{1}{2}\mathbf{a}_i(\Delta t)^2 - \tfrac{1}{6}\mathbf{b}_i(\Delta t)^3 + \cdots \tag{1.15}$$

Addition of these two equations gives a recipe for predicting the positions a time step Δt later from the current and previous positions, and the current acceleration, a method known as the *Verlet* algorithm:

$$\mathbf{r}_{i+1} = (2\mathbf{r}_i - \mathbf{r}_{i-1}) + \mathbf{a}_i(\Delta t)^2 + \cdots \tag{1.16}$$

Note that all odd terms in the Verlet algorithm disappear, that is the algorithm is correct to third order in the time step. The acceleration can be calculated from the force or, equivalently, the potential:

$$\mathbf{a} = \frac{\mathbf{F}}{m} = -\frac{1}{m}\frac{\partial \mathbf{V}}{\partial \mathbf{r}} \tag{1.17}$$

The time step Δt is an important control parameter for a simulation. The *largest* value of Δt is determined by the *fastest* process occurring in the system, typically being an order of magnitude smaller than the fastest process. For simulating nuclear motions, the fastest process is the motion of hydrogens, being the lightest particles. Hydrogen vibrations occur with a typical frequency of 3000 cm^{-1}, corresponding to $\sim 10^{14}$ s^{-1}, and therefore necessitating time steps of the order of one femtosecond (10^{-15} s).

1.6 Separation of Variables

As discussed in the previous section, the central problem is solving a differential equation with respect to either the position (classical) or wave function (quantum) for the particles in the system. The standard method of solving differential equations is to find a set of coordinates where the differential equation can be separated into less complicated equations. The first step is to introduce a *center of mass* coordinate system, defined as the mass-weighted sum of the coordinates of all particles, which allows the translation of the combined system with respect to a fixed coordinate system to be separated from the internal motion. For a two-particle system, the internal motion is then described in terms of a reduced mass moving relative to the center of mass, and this can be further transformed by introducing a coordinate system that reflects the symmetry of the interaction between the two particles. If the interaction only depends on the interparticle distance (e.g. Coulomb or gravitational interaction), the coordinate system of choice is normally a polar (two-dimensional) or spherical polar (three-dimensional) system. In these coordinate systems, the dynamical equation can be transformed into solving one-dimensional differential equations.

For more than two particles, it is still possible to make the transformation to the center of mass system. However, it is no longer possible to find a set of coordinates that allows a separation of the degrees of freedom for the internal motion, thus preventing an analytical solution. For many-body ($N > 2$) systems, the dynamical equation must therefore be solved by computational (numerical) methods. Nevertheless, it is often possible to achieve an approximate separation of variables based on physical properties, for example particles differing considerably in mass (such as nuclei and electrons). A two-particle system consisting of one nucleus and one electron can be solved exactly by introducing a center of mass system, thereby transforming the problem into a pseudo-particle with a reduced mass ($\mu = m_1 m_2/(m_1 + m_2)$) moving relative to the center of mass. In the limit of the nucleus being infinitely heavier than the electron, the center of mass system becomes identical to that of the nucleus. In this limit, the reduced mass becomes equal to that of the electron, which moves relative to the (stationary) nucleus. For large, but finite, mass ratios, the approximation $\mu \approx m_e$ is unnecessary but may be convenient for interpretative purposes. For many-particle systems, an exact separation is not possible, and the *Born–Oppenheimer* approximation corresponds to assuming that the nuclei are infinitely heavier than the electrons. This allows the electronic problem to be solved for a given set of stationary nuclei. Assuming that the electronic problem can be solved for a large set of nuclear coordinates, the electronic energy forms a parametric hypersurface as a function of the nuclear coordinates, and the motion of the nuclei on this surface can then be solved subsequently.

If an approximate separation is not possible, the many-body problem can often be transformed into a pseudo one-particle system by taking the average interaction into account. For quantum mechanics, this corresponds to the Hartree–Fock approximation, where the average electron–electron repulsion is incorporated. Such pseudo one-particle solutions often form the conceptual understanding of the system and provide the basis for more refined computational methods.

Molecules are sufficiently heavy that their motions can be described quite accurately by classical mechanics. In condensed phases (solution or solid state), there is a strong interaction between molecules, and a reasonable description can only be attained by having a large number of individual molecules moving under the influence of each other's repulsive and attractive forces. The forces in this case are complex and cannot be written in a simple form such as the Coulomb or gravitational interaction. No analytical solutions can be found in this case, even for a two-particle (molecular) system. Similarly, no approximate solution corresponding to a Hartree–Fock model can be constructed. The only method in this case is direct simulation of the full dynamical equation.

1.6.1 Separating Space and Time Variables

The time-dependent Schrödinger equation involves differentiation with respect to both time and position, the latter contained in the kinetic energy of the Hamiltonian operator:

$$\begin{aligned} \mathbf{H}(\mathbf{r},t)\Psi(\mathbf{r},t) &= i\frac{\partial \Psi(\mathbf{r},t)}{\partial t} \\ \mathbf{H}(\mathbf{r},t) &= \mathbf{T}(\mathbf{r}) + \mathbf{V}(\mathbf{r},t) \end{aligned} \tag{1.18}$$

For (bound) systems where the potential energy operator is time-independent ($\mathbf{V}(\mathbf{r},t) = \mathbf{V}(\mathbf{r})$), the Hamiltonian operator becomes time-independent and yields the total energy when acting on the wave function. The energy is a constant, independent of time, but depends on the space variables.

$$\begin{aligned} \mathbf{H}(\mathbf{r},t) &= \mathbf{H}(\mathbf{r}) = \mathbf{T}(\mathbf{r}) + \mathbf{V}(\mathbf{r}) \\ \mathbf{H}(\mathbf{r})\Psi(\mathbf{r},t) &= E(\mathbf{r})\Psi(\mathbf{r},t) \end{aligned} \tag{1.19}$$

Inserting this in the time-dependent Schrödinger equation shows that the time and space variables of the wave function can be separated:

$$\begin{aligned} \mathbf{H}(\mathbf{r})\Psi(\mathbf{r},t) = E(\mathbf{r})\Psi(\mathbf{r},t) &= i\frac{\partial \Psi(\mathbf{r},t)}{\partial t} \\ \Psi(\mathbf{r},t) &= \Psi(\mathbf{r})\mathrm{e}^{-iEt} \end{aligned} \tag{1.20}$$

The latter follows from solving the first-order differential equation with respect to time, and shows that the time dependence can be written as a simple phase factor multiplied by the spatial wave function. For time-*in*dependent problems, this phase factor is normally neglected, and the starting point is taken as the time-independent Schrödinger equation:

$$\mathbf{H}(\mathbf{r})\Psi(\mathbf{r}) = E(\mathbf{r})\Psi(\mathbf{r}) \tag{1.21}$$

1.6.2 Separating Nuclear and Electronic Variables

Electrons are very light particles and cannot be described by classical mechanics, while nuclei are sufficiently heavy that they display only small quantum effects. The large mass difference indicates that the nuclear velocities are much smaller than the electron velocities, and the electrons therefore adjust very fast to a change in the nuclear geometry.

For a general N-particle system, the Hamiltonian operator contains kinetic ($\mathbf{T}$) and potential ($\mathbf{V}$) energy for all particles:

$$\begin{aligned} \mathbf{H} &= \mathbf{T} + \mathbf{V} \\ \mathbf{T} = \sum_{i=1}^{N} \mathbf{T}_i \quad &; \quad \mathbf{V} = \sum_{i>j}^{N} \mathbf{V}_{ij} \end{aligned} \tag{1.22}$$

The potential energy operator is the Coulomb potential (Equation (1.3)). Denoting nuclear coordinates with **R** and subscript n, and electron coordinates with **r** and subscript e, this can be expressed as follows:

$$\begin{aligned} \mathbf{H}_{tot}\Psi_{tot}(\mathbf{R},\mathbf{r}) &= E_{tot}\Psi_{tot}(\mathbf{R},\mathbf{r}) \\ \mathbf{H}_{tot} = \mathbf{H}_e + \mathbf{T}_n \quad ; \quad \mathbf{H}_e &= \mathbf{T}_e + \mathbf{V}_{ne} + \mathbf{V}_{ee} + \mathbf{V}_{nn} \\ \Psi_{tot}(\mathbf{R},\mathbf{r}) &= \Psi_n(\mathbf{R})\Psi_e(\mathbf{R},\mathbf{r}) \\ \mathbf{H}_e\Psi_e(\mathbf{R},\mathbf{r}) &= E_e(\mathbf{R})\Psi_e(\mathbf{R},\mathbf{r}) \\ (\mathbf{T}_n + E_e(\mathbf{R}))\Psi_n(\mathbf{R}) &= E_{tot}\Psi_n(\mathbf{R}) \end{aligned} \tag{1.23}$$

The above approximation corresponds to neglecting the coupling between the nuclear and electronic velocities, that is the nuclei are stationary from the electronic point of view. The electronic wave function thus depends *parametrically* on the nuclear coordinates, since it only depends on the *position* of the nuclei, not on their *momentum*. To a good approximation, the electronic wave function thus provides a *potential energy surface* upon which the nuclei move, and this separation is known as the *Born–Oppenheimer* approximation.

The Born–Oppenheimer approximation is usually very good. For the hydrogen molecule (H_2) the error is of the order of 10^{-4} au, and for systems with heavier nuclei the approximation becomes better. As we shall see later, it is possible only in a few cases to solve the electronic part of the Schrödinger equation to an accuracy of 10^{-4} au, that is neglect of the nuclear–electron coupling is usually only a minor approximation compared with other errors.

1.6.3 Separating Variables in General

Assume that a set of variables can be found where the Hamiltonian operator **H** for two particles/variables can be separated into two independent terms, with each only depending on one particle/variable:

$$\mathbf{H} = \mathbf{h}_1 + \mathbf{h}_2 \tag{1.24}$$

Assume furthermore that the Schrödinger equation for one particle/variable can be solved (exactly or approximately):

$$\mathbf{h}_i\phi_i = \varepsilon_i\phi_i \tag{1.25}$$

The solution to the two-particle problem can then be composed of solutions of one-variable Schrödinger equations:

$$\Psi = \phi_1\phi_2 \quad ; \quad E = \varepsilon_1 + \varepsilon_2 \tag{1.26}$$

This can be generalized to the case of N particles/variables:

$$\begin{aligned} \mathbf{H} &= \sum_i \mathbf{h}_i \\ \Psi = \prod_i \phi_i \quad &; \quad E = \sum_i \varepsilon_i \end{aligned} \tag{1.27}$$

The properties in Equation (1.27) may be verified by inserting the entities in the Schrödinger Equation (1.21).

1.7 Classical Mechanics

1.7.1 The Sun–Earth System

The motion of the Earth around the Sun is an example of a two-body system that can be treated by classical mechanics. The interaction between the two "particles" is the gravitational force:

$$\mathbf{V}(\mathbf{r}_{12}) = -C_{\mathrm{grav}}\frac{m_1 m_2}{r_{12}} \tag{1.28}$$

The dynamical equation is Newton's second law, which in differential form can be written as

$$-\frac{\partial \mathbf{V}}{\partial \mathbf{r}} = m\frac{\partial^2 \mathbf{r}}{\partial t^2} \tag{1.29}$$

The first step is to introduce a center of mass system, and the internal motion becomes motion of a "particle" with a reduced mass given by

$$\mu = \frac{M_{\mathrm{Sun}} m_{\mathrm{Earth}}}{M_{\mathrm{Sun}} + m_{\mathrm{Earth}}} = \frac{m_{\mathrm{Earth}}}{(1 + m_{\mathrm{Earth}}/M_{\mathrm{Sun}})} \cong m_{\mathrm{Earth}} \tag{1.30}$$

Since the mass of the Sun is 3×10^5 times larger than that of the Earth, the reduced mass is essentially identical to the Earth's mass ($\mu = 0.999997 m_{\mathrm{Earth}}$). To a very good approximation, the system can therefore be described as the Earth moving around the Sun, which remains stationary.

The motion of the Earth around the Sun occurs in a plane, and a suitable coordinate system is a polar coordinate system (two-dimensional) consisting of r and θ (Figure 1.3).

The interaction depends only on the distance r, and the differential equation (Newton's equation) can be solved analytically. The bound solutions are elliptical orbits with the Sun (more precisely, the center of mass) at one of the foci, but for most of the planets, the actual orbits are close to circular. Unbound solutions corresponding to hyperbolas also exist and could, for example, describe the path of a (non-returning) comet (see Figure 1.4).

Each bound orbit can be classified in terms of the dimensions (largest and smallest distance to the Sun), with an associated total energy. In classical mechanics, there are no constraints on the energy and all sizes of orbits are allowed. If the zero point for the energy is taken as the two particles at rest infinitely far apart, positive values of the total energy correspond to unbound solutions (hyperbolas, with the kinetic energy being larger than the potential energy) while negative values correspond to bound orbits (ellipsoids, with the kinetic energy being less than the potential energy). Bound solutions are also called *stationary* orbits, as the particle position returns to the same value with well-defined time intervals.

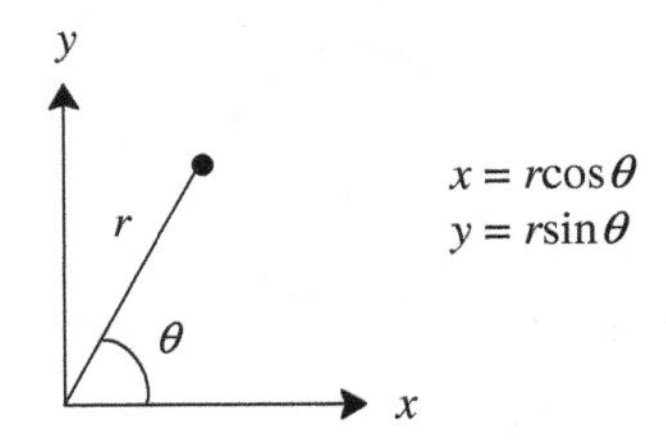

Figure 1.3 A polar coordinate system.

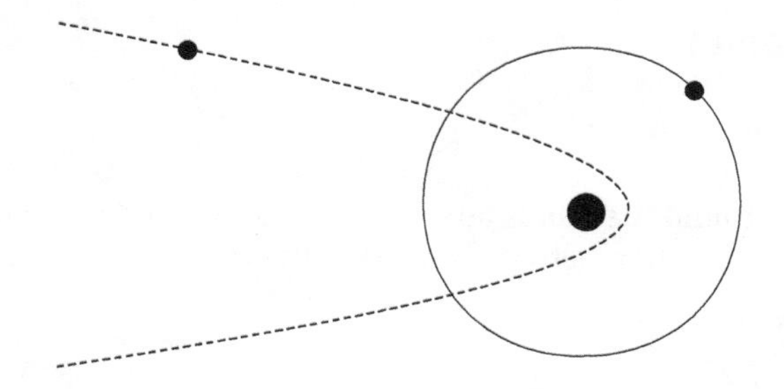

Figure 1.4 Bound and unbound solutions to the classical two-body problem.

1.7.2 The Solar System

Once we introduce additional planets in the Sun–Earth system, an analytical solution for the motions of all the planets can no longer be obtained. Since the mass of the Sun is so much larger than the remaining planets (the Sun is 1000 times heavier than Jupiter, the largest planet), the interactions between the planets can to a good approximation be neglected. For the Earth, for example, the second most important force is from the Moon, with a contribution that is 180 times smaller than that from the Sun. The next largest contribution is from Jupiter, being approximately 30 000 times smaller (on average) than the gravitational force from the Sun. In this *central field model*, the orbit of each planet is determined as if it was the only planet in the solar system, and the resulting computational task is a two-particle problem, that is elliptical orbits with the Sun at one of the foci. The complete solar system is the unification of eight such orbits and the total energy is the sum of all eight individual energies.

A formal refinement can be done by taking the *average* interaction between the planets into account, that is a Hartree–Fock type approximation. In this model, the orbit of one planet (e.g. the Earth) is determined by taking the average interaction with all the other planets into account. The average effect corresponds to spreading the mass of the other planets evenly along their orbits.

The Hartree–Fock model (Figure 1.5) represents only a very minute improvement over the independent orbit model for the solar system, since the planetary orbits do not cross. The effect of a planet inside the Earth's orbit corresponds to adding its mass to the Sun, while the effect of the spread-out mass of a planet outside the Earth's orbit is zero. The Hartree–Fock model for the Earth thus consists of increasing the Sun's effective mass with that of Mercury and Venus, that is a change of only 0.0003%. For the solar system there is thus very little difference between totally neglecting the planetary interactions and taking the average effect into account.

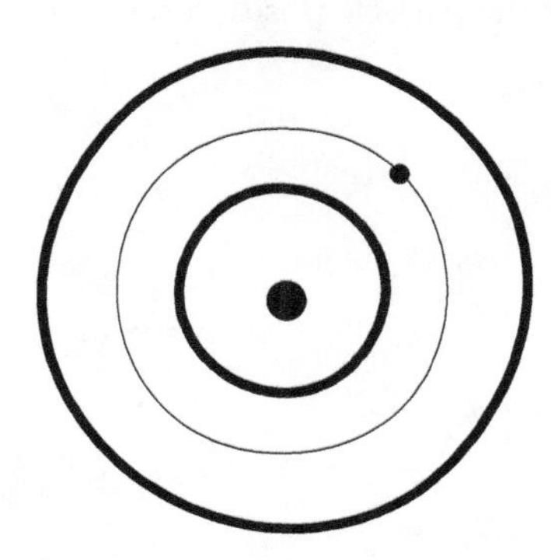

Figure 1.5 A Hartree–Fock model for the solar system.

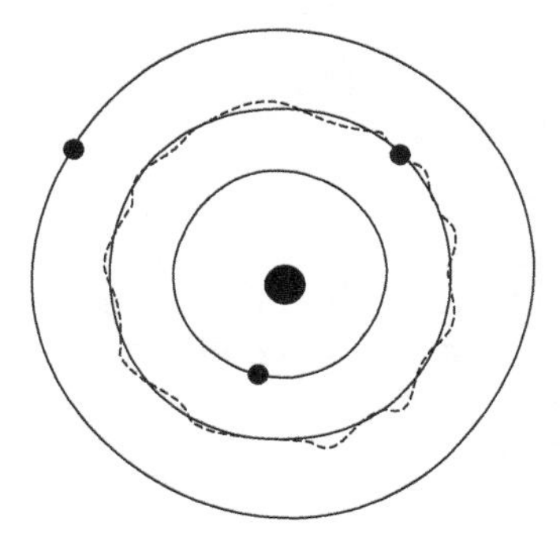

Figure 1.6 Modeling the solar system with actual interactions.

The real system, of course, includes all interactions, where each pair interaction depends on the actual distance between the planets (Figure 1.6). The resulting planetary motions cannot be solved analytically, but can be simulated numerically. From a given starting condition, the system is allowed to evolve for many small time steps, and all interactions are considered constant within each time step. By sufficiently small time steps, this yields a very accurate model of the real many-particle dynamics, and will display small wiggles of the planetary motion around the elliptical orbits calculated by either of the two independent-particle models.

Since the perturbations due to the other planets are significantly smaller than the interaction with the Sun, the "wiggles" are small compared with the overall orbital motion, and a description of the solar system as planets orbiting the Sun in elliptical orbits is a very good approximation to the true dynamics of the system.

1.8 Quantum Mechanics

1.8.1 A Hydrogen-Like Atom

A quantum analog of the Sun–Earth system is a nucleus and one electron, that is a hydrogen-like atom. The force holding the nucleus and electron together is the Coulomb interaction:

$$\mathbf{V}(\mathbf{r}_{12}) = \frac{q_1 q_2}{r_{12}} \tag{1.31}$$

The interaction again only depends on the distance, but owing to the small mass of the electron, Newton's equation must be replaced with the Schrödinger equation. For bound states, the time-dependence can be separated out, as shown in Section 1.6.1, giving the time-independent Schrödinger equation:

$$\mathbf{H}\Psi = E\Psi \tag{1.32}$$

The Hamiltonian operator for a hydrogen-like atom (nuclear charge of Z) can in Cartesian coordinates and atomic units be written as follows, with M being the nuclear and m the electron mass ($m = 1$ in atomic units):

$$\mathbf{H} = -\frac{1}{2M}\nabla_1^2 - \frac{1}{2m}\nabla_2^2 - \frac{Z}{\sqrt{(x_1 - x_2)^2 + (y_1 - y_2)^2 + (z_1 - z_2)^2}} \tag{1.33}$$

The Laplace operator is given by

$$\nabla_i^2 = \frac{\partial^2}{\partial x_i^2} + \frac{\partial^2}{\partial y_i^2} + \frac{\partial^2}{\partial z_i^2} \tag{1.34}$$

The two kinetic energy operators are already separated, since each only depends on the three coordinates for one particle. The potential energy operator, however, involves all six coordinates. The center of mass system is introduced by the following six coordinates:

$$\begin{aligned} X &= \frac{(Mx_1 + mx_2)}{(M+m)} \quad ; \quad x = x_1 - x_2 \\ Y &= \frac{(My_1 + my_2)}{(M+m)} \quad ; \quad y = y_1 - y_2 \\ Z &= \frac{(Mz_1 + mz_2)}{(M+m)} \quad ; \quad z = z_1 - z_2 \end{aligned} \tag{1.35}$$

Here the X, Y, Z coordinates define the center of mass system and the x, y, z coordinates specify the relative position of the two particles. In these coordinates the Hamiltonian operator can be rewritten as

$$\mathbf{H} = -\tfrac{1}{2}\nabla_{XYZ}^2 - \frac{1}{2\mu}\nabla_{xyz}^2 - \frac{Z}{\sqrt{x^2 + y^2 + z^2}} \tag{1.36}$$

The first term only involves the X, Y and Z coordinates, and the $\nabla^2{}_{XYZ}$ operator is obviously separable in terms of X, Y and Z. Solution of the XYZ part gives translation of the whole system in three dimensions relative to the laboratory-fixed coordinate system. The xyz coordinates describe the relative motion of the two particles in terms of a pseudo-particle with a reduced mass μ relative to the center of mass:

$$\mu = \frac{M_{\text{nuc}} m_{\text{elec}}}{M_{\text{nuc}} + m_{\text{elec}}} = \frac{m_{\text{elec}}}{\left(1 + {m_{\text{elec}}}/{M_{\text{nuc}}}\right)} \cong m_{\text{elec}} \tag{1.37}$$

For the hydrogen atom, the nucleus is approximately 1800 times heavier than the electron, giving a reduced mass of $0.9995 m_{\text{elec}}$. Similar to the Sun–Earth system, the hydrogen atom can therefore to a good approximation be considered as an electron moving around a stationary nucleus, and for heavier elements the approximation becomes better (with a uranium nucleus, for example, the nucleus/electron mass ratio is ~430 000). Setting the reduced mass equal to the electron mass corresponds to making the assumption that the nucleus is infinitely heavy and therefore stationary.

The potential energy again only depends on the distance between the two particles, but in contrast to the Sun–Earth system, the motion occurs in three dimensions, and it is therefore advantageous to transform the Schrödinger equation into a spherical polar set of coordinates (Figure 1.7).

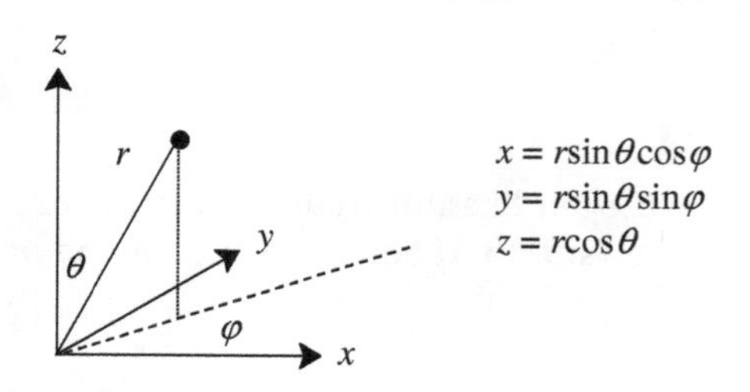

Figure 1.7 A spherical polar coordinate system.

The potential energy becomes very simple, but the kinetic energy operator becomes complicated:

$$\mathbf{H} = -\frac{1}{2\mu}\nabla^2_{r\theta\varphi} - \frac{Z}{r}$$
$$\nabla^2_{r\theta\varphi} = \frac{1}{r^2}\frac{\partial}{\partial r}r^2\frac{\partial}{\partial r} + \frac{1}{r^2\sin\theta}\frac{\partial}{\partial\theta}\sin\theta\frac{\partial}{\partial\theta} + \frac{1}{r^2\sin^2\theta}\frac{\partial^2}{\partial\varphi^2} \tag{1.38}$$

The kinetic energy operator, however, is almost separable in spherical polar coordinates, and the actual method of solving the differential equation can be found in a number of textbooks. The bound solutions (negative total energy) are called orbitals and can be classified in terms of three *quantum numbers*, n, l and m, corresponding to the three spatial variables r, θ and φ. The quantum numbers arise from the boundary conditions on the wave function, that is it must be periodic in the θ and φ variables and must decay to zero as $r \to \infty$. Since the Schrödinger equation is not completely separable in spherical polar coordinates, there exist the restrictions $n > l \geq |m|$. The n quantum number describes the *size* of the orbital, the l quantum number describes the *shape* of the orbital, while the m quantum number describes the *orientation* of the orbital relative to a fixed coordinate system. The l quantum number translates into names for the orbitals:

- $l = 0$: s-orbital
- $l = 1$: p-orbital
- $l = 2$: d-orbital
- $l = 3$: f-orbital, etc.

The orbitals can be written as a product of a *radial* function, describing the behavior in terms of the distance r between the nucleus and electron, and *spherical harmonic* functions Y_{lm}, representing the angular part in terms of the angles θ and φ. The orbitals can be visualized by plotting three-dimensional objects corresponding to the wave function having a specific value (e.g. $\Psi^2 = 0.10$) (see Table 1.2).

The orbitals for different quantum numbers are orthogonal and can be chosen to be normalized:

$$\langle \Psi_{n,l,m} | \Psi_{n',l',m'} \rangle = \delta_{n,n'}\delta_{l,l'}\delta_{m,m'} \tag{1.39}$$

Table 1.2 Hydrogenic orbitals obtained from solving the Schrödinger equation.

n	*l*	*m*	$\Psi_{n,l,m}(r,\theta,\varphi)$	Shape and size
1	0	0	$Y_{0,0}(\theta,\varphi)e^{-Zr}$	
2	0	0	$Y_{0,0}(\theta,\varphi)(2 - Zr)e^{-Zr/2}$	
	1	$\pm1, 0$	$Y_{1,m}(\theta,\varphi)Zre^{-Zr/2}$	
3	0	0	$Y_{0,0}(\theta,\varphi)(27 - 18Zr + 2Z^2r^2)e^{-Zr/3}$	
	1	$\pm1, 0$	$Y_{1,m}(\theta,\varphi)Zr(6 - Zr)e^{-Zr/3}$	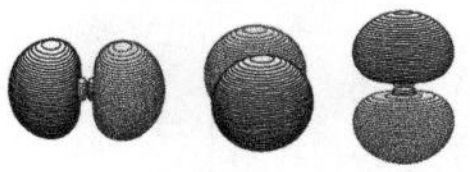
	2	$\pm2, \pm1, 0$	$Y_{2,m}(\theta,\varphi)Z^2r^2e^{-Zr/3}$	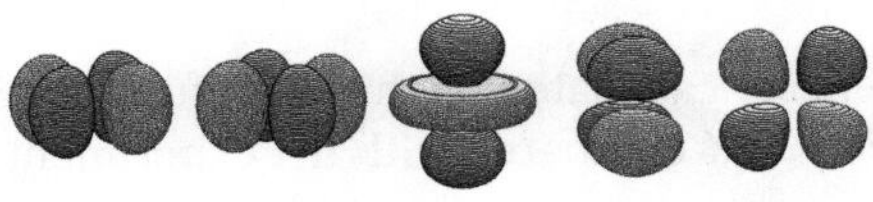

The orthogonality of the orbitals in the angular part (l and m quantum numbers) follows from the shape of the spherical harmonic functions, as these have l nodal planes (points where the wave function is zero). The orthogonality in the radial part (n quantum number) is due to the presence of $(n-l-1)$ radial nodes in the wave function.

In contrast to classical mechanics, where all energies are allowed, wave functions and associated energies are quantized, that is only certain values are allowed. The energy only depends on n for a given nuclear charge Z and is given by

$$E = -\frac{Z^2}{2n^2} \tag{1.40}$$

Unbound solutions have a positive total energy and correspond to scattering of an electron by the nucleus.

1.8.2 The Helium Atom

Like the solar system, it is not possible to find a set of coordinates where the Schrödinger equation can be solved analytically for more than two particles (i.e. for many-electron atoms). Owing to the dominance of the Sun's gravitational field, a central field approximation provides a good description of the actual solar system, but this is not the case for an atomic system. The main differences between the solar system and an atom such as helium are:

1. The interaction between the electrons is only a factor of two smaller than between the nucleus and electrons, compared with a ratio of at least 1000 for the solar system.
2. The electron–electron interaction is repulsive, compared with the attraction between planets.
3. The motion of the electrons must be described by quantum mechanics owing to the small electron mass, and the particle position is determined by an orbital, the square of which gives the probability of finding the electron at a given position.
4. Electrons are indistinguishable particles having a spin of $^1/_2$. This fermion character requires the total wave function to be antisymmetric, that is it must change sign when interchanging two electrons. The antisymmetry results in the so-called exchange energy, which is a non-classical correction to the Coulomb interaction.

The simplest atomic model would be to neglect the electron–electron interaction and only take the nucleus–electron attraction into account. In this model each orbital for the helium atom is determined by solving a hydrogen-like system with a nucleus and one electron, yielding hydrogen-like orbitals, 1s, 2s, 2p, 3s, 3p, 3d, etc., with $Z = 2$. The total wave function is obtained from the resulting orbitals subject to the aufbau and Pauli principles. These principles say that the lowest energy orbitals should be filled first and only two electrons (with different spin) can occupy each orbital, that is the electron configuration becomes $1s^2$. The antisymmetry condition is conveniently fulfilled by writing the total wave function as a Slater determinant, since interchanging any two rows or columns changes the sign of the determinant. For a helium atom, this would give the following (unnormalized) wave function, with the orbitals given in Table 1.2 with $Z = 2$:

$$\Phi = \begin{vmatrix} \phi_{1s\alpha}(1) & \phi_{1s\beta}(1) \\ \phi_{1s\alpha}(2) & \phi_{1s\beta}(2) \end{vmatrix} = \phi_{1s\alpha}(1)\phi_{1s\beta}(2) - \phi_{1s\beta}(1)\phi_{1s\alpha}(2) \tag{1.41}$$

The total energy calculated by this wave function is simply twice the orbital energy, −4.000 au, which is in error by 38% compared with the experimental value of −2.904 au. Alternatively, we can use the wave function given by Equation (1.41), but include the electron–electron interaction in the energy calculation, giving a value of −2.750 au.

A better approximation can be obtained by taking the *average* repulsion between the electrons into account when determining the orbitals, a procedure known as the Hartree–Fock approximation. If the orbital for one of the electrons was somehow known, the orbital for the second electron could be calculated in the electric field of the nucleus and the first electron, described by its orbital. This argument could just as well be used for the *second* electron with respect to the *first* electron. The goal is therefore to calculate a set of self-consistent orbitals, and this can be done by iterative methods.

For the solar system, the non-crossing of the planetary orbitals makes the Hartree–Fock approximation only a very minor improvement over a central field model. For a many-electron atom, however, the situation is different since the position of the electrons is described by three-dimensional probability functions (square of the orbitals), that is the electron "orbits" "cross". The average nucleus–electron distance for an electron in a 2s-orbital is larger than for one in a 1s-orbital, but there is a finite probability that a 2s-electron is closer to the nucleus than a 1s-electron. If the 1s-electrons in lithium were completely inside the 2s-orbital, the latter would experience an effective nuclear charge of 1.00, but owing to the 2s-electron penetrating the 1s-orbital, the effective nuclear charge for an electron in a 2s-orbital is 1.26. The 2s-electron in return screens the nuclear charge felt by the 1s-electrons, making the effective nuclear charge felt by the 1s-electrons less than 3.00. The mutual screening of the two 1s-electrons in helium produces an effective nuclear charge of 1.69, yielding a total energy of −2.848 au, which is a significant improvement relative to the model with orbitals employing a fixed nuclear charge of 2.00.

Although the effective nuclear charge of 1.69 represents the lowest possible energy with the functional form of the orbitals in Table 1.2, it is possible to further refine the model by relaxing the functional form of the orbitals from a strict exponential. Although the exponential form is the exact solution for a hydrogen-like system, this is not the case for a many-electron atom. Allowing the orbitals to adopt best possible form, and simultaneously optimizing the exponents ("effective nuclear charge"), gives an energy of −2.862 au. This represents the best possible independent-particle model for the helium atom, and any further refinement must include the instantaneous correlation between the electrons. By using the electron correlation methods described in Chapter 4, it is possible to reproduce the experimental energy of −2.904 au (see Table 1.3).

The equal mass of all the electrons and the strong interaction between them makes the Hartree–Fock model less accurate than desirable, but it is still a big improvement over an independent orbital model. The Hartree–Fock model typically accounts for ~99% of the total energy, but the remaining *correlation energy* is usually very important for chemical purposes. The correlation between the electrons describes the "wiggles" relative to the Hartree–Fock orbitals due to the instantaneous interaction between the electrons, rather than just the average repulsion. The goal of correlated methods

Table 1.3 Helium atomic energies in various approximations.

Wave function	Z_{eff}	Energy (au)
He^+ exponential orbital, no electron–electron repulsion	2.00	−4.000
He^+ exponential orbital, including electron–electron repulsion	2.00	−2.750
Optimum single exponential orbital	1.69	−2.848
Best orbital, Hartree–Fock limit		−2.862
Experimental		−2.904

for solving the Schrödinger equation is to calculate the remaining correction due to the electron–electron interaction.

1.9 Chemistry

The Born–Oppenheimer separation of the electronic and nuclear motions is a cornerstone in computational chemistry. Once the electronic Schrödinger equation has been solved for a large number of nuclear geometries (and possibly also for several electronic states), the *potential energy surface* (PES) is known. The motion of the nuclei on the PES can then be solved either classically (Newton) or by quantum (Schrödinger) methods. If there are N nuclei, the dimensionality of the PES is $3N$, that is there are $3N$ nuclear coordinates that define the geometry. Of these coordinates, three describe the overall translation of the molecule and three describe the overall rotation of the molecule with respect to three axes. For a linear molecule, only two coordinates are necessary for describing the rotation. This leaves $3N - 6(5)$ coordinates to describe the internal movement of the nuclei, which for small displacements may be chosen as "vibrational normal coordinates".

It should be stressed that nuclei are heavy enough that quantum effects are almost negligible, that is they behave to a good approximation as classical particles. Indeed, if nuclei showed significant quantum aspects, the concept of molecular structure (i.e. different configurations and conformations) would not have any meaning, since the nuclei would simply tunnel through barriers and end up in the global minimum. Dimethyl ether, for example, would spontaneously transform into ethanol. Furthermore, it would not be possible to speak of a molecular geometry, since the Heisenberg uncertainty principle would not permit a measure of nuclear positions with an accuracy smaller than the molecular dimension.

Methods aimed at solving the electronic Schrödinger equation are broadly referred to as "electronic structure calculations". An accurate determination of the electronic wave function is very demanding. Constructing a complete PES for molecules containing more than three or four atoms is virtually impossible. Consider, for example, mapping the PES by calculating the electronic energy for every 0.1 Å over, say, a 1 Å range (a very coarse mapping). With three atoms, there are three internal coordinates, giving 10^3 points to be calculated. Four atoms already produce six internal coordinates, giving 10^6 points, which is possible to calculate, but only with a determined effort. Larger systems are out of reach. Constructing global PESs for all but the smallest molecules is thus impossible. By restricting the calculations to the "chemically interesting" part of the PES, however, it is possible to obtain useful information. The interesting parts of a PES are usually nuclear arrangements that have low energies. For example, nuclear movements near a minimum on the PES, which corresponds to a stable molecule, are molecular vibrations. Chemical reactions correspond to larger movements, and may in the simplest approximation be described by locating the lowest energy path leading from one minimum on the PES to another.

These considerations lead to the following definition:

Chemistry is knowing the energy as a function of the nuclear coordinates.

The large majority of what are commonly referred to as molecular properties may similarly be defined as:

Properties are knowing how the energy changes upon adding a perturbation.

In the following chapters we will look at some aspects of solving the electronic Schrödinger equation or otherwise construct a PES, how to deal with the movement of nuclei on the PES and various technical points of commonly used methods. A word of caution here: although it is the nuclei that move, and the electrons follow "instantly" (according to the Born–Oppenheimer approximation), it is common also to speak of "atoms" moving. An isolated atom consists of a nucleus and some electrons, but in a molecule the concept of an atom is not well defined. Analogously to the isolated atom, an atom in a molecule should consist of a nucleus and some electrons. But how does one partition the total electron distribution in a molecule such that a given portion belongs to a given nucleus? Nevertheless, the words nucleus and atom are often used interchangeably.

Much of the following will concentrate on describing individual molecules. Experiments are rarely done on a single molecule; rather they are performed on macroscopic samples with perhaps 10^{20} molecules. The link between the properties of a single molecule, or a small collection of molecules, and the macroscopic observable is statistical mechanics. Briefly, macroscopic properties, such as temperature, heat capacity, entropy, etc., are the net effect of a very large number of molecules having a certain distribution of energies. If all the possible energy states can be determined for an individual molecule or a small collection of molecules, statistical mechanics can be used for calculating macroscopic properties.

References

1 T. Clark, *A Handbook of Computational Chemistry* (John Wiley & Sons, 1985).
2 D. M. Hirst, *A Computational Approach to Chemistry* (Blackwell, 1990).
3 A. Hinchcliffe, *Modelling Molecular Structure* (John Wiley & Sons, 1996).
4 D. Young, *Computational Chemistry* (Wiley-Interscience, 2001).
5 A. R. Leach, *Molecular Modelling. Principles and Applications* (Longman, 2001).
6 C. J. Cramer, *Essentials of Computational Chemistry* (John Wiley & Sons, 2002).
7 T. Schlick, *Molecular Modeling and Simulation* (Springer, 2002).

2

Force Field Methods

2.1 Introduction

As mentioned in Chapter 1, one of the major problems is calculating the electronic energy for a given nuclear configuration to give a potential energy surface. In *force field* (FF) methods, this step is bypassed by writing the electronic energy as a parametric function of the nuclear coordinates and fitting the parameters to experimental or higher level computational data. The "building blocks" in force field methods are atoms, that is electrons are not considered as individual particles. This means that bonding information must be provided *explicitly*, rather than being the *result* of solving the electronic Schrödinger equation.

In addition to bypassing the solution of the electronic Schrödinger equation, the quantum aspects of the nuclear motion are also neglected. This means that the dynamics of the atoms is treated by classical mechanics, that is Newton's second law. For time-independent phenomena, the problem reduces to calculating the energy at a given geometry. Often the interest is in finding geometries of stable molecules and/or different conformations, and possibly also interconversion between conformations. The problem is then reduced to finding energy minima (and possibly also some first-order saddle points) on the potential energy surface.

Molecules are described by a "ball and spring" model in force field methods, with atoms having different sizes and "softness" and bonds having different lengths and "stiffness".[1–5] Force field methods are also referred to as *molecular mechanics* (MM) methods. Many different force fields exist, and in this chapter we will use Allinger's MM2 and MM3 (Molecular Mechanics versions 2 and 3) to illustrate specific details.[6,7]

The foundation of force field methods is the observation that molecules tend to be composed of units that are structurally similar in different molecules. All C—H bond lengths, for example, are roughly constant in all molecules, being between 1.06 and 1.10 Å. The C—H stretch vibrations are also similar, between 2900 and 3300 cm^{-1}, implying that the C—H force constants are also comparable. If the C—H bonds are further divided into groups, for example those attached to single-, double- or triple-bonded carbon, the variation within each of these groups becomes even smaller. The same grouping holds for other features as well, for example all C=O bonds are approximately 1.22 Å long and have vibrational frequencies of approximately 1700 cm^{-1}, all double-bonded carbons are essentially planar, etc. The transferability also holds for energetic features. A plot of the heat of formation

Introduction to Computational Chemistry, Third Edition. Frank Jensen.

Companion Website: http://www.wiley.com/go/jensen/computationalchemistry3

for linear alkanes, that is $CH_3(CH_2)_nCH_3$, against the chain length n produces a straight line, showing that each CH_2 group contributes essentially the same amount of energy. (For a general discussion of estimating heat of formation from group additivities, see Benson.[8])

The picture of molecules being composed of structural units ("functional groups") that behave similarly in different molecules forms the very basis of organic chemistry. Molecular structure drawings, where alphabetic letters represent atoms and lines represent bonds, are used universally. Organic chemists often build ball and stick, or CPK space-filling, models of their molecules to examine their shapes. Force field methods are in a sense a generalization of these models, with the added feature that the atoms and bonds are not fixed at one size and length. Furthermore, force field calculations enable predictions of relative energies and barriers for interconversion of different conformations.

The idea of molecules being composed of atoms, which are structurally similar in different molecules, is implemented in force field models as atom *types*. The atom type depends on the atomic number and the type of chemical bonding it is involved in. The type may be denoted with either a number or a letter code. In MM2, for example, there are 71 different atom types (type 44 is missing). Type 1 is an sp^3-hybridized carbon, and an sp^2-hybridized carbon may be type 2, 3 or 50, depending on the neighbor atom(s). Type 2 is used if the bonding is to another sp^2-carbon (simple double bond), type 3 is used if the carbon is bonded to an oxygen (carbonyl group) and type 50 is used if the carbon is part of an aromatic ring with delocalized bonds. Table 2.1 gives a complete list of the MM2(91) atom types, where (91) indicates the year when the parameter set was released. The atom type numbers roughly reflect the order in which the corresponding functional groups were parameterized. The number of atom types represents a choice for how different two atoms in different bonding situations can be and still be considered identical, that is described by the same set of parameters. Results from electronic structure methods can be used for comparing similarities and thus deciding how many atoms types that should be used,[9] but most force fields rely on chemical intuition and calibration studies for the decision.

2.2 The Force Field Energy

The force field energy is written as a sum of terms (Figure 2.1), each describing the energy required for distorting a molecule in a specific fashion:

$$E_{FF} = E_{str} + E_{bend} + E_{tors} + E_{vdw} + E_{el} + E_{cross} \tag{2.1}$$

where E_{str} is the energy function for *stretching* a bond between two atoms, E_{bend} represents the energy required for *bending* an angle, E_{tors} is the *torsional* energy for rotation around a bond, E_{vdw} and E_{el} describe the *non-bonded* atom–atom interactions and finally E_{cross} describes coupling between the first three terms.

Given an energy function of the nuclear coordinates as in Equation (2.1), geometries and relative energies can be calculated by optimization. Stable molecules correspond to minima on the potential energy surface, and can be located by minimizing E_{FF} as a function of the nuclear coordinates. Conformational transitions can be described by locating transition structures on the E_{FF} surface. Exactly how such a multidimensional function optimization may be carried out is described in Chapter 13.

Table 2.1 MM2(91) atom types.

Type	Symbol	Description	Type	Symbol	Description
1	C	sp^3-carbon	28	H	Enol or amide
2	C	sp^2-carbon, alkene	48	H	Ammonium
3	C	sp^2-carbon, carbonyl, imine	36	D	Deuterium
4	C	sp-carbon	20	lp	Lone pair
22	C	sp^3-carbon, cyclopropane	15	S	Sulfide (R_2S)
29	C·	Radical	16	S^+	Sulfonium (R_3S^+)
30	C^+	Carbocation	17	S	Sulfoxide (R_2SO)
38	C	sp^2-carbon, cyclopropene	18	S	Sulfone (R_2SO_2)
50	C	sp^2-carbon, aromatic	42	S	sp^2-sulfur, thiophene
56	C	sp^3-carbon, cyclobutane	11	F	Fluoride
57	C	sp^2-carbon, cyclobutene	12	Cl	Chloride
58	C	Carbonyl, cyclobutanone	13	Br	Bromide
67	C	Carbonyl, cyclopropanone	14	I	Iodide
68	C	Carbonyl, ketene	26	B	Boron, trigonal
71	C	Ketonium carbon	27	B	Boron, tetrahedral
8	N	sp^3-nitrogen	19	Si	Silane
9	N	sp^2-nitrogen, amide	25	P	Phosphine
10	N	sp-nitrogen	60	P	Phosphor, pentavalent
37	N	Azo or pyridine (—N=)	51	He	Helium
39	N^+	sp^3-nitrogen, ammonium	52	Ne	Neon
40	N	sp^2-nitrogen, pyrrole	53	Ar	Argon
43	N	Azoxy (—N=N—O)	54	Kr	Krypton
45	N	Azide, central atom	55	Xe	Xenon
46	N	Nitro ($-NO_2$)	31	Ge	Germanium
72	N	Imine, oxime (=N—)	32	Sn	Tin
6	O	sp^3-oxygen	33	Pb	Lead
7	O	sp^2-oxygen, carbonyl	34	Se	Selenium
41	O	sp^2-oxygen, furan	35	Te	Tellurium
47	O^-	Carboxylate	59	Mg	Magnesium
49	O	Epoxy	61	Fe	Iron (II)
69	O	Amine oxide	62	Fe	Iron (III)
70	O	Ketonium oxygen	63	Ni	Nickel (II)
5	H	Hydrogen, except on N or O	64	Ni	Nickel (III)
21	H	Alcohol (OH)	65	Co	Cobalt (II)
23	H	Amine (NH)	66	Co	Cobalt (III)
24	H	Carboxyl (COOH)			

Note that special atom types are defined for carbon atoms involved in small rings, such as cyclopropane and cyclobutane. The reason for this will be discussed in Section 2.2.2.

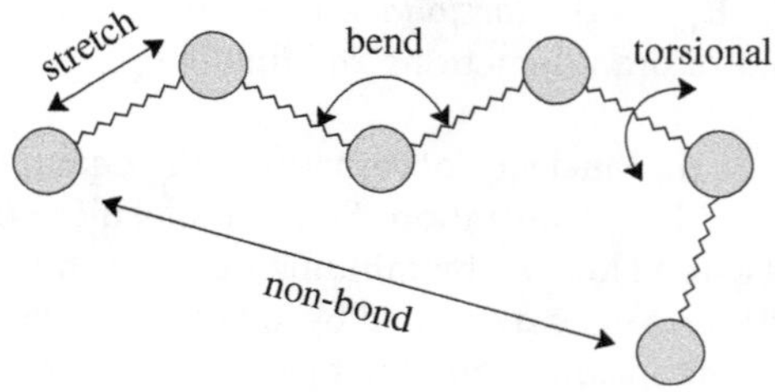

Figure 2.1 Illustration of the fundamental force field energy terms.

2.2.1 The Stretch Energy

E_{str} is the energy function for stretching a bond between two atom types A and B. In its simplest form, it is written as a Taylor expansion around a "natural", or "equilibrium", bond length, R_0. Terminating the expansion at second order gives

$$E_{str}\left(R^{AB}-R_0^{AB}\right)=E(0)+\frac{dE}{dR}\left(R^{AB}-R_0^{AB}\right)+\frac{1}{2}\frac{d^2E}{dR^2}\left(R^{AB}-R_0^{AB}\right)^2 \tag{2.2}$$

The derivatives are evaluated at $R=R_0$ and the $E(0)$ term is normally set to zero, since this is just the zero point for the energy scale. The second term is zero as the expansion is around the equilibrium value. In its simplest form the stretch energy can thus be written as

$$E_{str}\left(R^{AB}-R_0^{AB}\right)=k^{AB}\left(R^{AB}-R_0^{AB}\right)^2=k^{AB}\left(\Delta R^{AB}\right)^2 \tag{2.3}$$

Here k^{AB} is the "force constant" for the A—B bond. This is the form of a harmonic oscillator, with the potential being quadratic in the displacement from the minimum.

The harmonic form is the simplest possible, and sufficient for determining most equilibrium geometries. There are certain strained and crowded systems where the results from a harmonic approximation are significantly different from experimental values, and if the force field should be able to reproduce features such as vibrational frequencies, the functional form for E_{str} must be improved. The straightforward approach is to include more terms in the Taylor expansion:

$$E_{str}(\Delta R^{AB})=k_2^{AB}(\Delta R^{AB})^2+k_3^{AB}(\Delta R^{AB})^3+k_4^{AB}(\Delta R^{AB})^4+\cdots \tag{2.4}$$

This of course has a price: more parameters have to be assigned.

Polynomial expansions of the stretch energy do not have the correct limiting behavior. The cubic anharmonicity constant k_3 is normally negative, and if the Taylor expansion is terminated at third order, the energy will go toward $-\infty$ for long bond lengths. Minimization of the energy with such an expression can cause the molecule to fly apart if a poor starting geometry is chosen. The quartic constant k_4 is normally positive and the energy will go toward $+\infty$ for long bond lengths if the Taylor series is terminated at fourth order. The correct limiting behavior for a bond stretched to infinity is that the energy should converge towards the dissociation energy. A simple function that satisfies this criterion is the Morse potential:[10]

$$\begin{aligned} E_{Morse}(\Delta R) &= D(1-e^{-\alpha\Delta R})^2 \\ \alpha &= \sqrt{\frac{k}{2D}} \end{aligned} \tag{2.5}$$

Here D is the dissociation energy and α is related to the force constant. The Morse function reproduces the actual behavior quite accurately over a wide range of distances, as seen in Figure 2.2. There are, however, some difficulties with the Morse potential in actual applications, for example the restoring force is quite small for long bond lengths. Distorted structures, which may either be a poor starting geometry or one that develops during a simulation, will therefore display a slow convergence toward the equilibrium bond length. For minimization purposes and simulations at ambient temperatures (e.g. 300 K) it is sufficient that the potential is reasonably accurate up to ~40 kJ/mol above the minimum (the average kinetic energy is 3.7 kJ/mol at 300 K). In this energy range there is little difference between a Morse potential and a Taylor expansion, and most force fields therefore employ a simple polynomial for the stretch energy. The number of parameters is often reduced by taking the cubic, quartic, etc., constants as a predetermined fraction of the harmonic force constant. A popular method

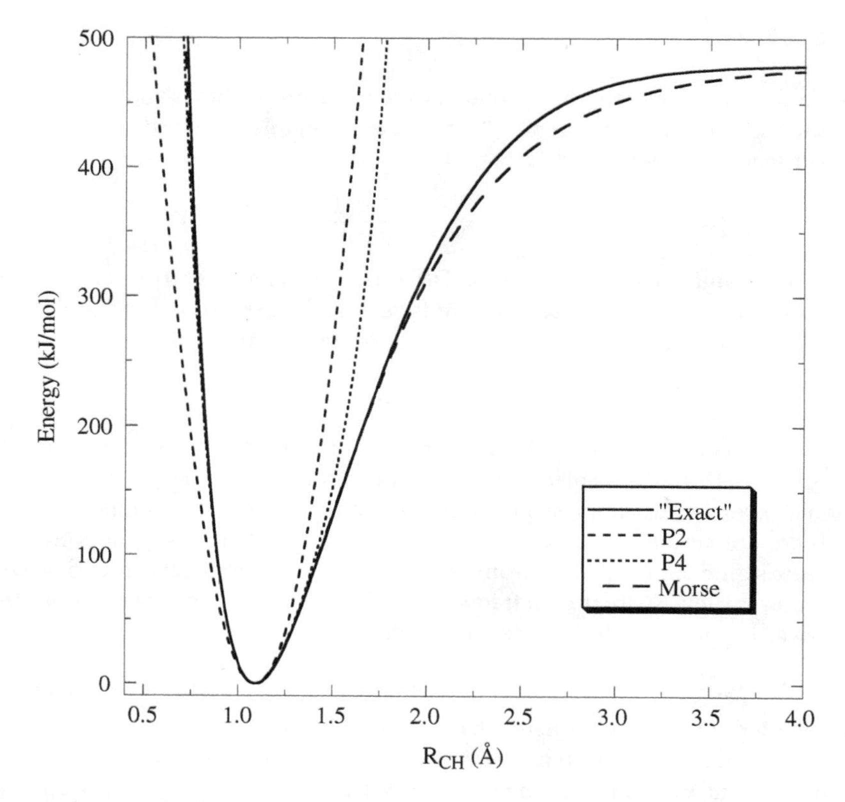

Figure 2.2 The stretch energy for CH_4.

is to require that the nth-order derivative at R_0 matches the corresponding derivative of the Morse potential. For a fourth-order expansion this leads to the following expression:

$$E_{\text{str}}(\Delta R^{\text{AB}}) = k_2^{\text{AB}}(\Delta R^{\text{AB}})^2 \left[1 - \alpha(\Delta R^{\text{AB}}) + \frac{7}{12}\alpha^2(\Delta R^{\text{AB}})^2\right] \tag{2.6}$$

The α constant is the same as that appearing in the Morse function, but may be taken as a fitting parameter. An alternative method for introducing anharmonicity is to use the harmonic form in Equation (2.3) but allow the force constant to depend on the bond distance.[11]

Figure 2.2 compares the performance of various functional forms for the stretch energy in CH_4. The "exact" form is taken from electronic structure calculations ([8,8]-CASMP2/aug-cc-pVTZ). The simple harmonic approximation (P2) is seen to be accurate to about ±0.1 Å from the equilibrium geometry and the quartic approximation (P4) up to ±0.3 Å. The Morse potential reproduces the real curve quite accurately up to an elongation of 0.8 Å, and becomes exact again in the dissociation limit.

For the large majority of systems, including simulations, the only important chemical region is within ~40 kJ/mol of the bottom of the curve. In this region, a fourth-order polynomial is essentially indistinguishable from either a Morse or the exact curve, as shown in Figure 2.3, and even a simple harmonic approximation does a quite good job.

Until now, we have used two different words for the R_0 parameter, the "natural" or the "equilibrium" bond length. The latter is slightly misleading. The R_0 parameter is *not* the equilibrium bond length for *any* molecule! Instead, it is the parameter which, when used to calculate the minimum energy structure of a molecule, will produce a geometry having the experimental equilibrium bond

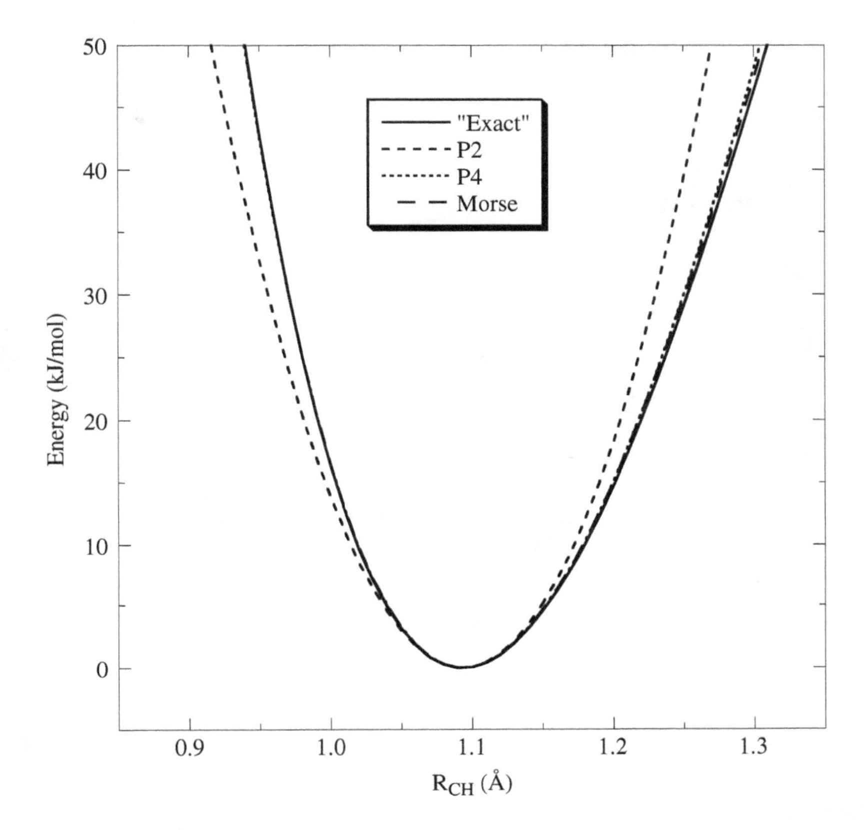

Figure 2.3 The stretch energy for CH_4.

length. If there were only one stretch energy in the whole force field energy expression (i.e. a diatomic molecule), R_0 would be the equilibrium bond length. However, in a polyatomic molecule the other terms in the force field energy will usually produce a minimum energy structure with bond lengths slightly longer than R_0. R_0 is the hypothetical bond length if no other terms are included, and the word "natural" bond length is a better description of this parameter than "equilibrium" bond length. Essentially all molecules have bond lengths that deviate very little from their "natural" values, typically by less than 0.03 Å. For this reason a simple harmonic is usually sufficient for reproducing experimental geometries.

For each bond type, that is a bond between two atom types A and B, there are at least two parameters to be determined, k^{AB} and R_0^{AB}. The higher-order expansions, and the Morse potential, have one additional parameter (α or D) that needs to be determined.

2.2.2 The Bending Energy

E_{bend} is the energy required for bending an angle formed by three atoms A—B—C, where there is a bond between A and B, and between B and C. Similarly to E_{str}, E_{bend} is usually expanded as a Taylor series around a "natural" bond angle and terminated at second order, giving the harmonic approximation:

$$E_{bend}\left(\theta^{ABC} - \theta_0^{ABC}\right) = k^{ABC}\left(\theta^{ABC} - \theta_0^{ABC}\right)^2 \tag{2.7}$$

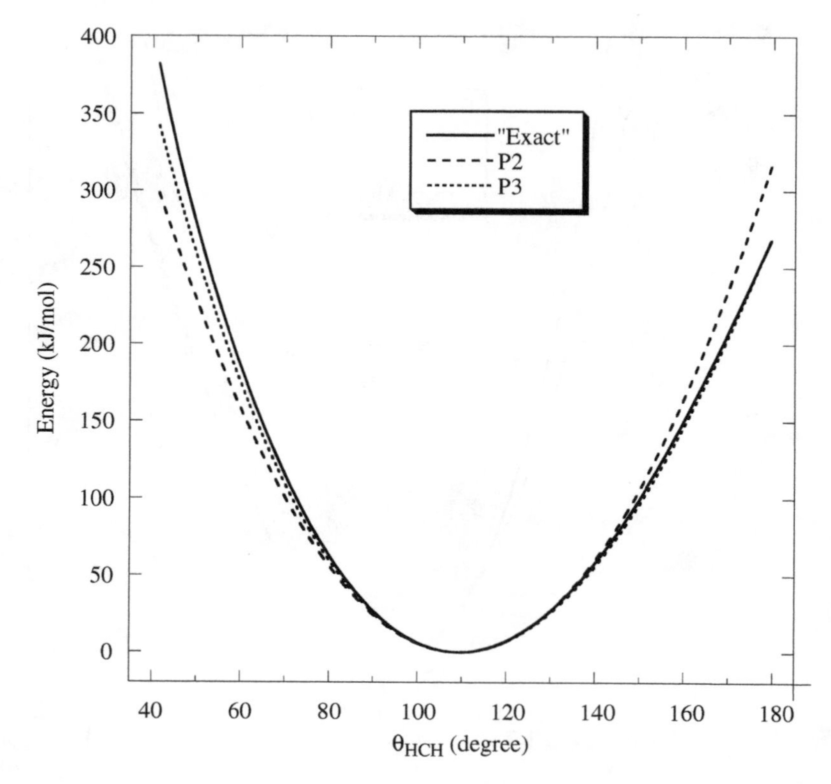

Figure 2.4 The bending energy for CH_4.

While the simple harmonic expansion is adequate for most applications, there may be cases where higher accuracy is required. The next improvement is to include a third-order term, analogous to E_{str}. This can give a very good description over a large range of angles, as illustrated in Figure 2.4 for CH_4. The "exact" form is again taken from electronic structure calculations (MP2/aug-cc-pVTZ). The simple harmonic approximation (P2) is seen to be accurate to about ±30° from the equilibrium geometry and the cubic approximation (P3) up to ±70°. Higher-order terms can be included in order to also reproduce vibrational frequencies. Analogous to E_{str}, the higher-order force constants are often taken as a fixed fraction of the harmonic constant. The constants beyond third order can rarely be assigned values with high confidence owing to insufficient experimental information. Fixing the higher-order constant in terms of the harmonic constant of course reduces the quality of the fit. While a third-order polynomial is capable of reproducing the actual curve very accurately if the cubic constant is fitted independently, the assumption that it is a fixed fraction (independent of the atom type) of the harmonic constant deteriorates the fit, but it still represents an improvement relative to a simple harmonic approximation. A second-order expansion is normally sufficient in the chemically important region below ~40 kJ/mol above the bottom of the energy curve.

Angles where the central atom is di- or trivalent (ethers, alcohols, sulfides, amines and enamines) present a special problem. In these cases, an angle of 180° corresponds to an energy maximum, that is the derivative of the energy with respect to the angle should be zero and the second derivative should be negative. This may be enforced by suitable boundary conditions on Taylor expansions of at least order three. A third-order polynomial fixes the barrier for linearity in terms of the harmonic

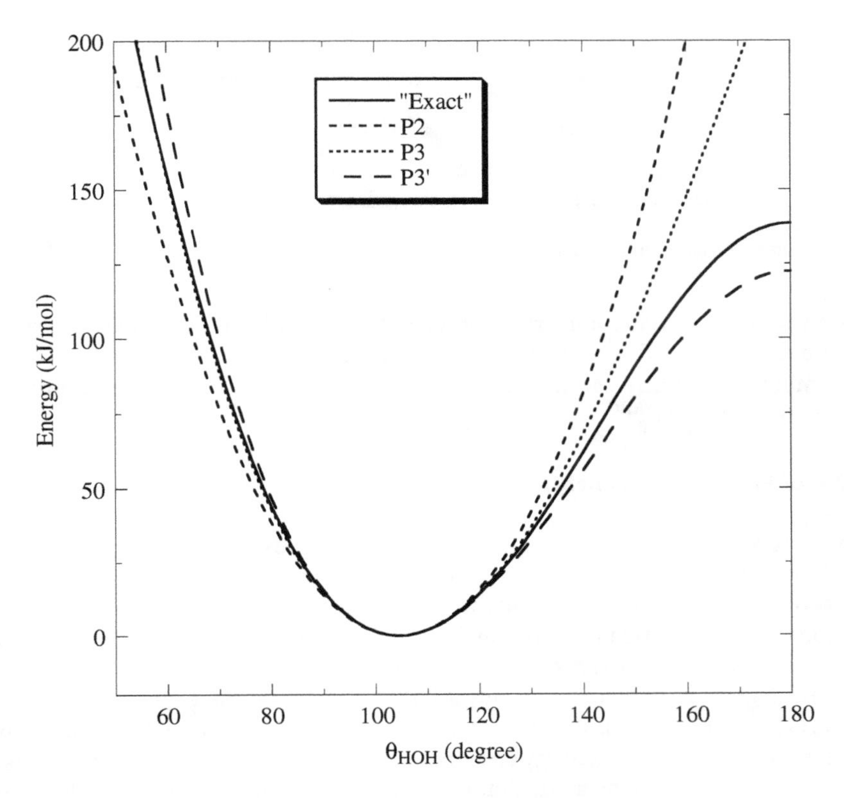

Figure 2.5 The bending energy for H_2O.

force constant and the equilibrium angle ($\Delta E^{\neq} = k(\theta - \theta_0)^2/6$). A fourth-order polynomial enables an independent fit of the barrier to linearity, but such constrained polynomial fittings are rarely done. Instead, the bending function is taken to be identical for all atom types, for example a fourth-order polynomial with cubic and quartic constants as a fixed fraction of the harmonic constant.

These features are illustrated for H_2O in Figure 2.5, where the "exact" form is taken from a parametric fit to a large amount of spectroscopic data.[12] The simple harmonic approximation (P2) is seen to be accurate to about ±20° from the equilibrium geometry and the cubic approximation (P3) up to ±40°. Enforcing the cubic polynomial to have a zero derivative at 180° (P3′) gives a qualitatively correct behavior, but reduces the overall fit, although it is still better than a simple harmonic approximation.

Although such refinements over a simple harmonic potential clearly improve the overall performance, they have little advantage in the chemically important region up to ~40 kJ/mol above the minimum. As for the stretch energy term, the energy cost for bending is so large that most molecules only deviate a few degrees from their natural bond angles. This again indicates that including only the harmonic term is adequate for most applications.

As noted above, special atom types are often defined for small rings, owing to the very different equilibrium angles for such rings. In cyclopropane, for example, the carbons are formally sp^3-hybridized, but have equilibrium CCC angles of 60°, in contrast to 110° in an acyclic system. With a low-order polynomial for the bend energy, the energy cost for such a deformation is large. For cyclobutane, for example, E_{bend} will dominate the total energy and cause the calculated structure to be planar, in contrast to the puckered geometry found experimentally. Introducing a special atom

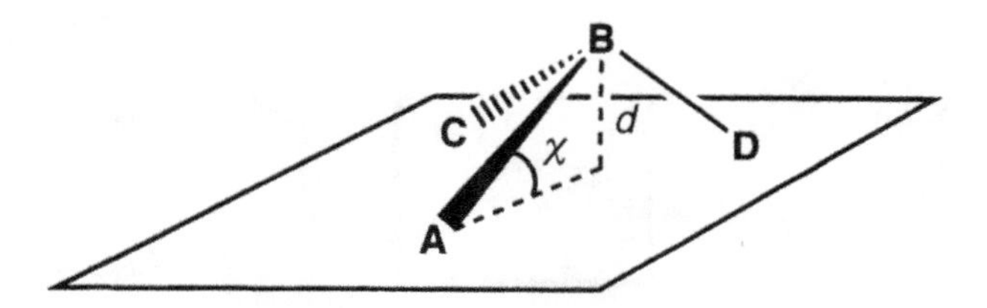

Figure 2.6 Out-of-plane variable definitions.

type for carbon atoms in a cyclobutane ring allows a natural bond angle parameter close to 90°, which in turn allows a description of low-energy ring puckering.

For each combination of three atom types, A, B and C, there are at least two bending parameters to be determined, k^{ABC} and θ_0^{ABC}.

2.2.3 The Out-of-Plane Bending Energy

If the central B atom in the angle ABC is sp^2-hybridized, there is a significant energy penalty associated with making the center pyramidal, since the four atoms prefer to be located in a plane. If the four atoms are exactly in a plane, the sum of the three angles with B as the central atom should be exactly 360°, but a quite large pyramidalization may be achieved without seriously distorting any of these three angles. Taking the bond distances to 1.5 Å, and moving the central atom 0.2 Å out of the plane, only reduces the angle sum to 354.8° (i.e. only a 1.7° decrease per angle). The corresponding out-of-plane angle, χ, is 7.7° for this case. Very large force constants must be used if the ABC, ABD and CBD angle distortions are to reflect the energy cost associated with the pyramidalization. This would have the consequence that the in-plane angle deformations for a planar structure would become unrealistically stiff. Thus a special *out-of-plane energy bend* term (E_{oop}) is usually added, while the in-plane angles (ABC, ABD and CBD) are treated as in the general case above. E_{oop} may be written as a harmonic term in the angle χ (the equilibrium angle for a planar structure is zero) or as a quadratic function in the distance d, as given below and shown in Figure 2.6:

$$E_{oop}(\chi) = k^B \chi^2 \quad \text{or} \quad E_{oop}(d) = k^B d^2 \tag{2.8}$$

Such energy terms may also be used for increasing the inversion barrier in sp^3-hybridized atoms (i.e. an extra energy penalty for being planar), and E_{oop} is also sometimes called E_{inv}. Inversion barriers are in most cases (e.g. in amines, NR_3) adequately modeled without an explicit E_{inv} term, the barrier arising naturally from the increase in bond angles upon inversion. The energy cost for non-planarity of sp^2-hybridized atoms may also be accounted for by an "improper" torsional energy, as described in Section 2.2.4, but the forms shown in Equation (2.8) have been found to produce better results.[13]

For each sp^2-hybridized atom there is one additional out-of-plane force constant to be determined, k^B.

2.2.4 The Torsional Energy

E_{tors} describes part of the energy change associated with rotation around a B—C bond in a four-atom sequence A—B—C—D, where A—B, B—C and C—D are bonded. Looking down the B—C bond, the torsional angle is defined as the angle formed by the A—B and C—D bonds, as shown in Figure 2.7. The angle ω may be taken to be in the range [0°, 360°] or [−180°, 180°].

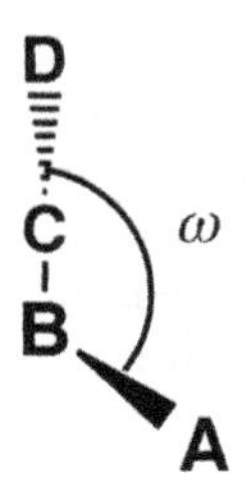

Figure 2.7 Torsional angle definition.

The torsional energy is fundamentally different from E_{str} and E_{bend} in three aspects:

1. A rotational barrier has contributions from both the non-bonded (van der Waals and electrostatic) terms, as well as the torsional energy, and the torsional parameters are therefore intimately coupled to the non-bonded parameters.
2. The torsional energy function must be periodic in the angle ω: if the bond is rotated 360° the energy should return to the same value.
3. The cost in energy for distorting a molecule by rotation around a bond is often low, that is large deviations from the minimum energy structure may occur and a Taylor expansion in ω is therefore not a good idea.

To encompass the periodicity, E_{tors} is written as a Fourier series:

$$E_{tors}(\omega) = \sum_{n=1} V_n \cos(n\omega) \tag{2.9}$$

The $n = 1$ term describes a rotation that is periodic by 360°, the $n = 2$ term is periodic by 180°, the $n = 3$ term is periodic by 120° and so on. The V_n constants determine the size of the barrier for rotation around the B—C bond. Depending on the situation, some of these V_n constants may be zero. In ethane, for example, the most stable conformation is one where the hydrogens are staggered relative to each other, while the eclipsed conformation represents an energy maximum. As the three hydrogens at each end are identical, it is clear that there are three energetically equivalent staggered, and three equivalent eclipsed, conformations. The rotational energy profile must therefore have three minima and three maxima. In the Fourier series only those terms that have $n = 3, 6, 9$, etc., can therefore have V_n constants different from zero.

For rotation around single bonds in substituted systems, other terms may be necessary. In the butane molecule, for example, there are still three minima, but the two *gauche* (torsional angle ~±60°) and *anti* (torsional angle =180°) conformations now have different energies. The barriers separating the two *gauche*, and the *gauche* and *anti*, conformations are also of different heights. This may be introduced by adding a term in Equation (2.9) corresponding to $n = 1$.

For the ethylene molecule, the rotation around the C=C bond must be periodic by 180°, and thus only $n = 2, 4$, etc., terms can enter. The energy cost for rotation around a double bond is much higher than that for rotation around a single bond in ethane, and this results in a larger value of the V_2 constant. For rotation around the C=C bond in a molecule such as 2-butene, there would again be a large V_2 constant, analogous to ethylene, but in addition there are now two different orientations of the two methyl groups relative to each other, *cis* and *trans*. The full rotation is periodic with a period of 360°, with deep energy minima at 0° and 180°, but slightly different energies of these two minima. This energy difference would show up as a V_1 constant, that is the V_2 constant essentially determines the barrier and location of the minima for rotation around the C=C bond, and the V_1 constant determines the energy difference between the *cis* and *trans* isomers.

Molecules that are composed of atoms having a maximum valence of four (essentially all organic molecules) are with a few exceptions found to have rotational profiles showing at most three minima. The first three terms in the Fourier series in Equation (2.9) are sufficient for qualitatively reproducing such profiles. Force fields that are aimed at large systems often limit the Fourier series to only one term, depending on the bond type (e.g. single bonds only have $\cos(3\,\omega)$ and double bonds only $\cos(2\,\omega)$).

Systems with bulky substituents on sp^3-hybridized atoms are often found to have four minima, the *anti* conformation being split into two minima with torsional angles of approximately ±170°. Other systems, notably polyfluoroalkanes, also split the *gauche* minima into two, often called *gauche* (angle of approximately ±50°) and *ortho* (angle of approximately ±90°) conformations, creating a rotational profile with six minima.[14] Rotations around a bond connecting sp^3- and sp^2-hybridized atoms (such as CH_3NO_2) also display profiles with six minima.[15] These exceptions from the regular three minima rotational profile around single bonds are caused by repulsive and attractive van der Waals interactions, and can still be modeled by having only terms up to $\cos(3\,\omega)$ in the torsional energy expression. Higher-order terms may be included to modify the detailed shape of the profile, and a few force fields employ terms with $n = 4$ and 6. Cases where higher-order terms are probably necessary are rotation around bonds to octahedral coordinated metals, such as Ru(pyridine)$_6$ or a dinuclear complex such as $Cl_4Mo{-}MoCl_4$. Here the rotation is periodic by 90° and thus requires a $\cos\,(4\,\omega)$ term.

It is customary to shift the zero point of the potential by adding a factor of one to each term. Most rotational profiles resemble either the ethane or ethylene examples above, and a popular expression for the torsional energy is given by

$$\begin{aligned} E_{\text{tors}}(\omega^{\text{ABCD}}) = {} & \tfrac{1}{2}V_1^{\text{ABCD}}[1+\cos(\omega^{\text{ABCD}})] + \tfrac{1}{2}V_2^{\text{ABCD}}[1-\cos(2\omega^{\text{ABCD}})] \\ & + \tfrac{1}{2}V_3^{\text{ABCD}}[1+\cos(3\omega^{\text{ABCD}})] \end{aligned} \tag{2.10}$$

The + and − signs are chosen such that the onefold rotational term has a minimum for an angle of 180°, the twofold rotational term has minima for angles of 0° and 180°, and the threefold rotational term has minima for angles of 60°, 180° and 300° (−60°). The factor $^1/_2$ is included such that the V_i parameters directly give the height of the barrier if only one term is present. A more general form for Equation (2.10) includes a phase factor, that is $\cos(n\omega - \tau)$, but for the most common cases of $\tau = 0°$ or 180°, this is completely equivalent to Equation (2.10). Figure 2.8 illustrates the functional behavior of the three individual terms in Equation (2.10).

The V_i parameters may also be negative, which corresponds to changing the minima on the rotational energy profile to maxima, and vice versa. Most commonly encountered rotational profiles can be obtained by combining the three V_i parameters. Figure 2.9 shows an example with one *anti* and two less stable *gauche* minima and with a significant *cis* barrier, corresponding to the combination $V_1 = 0.5$, $V_2 = -0.2$, $V_3 = 0.5$ in Equation (2.10).

As mentioned in Section 2.2.3, the out-of-plane energy may also be described by an "improper" torsional angle. For the example shown in Figure 2.6, a torsional angle ABCD may be defined, even though there is no bond between C and D. The out-of-plane E_{oop} may then be described by an angle ω^{ABCD}, for example as a harmonic function $(\omega - \omega_0)^2$ or Equation (2.10) with a large V_2 constant. Note that the definition of such improper torsional angles is not unique, the angle ω^{ABDC}, for example, is equally good. The main reason for using an improper torsional angle for describing E_{oop} is that no new functional forms need to be implemented, but the forms in Equation (2.8) have been shown to produce better results.[13]

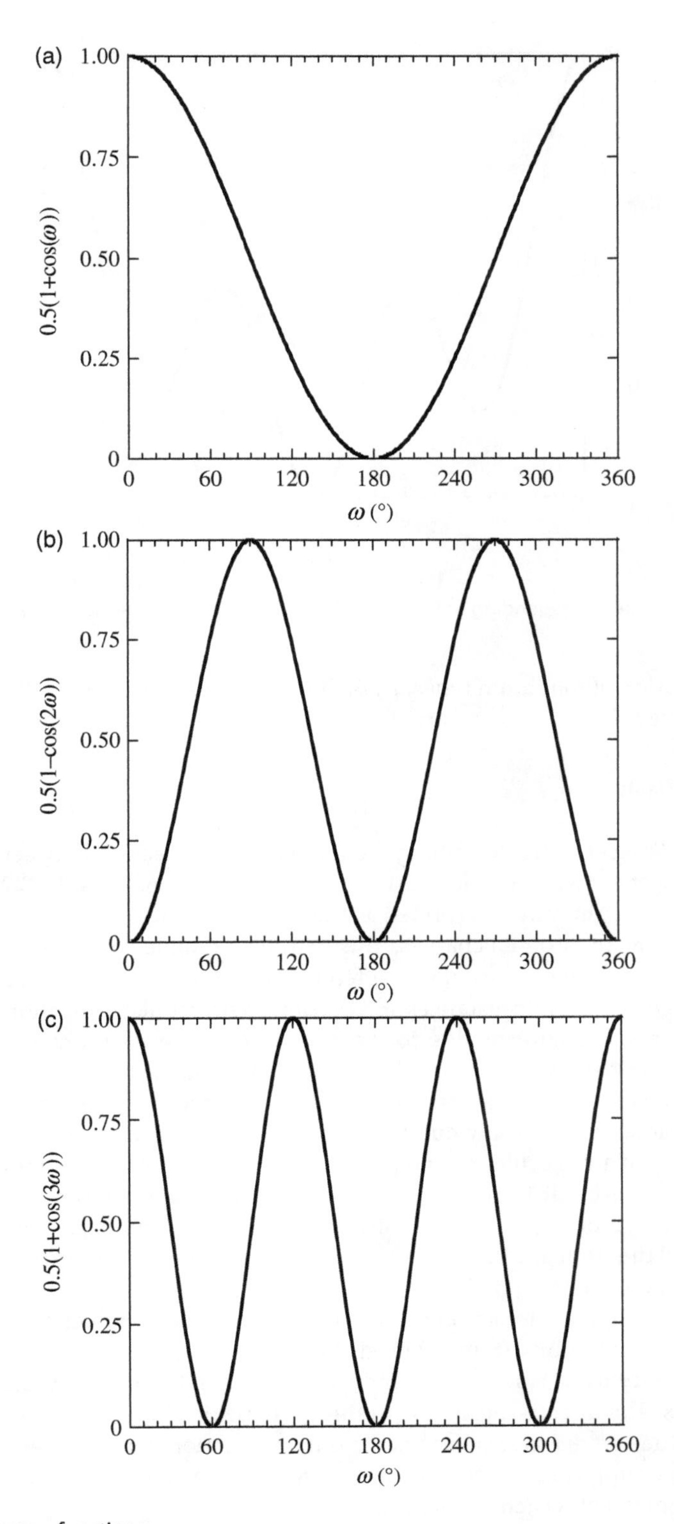

Figure 2.8 Torsional energy functions.

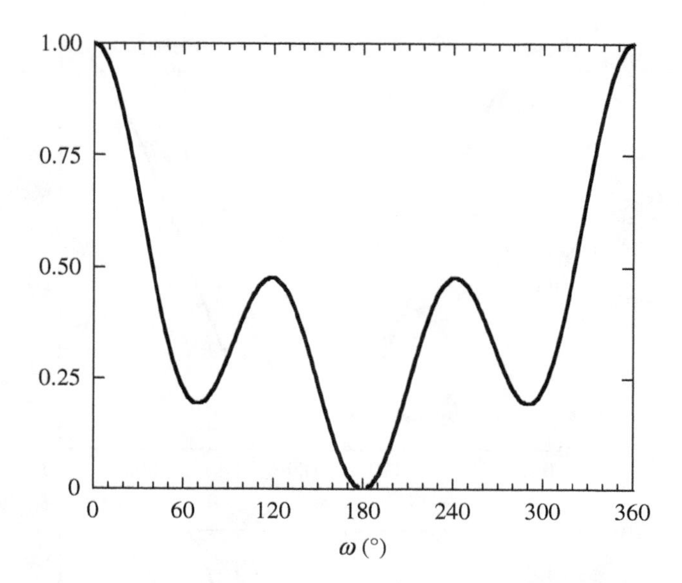

Figure 2.9 Rotational profile corresponding to Equation (2.10) with $V_1 = 0.5$, $V_2 = -0.2$, $V_3 = 0.5$.

For each combination of four atom types, A, B, C and D, there are generally three torsional parameters to be determined, V_1^{ABCD}, V_2^{ABCD} and V_3^{ABCD}.

2.2.5 The van der Waals energy

E_{vdw} is the van der Waals energy describing the repulsion or attraction between atoms that are not directly bonded. Together with the electrostatic term E_{el} (Sections 2.2.6 to 2.2.8), it describes the non-bonded energy. E_{vdw} may be interpreted as the non-polar part of the interaction not related to electrostatic energy due to (atomic) charges. This may, for example, be the interaction between two methane molecules or two methyl groups at different ends of the same molecule.

E_{vdw} is zero at large interatomic distances and becomes very repulsive for short distances. In quantum mechanical terms, the latter is due to the overlap of the electron clouds of the two atoms, which causes repulsion between electrons due to Coulomb and exchange interactions. At intermediate distances, however, there is a slight attraction between two such electron clouds from induced dipole–dipole interactions, physically due to electron correlation (discussed in Chapter 4). Even if the molecule (or part of a molecule) has no permanent dipole moment, the motion of the electrons will create a slightly uneven distribution at a given time. This dipole moment will induce a charge polarization in the neighbor molecule (or another part of the same molecule), creating an attraction, and it can be derived theoretically that this attraction varies as the inverse sixth power of the distance between the two fragments.

The induced dipole–dipole interaction is the leading term of such induced multipole interactions, but there are also contributions from induced dipole–quadrupole, quadrupole–quadrupole, etc., interactions. These vary as R^{-8}, R^{-10}, etc., and the R^{-6} dependence is only the asymptotic behavior at long distances. The force associated with this potential is often referred to as a "dispersion" or "London" force.[16] The van der Waals term is the only interaction between rare gas atoms (and thus the reason why, for example, argon can become a liquid and a solid) and it is the main interaction between non-polar molecules such as alkanes.

E_{vdw} is very positive at small distances, has a minimum that is slightly negative at a distance corresponding to the two atoms just "touching" each other and approaches zero as the distance becomes large. A general functional form that fits these conditions is given by

$$E_{vdw}(R^{AB}) = E_{repulsion}(R^{AB}) - \frac{C^{AB}}{(R^{AB})^6} \tag{2.11}$$

It is not possible to derive theoretically the functional form of the repulsive interaction; it is only required that it goes toward zero as R goes to infinity and it should approach zero faster than the R^{-6} term, as the energy should go towards zero from below.

A popular function that obeys these general requirements is the *Lennard-Jones* (LJ) potential,[17] where the repulsive part is given by an R^{-12} dependence (C_1 and C_2 are suitable constants):

$$E_{LJ}(R) = \frac{C_1}{R^{12}} - \frac{C_2}{R^6} \tag{2.12}$$

The Lennard-Jones potential can also be written as

$$E_{LJ}(R) = \varepsilon \left[\left(\frac{R_0}{R} \right)^{12} - 2\left(\frac{R_0}{R} \right)^{6} \right] \tag{2.13}$$

Here R_0 is the minimum energy distance and ε the depth of the minimum, and the connection to the form shown in Equation (2.12) is $C_1 = \varepsilon R_0{}^{12}$ and $C_2 = 2\varepsilon R_0{}^6$. It may alternatively be written in terms of a collision parameter σ, with the connection $R_0 = 2^{1/6}\sigma$, as

$$E_{LJ}(R) = 4\varepsilon \left[\left(\frac{\sigma}{R} \right)^{12} - \left(\frac{\sigma}{R} \right)^{6} \right] \tag{2.14}$$

There are no theoretical arguments for choosing the exponent in the repulsive part to be 12; this is purely a computational convenience and there is evidence that an exponent of 9 or 10 gives better results.

The *Merck Molecular Force Field* (MMFF) uses a generalized Lennard-Jones potential where the exponents and two empirical constants are derived from experimental data for rare gas atoms.[18] The resulting *buffered 14-7* potential is

$$E_{buf14\text{-}7}(R) = \varepsilon \left(\frac{1.07R_0}{R + 0.07R_0} \right)^7 \left(\frac{1.12R_0^7}{R^7 + 0.12R_0^7} - 2 \right) \tag{2.15}$$

From electronic structure theory it is known that the repulsion is due to overlap of the electronic wave functions, and furthermore that the electron density falls off exponentially with the distance from the nucleus (the exact wave function for the hydrogen atom is an exponential function). There is therefore some justification for choosing the repulsive part as an exponential function. The general form of the "*exponential* – R^{-6}" E_{vdw} function, also known as a "*Buckingham*" or "*Hill*" type potential,[19] is given by

$$E_{Hill}(R) = Ae^{-BR} - \frac{C}{R^6} \tag{2.16}$$

where A, B and C are here suitable constants. It is sometimes written in a slightly more convoluted form as

$$E_{Hill}(R) = \varepsilon \left[\frac{6}{\alpha - 6} e^{\alpha(1-R/R_0)} - \frac{\alpha}{\alpha - 6} \left(\frac{R_0}{R} \right)^6 \right] \tag{2.17}$$

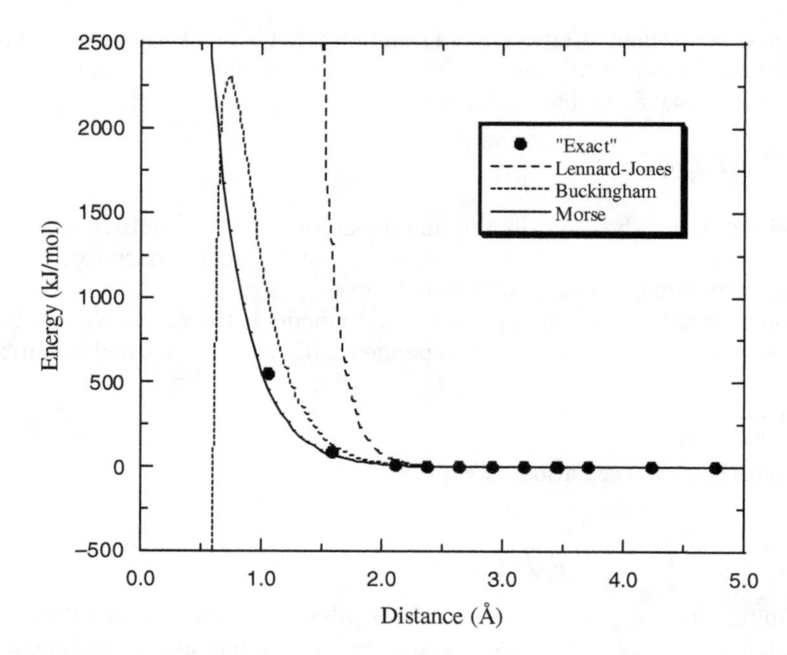

Figure 2.10 Comparison of E_{vdw} functionals for the H_2—He potential.

Here R_0 and ε have been defined in Equation (2.13) and α is a free parameter. Choosing an α value of 12 gives a long-range behavior identical to the Lennard-Jones potential, while a value of 13.772 reproduces the Lennard-Jones force constant at the equilibrium distance. The α parameter may also be taken as a fitting constant. The Buckingham potential has a problem for short interatomic distances where it "turns over". As R goes toward zero, the exponential becomes a constant while the R^{-6} term goes toward $-\infty$. Minimizing the energy of a structure that accidentally has a very short distance between two atoms will thus result in nuclear fusion! Special precautions therefore have to be taken for avoiding this when using Buckingham-type potentials.

A third functional form, which has an exponential dependence and the correct general shape, is the Morse potential (Equation (2.5)). It does not have the R^{-6} dependence at long range, but, as mentioned above, in reality there are also R^{-8}, R^{-10}, etc., terms. The D and α parameters of a Morse function describing E_{vdw} will of course be much smaller than for E_{str}, and R_0 will be longer.

For small systems, where accurate interaction energy profiles are available, it has been shown that the Morse function actually gives a slightly better description than a Buckingham potential, which again performs significantly better than a Lennard-Jones 12-6 potential.[20,21] This is illustrated for the H_2—He interaction in Figure 2.10, where the Buckingham and Morse parameters have been derived from the minimum energy and corresponding distance (ε and R_0) and by matching the force constant at the minimum.

The main difference between the three functions is in the repulsive part at short distances; the Lennard-Jones potential is much too hard and the Buckingham also tends to overestimate the repulsion. Furthermore, it has the problem of "inverting" at short distances. For chemical purposes, however, these "problems" are irrelevant, since energies in excess of 400 kJ/mol are sufficient to break most bonds and will never be encountered in actual calculations. The behavior in the attractive part of the potential, which is essential for intermolecular interactions, is very similar for the three functions, as shown in Figure 2.11.

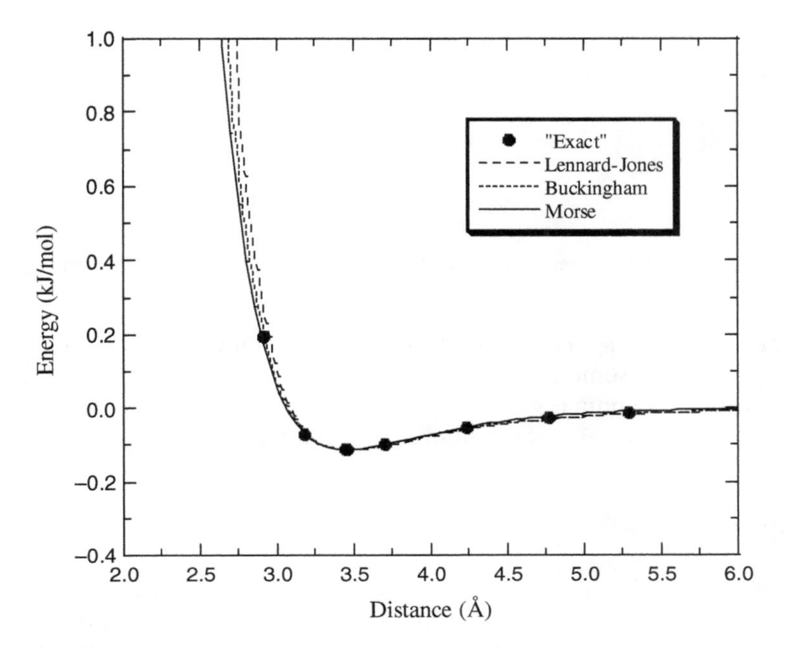

Figure 2.11 Comparison of E_{vdw} functionals for the attractive part of the H_2—He potential.

Part of the better description for the Morse and Buckingham potentials is due to the fact that they have three parameters, while the Lennard-Jones only employs two. Since the equilibrium distance and the well depth fix two constants, there is no additional flexibility in the Lennard-Jones function to fit the form of the repulsive interaction.

Most force fields employ the Lennard-Jones potential, despite the known inferiority to an exponential-type function. Let us examine the reason for this in a little more detail.

Essentially all force field calculations use atomic Cartesian coordinates as the variables in the energy expression. To obtain the distance between two atoms one needs to calculate the quantity shown by

$$R^{AB} = \sqrt{(x_A - x_B)^2 + (y_A - y_B)^2 + (z_A - z_B)^2} \tag{2.18}$$

In the exponential-type potentials, the distance is multiplied by a constant and used as the argument for the exponential. Computationally, it takes significantly more time (typical factor of ~5) to perform mathematical operations such as taking the square root and calculating exponential functions than to do simple multiplication and addition. The Lennard-Jones potential has the advantage that the distance itself is not needed, only R raised to *even* powers, and using square roots and exponential functions is thus avoided. The power of 12 in the repulsive part is chosen as it is simply the square of the power of 6 in the attractive part. Calculating E_{vdw} for an exponential-type potential is computationally more demanding than for the Lennard-Jones potential. For large molecules, the calculation of the non-bonded energy in the force field energy expression is by far the most time-consuming, as discussed in Section 2.7. The difference between the above functional forms is in the repulsive part of E_{vdw}, which is usually not very important. In actual calculations, the Lennard-Jones potential gives results comparable with the more accurate functions, and it is computationally more efficient.

The van der Waals distance, R_0^{AB}, and softness parameters, ε^{AB}, depend on both atom types A and B. These parameters are in essentially all force fields written in terms of parameters for the individual

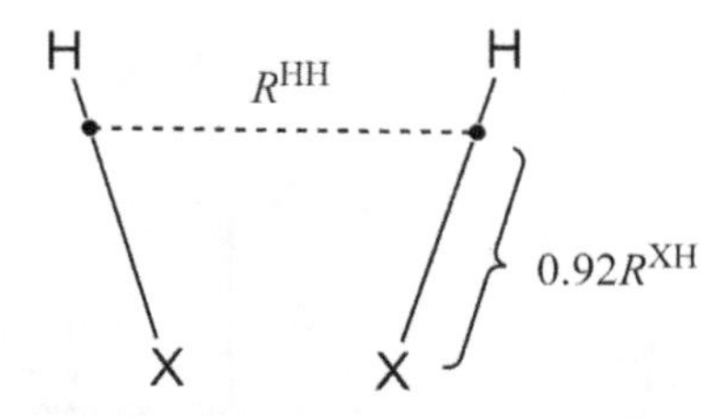

Figure 2.12 Illustration of the distance reduction that can be used for E_{vdw} involving hydrogens.

atom types (for an exception, see reference 22). There are several ways of combining atomic parameters to diatomic parameters, some of them being quite complicated.[18] A commonly used method is to take the van der Waals minimum distance as the sum of two van der Waals radii and the interaction parameter as the geometrical mean of the atomic "softness" constants (employed, for example, by the CHARMM and AMBER force fields):

$$R_0^{AB} = R_0^A + R_0^B \quad ; \quad \varepsilon^{AB} = \sqrt{\varepsilon^A \varepsilon^B} \tag{2.19}$$

In some force fields, especially those using the Lennard-Jones form in Equation (2.12), the R_0^{AB} parameter is defined as the geometrical mean of atomic radii, implicitly via the geometrical mean rule used for the C_1 and C_2 constants (employed, for example, by the OPLS and GROMOS force fields).

For each atom type there are two parameters to be determined, the van der Waals radius and atomic softness, R_0^A and ε^A. It should be noted that since the van der Waals energy is calculated between pairs of atoms, but parameterized against experimental data, the derived parameters represent an *effective* pair potential, which at least partly includes many-body contributions.

The van der Waals energy is the interaction between the electron clouds surrounding the nuclei. In the above treatment, the atoms are assumed to be spherical, but there are two instances where this may not be a good approximation. The first is when one (or both) of the atoms is hydrogen. Hydrogen has only one electron, which always is involved in bonding to the neighbor atom. For this reason the electron distribution around the hydrogen nucleus is not spherical but displaced toward the other atom. One way of modeling this anisotropy is to displace the position, which is used in calculating E_{vdw}, inwards along the bond. MM2 and MM3 use this approach with a scale factor of ~0.92, that is the distance used in calculating E_{vdw} is between points located 0.92 times the X—H bond distance, as shown in Figure 2.12.

The electron density around the hydrogen will also depend significantly on the nature of the X atom. Electronegative atoms such as oxygen or nitrogen will, for example, lead to a smaller effective van der Waals radius for the hydrogen than when it is bonded to carbon. Many force fields therefore have several different types of hydrogen, depending on whether they are bonded to carbon, nitrogen, oxygen, etc., and this may depend further on the type of the neighbor (e.g. alcohol or acid oxygen) (see Table 2.1).

The other case where the spherical approximation may be less than optimal is for atoms having lone pairs, such as oxygen and nitrogen. The lone pair electrons are more diffuse than the electrons involved in bonding, and the atom is thus "larger" in the lone pair direction. Some force fields choose to model lone pairs by assigning *pseudo-atoms* at the lone pair positions. Pseudo-atoms (type 20 in Table 2.1) behave as any other atom type with bond distances and angles, and have their own van der Waals parameters. They are significantly smaller than normal hydrogen atoms and thus make the oxygen or nitrogen atom "bulge" in the lone pair direction. In some cases, sulfur is also assigned lone pairs, although it has been argued that the third row atoms are more spherical owing to the increased number of electrons, and therefore should not need lone pairs.

It is unclear whether it is necessary to include these effects to achieve good models. The effects are small, and it may be that the error introduced by assuming spherical atoms can be absorbed in the other parameters. Introducing lone pair pseudo-atoms, and making special treatment for hydrogens, again make the time-consuming part of the calculation, the non-bonded energy, even more demanding. Most force fields thus neglect these effects.

Hydrogen bonds require special attention. Such bonds are formed between hydrogens attached to electronegative atoms such as oxygen and nitrogen, and lone pairs, especially on oxygen and nitrogen. They have bond strengths of typically 10–20 kJ/mol, where normal single bonds are 250–450 kJ/mol and van der Waals interactions are 0.5–1.0 kJ/mol. The main part of the hydrogen bond energy comes from electrostatic attraction between the positively charged hydrogen and negatively charged heteroatom (see Section 2.2.6). Additional stabilization may be modeled by assigning special deep and short van der Waals interactions (via large ε and small R_0 parameters). This does not mean that the van der Waals radius for a hydrogen bonded to an oxygen is especially short, since this would affect all van der Waals interactions involving this atom type. Only those *pairs* of interactions that are capable of forming hydrogen bonds are identified (by their atoms types) and the normal $E_{\rm vdw}$ parameters are replaced by special "hydrogen bonding" parameters. The functional form of $E_{\rm vdw}$ may also be different. One commonly used function is a modified Lennard-Jones potential of the form shown by

$$E_{\text{H-bond}}(R) = \varepsilon\left[5\left(\frac{R_0}{R}\right)^{12} - 6\left(\frac{R_0}{R}\right)^{10}\right] \tag{2.20}$$

In some cases $E_{\text{H-bond}}$ also includes a directional term such as $(1 - \cos\omega^{\rm XHY})$ or $(\cos\omega^{\rm XHY})^4$ multiplied with the distance-dependent part in Equation (2.20). The current trend seems to be that force fields are moving away from such specialized parameters and/or functional forms, and instead are accounting for hydrogen bonding purely by electrostatic interactions. The fixed partial atomic charge model described in the next section, however, is in general incapable of modeling the directionality, but it can be done using non-nuclear-centered charges or atomic multipoles.

2.2.6 The Electrostatic Energy: Atomic Charges

The other part of the non-bonded interaction is due to internal (re)distribution of the electrons, creating positive and negative parts of the molecule.[23] A carbonyl group, for example, has a negatively charged oxygen and a positively charged carbon. At the lowest approximation, this can be modeled by assigning (partial) charges to each atom. Alternatively, the bond may be assigned a bond dipole moment. These two descriptions give similar (but not identical) results. Only in the long-distance limit of interaction between such molecules do the two descriptions give identical results.

The interaction between point charges is given by the Coulomb potential, with ε being a dielectric constant:

$$E_{\rm el}(R^{\rm AB}) = \frac{Q^{\rm A}Q^{\rm B}}{\varepsilon R^{\rm AB}} \tag{2.21}$$

The atomic charges can be assigned by empirical rules,[24] but are more commonly assigned by fitting to the electrostatic potential calculated by electronic structure methods, as discussed in the next section. Since hydrogen bonding to a large extent is due to attraction between the electron-deficient

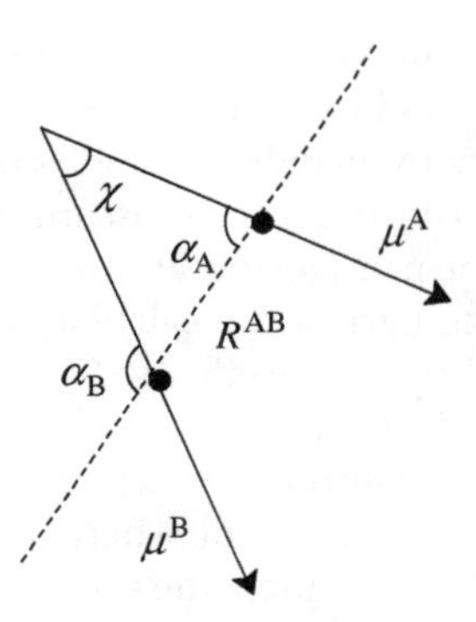

Figure 2.13 Definition of variables for a dipole–dipole interaction.

hydrogen and an electronegative atom such as oxygen or nitrogen, a proper choice of partial charges may adequately model this interaction.

The MM2 and MM3 force fields use a bond dipole description for E_{el}. The interaction between two dipoles is given by

$$E_{el}(R^{AB}) = \frac{\mu^A \mu^B}{\varepsilon (R^{AB})^3}(\cos\chi - 3\cos\alpha_A \cos\alpha_B) \tag{2.22}$$

The angles χ, α_A and α_B are defined as shown in Figure 2.13.

When properly parameterized, there is little difference in the performance of the two ways of representing E_{el}. There are exceptions where two strong bond dipoles are immediate neighbors (e.g. α-halogen ketones). The dipole model will here lead to a stabilizing electrostatic interaction for a *transoid* configuration (torsional angle of 180°), while the atomic charge model will be purely repulsive for all torsional angles (since all 1, 3-interactions are neglected). In either case, however, a proper rotational profile may be obtained by suitable choices of the constants in E_{tors} to compensate for incorrect behaviors in the electrostatic energy. The atomic charge model is easier to parameterize by fitting to an electronic wave function, and is preferred by almost all force fields.

The "effective" dielectric constant ε can be included to model the effect of surrounding molecules (solvent) and the fact that interactions between distant sites may be "through" part of the same molecule, that is a polarization effect. A value of 1 for ε corresponds to a vacuum, while a large ε reduces the importance of long-range charge–charge interactions. Often a value between 2 and 5 is used, although there is little theoretical justification for any specific value.[25] In some applications the dielectric constant is made distance-dependent (e.g. $\varepsilon = \varepsilon_0 R^{AB}$, changing the Coulomb interaction to $Q^A Q^B/\varepsilon_0 (R^{AB})^2$) to model the "screening" by solvent molecules. There is little theoretical justification for this, but it increases the efficiency of the calculation as a square root operation is avoided (discussed in Section 2.2.5), and it seems to provide reasonable results.

How far apart (in terms of number of bonds between them) should two atoms be before a non-bonded energy term contributes to E_{FF}? It is clear that two atoms directly bonded should not have an E_{vdw} or E_{el} term – their interaction is described by E_{str}. It is also clear that the interaction between two hydrogens at each end of, say, $CH_3(CH_2)_{50}CH_3$ is identical to the interaction between two hydrogens belonging to two different molecules, and they therefore should have an E_{vdw} and an E_{el} term. But where should the dividing line be? Most force fields included E_{vdw} and E_{el} for atom pairs that are separated by three bonds or more, although 1,4-interactions in many cases are scaled down by a factor between 1 and 2. This means that the rotational profile for an A—B—C—D sequence is determined both by E_{tors} and E_{vdw} and E_{el} terms for the A—D pair. In a sense, E_{tors} may be considered as a correction necessary for obtaining the correct rotational profile once the non-bonded contribution has been accounted for. Some force fields have chosen also to include a non-bonded energy term for atoms that are 1,3 with respect to each other, and these are called *Urey–Bradley* terms. In this case, the energy required to bend a three-atom sequence is a mixture of E_{bend} and the Urey–Bradley energy, where the latter serves to describe the stretch-bend coupling discussed in Section 2.2.9. Most modern force fields calculate E_{str} between all atoms pairs that are 1,2 with respect to each other in terms of bonding, E_{bend} for all pairs that are 1,3, E_{tors} between all pairs that are 1,4, and E_{vdw} and E_{el} between all pairs that are 1,4 or higher.

The electrostatic energy dominates the force field energy function for polar molecules, and an accurate representation is therefore important for obtaining good results. Within the partial charge model, the atomic charges are normally assigned by fitting to the molecular *ElectroStatic Potential* (ESP) calculated by an electronic structure method. The electrostatic potential ϕ_{esp} at a point $\mathbf{r}$ is given by the nuclear charges and electron density as shown in

$$\phi_{esp}(\mathbf{r}) = \sum_{A}^{N_{nuc}} \frac{Z_A}{|\mathbf{R}_A - \mathbf{r}|} - \int \frac{\rho(\mathbf{r}')}{|\mathbf{r}' - \mathbf{r}|} d\mathbf{r}' \tag{2.23}$$

The fitting is done by minimizing an error function of the form shown in

$$ErrF(\mathbf{Q}) = \frac{1}{N_{points}} \sum_{\mathbf{r}}^{N_{points}} \left(\phi_{esp}(\mathbf{r}) - \sum_{A}^{N_{atoms}} \frac{Q^A(\mathbf{R}_A)}{|\mathbf{R}_A - \mathbf{r}|} \right)^2 \tag{2.24}$$

under the constraint that the sum of the partial charges Q^A is equal to the total molecular charge. The electrostatic potential is typically sampled at a few thousand points in the near vicinity of the molecule, which is discussed in more detail in Section 10.2. The set of linear equations arising from minimizing the error function is often poorly conditioned, that is the calculated partial charges are sensitive to small details in the fitting data.[26] The physical reason for this is that the electrostatic potential is primarily determined by the atoms near the surface of the molecule, while the atoms buried within the molecule have very little influence on the external electrostatic potential. A straightforward fitting therefore often results in unrealistically large charges for the non-surface atoms. The problem can to some extent be avoided by adding a hyperbolic penalty term for having non-zero partial charges, since this ensures that only those charges that are important for the electrostatic potential have values significantly different from zero.[27] This *Restrained ElectroStatic Potential* (RESP) fitting scheme has been used in. for example, the AMBER force field. Other constraints are also often imposed, such as constraining the charges on the three hydrogens in a methyl group to be equal or the sum of all charges in a subgroup (such as a methyl group or an amino acid) to be zero.

Several papers have discussed how the sampling points should be selected, including surface-based,[28] grid-based,[29] randomly selected[30] and Boltzmann weighted by their van der Waals energy.[31] Tsiper and Burke have shown that the ESP on an isodensity surface that includes essentially all the electron density *uniquely* determines the ESP at *all* points outside the surface.[32] This implies that the exact procedure for selecting sampling points is largely irrelevant as long as a sufficiently dense set of points is selected, and different fitting schemes thus primarily differ in terms of an incomplete sampling and associated numerical instabilities.

One might anticipate that atomic charges based on fitting to the electrostatic potential would lead to well-defined values. This, however, is not the case. Besides the already mentioned dependence on the (incomplete) sampling points, another problem is that a straightforward fitting tends to give conformationally dependent charges.[26] The three hydrogens in a freely rotating methyl group, for example, may end up having significantly different charges, or two conformations may give two significantly different sets of fitted parameters. This is a problem in connection with force field methods that rely on the fundamental assumption that parameters are transferable between similar fragments, and atoms that are easily interchanged (e.g. by bond rotation) should have identical parameters. Conformationally dependent charges can be modeled in force field methods by fluctuating charge or polarization models (Section 2.2.8), but this leads to significantly more complicated force fields, and consequently loss of computational efficiency.

One way of eliminating the problem with conformationally dependent charges is to add additional constraints, for example forcing the three hydrogens in a methyl group to have identical charges[27] or averaging over different conformations.[33] A more fundamental problem is that the fitting procedure becomes statistically underdetermined for large systems, although the severity of this depends on how the fitting is done.[34,35] The difference between the true electrostatic potential and that generated by a set of atomic charges on, say, $^3/_4$ of the atoms is not significantly reduced by having fitting parameters on all atoms.[36] The electrostatic potential experienced outside the molecule is mainly determined by the atoms near the surface, and consequently the charges on atoms buried within a molecule cannot be assigned with any great confidence. Even for a medium-sized molecule, it may only be statistically valid to assign charges to perhaps half the nuclei. Having a full set of atomic charges thus forms a redundant set: many different sets of charges may be chosen, all of which are capable of reproducing the true electrostatic potential to almost the same accuracy. Although a very large number of sampling points (several thousand) may be chosen to be fitted by relatively few (perhaps 20–30) parameters, the fact that the sampling points are highly correlated makes the problem underdetermined. In practical applications, additional penalty terms are therefore often added, to ensure that only those atoms that contribute significantly in a statistical sense acquire non-zero charges.[27]

Another problem with atomic charges determined by fitting is related to the absolute accuracy. Although inclusion of charges on all atoms does not significantly improve the results over that determined from a reduced set of parameters, the *absolute* deviation between the true and fitted electrostatic potentials can be quite large. Interaction energies as calculated by an atom-centered charge model in a force field may be off by several kJ/mol per atom in certain regions of space just outside the molecular surface, an error of one or two orders of magnitude larger than the van der Waals interaction. In order to improve the description of the electrostatic interaction, additional non-nuclear-centered charges may be added,[37] or dipole, quadrupole, etc., moments may be added at nuclear or bond positions.[38] These descriptions are essentially equivalent since a dipole may be generated as two oppositely charged monopoles, a quadrupole as four monopoles, etc. The parameter redundancy problem increases significantly when adding atomic dipole and quadrupole moments, and $\sim ^1/_2$ of the parameters can be removed without affecting the ability to reproduce the electrostatic potential.[36]

The non-bonded energies are modeled by pair-interactions, but if the parameters are obtained by fitting to experimental data, they will include the average part of many-body effects. The most important is the two-body polarization, as discussed in the next section. Partial charges obtained by fitting to ESP do not include this effect, and will therefore lead to a systematic underestimation of the electrostatic energy. The three-body effect in the electrostatic energy may be considered as the interaction between two atomic charges being modified because a third atom or molecule polarizes the charges. The dipole moment of water, for example, increases from 1.8 Debye in the gas phase to an effective value of 2.5 Debye in the solid state.[39] The average polarization effect in condensed phases can be partly modeled by artificially increasing the partial charges by 10–20%, or by fitting to data that are known to overestimate the polarity (e.g. Hartree–Fock results, Section 4.3).

2.2.7 The Electrostatic Energy: Atomic Multipoles

Obtaining a good description of the electrostatic interaction between molecules (or between different parts of the same molecule) is one of the big challenges in force field developments.[23] Many commercial applications of force field methods are aimed at designing molecules that interact in a specific fashion. Such interactions are usually pure non-bonded, and for polar molecules such as proteins, the electrostatic interaction is very important.

Modeling the electrostatic energy by (fixed) partial atomic charges has five main deficiencies:

1. The fitting of atomic charges to electrostatic potentials focuses on reproducing *inter*molecular interactions, but the electrostatic energy also plays a strong role in the *intra*molecular energy, which determines conformational energies. For polar molecules the (relative) conformational energies are therefore often of significantly lower accuracy than for non-polar systems.
2. The partial charge model gives a rather crude representation of the electrostatic potential surrounding a molecule, with errors often being in the 10–20 kJ/mol range. For a given (fixed) geometry, the molecular electrostatic potential can be improved either by adding non-nuclear-centered partial charges or by including higher-order (dipole, quadrupole, etc.) atomic electric moments.
3. Charge penetration effects are neglected. This effect arises because two atoms at a typical van der Waals distance apart penetrate each others electron clouds and this leads to an attractive interaction. This cannot be described by a fixed partial charge model and the effect is instead included implicitly in the van der Waals energy term. An explicit description requires a screened partial charge model.[40]
4. The coupling of atomic charges and higher-order moments with the geometry is neglected. Analysis has shown that both partial atomic charges and higher-order electric moments depend significantly on the geometry, that is these quantities do not fulfill the requirement of "transferability".
5. Only two-body interactions are included, but for polar species the three-body contribution is quite significant, perhaps 10–20% of the two-body term.[41] A rigorous modeling of these effects requires inclusion of polarizability, but can be partly included in the two-body interaction by empirically increasing the interaction by 10–20%, although this leads to a systematic bias, as illustrated below.

The non-bonded terms together with the torsional energy determine the internal (conformational) degrees of freedom, and the torsional parameters are usually assigned as a residual correction to reproduce rotational barriers and energetic preferences after the assignment of the non-bonded parameters. The electrostatic energy is unimportant for non-polar systems such as hydrocarbons and is in these cases often completely neglected. The conformational space in such cases is consequently determined by E_{vdw} and E_{tors}. Since the van der Waals interaction is short-ranged, this means that the transferability assumption inherent in force field methods is valid, and torsional parameters determined for small model systems can also be used for predicting conformations for large systems.

For polar systems, however, the long-range electrostatic interaction is often the dominating energy term, and the transferability of torsional parameters determined for small model systems becomes more problematic. Clearly the variation of the electrostatic energy with the geometry must be accurately modeled in order for the torsional parameters to be sufficiently transferable, and this is difficult to achieve without including polarization.[42]

The representation of the electrostatic potential for a fixed geometry can be systematically improved by including non-atom-centered charges[43] or by including higher-order moments.[44] The *Distributed Multipole Analysis* (DMA) developed by A. Stone provides an exact method for expanding the electrostatic potential in terms of multipole moments distributed at a number of positions within the molecule, and these moments can be derived directly from the wave function without a fitting procedure (see Section 10.2).[45] By restricting the positions to only atoms, a quite accurate representation of the electrostatic potential can be obtained by including up to quadrupole moments at each site. The DMA-derived multipoles are sensitive to the reference data (i.e. wave function quality and basis set), and a better stability can be obtained by fitting a set of multipoles to the electrostatic potential[32] or by fitting atomic charges to match the atomic multipole moments.[46] Such fitted multipole methods typically reduce the required moments by one or two, that is fitted charges and dipoles can reproduce DMA results including up to quadrupoles or octupoles. The underdetermined nature of the fitting process (Equation (2.24)), however, becomes even more pronounced when atomic dipole, quadrupole, etc., moments are included as fitting parameters. Typically only ~$^1/_2$ of the atomic charges can be assigned statistically valid values for a medium-sized molecule, and even less of the higher-order moments can be considered non-redundant.

The interaction energy between atomic multipoles at positions $\mathbf{R}_A$ and $\mathbf{R}_B$ can conveniently be written in a polytensor form:[47]

$$E_{\text{el}}^{AB} = \mathbf{M}_A^{\text{t}} \mathbf{T}_{AB} \mathbf{M}_B \tag{2.25}$$

where $\mathbf{M}_A$ and $\mathbf{M}_B$ are vectors containing all the components of the electric moments and $\mathbf{T}_{AB}$ is a matrix describing the interaction between them, which only depends on the distance $R^{BA} = |\mathbf{R}_B - \mathbf{R}_A|$:

$$\mathbf{M}_A = \left(Q^A, \mu_x^A, \mu_y^A, \mu_z^A, \Theta_{xx}^A, \Theta_{xy}^A, \Theta_{xz}^A, \Theta_{yy}^A, \Theta_{yz}^A, \Theta_{zz}^A, \ldots \right)^{\text{t}} \tag{2.26}$$

$$\mathbf{T}_{AB} = \begin{pmatrix} 1 & \frac{\partial}{\partial x_B} & \frac{\partial}{\partial y_B} & \frac{\partial}{\partial z_B} & \cdots \\ \frac{\partial}{\partial x_A} & \frac{\partial^2}{\partial x_A \partial x_B} & \frac{\partial^2}{\partial x_A \partial y_B} & \frac{\partial^2}{\partial x_A \partial z_B} & \cdots \\ \frac{\partial}{\partial y_A} & \frac{\partial^2}{\partial y_A \partial x_B} & \frac{\partial^2}{\partial y_A \partial y_B} & \frac{\partial^2}{\partial y_A \partial z_B} & \cdots \\ \frac{\partial}{\partial z_A} & \frac{\partial^2}{\partial z_A \partial x_B} & \frac{\partial^2}{\partial z_A \partial y_B} & \frac{\partial^2}{\partial z_A \partial z_B} & \cdots \\ \vdots & \vdots & \vdots & \vdots & \ddots \end{pmatrix} \frac{1}{R^{BA}} \tag{2.27}$$

Inclusion of higher-order electric moments will improve the representation of the electrostatic potential of a given molecule, but it does not taken the geometry dependence into account, which is the topic of the next section.

2.2.8 The Electrostatic Energy: Polarizability and Charge Penetration Effects

The interaction between two polar molecules (or parts of the same molecule) can lead to either stabilization or destabilization depending on the orientation, as illustrated in Figure 2.14 for two dipolar molecules.

Inclusion of polarization will always lead to stabilization relative to the unpolarized situation, by increasing the dipole moments in the energetically favored orientation and by reducing the dipole moments in the energetically disfavored orientation. The stabilization of the energetically favored orientation can be modeled by enhancing the static dipole moments, but this leads to an incorrect destabilization of the energetically disfavored orientation. The common practice in fixed charge force fields of increasing partial atomic charges by 10–20% relative to the values obtained by fitting

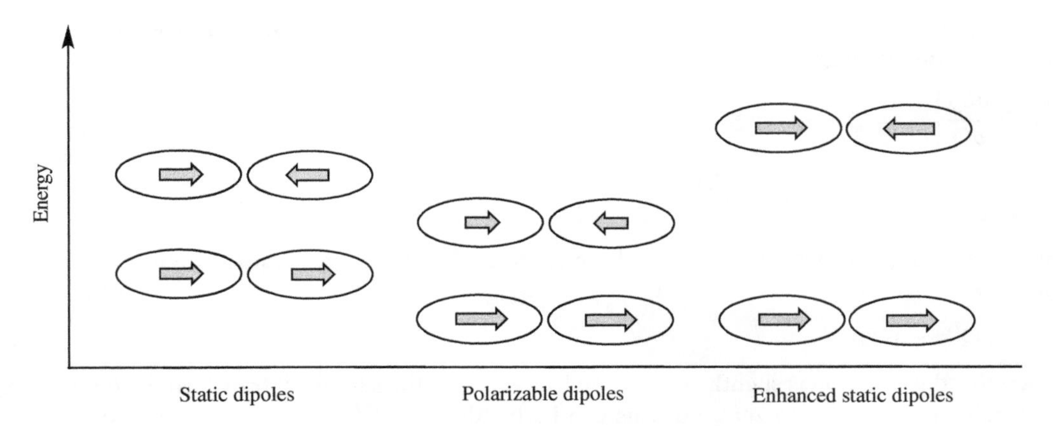

Figure 2.14 Interactions between static and polarizable dipolar molecules.

to gas-phase values (or fitting to data that overestimates the gas-phase polarity) corresponds to the "enhanced static dipole" part of Figure 2.14. The claim, that empirically increasing the partial atomic charges models the average polarization, is thus only correct for the energetically favorable orientations, and leads to a systematic bias against energetically disfavored orientations. The polarization energy is furthermore non-linear (quadratic in the lowest order) in the field strength, while the energy due to static enhanced dipoles is linear in the field strength, and the partial atomic charges model therefore also implicitly includes an average over field strengths.

The coupling of the electrostatic energy with the geometry requires explicit inclusion of electronic polarization, which can be modeled in several different ways, of which the *fluctuating charge* (FQ), the *Drude Oscillator* (DO) or *Charge-On-Spring* (COS), and induced *Point Dipole* (PD) models are the most popular approaches.[48] To differentiate between force fields that incorporate polarization and those that do not, the latter are sometimes denoted *additive force fields.*

In the *fluctuating charge* (FQ) model, the atomic charges are allowed to adjust to changes in the geometry based on electronegativity equalization.[49–51] Consider the following expansion of the energy as a function of the number of electrons N:

$$E = E_0 + \frac{\partial E}{\partial N}\Delta N + \frac{1}{2}\frac{\partial^2 E}{\partial N^2}(\Delta N)^2 + \ldots \tag{2.28}$$

The first derivative is the *electronegativity* χ, except for a change in sign ($\partial E/\partial N = -\chi$), while the second derivative is the *hardness* η. Although these have well-defined finite-value approximations in terms of ionization potentials and electron affinities (Section 16.2), they are usually treated as empirical parameters within a force field environment. For an atom in a molecule, the change in the number of electrons is equal to minus the change in the atomic charge ($-\Delta N = \Delta Q$). Taking the expansion point as the one with no atomic charges gives

$$E = \chi Q + \tfrac{1}{2}\eta Q^2 + \cdots \tag{2.29}$$

Terminating the expansion at second order, adding a term corresponding to the interaction of the charges with an external potential ϕ and summing over all sites gives an expression for the electrostatic energy:

$$E_{\text{el}} = \sum_A \phi_A Q^A + \sum_A \chi_A Q^A + \tfrac{1}{2}\sum_{AB} \eta_{AB} Q^A Q^B \tag{2.30}$$

Switching to a vector–matrix notation and requiring that the energy is stationary with and without an external potential gives

$$\left.\frac{\partial E_{\text{el}}}{\partial \mathbf{Q}}\right|_{\phi\neq 0} = \phi + \chi + \boldsymbol{\eta}\mathbf{Q} = 0$$
$$\left.\frac{\partial E_{\text{el}}}{\partial \mathbf{Q}}\right|_{\phi=0} = \chi + \boldsymbol{\eta}\mathbf{Q}^0 = 0 \tag{2.31}$$

Subtraction of these two equations leads to a recipe for the charge transfer due to the external potential

$$\Delta\mathbf{Q} = -\boldsymbol{\eta}^{-1}\phi \tag{2.32}$$

In practice the situation is slightly more complicated, since the sum of all transferred atomic charges must be constrained to be zero, but this can be handled by the Lagrange technique described in Section 13.5. Since the potential depends on the charges at all sites (Equation (2.33)), this must be solved self-consistently by iterative methods (see Equation (2.36) below):

$$\phi(\mathbf{r}) = \sum_A \frac{Q^A}{|\mathbf{r} - \mathbf{R}_A|} \tag{2.33}$$

Once the iterations have converged, the electrostatic energy is given by the simple Coulomb form in Equation (2.21). The diagonal elements of the hardness matrix can be associated with the atomic hardness parameters, while the off-diagonal elements introduce coupling between all pairs of atoms. In the original model these couplings where taken as simple Coulomb interactions, but this leads to an unphysical situation where atoms can transfer a fractional number of electrons at infinite distance. In more modern versions the coupling elements are replaced by screened Coulomb-type interactions to ensure that bond dissociation leads to uncharged fragments. Note that the FQ model treats through-bond and through-space polarization equivalently and is thus able to describe intermolecular charge transfer.

In the *Drude Oscillator* (DO) or *Charge-On-Spring* (COS) model, the atomic polarizability is modeled by transferring part of the atomic charge to a virtual site (Drude particle) connected to the atom by a harmonic spring,[52] similar to the use of pseudo-atoms for modeling lone pairs (Section 2.2.5). The polarizability α is related to the Drude particle charge Q_{D} and force constant k_{D} by

$$\alpha = \frac{Q_{\text{D}}^2}{k_{\text{D}}} \tag{2.34}$$

The force constant must be chosen large enough that the Drude particle always is much closer to the center of the atom ("nucleus") than the atomic dimension, thus representing an atomic dipole polarizability.[53] This in turn necessitates a large Drude charge, often several units of atomic charge. The Drude particles may be taken as massless, in which case their positions must be completely relaxed before calculating the atomic forces in an optimization or simulation. Alternatively, they can be assigned a small finite mass taken from the atom and their positions updated by extended Lagrangian methods (Section 15.2.5) for simulation purposes. This, as usual, requires two thermostats in order to keep the temperature of the Drude particles much smaller than the temperature of the atom. Normally only non-hydrogen atoms are assigned a Drude particle.

The induced *Point Dipole* (PD) model describes polarization by an atomic polarization tensor ($\boldsymbol{\alpha}$), which leads to an induced dipole moment ($\boldsymbol{\mu}_{\text{ind}}$) arising from the electric field ($\mathbf{F} = -\partial\phi/\partial\mathbf{r}$) created by the electric moments at other sites.[54] The stabilization energy due to the polarization is half the induced dipole moment times the field, where the sign is chosen such that an energy lowering corresponds to a negative polarization energy:

$$\begin{aligned} \mu_{ind} &= \alpha \mathbf{F} \\ E_{el}^{pol} &= -\tfrac{1}{2}\mu_{ind}\mathbf{F} = -\tfrac{1}{2}\alpha \mathbf{F}^2 \end{aligned} \quad (2.35)$$

Since the (atomic) hardness is inversely related to the average polarizability,[54] the charge transfer in Equation (2.32) is essentially the average polarizability times the potential. The FQ model can thus be considered as a polarizable monopole equivalent of the polarizable dipole model in Equation (2.35).

The induced dipole moment in Equation (2.35) depends on the total electric field **F**, which has static and induced components. The static field contains contributions from permanent multipole moments on all (other) atoms, while the induced field arises from the induced dipole moments on all other atoms:

$$\mu_{ind} = \alpha \mathbf{F}_{tot} = \alpha(\mathbf{F}_{static} + \mathbf{F}_{ind}) = \alpha(\mathbf{F}_{static} + \mathbf{T}^{Dipole}\mu_{ind}) \quad (2.36)$$

The dipole $\mathbf{T}^{Dipole}$ matrix elements are defined in Equation (2.27), and Equation (2.36) can be solved by combining the $\mathbf{T}^{Dipole}$ matrix with the atomic polarizabilities to form a *relay matrix* **R**:

$$\mu_{ind} = (\alpha^{-1} - \mathbf{T}^{Dipole})^{-1}\mathbf{F}_{static} = \mathbf{R}\mathbf{F}_{static} \quad (2.37)$$

The induced atomic dipoles can thus be calculated directly by matrix inversion, but since the **R** matrix has dimension of the order of the number of atoms in the system, this is computationally expensive even for medium-sized systems. Alternatively, the induced dipole moment can be calculated by iterating Equation (2.36) until self-consistency with a suitable initial guess. The self-consistent determination of atomic charges in the FQ model (Equations (2.32) and (2.33)) can be done completely analogously. For molecular dynamics simulations, the change in the induced dipoles or charges between each time step can be done by solving Equation (2.36) before the calculation of the new atomic forces, but can alternatively be treated by an extended Lagrange method (Section 15.2.5), where fictive masses are assigned to the dipoles or charges, and evolved along with the other variables in a simulation.[24] The self-consistent determination of the induced dipole moments in Equation (2.36) must be converged tightly when used for calculating forces for simulations, and this can be computationally expensive. Alternative versions where the relay matrix is approximated as simply the polarizability (denoted "direct" since this neglects the couplings between induced moments and thus avoids iterations) or estimated by perturbation theory from two iterations have also been proposed.[55]

The induced PD model in Equation (2.36) leads to infinite polarization for a distance of $(4\alpha_A\alpha_B)^{1/6}$ between two atoms with polarizabilities α_A and α_B, and this distance is often comparable to the sum of the two van der Waals radii. Thole suggested[56] that this unphysical situation could be avoided by introducing a short-range damping function, which corresponds to multiplying the dipole **T** matrix elements in Equation (2.27) by a damping function. Several choices are possible for the damping function with one example shown in Equation (2.38), which implies similar derived damping factors for monopole and higher-order multipoles:[57]

$$\begin{aligned} \mathbf{T}_{AB}^{Damp-Dipole} &= (1 - e^{-au^3})\mathbf{T}_{AB}^{Dipole} \\ u &= \frac{R^{AB}}{(\alpha_A\alpha_B)^{1/6}} \end{aligned} \quad (2.38)$$

The a parameter is a free variable controlling the damping.

It may be illustrative to put the above three approaches for introducing electronic polarization into force fields by the quantum mechanical treatment for polarization at the molecular level, discussed in

Section 11.1.1. The total energy of a (continuous) charge distribution subjected to an external electric potential ϕ can be written as a Taylor expansion:

$$E = Q\phi - \boldsymbol{\mu}\mathbf{F} - \tfrac{1}{2}\boldsymbol{\Theta}\mathbf{F}' - \cdots \tag{2.39}$$

Here Q is the total (molecular) charge, $\boldsymbol{\mu}$ is the dipole moment, $\boldsymbol{\Theta}$ is the quadrupole moment, etc., $\mathbf{F}$ is the electric field $(-\partial\phi/\partial\mathbf{r})$ and $\mathbf{F}'$ is the field gradient $(\partial\mathbf{F}/\partial\mathbf{r})$, etc. The dipole moment consists of a permanent term $\boldsymbol{\mu}_0$ (value in the absence of an external field) and induced components depending linearly, quadratically, etc., on the field strength, where $\boldsymbol{\alpha}$ and $\boldsymbol{\beta}$ are the polarizability and hyperpolarizability tensors, respectively:

$$\boldsymbol{\mu} = \boldsymbol{\mu}_0 + \boldsymbol{\alpha}\mathbf{F} + \tfrac{1}{2}\boldsymbol{\beta}\mathbf{F}^2 + \cdots \tag{2.40}$$

The quadrupole (and higher-order moments) can similarly be written as a Taylor expansion in terms of a static and field gradient induced moments. The FQ method can be considered as a polarization model where only monopole moments are considered and only including terms up to quadratic in the potential. Taking also mixed dipole–quadrupole polarization terms into account allows the energy to be written as in the following equation in terms of atomic distributed electric moments:

$$E = Q_0\phi - \boldsymbol{\mu}_0\mathbf{F} - \tfrac{1}{2}\boldsymbol{\Theta}_0\mathbf{F}' - \cdots - \eta^{-1}\phi^2 - \cdots - \tfrac{1}{2}\boldsymbol{\alpha}\mathbf{F}^2 - \cdots - \mathbf{A}\mathbf{F}\mathbf{F}' - \mathbf{B}\mathbf{F}^2\mathbf{F}' - \mathbf{C}(\mathbf{F}')^2 - \cdots \tag{2.41}$$

Here $\mathbf{A}$ is the mixed dipole–quadrupole polarizability tensor, $\mathbf{B}$ is the mixed dipole–quadrupole hyperpolarizability tensor and $\mathbf{C}$ is the quadrupole polarizability tensor, with implied tensor contraction of all terms. Equation (2.41) represents a systematic way of improving the electrostatic energy in a force field description by adding a sequence of higher-order static multipole moments and a sequence of multipole polarizabilities and hyperpolarizabilities at (all) atomic sites. The FQ method corresponds to including the first and fourth terms on the right-hand side of Equation (2.41). The PD method includes a polarization of the dipole moments and is usually combined with static multipoles up to quadrupole moments, which corresponds to the first three and the fifth terms on the right-hand side of Equation (2.41). The DO method can be considered as an approximate model for describing dipole polarization, and where only static monopole moments are included, that is the first term on the right-hand side of Equation (2.41) and an approximation of the fifth term.

The FQ method is the simplest way of introducing a coupling between the electrostatic energy and geometry, but it is, for example, unable to account for the charge polarization of a planar molecule when the external field is perpendicular to the molecule plane. This can be described by the DO method, but it is limited to describing polarization effects, which occur in the same direction as the external field. The same limitation is present in the PD method if the polarizability tensor is taken to be isotropic (i.e. only diagonal elements of equal magnitudes), but a full (anisotropic) tensor can also describe induced dipoles that are not along the external field direction.

A decision has to be made on which interactions to include and which to neglect when adding multipole and/or polarizability terms, as indicted by the expansion in Equation (2.41). The distance dependence on the interactions between multipoles is given in Table 2.2.

A model including up to quadrupole moments accounts for all interactions having up to R^{-3} distance dependencies, and in addition includes some of those having R^{-4} and R^{-5} distance dependencies. The charge-induced dipole interaction has a distance dependence of R^{-4}, while the dipole-induced dipole interaction is R^{-6}, suggesting that the former should be included when quadrupole moments are incorporated. Current wisdom suggests that the contributions from dipole hyperpolarizability and quadrupole polarizabilities (and mixed dipole–quadrupole terms) are at least an order of

Table 2.2 Distance dependence of multipole interactions.

	Q	μ	Θ	Ξ
Q	R^{-1}	R^{-2}	R^{-3}	R^{-4}
μ	R^{-2}	R^{-3}	R^{-4}	R^{-5}
Θ	R^{-3}	R^{-4}	R^{-5}	R^{-6}
Ξ	R^{-4}	R^{-5}	R^{-6}	R^{-7}

magnitude smaller than the dipole polarization for electric field strengths generated by the interaction of molecules. The most advanced general-purpose force fields thus employ up to (static) quadrupole moments on all atomic sites in addition to an atomic point dipole polarizability. Combinations of FQ and PD have also been proposed.[58]

Incorporation of electric multipole moments, fluctuating charges and atomic polarizabilities significantly increases the number of fitting parameters for each atom type or functional unit, and only electronic structure methods are capable of providing a sufficient number of reference data. Electronic structure calculations, however, automatically include all of the above effects, and also have higher-order terms. The data must therefore be "deconvoluted" in order to extract suitable multipole and polarization parameters for use in force fields.[59] A calculated set of distributed dipole moments, for example, must be decomposed into permanent and induced contributions, based on an assigned polarizability tensor. Furthermore, only the lowest non-vanishing multipole moment is independent of the origin of the coordinate system; that is for a non-centrosymmetric neutral molecule the dipole moment is unique, but the quadrupole moment depends on where the origin is placed.

Atomic charges and multipole moments are classical representations of the electrostatic potential generated by the electron distribution in a molecule, which can be calculated by quantum mechanical methods. The electron distribution is a continuous function that decays exponentially as a function of the distance to the nucleus. The van der Waals surface of a molecule is closely associated with an electronic isodensity surface with a value typically taken to be $\sim 4 \times 10^{-4}$. The closest approach between molecules during a simulation may correspond to a distance where the electron density has a value of $\sim 10^{-3}$. Although these are small values, it should be recognized that molecules at or below the van der Waals distance penetrate their mutual electron densities to a significant extent, and this leads to a net attraction. This charge penetration effect is a major component of so-called π–π stacking effects and the interaction between cations and aromatic systems. Piquemal *et al.* have suggested that the charge penetration effect can be modeled by an exponential screening factor for a fraction of the charge corresponding to the effective number of valence electrons N_{eff}. The electrostatic potential for an atom with a net charge of Q and $N_{\text{eff}} = N_{\text{val}} - Q$ can thus be written as in the following equation, where the screening parameter α has a typical value of ~ 3:[40]

$$\phi_{\text{ESP}}(\mathbf{r}) = \frac{1}{|\mathbf{R} - \mathbf{r}|} \left[Q + N_{\text{eff}} e^{-\alpha|\mathbf{R}-\mathbf{r}|} \right] \tag{2.42}$$

Other screening functions have also been proposed, as well as screening of higher-order moments.[60] It should be noted that standard force fields attempt to account for the charge penetration effect implicitly by the van der Waals energy term.

The addition of multipole moments increases the computational time for the electrostatic energy,[23] since there now are several components for each pair of sites, and for multipoles up to quadrupoles the evaluation time increases by almost an order of magnitude. It should be noted that addition of atomic multipoles is computationally more efficient than addition of (many) non-nuclear-centered charges, since the former has short-ranged interactions while the latter is long-ranged (Table 2.2). Inclusion of polarization further increases the computational complexity by adding an iterative procedure for evaluating the induced charges or dipole moments, although algorithmic advances have significantly reduced the computational overhead relative to a fixed charge model.[55] The DO model has the computational advantage that only existing methodologies for calculating charge–charge interactions are required, and the computational overhead is therefore limited to the increased number of particles, typically ~50% extra since only non-hydrogen atoms carry Drude particles, and the requirement of slightly smaller time step in simulations due to the low-mass Drude particles. These factors typically increase the computational time by a factor of ~four compared to a fixed charge force field.[61]

Advanced force fields with a description beyond fixed partial charges of the electrostatic energy are still at the experimental stage, with extensive testing and parameterization. It is clear that many physical important interactions require explicit consideration of higher-order multipoles, and polarization and charge penetration effects, as, for example, the description of halogen bonds and interactions between cations and π-systems of aromatic compounds. Several studies comparing the performance of polarized and fixed charge force fields have in many cases shown disappointing small improvements or even deterioration of the results.[61] These results are often related to the distribution of a macromolecule in the conformational phase space, and it should be realized that the energetics of the conformations space is a delicate balance of the electrostatic, van der Waals and torsional energy terms. Fixed charge force fields have been extensively reparameterized over several decades, and errors in one term, like electrostatic, can to some extent be absorbed in other terms, like van der Waals and torsional. Selectively improving the electrostatic term partly destroys the error cancellation, and thus is likely to require a complete reparameterization of the entire force field in order to achieve a systematic improvement.

2.2.9 Cross Terms

The first five terms in the general energy expression, Equation (2.1), are common to all force fields. The last term, E_{cross}, covers coupling between these fundamental, or diagonal, terms. Consider, for example, a molecule such as H_2O. It has an equilibrium angle of 104.5° and an O—H distance of 0.958 Å. If the angle is compressed to, say, 90°, and the optimal bond length is determined by electronic structure calculations, the equilibrium distance becomes 0.968 Å, that is slightly longer. Similarly, if the angle is widened, the lowest energy bond length becomes shorter than 0.958 Å. This may qualitatively be understood by noting that the hydrogens come closer together if the angle is reduced. This leads to an increased repulsion between the hydrogens, which can be partly alleviated by making the bonds longer. If only the first five terms in the force field energy are included, this coupling between bond distance and angle cannot be modeled. It may be taken into account by including a term that depends on both bond length and angle, or by including a Urey–Bradley non-bonded energy term between pairs of atoms that are bonded to a common atom. E_{cross} may in general include a whole series of terms that couple two (or more) of the bonded terms.

The components in E_{cross} are usually written as products of first-order Taylor expansions in the individual coordinates. The most important of these is the stretch/bend term, which for an A—B—C sequence may be written as

$$E_{str/bend} = k^{ABC}\left(\theta^{ABC} - \theta_0^{ABC}\right)\left[\left(R^{AB} - R_0^{AB}\right) - \left(R^{BC} - R_0^{BC}\right)\right] \tag{2.43}$$

Other examples of such cross terms are given by[13]

$$
\begin{aligned}
E_{\text{str/str}} &= k^{\text{ABC}}\left(R^{\text{AB}} - R_0^{\text{AB}}\right)\left(R^{\text{BC}} - R_0^{\text{BC}}\right) \\
E_{\text{bend/bend}} &= k^{\text{ABCD}}\left(\theta^{\text{ABC}} - \theta_0^{\text{ABC}}\right)\left(\theta^{\text{BCD}} - \theta_0^{\text{BCD}}\right) \\
E_{\text{str/tors}} &= k^{\text{ABCD}}\left(R^{\text{AB}} - R_0^{\text{AB}}\right)\cos(n\omega^{\text{ABCD}}) \\
E_{\text{bend/tors}} &= k^{\text{ABCD}}\left(\theta^{\text{ABC}} - \theta_0^{\text{ABC}}\right)\cos(n\omega^{\text{ABCD}}) \\
E_{\text{bend/tors/bend}} &= k^{\text{ABCD}}\left(\theta^{\text{ABC}} - \theta_0^{\text{ABC}}\right)\left(\theta^{\text{BCD}} - \theta_0^{\text{BCD}}\right)\cos(n\omega^{\text{ABCD}})
\end{aligned}
\tag{2.44}
$$

The constants involved in these cross terms are usually not taken to depend on all the atom types involved in the sequence. The stretch/bend constant, for example, in principle depends on all three atoms, A, B and C. However, it is usually taken to depend only on the central atom, that is $k^{\text{ABC}} = k^{\text{B}}$, or chosen as a universal constant independent of atom type. It should be noted that cross terms of the above type are inherently unstable if the geometry is far from equilibrium. Stretching a bond to infinity, for example, will make $E_{\text{str/bend}}$ go toward $-\infty$ if θ is less than θ_0. If the bond stretch energy itself is harmonic (or quartic) this is not a problem as it approaches $+\infty$ faster. However, if a Morse-type potential is used, special precautions will have to be made to avoid long bonds in geometry optimizations and simulations.

Another type of correction, which is related to cross terms, is modification of parameters based on atoms not directly involved in the interaction described by the parameter. Carbon–carbon bond lengths, for example, become shorter if there are electronegative atoms present at either end. Such electronegativity effects may be modeled by adding a correction to the natural bond length R_0^{AB} based on the atom C attached to the A—B bond:[62]

$$
R_0^{\text{AB-C}} = R_0^{\text{AB}} + \Delta R_0^{\text{C}} \tag{2.45}
$$

Other effects, such as hyperconjugation, can be modeled by allowing the natural bond length to depend on the adjacent torsional angle.[63] The hyperconjugation effect can be thought of as weakening of a σ-bond by donation of electron density into an adjacent empty π^*-bond, as illustrated in Figure 2.15.

Since the conjugation can only take place when the σ-orbital is aligned with the π-system, the resulting bond elongation will depend on the torsional angle, which can be modeled by an energy term such as in

$$
\begin{aligned}
R_0^{\text{AB}} &= R_0^{\text{AB}} + \Delta R_0^{\omega} \\
\Delta R_0^{\omega} &= k(1 - \cos 2\omega^{\text{ABCD}})
\end{aligned}
\tag{2.46}
$$

2.2.10 Small Rings and Conjugated Systems

It has already been mentioned that small rings present a problem as their equilibrium angles are very different from their acyclic cousins. One way of alleviating this problem is to assign new atom types. If a sufficient number of cross terms is included, however, the necessary number of atom types can

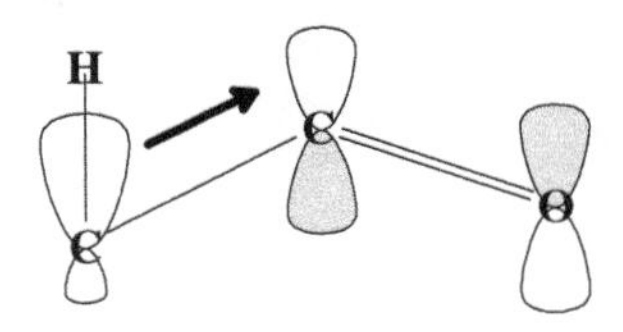

Figure 2.15 Illustrating the elongation of the C—H bond by hyperconjugation.

actually be reduced. Some force fields have only one sp^3-carbon atom type, covering bonding situations from cyclopropane to linear alkanes with the same set of parameters. The necessary flexibility in the parameter space is here transferred from the atom types to the parameters in the cross terms, that is the cross terms modify the diagonal terms such that a more realistic behavior is obtained for large deviations from the natural value.

One additional class of bonding that requires special consideration in force fields is conjugated systems. Consider, for example, 1,3-butadiene. According to the MM2 type convention (Table 2.1), all carbon atoms are of type 2. This means that the same set of parameters is used for the terminal and central C—C bonds. Experimentally, the bond lengths are 1.35 and 1.47 Å, that is very different, which is due to the partial delocalization of the π-electrons in the conjugated system.[64] The outer C═C bond is slightly reduced in double-bond character (and thus has a slightly longer bond length than in ethylene) while the central bond is roughly halfway between a single and a double bond. Similarly, without special precautions, the barriers for rotation around the terminal and central bonds are calculated to be the same, and assume a value characteristic of a localized double bond, ~230 kJ/mol. Experimentally, however, the rotational barrier for the central bond is only ~25 kJ/mol.[65]

There are two main approaches for dealing with conjugated systems. One is to identify certain bonding combinations and use special parameters for these cases, analogously to the treatment of hydrogen bonds in E_{vdw}. If four type 2 carbons are located in a linear sequence, for example, they constitute a butadiene unit and special stretch and torsional parameters should be used for the central and terminal bonds. Similarly, if six type 2 carbons are in a ring, they constitute an aromatic ring and a set of special aromatic parameters are used, or the atom type 2 may be changed to a type 50, identifying from the start that these carbons should be treated with a different parameter set. The main problem with this approach is that there are many such "special" cases requiring separate parameters. Three conjugated double bonds, for example, may either be linearly or cross-conjugated (1,3,5-hexatriene and 2-vinyl-1,3-butadiene), each requiring a set of special parameters different from those used for 1,3-butadiene. The central bond in biphenyl will be different from the central bond in 1,3-butadiene. Modeling the bond alterations in fused aromatics such as naphthalene or phenanthrene requires complicated bookkeeping to keep track of all the different bond lengths, etc.

The other approach, which is somewhat more general, is to perform a simple electronic structure calculation to determine the degree of delocalization within the π-system. This approach is used in the MM2 and MM3 force fields, often denoted MMP2 and MMP3.[66] The electronic structure calculation is of the Pariser–Pople–Parr (PPP) type (Section 7.3), which is only slightly more advanced than a simple Hückel calculation. From the calculated π-molecular orbitals, the π-bond order ρ (Section 10.1) for each bond can be calculated as given in the following equation, where n_i and c_i are the number of electrons and coefficient for the ith MO, respectively:

$$\rho_{AB} = \sum_{i}^{N_{occ}} n_i c_{Ai} c_{Bi} \tag{2.47}$$

Since the bond length, force constant and rotational energy depend on the π-bond order, these constants can be parameterized based on the calculated ρ. The connections used in MMP2 are as follows, with β_{BC} being a resonance parameter:

$$\begin{aligned} R_0^{AB} &= 1.503 - 0.166\rho_{AB} \\ k^{AB} &= 5.0 + 4.6\rho_{AB} \\ V_2^{ABCD} &= 62.5\rho_{BC}\beta_{BC} \end{aligned} \tag{2.48}$$

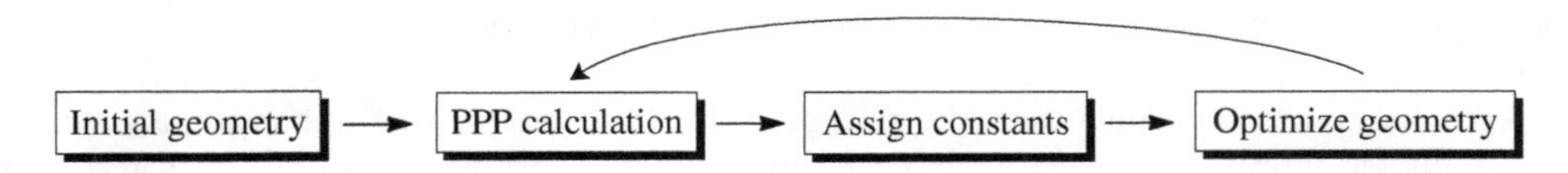

Figure 2.16 Illustration of the two-level optimization involved in an MMP2 calculation.

The natural bond length varies between 1.503 Å and 1.337 Å for bond orders between 0 and 1 – these are the values for pure single and double bonds between two sp^2-carbons. Similarly, the force constant varies between the values used for isolated single and double bonds. The rotational barrier for an isolated double bond is 250 kJ/mol, since there are four torsional contributions for a double bond.

This approach allows a general treatment of all conjugated systems, but requires the addition of a second level of iterations in a geometry optimization, as illustrated in Figure 2.16. At the initial geometry, a PPP calculation is performed, the π-bond orders are calculated and suitable bond parameters (R_0^{AB}, k^{AB} and V_2^{ABCD}) are assigned. These parameters are then used for optimizing the geometry. The optimized geometry will usually differ from the initial geometry; thus the parameters used in the optimization are no longer valid. At the "optimized" geometry, a new PPP calculation is performed and a new set of parameters derived. The structure is re-optimized and a new PPP calculation is carried out, etc. This is continued until the geometry change between two macro iterations is negligible.

For commonly encountered conjugated systems such as butadiene and benzene, the *ad hoc* assignment of new parameters is usually preferred as it is simpler than the computationally more demanding PPP method. For less common conjugated systems, the PPP approach is more elegant and has the definite advantage that the common user does not need to worry about assigning new parameters. If the system of interest contains conjugation and a force field that uses the parameter replacement method is chosen, the user should check that proper bond lengths and reasonable rotational barriers are calculated (i.e. that the force field has identified the conjugated moiety and contains suitable substitution parameters). Otherwise, very misleading results may be obtained without any indication of problems from the force field.

2.2.11 Comparing Energies of Structurally Different Molecules

The force field energy function has a zero point defined implicitly by the zero points chosen for each of the terms. For the three bonding terms, stretch, bend and torsion, this is at the bottom of the energy curve (natural bond lengths and angles), while for the two non-bonded terms, it is at infinite separation. The zero point for the total force field energy is therefore a hypothetical system, where all the bond distances, angles and torsional angles are at their equilibrium values, and at the same time all the non-bonded atoms are infinitely removed from each other. Except for small systems such as CH_4, where there are no non-bonded terms, this is a physically unattainable situation. The force field energy, E_{FF}, is often called the *steric energy*, as it is in some sense the excess energy relative to a hypothetical molecule with non-interacting fragments, but the *numerical value of the force field function has no physical meaning*!

Relative values, however, should ideally reflect conformational energies. If all atom and bond types are the same, as in cyclohexane and methyl-cyclopentane, the energy functions have the same zero point and relative stabilities can be directly compared. This is a rather special situation, however, and stabilities of different molecules can normally not be calculated by force field techniques. For comparing relative stabilities of chemically different molecules such as dimethyl ether and ethyl alcohol, or for comparing with experimental heat of formations, the zero point of the energy scale must be the same.

In electronic structure calculations, the zero point for the energy function has all particles (electrons and nuclei) infinitely removed from each other, and this common reference state allows energies for systems with different numbers of particles to be directly compared. If the same reference is used in force field methods, the energy function becomes an absolute measure of molecular stability. The difference relative to the normal reference state for force field functions is the sum of all bond dissociation energies, at least for a simple diagonal force field. If correction terms are added to the normal force field energy function based on average bond dissociation energies for each bond type, the energy scale become absolute, and can be directly compared with, for example, ΔH_f. Such bond dissociation energies again rest on the assumption of transferability, for example that all C—H bonds have dissociation energies close to 400 kJ/mol. In reality, the bond dissociation energy for a C—H bond depends on the environment: the value for the aldehyde C—H bond in CH_3CHO is 366 kJ/mol while it is 410 kJ/mol for C_2H_6.[67] This can be accounted for approximately by assigning an average bond dissociation energy to a C—H bond and a smaller correction based on larger structural units, such as CH_3 and CHO groups. The MM2 and MM3 force fields use an approach where such bond dissociation energies and structural factors are assigned based on fitting to experimental data, and this approach is quite successful for reproducing experimental ΔH_f values:

$$\Delta H_f = E_{FF} + \sum_{AB}^{bonds} \Delta H_{AB} + \sum_{G}^{groups} \Delta H_G \tag{2.49}$$

The heat of formation parameters may be considered as shifting the zero point of E_{FF} to a common origin. Since corrections from larger moieties are small, it follows that energy differences between systems having the same groups (e.g. methyl-cyclohexane and ethyl-cyclopentane) can be calculated directly from differences in steric energy.

If the heat of formation parameters are derived based on fitting to a large variety of compounds, a specific set of parameters is obtained. A slightly different set of parameters may be obtained if only certain "strainless" molecules are included in the parameterization. Typically molecules such as straight-chain alkanes and cyclohexane are defined to be strainless. By using these strainless heat of formation parameters, a strain energy may be calculated as illustrated in Figure 2.17.

Deriving such heat of formation parameters requires a large body of experimental ΔH_f values, and for many classes of compounds there are not sufficient data available. Only a few force fields, notably MM2 and MM3, also attempt to parameterize heats of formation. Most force fields are only concerned with reproducing geometries and possibly conformational relative energies, for which the steric energy is sufficient.

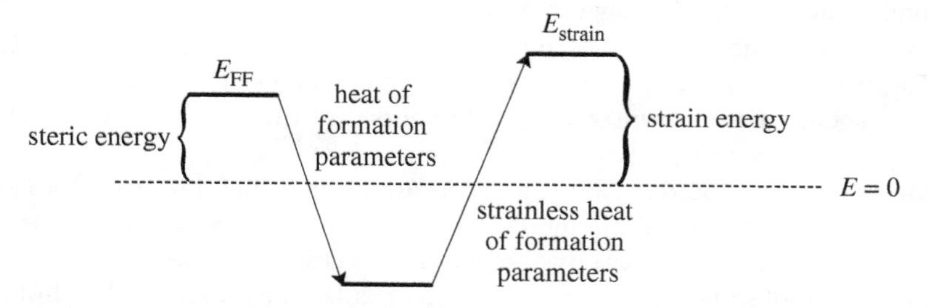

Figure 2.17 Illustrating the difference between steric energy and heat of formation.

2.3 Force Field Parameterization

Having settled on the functional description and a suitable number of cross terms, the problem of assigning numerical values to the parameters arises. This is by no means trivial.[68] Consider, for example, MM2(91) with 71 atom types. Not all of these can form stable bonds with each other; hydrogens and halogens can only have one bond, etc. For the sake of argument, however, assume that the effective number of atom types capable of forming bonds between each other is 30.

- Each of the 71 atom types has two van der Waals parameters, R_0^A and ε^A, giving 142 parameters.
- There are approximately $^1/_2 \times 30 \times 30 = 450$ possible different E_{str} terms, each requiring at least two parameters, k^{AB} and R_0^{AB}, for a total of at least 900 parameters.
- There are approximately $^1/_2 \times 30 \times 30 \times 30 = 13\,500$ possible different E_{bend} terms, each requiring at least two parameters, k^{ABC} and θ_0^{ABC}, for a total of at least 27 000 parameters.
- There are approximately $^1/_2 \times 30 \times 30 \times 30 \times 30 = 405\,000$ possible different E_{tors} terms, each requiring at least three parameters, V_1^{ABCD}, V_2^{ABCD} and V_3^{ABCD}, for a total of at least 1 215 000 parameters.
- Cross terms may add another million possible parameters.

To achieve just a rudimentary assignment of the value of one parameter, at least 3–4 independent data should be available. To parameterize MM2 for all molecules described by the 71 atom types would thus require of the order of 10^7 independent experimental data, not counting cross terms, which clearly is impossible. Furthermore, the parameters that are the most numerous, the torsional constants, are also the ones that are the hardest to obtain experimental data for. Experimental techniques normally probe a molecule near its equilibrium geometry. Getting energetic information about the whole rotational profile is very demanding and has only been done for a handful of small molecules. It has therefore become common to rely on data from electronic structure calculations to derive force field parameters. Calculating, for example, rotational energy profiles is computationally fairly easy. The so-called "Class II" and "Class III" force fields rely heavily on data from electronic structure calculations to derive force field parameters, especially the bonded parameters (stretch, bend and torsional).

While the non-bonded terms are relatively unimportant for the "local" structure, they are the only contributors to intermolecular interactions and the major factor in determining the global structure of a large molecule, such as protein folding. The electrostatic part of the interaction may be assigned based on fitting parameters to the electrostatic potential derived from an electronic wave function, as discussed in Sections 2.2.6 to 2.2.8. The van der Waals interaction, however, is difficult to calculate reliably by electronic structure methods, requiring a combination of electron correlation and very large basis sets, and these parameters are therefore usually assigned based on fitting to experimental data for either the solid or liquid state.[69]

For a system containing only a single atom type (e.g. liquid argon), the R_0 (atomic size) and ε (interaction strength) parameters can be determined, for example, by requiring that the experimental density and heat of evaporation are reproduced, respectively. Since the parameterization implicitly takes many-body effects into account, a (slightly) different set of van der Waals parameters will be obtained if the parameterization instead focuses on reproducing the properties of the crystal phase. For systems where several atom types are involved (e.g. water), there are two van der Waals parameters for each atom type, and the experimental density and heat of evaporation alone therefore give insufficient data for a unique assignment of all parameters. Although one may include additional experimental data, for example the variation of the density with temperature, this still provides insufficient data for a general system containing many atom types. Furthermore, it is possible that several combinations of van der Waals parameters for different atoms may be able to reproduce properties of a liquid, that is even if there are sufficient experimental data, the derived parameter set may not be unique. One approach for solving this problem is to use electronic structure methods to determine *relative* values for van der Waals parameters, for example using a neon atom as the probe, and determine the absolute values by fitting to experimental values.[70]

An alternative procedure is to derive the van der Waals parameters from other physical (atomic) properties. The interaction strength ε^{AB} between two atoms is related to the polarizabilities α_A and α_B, that is the ease with which the electron densities can be distorted by an electric field. The Slater–Kirkwood equation[71] provides an explicit relationship between these quantities, which has been found to give good results for the interaction of rare gas atoms:

$$\varepsilon^{AB} = C\frac{\alpha_A \alpha_B}{\sqrt{\frac{\alpha_A}{N_A^{\text{eff}}}} + \sqrt{\frac{\alpha_B}{N_B^{\text{eff}}}}} \tag{2.50}$$

Here C is a constant for converting between the units of ε and α, and N_A^{eff} is the effective number of electrons, which may be taken either as the number of valence electrons or treated as a fitting parameter. The R_0 parameter may similarly be taken from atomic quantities. One problem with this procedure is that the atomic polarizability will of course be modified by the bonding situation (i.e. the atom type), which is not taken into account by the Slater–Kirkwood equation.

The molecular (or atomic) van der Waals dispersion coefficients can also be obtained theoretically from an integration of the polarizability corresponding to imaginary frequencies (Section 11.10):[72]

$$\begin{aligned} E_{\text{dispersion}}(R^{AB}) &= -\frac{C_6^{AB}}{(R^{AB})^6} - \frac{C_8^{AB}}{(R^{AB})^8} - \cdots \\ C_6^{AB} &= \frac{3}{\pi}\int_0^\infty \alpha_{\text{A}}(\text{i}\omega)\alpha_{\text{B}}(\text{i}\omega)\text{d}\omega \end{aligned} \tag{2.51}$$

The above considerations illustrate the inherent contradiction in designing highly accurate force fields. To get a high accuracy for a wide variety of molecules and a range of properties, many functional complex terms must be included in the force field expression. For each additional parameter introduced in an energy term, the potential number of new parameters to be derived grows with the number of atom types to a power between 1 and 4. The higher the accuracy that is needed, the more finely the fundamental units must be separated, that is the more atom types must be used. In the extreme limit, each atom that is not symmetry related or conformationally easy to convert to another, in each new molecule is a new atom type. In this limit, each molecule will have its own set of parameters to be used just for this one molecule. To derive these parameters, the molecule must be subjected to many different experiments, or a large number of electronic structure calculations. This is the approach used in "inverting" spectroscopic data to produce a potential energy surface, and the resulting force field may be considered as a sophisticated interpolating function for reproducing the reference data. While this may be useful for parameterizing a small number of widely used systems (e.g. all natural nucleic acids or amino acids), it is unsuitable for the parameterization of, for example, a virtual screening data base containing millions of molecules. The latter case demands a parameterization strategy in terms of a limited number of atom types with transferable parameters. The number of atom types in each force field to a large extent represents a subjective choice based on chemical

Table 2.3 Comparison of possible and actual number of MM2(91) parameters.

Term	Estimated number of parameters	Actual number of parameters
E_{vdw}	142	142
E_{str}	900	290
E_{bend}	27 000	824
E_{tors}	1 215 000	2466

intuition and subsequent calibrations and these do not always agree with similarity measures based on results from electronic structure methods.[9]

The fundamental assumption of force fields is that structural units are transferable between different molecules. A compromise between accuracy and generality must thus be made. In MM2(91) the actual number of parameters compared with the theoretical estimated possible (based on the 30 effective atom types above) is shown in Table 2.3.

As seen from Table 2.3, there are a large number of possible compounds for which there are no parameters, and on which it is then impossible to perform force field calculations. Actually, the situation is not as bad as it would appear from Table 2.3. Although only ~0.2% of the possible combinations for the torsional constants has been parameterized, these encompass the majority of the chemically interesting compounds. It has been estimated that ~20% of all known compounds can be modeled by the parameters in MM2, the majority with a good accuracy. However, the problem of lacking parameters is very real, and anyone who has used a force field for all but the most rudimentary problems has encountered the problem. How does one progress if there are insufficient parameters for the molecule of interest?

There are two possible routes. The first is to estimate the missing parameters by comparison with force field parameters for similar systems. If, for example, there are missing torsional parameters for rotation around a H—X—Y—O bond in your molecule, but parameters exist for H—X—Y—C, then it is probably a good approximation to use the same values. In other cases, it may be less obvious what to choose. What if your system has an O—X—Y—O torsion, and parameters exist for O—X—Y—C and C—X—Y—O, but they are very different? What do you choose then, one or the other, or the average? After a choice has been made, the results should ideally be evaluated to determine how sensitive they are to the exact value of the guessed parameters. If the guessed parameters can be varied by ±50% without seriously affecting the final results, the property of interest is insensitive to the guessed parameters and can be trusted to the usual degree of the force field. If, on the other hand, the final results vary by a factor of two when the guessed parameters are changed by 10%, a better estimate of the critical parameters should be sought from external sources. If many parameters are missing from the force field, such an evaluation of the sensitivity to parameter changes becomes impractical, and one should consider either the second route described below or abandon force field methods altogether.

The second route to missing parameters is to use external information, experimental data or electronic structure calculations. If the missing parameters are bond length and force constant for a specific bond type, it is possible that an experimental bond distance may be obtained from an X-ray structure and the force constant estimated from measured vibrational frequencies, or missing torsional parameters may be obtained from a rotational energy profile calculated by electronic structure calculations. If many parameters are missing, this approach rapidly becomes very time-consuming, and may not give as good final results as you may have expected from the "rigorous" way of deriving the parameters. The reason for this is discussed below.

Assume now that the functional form of the force field has been settled. The next task is to select a set of reference data – for the sake of argument let us assume that they are derived from experiments, but they could also be taken from electronic structure calculations. The problem is then to assign numerical values to all the parameters such that the results from force field calculations match the reference data set as close as possible. The reference data may be of very different types and accuracy, containing bond distances, bond angles, relative energies, vibrational frequencies, dipole moments, etc. These data of course have different units, and a decision must be made as to how they should be weighted. How much weight should be put on reproducing a bond length of 1.532 Å relative to an energy difference of 10 kJ/mol? Should the same weight be used for all bond distances if, for example, one distance is determined to 1.532 ± 0.001 Å while another is known only to 1.73 ± 0.07 Å? The selection is further complicated by the fact that different experimental methods may give slightly different answers for, say, the bond distance, even in the limit of no experimental uncertainty. The reason for this is that different experimental methods do not measure the same property. X-ray diffraction, for example, determines the electron distribution, while microwave spectroscopy primarily depends on the nuclear position. The maximum in the electronic distribution may not be exactly identical to the nuclear position, and these two techniques will therefore give slightly different bond lengths.

Once the question of assigning weights for each reference data has been decided, the fitting process can begin. It may be formulated in terms of an error function:[73]

$$ErrF(\text{parameters}) = \sum_{i}^{\text{data}} weight_i \times (reference\ value - calculated\ value)_i^2 \tag{2.52}$$

The problem is to find the minimum of *ErrF* with the parameters as variables. From an initial set of guess parameters, force field calculations are performed for the *whole* set of reference molecules and the results compared with the reference data. The deviation is calculated and a new improved set of parameters can be derived. This is continued until a minimum has been found for the *ErrF* function. To find *the* best set of force field parameters corresponds to finding the *global* minimum for the multidimensional *ErrF* function. The simplest optimization procedure performs a cyclic minimization, reducing the *ErrF* value by varying one parameter at a time. More advanced methods rely on the ability to calculate the gradient (and possibly also the second derivative) of the *ErrF* with respect to the parameters. Such information may be used in connection with the optimization procedure, as described in Chapter 13.

The parameterization process may be done sequentially or in a combined fashion. In the sequential method, a certain class of compounds, such as hydrocarbons, is parameterized first. These parameters are held fixed and a new class of compounds, for example alcohols and ethers, are then parameterized. This method is in line with the basic assumption of a force field – that parameters are transferable. The advantage is that only a fairly small number of parameters are fitted at a time. The *ErrF* is therefore a relatively low-dimensional function, and one can be reasonably certain that a "good" minimum has been found (although it may not be the global minimum). The disadvantage is that the final set of parameters necessarily provides a poorer fit (as defined from the value of the *ErrF*) than if all the parameters are fitted simultaneously.

The combined approach tries to fit all the constants in a single parameterization step. Considering that the number of force field parameters may be many thousands, it is clear that the *ErrF* function will have a very large number of local minima. To find the global minimum of such a multivariable function is very difficult. It is thus likely that the final set of force field parameters derived by this procedure will in some sense be less than optimal, although it may still be "better" than that derived by the sequential procedure. Furthermore, many of the parameter sets that give low *ErrF* values (including

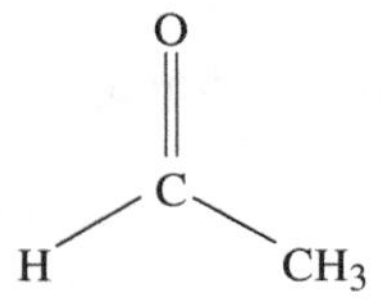

Figure 2.18 The structure of acetaldehyde.

the global minimum) may be "non-physical", for example force constants for similar bonds being very different. Due to the large dimensionality of the problem, such combined optimizations require the ability to calculate the gradient of the *ErrF* with respect to the parameters, and writing such programs is not trivial. There is also a more fundamental problem when new classes of compounds are introduced at a later time than the original parameterization. To be consistent, the whole set of parameters should be re-optimized. This has the consequence that (all) parameters change when a new class of compounds is introduced or whenever more data are included in the reference set. Such "time-dependent" force fields are clearly not desirable. Most parameterization procedures therefore employ a sequential technique, although the number of compound types parameterized in each step varies.

There is one additional point to be mentioned in the parameterization process that is also important for understanding why the addition of missing parameters by comparison with existing data or from external sources is somewhat problematic. This is the question of redundant variables, as can be exemplified by considering acetaldehyde (Figure 2.18).

In the energy bend expression there will be four angle terms describing the geometry around the carbonyl carbon, an HCC, an HCO, a CCO and an out-of-plane bend. Assuming the latter to be zero for the moment, it is clear that the other three angles are not independent. If the θ_{HCO} and θ_{CCO} angles are given, the θ_{HCC} angle must be $360° - \theta_{HCO} - \theta_{CCO}$. Nevertheless, there will be three natural angle parameters and three force constants associated with these angles. For the whole molecule there are six stretch terms, nine bending terms and six torsional terms (count them!) in addition to at least one out-of-plane term. This means that the force field energy expression has 22 degrees of freedom, in contrast to the 15 ($3N_{atom} - 6$) independent coordinates necessary to completely specify the system. The force field parameters, as defined by the E_{FF} expression, are therefore not independent.

The implicit assumption in force field parameterization is that, given sufficient amounts of data, this redundancy will cancel out. In the above case, additional data for other aldehydes and ketones may be used (at least partly) for removing this ambiguity in assigning angle bend parameters, but in general there are more force field parameters than required for describing the system. This clearly illustrates that force field parameters are just that – parameters. They do not necessarily have any direct connection with experimental force constants. Experimental vibrational frequencies can be related to a unique set of force constants, but only in the context of a non-redundant set of coordinates.

It is also clear that errors in the force field due to inadequacies in the functional forms used for each of the energy terms will to some extent be absorbed by the parameter redundancy. Adding new parameters from external sources, or estimating missing parameters by comparison with those for "similar" fragments, may partly destroy this cancellation of errors. This is also the reason why parameters are not transferable between different force fields, as the parameter values are dependent on the functional form of the energy terms and are mutually correlated. The energy profile for rotating around a bond, for example, contains contributions from the electrostatic, the van der Waals and the torsional energy terms. The torsional parameters are therefore intimately related to the atomic partial charges and cannot be transferred to another force field.

The parameter redundancy is also the reason that care should be exercised when trying to decompose energy differences into individual terms. Although it may be possible to rationalize the preference of one conformation over another by, for example, increased steric repulsion between certain atom pairs, this is intimately related to the chosen functional form for the non-bonded energy, and the balance between this and the angle bend/torsional terms. The rotational barrier in ethane, for example, may be reproduced solely by an HCCH torsional energy term, solely by an H—H van der Waals repulsion or solely by an H—H electrostatic repulsion. Different force fields will have (slightly) different balances of these terms, and while one force field may contribute a conformational difference primarily to steric interactions, another may have the major determining factor to be the torsional energy, and a third may "reveal" that it is all due to electrostatic interactions.

2.3.1 Parameter Reductions in Force Fields

The overwhelming problem in developing force fields is the lack of enough high-quality reference data. As illustrated above, there are literally millions of possible parameters in even quite simple force fields. The most numerous of these are the torsional parameters, followed by the bending constants. As force fields often are used for predicting properties of unknown molecules, it is inevitable that the problem of lacking parameters will be encountered frequently. Furthermore, many of the existing parameters may be based on very few reference data, and therefore be associated with substantial uncertainty.

Many modern force field programs are commercial. Having the program tell the user that his or her favorite molecule cannot be calculated owing to lack of parameters is not good for business. Making the user derive new parameters, and getting the program to accept them, may require more knowledge than the average user, who is just interested in the answer, has. Many force fields thus have "generic" parameters. This is just a fancy word for the program making more or less educated guesses for the missing parameters.

One way of reducing the number of parameters is to reduce the dependence on atom types. Torsional parameters, for example, can be taken to depend only on the types of the two central atoms. All C—C single bonds would then have the same set of torsional parameters. This does not mean that the rotational barrier for all C—C bonds is identical, since van der Waals and/or electrostatic terms also contribute. Such a reduction replaces all tetra-atomic parameters with diatomic constants, that is $V^{ABCD} \rightarrow V^{BC}$. Similarly, the triatomic bending parameters may be reduced to atomic constants by assuming that the bending parameters only depend on the central atom type ($k^{ABC} \rightarrow k^{B}, \theta_0^{ABC} \rightarrow \theta_0^{B}$). Generic constants are often taken from such reduced parameter sets. In the case of missing torsional parameters, they may also simply be omitted, that is setting the constants to zero. A good force field program informs the user of the quality of the parameters used in the calculation, especially if such generic parameters are used, and this is useful for evaluating the quality of the results. Some programs unfortunately use the necessary number of generic parameters to carry out the calculations without notifying the user. In extreme cases, one may perform calculations on molecules for which essentially no "good" parameters exist, and get totally useless results. *The ability to perform a calculation is no guarantee that the results can be trusted!*

The quality of force field parameters is essential for judging how much faith can be put in the results. If the molecule at hand only uses parameters that are based on many good-quality experimental results, then the computational results can be trusted to be almost of experimental quality. If, however, the employed parameters are based only on a few experimental data and/or many generic parameters have been used, the results should be treated with care. Using low-quality parameters for describing an "uninteresting" part of the molecule, such as a substituted aromatic ring in a distant side chain, is not problematic. In some cases, such uninteresting parts may simply be substituted by other simpler groups (e.g. a methyl group). However, if the low-quality parameters directly influence the property of interest, the results may potentially be misleading.

2.3.2 Force Fields for Metal Coordination Compounds

Coordination chemistry is an area that is especially plagued with the problems of assigning suitable functions for describing the individual energy terms and deriving good parameters.[74,75] The bonding around metals is much more varied than for organic molecules, where atoms form only two, three or four bonds. Furthermore, for a given number of ligands, more than one geometrical arrangement is usually possible. A four-coordinated metal, for example, may either be tetrahedral or square planar, and a five-coordinated metal may either have a square pyramidal or trigonal bipyramidal structure. This is in contrast to four-coordinated atoms such as carbon or sulfur that are always very close to tetrahedral. The increased number of ligands combined with the multitude of possible geometries significantly increases the problem of assigning suitable functional forms for each of the energy terms. Consider, for example, a "simple" compound such as $Fe(CO)_5$, which has a trigonal bipyramid structure (Figure 2.19).

It is immediately clear that a C—Fe—C angle bend must have three energy minima corresponding to 90°, 120° and 180°, indicating that a simple Taylor expansion around a (single) natural value is not suitable. Furthermore, the energy cost for a geometrical distortion (bond stretching and bending) is usually much smaller around a metal atom than for a carbon atom. This has the consequence that coordination compounds are much more dynamic, displaying phenomena such as pseudo-rotations, ligand exchange and large geometrical variations for changes in the ligands. In iron pentacarbonyl there exists a whole series of equivalent trigonal bipyramid structures that readily interconvert, that is the energy cost for changing the C—Fe—C angle from 90° to 120° and to 180° is small. Deviations up to 30° from the "natural" angle by introducing bulky substituents on the ligands are not uncommon. Furthermore, the bond distance for a given metal–ligand is often sensitive to the nature of the other ligands. An example there is the *trans* effect, where a metal–ligand bond distance can vary by perhaps 0.2 Å depending on the nature of the ligand on the opposite side.

Another problem encountered in metal systems is the lack of well-defined bonds. Consider, for example, an olefin coordinated to a metal. Should this be considered as a single bond between the metal and the center of the C—C bond or as a metallocyclopropane with two M—C bonds? A cyclopentadiene ligand may similarly be modeled either with a single bond to the center of the ring or with five M—C bonds. In reality, these represent limiting behaviors, and the structures on the left in Figure 2.20 correspond to a weak interaction while those on the right involve strong electron donation from the ligand to the metal (and vice versa). A whole range of intermediate cases is found in

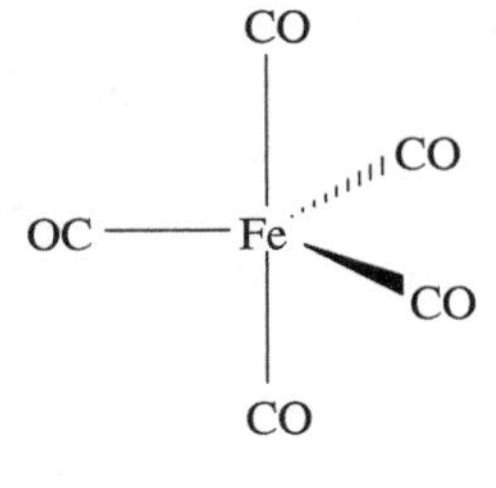

Figure 2.19 The structure of iron pentacarbonyl.

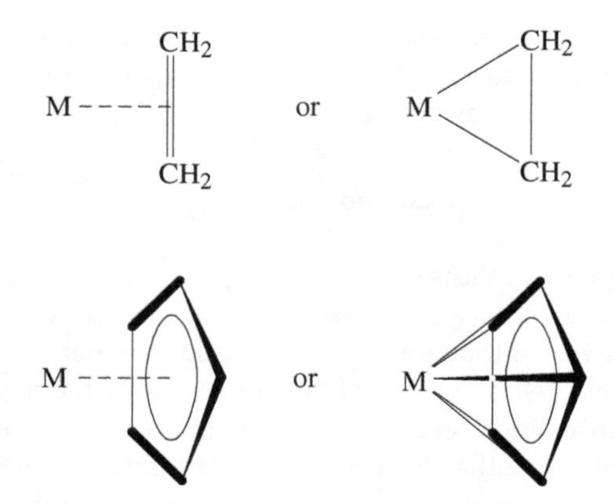

Figure 2.20 The ambiguity of modeling metal coordination.

coordination chemistry. The description with bonds between the metal and all the ligand atoms suffers from the lack of (free) rotation of the ligand. The coordination to the "center" of the ligand may be modeled by placing a *pseudo-atom* at that position and relating the ligand atoms to the pseudo-atom (a pseudo-atom is just a point in space, also sometimes called a dummy atom; see Appendix D). Alternatively, the coordination may be described entirely by non-bonded interactions (van der Waals and electrostatic).

One possible, although not very elegant, solution to these problems is to assign different atom types for each bonding situation. In the $Fe(CO)_5$ example, this would mean distinguishing between equatorial and axial CO units. There would then be three different C—Fe—C bending terms, C_{eq}—Fe—C_{eq}, C_{eq}—Fe—C_{ax} and C_{ax}—Fe—C_{ax}, with natural angles of 120°, 90° and 180°, respectively. This approach sacrifices the dynamics of the problem: interchanging an equatorial and axial CO no longer produces energetically equivalent structures. Similarly, the same metal atom in two different geometries (such as tetrahedral and square planar) would be assigned two different types, or in general a new type for each metal in a specific oxidation and spin state, and with a specific number of ligands. This approach encounters the parameter "explosion", as discussed above. It also biases the results in the direction of the user's expectations – if a metal atom is assigned a square planar atom type, the structure will end up close to square planar, even though the real geometry may be tetrahedral. The object of a computational study, however, is often a series of compounds that have similar bonding around the metal atom. In such cases, the specific parameterization may be quite useful, but the limitations should of course be kept in mind. Most force field modeling of coordination compounds to date have employed this approach, tailoring an existing method to also reproduce properties (most notably geometries) of a small set of reference systems.

Part of the problems may be solved by using more flexible functional forms for the individual energy terms, most notably the stretching and bending energies. The stretch energy may be chosen as a Morse potential (Equation (2.5)), allowing for quite large distortions away from the natural distance, and also being able to account for dissociation. However, phenomena such as the *trans* effect are inherently electronic in nature (similar to the delocalization in conjugated systems) and are not easily accounted for in a force field description.

The multiple minima nature of the bending energy, combined with the low barriers for interconversion, resembles the torsional energy for organic molecules. An expansion of E_{bend} in terms of cosine or sine functions to the angle is therefore more natural than a simple Taylor expansion in the angle. Furthermore, bending around a metal atom often has an energy maximum for an angle of 180°, with

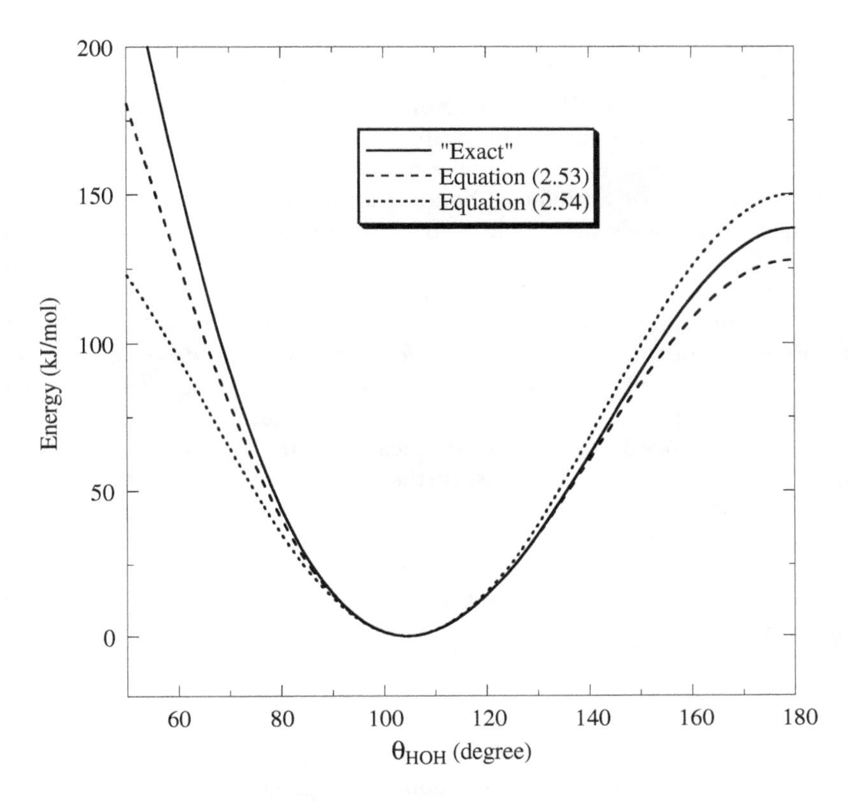

Figure 2.21 Comparing different E_{bend} functionals for the H_2O potential.

a low barrier. The following examples have a zero derivative for a linear angle and are reasonable for describing bond bending such as that encountered in the H_2O example (Figure 2.5):

$$E_{\mathrm{bend}}(\theta) = k\left(\sin^2\frac{\theta}{2} - \sin^2\frac{\theta_0}{2}\right)^2 \tag{2.53}$$

$$E_{\mathrm{bend}}(\theta) = k(1 + \cos(n\theta + \tau)) \tag{2.54}$$

The latter functional form contains a constant n that determines the periodicity of the potential (τ is a phase factor) and allows bending energies with multiple minima, analogously to the torsional energy. It does, however, have problems of unwanted oscillations if an energy minimum with a natural angle close to 180° is desired (this requires n to be large, creating many additional minima). It is also unable to describe situations where the minima are not regularly spaced, such as the $Fe(CO)_5$ system (minima for angles of 90°, 120° and 180°). The performance of Equations (2.53) and (2.54) for the H_2O case is given in Figure 2.21, which can be compared with Figure 2.5.

The barrier towards linearity is given implicitly by the force constant in both the potentials in Equations (2.53) and (2.54). A more general expression, which allows even quite complicated energy functionals to be fitted, is a Fourier expansion:

$$E_{\mathrm{bend}}(\theta) = \sum_n k_n \cos(n\theta) \tag{2.55}$$

An alternative approach consists of neglecting the L—M—L bending terms, and instead includes non-bonded 1,3-interactions. The geometry around the metal is then defined exclusively by the van der

Waals and electrostatic contributions (i.e. placing the ligands as far apart as possible), and this model is known as *Points-On-a-Sphere* (POS).[76] It is basically equivalent to the *Valence Shell Electron-Pair Repulsion* (VSEPR) model, with VSEPR focusing on the electron pairs that make up a bond and POS focusing on the atoms and their size.[77] For alkali, alkaline earth and rare earth metals, where the bonding is mainly electrostatic, POS gives quite reasonable results, but it is unable to model systems where the d-orbitals have a preferred bonding arrangement. Tetracoordinated metal atoms, for example, will in such models always end up being tetrahedral, although d^8-metals are normally square planar. An explicit coupling between the geometry and the occupancy of the d-orbitals can be achieved by adding a ligand field energy term to the force field energy function.[78,79]

The final problem encountered in designing force fields for metal complexes is the lack of sufficient numbers of experimental data. Geometrical data for metal compounds are much scarcer than for organic structures, and the soft deformation potentials mean that vibrational frequencies are often difficult to assign to specific modes. Deriving parameters from electron structure calculations is troublesome because the presence of multiple ligands means that the number of atoms is quite large and the metal atom itself contains many electrons. Furthermore, there are often many different low-lying electronic states owing to partly occupied d-orbitals, indicating that single reference methods (i.e. Hartree–Fock-type calculations) are insufficient for even a qualitative correct wave function. Finally, relativistic effects become important for some of the metals in the lower part of the periodic system. These effects have the consequence that electronic structure calculations of a sufficient quality are computationally expensive to carry out.

2.3.3 Universal Force Fields

The combination of many atom types and the lack of a sufficient number of reference data have prompted the development of force fields with reduced parameters sets, such as the *Universal Force Field* (UFF).[80] The idea is to derive di-, tri- and tetra-atomic parameters (E_{str}, E_{bend}, E_{tors}) from atomic constants (such as atom radii, ionization potentials, electronegativities, polarizabilities, etc.). Such force fields are in principle capable of describing molecules composed of elements from the whole periodic table, and these have been labeled as "all elements" in Table 2.4. They give less accurate results compared with conventional force fields, but geometries are often calculated qualitatively correctly. Relative energies, however, are much more difficult to obtain accurately, and conformational energies for organic molecules are generally quite poor. Another approach is to use simple valence bonding arguments (e.g. hybridization) to derive the functional form for the force field, as employed in the VALBOND approach.[81]

2.4 Differences in Atomistic Force Fields

There are many different force fields employing atoms as the fundamental particle in use. They differ in three main aspects:

1. What is the functional form of each energy term?
2. How many cross terms are included?
3. What types of information are used for fitting the parameters?

There are two general trends. If the force field is designed primarily to treat large systems, such as proteins or DNA, the functional forms are kept as simple as possible. This means that only harmonic functions are used for E_{str} and E_{bend} (or these term are omitted, forcing all bond lengths and angles

Table 2.4 Comparison of functional forms used in common force fields; the torsional energy, E_{tors}, is in all cases given as a Fourier series in the torsional angle. Note that several of the force fields denoted as polarizable (e.g. AMBER, CHARMM, GROMOS) are available also in an unpolarized version.

Force field	E_{str}	E_{bend}	E_{oop}	E_{vdw}	E_{el}	E_{cross}	Molecules
AMBER[82]	P2	P2	imp.	12–6	Charge, PD	None	Biomolecules
AMOEBA[83]	P2	P2		14–7	Quad, PD	None	Biomolecules
CFF91/93/95[84]	P4	P4	P2	9–6	Charge	ss, bb, st, sb, bt, btb	General
CHARMM[85]	P2	P2	P2(imp.)	12–6	Charge DO	None	Biomolecules
COMPASS[86]	P4	P4	P2	9–6	Charge	ss, sb, st, bb, bt, bbt	Organic
COSMIC[87]	P2	P2		Morse	Charge	None	General
CVFF[88]	P2 or Morse	P2	P2	12–6	Charge	ss, bb, sb, btb	General
DREIDING[89]	P2 or Morse	P2(cos)	P2(cos)	12–6 or Exp–6	Charge	None	General
EAS[90]	P2	P3	None	Exp–6	None	None	Alkanes
ECEPP[91]	Fixed	Fixed	Fixed	12–6 and 12–10	Charge	None	Proteins
EFF[92]	P4	P3	None	Exp–6	None	ss, bb, sb, st, btb	Alkanes
ENCAD[93]	P2	P2	imp.	12–6	Charge	None	Biomolecules
ESFF[94]	Morse	P2(cos)	P2	9–6	Charge	None	All elements
GAFF[95]	P2	P2	imp	12–6	Charge	None	Organic
GROMOS[96]	P2	P2	P2(imp.)	12–6	Charge, DO	None	Biomolecules
MM2[6]	P3	P2+P6	P2	Exp–6	Dipole	sb	General
MM3[7]	P4	P6	P2	Exp–6	Dipole or charge	sb, bb, st	General (all elements)
MM4[97]	P6	P6	imp.	Exp–6	Charge	ss, bb, sb, tt, st, tb, btb	General

(*Continue*)

Table 2.4 *(Continued)*

Force field	E_{str}	E_{bend}	E_{oop}	E_{vdw}	E_{el}	E_{cross}	Molecules
MMFF[98]	P4	P3	P2	14–7	Charge	sb	General
MOMEC[99]	P2	P2	P2	Exp–6	Charge, FQ	None	Metal coordination
NEMO[100]	Fixed	Fixed	None	Exp–6	Quad, PD	None	Special
OPLS[101]	P2	P2	imp.	12–6	Charge	None	Biomolecules
PFF[102]	P2	P2	imp.	12–6	Charge, FQ, PD	None	Proteins
PROSA[58]	P2	P2	imp.	12–6	Charge, FQ, PD	None	Proteins
QMFF[103]	P4	P4	P2	9–6	Charge	ss, sb, st, bb, bt, btb	General
SDFF[104]	P4	P4		9–6	Quad, PD	ss, st, tt	Hydrocarbons
SHAPES[105]	P2	cos($n\theta$)	P2(imp.)	12–6	Charge	None	Metal coordination
TraPPE[106]	Fixed	P2	Fixed	12–6	Charge	None	Organic
TRIPOS[107]	P2	P2	P2	12–6	Charge	None	General
UFF[80]	P2 or Morse	cos($n\theta$)	imp.	12–6	Charge	None	All elements
YETI[108]	P2	P2	imp.	12–6 and 12–10	Charge	None	Proteins

Notation. Pn: polynomial of order n; Pn(cos): polynomial of order n in cosine to the angle; Pn(imp.): polynomial of order n in the improper angle; cos($n\theta$): Fourier term(s) in cosine to the angle; Exp–6: exponential + R^{-6}; $n - m$: $R^{-n} + R^{-m}$ Lennard-Jones-type potential; Quad: electric moments up to quadrupoles; FQ: fluctuating charge, DO: Drude oscillator, PD: point dipole; Fixed: not a variable; imp.: improper torsional angle; ss: stretch–stretch; bb: bend–bend: sb: stretch–bend; st: stretch–torsional; bt: bend–torsional; tt: torsional–torsional; btb: bend–torsional–bend.

to be constant), no cross terms are included and the Lennard-Jones potential is used for E_{vdw}. Such force fields are often called "harmonic", "diagonal" or "Class I". The other branch concentrates on reproducing small- to medium-size molecules to a high(er) degree of accuracy. These force fields will include a number of cross terms, use at least cubic or quartic expansions of E_{str} and E_{bend}, and possibly an exponential-type potential for E_{vdw}. The small-molecule force fields strive to not only reproduce geometries and relative energies but also vibrational frequencies, and these are often called "Class II" force fields. Further refinements by allowing parameters to depend on neighboring atom types, for example for modeling hyperconjugation, and including electronic polarization effects have been denoted "Class III" force fields.

Table 2.4 gives a description of the functional forms used in some of the common force fields. The torsional energy is in all cases written as a Fourier series, typically of order 3. Many of the force fields undergo continuous developments, and Table 2.4 may thus be considered as a "snapshot" of the situation when the data were collected. The "universal"-type force fields, described in Section 2.3.3, are in principle capable of covering molecules composed of elements from the whole periodic table, and these have been labeled as "all elements".

Even for force fields employing the same mathematical form for an energy term there may be significant differences in the parameters. Table 2.5 shows the variability of the parameters for the stretch energy between different force fields. It should be noted that the stretching parameters are among those that vary the *least* between force fields.

It is perhaps surprising that force constants may differ by almost a factor of two, but this is of course related to the stiffness of the stretch and bending energies. Very few molecules have bond lengths deviating more than a few hundredths of an angstrom from the reference value, and the associated energy contribution will be small regardless of the force constant value. Stated another way, the minimum energy geometry is insensitive to the exact value of the force constant.

The torsional parameters are most important for determining geometries and relative energies of conformational structures, as well as activation energies for conformational transitions. When studies reveal that a given force field provides inaccurate or biased results for a given system, this often results in determining a new set of torsional parameters for a subset of atom types in the force field. These updated parameters are often denoted with a serial number or a year-label when they were released, as, for example, CHARMM-38 or OPLS-05. Torsional energy modifications have in some cases been applied by adding a look-up table value depending on two torsional angles, rather than modifying the torsional parameters of functional form, as has been done in the CMAP approach for protein backbone angles.[109]

Table 2.5 Comparison of stretch energy parameters for different force fields.

Force field	R_0 (Å)				k (mdyn/Å)			
	C—C	C—O	C—F	C=O	C—C	C—O	C—F	C=O
MM2	1.523	1.402	1.392	1.208	4.40	5.36	5.10	10.80
MM3	1.525	1.413	1.380	1.208	4.49	5.70	5.10	10.10
MMFF	1.508	1.418	1.360	1.222	4.26	5.05	6.01	12.95
AMBER	1.526	1.410	1.380	1.220	4.31	4.45	3.48	8.76
OPLS	1.529	1.410	1.332	1.229	3.73	4.45	5.10	7.92

2.5 Water Models

Water is arguably one of the most difficult molecules to model by classical force field potentials, but, as an essential solvent for biomolecular simulations, is also one that must be given special attention. The difficulties arise because water has unique hydrogen bonding capabilities, has large multibody contributions (polarization) as well as important quantum effects, since it contains two low-mass hydrogen nuclei (zero-point energies and hydrogen tunneling, leading to, for example, different melting points for H_2O and D_2O). The strong hydrogen bonding between water molecules has contributions of charge transfer and charge penetration effects, both of which are difficult to describe within a force field environment. These molecular properties lead to a number of unusual macroscopic properties for water, such as the non-monotonic density–temperature profile (maximum density at 4 °C), the volume expansion upon freezing, several different crystal phases, a large and temperature-dependent dielectric constant, the auto-dissociation into ions, and a large heat of evaporation. Parameterization for reproducing some of these properties are often quite successful for the properties included in the parameterization, but fails for properties not included. Since water often accounts for more than half of the total system size in a simulation, computational efficiency is of outmost importance. These conflicting requirements have led to a large number of proposed water models of varying degrees of accuracy and computational efficiencies, some of which will be described in this section.[110]

The water models can be classified by three criteria:

- Rigid or flexible geometry
- Number of interaction sites
- Static or polarizable electrostatic interactions.

The number of interaction sites refers to how many expansion points are used for representing the non-bonded interaction. These often coincide with the nuclear positions, but not necessarily, and additional off-nucleus sites are often included when fixed point charges are used for representing the electrostatic interaction. The non-bonded interactions must be calculated between all pairs of interaction sites for all water molecules, and the computational cost therefore increases quadratically with the number of interaction sites. The lower accuracy of models with few interaction sites can partly be compensated for by employing different parameters for the same underlying model for different situations, as, for example, different water parameters depending on the specific force field used for a protein or different water parameters depending on how the long-range electrostatic interaction is calculated. Alternatively, solute parameters can be tuned to a specific water model, as, for example, for metals ions.[111]

Some of the commonly used water models and selected properties are collected in Table 2.6.

The SPC (*Simple Point Charge*) and TIP3P (*Transferable Intermolecular Potential 3 Point*) model are three-site models where the nuclear positions are used as sites for partial charges. The TIP4P model moves the negative charge from the position of the oxygen atom to an off-center position along the HOH bisector, while the TIP5P model places partial negative charges at the oxygen lone pair positions. All of these models account for the non-polar van der Waals type interaction by a single interaction site located at the oxygen position. TIP3P thus have 9 interactions to be calculated for each pair of water molecules, while TIP5P has 17 interactions (16 charge–charge and 1 van der Waals), which makes the latter roughly a factor of two more expensive computationally. Six site models where the three nucleus positions are augmented with both a bisector and lone pair sites have also been proposed in order to accurately reproduce the molecular quadrupole moments within a fixed charge model.

Table 2.6 A selection of water models and associated properties.[112,113]

Model property	SPC/E	TIP3P	TIP4P	TIP4P/2005	TIP5P	BK3	AMOEBA	Exp.
T_m	215	146	232	252	274	250	261	273
ρ	0.994	0.98	0.988	0.993	0.979	0.997	1.000	0.997
ΔH_{vap}	49.3	42.0	44.6	50.2	43.8	45.8	43.5	44.0
C_p	88	79	84	88	121	92	88	75
ε	68	94	50	58	91	79	81	78

T_m: melting temperature (K), ρ: density at 298 °C (g/ml), ΔH_{vap}: heat of vaporization (kJ/mol), C_p: heat capacity (J/mol K), ε: dielectric constant.

The difference between SPC and SPC/E (E for "extended") is only that the latter has slightly larger partial charges on the oxygen/hydrogens to account for self-polarization. The original SPC model employed fixed bond lengths and angles, but a flexible SPC model has also been proposed. The TIP3P model in the CHARMM version has additional van der Waals interaction sites for the hydrogens. The TIP4P model similarly comes in several flavors. The TIP4P/Ew is a reparameterization for use in connection with Ewald summation techniques for calculating electrostatic interaction in periodic systems, in contrast to the original simple truncation scheme, while the TIP4P/2005 represent a reparameterization to better account for the whole phase diagram of water, including ice. They differ in the exact position of the off-nucleus site, the partial charges and the van der Waals parameters. In analogy with the SPC model, TIP4P also exist in a version with flexible geometry, but the difference in properties between these two models is very minor (e.g. a 2 °C higher melting temperature).

An isolated water molecule has a dipole moment of 1.85 Debye, but the solution value is ~2.85 Debye, which represents a ~50% increase due to polarization. Fixed charge force fields attempt to model this by increasing the atomic charges, but as noted in Section 2.2.8, this only accounts for the physical effect in an average fashion. The TIP4P-FQ model accounts for the electronic polarization by the fluctuation charge model within the TIP4P framework, while the BK3 (Baranyai and Kiss) model employs a charge-on-a-spring model using the three nuclei as sites. The AMOEBA water model employs up to quadrupole permanent electric moments and a dipole polarization at each nuclear site.

2.6 Coarse Grained Force Fields

Atomistic force fields employ a spherical atom as the fundamental particle which has a clear physical analogy. *Coarse Grained* (*CG*) force fields group several atoms together into a single particle, usually called a *bead*, in order to improve on the computational efficiency and thereby allow simulation of larger systems and longer simulation times. United atom force fields, where an atom with attached hydrogens is modeled as a single particle, can be considered as an intermediate model between an all-atom (AA) and a CG force field model.[114]

A typical CG force field employs a 4-to-1 mapping, meaning that four atoms with associated hydrogens are modeled as a single bead, and we will use the MARTINI protein force field as a representative example.[115,116] The beads are divided into four main groups: charged (Q), polar (P), nonpolar (N) and apolar (C), and each of these bead types are subdivided into 4–5 types depending on their (relative) polarity or hydrogen bond donor/acceptor capabilities for a total of 18 bead types. The bonded energy terms are parameterized by harmonic and Fourier potentials analogous to atomistic force fields. The non-bonded interactions are parameterized by pair-wise Lennard-Jones potentials and charge–charge interactions for the charge-type beads, where the electrostatic interaction employs the Coulomb form in Equation (2.20) with a dielectric constant of 15 to model implicit screening. All non-bonded terms furthermore employ a 12 Å cutoff distance (Section 2.7), beyond which the potentials are shifted to zero in order to improve the computational efficiency. The lower resolution has as a consequence that (slightly) different molecules are described as being identical; for example glycine and alanine and also cysteine and methionine are described identically in the MARTINI force field.

CG FFs sacrifice some molecular details in order to improve the computational efficiency, with the hope that a careful parameterization can absorb some of the deficiencies in the underlying model. Perhaps one of the most severe limitations is the inability to describe hydrogen bonds, which is an essential element in determining secondary protein structure and the interaction with water. CG FFs have so far been unable to solve the secondary protein structure problem by parameterization and rely on additional constraints for fixing specific parts of a protein to have, for example, an alpha-helixal or a beta-sheet structure. The MARTINI force field employs an environmental-dependent bead type, that is a given amino acid will be described by different bead types depending on the local secondary structure, but also employs constraints to keep the secondary structure.[117] The inability to model a protein secondary structure means that CG FFs are unable to simulate protein folding or dynamics that require secondary structure transformations, and the secondary structure must be defined by the user.

Water represents a special problem in a CG description. Simulations of biomolecular systems usually need to solvate the molecule of interest in a large box of water, and water molecules consequently often account for the majority of particles in a given simulation. Coarse graining of the water is therefore an essential requirement for achieving a significant computational improvement. The 4-to-1 mapping defines a water bead as a representation for four water molecules. Since the beads have no internal structure, this means that the water beads become non-polar, and the polar screening effect must instead be modeled by a large ($\varepsilon = 15$) dielectric constant in the Coulomb term. The non-polar nature of water beads has been addressed by introducing polarizable water beads, where the polarization is described by associating two Drude particles (Section 2.2.8) having charges of ± 0.46 within each bead,[118] in which case the dielectric constant is reduced to 2.5.

The 4-to-1 mapping of water molecules obviously leads to problems describing single water molecules within a protein, but also means that the model lacks part of the internal water dynamics. The reduced dynamics (entropy) implies that CG water beads have a too high freezing point (~290 K), which may cause problem in simulations. The freezing point can be lowered by adding a small amount (mole fraction ~0.1) of artificially larger (~20%) water beads, which disrupt the ordering and act as antifreeze particles. The use of the Drude particle polarizable water model has a slightly lower freezing point (~285 K). An implicit water model has also been proposed and implemented as a reparameterization of the non-bonded interaction between CG beads (nicknamed Dry MARTINI).[119]

The reduced number of particles and (internal) degrees of freedom in CG models means that the entropy described within the model is reduced. When parameterized against experimental free energies, the missing entropy leads to a decrease in the enthalpy terms defined by the parameters. This in turn leads to a poor description of the temperature dependence of the free energy, and one should therefore be careful of employing CG methods at temperatures different from that used in the parameterization. The same reservation holds to some extent also for AA models, but the problem is expected to be more severe for CG methods.

The simulation time with an AA force field is closely related to the actual time, such that simulations can be used to probe directly the time scale of various processes. This, however, is also a severe limitation, as computational resources rarely allow simulations of realistic sized systems for longer than a few microseconds (at best), while many interesting phenomena like protein folding and substrate binding/unbinding occur on millisecond time scales. A CG FF description of a given system contains significantly fewer particles than an AA description, and the lack of low-mass particles (hydrogens) allows for using larger time steps (typically an order of magnitude larger than for AA methods), resulting in significantly longer simulation times for a given computational cost. The less detailed description furthermore means that the CG FF energy surface is smoother and therefore intrinsically leads to faster dynamics. The connection between simulation time and "real" time is difficult to quantify, and depends both on the system and the property of interest, but a factor of four is often used as an estimate, that is one ns of simulations time equals four ns of "real" time.

The development of CG FF is still at the exploratory stages, where limitations are discovered and alternative parameterizations and functional forms are being explored. In analogy with the mixing of QM and MM methods, multiscale models consisting of mixing AA and CG FF have also been proposed, but such methods are in their infancy.[120]

2.7 Computational Considerations

Evaluation of the non-bonded energy is by far the most time-consuming step, which can be exemplified by the number of individual energy terms for the linear alkanes $CH_3(CH_2)_{n-2}CH_3$ shown in Table 2.7.

The number of bonded contributions, E_{str}, E_{bend} and E_{tors}, grow *linearly* with the system size, while the non-bonded contributions, E_{vdw} and E_{el}, grow as the *square* of the system size. This is fairly obvious as, for a large molecule, most of the atom pairs are not bonded, or are not bonded to a common atom, and thus contribute with E_{vdw}/E_{el} terms. For $CH_3(CH_2)_{98}CH_3$, which contains a mere 302 atoms, the non-bonded terms already account for ~96% of the computational effort. For a 1000 atom system, the percentage is 98.8%, and for 10 000 atoms it is 99.88%. *In the limit of large molecules, the computational time for calculating the force field energy grows approximately as the square of the number of atoms.* The majority of these non-bonded energy contributions are numerically very small, as the distance between the atom pairs is large. A considerable saving in computational time can be achieved by truncating the van der Waals potential at some distance, say 10 Å. If the distance is larger than this cutoff, the contribution is neglected. This is not quite as clean as it may sound at first. Although it is true that the contribution from a pair of atoms is very small if they are separated by 10 Å, there may be a large number of such atom pairs. The *individual* contribution falls off quickly, but the *number* of contributions also increases. Many force fields use cutoff distances around 10 Å, but it has been shown that the total van der Waals energy only converges if the cutoff distance is of the order of 20 Å. However, using a cutoff of 20 Å may significantly increase the computational time

Table 2.7 Number of terms for each energy contribution in $CH_3(CH_2)_{n-2}CH_3$.

n	N_{atoms}	E_{str}	E_{bend}	E_{tors}	$E_{non\text{-}bonded}$
10	32	31 (5%)	30 (10%)	81 (14%)	405 (70%)
20	62	61 (3%)	60 (6%)	171 (8%)	1710 (83%)
50	152	151 (1%)	300 (3%)	441 (4%)	11 025 (93%)
100	302	301 (1%)	600 (1%)	891 (2%)	44 550 (96%)
	N	$(N-1)$	$2(N-2)$	$3(N-5)$	$\frac{1}{2}N(N-1)-3N+5$

(by a factor of perhaps 5–10) relative to a cutoff of 10 Å. Alternatively, the exact van der Waals energy for a periodic system may be calculated by a variation of the particle mesh Ewald technique.[121]

The introduction of a cutoff distance does not by itself lead to a significant computational saving, since all the distances must be computed prior to the decision on whether to include the contribution or not. A substantial increase in computational efficiency can be obtained by keeping a *non-bonded* or *neighbor list* over atom pairs. From a given starting geometry, a list is prepared over the atom pairs that are within the cutoff distance plus a smaller buffer zone. During a minimization or simulation, only the contributions from the atom pairs on the list are evaluated, which avoids the calculation of distances between all pairs of atoms. Since the geometry changes during the minimization or simulation, the non-bonded list must be updated once an atom has moved more than the buffer zone or simply at (fixed) suitable intervals, for example every 10 or 20 steps.

The use of a cutoff distance reduces the scaling in the large system limit from N^2_{atom} to N_{atom} since the non-bonded contributions are then only evaluated within the local "sphere" determined by the cutoff radius. A cutoff distance of ~10 Å, however, is so large that the large system limit is not achieved in practical calculations. Furthermore, the updating of the neighbor list still involves calculating distances between all atom pairs. The actual scaling is thus N^n_{atom}, where n is between 1 and 2, depending on the details of the system. In most applications, however, it is not the energy of a single geometry that is of interest, but that of an optimized geometry. The larger the molecule, the more degrees of freedom, and the more complicated the geometry optimization is. The gain by introducing a non-bonded cutoff is partly offset by the increase in computational effort in the geometry optimization. Thus, as a rough guideline, the increase in computational time upon changing the size of the molecule can be taken as N^2_{atom}.

The introduction of a cutoff distance, beyond which E_{vdw} is set to zero, is quite reasonable as the neglected contributions rapidly become small for any reasonable cutoff distance. This is not true for the other part of the non-bonded energy, the Coulomb interaction. Contrary to the van der Waals energy, which falls off as R^{-6}, the charge–charge interaction varies as R^{-1}. This is actually only true for the interaction between molecules (or fragments) carrying a net charge. The charge distribution in neutral molecules or fragments makes the long-range interaction behave as a dipole–dipole interaction. Consider, for example, the interaction between two carbonyl groups. The carbons carry a positive and the oxygens a negative charge. Seen from a distance it looks like a bond dipole moment, not two net charges. The interaction between two dipoles behaves like R^{-3}, not R^{-1}, but an R^{-3} interaction still requires a significantly larger cutoff than the van der Waals R^{-6} interaction.

Table 2.8 shows the interaction energy between two carbonyl groups in terms of the MM3 E_{vdw} and E_{el}, the latter described either by an atomic point charge or a bond dipole model. The bond

Table 2.8 Comparing the distance behavior of non-bonded energy contributions (kJ/mol).

Distance (Å)	E_{vdw}	$E_{\text{dipole–dipole}}$	$E_{\text{point charges}}$	$E_{\text{net charges}}$
5	−0.92	1.665	1.598	28.5
10	−0.0060	0.208	0.206	14.2
15	−0.00054	0.0617	0.0614	9.5
20	-9.5×10^{-5}	0.0260	0.0259	7.1
30	-8.4×10^{-6}	0.00770	0.00770	4.7
50	-3.9×10^{-7}	0.00167	0.00167	2.8
100	-6.1×10^{-9}	0.000208	0.000208	1.4

dipole moment is 1.86 Debye, corresponding to atomic charges of ±0.32 separated by a bond length of 1.208 Å. For comparison, the interaction between two net charges of 0.32 is also given.

From Table 2.8 it is clearly seen that E_{vdw} becomes small (less than ~0.01 kJ/mol) beyond a distance of ~10 Å. The electrostatic interaction reaches the same level of importance at a distance of ~30 Å. The table also shows that the interaction between point charges behaves as a dipole–dipole interaction, that is an R^{-3} dependence. The interaction between net charges is very long-range – even at 100 Å separation there is a 1.4 kJ/mol energy contribution. The "cutoff" distance corresponding to a contribution of ~0.01 kJ/mol is of the order of 14 000 Å!

There are different ways of implementing a non-bonded cutoff. The simplest is to neglect all contributions if the distance is larger than the cutoff. This is in general not a very good method as the energy function becomes discontinuous. Derivatives of the energy function also become discontinuous, which causes problems in optimization procedures and when performing simulations. A better method is to use two cutoff distances between which a switching function connects the correct E_{vdw} or E_{el}, or the corresponding forces, smoothly with zero. Such interpolations solve the mathematical problems associated with optimization and simulation, but the chemical significance of the cutoff of course still remains. This is especially troublesome in simulation studies where the distribution of solvent molecules can be very dependent on the use of cutoffs. The modern approaches for evaluating the electrostatic contribution is the use of *fast multipole* or *Ewald* sum methods (see Section 15.3), both of which are able to calculate the electrostatic energy exactly (to within a specified numerical precision) with an effort that scales less than quadratic with the number of particles (linear for fast multipole, $N^{3/2}$ for Ewald and $N \ln N$ for particle mesh Ewald methods). These methods require only slightly more computer time than using a cutoff-based method, and give much better results in a mathematical sense. A caveat is that force field parameters often are optimized with a specific protocol for handling the non-bonded interactions and using, for example, particle mesh Ewald methods with a force field optimized with a cutoff-based protocol and a specific cutoff value may produce worse results, as compared with experimental results.

2.8 Validation of Force Fields

The quality of a force field calculation depends on two quantities: the appropriateness of the mathematical form of the energy expression and the accuracy of the parameters. If elaborate forms for the individual interaction terms have been chosen and a large number of experimental data is available for assigning the parameters, the results of a calculation may be as good as those obtained from experiments, but at a fraction of the cost. This is the case for simple systems such as hydrocarbons. Even a force field with complicated functional forms for each of the energy contributions contains only relatively few parameters when carbon and hydrogen are the only atom types, and experimental data exist for hundreds of such compounds. The parameters can therefore be assigned with a high degree of confidence. Other well-known compound types, such as ethers and alcohol, can achieve almost as good results. For less common species, such as sulfones or polyfunctional molecules, much less experimental information is available and the parameters are less well defined.

Force field methods are primarily geared to predicting two properties: geometries and relative energies. Structural features are in general much easier to predict than relative energies. Each geometrical feature depends only on a few parameters. For example, bond distances are essentially determined by R_0 and the corresponding force constant, bond angles by θ_0 and conformational minima by V_1, V_2 and V_3. It is therefore relatively easy to assign parameters that reproduce a given geometry. Relative energies of different conformations, however, are much more troublesome

Table 2.9 Average errors in heat of formation (kJ/mol) by MM2.[123–127]

Compound type	Average error in ΔH_f
Hydrocarbons	1.8
Ethers and alcohols	2.1
Carbonyl compounds	3.4
Aliphatic amines	1.9
Aromatic amines	12.1
Silanes	4.5

since they are a consequence of many small contributions, that is the exact functional form of the individual energy terms and the balance between them. The largest contributions to conformational energy differences are the non-bonded and torsional terms, and it is therefore important to have good representations of the whole torsional energy profile. Even though a given force field may be parameterized to reproduce rotational energy profiles for ethane and ethanol, and contains a good description of hydrogen bonding between two ethanol molecules, there is no guarantee that it will be successful in reproducing the relative energies of different conformations of, say, 1,2-dihydroxyethane. For large systems, it is inevitable that small inaccuracies in the functional forms for the energy terms and parameters will influence the shape of the whole energy surface to the point where minima may disappear or become saddle points. Essentially all force fields, no matter how elaborate the functional forms and parameterization, will have artificial minima and fail to predict real minima, even for quite small systems. The MM2 force field for cyclododecane predicts 122 different conformations, but the MM3 surface contains only 98 minima.[122] Given that cyclododecane belongs to a class of well-parameterized molecules, the saturated hydrocarbons, and that MM2 and MM3 are among the most accurate force fields, this clearly illustrates the point.

Validation of a force field is typically done by showing how accurately it reproduces a set of reference data, which may or may not have been used in the actual parameterization. Since different force fields employ different sets of reference data, it is difficult to compare their accuracies directly. Indeed, there is no single "best" force field; each has its advantages and disadvantages. They perform best for the type of compounds that have been used in the parameterization, but may give questionable results for other systems. Table 2.9 gives typical accuracies for ΔH_f that can be obtained with the MM2 force field.

The average error is the difference between the calculated and experimental ΔH_f. In this connection is should be noted that the average error in the *experimental* data for the hydrocarbons is 1.7 kJ/mol, that is MM2 essentially reproduce the experiments to within the experimental uncertainty.

There is one final thing that needs to be mentioned in connection with the validation of a force field, namely the reproducibility. The results of a calculation are determined completely by the mathematical expressions for the energy terms and the parameter set (assuming that the computer program is working correctly). A new force field is usually parameterized for a fairly small set of functional groups initially and may then evolve by addition of parameters for a larger diversity later. This sometimes has the consequence that some of the initial parameters must be modified to give an acceptable fit. Furthermore, new experimental data may warrant changes in existing parameters. In some cases, different sets of parameters are derived by different research groups for the same types of functional group. The result is that the parameter set for a given force field is not constant in time, and sometimes not in geographical location either. There may also be differences in the implementation details of the energy terms. The E_{oop} in MM2, for example, is defined as a harmonic term in the bending angle

(Figure 2.6), but may be substituted by an improper torsional angle in some computer programs. The consequence is that there often are several different "flavors" of a given force field, depending on the exact implementation, the original parameter set (which may not be the most recent) and any local additions to the parameters. A vivid example is the MM2 force field, which exists in several different implementations that do not give the exact same results but are nevertheless denoted as "MM2" results. Mainstream force fields such as CHARMM and AMBER have a fixed underlying mathematical form, which is used with periodically updated parameter lists containing a version number or release year to ensure reproducibility.

2.9 Practical Considerations

It should be clear that force field methods are models of the real quantum mechanical systems. The neglect of electrons as individual particles forces the user to define explicitly the bonding in the molecule prior to any calculations. The *user* must decide how to describe a given molecule in terms of the selected force field. The input to a calculation consists of three sets of information:

1. Which atom types are present?
2. How are they connected, that is which atoms are bonded to each other?
3. Make a start guess of the geometry.

The first two sets of information determine the functional form of E_{FF}, that is enable the calculation of the potential energy surface for the molecule. Normally the molecule will then be optimized by minimizing E_{FF}, which requires a starting guess of the geometry, and perhaps used as a starting point for a simulation. All the above three sets of information can be uniquely defined from a (three-dimensional) drawing of a molecule. Modern programs therefore have a graphical interface that allows the molecule simply to be drawn on the screen, constructed from preoptimized fragments or read from an external source, as, for example, an X-ray structure. The interface then automatically assigns suitable atom types based on the selected atomic symbols and the connectivity, and converts to Cartesian coordinates.

2.10 Advantages and Limitations of Force Field Methods

The main advantage of force field methods is the speed with which calculations can be performed, enabling large systems to be treated. Even with a desktop personal computer, molecules with several thousand atoms can be handled. This puts the applications in the region of modeling biomolecular macromolecules, such as proteins and DNA, and molecular modeling is used by most pharmaceutical companies. The ability to treat a large number of particles also makes force field models the only realistic method for performing simulations where solvent effects or crystal packing can be studied (Chapter 15).

For systems where good parameters are available, it is possible to make very good predictions of geometries and relative energies of a large number of molecules in a short time. It is also possible to determine barriers for interconversion between different conformations, although this is much less automated. One of the main problems is of course the lack of good parameters. If the molecule is slightly out of the ordinary, it is very likely that only poor-quality parameters exist, or none at all. Obtaining suitable values for these missing parameters can be a frustrating experience. Force field methods are good for predicting properties for classes of molecules where a lot of information already exists. For unusual molecules, their use is limited, as they often must rely on generic parameters.

Finally, force field methods are "zero-dimensional". It is not possible to assess the probable error of a given result *within* the method. The quality of the result can only be judged by comparing with other calculations on similar types of molecules for which relevant experimental data exist.

2.11 Transition Structure Modeling

Structural changes can be divided into two general types: those of a conformational nature and those involving bond breaking/forming. There are intermediate cases, such as bonds involving metal coordination, but since metal coordination is difficult to model anyway, we will neglect such systems at present. The bottleneck for structural changes is the highest energy point along the reaction path, called the *Transition State* or *Transition Structure* (TS) (Chapter 14). Conformational TSs have the same atom types and bonding for both the reactant and product, and can be located on the force field energy surface by standard optimization algorithms. Since conformational changes are often localized to rotation around a single bond, simply locating the maximum energy structure for rotation (so-called "torsional angle driving", see Section 13.4.1) around this bond often represents a good approximation of the real TS.

Modeling TSs for reactions involving bond breaking/forming within a force field methodology is much more difficult.[128] In this case, the reactant and product are *not* described by the same set of atom types and/or bonding. There may even be a different number of atoms at each end of the reaction (e.g. lone pairs disappearing). This means that there are two *different* force field energy functions for the reactant and product, that is the energy as a function of the reactant coordinate is not continuous. Nevertheless, methods have been developed for modeling *differences* in activation energies between similar reactions by means of force field techniques, and four approaches are described below.

2.11.1 Modeling the TS as a Minimum Energy Structure

One of the early applications of TS modeling was the work on steric effects in S_N2 reactions by DeTar and coworkers, and it has been further developed by Houk and coworkers.[129] The approach consists of first locating the TS for a typical example of the reaction with electronic structure methods, often at the Hartree–Fock or density functional theory level. The force field function is then modified such that an energy minimum is created with a geometry that matches the TS geometry found by the electronic structure method. The modification defines new parameters for all the energy terms involving the partly formed/broken bonds. The stretch energy terms have natural bond lengths taken from the electronic structure calculation, and force constants that are typically half the strength of normal bonds. Bond angle terms are similarly modified with respect to equilibrium values and force constants, the former taken from the electronic structure data and the latter usually estimated. These modifications often necessitate the definition of new "transition state" atom types. Once the force field parameters have been defined, the structure is minimized as usual. Sometimes a few cycles of parameter adjustments and re-optimizations are necessary for obtaining a set of parameters capable of reproducing the desired TS geometry. Norrby and coworkers have described a partly automated method for optimizing the parameters to reproduce the reference structure, with the acronym Q2MM.[130,131] When the modified force field is capable of reproducing the reference TS geometry, it can be used for predicting TS geometries and relative energies of reactions related to the model system. As long as the differences between the systems are purely "steric", it can be hoped that relative energy differences (between the reactant and the TS model) will correlate with relative activation energies. Purely electronic effects, such as Hammett-type effects due to *para*-substitution in aromatic systems, can of course not be modeled by force field techniques.

2.11.2 Modeling the TS as a Minimum Energy Structure on the Reactant/Product Energy Seam

There are two principal problems with the above modeling technique. First, the TS is modeled as a minimum on the energy surface, while it should be a first-order saddle point. This has the consequence that changes in the TS position along the reaction coordinate due to differences in the reaction energy will be in the wrong direction (Section 16.7). In many cases, this is probably not important. For reactions having a reasonable barrier, the TS geometry appears to be relatively constant, which may be rationalized in terms of the Marcus equation (Section 16.6). Comparing reactions that differ in terms of the steric hindrance at the TS, however, may be problematic, as the TS changes along the reaction coordinate will be in the wrong direction. The second problem is the more or less *ad hoc* assignment of parameters. Even for quite simple reactions, many new parameters must be added. Inventing perhaps 40 new parameters for reproducing maybe five relative activation energies raises the nagging question as to whether TS modeling is just a fancy way of describing five data points by 40 variables.[132]

Both of these problems are eliminated in the intersecting potential energy surface modeling technique called *SEAM*.[133] The force field TS is here modeled as the lowest point on the seam of the reactant and product energy functions, as shown in Figure 2.22. Locating the minimum energy structure on the seam is an example of a constrained optimization; the energy should be minimized subject to the constraint that the reactant and product energies are identical. Although this is computationally somewhat more complicated than the simple minimization required in the Houk approach, it can be handled in a quite efficient manner.

In the SEAM approach only the force field parameters for describing the reactant and products are necessary, alleviating the problem of assigning parameters specific for the TS. Furthermore, differences in reactivity due to differences in reaction energy or steric hindrance at the TS are automatically included. The question is how accurately the lowest energy point on the seam resembles the actual TS. This is difficult to evaluate rigorously as it is intimately connected with the accuracy of the force field used for describing the reactant and product structures. It is clear that the TS will have bond distances and angles significantly different from equilibrium structures. This method of TS modeling therefore

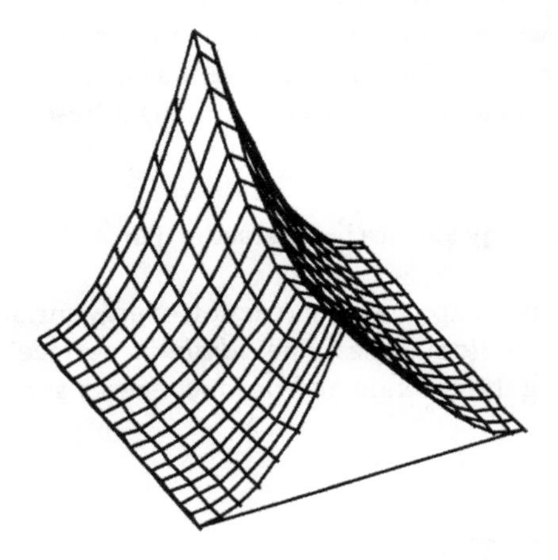

Figure 2.22 Modeling a transition structure as a minimum on the intersection of two potential energy surfaces.

requires a force field that is accurate over a much wider range of geometries than normal. Especially important is the stretch energy, which must be able to describe bond breaking. A polynomial expansion is therefore not suitable, and for example a Morse function is necessary. Similarly, the repulsive part of the van der Waals energy must be fairly accurate, which means that Lennard-Jones potentials are not suitable and should be replaced by, for example, Buckingham-type potentials. Furthermore, many of the commonly employed cross terms (Section 2.2.9) become unstable at long bonds lengths and must be modified. When such modifications are incorporated, however, the intersecting energy surface model appears to give surprisingly good results.

There are, of course, also disadvantages in this approach: these are essentially the same as the advantages! The SEAM method automatically includes the effect of different reaction energies, since a more exothermic reaction will move the TS toward the reactant and lower the activation energy (Section 16.6). This, however, requires that the force field be able to calculate relative energies of the reactant and product, that is the ability to convert steric energies to heat of formation. As mentioned in Section 2.2.11, there are only a few force fields that have been parameterized for this. In practice, this is not a major problem since the reaction energy for a prototypical example of the reaction of interest can be obtained from experimental data or estimated. Using the normal force field assumption of transferability of heat of formation parameters, the difference in reaction energy is thus equal to the difference in steric energy. Only the reaction energy for a single reaction of the given type therefore needs to be estimated and relative activation energies are not sensitive to the exact value used.

If the minimum energy seam structure does not accurately represent the actual TS (compared, for example, with that obtained from an electronic structure calculation) the lack of specific TS parameters becomes a disadvantage. In the Houk approach, it is fairly easy to adjust the relevant TS parameters to reproduce the desired TS geometry. In the intersecting energy surface method, the TS geometry is a complicated result of the force field parameters for the reactant and product, and the force field energy functions. Modifying the force field parameters, or the functional form of some of the energy terms, in order to achieve the desired TS geometry without destroying the description of the reactant/product, is far from trivial. A final disadvantage, which is inherent to the SEAM method, is the implicit assumption that all the geometrical changes between the reactant and product occur in a "synchronous" fashion, albeit weighted by the energy costs for each type of distortion. "Asynchronous" or "two-stage" reactions (as opposed to two-step reactions that involve an intermediate), where some geometrical changes occur mainly before the TS and others mainly after the TS, are difficult to model by this method.

Since the TS is given in terms of the diabatic energy surfaces for the reactant and product, it is also clear that activation energies will be too high. For evaluating relative activation energies of similar reactions this is not a major problem since the important aspect is the relative energies. The overestimation of the activation energy can be improved by adding a "resonance" term to the force field, as discussed in the next section.

2.11.3 Modeling the Reactive Energy Surface by Interacting Force Field Functions

Within a valence bond approach (Chapter 8), the reaction energy surface can be considered as arising from the interaction of two *diabatic* surfaces. The *adiabatic* surface can be generated by solving a 2 × 2 secular equation involving the reactant and product energy surfaces, E_r and E_p:

$$\begin{vmatrix} E_r - E & V \\ V & E_p - E \end{vmatrix} = 0$$
$$E = \frac{1}{2}\left[(E_r + E_p) - \sqrt{(E_r + E_p)^2 + 4V^2}\right] \tag{2.56}$$

Warshel has pioneered the *Extended Valence Bond* (EVB) method,[134,135] where the reactant and product surfaces are described by force field energy functions, and Truhlar has generalized the approach by the *MultiConfigurations Molecular Mechanics* (MCMM) method.[136] In either case, the introduction of the interaction term V generates a continuous energy surface for transforming the reactant into the product configuration, and the TS can be located analogously to energy surfaces generated by electronic structure methods. The main drawback of this method is the somewhat arbitrary interaction element, and the fact that the TS must be located as a first-order saddle point, which is significantly more difficult than locating minima or minima on seams. It can be noted that the SEAM method corresponds to the limiting case where $V \rightarrow 0$ in the EVB method.

2.11.4 Reactive Force Fields

Reactive force fields denote methods that are capable of breaking and forming bonds within a parameterized energy function model. A fundamental difference relative to standard force fields is that reactive force fields contain no explicit bonding information, and all energy terms must be defined entirely in terms of interatomic distances. Within the field of modeling reactions on surfaces, the energy function is often written in terms of two- and three-body potentials that are parameterized for a few atoms for specific purposes. The corresponding potentials are often associated with the names of Stillinger-Weber, Brenner and Tersoff. Reactive force fields for organic compounds are somewhat more involved, as the atoms (primarily C, N, O, S) can participate in several different types of bonding.

Reactive force fields for organic systems can be designed by modifying the energy terms such that the energy function can describe both end-points and all intermediate structures, and thus create a continuous surface connecting the reactant and product energy functions, which is the approach used in the *ReaxFF* method.[137] The necessary requirement for a reactive force field is that bonds should be able to break and form, that is the stretch energy should approach the dissociation energy as the bond distance goes towards infinity. This can, for example, be achieved by a Morse potential as shown in Equation (2.5), but this in addition requires that all the other energy terms are also modified. Breaking a bond A–B implies that bending and torsional terms involving this bond should disappear, while the bonded stretch interaction is gradually replaced by non-bonded electrostatic and van der Waals interactions. In order to create a smooth and realistic variation of the energy with geometry, these changes must be introduced by suitable interpolation functions and be properly parameterized.

The central feature in the ReaxFF method is the use of a bond order (Equation (10.10)) parameter to perform the interpolation of the various energy terms. The bond order ρ'_{AB} between two atoms A and B is parameterized as a sum of three terms accounting for a σ and up to two π bonds, where the p/q parameters model the distance behavior of each term:

$$\rho'_{AB} = \sum_{i=1}^{3} \exp\left(-p_i \left(\frac{R^{AB}}{R_0^{AB}}\right)^{q_i}\right) \tag{2.57}$$

Application of Equation (2.57) to a given atom will in general lead to overcoordination, that is the sum of the ρ' will exceed the valence. The directly "bonded" atoms (1–2 interactions in standard force fields) provide the main contribution, but longer-range contributions (primarily 1–3 interactions in standard force fields) increase the value, such that a carbon atom may end up having a total bond order significantly larger than its valence of 4. A number of multiplicative terms depending on the deviation of the sum of bond orders from the valence of the given atom are applied to the ρ' defined in Equation (2.57) to produce an (unprimed) bond order ρ_{AB}. Additional energy terms accounting for over- and undercoordination are also added. The stretch energy is modeled as shown below, where D_e is a dissociation energy parameter and p is a suitable parameter for a given bond:

$$E_{\text{str}}(\rho) = -D_{\text{e}}\rho \exp(p(1-\rho^p)) \tag{2.58}$$

Multiplicative interpolation functions of the type shown below are used to reduce bending and torsional terms to zero as a bond dissociates:

$$f(\rho) = 1 - \exp(-p\rho^q) \tag{2.59}$$

Since explicit bonding information is not present, non-bonded interactions are calculated between all pairs of atoms, in contrast to non-reactive force fields where 1–2 and 1–3 interactions are neglected. In order to avoid excessive repulsive van der Waals interactions and potentially infinite electrostatic attraction energies between atoms with opposite charges at short distances, the energy expressions in Equations (2.11) and (2.21) must be modified. The electrostatic energy is modeled by a screening Coulomb expression shown below, where ζ^{AB} is a suitable screening parameter:

$$E_{el}(R^{\text{AB}}) = \frac{Q^{\text{A}}Q^{\text{B}}}{((R^{\text{AB}})^3 + (\zeta^{\text{AB}})^3)^{1/3}} \tag{2.60}$$

An additional complication is that the atomic charge Q must be allowed to change as a function of bond distance, and this can be accomplished by a variation of the fluctuation charge method (Equation (2.31)).

2.12 Hybrid Force Field Electronic Structure Methods

Force field methods are inherently unable to describe the details of bond breaking/forming or electron transfer reactions, since there is an extensive rearrangement of the electrons. If the system of interest is too large to treat entirely by electronic structure methods, there are two possible approximate methods that can be used. In some cases, the system can be "pruned" to a size that can be treated by replacing "unimportant" parts of the molecule with smaller model groups, for example substitution of a hydrogen or methyl group for a phenyl ring. For studying enzymes, however, it is usually assumed that the whole system is important for holding the active size in the proper arrangement, and the "backbone" conformation may change during the reaction. Similarly, for studying solvation, it is not possible to "prune" the number of solvent molecules without severely affecting the accuracy of the model. Hybrid methods have been designed for modeling such cases, where the active size is calculated by electronic structure methods (usually semi-empirical, low-level *ab initio* or density functional methods), while the backbone is calculated by a force field method.[138] Such methods are often denoted *Quantum Mechanics–Molecular Mechanics* (QM/MM). The partition can formally be done by dividing the Hamiltonian and resulting energy into three parts:

$$\begin{aligned} \mathbf{H}_{\text{total}} &= \mathbf{H}_{\text{QM}} + \mathbf{H}_{\text{MM}} + \mathbf{H}_{\text{QM/MM}} \\ E_{\text{total}} &= E_{\text{QM}} + E_{\text{MM}} + E_{\text{QM/MM}} \end{aligned} \tag{2.61}$$

The QM and MM regions are described completely analogously to the corresponding isolated system, using the techniques discussed in Chapters 2 to 6. The main problem with QM/MM schemes is deciding how the two parts should interact (i.e. $\mathbf{H}_{\text{QM/MM}}$). The easiest situation is when the two regions are not connected by covalent bonds, as, for example, when using an MM description for modeling the effect of solvation on a QM system. If the two regions are connected by covalent bonding, as, for example, when using a QM model for the active site in an enzyme and describing the backbone by an MM model, the partitioning is somewhat more difficult.

The description of the QM/MM interaction can be divided into three levels of increasing accuracy:[138]

1. *Mechanical embedding.* Only the direct bonded and steric effects of the MM atom on the QM part are taken into account.
2. *Electrostatic embedding.* In addition, the electric field generated by the MM atoms, typically represented by partial atomic charges, is allowed to influence the QM region.
3. *Polarizable embedding.* The MM atoms are described by both static electric moments and polarizabilities, and the MM and QM regions are allowed to mutually polarize each other.

In the mechanical embedding approach only the bonded and steric energies of the two regions are included in the interaction term, that is QM atoms have additional forces generated by the MM framework, and vice versa, but there is no interaction between the electronic parts of the two regions. The QM atoms are assigned van der Waals parameters and included in an MM non-bonded energy expression, as illustrated by a Lennard-Jones potential:

$$\mathbf{H}_{QM/MM} = \sum_{A}^{N_{MM-Atoms}} \sum_{B}^{N_{QM-Atoms}} \varepsilon^{AB} \left[\left(\frac{R_0^{AB}}{R^{AB}} \right)^{12} - 2 \left(\frac{R_0^{AB}}{R^{AB}} \right)^{6} \right] \tag{2.62}$$

The QM atoms may also be assigned partial charges, for example from a population analysis, and charge–charge interactions between the QM and MM atoms included by a classical expression such as Equation (2.21). If the two regions are bonded there are additional terms corresponding to stretching and bending interactions. The mechanical embedding model is rarely a useful level of approximation, as the wave function of the QM region does not respond to changes in the MM region except for pure geometrical changes.

The next level of improvement is called electrostatic embedding, where the atoms in the MM regions are allowed to polarize the QM region. Partial charges on the MM atoms can be incorporated into the QM Hamiltonian analogously to nuclear charges (i.e. adding $\mathbf{V}_{ne}$-like terms to the one-electron matrix elements in Equation (3.59)), and the QM atoms thus feel the electric potential due to all the MM atoms:

$$\mathbf{V}_{QM/MM} = \sum_{A}^{N_{MM-Atoms}} \frac{Q^A}{|\mathbf{R}_A - \mathbf{r}_i|} \tag{2.63}$$

The non-bonded mechanical term in Equation (2.62) is still needed in order to prevent the MM atoms from drifting into the QM region. The electrostatic embedding allows the geometry of MM atoms to influence the QM region, that is the wave function in the QM region becomes coupled to the MM geometry. An interesting computational issue arises when the number of MM atoms is large and the QM region is small, since the calculation of the one-electron integrals associated with $\mathbf{V}_{QM/MM}$ may become a dominating factor, rather than the two-electron integrals associated with the QM region itself, but in most cases the inclusion of the $\mathbf{V}_{Q\,M/MM}$ term only marginally increases the computational effort over a mechanical embedding.

The third level, often called polarizable embedding, is made by allowing the QM atoms also to polarize the MM region, that is the electric field generated by the QM region influences the MM electric moments. This of course requires that a polarizable force field is employed (Section 2.2.8) and necessitates an iterative procedure for allowing the electric fields in both the QM and MM regions to be determined in a self-consistent fashion.

While electrostatic embedding typically accounts for the electric field generated by the MM atom by assigning a set of partial atomic charges to each atom, polarizable embedding usually assigns both partial charges, dipoles and quadrupole moments, as well as a dipole polarizability to each atom. This substantially increases the computational cost, and since polarizable force fields are not yet commonly used anyway, most QM/MM methods employ the electrostatic embedding approximation. An exception is the *Effective Fragment Potential* (EFP) method where the EFP energy is parameterized as a sum of five terms:[139]

$$E^{\mathrm{EFP}} = E_{\mathrm{el}} + E_{\mathrm{pol}} + E_{\mathrm{disp}} + E_{\mathrm{ex-rep}} + E_{\mathrm{CT}} \tag{2.64}$$

The electrostatic energy is described by distributed electric moments up to octupoles based on a DMA partitioning, while the polarization energy is taken into account by distributed dipole polarizabilities. The dispersion energy is parameterized as a two-term expansion with R^{-6} and R^{-8} distance dependence, where the latter is taken as a fixed fraction of the first term. The exchange-repulsion energy is parameterized in terms of overlap integrals between the fragments and the charge-transfer term is parameterized in terms of interactions between occupied and virtual orbitals on the two fragments.

The QM and MM regions in many cases belong to the same molecule, and the division between the two parts must be done by cutting one or more covalent bonds. This leaves one or more unpaired electrons in the QM part, which must be properly terminated. The dangling bonds are in most cases terminated by adding "link" atoms, typically a hydrogen. For semi-empirical methods, it can also be a pseudo-halogen atom with parameters adjusted to provide a special link atom.[140] Alternatively, the termination can be in the form of a localized molecular or generalized hybrid orbital.[141,142] Both approaches provide results of similar quality when care is taken in the selection of the two regions, but the link atom method is somewhat simpler to implement.[143] When the link atom procedure is used, the link atom(s) is only present in the QM calculation and is not seen by the MM framework. A number of choices must also be made for which and how many of the MM bend and torsional terms that involve one or more QM atoms are included. Bending terms involving two MM and one QM atoms are usually included, but those involving one MM and two QM atoms may be neglected. Similarly, the torsional terms involving only one QM atom are usually included, but those involving two or three QM atoms may or may not be neglected.

A general problem is that the QM wave function extends beyond the formal QM region. The presence of nearby positively/negatively charged MM atoms will thus attract/repel the electrons, which can lead to artificial distortions of the wave function. This is especially problematic with positively charged MM atoms in connection with QM wave functions expanded in a basis set with diffuse functions, where the MM atoms may attract significant electron density. If these MM atoms were described in a QM framework, there would be exchange–repulsion forces that would prevent the accumulation of electron density from the real QM region, but this is missing in the QM/MM description. For covalently bonded QM/MM systems, this dictates that the partitioning should be made through non-polar bonds. A heuristic fix is to limit the basis set in the QM region to typically of DZP quality, where the basis functions are sufficiently compact that the delocalization into the MM region is prevented. The concept of mixing methods of different accuracy has been generalized in the *ONIOM* (*Our own N-layered Integrated molecular Orbital Molecular mechanics*) method to include several (usually two or three) layers, for example using relatively high-level theory in the central part, a lower-level electronic structure theory in an intermediate layer and force field to treat the outer layer.[144] The original ONIOM method only employed mechanical embedding for the QM/MM interface, but more recent extensions have also included electronic embedding.

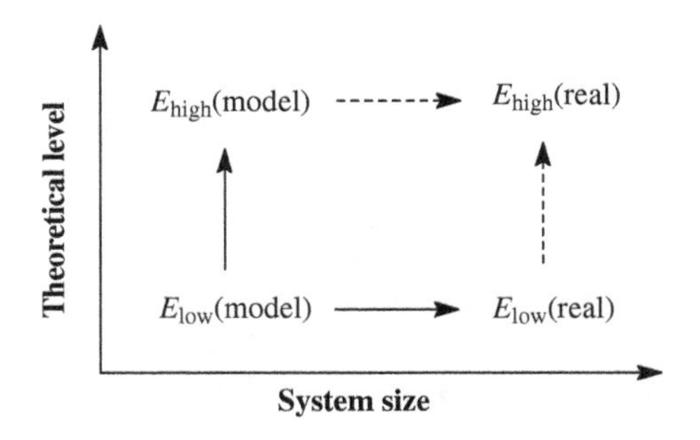

Figure 2.23 Illustration of the ONIOM extrapolation method.

The ONIOM method employs an extrapolation scheme based on assumed additivity, in analogy to the CBS, Gn and Wn methods discussed in Section 5.10. For a two-layer scheme, the small (model) system is calculated at both the low and high levels of theory, while the large (real) system is calculated at the low level of theory. The result for the real system at the high theoretical level is estimated by adding the change between the high and low levels of theory for the model system to the low-level results for the real system, as illustrated in Figure 2.23 and the following equation:

$$\begin{aligned} E_{\text{ONIOM}}(\text{real system, high level}) &= E_{\text{high level}}(\text{model system}) \\ &- E_{\text{low level}}(\text{model system}) + E_{\text{low level}}(\text{real system}) \end{aligned} \tag{2.65}$$

A similar extrapolation can be done for multilevel ONIOM models, although it requires several intermediate calculations. It should be noted that derivatives of the ONIOM model can be constructed straightforwardly from the corresponding derivative of the underlying methods, and it is thus possible to perform geometry optimizations and vibrational analysis using the ONIOM energy function.

QM/MM methods are often used for modeling solvent effects, with the solvent treated by MM methods, but in some cases the first solvation shell is included in the QM region. If such methods are used in connection with dynamical sampling of the configurational space, it is possible that MM solvent molecules can enter the QM regions, or QM solvent molecules can drift into the MM region. In order to handle such situations, there must be a procedure for allowing solvent molecules to switch between a QM and MM description. In order to ensure a smooth transition, a transition region can be defined between the two parts, where a switching function is employed to make a continuous transition between the two descriptions.[145]

The main problem with QM/MM methods is that there is no unique way of deciding which part should be treated by force field and which by quantum mechanics, and QM/MM methods are therefore not "black box" methods. The "stitching" together of the two regions is not unique and the many possible combinations of force field and QM methods make QM/MM methods very diverse. QM/MM methods for describing large systems, like a chemical reaction occurring within an enzyme, typically employ a fixed charge force field and a DFT method using a medium-sized basis set, while QM/MM methods aimed at, for example, investigating environment effects on spectra often employ a fairly small number of solvent molecules with an elaborate MM description including high-order static electric moments and polarizability and a high-level QM method, like CCSD, and a large basis set. Ideally a given choice of QM and MM partitioning should be calibrated by systematically enlarging the QM region and monitoring how the property of interest (hopefully) converges. When using a fixed charge force field, Sumowski and Ochsenfeld have shown that the energetic results converge when a QM region extends ~10 Å away from the reactive center, which means that the QM region should include ~1000 atoms, but this is prohibitively large for many applications.[146] Siegbahn and Himo, however, have argued that significantly smaller QM regions with ~100 atoms are sufficient for achieving results converged to chemical accuracy.[147] It is likely that the quality of the force field in the MM region influences the size of the QM region necessary for achieving a given accuracy; that is by the use of a polarizable force field instead of a simple fixed charge one may reduce the size of the required QM region, but no definitive investigations have been reported.

Another troublesome aspect of the QM/MM method is the sampling problem. The total QM/MM energy of a system with, say, 100 QM atoms and 10 000 MM atoms will be very sensitive to the exact configuration of the MM atoms. When constructing, for example, an energy profile for an enzymatic chemical reaction, where the position of the QM atoms are optimized in the presence of the full set of geometry relaxed MM atoms, it is essential to ensure that the MM atoms have the same configuration along the whole energy profile. If, for example, hydrogen bonds are broken or formed in the MM region as the QM region is changed, this will leads to discontinuous energy jumps and possible hysteresis (energy profiles in the forward and reverse directions are not identical). Since active sampling of the MM region rarely is possible due to the high computational cost of the QM region, the results may depend on the specific chosen MM configuration.

References

1 U. Burkert and N. L. Allinger, *Molecular Mechanics* (ACS, 1982).
2 U. Dinur and A. T. Hagler, *Reviews in Computational Chemistry* **2**, 99 (1991).
3 A. K. Rappe and C. J. Casewit, *Molecular Mechanics Across Chemistry* (University Science Books, 1997).
4 J. W. Ponder and D. A. Case, *Protein Simulations* **66**, 27 (2003).
5 A. D. Mackerell, *Journal of Computational Chemistry* **25** (13), 1584–1604 (2004).
6 N. L. Allinger, *Journal of the American Chemical Society* **99** (25), 8127–8134 (1977).
7 N. L. Allinger, Y. H. Yuh and J. H. Lii, *Journal of the American Chemical Society* **111** (23), 8551–8566 (1989).
8 S. W. Benson, *Thermochemical Kinetics* (John Wiley & Sons, 1976).
9 P. L. A. Popelier and F. M. Aicken, *Chemphyschem* **4** (8), 824–829 (2003).
10 P. M. Morse, *Physical Review* **34** (1), 57–64 (1929).
11 M. Mollhoff and U. Sternberg, *Journal of Molecular Modeling* **7** (4), 90–102 (2001).
12 P. Jensen, *Journal of Molecular Spectroscopy* **133** (2), 438–460 (1989).
13 A. T. Hagler, *Journal of Chemical Theory and Computation* **11** (12), 5555–5572 (2015).
14 B. Albinsson, H. Teramae, J. W. Downing and J. Michl, *Chemistry – A European Journal* **2** (5), 529–538 (1996).
15 P. C. Chen, *International Journal of Quantum Chemistry* **62** (2), 213–221 (1997).
16 F. London, *Zeitschrift Fur Physik* **63** (3–4), 245–279 (1930).
17 J. E. Jones, *Proceedings of the Royal Society of London A: Mathematical, Physical and Engineering Sciences* **106** (738), 463–477 (1924).
18 T. A. Halgren, *Journal of the American Chemical Society* **114** (20), 7827–7843 (1992).
19 T. L. Hill, *Journal of Chemical Physics* **16** (4), 399–404 (1948).
20 J. R. Hart and A. K. Rappe, *Journal of Chemical Physics* **97** (2), 1109–1115 (1992).
21 J. M. Hayes, J. C. Greer and D. A. Morton-Blake, *Journal of Computational Chemistry* **25** (16), 1953–1966 (2004).

22 C. M. Baker, P. E. M. Lopes, X. Zhu, B. Roux and A. D. MacKerell, *Journal of Chemical Theory and Computation* **6** (4), 1181–1198 (2010).
23 G. A. Cisneros, M. Karttunen, P. Ren and C. Sagui, *Chemical Reviews* **114** (1), 779–814 (2013).
24 J. Gasteiger and H. Saller, *Angewandte Chemie – International Edition in English* **24** (8), 687–689 (1985).
25 C. N. Schutz and A. Warshel, *Proteins – Structure Function and Bioinformatics* **44** (4), 400–417 (2001).
26 M. M. Francl and L. E. Chirlian, *Reviews in Computational Chemistry* **14**, 1–31 (2000).
27 C.I. Bayly, P. Cieplak, W. D. Cornell and P. A. Kollman, *Journal of Physical Chemistry* **97** (40), 10269–10280 (1993).
28 L. E. Chirlian and M. M. Francl, *Journal of Computational Chemistry* **8** (6), 894–905 (1987).
29 C. M. Breneman and K. B. Wiberg, *Journal of Computational Chemistry* **11** (3), 361–373 (1990).
30 R. J. Woods, M. Khalil, W. Pell, S. H. Moffat and V. H. Smith, *Journal of Computational Chemistry* **11** (3), 297–310 (1990).
31 J. C. White and E. R. Davidson, *Journal of Molecular Structure – Theochem* **101** (1–2), 19–31 (1993).
32 E. V. Tsiper and K. Burke, *Journal of Chemical Physics* **120** (3), 1153–1156 (2004).
33 P. Soderhjelm and U. Ryde, *Journal of Computational Chemistry* **30** (5), 750–760 (2009).
34 M. M. Francl, C. Carey, L. E. Chirlian and D. M. Gange, *Journal of Computational Chemistry* **17** (3), 367–383 (1996).
35 E. Sigfridsson and U. Ryde, *Journal of Computational Chemistry* **19** (4), 377–395 (1998).
36 S. Jakobsen and F. Jensen, *Journal of Chemical Theory and Computation* **10** (12), 5493–5504 (2014).
37 C. Aleman, M. Orozco and F. J. Luque, *Chemical Physics* **189** (3), 573–584 (1994).
38 Y. Yuan, M. J. L. Mills and P. L. A. Popelier, *Journal of Computational Chemistry* **35** (5), 343–359 (2014).
39 E. Whalley, *Chemical Physics Letters* **53** (3), 449–451 (1978).
40 Q. Wang, J. A. Rackers, C. He, R. Qi, C. Narth, L. Lagardere, N. Gresh, J. W. Ponder, J.-P. Piquemal and P. Ren, *Journal of Chemical Theory and Computation* **11** (6), 2609–2618 (2015).
41 G. R. Medders, V. Babin and F. Paesani, *Journal of Chemical Theory and Computation* **9** (2), 1103–1114 (2013).
42 T. D. Rasmussen, P. Ren, J. W. Ponder and F. Jensen, *International Journal of Quantum Chemistry* **107** (6), 1390–1395 (2007).
43 R. W. Dixon and P. A. Kollman, *Journal of Computational Chemistry* **18** (13), 1632–1646 (1997).
44 S. Cardamone, T. J. Hughes and P. L. A. Popelier, *Physical Chemistry Chemical Physics* **16** (22), 10367-10387 (2014).
45 A. J. Stone, *Journal of Chemical Theory and Computation* **1** (6), 1128–1132 (2005).
46 R. Anandakrishnan, C. Baker, S. Izadi and A. V. Onufriev, *PloS One* **8** (7), e67715 (2013).
47 C. E. Dykstra, *Chemical Reviews* **93** (7), 2339–2353 (1993).
48 H. B. Yu and W. F. van Gunsteren, *Computer Physics Communications* **172** (2), 69–85 (2005).
49 A. K. Rappe and W. A. Goddard, *Journal of Physical Chemistry* **95** (8), 3358–3363 (1991).
50 U. Dinur and A. T. Hagler, *Journal of Computational Chemistry* **16** (2), 154–170 (1995).
51 S. Patel and C. L. Brooks, *Journal of Computational Chemistry* **25** (1), 1–15 (2004).
52 G. Lamoureux, A. D. MacKerell and B. Roux, *Journal of Chemical Physics* **119** (10), 5185–5197 (2003).
53 G. Lamoureux and B. T. Roux, *Journal of Chemical Physics* **119** (6), 3025–3039 (2003).
54 P. Cieplak, F.-Y. Dupradeau, Y. Duan and J. Wang, *Journal of Physics – Condensed Matter* **21** (33), 333102 (2009).
55 A. C. Simmonett, F. C. Pickard, Y. Shao, T. E. Cheatham, III and B. R. Brooks, *Journal of Chemical Physics* **143** (7) 074115 (2015).
56 B. T. Thole, *Chemical Physics* **59** (3), 341–350 (1981).

57 P. Y. Ren and J. W. Ponder, *Journal of Physical Chemistry B* **107** (24), 5933–5947 (2003).

58 H. A. Stern, G. A. Kaminski, J. L. Banks, R. H. Zhou, B. J. Berne and R. A. Friesner, *Journal of Physical Chemistry B* **103** (22), 4730–4737 (1999).

59 P. Y. Ren and J. W. Ponder, *Journal of Computational Chemistry* **23** (16), 1497–1506 (2002).

60 M. A. Freitag, M. S. Gordon, J. H. Jensen and W. J. Stevens, *Journal of Chemical Physics* **112** (17), 7300–7306 (2000).

61 P. E. M. Lopes, J. Huang, J. Shim, Y. Luo, H. Li, B. Roux and A. D. MacKerell, *Journal of Chemical Theory and Computation* **9** (12), 5430–5449 (2013).

62 H. D. Thomas, K. H. Chen and N. L. Allinger, *Journal of the American Chemical Society* **116** (13), 5887–5897 (1994).

63 N. L. Allinger, K. S. Chen, J. A. Katzenellenbogen, S.R. Wilson and G.M. Anstead, *Journal of Computational Chemistry* **17** (5–6), 747–755 (1996).

64 K. Kveseth, R. Seip and D. A. Kohl, *Acta Chemica Scandinavica Series A – Physical and Inorganic Chemistry* **34** (1), 31–42 (1980).

65 R. Engeln, D. Consalvo and J. Reuss, *Chemical Physics* **160** (3), 427–433 (1992).

66 J. T. Sprague, J. C. Tai, Y. Yuh and N. L. Allinger, *Journal of Computational Chemistry* **8** (5), 581–603 (1987).

67 A. J. Gordon and R. A. Ford, *The Chemist's Companion* (John Wiley & Sons, 1972).

68 D. Reith and K. N. Kirschner, *Computer Physics Communications* **182** (10), 2184–2191 (2011).

69 S. W. I. Siu, K. Pluhackova and R. A. Boeckmann, *Journal of Chemical Theory and Computation* **8** (4), 1459–1470 (2012).

70 I. J. Chen, D. X. Yin and A. D. MacKerell, *Journal of Computational Chemistry* **23** (2), 199–213 (2002).

71 J. C. Slater and J. G. Kirkwood, *Physical Review* **37** (6), 682–697 (1931).

72 J. F. Stanton, *Physical Review A* **49** (3), 1698–1703 (1994).

73 L.-P. Wang, T. J. Martinez and V. S. Pande, *Journal of Physical Chemistry Letters* **5** (11), 1885–1891 (2014).

74 C. R. Landis, D. M. Root and T. Cleveland, *Rev. Comp. Chem.* **6**, 73 (1995).

75 P. Comba and M. Kerscher, *Coordination Chemistry Reviews* **253** (5–6), 564–574 (2009).

76 B. P. Hay, Coordination Chemistry Reviews **126** (1–2), 177–236 (1993).

77 I. H. R. J. Gillespie, *The VSEPR Model of Molecular Geometry* (Allyn and Bacon, Boston, 1991).

78 V. J. Burton, R. J. Deeth, C. M. Kemp and P. J. Gilbert, *Journal of the American Chemical Society* **117** (32), 8407–8415 (1995).

79 M. Foscato, R. J. Deeth and V. R. Jensen, *Journal of Chemical Information and Modeling* **55** (6), 1282–1290 (2015).

80 A. K. Rappe, C. J. Casewit, K. S. Colwell, W. A. Goddard and W. M. Skiff, *Journal of the American Chemical Society* **114** (25), 10024–10035 (1992).

81 D. M. Root, C. R. Landis and T. Cleveland, *Journal of the American Chemical Society* **115** (10), 4201–4209 (1993).

82 Y. Duan, C. Wu, S. Chowdhury, M. C. Lee, G. M. Xiong, W. Zhang, R. Yang, P. Cieplak, R. Luo, T. Lee, J. Caldwell, J. M. Wang and P. Kollman, *Journal of Computational Chemistry* **24** (16), 1999–2012 (2003).

83 Y. Shi, Z. Xia, J. Zhang, R. Best, C. Wu, J. W. Ponder and P. Ren, *Journal of Chemical Theory and Computation* **9** (9), 4046–4063 (2013).

84 M. J. Hwang, T. P. Stockfisch and A. T. Hagler, *Journal of the American Chemical Society* **116** (6), 2515–2525 (1994).

85 K. Vanommeslaeghe, E. Hatcher, C. Acharya, S. Kundu, S. Zhong, J. Shim, E. Darian, O. Guvench, P. Lopes, I. Vorobyov and A. D. MacKerell, Jr., *Journal of Computational Chemistry* **31** (4), 671–690 (2010).

86 H. Sun, *Journal of Physical Chemistry B* **102** (38), 7338–7364 (1998).

87 S. D. Morley, R. J. Abraham, I. S. Haworth, D. E. Jackson, M. R. Saunders and J. G. Vinter, *Journal of Computer-Aided Molecular Design* **5** (5), 475–504 (1991).

88 S. Lifson, A. T. Hagler and P. Dauber, *Journal of the American Chemical Society* **101** (18), 5111–5121 (1979).

89 S. L. Mayo, B. D. Olafson and W. A. Goddard, *Journal of Physical Chemistry* **94** (26), 8897–8909 (1990).

90 E. M. Engler, J. D. Andose and P. V. Schleyer, *Journal of the American Chemical Society* **95** (24), 8005–8025 (1973).

91 G. Nemethy, K. D. Gibson, K. A. Palmer, C. N. Yoon, G. Paterlini, A. Zagari, S. Rumsey and H. A. Scheraga, *Journal of Physical Chemistry* **96** (15), 6472–6484 (1992).

92 J. L. M. Dillen, *Journal of Computational Chemistry* **16** (5), 595–609 (1995).

93 M. Levitt, M. Hirshberg, R. Sharon and V. Daggett, *Computer Physics Communications* **91** (1–3), 215–231 (1995).

94 S. H. Shi, L. Yan, Y. Yang, J. Fisher-Shaulsky and T. Thacher, *Journal of Computational Chemistry* **24** (9), 1059–1076 (2003).

95 J.M. Wang, R.M. Wolf, J.W. Caldwell, P.A. Kollman and D.A. Case, *Journal of Computational Chemistry* **25** (9), 1157–1174 (2004).

96 W. Plazinski, A. Lonardi and P. H. Huenenberger, *Journal of Computational Chemistry* **37** (3), 354–365 (2016).

97 N. L. Allinger, K. S. Chen and J. H. Lii, *Journal of Computational Chemistry* **17** (5–6), 642–668 (1996).

98 T. A. Halgren, *Journal of Computational Chemistry* **17** (5–6), 490–519 (1996).

99 P. Comba, B. Martin and A. Sanyal, *Journal of Computational Chemistry* **34** (18), 1598–1608 (2013).

100 J. M. Hermida-Ramon, S. Brdarski, G. Karlstrom and U. Berg, *Journal of Computational Chemistry* **24** (2), 161–176 (2003).

101 W. Damm, A. Frontera, J. TiradoRives and W. L. Jorgensen, *Journal of Computational Chemistry* **18** (16), 1955–1970 (1997).

102 G. A. Kaminski, H. A. Stern, B. J. Berne, R. A. Friesner, Y. X. X. Cao, R. B. Murphy, R. H. Zhou and T. A. Halgren, *Journal of Computational Chemistry* **23** (16), 1515–1531 (2002).

103 C. S. Ewig, R. Berry, U. Dinur, J. R. Hill, M. J. Hwang, H. Y. Li, C. Liang, J. Maple, Z. W. Peng, T. P. Stockfisch, T. S. Thacher, L. Yan, X. S. Ni and A. T. Hagler, *Journal of Computational Chemistry* **22** (15), 1782–1800 (2001).

104 K. Palmo, B. Mannfors, N. G. Mirkin and S. Krimm, *Biopolymers* **68** (3), 383–394 (2003).

105 V. S. Allured, C. M. Kelly and C. R. Landis, *Journal of the American Chemical Society* **113** (1), 1–12 (1991).

106 N. Rai and J. I. Siepmann, *Journal of Physical Chemistry B* **111** (36), 10790–10799 (2007).

107 M. Clark, R. D. Cramer and N. Vanopdenbosch, *Journal of Computational Chemistry* **10** (8), 982–1012 (1989).

108 A. Vedani and D. W. Huhta, *Journal of the American Chemical Society* **112** (12), 4759–4767 (1990).

109 A. D. Mackerell, M. Feig and C. L. Brooks, *Journal of Computational Chemistry* **25** (11), 1400–1415 (2004).

110 B. Guillot, *Journal of Molecular Liquids* **101** (1–3), 219–260 (2002).

111 P. Li, L. F. Song and K. M. Merz, *Journal of Chemical Theory and Computation* **11** (4), 1645–1657 (2015).

112 C. Vega and J. L. F. Abascal, *Physical Chemistry Chemical Physics* **13** (44), 19663–19688 (2011).

113 O. Demerdash, E.-H. Yap and T. Head-Gordon, *Annual Review of Physical Chemistry* **65**, 149–174 (2014).

114 T. Hassinen and M. Perakyla, *Journal of Computational Chemistry* **22** (12), 1229–1242 (2001).

115 L. Monticelli, S. K. Kandasamy, X. Periole, R. G. Larson, D. P. Tieleman and S.-J. Marrink, *Journal of Chemical Theory and Computation* **4** (5), 819–834 (2008).

116 D. H. de Jong, G. Singh, W. F. D. Bennett, C. Arnarez, T. A. Wassenaar, L. V. Schafer, X. Periole, D. P. Tieleman and S. J. Marrink, *Journal of Chemical Theory and Computation* **9** (1), 687–697 (2013).

117 X. Periole, M. Cavalli, S.-J. Marrink and M. A. Ceruso, *Journal of Chemical Theory and Computation* **5** (9), 2531–2543 (2009).

118 S. O. Yesylevskyy, L. V. Schafer, D. Sengupta and S. J. Marrink, *PloS Computational Biology* **6** (6) (2010).

119 C. Arnarez, J. J. Uusitalo, M. F. Masman, H. I. Ingolfsson, D. H. de Jong, M. N. Melo, X. Periole, A. H. de Vries and S. J. Marrink, *Journal of Chemical Theory and Computation* **11** (1), 260–275 (2015).

120 T. A. Wassenaar, H. I. Ingolfsson, M. Priess, S. J. Marrink and L. V. Schaefer, *Journal of Physical Chemistry B* **117** (13), 3516–3530 (2013).

121 N. M. Fischer, P. J. van Maaren, J. C. Ditz, A. Yildirim and D. van der Spoel, *Journal of Chemical Theory and Computation* **11** (7), 2938–2944 (2015).

122 I. Kolossvary and W. C. Guida, *Journal of the American Chemical Society* **118** (21), 5011–5019 (1996).

123 N. L. Allinger, S. H. M. Chang, D. H. Glaser and H. Honig, *Israel Journal of Chemistry* **20** (1–2), 51–56 (1980).

124 S. Profeta and N. L. Allinger, *Journal of the American Chemical Society* **107** (7), 1907–1918 (1985).

125 J. P. Bowen, A. Pathiaseril, S. Profeta and N. L. Allinger, *Journal of Organic Chemistry* **52** (23), 5162–5166 (1987).

126 M. R. Frierson, M. R. Imam, V. B. Zalkow and N. L. Allinger, *Journal of Organic Chemistry* **53** (22), 5248–5258 (1988).

127 J. C. Tai and N. L. Allinger, *Journal of the American Chemical Society* **110** (7), 2050–2055 (1988).

128 F. Jensen and P. O. Norrby, *Theoretical Chemistry Accounts* **109** (1), 1–7 (2003).

129 J. E. Eksterowicz and K. N. Houk, *Chemical Reviews* **93** (7), 2439–2461 (1993).

130 E. Lime and P.-O. Norrby, *Journal of Computational Chemistry* **36** (4), 244–250 (2015).

131 E. Hansen, A. R. Rosales, B. Tutkowski, P.-O. Norrby and O. Wiest, *Accounts of Chemical Research* **49** (5), 996–1005 (2016).

132 F. M. Menger and M. J. Sherrod, *Journal of the American Chemical Society* **112** (22), 8071–8075 (1990).

133 P. T. Olsen and F. Jensen, *Journal of Chemical Physics* **118** (8), 3523–3531 (2003).

134 J. Aqvist and A. Warshel, *Chemical Reviews* **93** (7), 2523–2544 (1993).

135 S. C. L. Kamerlin and A. Warshel, *Wiley Interdisciplinary Reviews – Computational Molecular Science* **1** (1), 30–45 (2011).

136 Y. Kim, J. C. Corchado, J. Villa, J. Xing and D. G. Truhlar, *Journal of Chemical Physics* **112** (6), 2718–2735 (2000).

137 A. C. T. van Duin, S. Dasgupta, F. Lorant and W. A. Goddard, *Journal of Physical Chemistry A* **105** (41), 9396–9409 (2001).

138 H. M. Senn and W. Thiel, *Angewandte Chemie – International Edition* **48** (7), 1198–1229 (2009).

139 M. S. Gordon, Q. A. Smith, P. Xu and L. V. Slipchenko, in *Annual Review of Physical Chemistry*, edited by M. A. Johnson and T. J. Martinez (2013), Vol. **64**, pp. 553–578.

140 I. Antes and W. Thiel, *Journal of Physical Chemistry A* **103** (46), 9290–9295 (1999).

141 V. Thery, D. Rinaldi, J. L. Rivail, B. Maigret and G. G. Ferenczy, *Journal of Computational Chemistry* **15** (3), 269–282 (1994).

142 J. L. Gao, P. Amara, C. Alhambra and M. J. Field, *Journal of Physical Chemistry A* **102** (24), 4714–4721 (1998).

143 N. Reuter, A. Dejaegere, B. Maigret and M. Karplus, *Journal of Physical Chemistry A* **104** (8), 1720–1735 (2000).

144 L. W. Chung, W. M. C. Sameera, R. Ramozzi, A. J. Page, M. Hatanaka, G. P. Petrova, T. V. Harris, X. Li, Z. Ke, F. Liu, H.-B. Li, L. Ding and K. Morokuma, *Chemical Reviews* **115** (12), 5678–5796 (2015).

145 R. E. Bulo, B. Ensing, J. Sikkema and L. Visscher, *Journal of Chemical Theory and Computation* **5** (9), 2212–2221 (2009).

146 C. V. Sumowski and C. Ochsenfeld, *Journal of Physical Chemistry A* **113** (43), 11734–11741 (2009).

147 P. E. M. Siegbahn and F. Himo, *Wiley Interdisciplinary Reviews – Computational Molecular Science* **1** (3), 323–336 (2011).

3

Hartree–Fock Theory

If we are interested in describing the electron distribution in detail, there is no substitute for quantum mechanics. Electrons are very light particles and they cannot be described correctly, even qualitatively, by classical mechanics. We will in this chapter and in Chapter 4 concentrate on solving the time-independent Schrödinger equation, which in shorthand operator form is given by

$$\mathbf{H}\Psi = E\Psi \tag{3.1}$$

If solutions are generated without reference to experimental data, the methods are usually called *ab initio* (Latin: "from the beginning"), in contrast to semi-empirical models, which are described in Chapter 7.

An essential part of solving the Schrödinger equation is the Born–Oppenheimer approximation, where the coupling between the nuclei and electronic *motion* is neglected. This allows the electronic part to be solved with the nuclear positions as parameters, and the resulting *potential energy surface* (PES) forms the basis for solving the nuclear motion. The major computational effort is in solving the electronic Schrödinger equation for a given set of nuclear coordinates.

The dynamics of a many-electron system is very complex and consequently requires elaborate computational methods. A significant simplification, both conceptually and computationally, can be obtained by introducing *independent-particle* models, where the motion of one electron is considered to be independent of the dynamics of all other electrons. An independent-particle model means that the interactions between the particles is approximated, either by neglecting all but the most important one or by taking all interactions into account in an average fashion. Within electronic structure theory, only the latter has an acceptable accuracy, and is called *Hartree–Fock* (HF) theory. In the HF model, each electron is described by an orbital and the total wave function is given as a product of orbitals. Since electrons are indistinguishable fermions (particles with a spin of $^1/_2$), however, the overall wave function must be antisymmetric (change sign upon interchanging any two electrons), which is conveniently achieved by arranging the orbitals in a Slater determinant. The best set of orbitals is determined by the variational principle, that is the HF orbitals give the lowest energy within the restriction of the wave function being a single Slater determinant. The shape of a given molecular orbital describes the probability of finding an electron, where the attraction to all the nuclei and the average repulsion to all the other electrons are included. Since the other electrons are described by their respective orbitals, the HF equations depend on their own solutions, and must therefore be solved iteratively. When the molecular orbitals are expanded in a basis set, the resulting equations

Introduction to Computational Chemistry, Third Edition. Frank Jensen.

Companion Website: http://www.wiley.com/go/jensen/computationalchemistry3

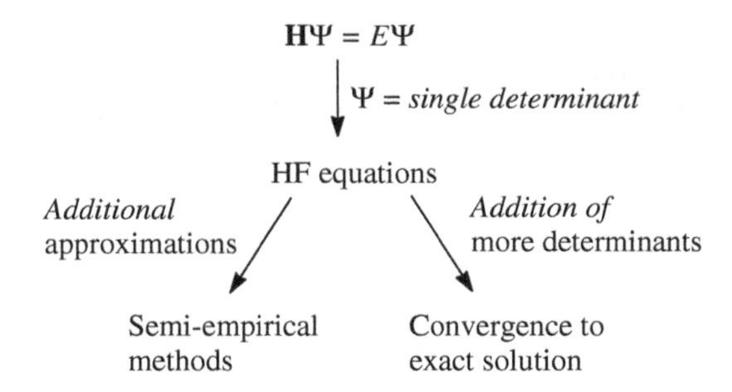

Figure 3.1 The HF model as a starting point for more approximate or more accurate treatments.

can be written as a matrix eigenvalue problem. The elements in the Fock matrix correspond to integrals of one- and two-electron operators over basis functions, multiplied by density matrix elements. The HF equations in a basis set can thus be obtained by repeated diagonalizations of a Fock matrix.

The HF model can be considered a branching point, where either additional approximations can be invoked, leading to semi-empirical methods, or it can be improved by adding additional determinants, thereby generating models that can be made to converge towards the exact solution of the electronic Schrödinger equation (Figure 3.1).[1–6]

Semi-empirical methods are derived from the HF model by neglecting all integrals involving more than two nuclei in the construction of the Fock matrix. Since the HF model by itself is only capable of limited accuracy, such approximations will by themselves lead to a poor model. The success of semi-empirical methods relies on turning the remaining integrals into parameters and fitting these to experimental data, especially molecular energies and geometries. Such methods are computationally much more efficient than the *ab initio* HF method, but are limited to systems for which parameters exist. These methods are described in Chapter 7.

HF theory only accounts for the average electron–electron interactions, and consequently neglects the correlation between electrons. Methods that include electron correlation require a multideterminant wave function, since HF is *the* best single-determinant wave function. Multideterminant methods are computationally much more involved than the HF model, but can generate results that systematically approach the exact solution of the Schrödinger equation. These methods are described in Chapter 4.

Density Functional Theory (DFT) in the Kohn–Sham version can be considered as an improvement on HF theory, where the many-body effect of electron correlation is modeled by a function of the electron density. DFT is, analogously to HF, an independent-particle model and is comparable to HF computationally, but provides significantly better results. The main disadvantage of DFT is that there is no systematic approach to improving the results towards the exact solution. These methods are described in Chapter 6.

We will also neglect relativistic effects in this chapter, which is justifiable for the first four rows in the periodic table (i.e. $Z < 36$) unless high accuracy is required, but the effects become important beyond the fourth row in the periodic table. A more detailed discussion can be found in Chapter 9. Spin-dependent effects are relativistic in origin (e.g. spin–orbit interaction), but can be introduced in an *ad hoc* fashion in non-relativistic theory, and calculated as corrections (e.g. by means of perturbation theory) after the electronic Schrödinger equation has been solved. This will be discussed in more detail in Chapter 11.

A word of caution before we start. A rigorous approach to many of the derivations requires keeping track of several different indices and validating why certain transformations are possible. The derivations will be performed less rigorously here, with the emphasis on illustrating the flow of the argument, rather than focusing on the mathematical details.

It is conventional to use the *bra-ket notation* for wave functions and multidimensional integrals in electronic structure theory in order to simplify the notation. The equivalences are defined as follows:

$$
\begin{aligned}
|\Psi\rangle &\equiv \Psi; \quad \langle\Psi| \equiv \Psi^* \\
\int \Psi^*\Psi d\mathbf{r} &= \langle\Psi|\Psi\rangle \\
\int \Psi^*\mathbf{H}\Psi d\mathbf{r} &= \langle\Psi|\mathbf{H}|\Psi\rangle
\end{aligned}
\tag{3.2}
$$

The *bra* $\langle n|$ denotes a *complex conjugate* wave function with quantum number n standing to the *left* of the operator, while the *ket* $|m\rangle$ denotes a wave function with quantum number m standing to the *right* of the operator, and the combined *bracket* denotes that the whole expression should be integrated over all coordinates. Such a bracket is often referred to as a *matrix element*, or as an *overlap element* when there is no operator involved.

3.1 The Adiabatic and Born–Oppenheimer Approximations

We will start by reviewing the Born–Oppenheimer approximation in more detail.[7–10] The total (non-relativistic) Hamiltonian operator can be written as kinetic and potential energies of the nuclei and electrons:

$$
\mathbf{H}_{\text{tot}} = \mathbf{T}_{\text{n}} + \mathbf{T}_{\text{e}} + \mathbf{V}_{\text{ne}} + \mathbf{V}_{\text{ee}} + \mathbf{V}_{\text{nn}} \tag{3.3}
$$

The Hamiltonian operator is first transformed to the center of mass system, where it may be written as (using atomic units, see Appendix C)

$$
\begin{aligned}
\mathbf{H}_{\text{tot}} &= \mathbf{T}_{\text{n}} + \mathbf{H}_{\text{e}} + \mathbf{H}_{\text{mp}} \\
\mathbf{H}_{\text{e}} &= \mathbf{T}_{\text{e}} + \mathbf{V}_{\text{ne}} + \mathbf{V}_{\text{ee}} + \mathbf{V}_{\text{nn}} \\
\mathbf{H}_{\text{mp}} &= -\frac{1}{2M_{\text{tot}}}\left(\sum_{i}^{N_{\text{elec}}} \nabla_i\right)^2
\end{aligned}
\tag{3.4}
$$

Here $\mathbf{H}_{\text{e}}$ is the *electronic Hamiltonian operator* and $\mathbf{H}_{\text{mp}}$ is called the *mass-polarization* (M_{tot} is the total mass of all the nuclei). The mass-polarization term arises because it is not possible to rigorously separate the center of mass motion from the internal motion for a system with more than two particles. We note that $\mathbf{H}_{\text{e}}$ only depends on the nuclear *positions* (via $\mathbf{V}_{\text{ne}}$ and $\mathbf{V}_{\text{nn}}$, see Equation (3.24)), but not on their *momenta*.

Assume for the moment that the full set of solutions to the electronic Schrödinger equation is available, where $\mathbf{R}$ denotes nuclear positions and $\mathbf{r}$ electronic coordinates:

$$
\mathbf{H}_{\text{e}}(\mathbf{R})\Psi_i(\mathbf{R},\mathbf{r}) = E_i(\mathbf{R})\Psi_i(\mathbf{R},\mathbf{r}); \quad i = 1, 2, \ldots, \infty \tag{3.5}
$$

The Hamiltonian operator is *Hermitian*:

$$
\int \Psi_i^*\mathbf{H}\Psi_j d\mathbf{r} = \int \Psi_j\mathbf{H}^*\Psi_i^* d\mathbf{r} \quad \leftrightarrow \quad \langle\Psi_i|\mathbf{H}|\Psi_j\rangle = \langle\Psi_j|\mathbf{H}|\Psi_i\rangle^* \tag{3.6}
$$

The Hermitian property means that the solutions can be chosen to be orthogonal and normalized (*orthonormal*):

$$\int \Psi_i^*(\mathbf{R},\mathbf{r})\Psi_j(\mathbf{R},\mathbf{r})\mathrm{d}\mathbf{r} = \delta_{ij} \quad \leftrightarrow \quad \langle \Psi_i | \Psi_j \rangle = \delta_{ij}$$

$$\begin{aligned} \delta_{ij} &= 1, \quad i = j \\ \delta_{ij} &= 0, \quad i \neq j \end{aligned} \tag{3.7}$$

Without introducing any approximations, the total (exact) wave function can be written as an expansion in the complete set of electronic functions, with the expansion coefficients being functions of the nuclear coordinates:

$$\Psi_{\text{tot}}(\mathbf{R},\mathbf{r}) = \sum_{i=1}^{\infty} \Psi_{\text{n}i}(\mathbf{R})\Psi_i(\mathbf{R},\mathbf{r}) \tag{3.8}$$

Inserting Equation (3.8) into the Schrödinger Equation (3.1) gives

$$\sum_{i=1}^{\infty} (\mathbf{T}_{\text{n}} + \mathbf{H}_{\text{e}} + \mathbf{H}_{\text{mp}})\Psi_{\text{n}i}(\mathbf{R})\Psi_i(\mathbf{R},\mathbf{r}) = E_{\text{tot}} \sum_{i=1}^{\infty} \Psi_{\text{n}i}(\mathbf{R})\Psi_i(\mathbf{R},\mathbf{r}) \tag{3.9}$$

The nuclear kinetic energy is a sum of differential operators:

$$\begin{aligned} \mathbf{T}_{\text{n}} &= -\sum_A \frac{1}{2M_A}\nabla_A^2 = \nabla_{\text{n}}^2 \\ \nabla_A &= \left(\frac{\partial}{\partial X_A}, \frac{\partial}{\partial Y_A}, \frac{\partial}{\partial Z_A}\right) \\ \nabla_A^2 &= \left(\frac{\partial^2}{\partial X_A^2} + \frac{\partial^2}{\partial Y_A^2} + \frac{\partial^2}{\partial Z_A^2}\right) \end{aligned} \tag{3.10}$$

We have here introduced the ∇_{n}^2 symbol, which implicitly includes the mass dependence, sign and summation. Expanding out (3.8) gives

$$\begin{aligned} \sum_{i=1}^{\infty} \left(\nabla_{\text{n}}^2 + \mathbf{H}_{\text{e}} + \mathbf{H}_{\text{mp}}\right)\Psi_{\text{n}i}\Psi_i &= E_{\text{tot}} \sum_{i=1}^{\infty} \Psi_{\text{n}i}\Psi_i \\ \sum_{i=1}^{\infty} \{\nabla_{\text{n}}^2\Psi_{\text{n}i}\Psi_i + \mathbf{H}_{\text{e}}\Psi_{\text{n}i}\Psi_i + \mathbf{H}_{\text{mp}}\Psi_{\text{n}i}\Psi_i\} &= E_{\text{tot}} \sum_{i=1}^{\infty} \Psi_{\text{n}i}\Psi_i \\ \sum_{i=1}^{\infty} \{\nabla_{\text{n}}(\Psi_i\nabla_{\text{n}}\Psi_{\text{n}i} + \Psi_{\text{n}i}\nabla_{\text{n}}\Psi_i) + \Psi_{\text{n}i}\mathbf{H}_{\text{e}}\Psi_i + \Psi_{\text{n}i}\mathbf{H}_{\text{mp}}\Psi_i\} &= E_{\text{tot}} \sum_{i=1}^{\infty} \Psi_{\text{n}i}\Psi_i \\ \sum_{i=1}^{\infty} \left\{ \begin{array}{c} \Psi_i\left(\nabla_{\text{n}}^2\Psi_{\text{n}i}\right) + 2(\nabla_{\text{n}}\Psi_i)(\nabla_{\text{n}}\Psi_{\text{n}i}) + \\ \Psi_{\text{n}i}\left(\nabla_{\text{n}}^2\Psi_i\right) + \Psi_{\text{n}i}E_i\Psi_i + \Psi_{\text{n}i}\mathbf{H}_{\text{mp}}\Psi_i \end{array} \right\} &= E_{\text{tot}} \sum_{i=1}^{\infty} \Psi_{\text{n}i}\Psi_i \end{aligned} \tag{3.11}$$

Here we have used the fact that $\mathbf{H}_{\text{e}}$ and $\mathbf{H}_{\text{mp}}$ only act on the electronic wave function and the fact that Ψ_i is an exact solution to the electronic Schrödinger Equation (3.5). We will now use the orthonormality of the Ψ_i by multiplying from the left by a *specific* electronic wave function $\Psi_j{}^*$ and integrate over the electron coordinates:

$$\nabla_{\text{n}}^2\Psi_{\text{n}j} + E_j\Psi_{\text{n}j} + \sum_{i=1}^{\infty} \left\{ \begin{array}{l} 2\langle\Psi_j|\nabla_{\text{n}}|\Psi_i\rangle(\nabla_{\text{n}}\Psi_{\text{n}i}) + \langle\Psi_j|\nabla_{\text{n}}^2|\Psi_i\rangle\Psi_{\text{n}i} + \\ \langle\Psi_j|\mathbf{H}_{\text{mp}}|\Psi_i\rangle\Psi_{\text{n}i} \end{array} \right\} = E_{\text{tot}}\Psi_{\text{n}j} \tag{3.12}$$

The electronic wave function has now been removed from the first two terms while the curly bracket contains terms that couple different electronic states. The first two of these are the first- and second-order *non-adiabatic coupling elements*, respectively, while the last is the mass polarization. The non-adiabatic coupling elements are important for systems involving more than one electronic surface, such as photochemical reactions.

In the *adiabatic* approximation the form of the total wave function is restricted to one electronic surface, that is all coupling elements in Equation (3.12) are neglected (only the terms with $i = j$ survive). Except for spatially degenerate wave functions, the diagonal first-order non-adiabatic coupling elements are zero:

$$\left(\nabla_{\mathrm{n}}^2 + E_j + \left\langle \Psi_j \middle| \nabla_{\mathrm{n}}^2 \middle| \Psi_j \right\rangle + \langle \Psi_j | \mathbf{H}_{\mathrm{mp}} | \Psi_j \rangle \right) \Psi_{\mathrm{n}j} = E_{\mathrm{tot}} \Psi_{\mathrm{n}j} \tag{3.13}$$

Neglecting the mass-polarization and reintroducing the kinetic energy operator gives

$$\left(\mathbf{T}_{\mathrm{n}} + E_j + \left\langle \Psi_j \middle| \nabla_{\mathrm{n}}^2 \middle| \Psi_j \right\rangle \right) \Psi_{\mathrm{n}j} = E_{\mathrm{tot}} \Psi_{\mathrm{n}j} \tag{3.14}$$

This can also be written as

$$(\mathbf{T}_{\mathrm{n}} + E_j(\mathbf{R}) + U(\mathbf{R}))\Psi_{\mathrm{n}j}(\mathbf{R}) = E_{\mathrm{tot}} \Psi_{\mathrm{n}j}(\mathbf{R}) \tag{3.15}$$

The $U(\mathbf{R})$ term is known as the *diagonal correction* and is smaller than $E_j(\mathbf{R})$ by a factor roughly equal to the ratio of the electronic and nuclear masses. It is usually a slowly varying function of $\mathbf{R}$, and the shape of the energy surface is therefore determined almost exclusively by $E_j(\mathbf{R})$.[11] In the *Born–Oppenheimer* approximation, the diagonal correction is neglected and the resulting equation takes on the usual Schrödinger form, where the electronic energy plays the role of a potential energy:

$$(\mathbf{T}_{\mathrm{n}} + E_j(\mathbf{R}))\Psi_{\mathrm{n}j}(\mathbf{R}) = E_{\mathrm{tot}} \Psi_{\mathrm{n}j}(\mathbf{R}) \tag{3.16}$$

In the Born–Oppenheimer picture, the nuclei move on a *potential energy surface* (PES), which is a solution to the *electronic* Schrödinger equation. The PES is independent of the nuclear masses (i.e. it is the same for isotopic molecules), but this is not the case when working in the adiabatic approximation since the diagonal correction (and mass-polarization) depends on the nuclear masses. Solving Equation (3.16) for the nuclear wave function leads to energy levels for molecular vibrations and rotations (Section 14.5), which in turn are the fundamentals for many forms of spectroscopy, such as infrared (IR), Raman, microwave, etc.

The Born–Oppenheimer (and adiabatic) approximation is usually a good approximation but breaks down when two (or more) solutions to the electronic Schrödinger equation come close together energetically.[12] Consider, for example, stretching the bond in the LiF molecule. Near the equilibrium distance the molecule is very polarized, that is described essentially by an ionic wave function, $\mathrm{Li^+F^-}$. The molecule, however, dissociates into neutral atoms (all bonds break homolytically in the gas phase), that is the wave function at long distance is of a covalent type, Li · F·. At the equilibrium distance, the covalent wave function is higher in energy than the ionic, but the situation reverses as the bond distance increases. At some point they must "cross". However, as they have the same symmetry, they do not actually cross, but make an *avoided crossing*. In the region of the avoided crossing, the wave function changes from being mainly ionic to covalent over a short distance, and the adiabatic, and therefore also the Born–Oppenheimer, approximation, breaks down. This is illustrated in Figure 3.2, where the two states have been calculated by a state average MCSCF procedure using the aug-cc-pVTZ basis set. The energy of the ionic state is given by the solid line, while the energy of the covalent state is shown by the dashed line. For bond distances near 6 Å, the lowest energy wave

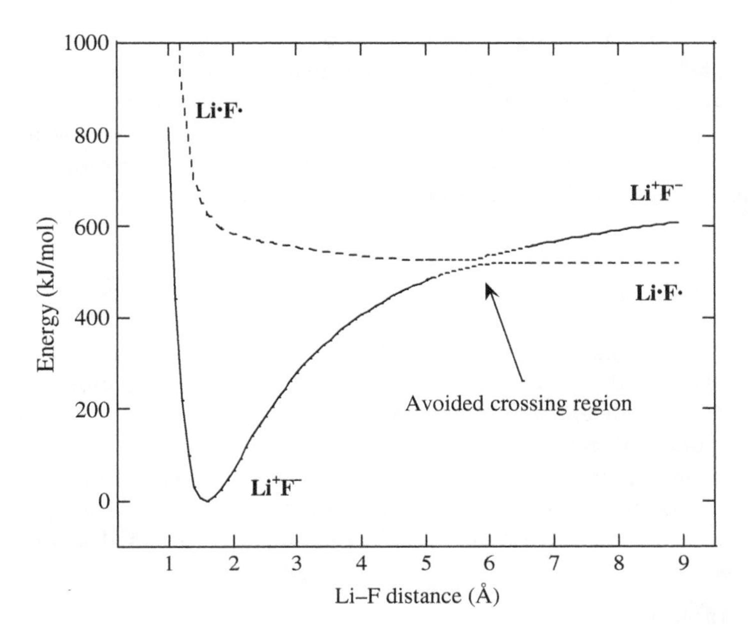

Figure 3.2 Avoided crossing of potential energy curves for LiF.

function suddenly switches from being almost ionic to being covalent, and the two states come within ~15 kJ mol^{-1} of each other. In this region the Born–Oppenheimer approximation becomes poor.

For the majority of systems the Born–Oppenheimer approximation introduces only very small errors. The *diagonal Born–Oppenheimer correction* (DBOC) can be evaluated relatively easy, as it is just the second derivative of the electronic wave function with respect to the nuclear coordinates and is therefore closely related to the nuclear gradient and second derivative of the energy (Section 11.9):

$$\Delta E_{\mathrm{DBOC}} = \sum_{A=1}^{N_{\mathrm{nuc}}} -\frac{1}{2M_A} \left\langle \Psi_{\mathrm{e}} \left| \nabla_A^2 \right| \Psi_{\mathrm{e}} \right\rangle \tag{3.17}$$

The largest effect is expected for hydrogen-containing molecules, since hydrogen has the lightest nucleus. The absolute magnitude of the DBOC for H_2O is ~7 kJ mol^{-1}, but the effect for the barrier towards linearity is only ~0.17 kJ/mol.[13] For the BH molecule, the equilibrium bond length elongates by ~0.0007 Å when the DBOC is included and the harmonic vibrational frequency changes by ~2 cm^{-1}. For systems with heavier nuclei, the effects are expected to be substantially smaller.[14]

When the Born–Oppenheimer approximation is expected to be poor, the non-adiabatic corrections will be large, and a better strategy in such cases may be to take the quantum nature of the nuclei into account directly. Starting from Equation (3.8), both the nuclear and electronic parts may be described by determinantal-based wave functions expanded in Gaussian basis sets. Each of the two wave functions (electronic and nuclear) can be described at different levels of approximations, with mean-field methods (i.e. Hartree–Fock) being the first step. Such methods are often denoted *Nuclear Orbital plus Molecular Orbital* (NOMO) and the energy spectrum arising from such methods directly gives both nuclear (e.g. vibrations) and electronic states.[10,15,16]

It should be noted that once methods beyond the Born–Oppenheimer approximation are employed, concepts such as molecular geometries become blurred and energy surfaces no longer exist. Nuclei are delocalized in a quantum description and a "bond length" is no longer a unique quantity, but must be defined according to the experiment that it is compared with. An X-ray structure, for example, measures the scattering of electromagnetic radiation by the electron density, neutron diffraction measures the scattering by the nuclei, while a microwave experiment measures the moments of inertia. With nuclei as delocalized wave packages, these quantities must be obtained as averages over the electronic and nuclear wave function components.

3.2 Hartree–Fock Theory

Having stated the limitations (non-relativistic Hamiltonian operator and the Born–Oppenheimer approximation), we are ready to consider the electronic Schrödinger equation. It can only be solved exactly for the H_2^+ molecule and similar one-electron systems. In the general case, we have to rely on approximate (numerical) methods. By neglecting relativistic effects, we also have to introduce electron spin as an *ad hoc* quantum effect. Each electron has a spin quantum number of $^1/_2$, and in the presence of a magnetic field there are two possible states, corresponding to alignment *along* or *opposite* to the field. The corresponding spin functions are denoted α and β, and obey the orthonormality conditions

$$\begin{aligned} \langle\alpha|\alpha\rangle &= \langle\beta|\beta\rangle = 1 \\ \langle\alpha|\beta\rangle &= \langle\beta|\alpha\rangle = 0 \end{aligned} \tag{3.18}$$

To generate approximate solutions we will employ the variational principle, which states that any approximate wave function has an energy above or equal to the exact energy (see Appendix B for a proof). The equality holds only if the wave function is the exact function. By constructing a trial wave function containing a number of parameters, we can generate the "best" trial function of the given form by minimizing the energy as a function of these parameters.

The energy of an approximate wave function can be calculated as the expectation value of the Hamiltonian operator, divided by the norm of the wave function:

$$E_e = \frac{\langle\Psi|\mathbf{H}_e|\Psi\rangle}{\langle\Psi|\Psi\rangle} \tag{3.19}$$

For a normalized wave function the denominator is 1, and therefore $E_e = \langle\Psi|\mathbf{H}_e|\Psi\rangle$. The total electronic wave function must be antisymmetric (change sign) with respect to interchange of any two electron coordinates (since electrons are fermions, having a spin of $^1/_2$). The *Pauli principle,* which states that two electrons cannot have all quantum numbers equal, is a direct consequence of this antisymmetry requirement. The antisymmetry of the wave function can be achieved by building it from *Slater Determinants* (SDs). The columns in a Slater determinant are single-electron wave functions, *orbitals,* while the electron coordinates are along the rows. Let us in the following assume that we are interested in solving the electronic Schrödinger equation for a molecule. The one-electron functions are thus *molecular orbitals* (MOs), which are given as the product of a spatial orbital and a spin function (α or β), also known as *spin-orbitals,* which may be taken as orthonormal. For the general case of N electrons and N spin-orbitals, the Slater determinant is given in

$$\Phi_{SD} = \frac{1}{\sqrt{N!}} \begin{vmatrix} \phi_1(1) & \phi_2(1) & \cdots & \phi_N(1) \\ \phi_1(2) & \phi_2(2) & \cdots & \phi_N(2) \\ \vdots & \vdots & \ddots & \vdots \\ \phi_1(N) & \phi_2(N) & \cdots & \phi_N(N) \end{vmatrix} \quad ; \quad \langle\phi_i|\phi_j\rangle = \delta_{ij} \tag{3.20}$$

We now make one further approximation, by taking the trial wave function to consist of a *single* Slater determinant. As will be seen later, this implies that electron correlation is neglected or, equivalently, the electron–electron repulsion is only included as an average effect. Having selected a single-determinant trial wave function the variational principle can be used to derive the *Hartree–Fock* (HF) equations, by minimizing the energy.

3.3 The Energy of a Slater Determinant

In order to derive the HF equations, we need an expression for the energy of a single Slater determinant. For this purpose, it is convenient to write it as an *antisymmetrizing operator* **A** working on the "diagonal" of the determinant, and expand **A** as a sum of permutations. We will denote the diagonal product by Π and use the symbol Φ to represent the determinant wave function:

$$\Phi = \mathbf{A}[\phi_1(1)\phi_2(2)\cdots\phi_N(N)] = \mathbf{A}\Pi \tag{3.21}$$

$$\mathbf{A} = \frac{1}{\sqrt{N!}}\sum_{p=0}^{N-1}(-1)^p\mathbf{P} = \frac{1}{\sqrt{N!}}\left[\mathbf{1} - \sum_{ij}\mathbf{P}_{ij} + \sum_{ijk}\mathbf{P}_{ijk} - \cdots\right] \tag{3.22}$$

The **1** operator is the identity, while the sum over $\mathbf{P}_{ij}$ generates all possible permutations of two electron coordinates, the sum over $\mathbf{P}_{ijk}$ all possible permutations of three electron coordinates, etc. It may be shown that **A** commutes with **H** and that **A** acting twice gives the same as **A** acting once, multiplied with the square root of N factorial:

$$\begin{aligned}\mathbf{AH} &= \mathbf{HA}\\ \mathbf{AA} &= \sqrt{N!}\mathbf{A}\end{aligned} \tag{3.23}$$

Consider now the Hamiltonian operator. The nuclear–nuclear repulsion does not depend on electron coordinates and is a constant for a given nuclear geometry. The nuclear–electron attraction is a sum of terms, each depending only on one electron coordinate. The same holds for the electron kinetic energy. The electron–electron repulsion, however, depends on two electron coordinates:

$$\begin{aligned}\mathbf{H}_\mathrm{e} &= \mathbf{T}_\mathrm{e} + \mathbf{V}_\mathrm{ne} + \mathbf{V}_\mathrm{ee} + \mathbf{V}_\mathrm{nn}\\ \mathbf{T}_\mathrm{e} &= -\sum_i^{N_\mathrm{elec}} \tfrac{1}{2}\nabla_i^2\\ \mathbf{V}_\mathrm{ne} &= -\sum_A^{N_\mathrm{nuclei}}\sum_i^{N_\mathrm{elec}} \frac{Z_A}{|\mathbf{R}_A - \mathbf{r}_i|}\\ \mathbf{V}_\mathrm{ee} &= \sum_i^{N_\mathrm{elec}}\sum_{j>i}^{N_\mathrm{elec}} \frac{1}{|\mathbf{r}_i - \mathbf{r}_j|}\\ \mathbf{V}_\mathrm{nn} &= \sum_A^{N_\mathrm{nuclei}}\sum_{B>A}^{N_\mathrm{nuclei}} \frac{Z_A Z_B}{|\mathbf{R}_A - \mathbf{R}_B|}\end{aligned} \tag{3.24}$$

We note that the zero point of the energy corresponds to the particles being at rest ($\mathbf{T}_\mathrm{e} = 0$) and infinitely removed from each other ($\mathbf{V}_\mathrm{ne} = \mathbf{V}_\mathrm{ee} = \mathbf{V}_\mathrm{nn} = 0$).

The operators may be collected according to the number of electron indices:

$$\begin{aligned}
\mathbf{h}_i &= -\tfrac{1}{2}\nabla_i^2 - \sum_A^{N_{\text{nuclei}}} \frac{Z_A}{|\mathbf{R}_A - \mathbf{r}_i|} \\
\mathbf{g}_{ij} &= \frac{1}{|\mathbf{r}_i - \mathbf{r}_j|} \\
\mathbf{H}_\text{e} &= \sum_i^{N_{\text{elec}}} \mathbf{h}_i + \sum_{j>i}^{N_{\text{elec}}} \mathbf{g}_{ij} + \mathbf{V}_{\text{nn}}
\end{aligned} \tag{3.25}$$

The one-electron operator $\mathbf{h}_i$ describes the motion of electron i in the field of all the nuclei, and $\mathbf{g}_{ij}$ is a two-electron operator giving the electron–electron repulsion.

The energy may be written in terms of the permutation operator as (using Equations (3.21) to (3.23))

$$\begin{aligned}
E &= \langle \Phi|\mathbf{H}|\Phi\rangle \\
&= \langle \mathbf{A}\Pi|\mathbf{H}|\mathbf{A}\Pi\rangle \\
&= \sqrt{N!}\langle \Pi|\mathbf{H}|\mathbf{A}\Pi\rangle \\
&= \sum_p (-1)^p \langle \Pi|\mathbf{H}|\mathbf{P}\Pi\rangle
\end{aligned} \tag{3.26}$$

The nuclear repulsion operator is independent of electron coordinates and can immediately be integrated to yield a constant:

$$\langle \Phi|\mathbf{V}_{\text{nn}}|\Phi\rangle = V_{\text{nn}}\langle \Phi|\Phi\rangle = V_{\text{nn}} \tag{3.27}$$

For the one-electron operator only the identity operator can give a non-zero contribution. For coordinate 1 this yields a matrix element over orbital 1:

$$\begin{aligned}
\langle \Pi|\mathbf{h}_1|\Pi\rangle &= \langle \phi_1(1)\phi_2(2)\cdots\ \phi_N(N)|\mathbf{h}_1|\phi_1(1)\phi_2(2)\cdots\ \phi_N(N)\rangle \\
&= \langle \phi_1(1)|\mathbf{h}_1|\phi_1(1)\rangle\langle \phi_2(2)|\phi_2(2)\rangle\cdots\ \langle \phi_N(N)|\phi_N(N)\rangle \\
&= \langle \phi_1(1)|\mathbf{h}_1|\phi_1(1)\rangle = h_1
\end{aligned} \tag{3.28}$$

This follows since all the MOs ϕ_i are normalized. All matrix elements involving a permutation operator gives zero. Consider, for example, the permutation of electrons 1 and 2:

$$\begin{aligned}
\langle \Pi|\mathbf{h}_1|\mathbf{P}_{12}\Pi\rangle &= \langle \phi_1(1)\phi_2(2)\cdots\ \phi_N(N)|\mathbf{h}_1|\phi_2(1)\phi_1(2)\cdots\ \phi_N(N)\rangle \\
&= \langle \phi_1(1)|\mathbf{h}_1|\phi_2(1)\rangle\langle \phi_2(2)|\phi_1(2)\rangle\cdots\ \langle \phi_N(N)|\phi_N(N)\rangle
\end{aligned} \tag{3.29}$$

This is zero as the integral over electron 2 is an overlap of two different MOs, which are orthogonal (Equation (3.20)).

For the two-electron operator, only the identity and $\mathbf{P}_{ij}$ operators can give non-zero contributions. A three-electron permutation will again give at least one overlap integral between two different MOs, which will be zero. The term arising from the identity operator is given by

$$\begin{aligned}
\langle \Pi|\mathbf{g}_{12}|\Pi\rangle &= \langle \phi_1(1)\phi_2(2)\cdots\ \phi_N(N)|\mathbf{g}_{12}|\phi_1(1)\phi_2(2)\cdots\ \phi_N(N)\rangle \\
&= \langle \phi_1(1)\phi_2(2)|\mathbf{g}_{12}|\phi_1(1)\phi_2(2)\rangle\cdots\ \langle \phi_N(N)|\phi_N(N)\rangle \\
&= \langle \phi_1(1)\phi_2(2)|\mathbf{g}_{12}|\phi_1(1)\phi_2(2)\rangle = J_{12}
\end{aligned} \tag{3.30}$$

The J_{12} matrix element is called a *Coulomb* integral. It represents the classical repulsion between two charge distributions described by $\phi_1^2(1)$ and $\phi_2^2(2)$. The term arising from the $\mathbf{P}_{ij}$ operator is given in

$$\begin{aligned}\langle\Pi|\mathbf{g}_{12}|\mathbf{P}_{12}\Pi\rangle &= \langle\phi_1(1)\phi_2(2)\cdots\ \phi_N(N)|\mathbf{g}_{12}|\phi_2(1)\phi_1(2)\cdots\ \phi_N(N)\rangle \\ &= \langle\phi_1(1)\phi_2(2)|\mathbf{g}_{12}|\phi_2(1)\phi_1(2)\rangle\cdots\ \langle\phi_N(N)|\phi_N(N)\rangle \\ &= \langle\phi_1(1)\phi_2(2)|\mathbf{g}_{12}|\phi_2(1)\phi_1(2)\rangle = K_{12}\end{aligned} \tag{3.31}$$

The K_{12} matrix element is called an *exchange* integral and has no classical analogy. Note that the order of the MOs in the J and K matrix elements is according to the electron indices. The Coulomb integral is non-zero for all electron pairs, regardless of spin, while the exchange integral is only non-zero when the two spin-orbitals ϕ_1 and ϕ_2 have the same spin. This follows from the spin orthogonality in Equation (3.18) by separating the spin dependence out in Equations (3.30 and (3.31). The energy can thus be written as

$$\begin{aligned}E &= \sum_{i=1}^{N_{\text{elec}}} \langle\phi_i(1)|\mathbf{h}_1|\phi_i(1)\rangle \\ &\quad + \sum_{i=1}^{N_{\text{elec}}}\sum_{j>i}^{N_{\text{elec}}} (\langle\phi_i(1)\phi_j(2)|\mathbf{g}_{12}|\phi_i(1)\phi_j(2)\rangle - \langle\phi_i(1)\phi_j(2)|\mathbf{g}_{12}|\phi_j(1)\phi_i(2)\rangle) + V_{\text{nn}} \\ E &= \sum_{i=1}^{N_{\text{elec}}} h_i + \sum_{i=1}^{N_{\text{elec}}}\sum_{j>i}^{N_{\text{elec}}} (J_{ij} - K_{ij}) + V_{\text{nn}}\end{aligned} \tag{3.32}$$

The minus sign for the exchange term comes from the factor of $(-1)^p$ in the antisymmetrizing operator, Equation (3.22). The energy may also be written in a more symmetrical form as

$$E = \sum_{i=1}^{N_{\text{elec}}} h_i + \frac{1}{2}\sum_{i=1}^{N_{\text{elec}}}\sum_{j=1}^{N_{\text{elec}}} (J_{ij} - K_{ij}) + V_{\text{nn}} \tag{3.33}$$

The factor of $^1/_2$ allows the double sum to run over all electrons since Equations (3.30) and (3.31) show that the Coulomb "self-interaction" J_{ii} is exactly cancelled by the corresponding "exchange" element K_{ii}. When the energy is written as in Equation (3.32), the exchange energy only includes contributions from pairs of electrons in different orbitals having the same spin (Equation (3.31)), but when the energy is written as in Equation (3.33), the exchange energy also includes the self-interaction correction from an electron in an orbital interacting with itself.

For the purpose of deriving the variation of the energy, it is convenient to express the energy in terms of Coulomb (**J**) and exchange (**K**) *operators*:

$$\begin{aligned}E &= \sum_{i}^{N_{\text{elec}}} \langle\phi_i|\mathbf{h}_i|\phi_i\rangle + \frac{1}{2}\sum_{ij}^{N_{\text{elec}}} (\langle\phi_j|\mathbf{J}_i|\phi_j\rangle - \langle\phi_j|\mathbf{K}_i|\phi_j\rangle) + V_{\text{nn}} \\ \mathbf{J}_i|\phi_j(2)\rangle &= \langle\phi_i(1)|\mathbf{g}_{12}|\phi_i(1)\rangle|\phi_j(2)\rangle \\ \mathbf{K}_i|\phi_j(2)\rangle &= \langle\phi_i(1)|\mathbf{g}_{12}|\phi_j(1)\rangle|\phi_i(2)\rangle\end{aligned} \tag{3.34}$$

Note that the **J** operator involves "multiplication" with a matrix element with the same orbital on both sides, while the **K** operator "exchanges" the two functions on the right-hand side of the $\mathbf{g}_{12}$ operator.

The objective is now to determine a set of MOs that makes the energy a minimum or at least stationary with respect to a change in the orbitals. The variation, however, must be carried out in such a way that the MOs remain orthogonal and normalized. This is a constrained optimization and can be handled by means of *Lagrange multipliers* (see Section 13.5). The condition is that a small change in the orbitals should not change the Lagrange function, that is the Lagrange function is stationary with respect to an orbital variation:

$$
\begin{aligned}
L &= E - \sum_{ij}^{N_{\text{elec}}} \lambda_{ij}(\langle \phi_i | \phi_j \rangle - \delta_{ij}) \\
\delta L &= \delta E - \sum_{ij}^{N_{\text{elec}}} \lambda_{ij}(\langle \delta\phi_i | \phi_j \rangle - \langle \phi_i | \delta\phi_j \rangle) = 0
\end{aligned}
\tag{3.35}
$$

The variation of the energy is given by

$$
\begin{aligned}
\delta E = &\sum_{i}^{N_{\text{elec}}} (\langle \delta\phi_i | \mathbf{h}_i | \phi_i \rangle + \langle \phi_i | \mathbf{h}_i | \delta\phi_i \rangle) \\
&+ \frac{1}{2} \sum_{ij}^{N_{\text{elec}}} \begin{pmatrix} \langle \delta\phi_i | \mathbf{J}_j - \mathbf{K}_j | \phi_i \rangle + \langle \phi_i | \mathbf{J}_j - \mathbf{K}_j | \delta\phi_i \rangle \\ + \langle \delta\phi_j | \mathbf{J}_i - \mathbf{K}_i | \phi_j \rangle + \langle \phi_j | \mathbf{J}_i - \mathbf{K}_i | \delta\phi_j \rangle \end{pmatrix}
\end{aligned}
\tag{3.36}
$$

The third and fifth terms are identical (since the summation is over all i and j), as are the fourth and sixth terms. They may be collected to cancel the factor of $^1/_2$ and the variation can be written in terms of a *Fock operator*, $\mathbf{F}_i$:

$$
\begin{aligned}
\delta E &= \sum_{i}^{N_{\text{elec}}} (\langle \delta\phi_i | \mathbf{h}_i | \phi_i \rangle + \langle \phi_i | \mathbf{h}_i | \delta\phi_i \rangle) + \sum_{ij}^{N_{\text{elec}}} (\langle \delta\phi_i | \mathbf{J}_j - \mathbf{K}_j | \phi_i \rangle + \langle \phi_i | \mathbf{J}_j - \mathbf{K}_j | \delta\phi_i \rangle) \\
\delta E &= \sum_{i}^{N_{\text{elec}}} (\langle \delta\phi_i | \mathbf{F}_i | \phi_i \rangle + \langle \phi_i | \mathbf{F}_i | \delta\phi_i \rangle) \\
\mathbf{F}_i &= \mathbf{h}_i + \sum_{j}^{N_{\text{elec}}} (\mathbf{J}_j - \mathbf{K}_j)
\end{aligned}
\tag{3.37}
$$

The Fock operator is an effective one-electron energy operator, describing the kinetic energy of an electron and the attraction to all the nuclei ($\mathbf{h}_i$), as well as the repulsion to all the other electrons (via the **J** and **K** operators). Note that the Fock operator is associated with the *variation* of the total energy, not the energy itself. The Hamiltonian operator (3.24) is *not* a sum of Fock operators.

The variation of the Lagrange function (Equation (3.35)) now becomes

$$
\delta L = \sum_{i}^{N_{\text{elec}}} (\langle \delta\phi_i | \mathbf{F}_i | \phi_i \rangle + \langle \phi_i | \mathbf{F}_i | \delta\phi_i \rangle) - \sum_{ij}^{N_{\text{elec}}} \lambda_{ij}(\langle \delta\phi_i | \phi_j \rangle + \langle \phi_i | \delta\phi_j \rangle)
\tag{3.38}
$$

The variational principle states that the desired orbitals are those that make $\delta L = 0$. Making use of the complex conjugate properties in Equation (3.39) gives Equation (3.40):

$$\begin{aligned}\langle\phi|\delta\phi\rangle &= \langle\delta\phi|\phi\rangle^* \\ \langle\phi|\mathbf{F}|\delta\phi\rangle &= \langle\delta\phi|\mathbf{F}|\phi\rangle^*\end{aligned} \tag{3.39}$$

$$\delta L = \sum_{i}^{N_{\text{elec}}} \langle\delta\phi_i|\mathbf{F}_i|\phi_i\rangle - \sum_{ij}^{N_{\text{elec}}} \lambda_{ij}\langle\delta\phi_i|\phi_j\rangle + \sum_{i}^{N_{\text{elec}}} \langle\delta\phi_i|\mathbf{F}_i|\phi_i\rangle^* - \sum_{ij}^{N_{\text{elec}}} \lambda_{ij}\langle\delta\phi_j|\phi_i\rangle^* = 0 \tag{3.40}$$

The variation of *either* $\langle\delta\phi|$ *or* $\langle\delta\phi|^*$ should make $\delta L = 0$, that is the first two terms in Equation (3.40) must cancel and the last two terms must cancel. Taking the complex conjugate of the last two terms and subtracting them from the first two gives

$$\sum_{ij}^{N_{\text{elec}}} \left(\lambda_{ij} - \lambda_{ji}^*\right)\langle\delta\phi_i|\phi_j\rangle = 0 \tag{3.41}$$

This means that the Lagrange multipliers are elements of a Hermitian matrix ($\lambda_{ij} = \lambda_{ji}^*$). The final set of *Hartree–Fock equations* may be written as

$$\mathbf{F}_i\phi_i = \sum_{j}^{N_{\text{elec}}} \lambda_{ij}\phi_j \tag{3.42}$$

The equations may be simplified by choosing a unitary transformation (Section 16.2) that makes the matrix of Lagrange multipliers diagonal, that is $\lambda_{ij} = 0$ and $\lambda_{ii} = \varepsilon_i$. This special set of molecular orbitals (ϕ') is called *canonical* MOs, and transforms Equation (3.42) into a set of pseudo-eigenvalue equations:

$$\mathbf{F}_i\phi_i' = \varepsilon_i\phi_i' \tag{3.43}$$

The Lagrange multipliers are seen to have the physical interpretation of MO energies, that is they are the expectation value of the Fock operator in the MO basis (multiply Equation (3.43) by ϕ'^*_i from the left and integrate):

$$\left\langle\phi_i'\left|\mathbf{F}_i\right|\phi_i'\right\rangle = \varepsilon_i\left\langle\phi_i'|\phi_i'\right\rangle = \varepsilon_i \tag{3.44}$$

The Hartree–Fock equations form a set of pseudo-eigenvalue equations as the Fock operator depends on all the occupied MOs (via the Coulomb and exchange operators, Equations (3.37) and (3.34)). A specific Fock orbital can only be determined if all the other occupied orbitals are known, and iterative methods must therefore be employed for solving the problem. A set of functions that is a solution to Equation (3.43) is called *Self-Consistent Field* (*SCF*) orbitals.

The canonical MOs may be considered as a convenient set of orbitals for carrying out the variational calculation. The total energy, however, depends only on the total wave function, which is a Slater determinant written in terms of the occupied MOs, Equation (3.20). The total wave function is unchanged by a unitary transformation of the occupied MOs among themselves (rows and columns in a determinant can be added and subtracted without affecting the determinant itself). After having determined the canonical MOs, other sets of MOs may be generated by forming linear combinations, such as localized MOs, or MOs displaying hybridization, which is discussed in more detail in Section 10.4.

The orbital energies can be considered as matrix elements of the Fock operator with the MOs (dropping the prime notation and letting ϕ be the canonical orbitals). The total energy can be written either as Equation (3.33) or in terms of MO energies (using the definition of **F** in Equations (3.37) and (3.44)):

$$E = \sum_{i}^{N_{\text{elec}}} \varepsilon_i - \frac{1}{2} \sum_{ij}^{N_{\text{elec}}} (J_{ij} - K_{ij}) + V_{\text{nn}} \tag{3.45}$$

$$\varepsilon_i = \langle \phi_i | \mathbf{F}_i | \phi_i \rangle = h_i + \sum_{j}^{N_{\text{elec}}} (J_{ij} - K_{ij}) \tag{3.46}$$

The total energy is *not* simply a sum of MO orbital energies. The Fock operator contains terms describing the repulsion to all other electrons (**J** and **K** operators), and the sum over MO energies therefore counts the electron–electron repulsion twice, which must be corrected for. It is also clear that the total energy cannot be exact, as it describes the repulsion between an electron and all the other electrons, assuming that their spatial distribution is described by a set of orbitals. The electron–electron repulsion is only accounted for in an average fashion and the HF method is therefore also referred to as a *mean-field* approximation. As mentioned previously, this is due to the approximation of a single Slater determinant as the trial wave function.

3.4 Koopmans' Theorem

The canonical MOs are convenient for the physical interpretation of the Lagrange multipliers. Consider the energy of an N-electron system and the corresponding system with one electron removed from orbital number k, and assume that the MOs are identical for the two systems (Equation (3.33)):

$$E_N = \sum_{i=1}^{N_{\text{elec}}} h_i + \frac{1}{2} \sum_{i=1}^{N_{\text{elec}}} \sum_{j=1}^{N_{\text{elec}}} (J_{ij} - K_{ij}) + V_{\text{nn}} \tag{3.47}$$

$$E_{N-1} = \sum_{i=1}^{N_{\text{elec}}-1} h_i + \frac{1}{2} \sum_{i=1}^{N_{\text{elec}}-1} \sum_{j=1}^{N_{\text{elec}}-1} (J_{ij} - K_{ij}) + V_{\text{nn}} \tag{3.48}$$

Subtracting the two total energies gives

$$E_N - E_{N-1} = h_k + \frac{1}{2} \sum_{i=1}^{N_{\text{elec}i}} (J_{ik} - K_{ik}) + \frac{1}{2} \sum_{j=1}^{N_{\text{elec}i}} (J_{kj} - K_{kj}) \tag{3.49}$$

The last two sums are identical and the energy difference becomes

$$E_N - E_{N-1} = h_k + \sum_{i=1}^{N_{\text{elec}i}} (J_{ik} - K_{ik}) = \varepsilon_k \tag{3.50}$$

As seen from Equation (3.46), this is exactly the orbital energy ε_k. The ionization energy within the "frozen MO" approximation is given simply as the orbital energy, a result known as *Koopmans' theorem*.[17] Similarly, the electron affinity of a neutral molecule is given as the orbital energy of the corresponding anion or, since the MOs are assumed constant, as the energy of the kth unoccupied orbital energy in the neutral species:

$$E_{N+1}^{k} - E_N = \varepsilon_k \tag{3.51}$$

Computationally, however, there is a significant difference between the eigenvalue of an occupied orbital for the anion and the eigenvalue corresponding to an unoccupied orbital in the neutral species when the orbitals are expanded in a set of basis functions (Section 3.5). Eigenvalues corresponding to occupied orbitals are well defined and they converge to a specific value as the size of the basis set is increased. In contrast, unoccupied orbitals in a sense are only the "left-over" functions in a given basis set and their number increases as the basis set is made larger. The lowest unoccupied eigenvalue usually converges to zero, corresponding to a solution for a free electron, described by a linear combination of the most diffuse basis functions. Equating ionization potentials to occupied orbital energies is therefore justified based on the frozen MO approximation, but taking unoccupied orbital energies as electron affinities is questionable, since continuum solutions are mixed in.

3.5 The Basis Set Approximation

For small highly symmetric systems, such as atoms and diatomic molecules, the Hartree–Fock equations can be solved by mapping the orbitals on a set of grid points, which are referred to as *numerical Hartree–Fock* methods.[18] However, essentially all calculations use a basis set expansion to express the unknown MOs in terms of a set of known functions. Any type of basis functions may in principle be used: exponential, Gaussian, polynomial, cube functions, wavelets, plane waves, etc. There are two guidelines for choosing the basis functions. One is that they should have a behavior that agrees with the physics of the problem, since this ensures that the convergence as more basis functions are added is reasonably rapid. For bound atomic and molecular systems, this means that the functions should go toward zero as the distance between the nucleus and the electron becomes large. The second guideline is a practical one: the chosen functions should make it easy to calculate all the required integrals.

The first criterion suggests the use of exponential functions located on the nuclei, since such functions are known to be exact solutions for the hydrogen atom. Unfortunately, exponential functions turn out to be computationally difficult. Gaussian functions are computationally much easier to handle, and although they are poorer at describing the electronic structure on a one-to-one basis, the computational advantages more than make up for this. For periodic systems, the infinite nature of the problem suggests the use of plane waves as basis functions, since these are the exact solutions for a free electron. We will return to the precise description of basis sets in Chapter 5, but for now simply assume that a set of M_{basis} basis functions located on the nuclei has been chosen.

Each MO ϕ is expanded in terms of the basis functions χ, conventionally called *atomic orbitals* (MO = LCAO, *Linear Combination of Atomic Orbitals*), although they are generally not solutions to the atomic HF problem:

$$\phi_i = \sum_{\alpha}^{M_{\text{basis}}} c_{\alpha i}\chi_\alpha \tag{3.52}$$

The Hartree–Fock Equations (3.43) may be written as

$$\mathbf{F}_i \sum_{\alpha}^{M_{\text{basis}}} c_{\alpha i}\chi_\alpha = \varepsilon_i \sum_{\alpha}^{M_{\text{basis}}} c_{\alpha i}\chi_\alpha \tag{3.53}$$

Multiplying from the left by a specific basis function and integrating yields the *Roothaan–Hall* equations (for a closed shell system).[19,20] These are the Hartree–Fock equations in the *atomic orbital basis*, and all the M_{basis} equations may be collected in a matrix notation:

$$\begin{gathered} \mathbf{FC} = \mathbf{SC}\boldsymbol{\varepsilon} \\ F_{\alpha\beta} = \langle \chi_\alpha | \mathbf{F} | \chi_\beta \rangle \quad ; \quad S_{\alpha\beta} = \langle \chi_\alpha | \chi_\beta \rangle \end{gathered} \tag{3.54}$$

The **S** matrix contains the overlap elements between basis functions and the **F** matrix contains the Fock matrix elements. Each $F_{\alpha\beta}$ element contains two parts from the Fock operator (Equation (3.37)), integrals involving the one-electron operators and a sum over occupied MOs of coefficients multiplied with two-electron integrals involving the electron–electron repulsion. The latter is often written as a product of a *density* matrix and two-electron integrals:

$$\begin{aligned} \langle \chi_\alpha | \mathbf{F} | \chi_\beta \rangle &= \langle \chi_\alpha | \mathbf{h} | \chi_\beta \rangle + \sum_{j}^{\text{occ.MO}} \langle \chi_\alpha | \mathbf{J}_j - \mathbf{K}_j | \chi_\beta \rangle \\ &= \langle \chi_\alpha | \mathbf{h} | \chi_\beta \rangle + \sum_{j}^{\text{occ.MO}} (\langle \chi_\alpha \phi_j | \mathbf{g} | \chi_\beta \phi_j \rangle - \langle \chi_\alpha \phi_j | \mathbf{g} | \phi_j \chi_\beta \rangle) \\ &= \langle \chi_\alpha | \mathbf{h} | \chi_\beta \rangle + \sum_{j}^{\text{occ.MO}} \sum_{\gamma\delta}^{M_{\text{basis}}} c_{\gamma j} c_{\delta_j} (\langle \chi_\alpha \chi_\gamma | \mathbf{g} | \chi_\beta \chi_\delta \rangle - \langle \chi_\alpha \chi_\gamma | \mathbf{g} | \chi_\delta \chi_\beta \rangle) \\ &= \langle \chi_\alpha | \mathbf{h} | \chi_\beta \rangle + \sum_{\gamma\delta}^{M_{\text{basis}}} D_{\gamma\delta} (\langle \chi_\alpha \chi_\gamma | \mathbf{g} | \chi_\beta \chi_\delta \rangle - \langle \chi_\alpha \chi_\gamma | \mathbf{g} | \chi_\delta \chi_\beta \rangle) \\ D_{\gamma\delta} &= \sum_{j}^{\text{occ.MO}} c_{\gamma j} c_{\delta_j} \end{aligned} \tag{3.55}$$

For use in Section 3.8, it can also be written in a more compact notation:

$$\begin{gathered} F_{\alpha\beta} = h_{\alpha\beta} + \sum_{\gamma\delta} G_{\alpha\beta\gamma\delta} D_{\gamma\delta} \\ \mathbf{F} = \mathbf{h} + \mathbf{G} \cdot \mathbf{D} \end{gathered} \tag{3.56}$$

Here $\mathbf{G} \cdot \mathbf{D}$ denotes the contraction of the **D** matrix with the four-dimensional **G** tensor.

The total energy (Equation (3.33)) in terms of integrals over basis functions is given as

$$\begin{aligned} E &= \sum_{i}^{N_{\text{elec}}} \langle \phi_i | \mathbf{h}_i | \phi_i \rangle + \frac{1}{2} \sum_{ij}^{N_{\text{elec}}} (\langle \phi_i \phi_j | \mathbf{g} | \phi_i \phi_j \rangle - \langle \phi_i \phi_j | \mathbf{g} | \phi_j \phi_i \rangle) + V_{\text{nn}} \\ &= \sum_{i}^{N_{\text{elec}}} \sum_{\alpha\beta}^{M_{\text{basis}}} c_{\alpha i} c_{\beta i} \langle \chi_\alpha | \mathbf{h} | \chi_\beta \rangle + \frac{1}{2} \sum_{ij}^{N_{\text{elec}}} \sum_{\alpha\beta\gamma\delta}^{M_{\text{basis}}} c_{\alpha i} c_{\gamma j} c_{\beta i} c_{\delta j} \begin{pmatrix} \langle \chi_\alpha \chi_\gamma | \mathbf{g} | \chi_\beta \chi_\delta \rangle - \\ \langle \chi_\alpha \chi_\gamma | \mathbf{g} | \chi_\delta \chi_\beta \rangle \end{pmatrix} + V_{\text{nn}} \\ &= \sum_{\alpha\beta}^{M_{\text{basis}}} D_{\alpha_\beta} h_{\alpha\beta} + \frac{1}{2} \sum_{\alpha\beta\gamma\delta}^{M_{\text{basis}}} D_{\alpha_\beta} D_{\gamma\delta} (\langle \chi_\alpha \chi_\gamma | \mathbf{g} | \chi_\beta \chi_\delta \rangle - \langle \chi_\alpha \chi_\gamma | \mathbf{g} | \chi_\delta \chi_\beta \rangle) + V_{\text{nn}} \end{aligned} \tag{3.57}$$

The latter expression may also be written as

$$E = \sum_{\alpha\beta}^{M_{\text{basis}}} D_{\alpha_\beta} h_{\alpha\beta} + \frac{1}{2} \sum_{\alpha\beta\gamma\delta}^{M_{\text{basis}}} (D_{\alpha_\beta} D_{\gamma\delta} - D_{\alpha\delta} D_{\gamma\beta}) \langle \chi_\alpha \chi_\gamma | \mathbf{g} | \chi_\beta \chi_\delta \rangle + V_{\text{nn}} \tag{3.58}$$

The one- and two-electron integrals in the atomic basis are given in Equations (3.59) and (3.60) using the operators in Equation (3.25):

$$\langle \chi_\alpha | \mathbf{h} | \chi_\beta \rangle = \int \chi_\alpha(1) \left(-\tfrac{1}{2}\nabla^2\right) \chi_\beta(1) d\mathbf{r}_1 + \sum_a^{N_{\text{nuclei}}} \int \chi_\alpha(1) \left(\frac{Z_a}{|\mathbf{R}_a - \mathbf{r}_1|}\right) \chi_\beta(1) d\mathbf{r}_1 \tag{3.59}$$

$$\langle \chi_\alpha \chi_\gamma | \mathbf{g} | \chi_\beta \chi_\delta \rangle = \int \chi_\alpha(1) \chi_\gamma(2) \left(\frac{1}{|\mathbf{r}_1 - \mathbf{r}_2|}\right) \chi_\beta(1) \chi_\delta(2) d\mathbf{r}_1 d\mathbf{r}_2 \tag{3.60}$$

The two-electron integrals are often written in a notation without electron coordinates or the **g** operator present:

$$\int \chi_\alpha(1) \chi_\gamma(2) \left(\frac{1}{|\mathbf{r}_1 - \mathbf{r}_2|}\right) \chi_\beta(1) \chi_\delta(2) d\mathbf{r}_1 d\mathbf{r}_2 = \langle \chi_\alpha \chi_\gamma | \chi_\beta \chi_\delta \rangle \tag{3.61}$$

This is known as the *physicist's* notation, where the ordering of the functions is given by the electron indices. They may also be written in an alternative order with both functions depending on electron 1 on the left and the functions depending on electron 2 on the right; this is known as the *Mulliken* or *chemist's* notation:

$$\int \chi_\alpha(1) \chi_\beta(1) \left(\frac{1}{|\mathbf{r}_1 - \mathbf{r}_2|}\right) \chi_\gamma(2) \chi_\delta(2) d\mathbf{r}_1 d\mathbf{r}_2 = (\chi_\alpha \chi_\beta | \chi_\gamma \chi_\delta) \tag{3.62}$$

The bra-ket notation has the electron indices $\langle 12|12\rangle$, while the parenthesis notation has the order $(11|22)$. In many cases the integrals are written with only the indices given, that is $\langle \chi_\alpha \chi_\beta | \chi_\gamma \chi_\delta \rangle = \langle \alpha\beta | \gamma\delta \rangle$. Since Coulomb and exchange integrals often are used as their difference, the following double-bar notations are also frequently used:

$$\begin{aligned} \langle \chi_\alpha \chi_\beta \| \chi_\gamma \chi_\delta \rangle &= \langle \chi_\alpha \chi_\beta | \chi_\gamma \chi_\delta \rangle - \langle \chi_\alpha \chi_\beta | \chi_\delta \chi_\gamma \rangle \\ (\chi_\alpha \chi_\beta \| \chi_\gamma \chi_\delta) &= (\chi_\alpha \chi_\beta | \chi_\gamma \chi_\delta) - (\chi_\alpha \chi_\gamma | \chi_\beta \chi_\delta) \end{aligned} \tag{3.63}$$

The Roothaan–Hall Equation (3.54) is a determination of the eigenvalues of the Fock matrix (see Section 17.2.3 for details). To determine the unknown MO coefficients $c_{\alpha i}$, the Fock matrix must be diagonalized. However, the Fock matrix is only known if all the MO coefficients are known (Equation (3.55)). The procedure therefore starts off by some guess of the coefficients, forms the **F** matrix and diagonalizes it. The new set of coefficients is then used for calculating a new Fock matrix, etc. This is continued until the set of coefficients used for constructing the Fock matrix is equal to those resulting from the diagonalization (to within a certain threshold), as illustrated in Figure 3.3. This set of coefficients determines a self-consistent field solution.

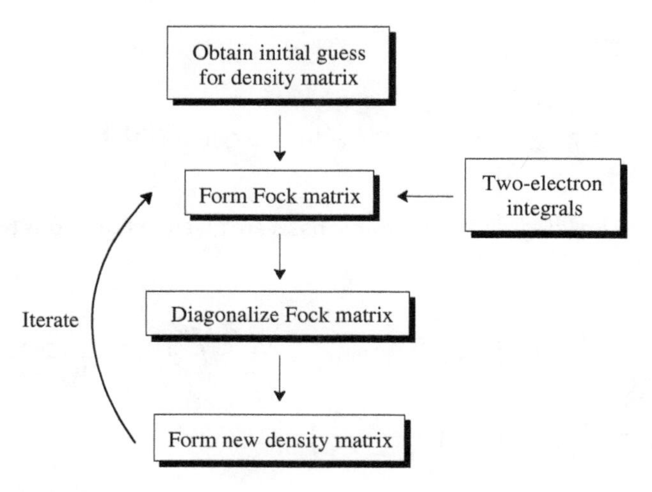

Figure 3.3 Illustration of the SCF procedure

The potential (or field) generated by the SCF electron density is identical to that produced by solving for the electron distribution. The Fock matrix, and therefore the total energy, only depends on the occupied MOs. Solving the Roothaan–Hall equations produces a total of M_{basis} MOs, that is there are N_{elec} occupied and $M_{\text{basis}} - N_{\text{elec}}$ unoccupied, or *virtual*, MOs. The virtual orbitals are orthogonal to all the occupied orbitals, but have no direct physical interpretation, except as electron affinities (via Koopmans' theorem).

In order to construct the Fock matrix in Equation (3.54), integrals between all pairs of basis functions and the one-electron operator **h** are needed. For M_{basis} functions there are of the order of M^2_{basis} such *one-electron integrals*. These one-electron integrals are also known as *core* integrals, as they describe the interaction of an electron with the whole frame of bare nuclei. The second part of the Fock matrix involves integrals over four basis functions and the **g** two-electron operator. There are of the order of M^4_{basis} of these *two-electron integrals*. In conventional HF methods, the two-electron integrals are calculated and saved before the SCF procedure is begun, and is then used in each SCF iteration. *Formally, in the large basis set limit the SCF procedure involves a computational effort that increases as the number of basis functions to the fourth power.* It will be shown below that the scaling may be substantially smaller in actual calculations.

For the two-electron integrals, the four basis functions may be located on one, two, three or four different atomic centers. It has already been mentioned that exponential-type basis functions ($\chi \propto \exp(-\alpha r)$) are fundamentally better suited for electronic structure calculations. However, it turns out that the calculation of especially three- and four-center two-electron integrals is very time-consuming for exponential functions. Gaussian functions ($\chi \propto \exp(-\alpha r^2)$) are much easier for calculating two-electron integrals. This is due to the fact that the product of two Gaussians located at two different positions ($\mathbf{R}_A$ and $\mathbf{R}_B$) with different exponents (α and β) can be written as a single Gaussian located at an intermediate position $\mathbf{R}_C$ between the two original integrals. This allows compact formulas for all types of one- and two-electron integrals to be derived:

$$
\begin{gathered}
G_A(\mathbf{r})G_B(\mathbf{r}) = K\mathrm{e}^{-\gamma(\mathbf{r}+\mathbf{R}_C)^2} \\
G_A(\mathbf{r}) = \left(\frac{2\alpha}{\pi}\right)^{3/4} \mathrm{e}^{-\alpha(\mathbf{r}+\mathbf{R}_A)^2} \quad ; \quad G_B(\mathbf{r}) = \left(\frac{2\beta}{\pi}\right)^{3/4} \mathrm{e}^{-\beta(\mathbf{r}+\mathbf{R}_B)^2} \\
\gamma = \alpha + \beta \quad ; \quad \mathbf{R}_C = \frac{\alpha\mathbf{R}_A + \beta\mathbf{R}_B}{\alpha+\beta} \quad ; \quad K = \left(\frac{2}{\pi}\right)^2 (\alpha\beta)^{3/4} \mathrm{e}^{-\frac{\alpha\beta}{\alpha+\beta}(\mathbf{R}_A - \mathbf{R}_B)^2}
\end{gathered}
\tag{3.64}
$$

As the number of basis functions increases, the accuracy of the MOs improves. In the limit of a complete basis set (infinite number of basis functions), the results are identical to those obtained by a numerical HF method, and this is known as the *Hartree–Fock limit*. This is *not* the exact solution to the Schrödinger equation, only the best single-determinant wave function that can be obtained. In practical calculations, the HF limit is never reached, and the term Hartree–Fock is normally used also to cover SCF solutions with an incomplete basis set. *Ab initio* HF methods, where all the necessary integrals are calculated from a given basis set, are one-dimensional. As the size of the basis set is increased, the variational principle ensures that the results become better (at least in an energetic sense). The quality of a result can therefore be assessed by running calculations with an increasingly larger basis set.

3.6 An Alternative Formulation of the Variational Problem

The objective is to minimize the total energy as a function of the molecular orbitals, subject to the orthogonality constraint. In the above formulation, this is handled by means of Lagrange multipliers. The final Fock matrix in the MO basis is diagonal, with the diagonal elements being the orbital energies. During the iterative sequence, that is before the orbitals have converged to an SCF solution, the Fock matrix is not diagonal. Starting from an initial set of molecular orbitals, the problem may also be formulated as a rotation of the orbitals (unitary transformation) in order to make the operator diagonal.[21] Since the operator depends on the orbitals, the procedure again becomes iterative. The orbital rotation is given by a unitary matrix **U**, which can be written as an exponential transformation of the orbitals in the determinant wave function (Section 17.2):

$$\phi' = \phi\mathbf{U} = \phi e^{X} \tag{3.65}$$

The **X** matrix contains the parameters describing the unitary transformation of the M_{basis} orbitals, being of the size of $M_{\text{basis}} \times M_{\text{basis}}$, and the orbital orthogonality is incorporated by requiring that **X** is anti-Hermitian.

It should be noted that the unoccupied orbitals do not enter the energy expression (Equation (3.33)), and a rotation between the virtual orbitals can therefore not change the energy. A rotation between the occupied orbitals corresponds to making linear combinations of these, but this does not change the total wave function or the total energy. The occupied–occupied and virtual–virtual blocks of the **X** matrix can therefore be chosen as zero. The variational parameters are the elements in the **X** matrix that describe the mixing of the occupied and virtual orbitals, that is there are a total of $N_{\text{occ}} \times (M_{\text{basis}} - N_{\text{occ}})$ parameters. The goal of the iterations is to make the off-diagonal elements in the occupied–virtual block of the Fock matrix zero. Alternatively stated, the off-diagonal elements are the gradients of the energy with respect to the orbitals and the stationary condition is that the gradient vanishes.

The exponential operator in Equation (3.65) can be written as a Taylor expansion, and the energy expanded in terms of the **X**-variables describing the occupied–virtual mixing of the orbitals:[21]

$$e^{\mathbf{X}} = 1 + \mathbf{X} + \tfrac{1}{2}\mathbf{X}\mathbf{X} + \cdots \tag{3.66}$$

$$E(\mathbf{X}) = E(\mathbf{0}) + \mathbf{E}'(\mathbf{0})\mathbf{X} + \tfrac{1}{2}\mathbf{X}^t\mathbf{E}''(\mathbf{0})\mathbf{X} + \cdots \tag{3.67}$$

The mixing of a specific occupied orbital i and a specific virtual orbital a can to first order be considered as making a linear combination of the reference and a singly excited Slater determinant, where orbital i is replaced with orbital a:

$$\Phi_0 \rightarrow \Phi_0 + x_i^a \Phi_i^a \tag{3.68}$$

The x_i^a notation here indicates an element x_{ai} in the **X** matrix, which implicitly defines x_{ia} as $-x_{ai}$. The first derivative of the energy with respect to the coefficient for mixing orbitals i and a can thus be calculated as the matrix elements of the reference and a singly excited Slater determinant:

$$\frac{\partial E}{\partial x_i^a} = 2\left\langle \Phi_0 | \mathbf{H} | \Phi_i^a \right\rangle \tag{3.69}$$

The second derivative can similarly be calculated as a sum of matrix elements between the reference and a doubly excited Slater determinant and between two singly excited Slater determinants:

$$\frac{1}{2}\frac{\partial^2 E}{\partial x_i^a \partial x_j^b} = \left\langle \Phi_0 \middle| \mathbf{H} \middle| \Phi_{ij}^{ab} \right\rangle + \left\langle \Phi_i^a \middle| \mathbf{H} \middle| \Phi_j^b \right\rangle - \delta_{ij}\delta_{ab}E_0 \tag{3.70}$$

The explicit expressions for the first and second derivatives of the energy with respect to the non-redundant **X**-variables can be written in terms of Fock matrix elements and two-electron integrals in the MO basis.[22] For an RHF-type wave function these are given in the following equations:

$$\frac{\partial E}{\partial x_i^a} = 4\langle \phi_i | \mathbf{F} | \phi_a \rangle \tag{3.71}$$

$$\frac{1}{2}\frac{\partial^2 E}{\partial x_i^a \partial x_j^b} = 2\begin{bmatrix} \delta_{ij}\langle \phi_a | \mathbf{F} | \phi_b \rangle - \delta_{ab}\langle \phi_i | \mathbf{F} | \phi_j \rangle \\ +4\langle \phi_i\phi_j | \phi_a\phi_b \rangle - \langle \phi_i\phi_j | \phi_b\phi_a \rangle - \langle \phi_i\phi_a | \phi_j\phi_b \rangle \end{bmatrix} \tag{3.72}$$

The gradient of the energy is an off-diagonal element of the molecular Fock matrix, which is easily calculated from the atomic Fock matrix. The second derivative, however, involves two-electron integrals that require an AO to MO transformation (see Section 4.2.1) and is therefore computationally expensive. Using the concepts from Chapter 17, the variational problem can be considered as a rotation of the coordinate system. In the original function space, the basis functions, the Fock operator depends on all the M_{basis} functions and the corresponding Fock matrix is non-diagonal. By performing a rotation of the coordinate system to the molecular orbitals, however, the matrix can be made diagonal, that is in this coordinate system the Fock operator only depends on N_{occ} functions.

3.7 Restricted and Unrestricted Hartree–Fock

So far there has not been any restriction on the MOs used to build the determinantal trial wave function. The Slater determinant has been written in terms of spin-orbitals, Equation (3.20), being products of a spatial orbital and a spin function (α or β). If there are no restrictions on the form of the spatial orbitals, the trial function is an *Unrestricted Hartree–Fock* (UHF) wave function.[23] The term *Different Orbitals for Different Spins* (DODS) is also sometimes used. If the interest is in systems with an even number of electrons and a singlet type of wave function (a *closed shell* system), the restriction that each spatial orbital should have two electrons, one with α and one with β spin, is normally made. Such wave functions are known as *Restricted Hartree–Fock* (RHF). Open-shell systems may also be described by restricted-type wave functions, where the spatial part of the doubly occupied orbitals is forced to be the same; this is known as *Restricted Open-shell Hartree–Fock* (ROHF). For open-shell species, a UHF treatment leads to well-defined orbital energies, which may be interpreted as

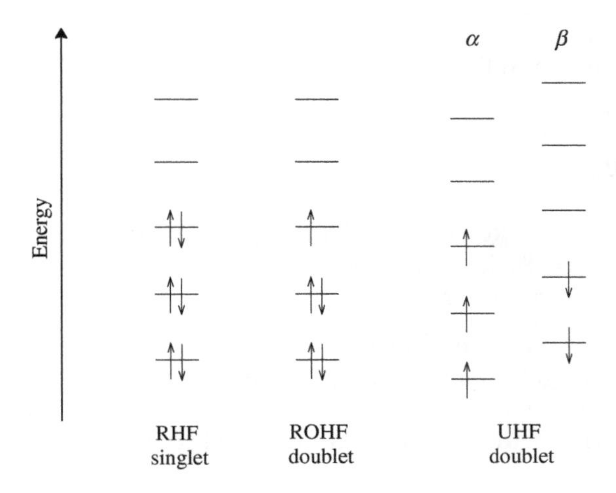

Figure 3.4 Illustrating an RHF singlet, and ROHF and UHF doublet states.

ionization potentials (Section 3.4). For an ROHF wave function, it is not possible to choose a unitary transformation that makes the matrix of Lagrange multipliers in Equation (3.42) diagonal, and orbital energies from an ROHF wave function are consequently not uniquely defined and cannot be equated to ionization potentials by a Koopmans-type argument.[24] The differences between these types of wave functions are illustrated in Figure 3.4.

The UHF wave function allows different spatial orbitals for the two electrons in an orbital. As restricted-type wave functions put constraints on the variation parameters, the energy of a UHF wave function is always lower than or equal to a corresponding R(O)HF-type wave function. For singlet states near the equilibrium geometry, it is usually not possible to lower the energy by allowing the α and β MOs to be different. For an open-shell system such as a doublet, however, it is clear that forcing the α and β MOs to be identical *is* a restriction. If the unpaired electron has α spin, it will interact differently with the other α electrons than with the β electrons due to the exchange operator, and consequently the optimum α and β orbitals will be different. The UHF description, however, has the disadvantage that the wave function is not an eigenfunction of the $\mathbf{S}^2$ operator (unless it is equal to the RHF solution), where the $\mathbf{S}^2$ operator evaluates the value of the total electron spin squared. This means that a "singlet" UHF wave function may also contain contributions from higher-lying triplet, quintet, etc., states. Similarly, a "doublet" UHF wave function will contain spurious (non-physical) contributions from higher-lying quartet, sextet, etc., states. This will be discussed in more detail in Section 4.4.

Semi-empirical methods (Chapter 7) sometimes employ the so-called *half-electron* method for describing open-shell systems, such as doublets and triplets. In this model a doublet state is described by putting two "half" electrons in the same orbitals with opposite spins, that is constructing an RHF-type wave function where all electron spins are paired. A triplet state may similarly be modeled as having two orbitals, each occupied by two half electrons with opposite spin. The main motivation behind this artificial construct is that open- and closed-shell systems (such as a triplet and singlet state) will have different amounts of electron correlation. Since semi-empirical methods perform the parameterization based on single-determinant wave functions, the half-electron method cancels the difference in electron correlations, and allows open- and closed-shell systems to be treated on an equal footing in terms of energy. It has the disadvantage that the open-shell nature is no longer present in the wave function; it is, for example, not possible to calculate spin densities (i.e. where the unpaired electron(s) is(are) most likely to be).

3.8 SCF Techniques

As discussed in Section 3.6, the Roothaan–Hall (or Pople–Nesbet for the UHF case) equations must be solved iteratively since the Fock matrix depends on its own solutions. The procedure illustrated in Figure 3.3 involves the following steps:

1. Calculate all one- and two-electron integrals.
2. Generate a suitable start guess for the MO coefficients.
3. Form the initial density matrix.
4. Form the Fock matrix as the core (one-electron) integrals + the density matrix times the two-electron integrals.
5. Diagonalize the Fock matrix. The eigenvectors contain the new MO coefficients.
6. Form the new density matrix. If it is sufficiently close to the previous density matrix, we are done; otherwise go to step 4.

There are several points hidden in this scheme. Will the procedure actually converge at all? Will the SCF solution correspond to the desired energy minimum (and not a maximum or saddle point)? Can the number of iterations necessary for convergence be reduced? Does the most efficient method depend on the type of computer and/or the size of the problem?

Let us look at some of the SCF techniques used in practice.

3.8.1 SCF Convergence

There is *no* guarantee that the above iterative scheme will converge. For geometries near equilibrium and using small basis sets, the straightforward SCF procedure often converges without problems. Distorted geometries (such as transition structures) and large basis sets containing diffuse functions, however, rarely converge, and metal complexes, where several states with similar energies are possible, are even more troublesome. There are various tricks that can be tried to help convergence:[25]

1. *Extrapolation.* This is a method for trying to make the convergence faster by extrapolating previous Fock matrices to generate a (hopefully) better Fock matrix than the one calculated directly from the current density matrix. Typically, the last three matrices are used in the extrapolation.
2. *Damping.* The reason for divergence, or very slow convergence, is often due to oscillations. A given density matrix $\mathbf{D}_n$ gives a Fock matrix $\mathbf{F}_n$, which, upon diagonalization, gives a density matrix $\mathbf{D}_{n+1}$. The Fock matrix $\mathbf{F}_{n+1}$ from $\mathbf{D}_{n+1}$ gives a density matrix $\mathbf{D}_{n+2}$ that is close to $\mathbf{D}_n$, but $\mathbf{D}_n$ and $\mathbf{D}_{n+1}$ are very different, as illustrated in Figure 3.5. The damping procedure tries to solve this by replacing the current density matrix with a weighted average, $\mathbf{D}'_{n+1} = \omega\mathbf{D}_n + (1-\omega)\mathbf{D}_{n+1}$. The weighting factor ω may be chosen as a constant or changed dynamically during the SCF procedure.
3. *Level shifting.* This technique[26] is perhaps best understood in the formulation of a rotation of the MOs that form the basis for the Fock operator (Section 3.6). At convergence, the Fock matrix elements in the MO basis between occupied and virtual orbitals are zero. The iterative procedure involves mixing (making linear combinations of) occupied and virtual MOs. During the iterative

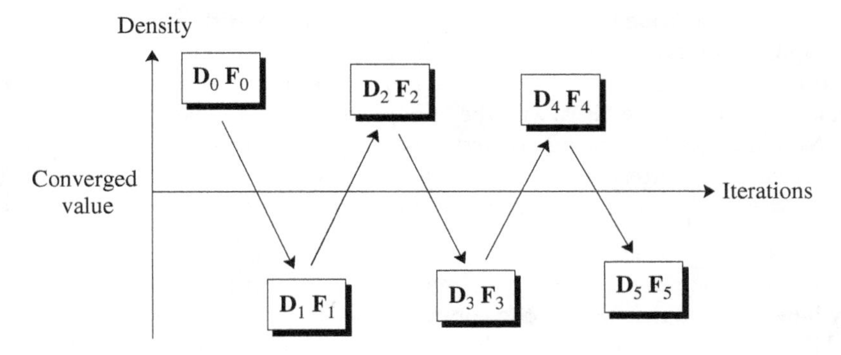

Figure 3.5 An oscillating SCF procedure.

procedure, these mixings may be large, causing oscillations or making the total energy increase. The degree of mixing may be reduced by artificially increasing the energy of the virtual orbitals. If a sufficiently large constant is added to the virtual orbital energies, it can be shown that the total energy is guaranteed to decrease, thereby forcing convergence. The more the virtual orbitals are raised in energy, the more stable is the convergence, but the *rate* of convergence also decreases with level shifting. For large enough shifts, convergence is guaranteed, but it is likely to occur very slowly, and may in some cases converge to a state that is not the ground state.

4. *Direct Inversion in the Iterative Subspace* (DIIS). This procedure was developed by P. Pulay and can be considered an extrapolation procedure (see Section 13.2.7 for more details).[27] It has proved to be very efficient in forcing convergence and in reducing the number of iterations at the same time, and it is one of the most commonly used methods for helping SCF convergence. The essence of DIIS is to replace the actual Fock matrix **F** in iteration n by a linear combination of all previous Fock matrices:

$$\mathbf{F}_n^* = \sum_{i=0}^{n} c_i \mathbf{F}_i \tag{3.73}$$

The coefficients in the linear combination are determined by minimizing a corresponding linear combination of "error" matrices, subject to a normalization condition for the sum of coefficients:

$$\mathbf{E}_n^* = \sum_{i=0}^{n} c_i \mathbf{E}_i \tag{3.74}$$

$$ErrF(\mathbf{c}) = \text{trace}(\mathbf{E}_n^* \cdot \mathbf{E}_n^*) \quad ; \quad \sum_{i=0}^{n} c_i = 1 \tag{3.75}$$

In the function space generated by the previous iterations we try to find an interpolation point with a lower error than any of the points actually calculated. Pulay suggested that the error estimate is taken as the **FDS** − **SDF** difference (**D** and **S** are the density and overlap matrices, respectively), which is related to the gradient of the SCF energy with respect to the MO coefficients; this has been found to work well in practice.

5. *"Direct minimization" techniques.* The variational principle indicates that we want to minimize the energy as a function of the MO coefficients or the corresponding density matrix elements, as given by Equation (3.57). In this formulation, the problem is no different from other types of non-linear

optimizations, and the same types of technique, such as steepest descent, conjugated gradient or Newton–Raphson methods can be used (see Chapter 17 for details).[28]

As mentioned in Section 3.6, the variational procedure can be formulated in terms of an exponential transformation of the MOs, with the (independent) variational parameters contained in an **X** matrix. Note that the **X**-variables are preferred over the MO coefficients in Equation (3.54) for optimization, since the latter are not independent (the MOs must be orthonormal). The first and second derivatives with respect to the non-redundant x-variables are given in Equations (3.71) and (3.72).

In a density matrix formulation, the energy depends on the density matrix elements as variables, and can formally be written as the trace of the contraction of the density matrix with the one-electron matrix **h** and the two-electron matrix **G**, with the latter depending implicitly on **D**:

$$E(\mathbf{D}) = \text{trace}(\mathbf{Dh}) + \text{trace}(\mathbf{DG}(\mathbf{D})) \tag{3.76}$$

The density matrix elements cannot be varied freely, however, as the orbitals must remain orthonormal, and this constraint can be formulated as the density matrix having to be *idempotent*, $\mathbf{DSD} = \mathbf{D}$. This is difficult to ensure during an optimization step, but the non-idempotent density matrix derived from taking an optimization step can be "purified" by the McWeeny procedure:[29]

$$\mathbf{D}_{\text{purified}} = 3\mathbf{D}^2 - 2\mathbf{D}^3 \tag{3.77}$$

The idempotency condition ensures that each orbital is occupied by exactly one electron. E. Cancès has shown that relaxing this condition to allow fractional occupancy during the optimization improves the convergence, a procedure named *relaxed constraint algorithm* (RCA)[30] and which was subsequently improved using ideas from the DIIS algorithm, leading to the *EDIIS* (*Energy DIIS*) method.[31] The optimization in terms of density matrix elements has the potential advantage that the matrix becomes sparse for large systems and can therefore be solved by techniques that scale linearly with the system's size.[28,32]

The Newton–Raphson method has the advantage of being quadratical convergent, that is sufficiently near the minimum that it converges very fast. The main problem in using Newton–Raphson methods for wave function optimization is computational efficiency. The exact calculation of the second derivative matrix is somewhat demanding, and each iteration in a Newton–Raphson optimization therefore takes longer than the simple Roothaan–Hall iterative scheme. Owing to the fast convergence near the minimum, a Newton–Raphson approach normally takes fewer iterations than, for example, DIIS, but the overall computational time is still a factor of ~two longer. Alternative schemes, where an approximation to the second derivative matrix is used (pseudo-Newton–Raphson), have also been developed, and they are often competitive with DIIS.[33] It should be kept in mind that the simple Newton–Raphson method is unstable and requires some form of stabilization, for example by using the augmented Hessian techniques discussed in Section 13.2.4.[34] Alternatively, for a large system (thousands of basis functions) the optimization may be carried out by conjugate gradient methods, but the convergence characteristic of these methods is significantly poorer.[28] Direct minimization methods have the advantage of a more stable convergence for difficult systems, where DIIS may display problematic behavior or converge to solutions that are not the global minimum.

3.8.2 Use of Symmetry

From group theory it may be shown that an integral can only be non-zero if the integrand belongs to the totally symmetric representation. Furthermore, the product of two functions can only be totally symmetric if they belong to the same irreducible representation. As both the Hamiltonian and Fock operators are totally symmetric (otherwise the energy would change by a rotation of the coordinate system), integrals of the following type can only be non-zero if the basis functions involving the same electron coordinate belong to the same representation:

$$\int \chi_\alpha(1)\chi_\beta(1)\mathbf{dr}_1 \quad ; \quad \int \chi_\alpha(1)\mathbf{F}\chi_\beta(1)\mathbf{dr}_1 \quad ; \quad \int \chi_\alpha(1)\mathbf{H}\chi_\beta(1)\mathbf{dr}_1 \tag{3.78}$$

Similar considerations hold for the two-electron integrals.

By forming suitable linear combinations of basis functions (*symmetry-adapted* functions), many one- and two-electron integrals need not be calculated as they are known to be exactly zero owing to symmetry. Furthermore, the Fock (in an HF calculation) or Hamiltonian matrix (in a configuration interaction (CI) calculation) will become block-diagonal, as only matrix elements between functions having the same symmetry can be non-zero. The saving depends on the specific system, but as a guideline the computational time is reduced by roughly a factor corresponding to the order of the point group (number of symmetry operations). Although the large majority of molecules do not have any symmetry, a sizeable proportion of the small molecules for which *ab initio* electronic structure calculations are possible are symmetric. Almost all *ab initio* programs employ symmetry as a tool for reducing the computational effort.

3.8.3 Ensuring that the HF Energy Is a Minimum, and the Correct Minimum

The standard iterative procedure produces a solution where the variation of the HF energy is *stationary* with respect to all orbital variations, that is the first derivatives of the energy with respect to the MO coefficients are zero. In order to ensure that this corresponds to an energy minimum, the second derivatives should also be calculated.[35,36] This is a matrix the size of the number of occupied MOs multiplied by the number of virtual MOs (identical to that arising in quadratic convergent SCF methods (Section 3.8.1)), and the eigenvalues of this matrix should all be positive in order to be an energy minimum (see also Section 6.9.1). Of course only the lowest eigenvalue is required to probe whether the solution is a minimum. A negative eigenvalue means that it is possible to get to a lower energy state by "exciting" an electron from an occupied to an unoccupied orbital, that is the solution is unstable. In practice, the stability is rarely checked – it is assumed that the iterative procedure has converged to a minimum. It should be noted that a positive definite second-order matrix only ensures that the solution is a *local* minimum; there may be other minima with lower energies.

The problem of convergence to saddle points in the wave function parameter space and the existence of multiple minima is rarely a problem for systems composed of elements from the first three rows in the periodic table. For systems having more than one metal atom with several partially filled d-orbitals, however, care must be taken to ensure that the iterative procedure converges to the desired solution. Consider, for example, the Fe_2S_2 system in Figure 3.6, where the d-electrons of two Fe atoms are coupled through the sulfur bridge atoms.

Each of the two Fe atoms is formally in the +III oxidation state, and therefore has a d^5 configuration. A high-spin state corresponding to all the 10 d-electrons being aligned can readily be described by a single determinant wave function, but the situation is more complicated for a low-spin singlet state. A singlet HF wave function must have an equal number of orbitals with α and β electron spin, but this can be obtained in several different ways. If each metal atom is in a high-spin state, an overall singlet state must have all the d-orbitals on one Fe atom occupied by electrons with α spin, while all the d-orbitals on the other Fe atom must be occupied by electrons with β spin. An alternative singlet

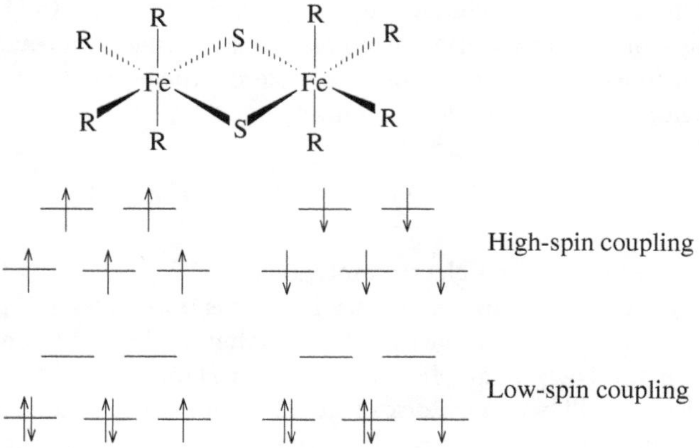

High-spin coupling

Low-spin coupling

Figure 3.6 Two different singlet states generated by coupling either two high-spin or two low-spin states.

state, however, can be generated by coupling the single unpaired electron from the two Fe centers in a low-spin configuration. Each of these two wave functions will be valid minima in the orbital parameter space, but clearly describe complexes with different properties. Note also that neither of these two singlet wave functions can be described by an RHF-type wave function. UHF-type wave functions with the above two types of spin coupling can be generated, but will often be severely spin contaminated. One can consider other spin coupling schemes to generate an overall singlet wave function, and the situation becomes more complicated if intermediate (triplet, pentet, etc.) spin states are desired, and for mixed valence states (Fe^{2+}/Fe^{3+}). The complications further increase when larger clusters are considered, as, for example, with the Fe_4S_4 moiety involved in electron transfer in the photosystem I and nitrogenase enzymes.

The question as to whether the energy is a minimum is closely related to the concept of wave function stability (see also Section 6.9.1). If a lower energy RHF solution can be found, the wave function is said to possess a *singlet instability*. It is also possible that an RHF-type wave function is a minimum in the coefficient space, but is a saddle point if the constraint of double occupancy of each MO is relaxed. This indicates that a lower energy wave function of the UHF type can be constructed, and this is called a *triplet instability*. It should be noted that in order to generate such UHF wave functions for a singlet state, an initial guess of the SCF coefficients must be specified that has the spatial parts of at least one set of α and β MOs different. There are other types of such instabilities, such as relaxing the constraint that the MOs should be real (allowing complex orbitals) or the constraint that an MO should only have a single spin function. Relaxing the latter produces the "*general*" HF method, where each MO is written as a spatial part having α spin plus another spatial part having β spin.[37] Such wave functions are no longer eigenfunctions of the $\mathbf{S}_z$ operator and are rarely used.

Another aspect of wave function instability concerns *symmetry breaking*, that is the wave function has a lower symmetry than the nuclear framework.[38] It occurs, for example, for the allyl radical with an ROHF-type wave function. The nuclear geometry has C_{2v} symmetry, but the C_{2v} symmetric wave function corresponds to a (first-order) saddle point. The lowest energy ROHF solution has only C_s symmetry and corresponds to a localized double bond and a localized electron (radical). Relaxing the double occupancy constraint and allowing the wave function to become UHF re-establishes the correct C_{2v} symmetry. Such symmetry breaking phenomena usually indicate that the type of wave function used is not flexible enough for even a qualitatively correct description.

3.8.4 Initial Guess Orbitals

The quality of the initial guess orbitals influences the number of iterations necessary for achieving convergence. As each iteration involves a computational effort proportional to M^4_{basis}, it is of course desirable to generate as good a guess as possible. Different start orbitals may in some cases result in convergence to different SCF solutions or make the difference between convergence and divergence. One possible way of generating a set of start orbitals is to diagonalize the Fock matrix consisting only of the one-electron contributions, the "core" matrix. This corresponds to initializing the density matrix as a zero matrix, totally neglecting the electron–electron repulsion in the first step. This is generally a poor guess, but it is available for all types of basis set and is easily implemented. Essentially all programs therefore have it as an option.

More sophisticated procedures involve taking the start MO coefficients from a semi-empirical calculation, such as Extended Hückel Theory (EHT) or Intermediate Neglect of Differential Overlap (INDO) (Chapter 7). The EHT method has the advantage that it is readily parameterized for all elements and it can provide start orbitals for systems involving elements from essentially the whole periodic table. An INDO calculation normally provides better start orbitals, but at a price. The INDO calculation itself is iterative and may suffer from convergence problems, just as the *ab initio* SCF itself.

Many systems of interest are symmetric. The MOs will transform as one of the irreducible representations in the point group, and most programs use this to speed up calculations. The initial guess for the start orbitals involves selecting how many MOs of each symmetry should be occupied, that is the electron configuration. Different start configurations produce different final SCF solutions. Many programs automatically select the start configuration based on the orbital energies of the starting MOs, which may be "wrong" in the sense that it does not produce the desired solution. Of course, a given solution may be checked to see if it actually corresponds to an energy minimum, but, as stated above, this is rarely done. Furthermore, there may be several (local) minima; thus the verification that the found solution is an energy minimum is no guarantee that it is the global minimum. A particular case is open-shell systems having at least one element of symmetry, as the open-shell orbital(s) determine the overall wave function symmetry. An example is the N_2^+ radical cation, where two states of Σ_g and Π_u symmetry exist with a difference of only ~ 70 kJ/mol in energy.

The reason different initial electron configurations may generate different final solutions is because matrix elements between orbitals belonging to different representations are exactly zero; thus only orbitals belonging to the same representation can mix. Forcing the program to run the calculation without symmetry usually does not help. Although turning the symmetry off will make the program actually calculate all matrix elements, those between MOs of different symmetry will still be zero (except for numerical inaccuracies). It is therefore often necessary to specify manually which orbitals should be occupied initially to generate the desired solution.

3.8.5 Direct SCF

The number of two-electron integrals formally grows as the fourth power of the size of the basis set. Owing to permutation symmetry (the following integrals are identical: $\langle \chi_1\chi_2|\chi_3\chi_4\rangle = \langle \chi_3\chi_2|\chi_1\chi_4\rangle = \langle \chi_1\chi_4|\chi_3\chi_2\rangle = \langle \chi_3\chi_4|\chi_1\chi_2\rangle = \langle \chi_2\chi_1|\chi_4\chi_3\rangle = \langle \chi_4\chi_1|\chi_2\chi_3\rangle = \langle \chi_2\chi_3|\chi_4\chi_1\rangle = \langle \chi_4\chi_3|\chi_2\chi_1\rangle$) the total number is approximately $^1/_8 M^4_{\text{basis}}$. Each integral is a floating point number associated with four indices indicating which basis functions are involved in the integral. Storing a floating point number in double precision (which is necessary for calculating the energy with an accuracy of ~14 digits) requires 64 bits = 8 bytes. A basis set with 100 functions thus generates ~12×10^6 integrals, requiring ~100 Mbytes of disk space or memory. The disk space required for storing the integrals rises rapidly; thus a basis set with 1000 functions requires ~1000 Gbytes of disk space (or memory). In practice, the storage requirement is somewhat less, since many of the integrals are small and can be ignored. Typically, a cutoff around $\sim 10^{-10}$ is employed; if the integral is less than this value it is not stored, and consequently makes a zero contribution to the construction of the Fock matrix in the iterative procedure. However, the disk space requirement effectively limits conventional HF methods to basis sets smaller than ~1000 functions.

Older computers had only very limited amounts of memory and disk storage of the integrals was the only option. Modern machines often have quite significant amounts of memory – a few hundred Gbytes is not uncommon. For small- and medium-sized systems, it may be possible to store all the integrals in memory instead of on disk. Such "in-core" methods are very efficient for performing an HF calculation. The integrals are only calculated once, and each SCF iteration is just a contraction of the integral tensor with a density matrix to form the Fock matrix (Equation (3.56)). Essentially all machines have optimized routines for doing matrix contraction efficiently. The only limitation is the quartic (M^4_{basis}) growth of the memory requirement with basis set size, which in practice restricts such in-core methods to basis sets with less than a few hundred functions.

The disk space (or memory) requirement can be reduced dramatically by performing the SCF in a *direct* fashion.[39] In the direct SCF method, the integrals are calculated from scratch in each iteration. At first this would appear to involve a computational effort that is larger than a conventional HF calculation by a factor close to the number of iterations. There are, however, a number of considerations that often makes direct SCF methods computationally quite competitive or even advantageous.

In disk-based methods, all the integrals are first calculated and written to disk. To reduce the disk space requirement, the four indices associated with each integral are "packed" into a single number and written to disk. The whole set of integrals must be read in each iteration, and the indices "unpacked" before the integrals are multiplied with the proper density matrix elements and added to the Fock matrix. Typically, half the time in an SCF procedure is spent calculating the integrals and writing them to disk; the other half is spent reading, unpacking and forming the Fock matrix maybe 20 times. In a direct approach, there is no overhead due to packing/unpacking of indices or writing/reading of integrals.

In disk-based methods, only integrals larger than a certain cutoff are saved. In direct methods, it is possible to ignore additional integrals. The contribution to a Fock matrix element is a product of density matrix elements and two-electron integrals. In disk-based methods, the density matrix is not known when the integrals are calculated and all integrals above the cutoff must be saved and processed in each iteration. In direct methods, however, the density matrix is known at the time when the integrals are calculated. Thus if the *product* of the density matrix elements and the integral is less than the cutoff, the integral can be ignored. Of course, this is only a saving if an estimate of the size of the integral is available before it is actually calculated. One such estimate is the *Schwarz inequality*:

$$|\langle \chi_\alpha \chi_\beta | \chi_\gamma \chi_\delta \rangle| \leq \sqrt{\langle \chi_\alpha \chi_\alpha | \chi_\beta \chi_\beta \rangle} \sqrt{\langle \chi_\gamma \chi_\gamma | \chi_\delta \chi_\delta \rangle} \tag{3.79}$$

but more advanced screening methods have also been developed.[40] The number of two-center integrals on the right-hand side is quite small (of the order of M^2_{basis}) and can easily be calculated beforehand. Thus if the product of the density matrix elements and the upper limit of the integral is less than the cutoff, the integral does not need to be calculated. In practice, integrals are calculated in *batches*, where a batch is a collection of integrals having the same exponent. For a $\langle pp|pp \rangle$-type batch there are thus 81 individual integrals, a $\langle dd|dd \rangle$-type batch has 625 individual integrals, etc. The integral screening is normally done at the batch level, that is if the largest term is smaller than a given cutoff, the whole batch can be neglected.

The above integral screening is even more advantageous if the Fock matrix is formed incrementally. Consider two sequential density and Fock matrices in the iterative procedure (Equation (3.56)):

$$\begin{aligned}
\mathbf{F}_n &= \mathbf{h} + \mathbf{G} \cdot \mathbf{D}_n \\
\mathbf{F}_{n+1} &= \mathbf{h} + \mathbf{G} \cdot \mathbf{D}_{n+1} \\
\mathbf{F}_{n+1} - \mathbf{F}_n &= \mathbf{G} \cdot (\mathbf{D}_{n+1} - \mathbf{D}_n) \\
\Delta\mathbf{F}_{n+1} &= \mathbf{G} \cdot \Delta\mathbf{D}_{n+1}
\end{aligned} \tag{3.80}$$

The *change* in the Fock matrix depends only on the *change* in the density matrix. Combined with the above screening procedure, it is thus only necessary to calculate those integrals to be multiplied with density matrix elements that have changed significantly since the last iteration. As the SCF converges, there are fewer and fewer integrals that need to be calculated.

The formal scaling of HF methods is M_{basis}^4, since the total number of two-electron integrals increases as M_{basis}^4. As just seen, however, we do not need to calculate all the two-electron integrals – many can be neglected without affecting the final results. The observed scaling is therefore less than the quartic dependence, but the exact power depends on how the size of the problem is increased. If the number of atoms is increased for a fixed basis set per atom, the scaling depends on the dimensionality of the atomic arrangement and the size of the atomic basis. The most favorable case is a small compact basis set (such as a minimum basis) and an essential one-dimensional system, such as polyacetylene, H—(C≡C)$_n$—H, or linear alkanes. In this case, the scaling is close to M_{basis}^2 once the number of functions exceeds ~100. A two-dimensional arrangement of atoms (such as a slab of graphite) has a slightly larger exponent dependence, while a three-dimensional system (such as a diamond structure) has a power dependence close to $M_{\text{basis}}^{2.3}$.[41] It should be noted that most molecular systems have a dimensionality between two and three – the presence of "holes" in the structure reduces the effective dimensionality to below three. With a larger basis set, especially if diffuse functions are present, the screening of integrals becomes much less efficient or, equivalently, the molecular system must be significantly larger to achieve the limiting scaling. In practice, however, the increase in the total number of basis functions is often not due to an enlargement of the molecular system, but rather to the use of an increasingly larger basis set per atom for a fixed sized molecule. For such cases, the observed scaling is often *worse* than the theoretical M_{basis}^4 dependence, since the integral screening becomes less and less efficient.

The combination of these effects means that the increase in computational time for a direct SCF calculation compared with a disk-based method is less than initially expected. For a medium-sized SCF calculation that requires, say, 20 iterations, the increase in CPU time may only be a factor of 2 or 3. Due to the more efficient screening, however, the direct method actually becomes more and more advantageous relative to disk-based methods as the size of the system increases. At some point, direct methods will therefore require *less* CPU time than a conventional method. Exactly where the cross-over point occurs depends on the way the number of basis functions is increased, the machine type and the efficiency of the integral code. Small compact basis sets in general experience the cross-over point quite early (perhaps around 100 functions) while it occurs later for large extended basis sets. Since conventional disk-based methods are limited to a few hundred basis functions, direct methods are normally the only choice for large calculations. Direct methods are essentially only limited by the available CPU time, and calculations involving up to several thousand basis functions have been reported.

Although direct methods for small- and medium-size systems require more CPU time than disk-based methods, this is in many cases irrelevant. For the user the determining factor is the time from submitting the calculation to the results being available. Over the years the speed of CPUs has increased much more rapidly than the speed of data transfer to and from disk. Most modern machines have very slow data transfer to disk compared with CPU speed. Measured by the elapsed wall clock time, disk-based HF methods are often the slowest in delivering the results, despite the fact that they require the least CPU time. Simply speaking, the CPU may be spending most of its time waiting for data to be transferred from disk. Direct methods, on the other hand, use the CPU with a near 100% efficiency. For machines without fast disk transfer the cross-over point for direct versus conventional methods in terms of wall clock time is often so low that direct methods are always preferred.

Modern computers are designed to run calculations in parallel on many (sometimes very many) individual computing units ("cores"). Making direct SCF calculations run efficiently in a parallel fashion is fairly easy: each core is given the task of calculating a certain batch of integrals and the total Fock matrix is simply the sum of contributions from each individual core.

3.8.6 Reduced Scaling Techniques

The computational bottleneck in HF methods is the calculation of the two-electron Coulomb and exchange terms arising from the electron–electron repulsion. In non-metallic systems, the exchange term is quite short-ranged, while the Coulomb interaction is long-ranged. In the large system limit, the Coulomb integrals thus dominate the computational cost. By using the screening techniques described in the previous section, the scaling in the large system limit will eventually be reduced from M_{basis}^4 to M_{basis}^2. Similar considerations hold for DFT methods (Chapter 6). Although an M_{basis}^2 scaling is quite modest, it is clear that a reduction down to linear scaling will be advantageous in order to move the calculations into the thousand atoms regime.[42]

The *Fast Multipole Moment* (FMM) method (Section 15.3) was originally developed for calculating interactions between point charges. A direct calculation involves a summation over all pairs, that is a computational effort that increases with M_{basis}^2. The idea in FMM is to split the total interaction into a near- and a far-field. The near-field is evaluated directly, while the far-field is calculated by dividing the physical space into boxes, and the interaction between all the charges in one box and all the charges in another is approximated as interactions between multipoles located at the center of the boxes. The further away from each other two boxes are, the larger the boxes can be for a given accuracy, thereby reducing the formal M_{basis}^2 behavior to linear scaling, that is proportional to M_{basis}.

The original FMM has been refined by also adjusting the accuracy of the multipole expansion as a function of the distance between boxes, producing the *very Fast Multipole Moment* (vFMM) method.[43] Both of these have been generalized to continuous charge distributions, as is required for calculating the Coulomb interaction between electrons in a quantum description.[44,45] The use of FMM methods in electronic structure calculations enables the Coulomb part of the electron–electron interaction to be calculated with a computational effort that depends linearly on the number of basis functions, once the system becomes sufficiently large.

Instead of dividing the physical space into a near- and far-field, the Coulomb operator itself may be partitioned into a short- and long-ranged part.[46] The short-ranged operator is evaluated exactly, while the long-ranged part is evaluated, for example, by means of a Fourier transformation. The net effect is again that the total Coulomb interaction can be calculated with a computational effort that only scales linearly with system size.

Although the exchange term in principle is short-ranged, and thus should benefit significantly from integral screening, this is normally not observed in practical calculations. This has been attributed to basis set incompleteness,[47] and this insight allowed a formulation of a more aggressive screening technique that enables the exchange part of the electron–electron interaction also to be reduced to an order M_{basis} method.

An approach called the *Auxiliary Density Matrix Method* (ADMM) approximates the HF exchange in a large basis set by its value in a small auxiliary basis set and a correction term calculated as the exchange energy difference between the large and small basis sets calculated by a DFT exchange functional.[48]

Another approach for achieving linear scaling is to break the system into smaller parts, perform a calculation on each subsection and subsequently piece these results together, as discussed in Section 4.12.3.[49]

3.8.7 Reduced Prefactor Methods

The use of methods with a reduced scaling does not necessarily lead to a reduced computational cost for systems that can be studied by the available resources, as the computational prefactor may be larger for reduced scaling methods. The cross-over point for when the linear scaling methods becomes competitive with traditional methods may be so high that is it of little practical use. An alternative approach is to accept a relatively large formal scaling and focus on reducing the prefactor in the computationally expensive step. These methods use information compression to reduce the computational expensive step.

The large number of two-electron integrals over basis functions can be considered as elements in a four-dimensional tensor, and since the basis functions are non-orthogonal, this tensor contains many near-redundancies, and these increase as the size of the basis set is increased. Tensor decomposition methods (Section 17.6.3) can be used to extract the non-redundant information (to within a suitable threshold), and only use this information in an HF calculation. This will significantly decrease the prefactor, often by one or two orders of magnitude, and the improvement increases as the basis set becomes larger. Several (slightly) different approaches are in common use for extracting the essential information, but they all share the tensor decomposition feature of replacing four-index integrals by lower-index (three and two) quantities.

Perhaps the easiest to understand of these methods is "*Resolution of the Identity*" (RI) techniques, where the product of two basis functions is expanded in an auxiliary basis set:

$$|\chi_\alpha \chi_\beta\rangle \approx \sum_{k}^{M_{aux}} c_k \chi_k \tag{3.81}$$

When the two basis functions depend on the same electron coordinate, the product on the left-hand side corresponds to an electron density and the procedure is then often referred to as *density fitting*. This allows replacing four-index two-electron integrals by products of three-index integrals:[50]

$$\langle \chi_\alpha \chi_\beta | \chi_\gamma \chi_\delta \rangle \approx \sum_{kl}^{M_{aux}} c_k c_l \langle \chi_\alpha \chi_\beta | \chi_k \rangle \langle \chi_k | \chi_l \rangle^{-1} \langle \chi_l | \chi_\gamma \chi_\delta \rangle \tag{3.82}$$

The expansion coefficients are determined by minimizing the fitting error, which can be done using either an overlap or a Coulomb metric, where the latter has been shown to give better results.[51] RI methods reduce the formal scaling from M_{basis}^4 to (only) M_{basis}^3, but actual timings show that the total computational cost is reduced by roughly an order of magnitude, without compromising the accuracy.[52] Versions where plane waves are used as the auxiliary basis have also been proposed and, properly implemented, these achieve linear scaling even for small systems and for large basis sets.[53]

Pseudospectral methods are very similar to RI techniques, except that the product of functions is approximated by a set of delta-functions (grid) rather than an auxiliary set of continuous functions.[54]

Cholesky decomposition methods are based on the fact that a positive definite matrix can be written as a sum of outer products of vectors (the same as (17.122), except that the **u** vectors are not normalized):[55]

$$\mathbf{A} = \sum_{k=1}^{N} \mathbf{u}_k \mathbf{u}_k^{\mathrm{t}} \tag{3.83}$$

The **u** vectors can be extracted by an iterative algorithm until their magnitudes fall below a suitable threshold, in which case the original matrix is represented with an accuracy comparable to this threshold. The four-dimensional tensor containing the two-electron integrals can be converted into a two-dimensional matrix by combining the indices two-by-two, and this super-matrix can be decomposed into vectors of length M^2:

$$\langle \chi_\alpha \chi_\beta | \chi_\gamma \chi_\delta \rangle \approx \sum_{k}^{M_{\text{Cholesky}}} \langle \chi_\alpha \chi_\beta | u_k \rangle \langle u_k | \chi_\gamma \chi_\delta \rangle \tag{3.84}$$

Cholesky decomposition methods can thus be viewed as an RI method where the auxiliary basis set is generated from the data, rather than predetermined.[56]

The non-redundant information contained in the two-electron integrals increases slower than the size of the basis set raised to the fourth power, and the efficiency gain by employing RI or the Cholesky method therefore increases with the basis set size, that is these methods become favorable even for small systems when large basis sets are used for achieving high accuracy.

Since the HF method may be formulated and implemented in several different ways, a practical question is which of these methods will be the fastest computationally for a given problem. The scaling only determines which method will be the fastest for large systems, that is for $N \to \infty$, while the prefactor determines the computational time for a given sized system. Figure 3.7 illustrates a quartic, quadratic and linear scaling algorithm with different prefactors.

For systems smaller than N_1, the most efficient method is the quartic one, the quadratic algorithm is the most efficient for systems sizes between N_1 and N_2, while the linear scaling method becomes the most efficient beyond N_2. Note that N_2 may be so large that the total computational resources may be exhausted before the cross-over point is reached.

With the advent of methods that enable the construction of the Fock matrix to be done with a computational effort that scales linearly with system size, the diagonalization step for solving the HF equations eventually becomes the computational bottleneck, since matrix diagonalization depends on the third power of the problem size, and this cross-over occurs for a few thousand basis functions. As discussed in Section 3.8.1, however, it is possible to reformulate the SCF problem in terms of a minimization of an energy functional that depends directly on the density matrix elements or orbital rotation parameters. This functional can then be minimized, for example, by conjugate gradient methods (Section 13.2.2), taking advantage of the fact that the density matrix becomes sparse for large systems.

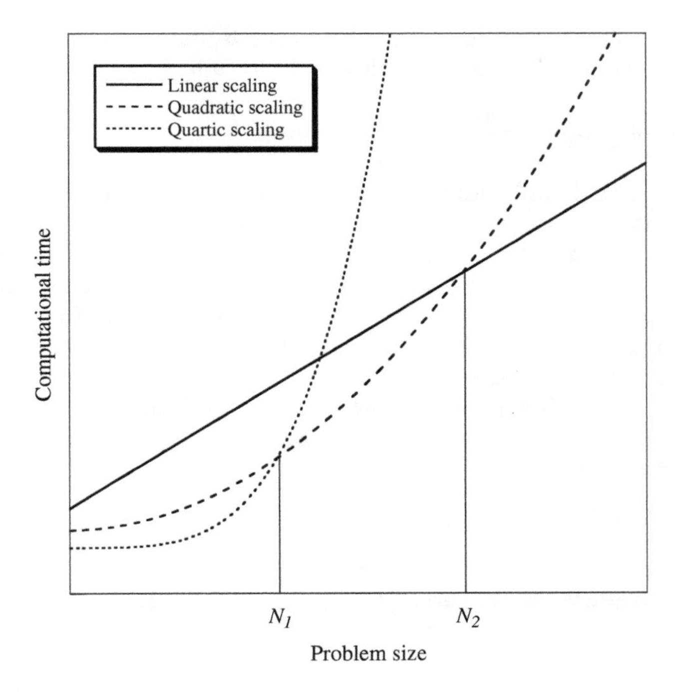

Figure 3.7 Method scaling with system size.

3.9 Periodic Systems

Periodic systems can be described as a fundamental *unit cell* being repeated to form an infinite system. The periodicity can be in one dimension (e.g. a polymer), two dimensions (e.g. a surface) or three dimensions (e.g. a crystal), with the latter being the most common. The unit cell in three dimensions can be characterized by three vectors $\mathbf{a}_1$, $\mathbf{a}_2$ and $\mathbf{a}_3$ spanning the physical space, with the length and the angles between them defining the shape (Figure 3.8).[57] There are seven possible shapes, the simplest of which is cubic, where all vector lengths are equal and all angles are 90°.

A unit cell can have atoms (or molecules) occupying various positions within the cell (corners, sides, center), and the combination of a unit cell and its occupancy is called a *Bravais lattice*, of which there are fourteen possible forms. The periodic (infinite) system can then be generated by translation of the unit cell (Bravais lattice) by lattice vectors **t**.

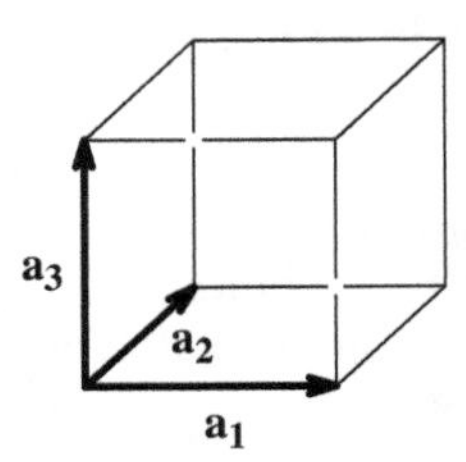

Figure 3.8 A cubic unit cell defined by three vectors.

The *reciprocal cell* is defined by three vectors $\mathbf{b}_1$, $\mathbf{b}_2$ and $\mathbf{b}_3$ derived from the $\mathbf{a}_1$, $\mathbf{a}_2$ and $\mathbf{a}_3$ vectors of the direct cell, and obeying the orthonormality condition $\mathbf{a}_i\mathbf{b}_j = 2\pi\delta_{ij}$:

$$\mathbf{b}_1 = 2\pi\frac{\mathbf{a}_2 \times \mathbf{a}_3}{L^3}, \quad \mathbf{b}_2 = 2\pi\frac{\mathbf{a}_3 \times \mathbf{a}_1}{L^3}, \quad \mathbf{b}_3 = 2\pi\frac{\mathbf{a}_1 \times \mathbf{a}_2}{L^3} \tag{3.85}$$

The reciprocal cell of a cubic cell with side length L is also a cube, with the side length $2\pi/L$. The equivalent of a unit cell in reciprocal space is called the (first) *Brillouin zone*. Just as a point in real space may be described by a vector **r**, a "point" in reciprocal space may be described by a vector **k**. Since **k** has units of inverse length, it is often called a wave vector. It is also closely related to the momentum and energy, for example the momentum and kinetic energy of a (free) particle described by a plane wave of the form $\mathrm{e}^{i\mathbf{k}\cdot\mathbf{r}}$ is k and $^1/_2k^2$, respectively.

The periodicity of the nuclei in the system means that the square of the wave function must display the same periodicity. This is inherent in the *Bloch theorem*, which states that the wave function value at equivalent positions in different cells are related by a complex phase factor involving the lattice vector **t** and a vector in the reciprocal space:

$$\phi(\mathbf{r}+\mathbf{t}) = \mathrm{e}^{i\mathbf{k}\cdot\mathbf{t}}\phi(\mathbf{r}) \tag{3.86}$$

Alternatively stated, the Bloch theorem indicates that a *crystalline orbital* (ϕ) for the nth band in the unit cell can be written as a wave-like part and a cell-periodic part (φ), called a *Bloch orbital*:

$$\phi_{n,k}(\mathbf{r}) = \mathrm{e}^{i\mathbf{k}\cdot\mathbf{r}}\varphi_n(\mathbf{r}) \tag{3.87}$$

The Bloch orbital can be expanded into a basis set of plane wave functions (χ^{PW}):

$$\begin{aligned}\varphi_n(\mathbf{r}) &= \sum_{\alpha}^{M_{\mathrm{basis}}} c_{n\alpha}\chi_\alpha^{\mathrm{PW}}(\mathbf{r}) \\ \varphi_{n,k}(\mathbf{r}) &= \mathrm{e}^{i\mathbf{k}\cdot\mathbf{r}}\sum_{\alpha}^{M_{\mathrm{basis}}} c_{n\alpha}\chi_\alpha^{\mathrm{PW}}(\mathbf{r})\end{aligned} \tag{3.88}$$

Alternatively, the basis set can be chosen as a set of nuclear-centered (Gaussian) basis functions, from which a set of Bloch orbitals can be constructed:

$$\begin{aligned}\varphi_{k\alpha}(\mathbf{r}) &= \sum_{\mathbf{t}} \mathrm{e}^{i\mathbf{k}\cdot\mathbf{t}}\chi_\alpha^{\mathrm{GTO}}(\mathbf{r}+\mathbf{t}) \\ \varphi_{n,k}(\mathbf{r}) &= \sum_{\alpha}^{M_{\mathrm{basis}}} c_{n\alpha}\varphi_{k\alpha}(\mathbf{r}) = \sum_{\alpha}^{M_{\mathrm{basis}}}\sum_{\mathbf{t}} c_{n\alpha}\mathrm{e}^{i\mathbf{k}\cdot\mathbf{t}}\chi_\alpha^{\mathrm{GTO}}(\mathbf{r}+\mathbf{t})\end{aligned} \tag{3.89}$$

The problem has now been transformed from treating an infinite number of orbitals (electrons) to only treating those within the unit cell. The price is that the solutions become a function of the reciprocal space vector **k** within the first Brillouin zone. For a system with M_{basis} functions, the variation problem can be formulated as a matrix equation analogous to Equation (3.54):

$$\mathbf{F}^{\mathbf{k}}\mathbf{C}^{\mathbf{k}} = \mathbf{S}^{\mathbf{k}}\mathbf{C}^{\mathbf{k}}\varepsilon^{\mathbf{k}} \tag{3.90}$$

The **k** appears as a parameter in the equation similarly to the nuclear positions in molecular Hartree–Fock theory. The solutions are continuous as a function of **k** and provide a range of energies called a *band*, with the total energy per unit cell being calculated by integrating over **k** space. Fortunately, the variation with **k** is rather slow for non-metallic systems, and the integration can be done numerically by including relatively few points.[58] Note that the presence of the phase factors in Equation (3.87) means that the matrices in Equation (3.90) are complex quantities.

For a given value of $\mathbf{k}$, the solution of Equation (3.90) provides M_{basis} orbitals. In molecular systems, the molecular orbitals are filled with electrons according to the *aufbau* principle, that is according to energy. The same principle is used for periodic systems, and the equivalent of the molecular HOMO (highest occupied molecular orbital) is the *Fermi* energy level. Depending on the system, two situations can occur:

- The number of electrons is such that a certain number of (non-overlapping) bands are completely filled, while the rest are empty.
- The number of electrons is such that one (or more) band(s) are only partly filled.

The first situation is analogous to that for molecular systems having a closed-shell singlet state. The energy difference between the "top" of the highest filled band and the "bottom" of the lowest empty band is called the *band gap*, and is equivalent to the HOMO–LUMO gap in molecular systems. The second situation is analogous to an open-shell electronic structure for a molecular system and corresponds to a band gap of zero. Systems with a band gap of zero are metallic, while those with a finite band gap are either insulators or semiconductors, depending on whether the band gap is large or small compared with the thermal energy kT.

As mentioned above, the basis functions within a unit cell can be either localized (e.g. Gaussian) or delocalized (plane wave) functions. For a Gaussian basis set, the computational problem of constructing the $\mathbf{F}^{\mathbf{k}}$ matrix is closely related to the molecular cases, involving multidimensional integrals over kinetic and potential energy operators. The periodic boundary condition means that the terms involving the potential energy operators in Equation (3.90) become infinite sums over $\mathbf{t}$ vectors. Since the operators involve both positive and negative quantities and only decay as r^{-1}, they require special care to ensure convergence to a definite quantity, as, for example, Ewald sum[59] or fast multipole methods.[60] For a plane wave basis, the construction of the energy matrix can be done efficiently by using fast Fourier transform (FFT) methods for switching between the real and reciprocal space. All local potential operators are easily evaluated in real space, while the kinetic energy is just the square operator in reciprocal space. FFT methods have the big advantage that the computational cost only increases as $N \ln N$, with N being the number of grid points in the Fourier transform.

The solution of Equation (3.90) can be done by repeated diagonalization of the $\mathbf{F}^{\mathbf{k}}$ matrix, analogously to the situation for non-periodic systems. A plane wave basis, however, often involves several thousand functions, which means that alternative methods are used for solving the equation.

References

1 A. Szabo and N. S. Ostlund, *Modern Quantum Chemistry* (McGraw-Hill, 1982).
2 W. J. Hehre, L. Radom, J. A. Pople and P. v. R. Schleyer, *Ab Initio Molecular Orbital Theory* (John Wiley & Sons, 1986).
3 R. McWeeny, *Methods of Molecular Quantum Mechanics* (Academic Press, 1992).
4 J. Simons and J. Nichols, *Quantum Mechanics in Chemistry* (Oxford University Press, 1997).
5 T. Helgaker, P. Jørgensen and J. Olsen, *Molecular Electronic Structure Theory* (John Wiley & Sons, 2000).
6 J. Simons, *Journal of Physical Chemistry* **95** (3), 1017–1029 (1991).
7 W. Kolos and L. Wolniewicz, *Journal of Chemical Physics* **41** (12), 3663–3673 (1964).
8 B. T. Sutcliffe, in *Advances in Quantum Chemistry, Recent Advances in Computational Chemistry*, edited by P. O. Lowdin, J. R. Sabin, M. C. Zerner, J. Karwowski and M. Karelson (Academic Press, 1997), Vol. 28, pp. 65–80.
9 B. Sutcliffe, in *Advances in Chemical Physics* (John Wiley & Sons, Inc., 2007), pp. 1–121.

10 S. Bubin, M. Pavanelo, W.-C. Tung, K. L. Sharkey and L. Adamowicz, *Chemical Reviews* **113** (1), 36–79 (2013).
11 N. C. Handy and A. M. Lee, *Chemical Physics Letters* **252** (5–6), 425–430 (1996).
12 L. J. Butler, *Annual Review of Physical Chemistry* **49**, 125–171 (1998).
13 E. F. Valeev and C. D. Sherrill, *Journal of Chemical Physics* **118** (9), 3921–3927 (2003).
14 S. Hirata, E. B. Miller, Y.-y. Ohnishi and K. Yagi, *Journal of Physical Chemistry A* **113** (45), 12461–12469 (2009).
15 H. Nakai, *International Journal of Quantum Chemistry* **107** (14), 2849–2869 (2007).
16 A. Sirjoosingh, M. V. Pak, C. Swalina and S. Hammes-Schiffer, *Journal of Chemical Physics* **139** (3), 034102 (2013).
17 T. Koopmans, *Physica* **1** (1–6), 104–113 (1934).
18 J. Kobus, *Computer Physics Communications* **184** (3), 799–811 (2013).
19 C. C. J. Roothaan, *Reviews of Modern Physics* **23** (2), 69–89 (1951).
20 G. G. Hall, *Proceedings of the Royal Society of London Series A – Mathematical and Physical Sciences* **205** (1083), 541–552 (1951).
21 M. Head-Gordon and J. A. Pople, *Journal of Physical Chemistry* **92** (11), 3063–3069 (1988).
22 J. Douady, Y. Ellinger, R. Subra and B. Levy, *Journal of Chemical Physics* **72** (3), 1452–1462 (1980).
23 J. A. Pople and R. K. Nesbet, *Journal of Chemical Physics* **22** (3), 571–572 (1954).
24 B. N. Plakhutin and E. R. Davidson, *Journal of Chemical Physics* **140** (1), 014102 (2014).
25 C. Kollmar, *International Journal of Quantum Chemistry* **62** (6), 617–637 (1997).
26 V. R. Saunders and I. H. Hillier, *International Journal of Quantum Chemistry* **7** (4), 699–705 (1973).
27 P. Pulay, *Journal of Computational Chemistry* **3** (4), 556–560 (1982).
28 H. Larsen, J. Olsen, P. Jorgensen and T. Helgaker, *Journal of Chemical Physics* **115** (21), 9685–9697 (2001).
29 R. McWeeny, *Reviews of Modern Physics* **32** (2), 335–369 (1960).
30 E. Cancès, *Journal of Chemical Physics* **114** (24), 10616–10622 (2001).
31 K. N. Kudin, G. E. Scuseria and E. Cances, *Journal of Chemical Physics* **116** (19), 8255–8261 (2002).
32 J. M. Millam and G. E. Scuseria, *Journal of Chemical Physics* **106** (13), 5569–5577 (1997).
33 G. Chaban, M. W. Schmidt and M. S. Gordon, *Theoretical Chemistry Accounts* **97** (1–4), 88–95 (1997).
34 S. Host, J. Olsen, B. Jansik, L. Thogersen, P. Jorgensen and T. Helgaker, *Journal of Chemical Physics* **129** (12), 124106 (2008).
35 R. Seeger and J. A. Pople, *Journal of Chemical Physics* **66** (7), 3045–3050 (1977).
36 R. Bauernschmitt and R. Ahlrichs, *Journal of Chemical Physics* **104** (22), 9047–9052 (1996).
37 C. A. Jimenez-Hoyos, T. M. Henderson and G. E. Scuseria, *Journal of Chemical Theory and Computation* 7 (9), 2667–2674 (2011).
38 E. R. Davidson and W. T. Borden, *Journal of Physical Chemistry* **87** (24), 4783–4790 (1983).
39 J. Almlof, K. Faegri and K. Korsell, *Journal of Computational Chemistry* **3** (3), 385–399 (1982).
40 D. S. Lambrecht and C. Ochsenfeld, *Journal of Chemical Physics* **123** (18), 184101 (2005).
41 D. L. Strout and G. E. Scuseria, *Journal of Chemical Physics* **102** (21), 8448–8452 (1995).
42 J. Kussmann, M. Beer and C. Ochsenfeld, *Wiley Interdisciplinary Reviews – Computational Molecular Science* **3** (6), 614–636 (2013).
43 H. G. Petersen, D. Soelvason, J. W. Perram and E. R. Smith, *Journal of Chemical Physics* **101** (10), 8870–8876 (1994).
44 C. A. White and M. Head-Gordon, *Journal of Chemical Physics* **104** (7), 2620–2629 (1996).
45 M. C. Strain, G. E. Scuseria and M. J. Frisch, *Science* **271** (5245), 51–53 (1996).
46 A. M. Lee, S. W. Taylor, J. P. Dombroski and P. M. W. Gill, *Physical Review A* **55** (4), 3233–3235 (1997).

47 E. Schwegler, M. Challacombe and M. Head-Gordon, *Journal of Chemical Physics* **106** (23), 9708–9717 (1997).

48 M. Guidon, J. Hutter and J. Van de Vondele, *Journal of Chemical Theory and Computation* **6** (8), 2348–2364 (2010).

49 T. Akama, M. Kobayashi and H. Nakai, *Journal of Computational Chemistry* **28** (12), 2003–2012 (2007).

50 F. Weigend, A. Kohn and C. Hattig, *Journal of Chemical Physics* **116** (8), 3175–3183 (2002).

51 P. Merlot, T. Kjaergaard, T. Helgaker, R. Lindh, F. Aquilante, S. Reine and T. B. Pedersen, *Journal of Computational Chemistry* **34** (17), 1486–1496 (2013).

52 F. Weigend, *Physical Chemistry Chemical Physics* **4** (18), 4285–4291 (2002).

53 J. Van de Vondele, M. Krack, F. Mohamed, M. Parrinello, T. Chassaing and J. Hutter, *Computer Physics Communications* **167** (2), 103–128 (2005).

54 Y. X. Cao, M. D. Beachy, D. A. Braden, L. Morrill, M. N. Ringnalda and R. A. Friesner, *Journal of Chemical Physics* **122** (22), 224116 (2005).

55 H. Koch, A. S. de Meras and T. B. Pedersen, *Journal of Chemical Physics* **118** (21), 9481–9484 (2003).

56 T. B. Pedersen, F. Aquilante and R. Lindh, *Theoretical Chemistry Accounts* **124** (1–2), 1–10 (2009).

57 R. Dovesi, B. Civalleri, R. Orlando, C. Roetti and V. R. Saunders, in *Reviews in Computational Chemistry*, edited by K. B. Lipkowitz, R. Larter and T. R. Cundari (John Wiley & Sons, 2005), Vol. 21, pp. 1–125.

58 H. J. Monkhorst and J. D. Pack, *Physical Review B* **13** (12), 5188–5192 (1976).

59 J. E. Jaffe and A. C. Hess, *Journal of Chemical Physics* **105** (24), 10983–10998 (1996).

60 K. N. Kudin and G. E. Scuseria, *Physical Review B* **61** (24), 16440–16453 (2000).

4

Electron Correlation Methods

The Hartree–Fock method generates solutions to the Schrödinger equation where the real electron–electron interaction is replaced by an average interaction (Chapter 3). In a sufficiently large basis set, the HF wave function is able to account for ~99% of the total energy, but the remaining ~1% is often very important for describing chemical phenomena. The difference in energy between the HF and the lowest possible energy in the given basis set is called the *electron correlation* (EC) energy.[1–6] Physically, it corresponds to the motion of the electrons being correlated, that is on average they are further apart than described by the HF wave function. As shown below, an unrestricted Hartree–Fock (UHF) type of wave function is, to a certain extent, able to include electron correlation. The proper reference for discussing electron correlation is therefore a restricted (RHF) or restricted open-shell (ROHF) wave function, although many authors use a UHF wave function for open-shell species. In the RHF case, all the electrons are paired in molecular orbitals. The two electrons in an MO occupy the same physical space and differ only in the spin function. The spatial overlap between the orbitals of two such "pair"-electrons is (exactly) one, while the overlap between two electrons belonging to different pairs is (exactly) zero, owing to the orthonormality of the MOs. The latter is not the same as saying that there is no repulsion between electrons in different MOs, since the electron–electron repulsion integrals involve products of MOs ($\langle\phi_i|\phi_j\rangle = 0$ for $i \neq j$, but $\langle\phi_i\phi_j|\mathbf{g}|\phi_i\phi_j\rangle$ and $\langle\phi_i\phi_j|\mathbf{g}|\phi_j\phi_i\rangle$ are not necessarily zero).

Naively it may be expected that the correlation between pairs of electrons belonging to the same spatial MO would be the major part of the electron correlation. However, as the size of the molecule increases, the number of electron pairs belonging to different spatial MOs grows faster than those belonging to the same MO. Consider, for example, the valence orbitals for CH_4. There are four *intra*-orbital electron pairs of opposite spins, but there are twelve *inter*-orbital pairs of *opposite* spins, and twelve *inter*-orbital pairs of *same* spin. A typical value for the intra-orbital pair correlation of a single bond is ~80 kJ/mol, while that of an inter-orbital pair (where the two MOs are spatially close, as in CH_4) is ~8 kJ/mol. The interpair correlation is therefore often comparable to the intrapair contribution.

Since the correlation between opposite spins has both intra- and inter-orbital contributions, it will be larger than the correlation between electrons having the same spin. The Pauli principle (or, equivalently, the antisymmetry of the wave function) has the consequence that there is no intra-orbital correlation from electron pairs with the same spin. The opposite spin correlation is sometimes called the *Coulomb correlation*, while the same spin correlation is called the *Fermi correlation*, that is the

Introduction to Computational Chemistry, Third Edition. Frank Jensen.

Companion Website: http://www.wiley.com/go/jensen/computationalchemistry3

Coulomb correlation is the largest contribution. Another way of looking at electron correlation is in terms of the electron density. In the immediate vicinity of an electron, there is a reduced probability of finding another electron. For electrons of opposite spin, this is often referred to as the *Coulomb hole* and the corresponding phenomenon for electrons of same spin is the *Fermi hole*. This hole picture is discussed in more detail in connection with density functional theory in Chapter 6.

Another distinction is between *dynamic* and *non-dynamical*, often also called *static*, electron correlation. The dynamic contribution is associated with the "instant" correlation between electrons, such as between those occupying the same spatial orbital. The non-dynamical part is associated with electrons avoiding each other on a more "permanent" basis, such as those occupying different spatial orbitals. The latter is also sometimes called a near-degeneracy effect, as it becomes important for systems where different orbitals (configurations) have similar energies. The electron correlation in a helium atom is almost purely dynamic, while the correlation in the H_2 molecule at the dissociation limit is purely non-dynamical (here the bonding and antibonding MOs become degenerate). At the equilibrium distance for H_2 the correlation is mainly dynamic (resembles the He atom), but this gradually changes to non-dynamical correlation as the bond distance is increased. Similarly, the Be atom contains both non-dynamical (near degeneracy of the $1s^22s^2$ and $1s^22p^2$ configurations) and dynamical correlation. There is therefore no clear-cut way of separating the two types of correlation, although they form a conceptually useful way of thinking about correlation effects. A UHF-type wave function can, compared to RHF, to a certain extent account for non-dynamical, but not dynamical, correlation, as discussed in Section 4.4.

The HF method determines *the* energetically best one-determinant trial wave function (within the given basis set). It is therefore clear that, in order to improve on HF results, the starting point must be a trial wave function that contains more than one Slater determinant (SD) Φ. This also means that the mental picture of electrons residing in orbitals has to be abandoned and the more fundamental property, the electron density, should be considered. As the HF solution usually gives ~99% of the correct answer, electron correlation methods normally use the HF wave function as a starting point for improvements.

A generic multideterminant trial wave function can be written as in Equation (4.1), where a_0 is usually close to one:

$$\Psi = a_0\Phi_{HF} + \sum_{i=1} a_i\Phi_i \tag{4.1}$$

Electron correlation methods differ in how they calculate the coefficients in front of the other determinants, with a_0 being determined by the normalization condition.

As discussed in Chapter 5, one can think of the expansion of an unknown MO in terms of basis functions as describing the MO "function" in the "coordinate system" of the basis functions. The multideterminant wave function (Equation (4.1)) can similarly be considered as describing the total wave function in a "coordinate" system of Slater determinants. The *basis set* determines the size of the *one-electron basis* (and thus limits the description of the one-electron functions, the MOs), while the *number of determinants* included determines the size of the *many-electron basis* (and thus limits the description of electron correlation), as illustrated in Figure 4.1.

4.1 Excited Slater Determinants

The starting point is usually an RHF calculation, where a solution of the Roothaan–Hall equations for a system with N electrons and M basis functions will yield $^1/_2\,N_{elec}$ occupied MOs and $M_{basis} - {}^1/_2\,N_{elec}$ unoccupied (virtual) MOs. Except for a minimum basis, there will always be more virtual

$$\chi \rightarrow \phi \rightarrow \Phi \rightarrow \Psi$$
$$\text{AO} \rightarrow \text{MO} \rightarrow \text{SD} \rightarrow \text{ME}$$
$$\phi = \sum_{\alpha=1} c_\alpha \chi_\alpha$$
$$\Psi = \sum_{i=1} a_i \Phi_i$$

Figure 4.1 Progression from atomic orbitals (AO) (basis functions), to molecular orbitals (MO), to Slater determinants (SD) and to a many-electron (ME) wave function.

than occupied MOs. A Slater determinant is constructed from $^1/_2\ N_{\text{elec}}$ spatial MOs multiplied by the two spin functions to yield N_{elec} spin-orbitals. A whole series of determinants may be generated by replacing MOs that are occupied in the HF determinant by MOs that are unoccupied. These can be denoted according to how many occupied HF MOs have been replaced by unoccupied MOs, that is Slater determinants that are *singly, doubly, triply, quadruply*, etc., *excited* relative to the HF determinant, up to a maximum of N_{elec} excited electrons (see Figure 4.2). These determinants are often referred to as *Singles* (S), *Doubles* (D), *Triples* (T), *Quadruples* (Q), etc.

The total number of determinants that can be generated depends on the size of the basis set: the larger the basis, the more virtual MOs, and the more excited determinants can be constructed. If all possible determinants in a given basis set are included, all the electron correlation (in the given basis) is (or can be) recovered. For an infinite basis, the Schrödinger equation is then solved exactly. Note that "exact" is this context is not the same as the experimental value, as the nuclei are assumed to have infinite masses (Born–Oppenheimer approximation) and relativistic effects are neglected. *Methods that include electron correlation are thus two-dimensional*, the larger the one-electron expansion (basis set size) and the larger the many-electron expansion (number of determinants), the better the results. This is illustrated in Figure 4.3.

In order to calculate total energies with a "chemical accuracy" of ~4 kJ/mol (~1 kcal/mol), it is necessary to use sophisticated methods for including electron correlation and large basis sets, which is only computationally feasible for small systems. The focus is therefore on calculating relative energies, where error cancellation can improve the accuracy of the calculated results. The important chemical changes take place in the valence orbitals, with the core orbitals being almost independent of the molecular environment. In many cases, the interest is therefore only in calculating the correlation energy associated with the valence electrons. Limiting the number of determinants to only those that can be generated by exciting the valence electrons is known as the *frozen-core* approximation. In some

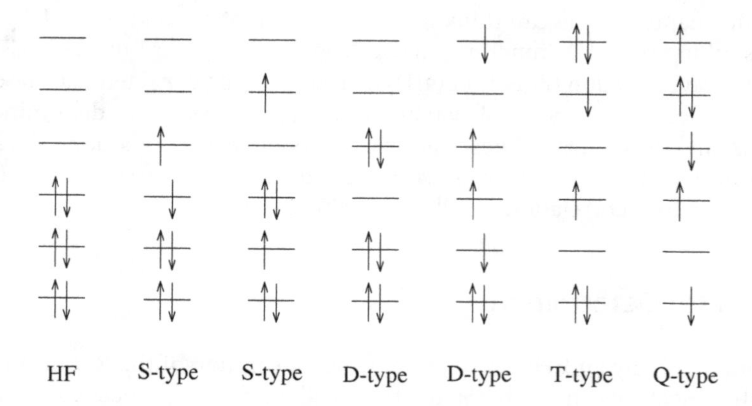

Figure 4.2 Excited Slater determinants generated from an HF reference.

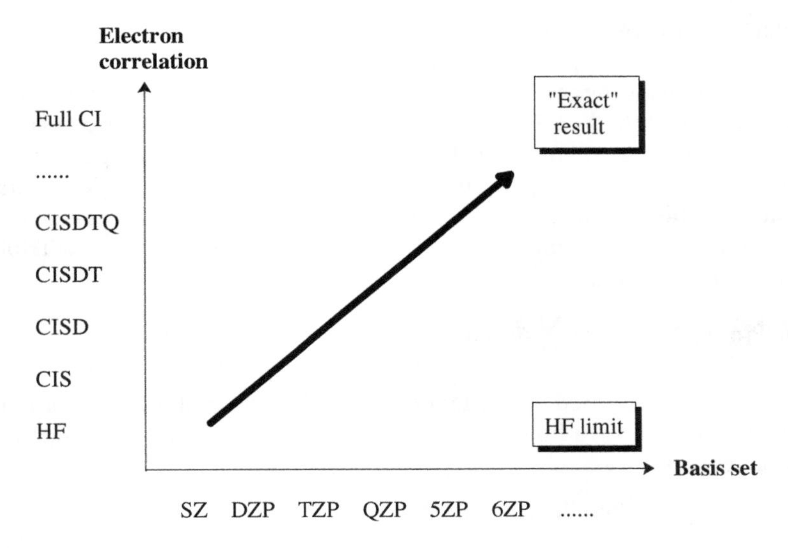

Figure 4.3 Convergence to the exact solution.

cases, the highest virtual orbitals corresponding to the antibonding combinations of the core orbitals are also removed from the correlation treatment (*frozen virtuals*). The frozen-core approximation is not justified in terms of total energy; the correlation of the core electrons gives a substantial energy contribution. However, it is a near-constant factor, which drops out when calculating relative energies. Furthermore, if we really want to calculate the core electron correlation, the standard basis sets are insufficient. In order to represent the angular correlation, higher angular momentum functions with the same radial size as the filled orbitals are needed, for example *p*- and *d*-functions with large exponents for correlating the 1s-electrons, as discussed in Section 5.4.6. Just allowing excitations of the core electrons in a standard basis set does not "correlate" the core electrons.

There are three main methods for calculating electron correlation: *Configuration Interaction* (CI), *Many-Body Perturbation Theory* (MBPT) and *Coupled Cluster* (CC). A word of caution before we describe these methods in more detail. The Slater determinants are composed of spin-MOs, but since the Hamiltonian operator is independent of spin, the spin dependence can be factored out. Furthermore, to facilitate notation, it is often assumed that the HF determinant is of the RHF type, rather than the more general UHF type. Finally, many of the expressions below involve double summations over identical sets of functions. To ensure only the unique terms are included, one of the summation indices must be restricted. Alternatively, both indices can be allowed to run over all values and the overcounting corrected by a factor of $^1/_2$. Various combinations of these assumptions result in final expressions that differ by factors of $^1/_2$, $^1/_4$, etc., from those given here. In the present chapter, the MOs are always spin-MOs and conversion of a restricted summation to unrestricted is always noted explicitly.

Finally, a comment on notation. The quality of a calculation is given by the level of theory (i.e. how much electron correlation is included) and the size of the basis set. In a commonly used "/" notation, introduced by J. A. Pople, this is denoted as "*level/basis*". If nothing further is specified, this implies that the geometry is optimized at this level of theory. As discussed in Section 5.7, the geometry is usually much less sensitive to the theoretical level than relative energies and high-level calculations are therefore often carried out using geometries optimized at a lower level. This is denoted as "*level2/basis2//level1/basis1*", where the notation after the "//" indicates the level at which the geometry is optimized.

4.2 Configuration Interaction

This is the oldest and perhaps the easiest method to understand and is based on the variational principle (Appendix B), analogous to the HF method. The trial wave function is written as a linear combination of determinants with the expansion coefficients determined by requiring that the energy should be a minimum (or at least stationary), a procedure known as *Configuration Interaction* (CI).[7,8] The MOs used for building the excited Slater determinants are taken from a Hartree–Fock calculation and held fixed. Subscripts S, D, T, etc., indicate determinants that are Singly, Doubly, Triply, etc., excited relative to the HF configuration:

$$\Psi_{\text{CI}} = a_0\Phi_{\text{HF}} + \sum_{\text{S}} a_{\text{S}}\Phi_{\text{S}} + \sum_{\text{D}} a_{\text{D}}\Phi_{\text{D}} + \sum_{\text{T}} a_{\text{T}}\Phi_{\text{T}} + \cdots = \sum_{i=0} a_i\Phi_i \tag{4.2}$$

This is an example of a constrained optimization, where the energy should be minimized under the constraint that the total CI wave function is normalized. Introducing a Lagrange multiplier (Section 13.5), this can be written as

$$L = \langle \Psi_{\text{CI}}|\mathbf{H}|\Psi_{\text{CI}}\rangle - \lambda(\langle \Psi_{\text{CI}}|\Psi_{\text{CI}}\rangle - 1) \tag{4.3}$$

The first bracket is the energy of the CI wave function and the second bracket is the norm of the wave function. In terms of determinants (Equation (4.2)), these can be written as in the following equations:

$$\langle \Psi_{\text{CI}}|\mathbf{H}|\Psi_{\text{CI}}\rangle = \sum_{i=0}\sum_{j=0} a_i a_j \langle \Phi_i|\mathbf{H}|\Phi_j\rangle = \sum_{i=0} a_i^2 E_i + \sum_{i=0}\sum_{j\neq i} a_i a_j \langle \Phi_i|\mathbf{H}|\Phi_j\rangle \tag{4.4}$$

$$\langle \Psi_{\text{CI}}|\Psi_{\text{CI}}\rangle = \sum_{i=0}\sum_{j=0} a_i a_j \langle \Phi_i|\Phi_j\rangle = \sum_{i=0} a_i^2 \langle \Phi_i|\Phi_i\rangle = \sum_{i=0} a_i^2 \tag{4.5}$$

The diagonal elements in the sum involving the Hamiltonian operator (Equation (4.4)) are energies of the corresponding determinants. The overlap elements between different determinants (Equation (4.5)) are zero as they are built from orthogonal MOs (Equation (3.20)). The variational procedure corresponds to setting all the derivatives of the Lagrange function (Equation (4.3)) with respect to the a_i expansion coefficients equal to zero:

$$\begin{aligned} \frac{\partial L}{\partial a_i} &= 2\sum_{j} a_j \langle \Phi_i|\mathbf{H}|\Phi_j\rangle - 2\lambda a_i = 0 \\ a_i(E_i - \lambda) &+ \sum_{j\neq i} a_j \langle \Phi_i|\mathbf{H}|\Phi_j\rangle = 0 \end{aligned} \tag{4.6}$$

If there is only one determinant in the expansion ($a_0 = 1, a_{i\neq 0} = 0$), the latter equation shows that the Lagrange multiplier λ is the (CI) energy.

As there is one Equation (4.6) for each i, the variational problem is transformed into solving a set of CI *secular equations*. Introducing the notation $H_{ij} = \langle \Phi_i|\mathbf{H}|\Phi_j\rangle$, the stationary conditions in Equation (4.6) can be written as

$$\begin{pmatrix} H_{00} - E & H_{01} & \cdots & H_{0j} & \cdots \\ H_{10} & H_{11} - E & \cdots & H_{1j} & \cdots \\ \vdots & \vdots & \ddots & \vdots & \cdots \\ H_{j0} & \vdots & \cdots & H_{jj} - E & \cdots \\ \vdots & \vdots & \cdots & \vdots & \ddots \end{pmatrix} \begin{pmatrix} a_0 \\ a_1 \\ \vdots \\ a_j \\ \vdots \end{pmatrix} = \begin{pmatrix} 0 \\ 0 \\ \vdots \\ 0 \\ \vdots \end{pmatrix} \tag{4.7}$$

This can also be written as matrix equations:

$$\begin{aligned}(\mathbf{H} - E\mathbf{I})\mathbf{a} &= \mathbf{0} \\ \mathbf{H}\mathbf{a} &= E\mathbf{a}\end{aligned} \tag{4.8}$$

Solving the secular equations is equivalent to diagonalizing the CI matrix (see Section 17.2.4). The CI energy is obtained as the lowest eigenvalue of the CI matrix, and the corresponding eigenvector contains the a_i coefficients in front of the determinants in Equation (4.2). The second lowest eigenvalue corresponds to the first excited state, etc.

4.2.1 CI Matrix Elements

The CI matrix elements H_{ij} can be evaluated by the same strategy employed for calculating the energy of a single determinant used for deriving the Hartree–Fock equations (Section 3.3). This involves expanding the determinants in a sum of products of MOs, thereby making it possible to express the CI matrix elements in terms of MO integrals. There are, however, some general features that make many of the CI matrix elements equal to zero.

The Hamiltonian operator (Equation (3.24)) does not contain spin; thus if two determinants have different total spins the corresponding matrix element is zero. This situation occurs if an electron is excited from an α spin-MO to a β spin-MO, such as the second S-type determinant in Figure 4.2. When the HF wave function is a singlet, this excited determinant is (part of) a triplet. The corresponding CI matrix element can be written in terms of integrals over MOs, and the spin dependence can be separated out. If there is a different number of α and β spin-MOs, there will always be at least one integral $\langle\alpha|\beta\rangle = 0$. That matrix elements between different spin states are zero may be fairly obvious. If we are interested in a singlet wave function, only singlet determinants can enter the expansion with non-zero coefficients. However, if the Hamiltonian operator includes, for example, the spin-orbit operator, matrix elements between singlet and triplet determinants are not necessarily zero, and the resulting CI wave function will be a mixture of singlet and triplet determinants.

Consider now the case where an electron with α spin is moved from orbital i to orbital a. The first S-type determinant in Figure 4.2 is of this type. Alternatively, the electron with β spin could be moved from orbital i to orbital a. Both of these excited determinants will have an S_z value of 0, but neither are eigenfunctions of the $\mathbf{S}^2$ operator. The difference and sum of these two determinants describe a singlet state and the $S_z = 0$ component of a triplet, as illustrated in Figure 4.4.

Such linear combinations of determinants, which are proper spin eigenfunctions, are called *Spin-Adapted Configurations* (SACs) or *Configurational State Functions* (CSFs). The construction of

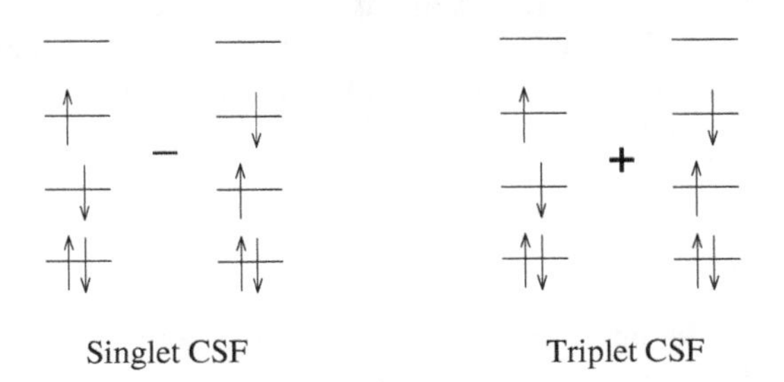

Figure 4.4 Forming configurational state functions from Slater determinants.

proper CSFs may involve several determinants for higher excited states. The first D-type determinant in Figure 4.2 is already a proper CSF, but the second D-type excitation must be combined with five other determinants corresponding to rearrangement of the electron spins to make a singlet CSF (actually there are two linearly independent CSFs that can be made). By making suitable linear combinations of determinants the number of non-zero CI matrix elements can therefore be reduced.

If the system contains symmetry, there are additional CI matrix elements that become zero. The symmetry of a determinant is given as the direct product of the symmetries of the MOs. The Hamiltonian operator always belongs to the totally symmetric representation; thus if two determinants belong to different irreducible representations, the CI matrix element is zero. This is again fairly obvious: if the interest is in a state of a specific symmetry, only those determinants that have the correct symmetry can contribute.

The Hamiltonian operator consists of a sum of one-electron and two-electron operators (Equation (3.25)). If two determinants differ by more than two (spatial) MOs there will always be an overlap integral between two different MOs that is zero (the same argument as in Equation (3.29)). CI matrix elements can therefore only be non-zero if the two determinants differ by 0, 1 or 2 MOs, and they may be expressed in terms of integrals of one- and two-electron operators over MOs. These connections are known as the *Slater–Condon* rules. If the two determinants are identical, the matrix element is simply the energy of a single-determinant wave function, as given by Equation (3.33). For matrix elements between determinants differing by 1 (exciting an electron from orbital i to a) or 2 (exciting two electrons from orbitals i and j to orbitals a and b) MOs, the results are given in the Equations (4.9) and (4.10) (compare with Equation (3.34), where the **g** operator is implicit in the notation for the two-electron integrals (Equation (3.61)):

$$\langle \Phi_0 | \mathbf{H} | \Phi_i^a \rangle = \langle \phi_i | \mathbf{h} | \phi_a \rangle + \sum_j (\langle \phi_i \phi_j | \phi_a \phi_j \rangle - \langle \phi_i \phi_j | \phi_j \phi_a \rangle) \tag{4.9}$$

$$\langle \Phi_0 | \mathbf{H} | \Phi_{ij}^{ab} \rangle = \langle \phi_i \phi_j | \phi_a \phi_b \rangle - \langle \phi_i \phi_j | \phi_b \phi_a \rangle \tag{4.10}$$

The matrix element between the HF and a singly excited determinant is a matrix element of the Fock operator (Equation (3.37)) between two different MOs:

$$\langle \phi_i | \mathbf{h} | \phi_a \rangle + \sum_j (\langle \phi_i \phi_j | \phi_a \phi_j \rangle - \langle \phi_i \phi_j | \phi_j \phi_a \rangle) = \langle \phi_i | \mathbf{F} | \phi_a \rangle \tag{4.11}$$

This is an occupied–virtual off-diagonal element of the Fock matrix in the MO basis, and is identical to the gradient of the energy with respect to an occupied–virtual mixing parameter (except for a factor of 4) (see Equation (3.71)). If the determinants are constructed from optimized canonical HF MOs, the gradient is zero, and the matrix element is zero. This may also be realized by noting that the MOs are eigenfunctions of the Fock operator in Equation (3.43):

$$\begin{aligned} \mathbf{F}\phi_a &= \varepsilon_a \phi_a \\ \langle \phi_i | \mathbf{F} | \phi_a \rangle &= \varepsilon_a \langle \phi_i | \phi_a \rangle = \varepsilon_a \delta_{ia} \end{aligned} \tag{4.12}$$

The disappearance of matrix elements between the HF reference and singly excited states is known as *Brillouin's theorem*. The HF reference state therefore only has non-zero matrix elements with doubly excited determinants, and the full CI matrix acquires a block diagonal structure (Figure 4.5).

$$
\begin{array}{c|ccccccc}
 & \Phi_{\mathrm{HF}} & \Phi_{\mathrm{S}} & \Phi_{\mathrm{D}} & \Phi_{\mathrm{T}} & \Phi_{\mathrm{Q}} & \Phi_{5} & \cdots \\
\hline
\Phi_{\mathrm{HF}} & E_{\mathrm{HF}} & 0 & x & 0 & 0 & 0 & 0 \\
\Phi_{\mathrm{S}} & 0 & E_{\mathrm{S}} & x & x & 0 & 0 & 0 \\
\Phi_{\mathrm{D}} & x & x & E_{\mathrm{D}} & x & x & 0 & 0 \\
\Phi_{\mathrm{T}} & 0 & x & x & E_{\mathrm{T}} & x & x & 0 \\
\Phi_{\mathrm{Q}} & 0 & 0 & x & x & E_{\mathrm{Q}} & x & x \\
\Phi_{5} & 0 & 0 & 0 & x & x & E_{5} & x \\
\vdots & 0 & 0 & 0 & 0 & x & x & \ddots
\end{array}
$$

Figure 4.5 Structure of the CI matrix.

In order to evaluate the CI matrix elements we need one- and two-electron integrals over MOs. These can be expressed in terms of the corresponding AO integrals and the MO coefficients:

$$\langle \phi_i | \mathbf{h} | \phi_a \rangle = \sum_{\alpha}^{M_{\text{basis}}} \sum_{\beta}^{M_{\text{basis}}} c_{\alpha i} c_{\beta j} \langle \chi_\alpha | \mathbf{h} | \chi_\beta \rangle \tag{4.13}$$

$$\langle \phi_i \phi_j | \phi_k \phi_l \rangle = \sum_{\alpha}^{M_{\text{basis}}} \sum_{\beta}^{M_{\text{basis}}} \sum_{\gamma}^{M_{\text{basis}}} \sum_{\delta}^{M_{\text{basis}}} c_{\alpha i} c_{\beta j} c_{\gamma k} c_{\delta l} \langle \chi_\alpha \chi_\beta | \chi_\gamma \chi_\delta \rangle \tag{4.14}$$

Such MO integrals are required for all electron correlation methods. The two-electron AO integrals are the most numerous and the above equation appears to involve a computational effect proportional to M_{basis}^8 (M_{basis}^4 AO integrals each multiplied by four sets of M_{basis} MO coefficients). However, by performing the transformation one index at a time, the computational effort can be reduced to M_{basis}^5:

$$\langle \phi_i \phi_j | \phi_k \phi_l \rangle = \sum_{\alpha}^{M_{\text{basis}}} c_{\alpha i} \left(\sum_{\beta}^{M_{\text{basis}}} c_{\beta j} \left(\sum_{\gamma}^{M_{\text{basis}}} c_{\gamma k} \left(\sum_{\delta}^{M_{\text{basis}}} c_{\delta l} \langle \chi_\alpha \chi_\beta | \chi_\gamma \chi_\delta \rangle \right) \right) \right) \tag{4.15}$$

Each step now only involves multiplication of M_{basis}^4 integrals with M_{basis} coefficients, that is the M_{basis}^8 dependence is reduced to four M_{basis}^5 operations. *In the large basis set limit, all electron correlation methods formally scale as at least* M_{basis}^5, since this is the scaling for the AO to MO integral transformation. The transformation is an example of a "rotation" of the "coordinate" system consisting of the AOs, to one where the Fock operator is diagonal, the MOs (see Section 17.2). The diagonal system allows a much more compact representation of the matrix elements needed for the electron correlation treatment. The coordinate change is also known as a *four-index transformation,* since it involves four indices associated with the basis functions.

4.2.2 Size of the CI Matrix

The excited Slater determinants are generated by removing electrons from occupied orbitals and placing them in virtual orbitals. The number of possible excited SDs is thus a combinatorial problem, and therefore increases *factorially* with the number of electrons and basis functions. Consider, for example, a system such as H_2O with a 6-31G(d) basis. For the purpose of illustration, let us for a moment return to the spin-orbital description. There are 10 electrons and 38 spin-MOs, of which 10 are occupied and 28 are empty. There are $K_{10,n}$ possible ways of selecting n electrons out of the 10

Table 4.1 Number of singlet CSFs as a function of excitation level for H_2O with a 6-31G(d) basis.

Excitation level n	Number of nth excited CSFs	Total number of CSFs
1	71	71
2	2 485	2 556
3	40 040	42 596
4	348 530	391 126
5	1 723 540	2 114 666
6	5 033 210	7 147 876
7	8 688 680	15 836 556
8	8 653 645	24 490 201
9	4 554 550	29 044 751
10	1 002 001	30 046 752

occupied orbitals and $K_{28,n}$ ways of distributing them in the 28 empty orbitals. The number of excited states for a given excitation level is thus $K_{10,n} \cdot K_{28,n}$ and the total number of excited determinants will be a sum over 10 such terms. This is also equivalent to $K_{38,10}$, the total number of ways 10 electrons can be distributed in 38 orbitals:

$$\text{Number of SDs} = \sum_{n=0}^{10} K_{10,n} \cdot K_{28,n} = K_{38,10} = \frac{38!}{10!(38-10)!} \tag{4.16}$$

Many of these excited determinants will of course have different spin multiplicity (triplet, pentet, etc., states for a singlet HF determinant) and can therefore be left out of the calculation. Generating only the singlet CSFs, the number of configurations at each excitation level is shown in Table 4.1.

The number of determinants (or CSFs) that can be generated grows wildly with the excitation level! Even if the C_{2v} symmetry of H_2O is employed, there is still a total of 7 536 400 singlet CSFs with A_1 symmetry. If all possible determinants are included, we have a *full CI* wave function and there is no truncation in the many-electron expansion besides that generated by the finite one-electron expansion (size of the basis set). This is the best possible wave function within the limitations of the basis set, that is it recovers 100% of the electron correlation in the given basis. For the water case with a medium basis set, this corresponds to diagonalizing a matrix of size 30 046 752 × 30 046 752, which is impossible. However, normally the interest is only in the lowest (or a few of the lowest) eigenvalue(s) and eigenvector(s), and there are special iterative methods (Sections 4.2.4 and 17.2.5) for determining one (or a few) eigenvector(s) of a large matrix.

In the general case of N electrons and M basis functions the total number of singlet CSFs that can be generated is given by

$$\text{Number of CSFs} = \frac{M!(M+1)!}{\left(\frac{N}{2}\right)!\left(\frac{N}{2}+1\right)!\left(M-\frac{N}{2}\right)!\left(M-\frac{N}{2}+1\right)!} \tag{4.17}$$

For H_2O with the above 6-31G(d) basis there are ~30 × 10^6 CSFs ($N = 10$, $M = 19$), and with the larger 6-311G(2d,2p) basis there are ~106 × 10^9 CSFs ($N = 10$, $M = 41$). For $H_2C{=}CH_2$ with the 6-31G(d) basis there are ~334 × 10^{12} CSFs ($N = 16$, $M = 38$).

One of the large-scale full CI calculations considered H_2O in a DZP type basis with 24 functions. Allowing all possible excitations of the 10 electrons generates 451 681 246 determinants.[9] The variational wave function thus contains roughly half a billion parameters, that is the formal size of the CI matrix is of the order of half a billion squared. Although a determination of the lowest eigenvalue of such a problem can be done in a matter of hours on a modern computer, the result is a single number, the ground state energy of the H_2O molecule. Due to basis set limitations, however, it is still some 0.2 au (~500 kJ/mol) larger than the experimental value. The computational effort for extracting a single eigenvalue and eigenvector scales essentially linearly with the number of CSFs, and it is possible to handle systems with up to a few billion determinants. The factorial growth of the number of determinants with the size of the basis set, however, makes the full CI method infeasible for all but the very smallest systems. Full CI calculations are thus *not* a routine computational procedure for including electron correlation, but they are a very useful reference for developing more approximate methods, as the full CI gives the best results that can be obtained in the given basis.

4.2.3 Truncated CI Methods

In order to develop a computationally tractable model, the number of excited determinants in the CI expansion (Equation (4.2)) must be reduced. Truncating the excitation level at one (*CI with singles* (CIS)) does not give any improvement over the HF result as all matrix elements between the HF wave function and singly excited determinants are zero. CIS is equal to HF for the ground state energy, although higher roots from the secular equations may be used as approximations to excited states. It has already been mentioned that only doubly excited determinants have matrix elements with the HF wave function different from zero; thus the lowest CI level that gives an improvement over the HF result is to include only doubly excited states, yielding the *CI with doubles* (CID) model. Compared with the number of doubly excited determinants, there are relatively few singly excited determinants (see, e.g., Table 4.1), and including these gives the CISD method. Computationally, this is only a marginal increase in effort over CID. Although the singly excited determinants have zero matrix elements with the HF reference, they enter the wave function indirectly as they have non-zero matrix elements with the doubly excited determinants. In the large basis set limit the CISD method scales as M_{basis}^6.

The next level in improvement is inclusion of the triply excited determinants, giving the CISDT method, which is an M_{basis}^8 method. Taking into account also quadruply excited determinants yields the CISDTQ method, which is an M_{basis}^{10} method. As shown below, the CISDTQ model in general gives results close to the full CI limit, but even truncating the excitation level at four produces so many configurations that it can only be applied to small molecules and small basis sets. The only CI method that is generally applicable for a large variety of systems is CISD. For computationally feasible systems (i.e. medium-size molecules and basis sets), it typically recovers 80–90% of the available correlation energy. The percentage is highest for small molecules; as the molecule gets larger the CISD method recovers less and less of the correlation energy, which is discussed in more detail in Section 4.5.

Since only doubly excited determinants have non-zero matrix elements with the HF state, these are the most important. This may be illustrated by considering a full CI calculation for the Ne atom in a [5s4p3d] basis, where the 1s-electrons are omitted from the correlation treatment.[10] The contribution to the full CI wave function from each level of excitation is given in Table 4.2.

The weight is the sum of a_i^2 coefficients at the given excitation level (Equation (4.2)). The CI method determines the coefficients from the variational principle; thus Table 4.2 shows that the doubly excited determinants are by far the most important in terms of energy. The singly excited determinants are the second most important, followed by the quadruples and triples. Excitations higher than four make

Table 4.2 Weights of excited configurations for the neon atom.

Excitation level	Weight
0	9.6×10^{-1}
1	9.8×10^{-4}
2	3.4×10^{-2}
3	3.7×10^{-4}
4	4.5×10^{-4}
5	1.9×10^{-5}
6	1.7×10^{-6}
7	1.4×10^{-7}
8	1.1×10^{-9}

only very small contributions, although there are actually many more of these highly excited determinants than the triples and quadruples, as illustrated in Table 4.1.

The relative importance of the different excitations may qualitatively be understood by noting that the doubles provide electron correlation for electron pairs. Quadruply excited determinants are important as they primarily correspond to products of double excitations. The singly excited determinants allow inclusion of multireference character in the wave function, that is they allow the orbitals to "relax" (they correspond to gradients of the HF energy with respect to orbital coefficients, Equation (4.11)). Although the HF orbitals are optimum for the single-determinant wave function, that is no longer the case when many determinants are included. The triply excited determinants are doubly excited relative to the singles and can then be viewed as providing correlation for the "multireference" part of the CI wave function.

While singly excited states make relatively small contributions to the correlation energy of a CI wave function, they are very important when calculating properties (Chapter 11). Molecular properties measure how the wave function changes when a perturbation, such as an external electric field, is added. The change in the wave function introduced by the perturbation makes the MOs no longer optimal in the variational sense. The first-order change in the MOs is described by the off-diagonal elements in the Fock matrix, as these are essentially the gradient of the HF energy with respect to the MOs. In the absence of a perturbation, these are zero, as the HF energy is stationary with respect to an orbital variation (Equation (3.40)). As shown in Equations (4.9) to (4.11), the Fock matrix off-diagonal elements are CI matrix elements between the HF and singly excited states. For molecular properties, the singly excited states thus allow the CI wave function to "relax" the MOs, that is letting the wave function respond to the perturbation.

4.2.4 Direct CI Methods

As illustrated above, even quite small systems at the CISD level result in millions of CSFs. The variational problem is to extract one or possibly a few of the lowest eigenvalues and eigenvectors of a matrix the size of millions squared. This cannot be done by standard diagonalization methods where all the eigenvalues are found. There are, however, iterative methods for extracting one, or a few, eigenvalues and eigenvectors of a large matrix. The CI problem of Equation (4.8) may be written as:

$$(\mathbf{H} - E\mathbf{I})\mathbf{a} = \mathbf{0} \tag{4.18}$$

The **H** matrix contains the matrix element between the CSFs in the CI expansion and the **a** vector of the expansion coefficients. The idea in iterative methods is to generate a suitable guess for the coefficient vector and calculate $(\mathbf{H} - E\mathbf{I})\mathbf{a}$. This will in general not be zero, and the deviation may be used for adding a correction to **a**, forming an iterative algorithm. If the interest is in the lowest eigenvalue, a suitable start eigenvector may be one that only contains the HF configuration, that is $\{1,0,0,0, \ldots\}$. Since the **H** matrix elements are essentially two-electron integrals in the MO basis (Equation (4.8)), the iterative procedure may be formulated as integral driven, that is a batch of integrals are read in (or generated otherwise) and used directly in the multiplication with the corresponding **a**-coefficients. The CI matrix is therefore not needed explicitly, only the effect of its multiplication with a vector containing the variational parameters, and storage of the entire matrix is avoided. This is the basis for being able to handle CI problems of almost any size, and is known as *direct CI*. Note that it is not "direct" in the sense used to describe the direct SCF method, where all the AO integrals are calculated as needed. The direct CI approach just assumes that the CI matrix elements (e.g. two-electron integrals in the MO basis) are available as required, traditionally stored in a file on a disk. There are several variations on how the **a**-vector is adjusted in each iteration, and the most commonly used versions are based on the *Davidson algorithm* discussed in Section 17.2.5.[11]

4.3 Illustrating how CI Accounts for Electron Correlation, and the RHF Dissociation Problem

Consider the H_2 molecule in a minimum basis consisting of one *s*-function on each center, χ_A and χ_B. An RHF calculation will produce two MOs, ϕ_1 and ϕ_2, being the sum and difference of the two AOs. The sum of the two AOs is a bonding MO, with increased probability of finding the electrons between the two nuclei, while the difference is an antibonding MO, with decreased probability of finding the electrons between the two nuclei, as illustrated in Figure 4.6:

$$\begin{aligned} \phi_1 &= (\chi_A + \chi_B) \\ \phi_2 &= (\chi_A - \chi_B) \end{aligned} \tag{4.19}$$

The HF wave function will have two electrons in the lowest energy (bonding) MO:

$$\Phi_0 = \begin{vmatrix} \phi_1(1) & \overline{\phi_1}(1) \\ \phi_1(2) & \overline{\phi_1}(2) \end{vmatrix} \tag{4.20}$$

We have here neglected the normalization constants for both the MOs and the determinantal wave function. The bar above the MO indicates that the electron has a β spin function, while no bar

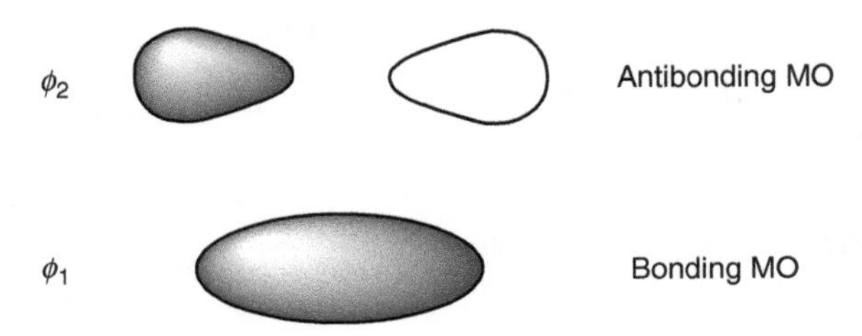

Figure 4.6 Molecular orbitals for H_2.

indicates an α spin function. In this basis, there are one doubly (Φ_1) and four singly excited Slater determinants (Φ_{2-5}):

$$\Phi_1 = \begin{vmatrix} \phi_2(1) & \overline{\phi_2}(1) \\ \phi_2(2) & \overline{\phi_2}(2) \end{vmatrix}$$
$$\Phi_2 = \begin{vmatrix} \phi_1(1) & \overline{\phi_2}(1) \\ \phi_1(2) & \overline{\phi_2}(2) \end{vmatrix}$$
$$\Phi_3 = \begin{vmatrix} \overline{\phi_1}(1) & \phi_1(1) \\ \overline{\phi_1}(2) & \phi_1(2) \end{vmatrix} \tag{4.21}$$
$$\Phi_4 = \begin{vmatrix} \phi_1(1) & \phi_2(1) \\ \phi_1(2) & \phi_2(2) \end{vmatrix}$$
$$\Phi_5 = \begin{vmatrix} \overline{\phi_1}(1) & \overline{\phi_2}(1) \\ \overline{\phi_1}(2) & \overline{\phi_2}(2) \end{vmatrix}$$

Configurations Φ_4 and Φ_5 are clearly the $S_z = 1$ and $S_z = -1$ components of a triplet state. The plus combination of Φ_2 and Φ_3 is the $S_z = 0$ component of the triplet and the minus combination is a singlet configuration (see Figure 4.4). The H_2 molecule belongs to the $D_{\infty h}$ point group and the two MOs transform as the σ_g (ϕ_1) and σ_u (ϕ_2) representations. The singly excited CSF ($\Phi_2 - \Phi_3$) has overall Σ_u symmetry, while the HF (Φ_0) and doubly excited determinant (Φ_1) have Σ_g. The full 6×6 CI problem therefore blocks into a 2×2 block of singlet Σ_g states, a 1×1 block of singlet Σ_u and a 3×3 block of triplet Σ_u states, as shown in Figure 4.7. Owing to the orthogonality of the spin functions, the triplet block is already diagonal.

The full CI for the $^1\Sigma_g$ states involves only two configurations, the reference HF and the doubly excited determinant:

$$\Phi_0 = \phi_1(1)\overline{\phi_1}(2) - \overline{\phi_1}(1)\phi_1(2) = \phi_1\phi_1(\alpha\beta - \beta\alpha) \tag{4.22}$$
$$\Phi_1 = \phi_2(1)\overline{\phi_2}(2) - \overline{\phi_2}(1)\phi_2(2) = \phi_2\phi_2(\alpha\beta - \beta\alpha) \tag{4.23}$$

The electron coordinate in Equations (4.22) and (4.23) is given implicitly by the order in which the orbitals are written, that is $\phi_1\phi_1[\alpha\beta - \beta\alpha] = \phi_1(1)\phi_1(2)[\alpha(1)\beta(2) - \beta(1)\alpha(2)]$. Ignoring the spin

$$\begin{pmatrix} {}^1\Phi_{0,\Sigma_g} & {}^1\Phi_{1,\Sigma_g} & {}^1(\Phi_2-\Phi_3)_{\Sigma_u} & {}^3\Phi_{4,\Sigma_u} & {}^3(\Phi_2+\Phi_3)_{\Sigma_u} & {}^3\Phi_{5,\Sigma_u} \end{pmatrix}$$
$$\begin{pmatrix} {}^1\Phi_{0,\Sigma_g} \\ {}^1\Phi_{1,\Sigma_g} \\ {}^1(\Phi_2-\Phi_3)_{\Sigma_u} \\ {}^3\Phi_{4,\Sigma_u} \\ {}^3(\Phi_2+\Phi_3)_{\Sigma_u} \\ {}^3\Phi_{5,\Sigma_u} \end{pmatrix} \begin{pmatrix} E_0 & H_{01} & 0 & 0 & 0 & 0 \\ H_{01} & E_1 & 0 & 0 & 0 & 0 \\ 0 & 0 & E_2 & 0 & 0 & 0 \\ 0 & 0 & 0 & E_3 & 0 & 0 \\ 0 & 0 & 0 & 0 & E_4 & 0 \\ 0 & 0 & 0 & 0 & 0 & E_5 \end{pmatrix}$$

Figure 4.7 Structure of the full CI matrix for the H_2 system in a minimum basis.

functions (which may be integrated out since **H** is spin-independent), the determinants can be expanded in AOs:

$$\Phi_0 = (\chi_A(1) + \chi_B(1))(\chi_A(2) + \chi_B(2)) = \chi_A\chi_A + \chi_B\chi_B + \chi_A\chi_B + \chi_B\chi_A \tag{4.24}$$

$$\Phi_1 = (\chi_A(1) - \chi_B(1))(\chi_A(2) - \chi_B(2)) = \chi_A\chi_A + \chi_B\chi_B - \chi_A\chi_B - \chi_B\chi_A \tag{4.25}$$

The first two terms on the right-hand side in Equations (4.24) and (4.25) have both electrons on the same nuclear center, and they describe *ionic* contributions to the wave function, H^+H^-. The latter two terms describe *covalent* contributions to the wave function, H·H·. The HF wave function thus contains equal amounts of ionic and covalent contributions.

The full CI wave function may be written in terms of AOs, with the optimum values of the a_0 and a_1 coefficients determined by the variational procedure:

$$\begin{aligned}\Psi_{CI} &= a_0\Phi_0 + a_1\Phi_1 \\ \Psi_{CI} &= (a_0 + a_1)(\chi_A\chi_A + \chi_B\chi_B) + (a_0 - a_1)(\chi_A\chi_B + \chi_B\chi_A)\end{aligned} \tag{4.26}$$

The HF wave function constrains both electrons to move in the same bonding orbital. By allowing the doubly excited state to enter the wave function, the electrons can better avoid each other, as the antibonding MO is now also available. The antibonding MO has a nodal plane (where $\phi_2 = 0$) perpendicular to the molecular axis (Figure 4.6), and the electrons are able to correlate their movements by being on opposite sides of this plane. This *left–right* correlation is a molecular equivalent of the atomic radial correlation discussed in Section 5.2.

Consider now the behavior of the HF wave function Φ_0 (Equation (4.24)) as the distance between the two nuclei is increased toward infinity. Since the HF wave function is an equal mixture of ionic and covalent terms, the dissociation limit is 50% H^+H^- and 50% H·H·. In the gas phase, all bonds dissociate homolytically and the ionic contribution should be 0%. The HF dissociation energy is therefore much too high. This is a general problem with RHF-type wave functions: the constraint of doubly occupied MOs is inconsistent with breaking bonds to produce radicals. In order for an RHF wave function to dissociate correctly, an even-electron molecule must break into two even-electron fragments, each being in the lowest electronic state. Furthermore, the orbital symmetries must match. There are only a few covalently bonded systems that obey these requirements (the simplest example is HHe^+). The wrong dissociation limit for RHF wave functions has several consequences:

1. The energy for stretched bonds is too high. Most transition structures have partly formed/broken bonds; thus activation energies are too high at the RHF level.
2. The too-steep increase in energy as a function of the bond length causes the minimum on a potential energy curve to occur too "early" for covalently bonded systems, and equilibrium bond lengths are too short at the RHF level.
3. The too-steep increase in energy as a function of the bond length causes the curvature of the potential energy surface near the equilibrium to be too large, and vibrational frequencies, especially those describing bond stretching, are in general too high.
4. The wave function contains too much "ionic" character, and RHF dipole moments (and also atomic charges) are in general too large.

It should be noted that dative bonds, such as in metal complexes and charge transfer species, in general have RHF wave functions that dissociate correctly, and the equilibrium bond lengths in these cases are normally too long.

The dissociation problem is solved in the case of a full CI wave function in this minimum basis. As seen from Equation (4.26), the ionic term can be made to disappear by setting $a_1 = -a_0$. The full CI wave function generates the lowest possible energy (within the limitations of the chosen basis set) at all distances, with the optimum weights of the HF and doubly excited determinant determined by the variational principle. In the general case of a polyatomic molecule and large basis set, correct dissociation of all bonds can be achieved if the CI wave function contains all determinants generated by a full CI in the valence orbital space. The latter corresponds to a full CI if a minimum basis is employed, but is much smaller than a full CI if an extended basis is used.

4.4 The UHF Dissociation and the Spin Contamination Problem

The dissociation problem can also be "solved" by using a wave function of the UHF type. Here the α and β bonding Mos are allowed to "localize", thereby reducing the MO symmetries to $C_{\infty v}$:

$$\phi_1 = (\chi_A + c\chi_B)\alpha \quad ; \quad \overline{\phi_1} = (c\chi_A + \chi_B)\beta \tag{4.27}$$

$$\Phi_0^{UHF} = \begin{vmatrix} \phi_1(1) & \overline{\phi_1}(1) \\ \phi_1(2) & \overline{\phi_1}(2) \end{vmatrix} \tag{4.28}$$

The optimum value of c is determined by the variational principle. If $c = 1$, the UHF wave function is identical to RHF. This will normally be the case near the equilibrium distance. As the bond is stretched, the UHF wave function allows each of the electrons to localize on a nucleus, causing c to go towards 0. The point where the RHF and UHF descriptions start to differ is often referred to as the RHF/UHF *instability* point.[12] This is an example of symmetry breaking, as discussed in Section 3.8.3. The UHF wave function correctly dissociates into two hydrogen atoms, but the symmetry breaking of the MOs has two other, closely connected, consequences: introduction of *electron correlation* and *spin contamination*. To illustrate these concepts, we need to look at the Φ_0 UHF determinant, and the six RHF determinants in Equations (4.20) and (4.21) in more detail. We will again ignore all normalization constants. The six RHF determinants can be expanded in terms of the AOs as in

$$\begin{aligned}
\Phi_0 &= [\chi_A\chi_A + \chi_B\chi_B + \chi_A\chi_B + \chi_B\chi_A](\alpha\beta - \beta\alpha) \\
\Phi_1 &= [\chi_A\chi_A + \chi_B\chi_B - \chi_A\chi_B - \chi_B\chi_A](\alpha\beta - \beta\alpha) \\
\Phi_2 &= [\chi_A\chi_A - \chi_B\chi_B](\alpha\beta - \beta\alpha) - [\chi_A\chi_B](\alpha\beta + \beta\alpha) \\
\Phi_3 &= [\chi_A\chi_A - \chi_B\chi_B](\alpha\beta - \beta\alpha) + [\chi_A\chi_B - \chi_B\chi_A](\alpha\beta + \beta\alpha) \\
\Phi_4 &= [\chi_A\chi_B - \chi_B\chi_A](\alpha\alpha) \\
\Phi_5 &= [\chi_A\chi_B - \chi_B\chi_A](\beta\beta)
\end{aligned} \tag{4.29}$$

Subtracting and adding Φ_2 and Φ_3 produces a pure singlet ($^1\Phi_-$) and the $S_z = 0$ component of the triplet ($^3\Phi_+$) wave function:

$$^1\Phi_- = \Phi_2 - \Phi_3 = [\chi_A\chi_A - \chi_B\chi_B](\alpha\beta - \beta\alpha) \tag{4.30}$$

$$^3\Phi_+ = \Phi_2 + \Phi_3 = [\chi_A\chi_B - \chi_B\chi_A](\alpha\beta + \beta\alpha) \tag{4.31}$$

Performing the expansion of the Φ_0^{UHF} determinant (Equation (4.28)) gives

$$\begin{aligned}
\Phi_0^{UHF} = & \; c[\chi_A\chi_A + \chi_B\chi_B](\alpha\beta - \beta\alpha) \\
& + [\chi_A\chi_B\alpha\beta - c^2\chi_B\chi_A\beta\alpha] \\
& + [c^2\chi_B\chi_A\alpha\beta - \chi_A\chi_B\beta\alpha]
\end{aligned} \tag{4.32}$$

Adding and subtracting factors of $\chi_A\chi_B\alpha\beta$ and $\chi_B\chi_A\beta\alpha$ allow this to be written as

$$\begin{aligned}\Phi_0^{\text{UHF}} &= [c(\chi_A\chi_A + \chi_B\chi_B) + (\chi_A\chi_B + \chi_B\chi_A)](\alpha\beta - \beta\alpha) \\ &\quad + (1-c^2)[\chi_A\chi_B\beta\alpha - \chi_B\chi_A\alpha\beta]\end{aligned} \tag{4.33}$$

Since $0 \leq c \leq 1$, the first term shows that UHF orbitals reduce the ionic contribution relative to the covalent structures, compared with the RHF case (Equation (4.24)). This is the same effect as for the CI procedure (Equation (4.26)), that is the first term shows that the UHF wave function partly includes electron correlation.

The first term in Equation (4.33) can be written as a linear combination of the Φ_0 and Φ_1 determinants, and describes a pure singlet state. The last part of the UHF determinant, however, has terms identical to two of those in the triplet $^3\Phi_+$ combination of Equation (4.31). If we had chosen the alternative set of UHF orbital with the α spin being primarily on center B in Equation (4.27), we would have obtained the other two terms in $^3\Phi_+$, that is the last term in Equation (4.33) breaks the symmetry. The UHF determinant is therefore not a pure spin state; it contains both singlet and triplet spin states. This feature is known as *spin contamination.* For $c = 1$, the UHF wave function is identical to RHF and Φ_0^{UHF} is a pure singlet. For $c = 0$, the UHF wave function only contains the covalent terms, which is the correct dissociation behavior, but also contains equal amounts of singlet and triplet character. When the bond distance is very large, the singlet and triplet states have identical energies, and the spin contamination has no consequence for the energy. In the intermediate region where the bond is not completely broken, however, spin contamination is important.

Compared with full CI, the UHF energy is too high as the higher lying triplet state is mixed into the wave function. The variational principle guarantees that the UHF energy is lower than or equal to the RHF energy since there are more variational parameters. The full CI energy is the lowest possible (for the given basis set) as it recovers 100% of the correlation energy. The UHF wave function thus lowers the energy by introducing some electron correlation, but at the same time raises the energy by including higher energy spin states. At the single-determinant level, the variational principle guarantees that the first effect dominates. If the second effect dominated, the UHF would collapse to the RHF solution. The correlation energy in general increases as a bond is stretched, and the instability point can thus be viewed as the geometry where the correlation effect becomes larger than the spin contamination. Pictorially, the dissociation curves appear as shown in Figure 4.8.

Another way of viewing spin contamination is to write the UHF wave function as a linear combination of pure R(O)HF determinants; for example for a singlet state,

$$^1\Phi^{\text{UHF}} = a_1\,{}^1\Phi^{\text{RHF}} + a_3\,{}^3\Phi^{\text{ROHF}} + a_5\,{}^5\Phi^{\text{ROHF}} + \cdots \tag{4.34}$$

Since the UHF wave function is multideterminantal in terms of R(O)HF determinants, it follows that it to some extent includes electron correlation (relative to the RHF reference).

The amount of spin contamination is given by the expectation value of the $\mathbf{S}^2$ operator, $\langle\mathbf{S}^2\rangle$. The theoretical value for a pure spin state is $S(S+1)$, where S is the maximum value for S_z, that is $\langle\mathbf{S}^2\rangle = 0$ for a singlet ($S = 0$), $\langle\mathbf{S}^2\rangle = 0.75$ for a doublet ($S = 1/2$), $\langle\mathbf{S}^2\rangle = 2.00$ for a triplet ($S = 1$), etc. A UHF "singlet" wave function will contain some amounts of triplet, quintet, etc., states, increasing the $\langle\mathbf{S}^2\rangle$ value from its theoretical value of zero for a pure spin state. Similarly, a UHF "doublet" wave function will contain some amounts of quartet, sextet, etc., states. Usually the contribution from the next higher spin state than the desired is the most important. The $\langle\mathbf{S}^2\rangle$ value for a UHF wave function is operationally calculated from the spatial overlap between all pairs of α and β spin-orbitals:

$$\langle\mathbf{S}^2\rangle = S(S+1) + N_\beta - \sum_{ij}^{N_{\text{MO}}} \left\langle \phi_i^\alpha \middle| \phi_j^\beta \right\rangle^2 \tag{4.35}$$

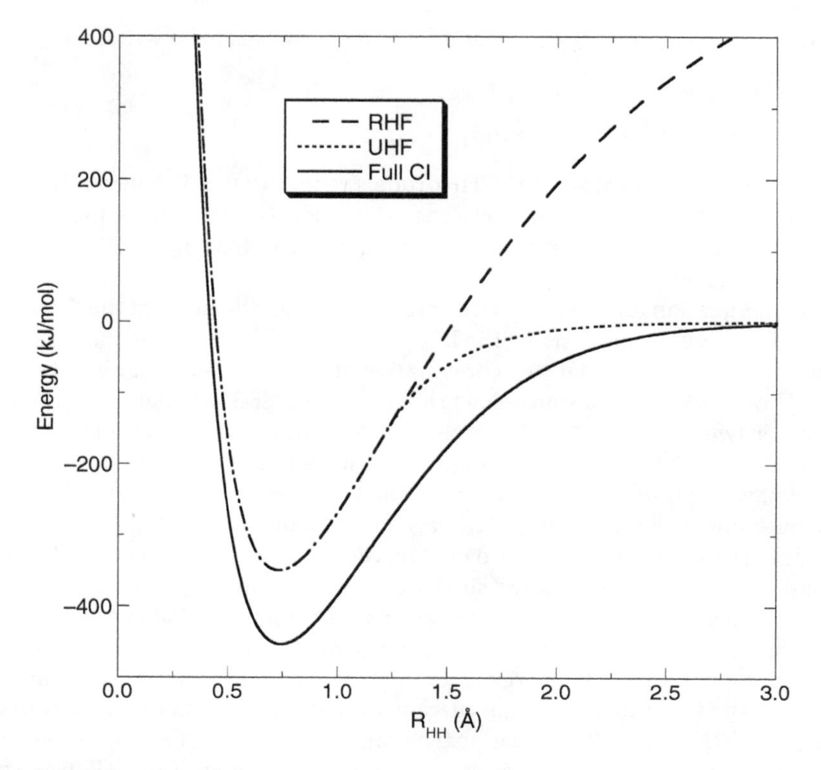

Figure 4.8 Bond dissociation curves for H_2.

If the α and β orbitals are identical, there is no spin contamination, and the UHF wave function is identical to RHF.

By including electron correlation in the wave function, the UHF method introduces more biradical character into the wave function than RHF. The spin contamination part is also purely biradical in nature, that is a UHF treatment in general will overestimate the biradical character of the wave function. Most singlet states are well described by a closed-shell wave function near the equilibrium geometry and, in those cases, it is not possible to generate a UHF solution that has a lower energy than the RHF. There are systems, however, for which this does not hold. An example is the ozone molecule, where two types of resonance structures can be drawn, as shown in Figure 4.9.

The biradical resonance structure for ozone requires two singly occupied MOs, and it is clear that an RHF-type wave function, which requires all orbitals to be doubly occupied, cannot describe this. A UHF-type wave function, however, allows the α and β orbitals to be spatially different, and can to a certain extent incorporate both resonance structures. Systems with biradical character will in general have a (singlet) UHF wave function different from an RHF wave function.

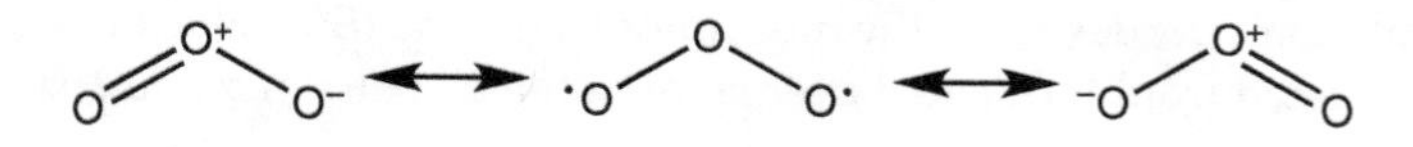

Figure 4.9 Resonance structures for ozone.

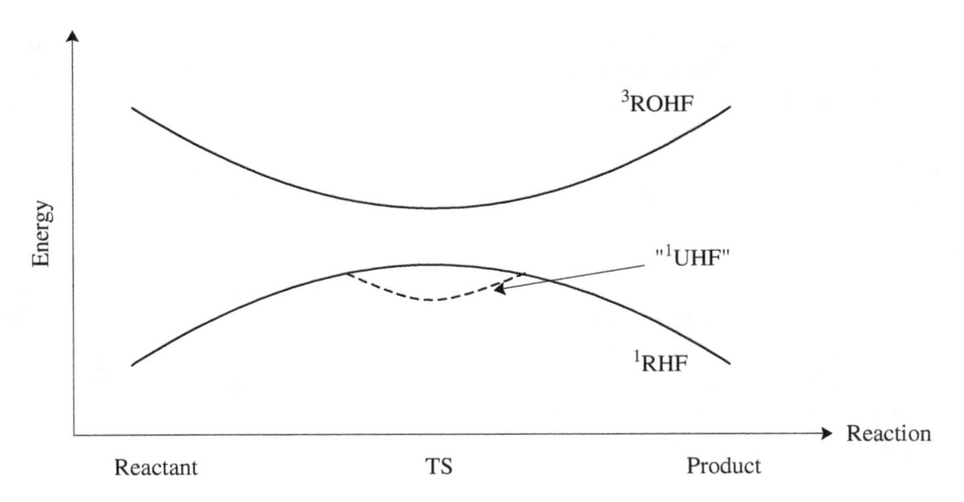

Figure 4.10 Mixing of pure singlet and triplet states may generate artificial minima on the UHF energy surface.

As mentioned above, spin contamination in general increases as a bond is stretched. This has important consequences for transition structures, which contain elongated bonds. While most singlet systems have identical RHF and UHF descriptions near the equilibrium geometry, it will often be possible to find a lower energy UHF solution in the TS region. However, since the spin contamination is not constant along the reaction coordinate, and since the UHF overestimates the biradical character, it is possible that the TS actually becomes a minimum on the UHF energy surface. In other words, the spin contamination may severely distort the shape of the potential energy surface. This may qualitatively be understood by considering the "singlet" UHF wave function as a linear combination of a singlet and a triplet states, as shown in Figure 4.10.

The degree of mixing is determined by the energy difference between the pure singlet and triplet states (as shown, for example, by second-order perturbation theory, see Section 4.8), which in general decreases along the reaction coordinate. Even if the mixing is not large enough to actually transform a TS to a minimum, it is clear that the UHF energy surface will be much too flat in the TS region. Activation energies calculated at the UHF level will always be lower than the RHF value, but may be either higher or lower than the "correct" value, depending on the amount of spin contamination, since RHF normally overestimates activation energies.

From the above it should be clear that UHF wave functions that are spin contaminated (more than a few percent deviation of $\langle \mathbf{S}^2 \rangle$ from the theoretical value of $S(S+1)$) have disadvantages. For closed-shell systems, an RHF procedure is therefore normally preferred. For open-shell systems, however, the UHF method has been widely used. It is possible to use an ROHF-type wave function for open-shell systems, but this leads to computational procedures that are somewhat more complicated than for the UHF case when electron correlation is introduced.

The main problem with the UHF method is the spin contamination, and there have been several proposals on how to remove these unwanted states. There are three strategies that can be considered for removing the contamination:

- During the SCF procedure.
- After the SCF has converged.
- After electron correlation has been added to the UHF solution.

A popular method of removing unwanted states is to project them out with a suitable projection operator (in the picture of the wave function being described in the coordinate system consisting of determinants, the components of the wave function along the higher spin states is removed). As mentioned above, the next higher spin state is usually the most important, and in many cases it is a quite good approximation to only remove this state. After projection, the wave function is then renormalized. If only the first contaminant is removed, this may in extreme cases actually increase the $\langle \mathbf{S}^2 \rangle$ value.

Performing the projection *during* the SCF procedure produces a wave function for which it is difficult to formulate a satisfactory theory for including electron correlation by means of perturbation or coupled cluster methods (Sections 4.8 and 4.9). Projections of the *converged* UHF wave function will lower the energy (although the projected UHF (PUHF) energy is no longer variational), since the contributions of the higher lying states are removed, and only the correlation effect remains. However, the problem of artificial distortion of the energy surface is even more pronounced at the PUHF level than with the UHF method itself. For example, it is often found that a false minimum is generated just after the RHF/UHF instability point on a bond dissociation curve. Furthermore, the derivatives of the PUHF energy are not continuous at the RHF/UHF instability point (Section 12.5). Projection of the wave function *after* electron correlation has been added, however, turns out to be a viable pathway. This has mainly been used in connection with perturbation methods, to be described in Section 4.8.2.

An approximate spin correction procedure has been proposed by K. Yamaguchi where the energy of the pure spin state is estimated by subtracting a fraction of the energy of the higher spin state, where the fraction is estimated from the $\langle \mathbf{S}^2 \rangle$ values.[13] For the typical case of a singlet state contaminated by a triplet state, the energy of the pure singlet state is estimated as shown by

$$\begin{aligned} {}^1E_{\text{pure}} &= {}^1E_{\text{impure}} + f\left({}^1E_{\text{impure}} - {}^3E_{\text{pure}}\right) \\ f &= \frac{{}^1\langle \mathbf{S}^2 \rangle_{\text{impure}}}{{}^3\langle \mathbf{S}^2 \rangle_{\text{pure}} - {}^1\langle \mathbf{S}^2 \rangle_{\text{impure}}} \end{aligned} \tag{4.36}$$

The Yamaguchi correction procedure assumes that only one higher spin state contributes to the spin contamination and that the spin contamination of the higher spin state is negligible. The procedure can be extended to also include systems with spin contamination from several different spin states and has the advantage that it can be applied for highly correlated wave functions and only requires one (or a few) additional energy calculations.

4.5 Size Consistency and Size Extensivity

As mentioned in Section 4.2.2, full CI is impossible except for very small systems. The only generally applicable method is CISD. Consider now a series of CISD calculations in order to construct the interaction potential between two H_2 molecules as a function of the distance between them. Relative to the HF wave function, there will be determinants that correspond to single excitations on only one of the H_2 fragments (S-type determinants), single excitations on both (D-type determinants) and double excitations only on one of the H_2 fragments (also D-type determinants). This will be the case at all intermolecular distances, even when the separation is very large. In that case, however, the system is just two H_2 molecules, and we could consider calculating the energy instead as twice the energy of one H_2 molecule. A CISD calculation on one H_2 molecule would generate singly and doubly excited determinants, and multiplying this by two would generate determinants that are triply and quadruply excited for the combined H_4 system. A CISD calculation of two H_2 molecules separated by, say, 100 Å will therefore not give the same energy as twice the results from a CISD calculation on one H_2 molecule (the latter will be lower). This problem is referred to as a *Size Inconsistency*. A very similar, but not identical concept, is *Size Extensivity*. Size consistency is only defined if the two fragments are non-interacting (separated by, say, 100 Å), while size extensivity implies that the method scales properly with the number of particles, that is the fragments can be interacting (separated by, say, 5 Å). Full CI *is* size consistent (and extensive), but all forms of truncated CI are not. The lack of size extensivity is the reason why CISD recovers less and less electron correlation as the systems grow larger.

4.6 Multiconfiguration Self-Consistent Field

The *MultiConfiguration Self-Consistent Field* (MCSCF) method can be considered as a CI where not only the coefficients in front of the determinants (Equation (4.2)) are optimized by the variational principle but the MOs used for constructing the determinants are also optimized.[14] The MCSCF optimization is iterative like the SCF procedure (if the "multiconfiguration" is only one, it is simply HF). Since the number of MCSCF iterations required for achieving convergence tends to increase with the number of configurations included, the size of MCSCF wave functions that can be treated is somewhat smaller than for CI methods.

When deriving the HF equations only the *variation* of the energy with respect to an orbital variation was required to be zero, which is equivalent to the first derivatives of the energy with respect to the MO expansion coefficients being equal to zero. The HF equations can be solved by an iterative SCF method, and there are many techniques for helping the iterative procedure to converge (Section 3.8). There is, however, no guarantee that the solution found by the SCF procedure is a *minimum* of the energy as a function of the MO coefficients. In order to ensure that a minimum has been found, the matrix of second derivatives of the energy with respect to the MO coefficients can be calculated and diagonalized, with a minimum having only positive eigenvalues (Section 6.9.1). This is rarely checked for SCF wave functions; in the large majority of cases the SCF procedure converges to a minimum without problems. MCSCF wave functions, on the other hand, are much harder to converge and much more prone to converge on solutions that are not minima. MCSCF wave function optimizations are therefore normally carried out by expanding the energy to second order in the variational parameters (orbital and configurational coefficients), analogously to the second-order SCF procedure described in Section 3.8.1, and using the Newton–Raphson-based methods described in Section 13.2.3 to force convergence to a minimum.

MCSCF methods are rarely used for calculating large fractions of the correlation energy. The orbital relaxation usually does not recover much electron correlation, and it is more efficient to include additional determinants and keep the MOs fixed (CI) if the interest is just in obtaining a large fraction of the correlation energy. Single-determinant HF wave functions normally give a qualitatively correct description of the electron structure, but there are many examples where this is not the case. MCSCF methods can be considered as an extension of single-determinant methods to give a qualitatively correct description.

Consider again the ozone molecule with the two resonance structures shown in Figure 4.9. Each type of resonance structure essentially translates into a different determinant. If more than one *non-equivalent* resonance structure is important, this means that the wave function cannot be described even qualitatively correctly at the RHF single-determinant level (benzene, for example, has two *equivalent* cyclohexatriene resonance structures and is adequately described by an RHF wave function). A UHF wave function allows some biradical character, with the disadvantages discussed in Section 4.4.

Alternatively, a second restricted type of CSF (consisting of two determinants) with two singly occupied MOs may be included in the wave function. The simplest MCSCF for ozone contains two configurations (often denoted TCSCF), with the optimum MOs and configurational weights determined by the variational principle. The CSFs entering an MCSCF expansion are pure spin states, and MCSCF wave functions therefore do not suffer from the problem of spin contamination.

Our definition of electron correlation uses the RHF energy as the reference. For ozone, both the UHF and the TCSCF wave functions have lower energies and include some electron correlation. This type of "electron correlation" is somewhat different from the picture presented at the start of this chapter. In a sense it is a consequence of our chosen zero point for the correlation energy, the RHF energy. The energy lowering introduced by adding enough flexibility in the wave function to be able to qualitatively describe the system is sometimes called the *static* electron correlation. This is essentially the effect of allowing orbitals to become (partly) singly occupied instead of forcing double occupation, that is describing near-degeneracy effects (two or more configurations having almost the same energy). The remaining energy lowering by correlating the motion of the electrons is called *dynamical* correlation. The problem is that there is no rigorous way of separating these effects. In the ozone example the energy lowering by going from RHF to UHF, or to a TCSCF, is almost pure static correlation. Increasing the number of configurations in an MCSCF will recover more and more of the dynamical correlation, until, at the full CI limit, the correlation treatment is exact. As mentioned above, MCSCF methods are mainly used for generating a qualitatively correct wave function, that is recovering the "static" part of the correlation.

An RHF wave function cannot describe systems with partial orbital occupancy, as, for example, a singlet biradical where two (or more) orbitals are near-degenerate. A UHF wave function can formally describe this situation, but it is often severely spin contaminated by the triplet state. An MCSCF-type wave function with a proper chosen set of configurations is sufficient, but is also computationally demanding. An interesting idea is to instead describe such systems as a *spin-flip* excitation from a higher spin-multiplicity state, where an electron with α spin in an orbital is transferred to another orbital with β spin. The trick is that the higher multiplicity state often can be adequately described by a single determinant wave function, and by using the linear response formalism in Sections 6.9.1 and 11.10, this allows a description of multiconfigurational states by single reference methods, such as EOM-CC[15] or TDDFT.[16] The singlet states of *o*-, *m*-, *p*-benzyne, for example, are strongly multiconfigurational, but the corresponding triplet states are adequately described by a single configuration. Note that the reference higher multiplicity state can be either higher or lower in energy than the desired low-multiplicity states. The major problem with MCSCF methods is selecting which configurations are necessary to include for the property of interest. One of the most popular approaches is the *Complete Active Space Self-Consistent Field* (CASSCF) method (also called *Full Optimized Reaction Space* (FORS)). Here the selection of configurations is done by partitioning the MOs into *active* and *inactive* spaces. The active MOs will typically be some of the highest occupied and some of the lowest unoccupied MOs from an RHF calculation. The inactive MOs have either 2 or 0 electrons, that is always either doubly occupied or empty. Within the active MOs a full CI is performed and all the proper symmetry-adapted configurations are included in the MCSCF optimization. Which MOs to include in the active space must be decided manually, by considering the problem at hand and the computational expense. If several points on the potential energy surface are desired, the MCSCF active space should include all those orbitals that change significantly, or for which the electron correlation is expected to change. A common notation is $[n,m]$-CASSCF, which indicates that n electrons are distributed in all possible ways in m orbitals.

As for any full CI expansion, the CASSCF becomes unmanageably large even for quite small active spaces. A variation of the CASSCF procedure is the *Restricted Active Space Self-Consistent Field*

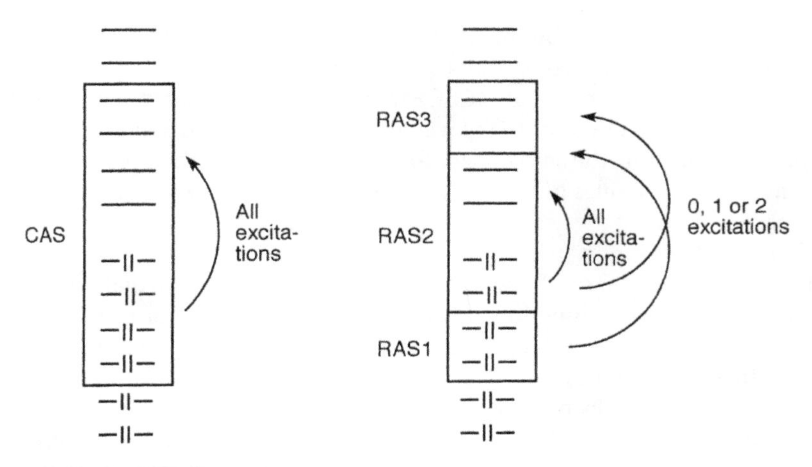

Figure 4.11 Illustrating the CAS and RAS orbital partitions.

(RASSCF) method.[17] Here the active MOs are divided into three sections, RAS1, RAS2 and RAS3, each having restrictions on the occupation numbers (excitations) allowed. A typical model consists of the configurations in the RAS2 space being generated by a full CI (analogously to CASSCF), or perhaps limited to SDTQ excitations. The RAS1 space consists of MOs that are doubly occupied in the HF reference determinant and the RAS3 space consists of MOs that are empty in the HF. Configurations additional to those from the RAS2 space are generated by allowing, for example, a maximum of two electrons to be excited *from* the RAS1 and a maximum of two electrons to be excited *to* the RAS3 space. In essence, a typical RASSCF procedure thus generates configurations by a combination of a full CI in a small number of MOs (RAS2) and a CISD in a somewhat larger MO space (RAS1 and RAS3), as illustrated in Figure 4.11. The RASSCF concept can be generalized to multiple active spaces with different restrictions of electron excitations to and from each space, an approach denoted *Generalized Active Space Self-Consistent Field* (GASSCF).[18] A similar idea is employed in the *Occupation Restricted Maximum Orbital Spaces* (ORMAS) method, where the active orbitals are subdivided into smaller active spaces with restrictions on the number of excitations within each subspace.[19]

The full CI expansion within the active space severely restricts the number of orbitals and electrons that can be treated by CASSCF methods. Table 4.3 shows how many singlet CSFs are generated for an $[n,n]$-CASSCF wave function (Equation (4.16)), without reductions arising from symmetry.

Table 4.3 Number of configurations generated in an $[n,n]$-CASSCF wave function.

n	Number of CSFs
2	3
4	20
6	175
8	1 764
10	19 404
12	226 512
14	2 760 615

The factorial increase in the number of CSFs effectively limits the active space for CASSCF wave functions to fewer than 12–14 electrons/orbitals. Selecting the "important" orbitals to correlate therefore becomes very important. The goal of MCSCF methods is usually not to recover a large fraction of the total correlation energy, but rather to recover all the *changes* that occur in the correlation energy for the given process. Selecting the active space for an MCSCF calculation requires some insight into the problem. There are a few rules of thumb that may be of help in selecting a proper set of orbitals for the active space:

1. For each occupied orbital, there will typically be one corresponding virtual orbital. This leads naturally to $[n,m]$-CASSCF wave functions where n and m are identical or nearly so.
2. Including all the valence orbitals, that is the space spanned by a minimum basis set, leads to a wave function that can correctly describe all dissociation pathways. Unfortunately, a full valence CASSCF wave function rapidly becomes unmanageably large for realistic-sized systems.
3. The orbital energies from an RHF calculation may be used for selecting the important orbitals. The highest occupied and lowest unoccupied are usually the most important orbitals to include in the active space. This can be partly justified by the formula for the second-order perturbation energy correction (Section 4.8.1, Equation (4.58)): the smaller the orbital energy difference, the larger contribution to the correlation energy. The numerator in Equation (4.58) should also be sizable and this implies that the occupied and virtual orbitals should occupy the same spatial region. Using RHF orbital energies for selecting the active space may be problematic in two situations. The first is when extended basis sets are used, where there will be many virtual orbitals with low energies and the exact order is more or less accidental. Furthermore, RHF virtual orbitals basically describe electron attachment (via Koopmans' theorem, Section 3.4), and are therefore not particularly well suited for describing electron correlation. An inspection of the form of the orbitals may reveal which to choose; they should be the ones that resemble the occupied orbitals in terms of basis function contribution. A commonly used trick is to generate improved virtual orbitals by running a calculation for the system of interest with a few electrons removed, since the lowest virtual orbitals then become much more valence-like. The second problem is more fundamental. If the real wave function has significant multiconfigurational character, then RHF may be qualitatively wrong, and selecting the active orbitals based on a qualitatively wrong wave function may lead to erroneous results. The problem is that we wish to include the important orbitals for describing the multideterminant nature, but these are not known until the final wave function is known.
4. An attempt to overcome this self-referencing problem is to use the concept of natural orbitals. The natural orbitals are those that diagonalize the density matrix (Section 10.5) and the eigenvalues are the occupation numbers. Orbitals with occupation numbers significantly different from 0 or 2 (for a closed-shell system) are usually those that are the most important to include in the active space. An RHF wave function will have occupation numbers of exactly 0 or 2, and some electron correlation must be included to obtain orbitals with non-integer occupation numbers. This may, for example, be done by running a preliminary MP2 or CISD calculation prior to the MCSCF. Alternatively, a UHF (when different from RHF)-type wave function may also be used. The total UHF density, which is the sum of the α and β density matrices, will also provide fractional occupation numbers since UHF includes some electron correlation. The procedure may still fail. If the underlying RHF wave function is poor, the MP2 correction may also give poor results, and selecting the active MCSCF orbitals based on the MP2 occupation number may again lead to erroneous results. In practice, however, selecting active orbitals based on, for example, MP2 occupation numbers appears to be quite efficient, and better than using RHF orbital energies.

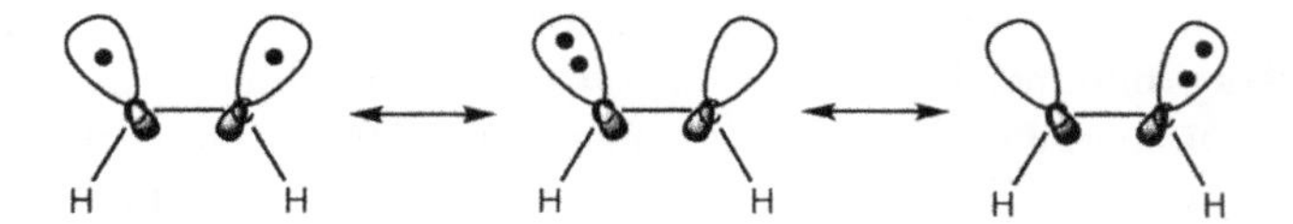

Figure 4.12 Important configurations for a bend acetylene model.

In a CASSCF-type wave function the CI coefficients do not have the same significance as for a single-reference CI based on HF orbitals. In a full CI (as in the active space of the CASSCF), the orbitals may be rotated among themselves without affecting the total wave function. A rotation of the orbitals, however, influences the magnitude of the coefficients in front of each CSF. While the HF coefficient in a single-reference CISD gives some indication of the "multireference" nature of the wave function, this is not the case for a CASSCF wave function, where the corresponding CI coefficient is arbitrary.

It should be noted that CASSCF methods inherently tend to give an unbalanced description, since all the electron correlation recovered is in the active space, with none in the inactive space, or between the active and inactive electrons.[20] This is not a problem if all the valence electrons are included in the active space, but this is only possible for small systems. If only some of the valence electrons are included in the active space, the CASSCF method tends to overestimate the importance of "biradical" structures. Consider, for example, acetylene where the hydrogens have been bent 60° away from linearity (this may be considered a model for orthobenzyne). The in-plane "π-orbital" now acquires significant biradical character. The true structure may be described as a linear combination of the following three configurations.

The structure on the left is biradical, while the two others are ionic, corresponding to both electrons being at the same carbon. The simplest CASSCF wave function that can qualitatively describe this system has two electrons in two orbitals, giving the three configurations shown above. The dynamical correlation between the two active electrons will tend to keep them as far apart as possible, that is favoring the biradical structure. Consider now a full valence CASSCF wave function with ten electrons in ten orbitals. This will analogously tend to separate the two electrons in each bond with one being at each end. The correlation of the electrons in the C—H bonds, for example, will place more electron density on the carbon atoms. This in turn favors the ionic structures in Figure 4.12 and disfavors the biradical, that is the dynamical correlation of the other electrons may take advantage of the empty orbital in the ionic structures but not in the biradical structure. These general considerations may be quantified by considering the natural orbital occupancies for increasingly large CASSCF wave functions, as shown in Table 4.4 with the 6-31G(d,p) basis.

The [4,4]-CASSCF also includes the two out-of-plane π-orbitals in the active space, while the [10,10]-CASSCF generates a full-valence CI wave function. The unbalanced description for the

Table 4.4 Natural orbital occupation numbers for the distorted acetylene model in Figure 4.12; only the occupation numbers for the six "central" orbitals are shown.

	n_5	n_6	n_7	n_8	n_9	n_{10}
RHF	2.00	2.00	2.00	0.00	0.00	0.00
UHF	2.00	1.72	1.30	0.70	0.28	0.01
[2,2]-CASSCF	2.00	2.00	1.62	0.38	0.00	0.00
[4,4]-CASSCF	2.00	1.85	1.67	0.33	0.14	0.00
[10,10]-CASSCF	1.97	1.87	1.71	0.30	0.13	0.02

[2,2]-CASSCF is reminiscent of the spin contamination problem for UHF wave functions, although the effect is much less pronounced. Nevertheless, the overestimation may be severe enough to alter the qualitative shape of energy surfaces, for example turning transition structures into minima, as illustrated in Figure 4.10. MCSCF methods are therefore not "black box" methods such as, for example, HF and MP (Section 4.8.1); selecting a proper number of configurations, and the correct orbitals, to give a balanced description of the problem at hand requires some experimentation and insight.

4.7 Multireference Configuration Interaction

The CI methods described so far consider only CSFs generated by exciting electrons from a single determinant. This corresponds to having an HF-type wave function as the reference. However, an MCSCF wave function may also be chosen as the reference. In that case, a CISD involves excitations of one or two electrons out of *all* the determinants that enter the MCSCF, defining the *MultiReference Configuration Interaction* (MRCI) method.[14] Compared with the single-reference CISD, the number of configurations is increased by a factor roughly equal to the number of configurations included in the MCSCF. Large-scale MRCI wave functions (many configurations in the MCSCF) can generate very accurate wave functions, but are also computationally very intensive. Since MRCI methods truncate the CI expansion, they are not size extensive.

Even truncating the (MR) CI expansion at the singles and doubles level frequently generates more configurations than can be handled readily. A further truncation is sometimes performed by selecting only those configurations that have an "interaction" with the reference configuration(s) above a selected threshold, where the "interaction" is evaluated by second-order perturbation theory (Section 4.8). Such *state-selected* CI (or MCSCF) methods all involve a preset cutoff below which configurations are neglected. This may cause problems for comparing energies of different geometries, since the potential energy surface may become discontinuous; that is at some point the importance of a given configuration drops below the threshold and the contribution suddenly disappears.

4.8 Many-Body Perturbation Theory

The idea in perturbation methods is that the problem at hand only differs slightly from a problem that has already been solved (exactly or approximately). The solution to the given problem should therefore in some sense be close to the solution to the already known system. This is described mathematically by defining a Hamiltonian operator that consists of two parts, a reference ($\mathbf{H}_0$) and a perturbation ($\mathbf{H}'$). The premise of perturbation methods is that the $\mathbf{H}'$ operator in some sense is "small" compared with $\mathbf{H}_0$. Perturbation methods can be used in quantum mechanics for adding corrections to solutions that employ an independent-particle approximation, and the theoretical framework is then called *Many-Body Perturbation Theory* (MBPT).

Let us assume that the Schrödinger equation for the reference Hamiltonian operator is solved:

$$\begin{aligned} \mathbf{H} &= \mathbf{H}_0 + \lambda\mathbf{H}' \\ \mathbf{H}_0\Phi_i &= E_i\Phi_i, \quad i = 0, 1, 2, \ldots, \infty \end{aligned} \tag{4.37}$$

The solutions for the unperturbed Hamiltonian operator form a complete set (since $\mathbf{H}_0$ is Hermitian), which can be chosen to be orthonormal, and λ is a (variable) parameter determining the strength of the perturbation. At present, we will only consider cases where the perturbation is time-independent and the reference wave function is non-degenerate. To keep the notation simple, we will furthermore only consider the lowest energy state. The perturbed Schrödinger equation is given by

$$\mathbf{H}\Psi = W\Psi \tag{4.38}$$

As the perturbation is increased from zero to a finite value, the energy and wave function must change continuously, and they can be written as an expansion in powers of the perturbation parameter λ:

$$\begin{aligned} W &= \lambda^0 W_0 + \lambda^1 W_1 + \lambda^2 W_2 + \lambda^3 W_3 + \cdots \\ \Psi &= \lambda^0 \Psi_0 + \lambda^1 \Psi_1 + \lambda^2 \Psi_2 + \lambda^3 \Psi_3 + \cdots \end{aligned} \tag{4.39}$$

For $\lambda = 0$, $\mathbf{H} = \mathbf{H}_0$ and it is seen that $\Psi_0 = \Phi_0$ and $W_0 = E_0$, and this is the *unperturbed* or *zeroth-order* wave function and energy. The Ψ_1, Ψ_2, ... and W_1, W_2, ... are the *first-order, second-order*, etc., *corrections.* The λ parameter will eventually be set equal to 1 and the nth-order energy or wave function becomes a sum of all terms up to order n.

It is convenient to choose the perturbed wave function to be *intermediately normalized,* that is the overlap with the unperturbed wave function should be 1. This has the consequence that all correction terms are orthogonal to the reference wave function:

$$\begin{aligned} \langle \Psi | \Phi_0 \rangle &= 1 \\ \langle \Psi_0 + \lambda \Psi_1 + \lambda^2 \Psi_2 + \cdots | \Phi_0 \rangle &= 1 \\ \langle \Phi_0 | \Phi_0 \rangle + \lambda \langle \Psi_1 | \Phi_0 \rangle + \lambda^2 \langle \Psi_2 | \Phi_0 \rangle + \cdots &= 1 \\ \langle \Psi_{i \neq 0} | \Phi_0 \rangle &= 0 \end{aligned} \tag{4.40}$$

Once all the correction terms have been calculated, it is trivial to normalize the total wave function.

With the expansions in Equation (4.39), the Schrödinger Equation (4.38) becomes

$$\begin{aligned} &(\mathbf{H}_0 + \lambda \mathbf{H}')(\lambda^0 \Psi_0 + \lambda^1 \Psi_1 + \lambda^2 \Psi_2 + \cdots) \\ &\quad = (\lambda^0 W_0 + \lambda^1 W_1 + \lambda^2 W_2 + \cdots)(\lambda^0 \Psi_0 + \lambda^1 \Psi_1 + \lambda^2 \Psi_2 + \cdots) \end{aligned} \tag{4.41}$$

Since this holds for any value of λ, we can collect terms with the same power of λ to give

$$\begin{aligned} &\lambda^0 : \mathbf{H}_0 \Psi_0 = W_0 \Psi_0 \\ &\lambda^1 : \mathbf{H}_0 \Psi_1 + \mathbf{H}' \Psi_0 = W_0 \Psi_1 + W_1 \Psi_0 \\ &\lambda^2 : \mathbf{H}_0 \Psi_2 + \mathbf{H}' \Psi_1 = W_0 \Psi_2 + W_1 \Psi_1 + W_2 \Psi_0 \\ &\lambda^n : \mathbf{H}_0 \Psi_n + \mathbf{H}' \Psi_{n-1} = \sum_{i=0}^{n} W_i \Psi_{n-i} \end{aligned} \tag{4.42}$$

These are the zero-, first-, second-, nth-order perturbation equations. The zeroth-order equation is just the Schrödinger equation for the unperturbed problem. The first-order equation contains two unknowns, the first-order correction to the energy, W_1, and the first-order correction to the wave function, Ψ_1. The nth-order energy correction can be calculated by multiplying from the left by Φ_0 and integrating, and using the "*turnover rule* " $\langle \Phi_0 | \mathbf{H}_0 | \Psi_i \rangle = \langle \Psi_i | \mathbf{H}_0 | \Phi_0 \rangle^*$:

$$\begin{aligned} \langle \Phi_0 | \mathbf{H}_0 | \Psi_n \rangle + \langle \Phi_0 | \mathbf{H}' | \Psi_{n-1} \rangle &= \sum_{i=0}^{n-1} W_i \langle \Phi_0 | \Psi_{n-i} \rangle + W_n \langle \Phi_0 | \Psi_0 \rangle \\ E_0 \langle \Psi_n | \Phi_0 \rangle + \langle \Phi_0 | \mathbf{H}' | \Psi_{n-1} \rangle &= W_n \langle \Phi_0 | \Psi_0 \rangle \\ W_n &= \langle \Phi_0 | \mathbf{H}' | \Psi_{n-1} \rangle \end{aligned} \tag{4.43}$$

From this it would appear that the $(n-1)$th-order wave function is required for calculating the nth-order energy. However, by using the turnover rule and the nth- and lower-order perturbation Equations (4.42), it can be shown that knowledge of the nth-order wave function actually allows a calculation of the $(2n+1)$th-order energy:

$$W_{2n+1} = \langle \Psi_n | \mathbf{H}' | \Psi_n \rangle - \sum_{k,l=0}^{n} W_{2n+1-k-l} \langle \Psi_k | \Psi_l \rangle \tag{4.44}$$

Up to this point, we are still dealing with undetermined quantities, energy and wave function corrections at each order. The first-order equation is one equation with two unknowns. Since the solutions to the unperturbed Schrödinger equation generate a complete set of functions, the unknown first-order correction to the wave function can be expanded in these functions. This is known as *Rayleigh–Schrödinger* perturbation theory, and the λ^1 equation in Equation (4.42) becomes

$$\begin{aligned} \Psi_1 &= \sum_i a_i \Phi_i \\ (\mathbf{H}_0 - W_0)\left(\sum_i a_i \Phi_i\right) &+ (\mathbf{H}' - W_1)\Phi_0 = 0 \end{aligned} \tag{4.45}$$

Multiplying from the left by Φ_0^* and integrating yields the following equation, where the orthonormality of the Φ_i is used (this also follows directly from Equation (4.44)):

$$\begin{aligned} \sum_i a_i \langle \Phi_0 | \mathbf{H}_0 | \Phi_i \rangle - W_0 \sum_i a_i \langle \Phi_0 | \Phi_i \rangle + \langle \Phi_0 | \mathbf{H}' | \Phi_0 \rangle - W_1 \langle \Phi_0 | \Phi_0 \rangle &= 0 \\ \sum_i a_i E_i \langle \Phi_0 | \Phi_i \rangle - a_0 E_0 + \langle \Phi_0 | \mathbf{H}' | \Phi_0 \rangle - W_1 &= 0 \\ a_0 E_0 - a_0 E_0 + \langle \Phi_0 | \mathbf{H}' | \Phi_0 \rangle - W_1 &= 0 \end{aligned} \tag{4.46}$$

$$W_1 = \langle \Phi_0 | \mathbf{H}' | \Phi_0 \rangle$$

The last equation shows that the first-order correction to the energy is an average of the perturbation operator over the unperturbed wave function.

The first-order correction to the wave function can be obtained by multiplying Equation (4.42) from the left by a function other than Φ_0 (Φ_j) and integrating to give

$$\begin{aligned} \sum_i a_i \langle \Phi_j | \mathbf{H}_0 | \Phi_i \rangle - W_0 \sum_i a_i \langle \Phi_j | \Phi_i \rangle + \langle \Phi_j | \mathbf{H}' | \Phi_0 \rangle - W_1 \langle \Phi_j | \Phi_0 \rangle &= 0 \\ \sum_i a_i E_i \langle \Phi_j | \Phi_i \rangle - a_j E_0 + \langle \Phi_0 | \mathbf{H}' | \Phi_0 \rangle &= 0 \\ a_j E_j - a_j E_0 + \langle \Phi_j | \mathbf{H}' | \Phi_0 \rangle &= 0 \end{aligned} \tag{4.47}$$

$$a_j = \frac{\langle \Phi_j | \mathbf{H}' | \Phi_0 \rangle}{E_0 - E_j}$$

The expansion coefficients determine the first-order correction to the perturbed wave function (Equation (4.36)), and they can be calculated from the known unperturbed wave functions and energies. The coefficient in front of Φ_0 for Ψ_1 cannot be determined from the above formula, but the assumption of intermediate normalization (Equation (4.40)) makes $a_0 = 0$. Starting from the second-order perturbation Equation (4.42), analogous formulas can be generated for the second-order corrections. Using intermediate normalization ($a_0 = b_0 = 0$), the second-order energy correction is given by Equation (4.49):

$$\Psi_2 = \sum_i b_i \Phi_i \tag{4.48}$$

$$\begin{aligned}
(\mathbf{H}_0 - W_0)\left(\sum_i b_i \Phi_i\right) + (\mathbf{H}' - W_1)\left(\sum_i a_i \Phi_i\right) - W_2 \Phi_0 = 0 \\
\sum_i b_i \langle \Phi_0 | \mathbf{H}_0 | \Phi_i \rangle - W_0 \sum_i b_i \langle \Phi_0 | \Phi_i \rangle + \\
\sum_i a_i \langle \Phi_0 | \mathbf{H}' | \Phi_i \rangle - W_1 \sum_i a_i \langle \Phi_0 | \Phi_i \rangle - W_2 \langle \Phi_0 | \Phi_0 \rangle = 0 \\
\sum_i b_i E_i \langle \Phi_0 | \Phi_i \rangle - b_0 E_0 + \sum_i a_i \langle \Phi_0 | \mathbf{H}' | \Phi_i \rangle - a_0 W_1 - W_2 = 0 \\
b_0 E_0 - b_0 E_0 + \sum_i a_i \langle \Phi_0 | \mathbf{H}' | \Phi_i \rangle - W_2 = 0 \\
W_2 = \sum_i a_i \langle \Phi_0 | \mathbf{H}' | \Phi_i \rangle = \sum_{i \neq 0} \frac{\langle \Phi_0 | \mathbf{H}' | \Phi_i \rangle \langle \Phi_i | \mathbf{H}' | \Phi_0 \rangle}{E_0 - E_i}
\end{aligned} \tag{4.49}$$

The last equation shows that the second-order energy correction may be written in terms of the first-order wave function (c_i) and matrix elements over unperturbed states. The second-order wave function correction is given by

$$\begin{aligned}
\sum_i b_i \langle \Phi_j | \mathbf{H}_0 | \Phi_i \rangle - W_0 \sum_i b_i \langle \Phi_j | \Phi_i \rangle \\
+ \sum_i a_i \langle \Phi_j | \mathbf{H}' | \Phi_i \rangle - W_1 \sum_i a_i \langle \Phi_j | \Phi_i \rangle - W_2 \langle \Phi_j | \Phi_0 \rangle = 0 \\
\sum_i b_i E_i \langle \Phi_j | \Phi_i \rangle - b_j E_0 + \sum_i a_i \langle \Phi_j | \mathbf{H}' | \Phi_i \rangle - a_j W_1 = 0 \\
b_j E_j - b_j E_0 + \sum_i a_i \langle \Phi_j | \mathbf{H}' | \Phi_i \rangle - a_j \langle \Phi_0 | \mathbf{H}' | \Phi_0 \rangle = 0 \\
b_j = \sum_{i \neq 0} \frac{\langle \Phi_j | \mathbf{H}' | \Phi_i \rangle \langle \Phi_i | \mathbf{H}' | \Phi_0 \rangle}{(E_0 - E_j)(E_0 - E_i)} - \frac{\langle \Phi_j | \mathbf{H}' | \Phi_0 \rangle \langle \Phi_0 | \mathbf{H}' | \Phi_0 \rangle}{(E_0 - E_j)^2}
\end{aligned} \tag{4.50}$$

The formulas for higher-order corrections become increasingly complex; the third-order energy correction, for example, is given by

$$W_3 = \sum_{i,j \neq 0} \frac{\langle \Phi_0 | \mathbf{H}' | \Phi_i \rangle [\langle \Phi_i | \mathbf{H}' | \Phi_j \rangle \langle \Phi j | \mathbf{H}' | \Phi_0 \rangle - \delta_{ij} \langle \Phi_0 | \mathbf{H}' | \Phi_0 \rangle \langle \Phi_j | \mathbf{H}' | \Phi_0 \rangle]}{(E_0 - E_i)(E_0 - E_j)} \tag{4.51}$$

The main point, however, is that all corrections can be expressed in terms of matrix elements of the perturbation operator over unperturbed wave functions and the unperturbed energies.

4.8.1 Møller–Plesset Perturbation Theory

So far, the theory has been completely general. In order to apply perturbation theory to the calculation of correlation energy, the unperturbed Hamiltonian operator must be selected. The most common choice is to take this as a sum over Fock operators, leading to *Møller–Plesset* (MP) perturbation theory.[21,22] The sum of Fock operators counts the (average) electron–electron repulsion twice (Equation (3.44)), and the perturbation becomes the exact $\mathbf{V}_{ee}$ operator minus twice the $\langle \mathbf{V}_{ee} \rangle$ operator. The operator associated with this difference is often referred to as the *fluctuation potential*. This choice is not really consistent with the basic assumption that the perturbation should be small compared with $\mathbf{H}_0$. However, it does fulfill the other requirement that solutions to the unperturbed Schrödinger equation should be known. Furthermore, this is the only choice that leads to a size extensive method, which is a desirable feature:

$$\begin{aligned}\mathbf{H}_0 &= \sum_{i=1}^{N_{elec}} \mathbf{F}_i = \sum_{i=1}^{N_{elec}} \left(\mathbf{h}_i + \sum_{j=1}^{N_{elec}} (\mathbf{J}_j - \mathbf{K}_j) \right) \\ &= \sum_{i=1}^{N_{elec}} \mathbf{h}_i + \sum_{i=1}^{N_{elec}} \sum_{j=1}^{N_{elec}} \langle \mathbf{g}_{ij} \rangle = \sum_{i=1}^{N_{elec}} \mathbf{h}_i + 2\langle \mathbf{V}_{ee} \rangle \\ \mathbf{H}' = \mathbf{H} - \mathbf{H}_0 &= \sum_{i=1}^{N_{elec}} \sum_{j>i}^{N_{elec}} \mathbf{g}_{ij} - \sum_{i=1}^{N_{elec}} \sum_{j=1}^{N_{elec}} \langle \mathbf{g}_{ij} \rangle = \mathbf{V}_{ee} - 2\langle \mathbf{V}_{ee} \rangle\end{aligned} \tag{4.52}$$

The zeroth-order wave function is the HF determinant and the zeroth-order energy is just a sum of MO energies:

$$W_0 = \langle \Phi_0 | \mathbf{H}_0 | \Phi_0 \rangle = \left\langle \Phi_0 \left| \sum_{i=1}^{N_{elec}} \mathbf{F}_i \right| \Phi_0 \right\rangle = \sum_{i=1}^{N_{elec}} \varepsilon_i \tag{4.53}$$

Recall that the orbital energy is the energy of an electron in the field of all the nuclei and includes the repulsion to all other electrons (Equation (3.46)), and therefore counts the electron–electron repulsion twice. The first-order energy correction is the average of the perturbation operator over the zeroth-order wave function (Equation (4.46)):

$$W_1 = \langle \Phi_0 | \mathbf{H}' | \Phi_0 \rangle = \langle \mathbf{V}_{ee} \rangle - 2\langle \mathbf{V}_{ee} \rangle = -\langle \mathbf{V}_{ee} \rangle \tag{4.54}$$

This yields a correction for the overcounting of the electron–electron repulsion at zeroth order. Comparing Equation (4.54) with the expression for the total energy in Equation (3.33), it is seen that the first-order energy (sum of W_0 and W_1) is exactly the HF energy. Using the notation E(MPn) to indicate the correction at order n and MPn to indicate the total energy up to order n, we have

$$\begin{aligned}\mathrm{MP0} &= E(MP0) = \sum_{i=1}^{N_{elec}} \varepsilon_i \\ \mathrm{MP1} &= E(MP0) + E(MP1) = E(HF)\end{aligned} \tag{4.55}$$

Electron correlation energy thus starts at order two with this choice of $\mathbf{H}_0$.

In developing the perturbation theory, it was assumed that the solutions to the unperturbed problem formed a complete set. This in general means that there must be an infinite number of functions, which is impossible in actual calculations. The lowest energy solution to the unperturbed problem is the HF wave function and additional higher energy solutions are excited Slater determinants, analogous to the CI method. When a finite basis set is employed, it is only possible to generate a finite number of excited determinants. The expansion of the many-electron wave function is therefore truncated.

Let us look at the expression for the second-order energy correction, Equation (4.49). This involves matrix elements of the perturbation operator between the HF reference and all possible excited states. Since the perturbation is a two-electron operator, all matrix elements involving triple, quadruple, etc., excitations are zero. When canonical HF orbitals are used, matrix elements with singly excited states are also zero, as indicated in

$$\begin{aligned}\langle \Phi_0 | \mathbf{H}' | \Phi_i^a \rangle &= \left\langle \Phi_0 \middle| \mathbf{H} - \sum_j^{N_{\rm elec}} \mathbf{F}_j \middle| \Phi_i^a \right\rangle \\ &= \langle \Phi_0 | \mathbf{H} | \Phi_i^a \rangle - \left\langle \Phi_0 \middle| \sum_j^{N_{\rm elec}} \mathbf{F}_j \middle| \Phi_i^a \right\rangle \\ &= \langle \Phi_0 | \mathbf{H} | \Phi_i^a \rangle - \left(\sum_j^{N_{\rm elec}} \varepsilon_j \right) \langle \Phi_0 | \Phi_i^a \rangle = 0\end{aligned} \tag{4.56}$$

The first bracket is zero owing to Brillouin's theorem (Section 4.2.1) and the second set of brackets is zero owing to the orbitals being eigenfunctions of the Fock operators and orthogonal to each other. The second-order correction to the energy, which is the first contribution to the correlation energy, thus only involves a sum over doubly excited determinants. These can be generated by promoting two electrons from occupied orbitals *i* and *j* to virtual orbitals *a* and *b*. The summation must be restricted such that each excited state is only counted once:

$$W_2 = \sum_{i<j}^{\rm occ} \sum_{a<b}^{\rm vir} \frac{\langle \Phi_0 | \mathbf{H}' | \Phi_{ij}^{ab} \rangle \langle \Phi_{ij}^{ab} | \mathbf{H}' | \Phi_0 \rangle}{E_0 - E_{ij}^{ab}} \tag{4.57}$$

The matrix elements between the HF and a doubly excited state are given by two-electron integrals over MOs (Equation (4.10)). The difference in total energy between two Slater determinants becomes a difference in MO energies (essentially Koopmans' theorem), and the explicit formula for the second-order Møller–Plesset correction is given by

$$E(\mathrm{MP2}) = \sum_{i<j}^{\rm occ} \sum_{a<b}^{\rm vir} \frac{(\langle \phi_i \phi_j | \phi_a \phi_b \rangle - \langle \phi_i \phi_j | \phi_b \phi_a \rangle)^2}{\varepsilon_i + \varepsilon_j - \varepsilon_a - \varepsilon_b} \tag{4.58}$$

Once the two-electron integrals over MOs are available, the second-order energy correction can be calculated as a sum over such integrals. There are of the order of $M_{\rm basis}^4$ integrals; thus the calculation of the energy (only) increases as $M_{\rm basis}^4$ with the system size. However, the transformation of the integrals from the AO to the MO basis grows as $M_{\rm basis}^5$ (Section 4.2.1). MP2 is an $M_{\rm basis}^5$ method, but fairly inexpensive as not all two-electron integrals over MOs are required. Only those corresponding to the combination of two occupied and two virtual MOs are needed. In practical calculations, this means that the MP2 energy for systems with a few hundred basis functions can be calculated at a cost similar to or less than what is required for calculating the HF energy. MP2 typically accounts for 80–90% of the correlation energy and it is the most economical method for including electron correlation.

The HF wave function includes Fermi correlation (between electrons having the same spin) due to the antisymmetry of the wave function, and it is thus expected that MP2 will describe the same-spin (SS) and opposite-spin (OS) correlation with different accuracy. This forms the basis for the *Spin Component Scaled* MP2 (SCS-MP2) methods where the SS and SO components of the MP2 correlation energy are scaled separately by empirical constants.[23] If the constant for the SO contribution is set to zero, the method is denoted *Spin Opposite Scaled* MP2 (SOS-MP2) which has the advantage that it can be evaluated with a computational effort that only scales as $M_{\rm basis}^4$. The optimum scaling constants depend on the application, but the results are usually significantly better than the regular MP2 methods.

The formula for the first-order correction to the wave function (Equation (4.47)) similarly only contains contributions from doubly excited determinants. Since knowledge of the first-order wave function allows calculation of the energy up to third order ($2n + 1 = 3$, Equation (4.44)), it is immediately clear that the third-order energy also only contains contributions from doubly excited determinants. Qualitative speaking, the MP2 contribution describes the correlation between pairs of electrons while MP3 describes the interaction between pairs. The formula for calculating this contribution is given in Equation (4.51) and involves a computational effort that formally increases as M_{basis}^6. The third-order energy typically accounts for 90–95% of the correlation energy.

The formula for the second-order correction to the wave function (Equation (4.50)) contains products of the type $\langle \Phi_j | \mathbf{H}' | \Phi_i \rangle \langle \Phi_i | \mathbf{H}' | \Phi_0 \rangle$. The Φ_0 is the HF determinant and the *last* bracket can only be non-zero if Φ_i is a doubly excited determinant. This means that the *first* bracket only can be non-zero if Φ_j is either a singly, doubly, triply or quadruply excited determinant (since $\mathbf{H}'$ is a two-electron operator). The second-order wave function allows calculation of the fourth- and fifth-order energies, and these terms therefore have contributions from determinants that are singly, doubly, triply or quadruply excited. The computational cost of the fourth-order energy without the contribution from the triply excited determinants, MP4(SDQ), increases as M_{basis}^6, while the triples contribution increases as M_{basis}^7. MP4 is still a computationally feasible model for many molecular systems, requiring a time similar to CISD. In typical calculations, the T contribution to MP4 will take roughly the same amount of time as the SDQ contributions, but the triples are often the most important at fourth order. The full fourth-order energy typically accounts for 95–98% of the correlation energy.

The fifth-order correction to the energy also involves S, D, T and Q contributions, and the sixth-order term introduces quintuple and sextuple excitations.[22] The working formulas for the MP5 and MP6 contributions are so complex that actual calculations are only possible for small systems. The computational effort for MP5 increases as M_{basis}^8 and for MP6 as M_{basis}^9. There is little experience with the performance of MPn beyond MP4.

As shown in Table 4.2, the most important contribution to the energy in a CI procedure comes from doubly excited determinants. This is also shown by the perturbation expansion; the second- and third-order energy corrections only involve doubles. At fourth order the singles, triples and quadruples enter the expansion for the first time. This is again consistent with Table 4.2, which shows that these types of excitations are of similar importance.

CI methods determine the energy by a variational procedure and the energy is consequently an upper bound to the exact energy. There is no such guarantee for perturbation methods and it is possible that the energy will be lower than the exact energy. This is rarely a problem and may in fact be advantageous. Limitations in the basis set often mean that the error in total energy is several au (thousands of kJ/mol) anyway. In the large majority of cases, the interest is not in total energies but in energy differences. Having a variational upper bound for two energies does not give any bound for the difference between these two numbers. The main interest is therefore that the error remains relatively constant for different systems, and the absence of a variational bound can allow for error cancellations. The lack of size extensivity of CI methods, on the other hand, is disadvantageous in this respect. The MP perturbation method *is* size extensive, but other forms of MBPT are not. It is now generally recognized that size extensivity is an important property and the MP form of MBPT is used almost exclusively.

The main limitation of perturbation methods is the assumption that the zeroth-order wave function is a reasonable approximation to the real wave function, that is the perturbation operator is sufficiently "small". The more poorly the HF wave function describes the system, the larger are the correction terms and the more terms must be included to achieve a given level of accuracy. If the reference state is a poor description of the system, the convergence may be so slow or erratic that perturbation methods cannot be used. Actually, it is difficult to assess whether the perturbation expansion is convergent or not, although the first few terms for many systems show a behavior that suggests that it is the case. This may to some extent be deceptive, as it has been demonstrated that the convergence properties depend on the size of the basis set[24] and the majority of studies have employed small- or medium-sized basis sets. A convergent series in, for example, a DZP-type basis may become divergent or oscillating in a larger basis, especially if diffuse functions are present.

The convergence properties for the perturbation series can be analyzed by considering the partitioned Hamiltonian in Equation (4.37), with λ being a parameter connecting the reference system ($\lambda = 0$) with the real physical system ($\lambda = 1$).[25,26] For analyzing the convergence behavior, we must allow λ also to have complex values:

$$\mathbf{H}(\lambda) = \mathbf{H}_0 + \lambda \mathbf{H}' \tag{4.59}$$

For a given λ value, the (exact) energy of the ground state can be written as an infinite summation of all perturbation terms:

$$E(\lambda) = \sum_{i=0}^{\infty} W_i \lambda^i \tag{4.60}$$

The mathematical theory of infinite series states that this summation is only convergent within a given radius R, that is the infinite series in Equation (4.60) only has a well-defined value if $|\lambda| < R$. Since we are interested in the situation where $\lambda = 1$, this translates into the condition $R > 1$. The convergence radius is determined by the smallest value of λ where another state becomes degenerate with the ground state; that is the MP perturbation series is only convergent if there are no excited states that become degenerate with the ground state within the circle in the complex plane corresponding to $|\lambda| = 1$. This includes non-physical situations where λ is negative, that is where the perturbation corresponds to the electron–electron interaction being *attractive*. In MP theory the zeroth-order energy is the sum of orbital energies, which includes the average electron–electron interaction twice, and the first-order energy correction W_1 is the negative of the average electron–electron interaction (Equation (4.54)). Since W_1 is smaller for excited states than for the ground state (more diffuse orbitals), this means that a negative λ value will raise the ground state more in energy than an excited state, and this may be sufficient to overcome the energy separation at $\lambda = 0$. This is especially true when diffuse basis functions are included, since they preferentially improve the description of excited states and such *intruder states* are the reason for the non-convergent behavior of the MP perturbation series. A complete search for intruder states within the complex plane corresponding to $|\lambda| = 1$ is difficult even for simple systems and a less rigorous search for avoided crossings along the real axis is also demanding. Establishing the convergence or divergence of the MP expansion on a case-by-case basis is unmanageable and one is therefore limited to observing the behavior for the first few terms.

In the ideal case, the HF, MP2, MP3 and MP4 results show a monotonic convergence toward a limiting value, with the corrections being of the same sign and numerically smaller as the order of perturbation increases. Unfortunately, this is not the typical behavior. Even in systems where the reference is well described by a single determinant, oscillations in a given property as a function of perturbation order are often observed. An analysis by Cremer indicates that a smooth convergence (of the total energy) is only expected for systems containing well-separated electron pairs and that oscillations occur when this is not that case.[22] The latter encompass systems containing lone pairs and/or multiple bonds, covering the large majority of molecules. It should be noted that one cannot

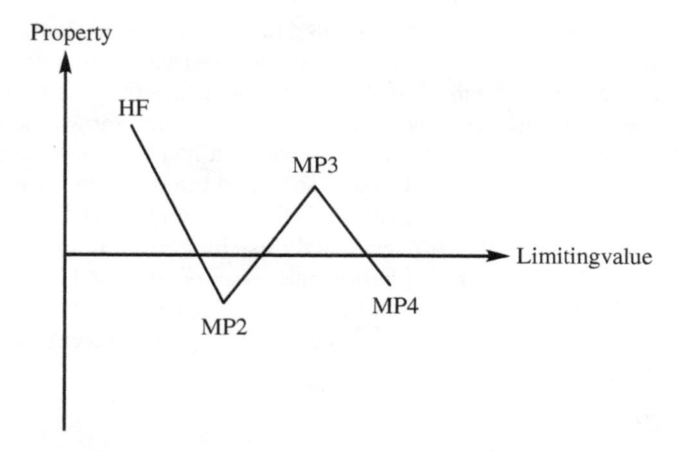

Figure 4.13 Typical oscillating behavior of results obtained with the MP method.

conclude anything about the convergence properties of the whole perturbation series from either the monotonic or oscillating behavior of the first few terms.

In practice, only low orders of perturbation theory can be carried out, and it is often observed that the HF and MP2 results differ considerably, the MP3 result moves back towards the HF and the MP4 away again (Figure 4.13). For "well-behaved" systems the correct answer is often somewhere between the MP3 and MP4 results. MP2 typically overshoots the correlation effect, but often gives a better answer than MP3, at least if medium-sized basis sets are used. Just as the first term involving doubles (MP2) tends to overestimate the correlation effect, it is often observed that MP4 overestimates the effect of the singles and triples contributions, since they enter the series for the first time at fourth order.

When the reference wave function contains substantial multireference character, a perturbation expansion based on a single determinant will display poor convergence. If the reference wave function suffers from symmetry breaking (Section 3.8.3), the MP method is almost guaranteed to give absurd results. The questionable convergence of the MP method has caused it to be significantly less popular, although MP2 continues to be a computationally inexpensive way of including the majority of the electron correlation effect.

4.8.2 Unrestricted and Projected Møller–Plesset Methods

When the reference is an RHF-type wave function the dissociation limit will normally be incorrect. As a bond is stretched, RHF gives an increasingly poorer description of the wave function, and consequently causes the perturbation series to break down. The use of a UHF wave function allows a correct dissociation limit in terms of energy but at the cost of introducing spin contamination (Sections 4.3 and 4.4). It is straightforward to derive an MP method based on a UHF reference wave function (UMP): in this case the unperturbed Hamiltonian operator is a sum of the α and β Fock operators. The addition of electron correlation decreases the spin contamination of the wave function (in the full CI limit the spin contamination is zero) but the improvement is usually small at low orders (2–4) of perturbation theory. As illustrated in Section 4.4, the UHF energy is lower than that of RHF owing to the inclusion of some electron correlation (mainly static), but it also contains some amounts of higher energy spin states. Since MP methods recover a large part of the electron correlation (both static and dynamical), the net effect at the UMP level is an increase in energy due to spin contamination. In the dissociation limit, this has no consequence, as the different spin states have equal energies. In the intermediate region, where the bond is not completely broken, it is usually observed that the RMPn energy is *lower* than the UMPn energy, although the RHF energy is *higher* than the UHF energy (see also Section 12.5). The spin contamination in UHF wave functions causes an UMPn expansion to converge more slowly than RMPn.[27] For open-shell systems, where RHF cannot be used, this would suggest that the reference wave function should be of the ROHF type, instead of UHF. Formulation of ROHF-based perturbation methods, however, is somewhat more difficult than for the UHF case. The reason is that for an ROHF wave function it is not possible to choose a set of MOs that makes the matrix of Lagrange multipliers diagonal (Equations (3.41) and (3.42)). There is thus not a unique set of canonical MOs to be used in the perturbation expansion, which again has the consequence that several choices of the unperturbed Hamiltonian operator are possible.[28] Different ROMP methods therefore give different energies, and there are no firm theoretical grounds for choosing one over the other.[29,30] In practice, however, different choices of the unperturbed Hamiltonian operator lead to similar results.

While projection methods for removing spin contamination are not recommended at the HF level, they work quite well at the UMP level. Formulas have been derived for removing all contaminants at the UMP2 level and also the first few states at the UMP3 and UMP4 levels.[31,32] The associated acronyms are PUMP and PMP, denoting slightly different methods, although in practice they give similar results. For singlet wave functions with bond lengths only slightly longer than the RHF/UHF instability point, such PUMP methods tend to give results very similar to those based on an RHF wave function. At longer bond lengths the RMP perturbation series eventually breaks down, while the PUMP methods approach the correct dissociation limit. It would therefore appear that PUMP methods should always be preferred. There are, however, also some computational factors to consider. First, UMP methods are by nature a factor of ~two more expensive since there are twice as many MO coefficients. Second, the projection itself also requires CPU time. This is especially true if many of the higher spin states need to be removed, or for projection at the MP4 level. Third, it is difficult to formulate derivatives of projected wave functions, which limits PUMP methods to the calculation of energies. A rule of thumb says that for uncomplicated systems the RMP4 treatment gives acceptable accuracy (relative errors of the order of a ~10 kJ/mol) up to bond lengths ~1.5 times the equilibrium length. Longer bonds are better treated by PUMP methods (see also Section 12.5). Most transition structures have bond lengths shorter than ~1.5 times the equilibrium length and RMP4 often gives quite accurate activation energies.

Just as single-reference CI can be extended to MRCI, it is also possible to use perturbation methods with a multideterminant reference wave function. A formulation of MR-MBPT methods, however, is not straightforward. The main problem here is similar to that with ROMP methods: the choice of the unperturbed Hamiltonian operator. Several different choices are possible, which will give different answers when the theory is carried out only to low order. Nevertheless, there are several different implementations of MP2-type expansions based on a MCSCF reference, often denoted CASMP2, CASPT2 or MRPT2.[33,34] An extension to third order has also been reported, with the acronym CASPT3.[35]

4.9 Coupled Cluster

Perturbation methods add *all types of corrections* (S, D, T, Q, etc.) to the reference wave function *to a given order* (2, 3, 4, etc.). The idea in *Coupled Cluster* (CC) methods is to include *all corrections of a given type to infinite order*.[36] Let us start by defining an excitation operator $\mathbf{T}$ as in

$$\mathbf{T} = \mathbf{T}_1 + \mathbf{T}_2 + \mathbf{T}_3 + \cdots + \mathbf{T}_{N_{\text{elec}}} \tag{4.61}$$

The $\mathbf{T}_i$ operator acting on an HF reference wave function Φ_0 generates all ith excited Slater determinants:

$$\mathbf{T}_1\Phi_0 = \sum_i^{\text{occ}}\sum_a^{\text{vir}} t_i^a \Phi_i^a \tag{4.62}$$

$$\mathbf{T}_2\Phi_0 = \sum_{i<j}^{\text{occ}}\sum_{a<b}^{\text{vir}} t_{ij}^{ab} \Phi_{ij}^{ab} \tag{4.63}$$

It is customary in coupled cluster theory to use the term *amplitudes* for the expansion coefficients t, which are equivalent to the a coefficients in Equation (4.1).

Using intermediate normalization, a CI wave function can be generated by allowing the excitation operator to work on an HF wave function:

$$\Psi_{\text{CI}} = (\mathbf{1}+\mathbf{T})\Phi_0 = (\mathbf{1}+\mathbf{T}_1+\mathbf{T}_2+\mathbf{T}_3+\mathbf{T}_4+\cdots)\Phi_0 \tag{4.64}$$

The corresponding coupled cluster wave function, on the other hand, is defined in

$$\Psi_{\text{CC}} = \mathrm{e}^{\mathbf{T}}\Phi_0 \tag{4.65}$$

The exponential operator can be written as a Taylor expansion:

$$\mathrm{e}^{\mathbf{T}} = \mathbf{1}+\mathbf{T}+\tfrac{1}{2}\mathbf{T}^2+\tfrac{1}{6}\mathbf{T}^3+\cdots = \sum_{k=0}^{\infty}\frac{1}{k!}\mathbf{T}^k \tag{4.66}$$

From Equations (4.61) and (4.65) the exponential operator may be written as

$$\begin{aligned}\mathrm{e}^{\mathbf{T}} &= \mathbf{1}+\mathbf{T}_1+\left(\mathbf{T}_2+\tfrac{1}{2}\mathbf{T}_1^2\right)+\left(\mathbf{T}_3+\mathbf{T}_2\mathbf{T}_1+\tfrac{1}{6}\mathbf{T}_1^3\right)\\ &\quad+\left(\mathbf{T}_4+\mathbf{T}_3\mathbf{T}_1+\tfrac{1}{2}\mathbf{T}_2^2+\tfrac{1}{2}\mathbf{T}_2\mathbf{T}_1^2+\tfrac{1}{24}\mathbf{T}_1^4\right)+\cdots\end{aligned} \tag{4.67}$$

The first term generates the reference HF and the second all singly excited states. The first parenthesis generates all doubly excited states, which may be considered as *connected* ($\mathbf{T}_2$) or *disconnected* ($\mathbf{T}_1^2$). The second parenthesis generates all triply excited states, which again may be either "true" ($\mathbf{T}_3$) or "product" triples ($\mathbf{T}_2\mathbf{T}_1$, $\mathbf{T}_1^3$). The quadruply excited states can similarly be viewed as composed of five terms, a true quadruple and four product terms. Physically, a connected type such as $\mathbf{T}_4$ corresponds to four electrons interacting simultaneously, while a disconnected term such as $\mathbf{T}_2^2$ corresponds to two non-interacting pairs of interacting electrons. By comparison with the CI wave function in Equation (4.64), it is seen that the CC wave function at each excitation level contains additional terms arising from products of excitations.

With the coupled cluster wave function in Equation (4.65) the Schrödinger equation becomes

$$\mathbf{H}\mathrm{e}^{\mathbf{T}}\Phi_0 = E_{\text{CC}}\mathrm{e}^{\mathbf{T}}\Phi_0 \tag{4.68}$$

At this point, one could proceed analogously to CI and evaluate the energy as an expectation value of the CC wave function and use the variational principle to determine the amplitudes:

$$E_{\text{CC}}^{\text{var}} = \frac{\langle\Psi_{\text{CC}}|\mathbf{H}|\Psi_{\text{CC}}\rangle}{\langle\Psi_{\text{CC}}|\Psi_{\text{CC}}\rangle} = \frac{\langle\mathrm{e}^{\mathbf{T}}\Phi_0|\mathbf{H}|\mathrm{e}^{\mathbf{T}}\Phi_0\rangle}{\langle\mathrm{e}^{\mathbf{T}}\Phi_0|\mathrm{e}^{\mathbf{T}}\Phi_0\rangle} \tag{4.69}$$

Expansion of the numerator and denominator according to Equation (4.66) unfortunately leads to a series of non-vanishing terms all the way up to order N_{elec}, which makes a variational coupled cluster approach unmanageable for all but the smallest systems:[37]

$$E_{CC}^{var} = \frac{\left\langle \left(1 + \mathbf{T} + \frac{1}{2}\mathbf{T}^2 \cdots \frac{1}{N!}\mathbf{T}^N\right)\Phi_0 \middle| \mathbf{H} \middle| \left(1 + \mathbf{T} + \frac{1}{2}\mathbf{T}^2 \cdots \frac{1}{N!}\mathbf{T}^N\right)\Phi_0 \right\rangle}{\left\langle \left(1 + \mathbf{T} + \frac{1}{2}\mathbf{T}^2 \cdots \frac{1}{N!}\mathbf{T}^N\right)\Phi_0 \middle| \left(1 + \mathbf{T} + \frac{1}{2}\mathbf{T}^2 \cdots \frac{1}{N!}\mathbf{T}^N\right)\Phi_0 \right\rangle} \tag{4.70}$$

The standard formulation of coupled cluster theory instead proceeds by *projecting* the coupled cluster Schrödinger Equation (4.68) on to the reference wave function. Multiplying from the left by $\Phi_0{}^*$ and integrating gives

$$\begin{aligned} \langle \Phi_0 | \mathbf{H}\mathrm{e}^{\mathbf{T}} | \Phi_0 \rangle &= E_{CC} \langle \Phi_0 | \mathrm{e}^{\mathbf{T}} \Phi_0 \rangle \\ \langle \Phi_0 | \mathbf{H}\mathrm{e}^{\mathbf{T}} | \Phi_0 \rangle &= E_{CC} \langle \Phi_0 | (1 + \mathbf{T}_1 + \mathbf{T}_2 + \cdots) \Phi_0 \rangle \\ E_{CC} &= \langle \Phi_0 | \mathbf{H}\mathrm{e}^{\mathbf{T}} | \Phi_0 \rangle \end{aligned} \tag{4.71}$$

Expanding out the exponential in Equation (4.66) and using the fact that the Hamiltonian operator contains only one- and two-electron operators (Equation (3.25)) we get

$$\begin{aligned} E_{CC} &= \left\langle \Phi_0 \middle| \mathbf{H}\left(1 + \mathbf{T}_1 + \mathbf{T}_2 + \tfrac{1}{2}\mathbf{T}_1^2\right) \middle| \Phi_0 \right\rangle \\ E_{CC} &= \langle \Phi_0 | \mathbf{H} | \Phi_0 \rangle + \langle \Phi_0 | \mathbf{H} | \mathbf{T}_1 \Phi_0 \rangle + \langle \Phi_0 | \mathbf{H} | \mathbf{T}_2 \Phi_0 \rangle + \tfrac{1}{2} \left\langle \Phi_0 | \mathbf{H} | \mathbf{T}_1^2 \Phi_0 \right\rangle \\ E_{CC} &= E_0 + \sum_i^{occ} \sum_a^{vir} t_i^a \left\langle \Phi_0 | \mathbf{H} | \Phi_i^a \right\rangle + \sum_{i<j}^{occ} \sum_{a<b}^{vir} \left(t_{ij}^{ab} + t_i^a t_j^b - t_i^b t_j^a \right) \left\langle \Phi_0 \middle| \mathbf{H} \middle| \Phi_{ij}^{ab} \right\rangle \end{aligned} \tag{4.72}$$

Note that the infinite expansion of the exponential operator in Equations (4.71) and (4.72) terminates at the **1** and $\mathbf{T}_2$ levels, in contrast to $\mathbf{T}_N$ in Equation (4.70). Furthermore, when using HF orbitals for constructing the Slater determinants, the matrix elements in Equation (4.72) involving the singly excited Slater determinant are zero (Brillouin's theorem) and the second matrix elements are just two-electron integrals over MOs (Equation (4.10)):

$$\begin{aligned} E_{CC} &= E_0 + \sum_{i<j}^{occ} \sum_{a<b}^{vir} \left(t_{ij}^{ab} + t_i^a t_j^b - t_i^b t_j^a \right) \left\langle \Phi_0 \middle| \mathbf{H} \middle| \Phi_{ij}^{ab} \right\rangle \\ E_{CC} &= E_0 + \sum_{i<j}^{occ} \sum_{a<b}^{vir} (t_{ij}^{ab} + t_i^a t_j^b - t_i^b t_j^a)(\langle \phi_i \phi_j | \phi_a \phi_b \rangle - \langle \phi_i \phi_j | \phi_b \phi_a \rangle) \end{aligned} \tag{4.73}$$

The coupled cluster correlation energy is therefore determined completely by the singles and doubles amplitudes and the two-electron MO integrals.

Equations for the amplitudes can be obtained by projecting the Schrödinger Equation (4.68) on to the space of singly, doubly, triply, etc., excited determinants:

$$\begin{aligned} \left\langle \Phi_m^e | \mathbf{H}\mathrm{e}^{\mathbf{T}} | \Phi_0 \right\rangle &= E_{CC} \left\langle \Phi_m^e | \mathrm{e}^{\mathbf{T}} \Phi_0 \right\rangle \\ \left\langle \Phi_{mn}^{ef} \middle| \mathbf{H}\mathrm{e}^{\mathbf{T}} \middle| \Phi_0 \right\rangle &= E_{CC} \left\langle \Phi_{mn}^{ef} \middle| \mathrm{e}^{\mathbf{T}} \Phi_0 \right\rangle \\ \left\langle \Phi_{mnl}^{efg} \middle| \mathbf{H}\mathrm{e}^{\mathbf{T}} \middle| \Phi_0 \right\rangle &= E_{CC} \left\langle \Phi_{mnl}^{efg} \middle| \mathrm{e}^{\mathbf{T}} \Phi_0 \right\rangle \end{aligned} \tag{4.74}$$

An alternative formulation is in terms of a *similarity transformation* of the Hamiltonian operator (Section 17.3). Consider Equation (4.56) where we multiply from the left by $\mathrm{e}^{-\mathbf{T}}$:

$$\mathrm{e}^{-\mathbf{T}}\mathbf{H}\mathrm{e}^{\mathbf{T}}\Phi_0 = E_{CC}\Phi_0 \tag{4.75}$$

Multiplying with $\Phi_0{}^*$ from the left and integrating leads directly to the energy equation

$$E_{CC} = \langle \Phi_0 | e^{-\mathbf{T}} \mathbf{H} e^{\mathbf{T}} | \Phi_0 \rangle \tag{4.76}$$

The coupled cluster energy can thus be considered as the expectation value of a similarity transformed (non-Hermitian) Hamiltonian. Equations for the amplitudes can be obtained by multiplying with an excited state, analogous to Equation (4.74):

$$\begin{aligned} \left\langle \Phi_m^e \middle| e^{-\mathbf{T}} \mathbf{H} e^{\mathbf{T}} \middle| \Phi_0 \right\rangle &= 0 \\ \left\langle \Phi_{mn}^{ef} \middle| e^{-\mathbf{T}} \mathbf{H} e^{\mathbf{T}} \middle| \Phi_0 \right\rangle &= 0 \\ \left\langle \Phi_{mnl}^{efg} \middle| e^{-\mathbf{T}} \mathbf{H} e^{\mathbf{T}} \middle| \Phi_0 \right\rangle &= 0 \end{aligned} \tag{4.77}$$

The similarity transformed Hamiltonian can be written as a series of nested commutators (*Baker–Campbell–Hausdorff expansion*):

$$\begin{aligned} e^{-\mathbf{T}} \mathbf{H} e^{\mathbf{T}} = \mathbf{H} + [\mathbf{H}, \mathbf{T}] + \tfrac{1}{2}[[\mathbf{H}, \mathbf{T}], \mathbf{T}] + \tfrac{1}{6}[[[\mathbf{H}, \mathbf{T}], \mathbf{T}], \mathbf{T}] \\ + \tfrac{1}{24}[[[[\mathbf{H}, \mathbf{T}], \mathbf{T}], \mathbf{T}], \mathbf{T}] \end{aligned} \tag{4.78}$$

Since the Hamilton operator only contains one- and two-electron operators, the above expansion is exact, and shows that the coupled cluster equations at most contain terms that are quartic in the unknown amplitudes. The similarity transform formulation and its commutation expansion is especially advantageous for deriving equations within the second quantization formalism (Appendix E).

4.9.1 Truncated coupled cluster methods

So far, everything has been exact. If all cluster operators up to $\mathbf{T}_N$ are included in $\mathbf{T}$, all possible excited determinants are generated and the coupled cluster wave function is equivalent to full CI. This is, as already stated, impossible for all but the smallest systems. The cluster operator must therefore be truncated at some excitation level. When the $\mathbf{T}$ operator is truncated, some of the terms in the amplitude equations will become zero and the amplitudes derived from these approximate equations will no longer be exact. The energy calculated from these approximate singles and doubles amplitudes (Equation (4.73)) will therefore also be approximate. How severe the approximation is depends on how many terms are included in $\mathbf{T}$. Including only the $\mathbf{T}_1$ operator does not give any improvement over HF, as matrix elements between the HF and singly excited states are zero. The lowest level of approximation is therefore $\mathbf{T} = \mathbf{T}_2$, referred to as *Coupled Cluster Doubles* (CCD). Compared with the number of doubles, there are relatively few singly excited states. Using $\mathbf{T} = \mathbf{T}_1 + \mathbf{T}_2$ gives the CCSD model, which is only slightly more demanding than CCD and yields a more complete model. Both CCD and CCSD involve a computational effort that scales as M_{basis}^6 in the limit of a large basis set. The next higher level has $\mathbf{T} = \mathbf{T}_1 + \mathbf{T}_2 + \mathbf{T}_3$, giving the CCSDT model. This involves a computational effort that scales as M_{basis}^8 and is more demanding than CISDT. It (and higher-order methods such as CCSDTQ) can consequently only be used for small systems, and CCSD is the only generally applicable coupled cluster method.

Let us look in a bit more detail at the CCSD method. In this case, we have, from Equation (4.67),

$$e^{\mathbf{T}_1+\mathbf{T}_2} = \mathbf{1} + \mathbf{T}_1 + \left(\mathbf{T}_2 + \tfrac{1}{2}\mathbf{T}_1^2\right) + \left(\mathbf{T}_2\mathbf{T}_1 + \tfrac{1}{6}\mathbf{T}_1^3\right) + \left(\tfrac{1}{2}\mathbf{T}_2^2 + \tfrac{1}{2}\mathbf{T}_2\mathbf{T}_1^2 + \tfrac{1}{24}\mathbf{T}_1^4\right) + \cdots \tag{4.79}$$

The CCSD energy is given by the general CC Equation (4.73), and amplitude equations can be derived by projecting against a singly excited Slater determinant (Equation (4.74)):

$$
\begin{aligned}
&\left\langle \Phi_m^e \middle| \mathbf{H} \middle| \left(1 + \mathbf{T}_1 + \left(\mathbf{T}_2 + \tfrac{1}{2}\mathbf{T}_1^2\right) + \left(\mathbf{T}_2\mathbf{T}_1 + \tfrac{1}{6}\mathbf{T}_1^3\right)\right)\Phi_0 \right\rangle = E_{CC}\left\langle \Phi_m^e \middle| (1 + \mathbf{T}_1)\Phi_0 \right\rangle \\
&t_i^a \left\langle \Phi_m^e \middle| \mathbf{H} \middle| \Phi_i^a \right\rangle + \left(t_{ij}^{ab} + t_i^a t_j^b - t_i^b t_j^a\right)\left\langle \Phi_m^e \middle| \mathbf{H} \middle| \Phi_{ij}^{ab} \right\rangle \\
&+ \left(t_{ij}^{ab} t_k^c + \cdots + t_i^a t_j^b t_k^c + \cdots\right)\left\langle \Phi_m^e \middle| \mathbf{H} \middle| \Phi_{ijk}^{abc} \right\rangle = E_{CC} t_m^e
\end{aligned}
\tag{4.80}
$$

Here the Brillouin theorem and the orthonormality of the Slater determinants have been employed and the notation $(t_i^a t_j^b t_k^c + \cdots)$ indicates that the terms involving permutations of the indices are omitted. Summation over indices has been omitted. Projecting against a doubly excited Slater determinant yields

$$
\begin{aligned}
&\left\langle \Phi_{mn}^{ef} \middle| \mathbf{H} \middle| \begin{pmatrix} 1 + \mathbf{T}_1 + \left(\mathbf{T}_2 + \frac{1}{2}\mathbf{T}_1^2\right) + \left(\mathbf{T}_2\mathbf{T}_1 + \frac{1}{6}\mathbf{T}_1^3\right) + \\ \left(\frac{1}{2}\mathbf{T}_2^2 + \frac{1}{2}\mathbf{T}_2\mathbf{T}_1^2 + \frac{1}{24}\mathbf{T}_1^4\right) \end{pmatrix} \Phi_0 \right\rangle = E_{CC}\left\langle \Phi_{mn}^{ef} \middle| \begin{pmatrix} 1 + \mathbf{T}_1 + \\ \left(\mathbf{T}_2 + \frac{1}{2}\mathbf{T}_1^2\right) \end{pmatrix} \Phi_0 \right\rangle \\
&\left\langle \Phi_{mn}^{ef} \middle| \mathbf{H} \middle| \Phi_0 \right\rangle + t_i^a \left\langle \Phi_{mn}^{ef} \middle| \mathbf{H} \middle| \Phi_i^a \right\rangle + \left(t_{ij}^{ab} + t_i^a t_j^b - t_i^b t_j^a\right)\left\langle \Phi_{mn}^{ef} \middle| \mathbf{H} \middle| \Phi_{ij}^{ab} \right\rangle \\
&+ \left(t_{ij}^{ab} t_k^c + \cdots + t_i^a t_j^b t_k^c + \cdots\right)\left\langle \Phi_{mn}^{ef} \middle| \mathbf{H} \middle| \Phi_{ijk}^{abc} \right\rangle \\
&+ \left(t_{ij}^{ab} t_{kl}^{cd} + \cdots + t_{ij}^{ab} t_k^c t_l^d + \cdots + t_i^a t_j^b t_k^c t_l^d + \cdots\right)\left\langle \Phi_{mn}^{ef} \middle| \mathbf{H} \middle| \Phi_{ijkl}^{abcd} \right\rangle = E_{CC}\left(t_{mn}^{ef} + t_m^e t_n^f - t_m^f t_n^e\right)
\end{aligned}
\tag{4.81}
$$

Note that the coupled cluster energy from Equation (4.73) must be inserted and also contains amplitudes.

Equations (4.80) and (4.81) involve matrix elements between singles and triples, and between doubles and quadruples. However, since the Hamiltonian operator only contains one- and two-electron operators, these are actually identical to matrix elements between the reference and a doubly excited state. Consider, for example, $\langle \Phi_m^e | \mathbf{H} | \Phi_{ijk}^{abc} \rangle$. Unless m equals either i, j or k, and e equals either a, b or c, there will be one overlap integral between different MOs that makes the matrix element zero. If, for example, $m = k$ and $e = c$, then the MO integral over these indices factor out as 1, and the rest is equal to a matrix element $\langle \Phi_0 | \mathbf{H} | \Phi_{ij}^{ab} \rangle$. Similarly, the matrix element $\langle \Phi_{mn}^{ef} | \mathbf{H} | \Phi_{ijkl}^{abcd} \rangle$ between a doubly and a quadruply excited determinant is only non-zero if *mn* matches up with two of the *ijkl* indices and *ef* matches up with *abcd*. Again, such non-zero matrix elements are equal to matrix elements between the reference and a doubly excited determinant (Equation (4.10)).

All the matrix elements can be evaluated in terms of MO integrals, and the expressions in Equations (4.80) and (4.81) form coupled non-linear equations for the singles and doubles amplitudes. The equations contain terms up to quartic in the amplitudes, for example $(t_i^a)^4$, and must be solved by iterative techniques. Once the amplitudes are known, the energy and wave function can be calculated. The important aspect in coupled cluster methods is that excitations of higher order than the truncation of the **T** operator enter the amplitude equation. Quadruply excited states, for example, are generated by the $\mathbf{T}_2^2$ operator in CCSD, and they enter the amplitude equations with a weight given as a *product* of doubles amplitudes. Quadruply excited states influence the doubles amplitudes, and thereby also the CCSD energy. It is the inclusion of these products of excitations that makes coupled cluster theory size extensive. For the case of a single H_2 molecule, a CISD calculation is equivalent to CCSD, and is also equivalent to a full CI calculation. For two H_2 molecules separated by 100 Å, however, a CISD is not equivalent to a full CI (it is missing the T and Q excitations), but a CCSD calculation is still equivalent to a full CI.

The above single reference coupled cluster methods can, analogous to CI and MP, be extended to the multireference case with the acronym MRCC, but there are several variations on the exact formalism.[38]

4.10 Connections between Coupled Cluster, Configuration Interaction and Perturbation Theory

The general cluster operator is given by the following equation, where terms have been collected according to the excitation they generate:

$$\begin{aligned} e^{\mathbf{T}} &= \mathbf{1} + \mathbf{T}_1 + \left(\mathbf{T}_2 + \tfrac{1}{2}\mathbf{T}_1^2\right) + \left(\mathbf{T}_3 + \mathbf{T}_2\mathbf{T}_1 + \tfrac{1}{6}\mathbf{T}_1^3\right) \\ &+ \left(\mathbf{T}_4 + \mathbf{T}_3\mathbf{T}_1 + \tfrac{1}{2}\mathbf{T}_2^2 + \tfrac{1}{2}\mathbf{T}_2\mathbf{T}_1^2 + \tfrac{1}{24}\mathbf{T}_1^4\right) + \cdots \end{aligned} \tag{4.82}$$

Each of the operators in a given parentheses generates *all* the excited determinants of the given type and the terms in Equation (4.82) generate all determinants that are included in a CISDTQ calculation. In terms of parameterization, the following connections can be made between CI and CC:

$$\begin{aligned} \mathbf{A}_S &\Leftrightarrow \mathbf{T}_1 \\ \mathbf{A}_D &\Leftrightarrow \mathbf{T}_2 + \tfrac{1}{2}\mathbf{T}_1^2 \\ \mathbf{A}_T &\Leftrightarrow \mathbf{T}_3 + \mathbf{T}_2\mathbf{T}_1 + \tfrac{1}{6}\mathbf{T}_1^3 \\ \mathbf{A}_Q &\Leftrightarrow \mathbf{T}_4 + \mathbf{T}_3\mathbf{T}_1 + \tfrac{1}{2}\mathbf{T}_2^2 + \tfrac{1}{2}\mathbf{T}_2\mathbf{T}_1^2 + \tfrac{1}{24}\mathbf{T}_1^4 \end{aligned} \tag{4.83}$$

The cluster expansion can be viewed as a method of partitioning the contributions from each excitation type. The total contribution from double excitations is the sum of two terms, one that is the square of the singles contributions $\mathbf{T}_1^2$ and the remaining is (by definition) the connected doubles $\mathbf{T}_2$. Similarly, the total contribution from triple excitations is a sum of three terms, the cube of the singles contributions, the product of the singles and doubles contribution, and the remaining is the connected triples. In the CI parameterization these are treated equivalently as triply excited determinants by $\mathbf{A}_T$.

The $\mathbf{T}_1$ effect is small when canonical HF orbitals are used, although not zero since singles enter indirectly via the doubly excited states (note that if non-canonical orbitals are used, the $\mathbf{T}_1$ term can be large). From CI we know that the effect of doubles is the most important (Section 4.2.3), and this contribution is in coupled cluster theory divided into $\mathbf{T}_1^2$ and $\mathbf{T}_2$. If $\mathbf{T}_1$ is small, then $\mathbf{T}_1^2$ must also be small, and the most important term is therefore $\mathbf{T}_2$. For the triple excitations, $\mathbf{T}_1^3$ must be negligible and $\mathbf{T}_1\mathbf{T}_2$ is small owing to $\mathbf{T}_1$. The most important contribution is therefore from connected triples $\mathbf{T}_3$. For the quadruple excitations, all the terms involving $\mathbf{T}_1$ must again be small, and since $\mathbf{T}_2$ is large, we expect the disconnected quadruples $\mathbf{T}_2^2$ to be the dominant term. This again suggests that the connected quadruples term $\mathbf{T}_4$ is small, which is reasonable since it corresponds to a simultaneous correlation of four electrons. Higher-order excitations will always contain terms appearing as powers and/or products of $\mathbf{T}_2$ and $\mathbf{T}_3$, which will normally dominate. Higher-order connected terms, $\mathbf{T}_n$ with $n > 4$, are therefore expected to have small effects. This is consistent with the physical picture that connected $\mathbf{T}_n$ operators correspond to n electrons interacting simultaneously. As n becomes large, this is increasingly improbable. It should be noted, however, that the higher-order cluster operators ($\mathbf{T}_4$, $\mathbf{T}_5$, ...) are expected to become more and more important as the number of electrons increases.

The principal deficiency of CISD is the lack of the $\mathbf{T}_2^2$ term which in the CI parameterization only enters at the CISDTQ level, and this is the main reason for CISD not being size extensive. Furthermore, this term becomes more and more important as the number of electrons increases, and CISD therefore recovers a smaller and smaller percentage of the correlation energy as the system increases. There are various approximate corrections for this lack of size extensivity that can be added to standard CISD. The most widely known of these is the *Davidson correction*, sometimes denoted CISD+Q(Davidson), where the quadruples contribution is approximated as in Equation (4.84), with a_0 being the coefficient for the HF reference wave function:

$$\Delta E_{\text{Q}} = \left(1 - a_0^2\right) \Delta E_{\text{CISD}} \tag{4.84}$$

If the renormalization of the wave function is also taken into account, the $(1 - a_0^2)$ quantity is divided by a_0^2 and the corresponding correction is called the *renormalized Davidson correction*. The effect of higher-order excitations is thus estimated from the correlation energy obtained at the CISD level times a factor that measures how important the single-determinant reference is at the CISD level. The Davidson correction does not yield zero for two-electron systems, where CISD is equivalent to a full CI, and it is likely that it overestimates the higher-order corrections for systems with few electrons. More complicated correction schemes have also been proposed,[39] but are rarely used.

Coupled cluster is closely connected with Møller–Plesset perturbation theory, as mentioned at the start of this section. The infinite Taylor expansion of the exponential operator (Equation (4.66)) ensures that the contributions from a given excitation level are included to infinite order. Perturbation theory indicates that doubles are the most important, since they are the only contributors to MP2 and MP3. At fourth order, there are contributions from singles, doubles, triples and quadruples. The MP4 quadruples contribution is actually the disconnected $\mathbf{T}_2^2$ term in the coupled cluster language, and the triples contribution corresponds to $\mathbf{T}_3$. This is consistent with the above analysis, the most important being $\mathbf{T}_2$ (and products thereof) followed by $\mathbf{T}_3$. The CCD energy is equivalent to MP∞(D) where all disconnected contributions of products of doubles are included. If the perturbation series is reasonably converged at fourth order, we expect that CCD will be comparable to MP4(DQ) and CCSD will be comparable to MP4(SDQ). The MP2, MP3 and MP4(SDQ) results may be obtained in the first iteration for the CCSD amplitudes, allowing a direct test of the convergence of the MP series. This also points out the principal limitation of the CCSD method: the neglect of the connected triples. Including $\mathbf{T}_3$ in the $\mathbf{T}$ operator leads to the CCSDT method, which, as mentioned above, is too demanding computationally for all but the smallest systems. Alternatively, the triples contribution may be evaluated by perturbation theory and *added* to the CCSD results. Several such hybrid methods have been proposed, but only the method with the acronym CCSD(T) is commonly used.[40] In this case, the triples contribution is calculated from the formula given by MP4, but using the CCSD amplitudes instead of the perturbation coefficients for the wave function corrections and adding a term arising from fifth-order perturbation theory, describing the coupling between singles and triples. Higher-order hybrid methods such as CCSD(TQ) and CCSDT(Q), where the connected quadruples contribution is estimated by fifth-order perturbation theory, are also possible, but they are again so demanding that they can only be used for small systems.[41]

The singles make, as mentioned, a fairly small contribution to the correlation energy when canonical HF orbitals are used. *Brueckner theory* is a variation of coupled cluster where the orbitals used for constructing the Slater determinants are optimized such that the contribution from singles is exactly zero, that is $t_i^a = 0$.[42,43] The lowest level of Brueckner theory includes only doubles, giving the acronym BD. Although BD in theory should be slightly better than CCSD, since it includes orbital relaxation, they give in practice essentially identical results (differences between BD and CCSD are of fifth order or higher in terms of perturbation theory). This is presumably rooted in the fact that the singles in CCSD introduce orbital relaxation.[44] The computational cost for BD is slightly higher than for CCSD, since the orbital relaxation necessitates additional iterations for solving the amplitudes.[45] Similarly, BD(T) is essentially equivalent to CCSD(T)[46] and BD(TQ) to CCSD(TQ).

Since the singly excited determinants effectively relax the orbitals in a CCSD calculation, non-canonical HF orbitals can also be used in coupled cluster methods. This allows, for example, the use of open-shell singlet states (which require two Slater determinants) as reference for a coupled cluster calculation.[47]

Another commonly used method is *Quadratic* CISD (QCISD). It was originally derived from CISD by including enough higher-order terms to make it size extensive.[48] It has since been shown that the resulting equations are identical to CCSD where some of the terms have been omitted.[8,49] The omitted terms are computationally inexpensive, and there appears to be no reason for using the less complete QCISD over CCSD (or QCISD(T) in place of CCSD(T)). In practice they normally give very similar results,[50] but exceptions have been reported.[51] There are a few other methods that may be considered either as CISD with the addition of extra terms to make them approximately size extensive or as approximate versions of CCSD. Some of the methods falling into this category are *Averaged Coupled-Pair Functional* (ACPF) and *Coupled Electron Pair Approximation* (CEPA). The simplest form of CEPA, CEPA-0, is also known as *Linear Coupled Cluster Doubles* (LCCD).

In connection with the calculation of molecular properties, it is useful to define some intermediate coupled cluster levels (see also Section 4.14). As already mentioned, the single excitations allow the MOs to relax from their HF form but do not give any direct contribution to the energy due to Brillouin's theorem. For studying properties that measure the response of the energy to a perturbation, the HF orbitals are no longer optimum, and the singles are at least as important as the doubles. The CC2 method is derived from CCSD by only including the doubles contribution arising from the lowest (non-zero) order in perturbation theory, where the perturbation is defined as in MP theory (i.e. as the true electron–electron potential minus the average repulsion).[52] The amplitude equations corresponding to multiplication of a doubly excited determinant in the CCSD equations (Equation (4.81)) thereby reduce to an MP2-like expression, and the t_2 amplitudes can be expressed directly in terms of the t_1 amplitudes and MO integrals. The iterative procedure therefore only involves the t_1 amplitudes. CC2 may loosely be defined as MP2 with the added feature of orbital relaxation arising from the singles. Similarly, CC3 is an approximation to the full CCSDT model, where the triples contribution is approximated by the expression arising from the lowest non-vanishing order in perturbation theory.[53] The triples amplitudes can then be expressed directly in terms of the singles and doubles amplitudes, and MO integrals. Both in terms of computational cost and accuracy, the following progression is expected:

$$\text{HF} \ll \text{CC2} < \text{CCSD} < \text{CC3} < \text{CCSDT} \tag{4.85}$$

Analogously to MP methods, coupled cluster theory can also be based on a UHF reference wave function. The resulting UCC methods again suffer from spin contamination of the underlying UHF, but the infinite nature of coupled cluster methods is substantially better at reducing spin contamination relative to UMP.[54] Projection methods analogous to the PUMP case have been considered but are not commonly used. ROHF-based coupled cluster methods have also been proposed but appear to give results very similar to UCC, especially at the CCSD(T) level.[40]

Standard coupled cluster theory is based on a single-determinant reference wave function. It suffers from the same problem as MP, in that it works best if the zeroth-order wave function is sufficiently "good". Owing to the summation of contributions to infinite order, however, coupled cluster is somewhat more tolerant to a poor reference wave function than MP methods. Since the singly excited determinants allow the MOs to relax in order to describe the multireference character in the wave function, the magnitude of the singles amplitude can be taken as an indication of how good the HF single determinant is as the reference. The T_1*-diagnostic* defined as the norm of the singles amplitude vector divided by the square root of the number of electrons has been suggested as an internal evaluation of the quality of a CCSD wave function:[55]

$$T_1 = \frac{1}{\sqrt{N_{elec}}} |\mathbf{t}_1| \tag{4.86}$$

Specifically, if $T_1 < 0.02$, the CCSD(T) method is expected to give results close to the full CI limit for the given basis set. If T_1 is larger than 0.02, it indicates that the reference wave function has significant multideterminant character, and multireference coupled cluster should preferentially be employed. Such methods are being developed,[56] but have not yet seen any extensive use. The T_1-diagnostic in Equation (4.86) is not independent of the system size and assumes that the orbitals are obtained from an HF calculation (the BD method, for example, has $T_1 = 0$ for all systems), and other diagnostics have also been proposed.[57]

4.10.1 Illustrating Correlation Methods for the Beryllium Atom

The beryllium atom has four electrons ($1s^2 2s^2$ electron configuration) and the ground-state wave function contains significant multireference character owing to the presence of the low-lying 2p-orbital. The correlation energies calculated with MP, CI and CC methods in a 4s2p basis set (cc-pVDZ basis augmented with one set of tight *s*- and *p*-functions for correlating the 1s-electrons) are given in Table 4.5.

Since beryllium only has four electrons, CISDTQ is a full CI treatment and completely equivalent to a CCSDTQ calculation. The multireference character displays itself as a relatively slow convergence of the perturbation series, with millihartree accuracy being attained at the MP6 level, and inclusion of terms up to MP20 is required in order to converge the energy to within 10^{-6} au of the exact answer. Note also that the correlation energy is overestimated at order seven, that is the perturbation series oscillates at higher orders. The contribution from triply excited states is minute, as expected for a system with two well-separated electron pairs, that is CISDT is only a marginal improvement over CISD.

Table 4.5 Correlation energies for the beryllium atom in a 4s2p basis set.

Level	ΔE_{corr} (au)	%	Level	ΔE_{corr} (au)	%	Level	ΔE_{corr} (au)	%
MP2	0.05317	67.85						
MP3	0.06795	86.70	CISD	0.07528	96.05	CCSD	0.07818	99.75
MP4	0.07412	94.58				CCSD(T)	0.07836	99.99
MP5	0.07692	98.15	CISDT	0.07547	96.29	CCSDT	0.07836	99.99
MP6	0.07809	99.64						
MP7	0.07849	100.15	CISDTQ	0.07837	100	CCSDTQ	0.07837	100

Results are courtesy of Professor Jeppe Olsen.

Table 4.6 Coefficients for dominating excited states.

Excitation	Type	a_{CISDTQ}	t_{CCSDTQ}	$t_{CCSDTQ} \cdot t_{CCSDTQ}$
$2s^2 \rightarrow 2p^2$	D	−0.18612	−0.18523	
$2s^2 \rightarrow 2s'^2$	D	−0.04341	−0.04376	
$1s^2 \rightarrow 1s'^2$	D	−0.02171	−0.02170	
$1s^2 2s^2 \rightarrow 1s'^2 2p^2$	Q	0.00407	4.10^{-7}	0.00402
$1s^2 2s^2 \rightarrow 1s'^2 2s'^2$	Q	0.00092		0.00095

The coefficients in the CISDTQ and CCSDTQ wave functions (using intermediate normalization) for the dominating excitations are given in Table 4.6.

There is little difference between the full CI and CCSD coefficients for the three most important doubly excited states. The quadruply excited states enter the CI wave function with non-negligible weights, contributing ~4% of the correlation energy (Table 4.5), but as shown in the last column of Table 4.6, these contributions are estimated very well by the product terms in the CC wave function. The CCSD energy in Table 4.5 and the t_4 amplitude in Table 4.6 show that the quadruply excited states in the CI wave function are mainly of the product type and not a true quadruply excited state. It is this feature that makes CC superior to CI-based methods.

4.11 Methods Involving the Interelectronic Distance

The necessity of going beyond the HF approximation is due to the fact that electrons are further apart than described by the product of their orbital densities, that is their motions are correlated. This arises from the electron–electron repulsion operator, which is a sum of terms of the type shown by

$$\frac{1}{|\mathbf{r}_1 - \mathbf{r}_2|} = \frac{1}{r_{12}} \tag{4.87}$$

Without these terms, the Schrödinger equation can be solved exactly, with the solution being a Slater determinant composed of orbitals.

The electron–electron repulsion operator has a *singularity* for $r_{12} = 0$, which results in the exact wave function having a *cusp* (discontinuous derivative),[58] since an infinite kinetic energy must cancel the infinity of the potential energy to give a finite result:

$$\left(\frac{\partial \Psi}{\partial r_{12}}\right)_{r_{12}=0} = \frac{1}{2}\Psi(r_{12}) \tag{4.88}$$

The cusp condition in Equation (4.88) implies that the exact wave function must be linear in the interelectronic distance for small values of r_{12}:

$$\Psi_{\text{exact}}(r_{12}) = \text{constant} + \frac{1}{2}r_{12} + \cdots \tag{4.89}$$

It would therefore seem logical that the interelectronic distance should be a necessary variable for describing electron correlation. For two-electron systems, extremely accurate wave functions may be generated by taking a trial wave function consisting of an orbital product times an expansion in electron coordinates, as given in the following equation, and variationally optimizing the α_i and C_{klm} parameters:

$$\Psi(r_1, r_2) = e^{-\alpha_1 r_1} e^{-\alpha_2 r_2} \sum_{klm} C_{klm}(r_1 + r_2)^k (r_1 - r_2)^l r_{12}^m \tag{4.90}$$

Expansions such as Equation (4.90) are known as *Hylleraas* -type wave functions.[59,60] For the hydrogen molecule, it is possible with a wave function containing ~1000 terms to converge the total energy to ~10^{-9} au, which is more accurate than what can be determined experimentally. In fact, the prediction that the experimental dissociation energy for H_2 was wrong, based on calculations, was one of the first hallmarks of quantum chemistry.[61] The total energy of the helium atom has been converged to more than 40 digits by Hylleraas-type wave functions containing ~10 000 terms. Such wave functions unfortunately become impractical for more than 3–4 electrons due to the large number of many-electron integrals and the non-linear optimization parameters.

All electron correlation methods based on expanding the N-electron wave function in terms of Slater determinants built from orbitals (one-electron functions) suffer from an agonizingly slow convergence. Literally millions or billions of determinants are required for obtaining results that in an absolute sense are close to the exact results. This is due to the fact that products of one-electron functions are poor at describing the cusp behavior of the wave function when two electrons are close together. At the second-order perturbation level (i.e. MP2) it may be shown that the *error* in the correlation energy for a singlet-coupled electron pair behaves asymptotically as $(l + 1/2)^{-4}$, where l is the highest angular momentum in the basis set. For a general wave function the convergence is $(l + 1/2)^{-4} + (l + 1/2)^{-5} + (l + 1/2)^{-6} + \cdots$. This means that the total energy will converge as $(L + 1)^{-3} + (L + 1)^{-4} + (L + 1)^{-5} + \cdots$, if the basis set is saturated up to angular momentum L.[62] For sufficiently large values of L the convergence is thus $\sim(L + 1)^{-3}$, which is quite slow. The corresponding convergence for a triplet-coupled electron pair is $(l + 1/2)^{-6} + (l + 1/2)^{-7} + (l + 1/2)^{-8} + \cdots$, leading to a $(L + 1)^{-5} + (L + 1)^{-6} + (L + 1)^{-7} + \cdots$ convergence for a basis set saturated up to angular momentum L. In order to achieve a high accuracy, it would seem desirable to explicitly include terms in the wave functions that are linear in the interelectronic distance. This is the idea in the R12 methods proposed by Kutzelnigg and Klopper.[63] The first-order correction to the HF wave function only involves doubly excited determinants (Equation (4.47)). In R12 methods additional terms are included, which are essentially the HF determinant multiplied with r_{ij} factors:

$$\Psi_{\text{R12}} = \Phi_{\text{HF}} + \sum_{ijab} a_{ijab} \Phi_{ij}^{ab} + \sum_{ij} b_{ij} r_{ij} \Phi_{\text{HF}} \tag{4.91}$$

The exact definition is slightly more complicated, since the wave function has to be properly antisymmetrized and projected on to the actual basis but, for illustration, the above form is sufficient. Such R12 wave functions may then be used in connection with the CI, MBPT or CC methods described in the previous sections. Consider, for example, a CI calculation with an R12-type wave function. The energy is given by the following equation, where the a_{ijab} and b_{ij} parameters in Equation (4.91) are optimized variationally:

$$E_{\text{CI-R12}} = \langle \Psi_{\text{R12}} | \mathbf{H} | \Psi_{\text{R12}} \rangle \tag{4.92}$$

The overwhelming problem is that matrix elements from Equation (4.92) now involve integrals depending on three and four electron coordinates. Consider, for example, the following terms arising from the r_{ij} operator written out in terms of the one- and two-electron operators (**h** and **g**, Equation (3.25)):

$$\begin{aligned} \langle \Phi_{HF} | \mathbf{H} | r_{ij} \Phi_{HF} \rangle &= \langle \Phi_{HF} | \mathbf{h} | r_{ij} \Phi_{HF} \rangle + \langle \Phi_{HF} | \mathbf{g} | r_{ij} \Phi_{HF} \rangle \\ \langle r_{ij} \Phi_{HF} | \mathbf{H} | r_{ij} \Phi_{HF} \rangle &= \langle r_{ij} \Phi_{HF} | \mathbf{h} | r_{ij} \Phi_{HF} \rangle + \langle r_{ij} \Phi_{HF} | \mathbf{g} | r_{ij} \Phi_{HF} \rangle \end{aligned} \tag{4.93}$$

The **g** operator leads to integrals over molecular orbitals of the type shown below:

$$\begin{aligned} &\left\langle \phi_i(1)\phi_j(2)\phi_k(3) \left| \frac{r_{12}}{r_{13}} \right| \phi_i(1)\phi_j(2)\phi_k(3) \right\rangle \\ &\left\langle \phi_i(1)\phi_j(2)\phi_k(3) \left| \frac{r_{12}r_{23}}{r_{13}} \right| \phi_i(1)\phi_j(2)\phi_k(3) \right\rangle \\ &\left\langle \phi_i(1)\phi_j(2)\phi_k(3)\phi_l(4) \left| \frac{r_{12}r_{34}}{r_{23}} \right| \phi_i(1)\phi_j(2)\phi_k(3)\phi_l(4) \right\rangle \end{aligned} \tag{4.94}$$

Not only are such integrals difficult to calculate but, when the MOs are expanded in a basis set consisting of M_{basis} AOs, there will be of the order of M_{basis}^6 three-electron integrals and of the order of M_{basis}^8 four-electron integrals.[64] Such methods are therefore inherently more expensive than, for example, the full CCSDT model.

The trick for turning the R12 method into a viable computational tool is to avoid calculating the three- and four-electron integrals without jeopardizing the accuracy. In a complete basis, a three-electron integral may be written in terms of products of two-electron integrals by inserting a "resolution of the identity" between the two operators (see Section 17.4):

$$\begin{aligned} \mathbf{1} &= \sum_{p}^{\infty} |\phi_p\rangle\langle\phi_p| = \sum_{pqr}^{\infty} |\phi_p\phi_q\phi_r\rangle\langle\phi_p\phi_q\phi_r| \\ \left\langle \phi_i\phi_j\phi_k \left| \frac{r_{12}}{r_{13}} \right| \phi_i\phi_j\phi_k \right\rangle &= \sum_{pqr}^{\infty} \langle \phi_i\phi_j\phi_k | r_{12} | \phi_p\phi_q\phi_r \rangle \left\langle \phi_p\phi_q\phi_r \left| \frac{1}{r_{13}} \right| \phi_i\phi_j\phi_k \right\rangle \\ &= \sum_{pqr}^{\infty} (\delta_{kr} \langle \phi_i\phi_j | r_{12} | \phi_p\phi_q \rangle) \left(\delta_{qj} \left\langle \phi_p\phi_r \left| \frac{1}{r_{13}} \right| \phi_i\phi_k \right\rangle \right) \\ &= \sum_{p}^{\infty} \langle \phi_i\phi_j | r_{12} | \phi_p\phi_j \rangle \left\langle \phi_p\phi_k \left| \frac{1}{r_{13}} \right| \phi_i\phi_k \right\rangle \end{aligned} \tag{4.95}$$

The first reduction occurs since the r_{12} and r_{13}^{-1} operators only involve two electron coordinates; the second reduction is due to the two delta functions. Three- and four-electron integrals can therefore be written as a sum over products of integrals involving only two electron coordinates. In a finite basis set, the resolution is not exact and the identities in Equation (4.95) become approximations. The beauty of the R12 methods is that this error can be controlled, albeit at the price of calculating and handling a significantly larger number (and different types) of two-electron integrals. The original R12 method included a linear r_{12} term, which is the correct behavior in the limit of a vanishing small r_{12}, but physically incorrect in the large r_{12} limit, since the Coulomb hole must integrate to zero. A better representation of the Coulomb hole can be obtained by replacing r_{12} with a function of r_{12} such as in the equation below, and these methods are denoted F12.[65] Note that the lowest-order Taylor expansion of $f(r_{12})$ is r_{12}:

$$f(r_{12}) = \gamma^{-1}(1 - e^{-\gamma r_{12}}) \tag{4.96}$$

The γ factor can be considered as a length scale parameter for the electron correlation, and depends on the actual system and electron pairs being correlated, but only weakly. A value of 1 Bohr^{-1} has been found to be close to optimum for a variety of systems and is often used in actual calculations. For practical applications the exponential function in Equation (4.96) is often replaced by a fixed finite expansion in Gaussian functions, analogous to the fitting of Slater-type orbitals by Gaussian-type orbitals (Section 5.4.1), since this facilitates the evaluation of the required integrals:

$$e^{-\gamma r_{12}} \simeq \sum_{i}^{n} c_i e^{-\alpha_i r_{12}^2} \tag{4.97}$$

Typically 3–6 Gaussian functions are employed in the expansion.

The above provides the gist of F12 methods, but the actual working equations are quite technical with a number of variations related to the projection on to occupied and virtual spaces.[65]

The significance of F12/R12 methods is that the energy error in terms of angular momentum of the basis set now behaves approximately as $(L + 1)^{-7}$, which is a significant improvement over standard methods. It should be noted that in the limit of a complete basis set the MP2-F12, for example, will give the same result as a traditional MP2 calculation, that is the F12 approach speeds up the basis set convergence, but does not change the fundamental characteristics of the MP2 method. The drawback is that F12 methods require a number of additional types of electron integrals as well as basis sets capable of controlling the errors due to the resolution of identity approximation. The explicit inclusion of the short-range electron correlation cusp condition also changes the orbital basis set requirement since the faster convergence means that fewer high angular momentum functions are required and construction of modified cc-pVXZ basis sets, denoted cc-pVXZ-F12, have been developed to provide a faster basis set convergence (Section 5.4.7). The auxiliary basis set required for the resolution of identity requires saturation with basis functions having angular momentum up to $\max(L_{orb}, 3L_{occ})+1$, where L_{orb} is the maximum angular momentum functions in the orbital basis set and L_{occ} is the maximum angular momentum function occupied in the atom(s). In the original method, the basis set for the resolution of the identity was the same as for expanding the orbitals and therefore needed to be large for the resolution of identity to be reasonably fulfilled.[66] Subsequently, an auxiliary basis set was used for the resolution of identity,[67] and the most recent improvement is to employ an auxiliary basis set composed of the orbital basis set and a *Complementary Auxiliary Basis Set* (CABS), where only the latter is used for the projection operators involved in the method.[68]

4.12 Techniques for Improving the Computational Efficiency

Performing an HF calculation can be divided into the following fundamental steps:

1. Calculate the one-electron integrals in a set of basis functions (atomic orbitals).
2. Calculate the two-electron integrals in a set of basis functions.
3. Contract the two-electron integrals with the density matrix and add the one-electron integrals to form the Fock matrix.
4. Form a new density matrix, either by diagonalization of the Fock matrix or by minimizing an energy function.

In a straightforward implementation of this scheme, each of the above four steps scales with the system size as N^2, N^4, N^4 and N^3. The calculation of the electron correlation energy can be formulated as involving a transformation of the two-electron integrals from atomic to molecular orbitals, which scales with the system size as N^5. Calculating the electron correlation energy using these integrals may involve an effort that scales with the system size as $N^{5\text{-}8}$ or higher. The scaling behavior, however, only indicates *how* the computational time increases with system size, not how *much* computational time each step requires. The latter depends on the *prefactor*, that is an N^5 step with a small prefactor may require less time than an N^4 step with a large prefactor for calculations that are actually feasible within a given computational time. Calculating the MP2 energy (an N^5 process), for example, often requires less computational time than calculating the HF energy (an N^4 process) for systems that can be calculated within a reasonable computational time. A number of technical techniques have been developed over the years that seek to reduce the computational time for a given type of calculation by reducing either the prefactor or the scaling, or both. The following sections give an overview of some of these techniques.

4.12.1 Direct Methods

Conventional HF methods rely on storing the two-electron integrals over atomic orbitals on disk and reading them in each SCF iteration, while direct methods generate the integrals as they are needed (Section 3.8.5) and then discarded. Since CPU speeds are much higher than data transfer to/from disk, and with the implementation of massive parallel algorithms for calculating integrals, modern electronic structure calculations are normally performed by integral direct algorithms. This is an easy change in algorithm since the HF energy is expressed directly in terms of AO integrals. Although this formally requires generating all two-electron integrals in each SCF iteration, rather than only once in conventional methods, the use of integral screening methods[69] means that only the fraction of integrals that makes a significant contribution in a given iteration is required, and this fraction decreases as the density matrix converges (Section 3.8.5). A direct SCF calculation usually requires more CPU time than a conventional one, but this is irrelevant, because a direct calculation will (usually) require less wall time than a conventional one. In a conventional disk-based calculation, the CPU will be idle a large percentage of the time while waiting for data transfer to/from the disk.

Methods involving electron correlation require matrix elements between Slater determinants, which can be expressed in terms of integrals over MOs (Equations (4.9) and (4.10)). Conventional methods for the integral transformation (Section 4.2.1) read the AOs, perform the multiplications with the MO coefficients (Equation (4.15)) and write the MO integrals to disk. These can then be read and used in the correlation treatment. Although the number of MO integrals typically is somewhat smaller than the number of AO integrals (e.g. MO integrals involving four virtual orbitals may not be needed), the disk space requirements are still significant if more than a few hundred basis functions are used. To eliminate the disk space requirements and remove the inefficient data transfer step for reading/writing to disk, it is desirable also to have direct algorithms for electron correlation methods. The need for integrals over MOs instead of AOs, however, makes the development of direct methods in electron correlation somewhat more complicated than at the HF level.

Consider, for example, the MP2 energy expression given in[70]

$$E(MP2) = \sum_{i<j}^{\text{occ}} \sum_{a<b}^{\text{vir}} \frac{(\langle \phi_i \phi_j | \phi_a \phi_b \rangle - \langle \phi_i \phi_j | \phi_b \phi_a \rangle)^2}{\varepsilon_i + \varepsilon_j - \varepsilon_a - \varepsilon_b} \tag{4.98}$$

The MO integrals are given in

$$\langle \phi_i \phi_j | \phi_k \phi_l \rangle = \sum_{\alpha}^{M_{\text{basis}}} \sum_{\beta}^{M_{\text{basis}}} \sum_{\gamma}^{M_{\text{basis}}} \sum_{\delta}^{M_{\text{basis}}} c_{\alpha i} c_{\beta j} c_{\gamma k} c_{\delta l} \langle \chi_\alpha \chi_\beta | \chi_\gamma \chi_\delta \rangle \tag{4.99}$$

Since each MO integral in principle contains contributions from *all* the AO integrals, a straightforward calculation of an MO integral each time it is needed will involve a generation of all the AO integrals. In other words, it would be necessary to recalculate the AO integrals $\sim O^2V^2$ times (O and V being the number of occupied and virtual orbitals, respectively), compared with the 15–20 times in an SCF calculation. The MP2 method would therefore change from being an M_{basis}^5 to an M_{basis}^8 method, which clearly is an unacceptably large penalty for a direct method.

The M_{basis}^8 dependence is a consequence of performing the four index transformation with all four indices at once. As shown in Section 4.2.1, it is advantageous to perform the transformation one index at a time:

$$
\begin{aligned}
\langle \phi_i\phi_j|\phi_a\phi_b\rangle &= \sum_{\delta}^{M_{\text{basis}}} c_{\delta b}\langle \phi_i\phi_j|\phi_a\chi_\delta\rangle \\
\langle \phi_i\phi_j|\phi_a\chi_\delta\rangle &= \sum_{\gamma}^{M_{\text{basis}}} c_{\gamma a}\langle \phi_i\phi_j|\chi_\gamma\chi_\delta\rangle \\
\langle \phi_i\phi_j|\chi_\gamma\chi_\delta\rangle &= \sum_{\beta}^{M_{\text{basis}}} c_{\beta j}\langle \phi_i\chi_\beta|\chi_\gamma\chi_\delta\rangle \\
\langle \phi_i\chi_\beta|\chi_\gamma\chi_\delta\rangle &= \sum_{\alpha}^{M_{\text{basis}}} c_{\alpha i}\langle \chi_\alpha\chi_\beta|\chi_\gamma\chi_\delta\rangle
\end{aligned}
\tag{4.100}
$$

By choosing the correct order of the transformation the scaling can be reduced considerably. In Equation (4.100) the indices corresponding to the occupied orbitals may be transformed before the virtuals. There are of the order of M_{basis}^4 of the AO integrals, $\langle \chi_\alpha\chi_\beta|\chi_\gamma\chi_\delta\rangle$, but only OM_{basis}^3 of the quarter transformed integrals, $\langle \phi_i\chi_\beta|\chi_\gamma\chi_\delta\rangle$. Instead of storing and reading the AO integrals from the SCF step, they can be recalculated in the transformation step, reducing the storage from M_{basis}^4 to OM_{basis}^3. The subsequent quarter transformations require less storage; that is the next transformation with an occupied index reduces the number of integrals to $O^2M_{\text{basis}}^2$, the third to O^2VM_{basis} and the last to O^2V^2. Since the MP2 energy can be written as a sum of contributions from each occupied orbital, the occupied orbitals can be treated one at a time, that is first sum all contributions of $\langle \phi_1\chi_\beta|\chi_\gamma\chi_\delta\rangle$, then $\langle \phi_2\chi_\beta|\chi_\gamma\chi_\delta\rangle$, etc. This reduces the necessary storage to only order M_{basis}^3. It may be further reduced to OVM_{basis} by proper scheduling of the evaluation order of the remaining three indices. The OVM_{basis} number of integrals is much less than the original M_{basis}^4, and will in many cases fit into memory. The net result is that disk storage is effectively eliminated, or at least greatly reduced. If only one occupied orbital is treated at a time, O integral evaluations are required, but the more memory that is available, the more occupied orbitals can be treated in a single sweep, decreasing the number of integral evaluations.

The whole MP2 expression in Equation (4.98) can be formulated directly in the AO basis by rewriting the inverse orbital difference by means of a Laplace transformation and evaluating the integral as a finite sum with suitable weighting factors:[71]

$$
\frac{1}{\varepsilon_i+\varepsilon_j-\varepsilon_a-\varepsilon_b} = \frac{1}{\Delta_{ijab}} = \int_0^\infty \mathrm{e}^{-\Delta_{ijab}t}\mathrm{d}t \simeq \sum_{\alpha=1}^{\tau} w_\alpha \mathrm{e}^{-\Delta_{ijab}t_\alpha}
\tag{4.101}
$$

The exponential orbital energy differences can be split into components and incorporated directly into the two-electron integrals. The MP2 energy expression can thus be written as a sum of exponentially energy weighted two-electron integrals in the AO basis, and 5–10 points in the numerical integration are usually sufficient for attaining a useful accuracy. The key point is that this formulation in localized basis functions, rather than delocalized canonical molecular orbitals, allows the use of integral screening techniques that leads to a linear scaling in the large-system limit.[72]

These are examples of how direct algorithms may be formulated for methods involving electron correlation. They illustrate that it is not as straightforward to apply direct methods at the correlated level as at the SCF level. However, the steady increase in CPU performance, and especially the evolution of multiprocessor machines, favors direct (and semidirect where some intermediate results are stored on disk) algorithms.[73]

4.12.2 Localized Orbital Methods

Ab initio calculations involving electron correlation traditionally employ a set of canonical HF orbitals and this leads to a computational effort that increases as a rather high power of the system size, that is N^5–N^8. Considering that the fundamental physical force is only between pairs of particles, this scaling is "non-physical". One of the reasons for the high scaling is the fact that canonical orbitals are delocalized over the whole molecule, that is essentially all orbitals make a (small) contribution to the wave function for a specific part of the molecule. This suggests that a set of localized orbitals (Section 10.4) may be a better starting point, since only a few orbitals would then contribute the large majority at a given point and the remaining contributions could simply be neglected. Alternatively, the problem may be formulated directly in the atomic orbital basis, since the basis functions are naturally localized on a single atom. Such local MP2 and local CC methods are major focus areas and they hold the promise of linear scaling, that is the computational effort only increases as N^1.[74] These methods are somewhat more complicated to formulate as the Fock matrix is only diagonal in the canonical orbitals. Methods based on localized orbitals will display a near-linear scaling with problem size in the large-scale limit, and the cross-over point in terms of computational resources for entering the large-scale regime depends on the specific formulation, but will typically be in the few tenths to a few hundred atoms region.

The use of localized orbitals will only lead to a computational saving if it is combined with criteria for neglecting a (large) fraction of the terms. Many localized orbital schemes employ one or more predetermined cutoff distances, or equivalent orbital or energy thresholds, which imply the risk of producing non-continuous energy surfaces, for example during a geometry optimization a given atomic distance or orbital contribution may cross one of the threshold values, and some of the correlation contributions suddenly change between zero and a finite value. The energy will thus experience a discontinuity upon increasing an interatomic distance by an infinitesimal amount, leading to corresponding discontinuities in the gradient and second derivatives. This in turn may lead to problems in the geometry optimization and/or vibrational frequencies. These errors can of course be controlled by choosing threshold values such that the discontinuities are well below chemical significance, but tight threshold values are counterproductive from a computational efficiency point of view and a compromise is necessary.[75] Another aspect is that the total energy is often only accurate to a few millihartrees (~1 kJ/mol) and different implementations may thus give slightly different results. Finally, it should be noted that there are certain systems, typically having stretched bonds (TS) or being aromatic, where it may be difficult to obtain localized orbitals or where more than one set of localized orbitals are possible, and a small geometry change may in such cases suddenly switch from one localization to another.

The *Divide–Expand–Consolidate* (DEC) approach employs a dynamical cutoff parameter based on an absolute energy criterion.[76] The total correlation energy can be written as a sum of terms for each (localized) occupied orbital correlated in a set of (localized) virtual orbitals. The size of the virtual space is gradually enlarged until the orbital correlation energy is converged to the chosen threshold. The error in the total correlation energy is thus completely determined by a single accuracy parameter, and this can be chosen as required by the application. A small threshold value necessarily implies more virtual orbitals for each occupied orbital, and thus an increased computational time. The DEC method can be employed for both MP2 and coupled cluster methods, and it is trivially parallel for each occupied orbital and thus well suited for massive parallelization. The DEC approach is independent (to within the specified energy threshold) on the specific form of the localized orbitals, but the computational efficiency depends strongly on having molecular orbitals that are as spatially localized as possible, and this was the main incitement for developing the modifying Forster–Boys localization methods discussed in Section 9.4.

4.12.3 Fragment-Based Methods

Localized orbital methods can be considered as a method for dividing the whole system into fragments that can be treated separately, where the fragmentation is determined by the orbital localization criterion. Alternatively, the fragmentation can be done *a priori* by a rule-based method.[77,78] Several such methods are available with slightly different techniques, such as *Fragment Molecular Orbital* (FMO)[79] and *Divide-and-Conquer* methods.[80] In contrast to localized orbital approaches, these methods rely on breaking the system into molecular fragments that can be treated independently and properties of the whole system are written as a sum of terms for each fragment.

A typical example is a protein, where the FMO approach defines fragments as individual amino acids. The bonds being broken are replaced by frozen localized orbitals, and the wave function for each fragment is optimized in the presence of the electrostatic potential for all the other fragments. This necessitates an iterative procedure over fragments, as changes in the wave function for one fragment changes the electrostatic potential for all other fragments, which then needs to be re-optimized, etc. When this procedure has converged, the interfragment polarization is accounted for. When properties for the whole system are taken as the sum of fragment terms, this is denoted as the FMO1 approximation. A limitation of FMO1 is that each fragment must necessarily have an integer number of electrons, which implies that charge transfer between fragments is not accounted for. This can be partly remedied by performing explicit calculations for all pairs of fragments, again in a self-consistent way, leading to the FMO2 approximation, where correction terms based on the dimer calculations are added to the estimated property for the whole system. This procedure can continue to the FMO3 approximation, where explicit calculations are performed for all timer fragment combinations, etc., but the number of dimers and trimers rises rapidly, and the size of each combined fragment also increases. The total computational time is thus a balance between defining small fragments requiring perhaps up to FMO4 and large(r) fragments requiring perhaps only FMO2. If care is taken to divide the system into fragments where the charge transfer between fragments is small, the FMO2 approximation often gives acceptable accuracy. For proteins, for example, the fragmentation should be done between the carbonyl carbon and C_α atoms in the backbone.

The advantage of the FMO approach is that each fragment calculation is trivially parallel, and thus scalable to a large number of processors. It can furthermore readily be extended to, for example, DFT or MP2 methods, as well as calculating molecular properties, like molecular gradients, excitation energies or nuclear magnetic shielding constants.

4.12.4 Tensor Decomposition Methods

The tensor decomposition ideas used for basis function integrals (Section 3.8.7) can similarly be used for decomposing the molecular orbital integrals used in the calculation of electron correlation, as well as the coefficients (e.g. coupled cluster amplitudes).[81] The general idea is to compress the information contained in the large number of two-electron integrals and the large number of coefficients for the Slater determinants into more compact representations. Two-electron integrals are four-index quantities and can thus be considered as elements in a four-dimensional tensor. Coupled cluster double amplitudes can similarly be considered as elements in a four-dimensional tensor, while, for example, triple amplitudes can be considered as elements in a six-dimensional tensor. Tensor decomposition methods attempt to approximate these multidimensional tensors into linear combinations of products of lower-order tensors, which in the limiting case are just one-dimensional vectors (Section 17.6.3). Resolution of identity and Cholesky decomposition methods can be viewed as special cases of such tensor decompositions.[82] Since electron correlation methods have a tensor contraction of expansion coefficients with two-electron integrals at the core, this can potentially lead to a large reduction in computational cost, but it must be balanced by the cost of performing the actual tensor decompositions. Methods such as *Pair Natural Orbital* (PNO), *Orbital Specific Virtual* (OSV) and *Optimized Virtual Orbital Space* (OVOS) can be considered as belonging to this family of methods, where the actual tensor decomposition is bypassed by constructing reduced orbital spaces motivated by physical arguments.

4.13 Summary of Electron Correlation Methods

The only generally applicable methods are CISD, MP2, MP3, MP4, CCSD and CCSD(T). CISD is variational, but not size extensive, while MP and CC methods are non-variational but size extensive. CISD and MP are in principle non-iterative methods, although the matrix diagonalization involved in CISD is usually so large that it has to be done iteratively. Solution of the coupled cluster equations must be done by an iterative technique since the parameters enter in a non-linear fashion. In terms of the most expensive step in each of the methods, they may be classified according to how they formally scale in the large system limit, as shown in Table 4.7.

We have so far been careful to use the wording "formal scaling". As already discussed, HF is formally an M^4 method but in practice the scaling may be reduced all the way down to M^1. Similarly, MP2 is formally an M^5 method. However, an MP2 calculation consists of three main parts: the HF calculation, the AO to MO integral transformation and the MP2 energy calculation. Only the second part has a formal scaling of M^5; the others are (formal) M^4 steps. In the large-system limit, the transformation required for the MP2 procedure *will* become the most expensive step, but in practice, where calculations may be restricted to a few hundred or thousand basis functions, it is often observed that the MP2 step takes less time than the HF step. The formal scaling only indicates what the rate-limiting

Table 4.7 Limiting scaling in terms of basis set size M for various methods.

Scaling	CI methods	MP methods	CC methods (iterative)
M^5	CIS	MP2	CC2
M^6	CISD	MP3	CCSD
M^7		MP4	CC3, CCSD(T)
M^8	CISDT	MP5	CCSDT
M^9		MP6	
M^{10}	CISDTQ	MP7	CCSDTQ

step will be in the large-system limit. Whether this limit actually is reached in practical calculations is another matter.

The lower value of M^5 scaling for methods involving electron correlation arises from the transformation of the two-electron integrals from the AO to MO basis, but if the transformation is carried out with one of the indices belonging to an occupied MO first, the scaling is actually the number of occupied orbitals (O) times M^4. If we consider making the system larger by doubling the fundamental unit (e.g. calculations on a series of increasingly larger water clusters), keeping the basis set per atom constant, O scales linearly with M, and we arrive at the M^5 scaling. This assumption (increasing system size) is the basis for Table 4.7. More often, however, a series of calculations is performed on the same system with increasingly larger basis sets. In this case, the number of electrons (occupied orbitals) is constant and the scaling is M^4. Many of the commonly employed methods for electron correlation (including, for example, MP2, MP3, MP4, CISD, CCSD and CCSD(T)) scale in fact as M^4 when the number of occupied orbitals is constant.

In terms of accuracy with a medium-sized basis set the following order is often observed:

$$\text{HF} \ll \text{MP2} < \text{CISD} < \text{MP4(SDQ)} \sim \text{CCSD} < \text{MP4} < \text{CCSD(T)} \tag{4.102}$$

All of these are single-determinant-based methods. Multireference methods cannot easily be classified as the quality of the results depends heavily on the size of the reference. A two-configurational reference is only a slight improvement over HF, but including all configurations generates a full CI. The ordering above is only valid when the HF reference is a "good" zeroth-order description of the system. The more multireference character in the wave function, the better the "infinite"-order coupled cluster performs relative to perturbation methods.

MP3 has not been included in the above comparison. As already mentioned, MP3 results are often inferior to those at MP2. In fact, MP2 often gives surprisingly good results, especially if large basis sets are used.[83] Furthermore, it should be kept in mind that the MP perturbation series in many cases may actually be divergent, although corrections carried out to low order (i.e. 2–4) rarely display excessive oscillations.

HF results should by modern standards be considered as model calculations, like semi-empirical methods such as PM6/7 (Chapter 7). Minimal basis HF calculations often give results that are worse than PM6/7, but at a computational cost several orders of magnitude larger. Medium and large basis set HF calculations usually do not give absolute results that are particularly close to experimental values, but since the errors to a certain degree are systematic (such as all vibrational frequencies being overestimated by ~10%), they can be used with more or less "empirical" corrections to treat systems for which correlated calculations are not possible. The distinct advantage of *ab initio* methods is the ability to treat all systems at an equal level of accuracy, independent of whether experimental data exist or not. A detailed assessment of the level of accuracy that can be expected at a given level of theory is difficult to establish as it is heavily dependent on the quality of the basis set. Given a sufficiently large basis set, however, the CCSD(T) method is able to meet the goal of an accuracy of ~4 kJ/mol (~1 kcal/mol) for most systems. Even with less complete methods (such as MP4) and medium-size basis sets such as DZP or TZP, it is often possible to get accuracies of the order of a few tens of kJ/mol.

The uses of CI methods has been declining at the expense of MP and especially CC methods. It is now recognized that size extensivity is important for obtaining accurate results. Excited states, however, are somewhat difficult to treat by perturbation or coupled cluster methods, and CI- or MCSCF-based methods have been the preferred methods here. Linear response methods (Section 11.10) have been developed for coupled cluster wave functions, and this allows calculation of excited state properties.

Finally, a few words on the size of systems that can be treated. The limiting parameters will again be taken as the number of basis functions although, as noted above, a more detailed breakdown in terms of occupied and virtual MOs can be done. Note also that a given limit in terms of basis functions may translate either into a large molecular system with a small basis per atom or a small molecular system with a very large basis set on each atom. The ordering in Equation (4.102) suggests three levels of electron correlation: none (HF), MP2 or extended (CCSD(T)). HF methods are in general possible with up to ~several thousand basis functions and MP2 is often less expensive than the HF calculation, even for thousands of functions. Advanced correlation methods are usually limited to ~thousand basis functions. With a DZP basis set these values translate into roughly 300 and 50 CH_2 fragments, respectively, while the values drop to ~100 and ~20 with a TZP basis set. These limits hold for just calculating the energy at a single geometry. If more advanced features are desired, such as optimizing the geometry or calculating frequencies, the limits drop to roughly half of the above.

With the continuing advances of computer hardware and more efficient algorithms, these limits are gradually being shifted upwards. Owing to the rather steep scaling with system size, however, they will (barring a fundamental breakthrough) give a rough idea of the size of systems that can be handled also in the future. The performance of computer hardware continues to improve by a factor of two in a timespan of about 18 months. In other words, a factor of 10 in terms of performance for the same price is gained roughly every 5 years. Owing to the scaling between 4 and 7 of the various methods, however, a factor of 10 increase in performance only translates into an increase of system size of 1.7 (M^4 scaling) or 1.4 (M^7 scaling). Linear scaling methods will of course benefit fully from increased computational performance, but these methods often come with a significantly larger prefactor in terms of computational effort.

4.14 Excited States

The description of HF and correlated methods in the previous chapter/sections has focused on the electronic ground state. In some cases it is also of interest to consider electronically excited states. It is useful to distinguish between two cases, depending on whether the excited state has the *same* or a *different* symmetry than the lower state(s), where symmetry includes both the spatial (irreducible representation in the point group) and spin (multiplicity) part of the wave function. The different symmetry case is easy to handle, as the lowest energy state of a given symmetry can be handled completely analogously to the ground state. An HF wave function may be obtained by a proper specification of the occupied orbitals, and the resulting wave function can be improved by adding electron correlation by, for example, CI, MP or CC methods. The only caveat may be that the state is an open shell, which often requires a (small) MCSCF wave function for an adequate zeroth-order description.

Excited states having lower energy solutions of the same symmetry are somewhat more difficult to treat. It is difficult to generate an HF-type wave function for such states, as the variational optimization tends to collapse to the lowest energy solution of the given symmetry. Special techniques where the orbital occupancy in each SCF iteration is chosen by a maximum orbital overlap criterion, rather than the aufbau principle, can be used to converge a single determinant wave function to excited state solutions.[84] It should be noted that excited state wave functions obtained in this fashion are not orthogonal to the ground state. The same technique can be employed for DFT methods, where electron correlation is included implicitly by the exchange-correlation functional. Excitation energies calculated by subtraction of two total energies are often referred to as ΔSCF methods.

The lack of a proper HF reference wave function means that perturbation and coupled cluster methods are not well suited for calculating excited state wave functions, although excited state *properties* (e.g. excitation energies) may be calculated *directly* with response methods (Section 11.10). It is relatively easy to generate higher-energy states by CI methods as this simply corresponds to using the $(n + 1)$th eigenvalue from the diagonalization of the CI matrix as a description of the nth excited state (the second root is the first excited state, etc.). Such a CI procedure will normally employ a set of HF orbitals from a calculation on the lowest energy state, and the CI procedure is therefore biased against the excited states.

The language for discussing excited states is complicated by the convention of also using the word excited for Slater determinants having different orbital occupancy than the reference HF determinant. An excited state (a wave function corresponding to a higher energy solution of the Schrödinger equation) can be written as a linear combination of excited (relative to the HF reference) Slater determinants. An excited state with large coefficients for S-type Slater determinants, corresponding to "exciting" a single electron from the HF reference, is denoted as a single-electron excitation, or singly excited state for short. Similarly, an excited state with large coefficients for D-type Slater determinants is denoted as a double-electron excitation, or doubly excited state for short. Note the wording "large coefficients" as both singly and doubly excited states may contain Slater determinants of S-, D-, T-, Q-, etc., types for describing orbital relaxation and electron correlation. A singly excited state may thus contain contributions from singly, doubly, triply, etc., excited Slater determinants. The excited state may in some cases have large contributions from both S- and D-type Slater determinants, which makes it ambiguous to classify it as either singly or doubly excited. The next section describes methods for quantifying the number of electrons involved in a given excitation, but for providing the basic understanding, we will in the following assume that excitations are pure single- or double-electron excitations.

The simplest description of an excited state is the orbital picture where one electron has been moved from an occupied to an unoccupied orbital, that is an S-type Slater determinant as illustrated in Figure 4.2. The lowest level of theory for a qualitative description of excited states is therefore CI including only the singly excited determinants, denoted CIS. CIS gives wave functions of roughly HF quality for singly excited states, since no orbital optimization is involved and no electron correlation is included. Time-dependent HF or DFT methods within the adiabatic approximation (Section 6.9.1) provide a similar quality description for such singly excited states. Doubly excited states are significantly more difficult to describe, and methods like CIS or standard TD-HF/DFT completely miss these states.

The ability of different methods for describing excited states can be illustrated for singlet methylene (CH_2) with a cc-pVDZ basis set. The valence orbitals and orbital occupancy of the dominating Slater determinant for the six lowest singlet excited states are shown in Figure 4.14, while the excitation energies calculated by a sequence of increasingly sophisticated CI wave functions are shown in Figure 4.15. The HOMO in the reference wave function has two electrons in a σ-type lone pair, while the LUMO is a π-type orbital. The lowest excited state S1 corresponds to a single-electron HOMO-LUMO excitation, but with a sufficient flexible CI wave function (CISDTQ) the excited state D1 corresponding to transferring both HOMO electrons to the LUMO is the second lowest in terms of energy. Such doubly excited states only appear when a method including doubly excited Slater determinants, like CISD, is employed, but a balanced description of the various states requires quite sophisticated methods. Remembering that doubly excited Slater determinants account for the majority of electron correlations, it is clear that CIS is equally poor (no electron correlation) for both the ground and (singly) excited states. CISD provides electron correlation for the ground state, but essentially none for the excited states, and thus leads to a large systematic overestimation of the excitation energies. CISDT has T-type Slater determinants that are doubly excited relative to the singly excited states and thus provide a balanced description of the ground and excited states that are dominated by

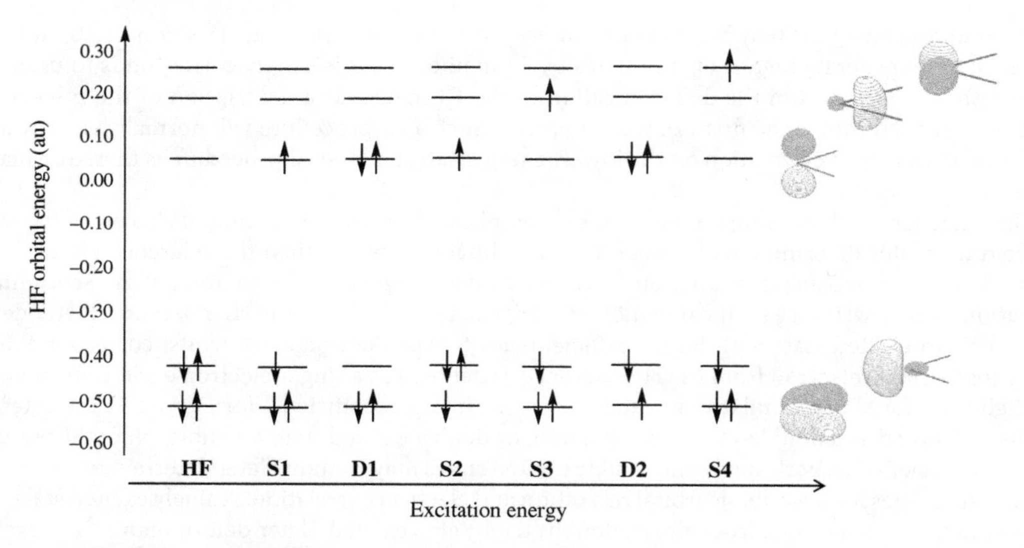

Figure 4.14 HF orbital energies and occupancies for the first six excited states in methylene (CH_2) with a cc-pVDZ basis set.

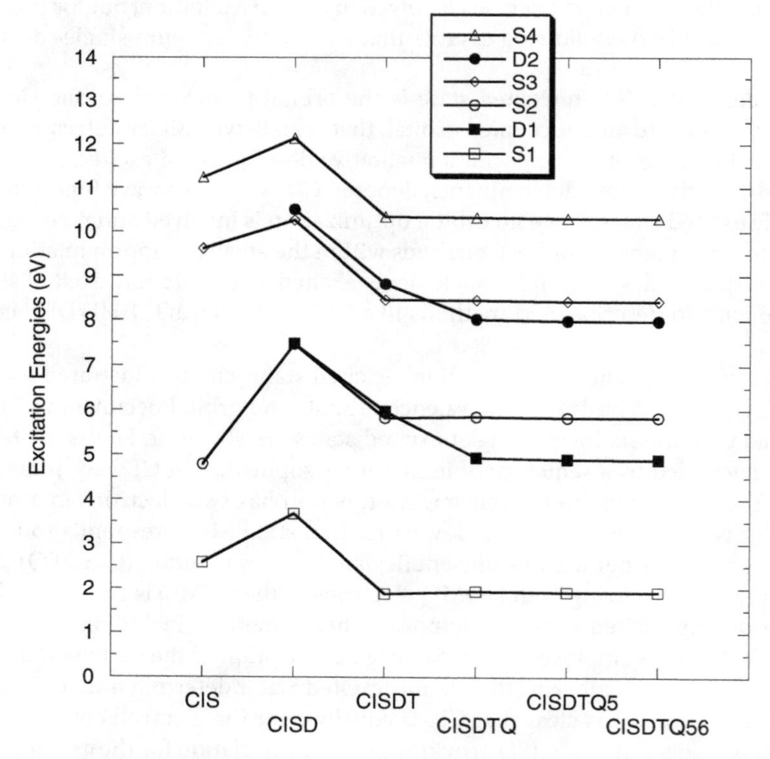

Figure 4.15 Energies for the first six singlet excited states in methylene (CH_2) with the cc-pVDZ basis set and increasing CI excitations levels. Open and closed symbols indicate excited states dominated by single (S)- and double (D)-electron transfer.

single-electron transfer (S1–S4). CISDTQ is required for including electron correlation for states dominated by double-electron transfer (D1, D2), since Q-type Slater determinants are necessary for correlating the electrons in doubly excited states. The results at the CISDTQ56 (full CI with the carbon 1s core-orbital frozen) level show that the description of the six lowest excited states is converged at the CISDTQ level.

Figure 4.15 shows that CISDTQ is required to give a balanced description of singly and doubly excited states, but this level of theory is feasible only for small systems. Singly excited states are relatively easy to describe and many single-determinant-based methods like CIS and TDHF methods often provide good results; however, it should be recognized that they fail to describe electron correlation effects. This usually leads to a systematic overestimation of the excitation energy in the 0.5–2 eV range for excited states of the same spin-multiplicity as the ground state (usually a singlet state), but a systematic underestimation of the excitation energy for states of higher spin-multiplicity (e.g. triplet states for a singlet ground state). The CIS(D) method[85] includes electron correlation for the CIS excited states by an MP2 like expression, which normally improves the excitation energies but, in analogy with MP2 for ground states, often overestimates the electron correlation effect. TDDFT includes electron correlation effects in the exchange-correlation term and often improves the performance to deviations in the ~0.4 eV range, but the performance depends on the exact exchange-correlation functional employed.

Coupled cluster methods are more effective than CI for including electrons and can also be used for calculating excited states by response methods (Section 11.10). The CCS model includes only S-type determinants and is identical to CIS. CC2 is an approximation to CCSD where the D-type excitations are included only to lowest order in the perturbation series. Including them non-iteratively leads to the CC(2) model, which is identical to CIS(D). CC3 is an approximation to CCSDT where the T-type excitations are included only to lowest order in the perturbation series. CC(3) (also denoted CCSDR(3)) is in analogy with CC(2), a model where the T-type excitations are included non-iteratively, and thus is similar to CCSD(T) for ground state energies. The sequence CCS, CC(2), CC2, CCSD, CC(3), CC3, CCSDT is thus expected to provide increasingly better results, but it should be recognized that they are all based on a single determinant HF reference wave function. In analogy with the CI methods described above, this means that the convergence with respect to increasing amounts of electron correlation is significantly slower for doubly excited states than for a singly excited state, and doubly excited states are missing in the CCS, CC(2) and CC2 models. This is illustrated in Figure 4.16 for the same excited states for CH_2 as in Figure 4.15.

A common caveat of CI and CC methods is the use of HF optimized orbitals. Excited states furthermore often require a linear combination of a (small) number of configurations for a qualitative description, and this implies the use of MCSCF methods. MCSCF employs an orbital optimization, which can be either *state-specific*, where the orbitals are optimized for each particular state, or *state-averaged*, where they are optimized for a suitable average of the desired states. The latter has the advantage that the different states are mutually orthogonal, which is important if, for example, transition moments are desired. The weight factors for the state averaging, however, are free variables that must be decided by the user. It should be noted that excited state MCSCF wave functions often are difficult to converge numerically, and second-order optimization techniques are therefore almost mandatory. A typical MCSCF wave function is of the CASSCF type, where all excitations are allowed within a restricted orbital space, and if the latter is chosen properly, it may provide a balanced zeroth-order description of both singly and doubly excited states. In order to obtain accurate excitation energies it is normally necessary also to include dynamical correlation, for example by a multireference second-order perturbation method approach. Figure 4.16 shows the results from CASSCF and CAS-MR2 calculations using a 4-electron-5-orbital active space, corresponding to the set of orbitals shown

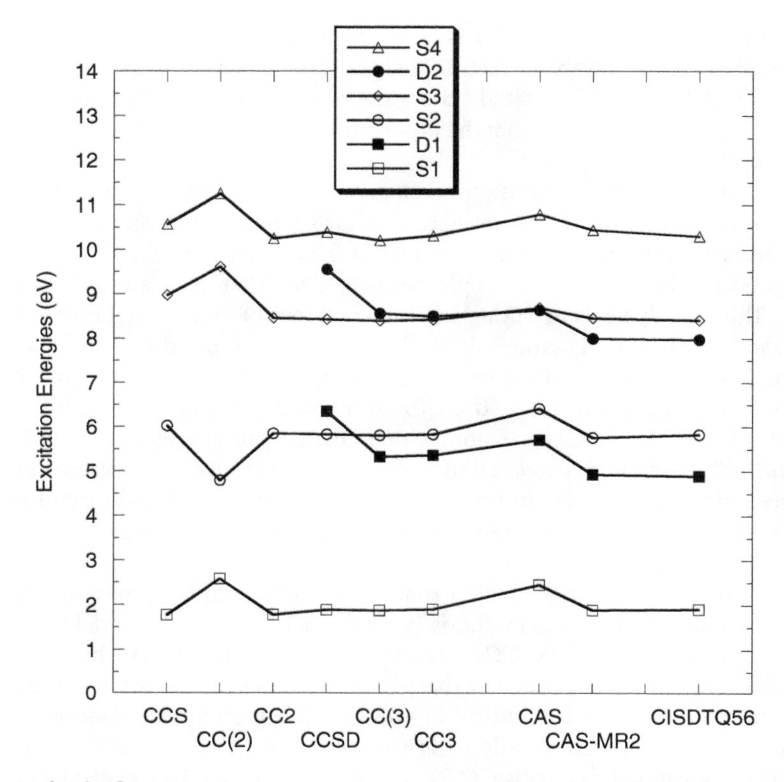

Figure 4.16 Energies for the first six singlet excited states in methylene (CH_2) with the cc-pVDZ basis set and increasing CC excitation levels, as well as CASSCF and CAS-MR2 levels. Open and closed symbols indicate excited states dominated by single (S)- and double (D)-electron transfer.

in Figure 4.14. The [4,5]-CASSCF wave function provides a good balance of the various excited states and is capable of predicting the correct order; including dynamical correlation by second-order perturbation theory in the full orbital space gives (not surprisingly) essentially complete agreement with the reference CISDTQ56 results.

The six lowest singlet excited states in the CH_2 example can quite clearly be characterized as four singly and two doubly excited states. Excited states for larger systems, however, often have characteristics intermediate between single- and double-electron excitation.[86] The use of methods that only account for singly excited states (TD-HF/DFT, CIS, CIS(D), CC2) will lead to a systematic overestimation of the excitation energies, with the error increasing with the amount of doubly excited character. Truly doubly excited states will simply be missing. As illustrated in Figures 4.15 and 4.16, it is necessary to include up to Q-type Slater determinants in CI methods and up to T-type Slater determinants in CC response methods, in order to provide a balanced description of states with different degrees of singly and doubly excitation character, but this is often computationally infeasible. A CASSCF combined with inclusion of dynamical correlation is often the only computational viable approach, but selecting the active orbital space may be challenging for large systems.

Excited states involve electrons that are more loosely bound than in the ground state, and they therefore usually require basis sets with diffuse functions for a proper description. This is especially true for so-called *Rydberg* states, which may be considered as an electron orbiting a positively charged molecule. Such states resemble a hydrogenic atomic system, with the molecular cation playing the rule of the proton, and can be characterized as having s-, p-, d-, etc., character. Rather than using a regular basis set with diffuse functions on each nucleus, such Rydberg states can be modeled by having a single set of diffuse functions located at the molecular center of mass.[87]

A general feature is that a few of the lowest excited states typically are dominated by single excitations within the valence orbitals and can be reasonably described by simple approaches such as CIS or TDDFT, but it becomes progressively more difficult to describe higher-lying states. Doubly excited and Rydberg states require sophisticated wave functions and/or extended basis sets for an adequate description, and are often surprisingly low in energy. In the case of polyenes, for example, even the lowest excited state has a very large doubly excited character. Unfortunately it is not possible to detect these states by a computationally inexpensive screening method, like CIS or TDDFT, since they cannot provide even a rudimentary description. Calculating ten or more excited states by, for example, CIS and a medium-sized basis set is just a waste of computational resources, as only the few lowest states may reflect the real excited state spectrum.

4.14.1 Excited State Analysis

Excited states are often conceptually visualized as moving a single electron from an occupied to a virtual (HF or DFT) orbital, but this may be an inadequate description. Excited states correspond to higher-energy solutions of the full Hamiltonian and involve orbital relaxation relative to the ground state orbitals, and usually also need a description in terms of a linear combination of several (many) excited Slater determinants, which may be singly, doubly, triply, etc., excited relative to the ground state. The degree of singly and doubly excited character can be evaluated based on the weights of the S- and D-type Slater determinants in a CI approach or by the t_1 and t_2 amplitudes in CC methods. Even in the cases where the excited state can be described by only a few excited Slater determinants, it may be difficult to analyze since the canonical orbitals used in the construction of the Slater determinants are delocalized over the whole system. The orbital excitation picture, however, can be reintroduced by describing the *change* between the ground and excited states by a set of natural orbitals. The change can be defined either by a difference in the total electron density or from the transition matrix elements.[88]

The change in the total electron density matrix between the excited and reference states can be obtained by a simple subtraction to produce a density difference matrix $\mathbf{\Delta}$:[89]

$$\delta = \mathbf{D}_{ex} - \mathbf{D}_0 \tag{4.103}$$

Diagonalization of the $\mathbf{\Delta}$ matrix yields a diagonal matrix $\boldsymbol{\delta}$ containing the eigenvalues with the unitary transformation matrix $\mathbf{U}$ containing the eigenvectors:

$$\delta = \mathbf{U}\Delta\mathbf{U}^{-1} \tag{4.104}$$

The sum of all eigenvalues must be zero for an electronic transition since the number of electrons is conserved (an ionization or electron attachment process would have eigenvalue sums of −1 and +1, respectively). The sum of the positive (or negative) eigenvalues directly provides the number of electrons involved in the transition, that is if it can be described as a single electron transition or if it contains a substantial double excitation nature. The eigenvectors corresponding to negative eigenvalues ($\boldsymbol{\delta}_-$) can be back-transformed using the absolute $\boldsymbol{\delta}_-$ values to yield a *detachment density matrix* $\mathbf{D}_-$ and the eigenvectors corresponding to positive eigenvalues ($\boldsymbol{\delta}_+$) can be back-transformed to yield an *attachment density matrix* $\mathbf{D}_+$:

$$\begin{aligned} \mathbf{D}_- &= \mathbf{U}^{-1}|\boldsymbol{\delta}_-|\mathbf{U} \\ \mathbf{D}_+ &= \mathbf{U}^{-1}\boldsymbol{\delta}_+\mathbf{U} \end{aligned} \tag{4.105}$$

The $\mathbf{D}_{-/+}$ matrices correspond to hole and particle density matrices and can be diagonalized to yield natural orbitals and occupation numbers, which in a compact form describe the change in the excited state electron density relative to the ground state reference.

Instead of the density difference matrix, the one-electron transition density matrix $\mathbf{T}$ can be used for analysis.[90] The $\mathbf{T}$ matrix is defined as the outer product of the excited and reference states integrated over $N-1$ electron coordinates:

$$\mathbf{T} = N_{\text{elec}} \int |\Psi_{\text{ex}}\rangle\langle\Psi_0| \mathrm{d}r_2 \cdots \mathrm{d}r_N \tag{4.106}$$

For the simplest case of a single determinant reference wave function and an excited state described by a linear combination of singly excited Slater determinants (CISs), the T_{ia} matrix element (i denoting an occupied and a denoting a virtual orbital) is given as a product of the ϕ_i and ϕ_a orbitals multiplied by the coefficient c_{ia} in the Slater determinant expansion. The $\mathbf{T}$ matrix is rectangular with dimension $N_{\text{occ}} \times N_{\text{virt}}$ and can be analyzed by singular value decomposition (Section 17.6.3):

$$\mathbf{T} = \mathbf{U}\mathbf{\Lambda}\mathbf{V}^{\text{t}} \tag{4.107}$$

The $\mathbf{U}$ and $\mathbf{V}$ matrices contain the eigenvectors of the $\mathbf{T}\mathbf{T}^{\text{t}}$ and $\mathbf{T}^{\text{t}}\mathbf{T}$ matrices, respectively, and $\mathbf{\Lambda}$ contains the square root of the corresponding eigenvalues. The sum of diagonal elements in $\mathbf{\Lambda}$ provides a measure of how many electrons are involved in the transition. Transformation of the occupied and virtual orbitals by the $\mathbf{U}$ and $\mathbf{V}$ matrices, respectively, produces a set of *Natural Transition Orbitals* (*NTOs*):

$$\begin{aligned} \phi_i^{\text{NTO}} &= \phi_i \mathbf{U} \\ \phi_a^{\text{NTO}} &= \phi_a \mathbf{V} \end{aligned} \tag{4.108}$$

The set of ϕ_i^{NTO} describes the hole orbitals in the occupied space (removal of electrons) while the set of ϕ_a^{NTO} describes the particle orbitals in the virtual space (addition of electrons). These form particle–hole pairs of orbitals associated with a singular value in the $\mathbf{\Lambda}$ matrix, which measures the fraction of electrons transferred. The $\mathbf{\Lambda}$ matrix often has one value close to one, while the others are close to zero, and the ϕ_i^{NTO} and ϕ_a^{NTO} hole and particle natural transition orbitals associated with the dominating singular value in such cases form a compact description of the excited state corresponding to a single-electron excitation.

Excited states can also be characterized in terms of their *charge-transfer* (CT) character, that is to what extent the ground and excited states have differences in the spatial electron distribution over the molecular structure. In the simple picture of exciting an electron from the HOMO to the LUMO, a CT excitation corresponds to the situation where the HOMO and LUMO are localized on different parts of the molecule. A quantitative measure of the CT character can be defined in terms of an index varying between 0 and 1, where a value of 0 indicates a pure CT excitation. In the simple HOMO–LUMO picture, the index can be defined as the spatial overlap of the absolute values of the HOMO and LUMO. In the more general case the CT index can be defined as the (normalized) overlap between the attachment and detachment density matrices (Equation (4.105)), or in terms of the overlap between absolute values of the NTO orbitals in the dominating pair, as defined by[91]

$$\mathit{CT\ index} = \left\langle \left|\phi_i^{\text{NTO}}\right| \middle| \left|\phi_a^{\text{NTO}}\right| \right\rangle \tag{4.109}$$

Peach *et al.* have defined a very similar Λ index as a weighted sum of overlaps between absolute values of occupied (ϕ_i) and virtual (ϕ_a) orbitals, where the κ_{ia} weighting factor is the contribution of the particular orbital pair to the particular excitation.[92] The Λ index is essentially just the CT index in Equation (4.109) formulated in terms of canonical orbitals instead of natural transition orbitals:

$$\Lambda = \frac{\sum_{ia} \kappa_{ia}^2 \langle |\phi_i| \, | |\phi_a| \rangle}{\sum_{ia} \kappa_{ia}^2} \tag{4.110}$$

The various CT indices are in practice very similar, and values less than ~0.5 are often taken as an indicator of a CT excitation. The difference in the dipole moments of the ground and excited states, which can be considered as the integrated charge of the attatchment/detatchment matrices times the distance between the two center-of-charge points for these two matrices, have also been suggested as an indicator for CT character, but this is not constrained to the interval from 0 to 1.[93]

4.15 Quantum Monte Carlo Methods

Monte Carlo methods refer to techniques for obtaining the value of a multidimensional integral of a function by randomly probing its value within the whole variable space and estimating the integral by statistical averaging. In the limit of an infinite number of sampling points, the result is identical to that obtained from an analytical integration, but for a finite number of points, the calculated value is given as an average with an associated standard deviation. The standard deviation, the uncertainty, depends inversely on the square root of the number of sampling points (Section 18.2).

Since the square of the wave function represent a probability function, the associated energy can be calculated by *Quantum Monte Carlo* (QMC) methods.[94,95] For a (approximate) variational wave function, the energy can be rewritten as

$$\begin{aligned} E &= \frac{\langle \Phi | \mathbf{H} | \Phi \rangle}{\langle \Phi | \Phi \rangle} = \frac{\int \Phi^* \mathbf{H} \Phi d\mathbf{r}}{\int \Phi^* \Phi d\mathbf{r}} \\ E &= \frac{\int \Phi^* \Phi (\Phi^{-1} \mathbf{H} \Phi) d\mathbf{r}}{\int \Phi^* \Phi d\mathbf{r}} = \frac{\int |\Phi(\mathbf{r})|^2 (\Phi^{-1} \mathbf{H} \Phi) d\mathbf{r}}{\int |\Phi(\mathbf{r})|^2 d\mathbf{r}} \\ E &= \int E_{\text{local}}(\mathbf{r}) P(\mathbf{r}) d\mathbf{r} \quad ; \quad P(\mathbf{r}) = \frac{|\Phi(\mathbf{r})|^2}{\int |\Phi(\mathbf{r})|^2 d\mathbf{r}} \quad ; \quad E_{\text{local}}(\mathbf{r}) = \Phi^{-1} \mathbf{H} \Phi \end{aligned} \tag{4.111}$$

The last equation shows that the energy can be calculated as an integral of the local energy function $\Phi^{-1}\mathbf{H}\Phi$ weighted with the probability density P. In principle, this integral could be calculated by numerical quadrature methods, such as the Simpson trapezoidal rule, but this becomes very inefficient when the number of variables is large. For a system with N electrons, the dimensionality of the problem is $3N_{\text{elec}}$, and the integral can be estimated much more efficiently by sampling the function point-wise within the whole function space. Estimating the functional value by a random sampling of points within the integration limits, weighted by the probability factors, is called *variational* QMC. The generation of points is done using a Metropolis algorithm, as discussed in more detail in Section 15.1, and the calculated energy is simply the average of the local energies over the sampling points:

$$E = \frac{1}{M_{\text{point}}} \sum_{i=1}^{M_{\text{point}}} E_{\text{local}}(\mathbf{r}_i) \tag{4.112}$$

An improvement of the variational QMC can be obtained by the *diffusion* QMC approach. Consider the time-dependent Schrödinger equation, where the time is replaced with an imaginary time variable $\tau = it$:

$$
\begin{aligned}
i\frac{\partial \Phi(\mathbf{r},t)}{\partial t} &= \mathbf{H}\Phi(\mathbf{r},t) \\
-\frac{\partial \Phi(\mathbf{r},\tau)}{\partial \tau} &= \mathbf{H}\Phi(\mathbf{r},\tau)
\end{aligned}
\tag{4.113}
$$

For a free electron, the Hamiltonian is only kinetic energy, and the resulting equation is identical to that describing a diffusion process:

$$
\frac{\partial \Phi(\mathbf{r},\tau)}{\partial \tau} = \tfrac{1}{2}\nabla^2\Phi(\mathbf{r},\tau) \tag{4.114}
$$

Addition of a potential energy results in a generalized diffusion equation:

$$
\frac{\partial \Phi(\mathbf{r},\tau)}{\partial \tau} = \tfrac{1}{2}\nabla^2\Phi(\mathbf{r},\tau) - \mathbf{V}(\mathbf{r})\Phi(\mathbf{r},\tau) \tag{4.115}
$$

The generalized diffusion equation can be solved by a random walk procedure and, in the long time limit, the resulting distribution converges to the ground state wave function. This can be seen by expanding an approximate wave function in terms of the exact wave functions (Equation (1.20)):

$$
\Phi(\mathbf{r},\tau) = \sum_k c_k \Psi_k(\mathbf{r},\tau) = \sum_k c_k \Psi_k(\mathbf{r}) e^{-E_k \tau} \tag{4.116}
$$

The exponential dependence on the energy means that the high-energy states decay faster than the low-energy ones and, in the long (imaginary) time limit, only the ground state wave function survives.

The main problem with QMC methods is the requirement of an antisymmetric wave function, since the electrons are fermions. The antisymmetry means that the wave function has both positive and negative regions, and consequently $3N_{\text{elec}} - 1$-dimensional surfaces where the wave function is zero. These surfaces are called *nodes* and correspond to zero-probability regions of space. Clearly, a procedure that indiscriminately samples the nodal regions will yield inaccurate answers. QMC methods thus require a guiding function, a trial wave function, for determining how to sample the huge phase space most efficiently and in agreement with the fermion nature of the electrons.

A suitable trial wave function can be constructed from a Hartree–Fock wave function multiplied with a suitable correlation function, often taken as a *Jastrow* factor $J(\mathbf{r})$:

$$
J(\mathbf{r}) = \sum_{i}^{N_{\text{elec}}} \chi(\mathbf{r}_i) - \sum_{i>j}^{N_{\text{elec}}} u(\mathbf{r}_i, \mathbf{r}_j) \tag{4.117}
$$

The functional forms of the one- and two-electron terms χ and u are chosen such that they model the nuclear–electron and electron–electron cusp conditions, respectively, and the parameters inherent in these functions are variationally optimized by the QMC procedure.

In order to maintain the wave function antisymmetry, the diffusion QMC is normally used within the *fixed node* approximation, that is the nodes are fixed by the initial trial wave function. Unfortunately, the location of nodes for the exact wave function is far from trivial to determine, although simple approximations such as HF can give quite reasonable estimates.[96] The fixed node diffusion QMC thus determines the best wave function with the nodal structure of the initial trial wave function. If the trial wave function has the correct nodal structure, the QMC will provide the exact solution to the Schrödinger equation, including the electron correlation energy. It should be noted that the region near the nuclei contributes most to the statistical error in QMC methods, and in many applications the core electrons are therefore replaced by a pseudo-potential (Section 5.12).

The scaling of QMC methods is $N_{\text{occ}}^2 N_{\text{basis}}$, but the prefactor makes these methods roughly two orders of magnitude more expensive than independent-particle models such as HF and DFT. The relatively low-order scaling, however, makes QMC competitive with, for example, coupled cluster methods, even for relatively small systems. The main disadvantage of QMC is the statistical error in the calculated results, which only decays as the inverse square root of the number of sampling points. Generating highly accurate results is thus computationally expensive, although the calculations are well suited for running on large parallel computers. The statistical uncertainty furthermore makes it difficult to calculate nuclear forces and second derivatives, which are essential for optimizing structures and calculating vibrational frequencies. Finally, the accuracy of the results is tightly coupled to the form of the trial wave function, and a poor trial wave function can generate poor-quality results.

References

1 A. Szabo and N. S. Ostlund, *Modern Quantum Chemistry* (McGraw-Hill, 1982).
2 W. J. Hehre, L. Radom, J. A. Pople and P. v. R. Schleyer, *Ab Initio Molecular Orbital Theory* (John Wiley & Sons, 1986).
3 R. McWeeny, *Methods of Molecular Quantum Mechanics* (Academic Press, 1992).
4 J. Simons, *Journal of Physical Chemistry* **95** (3), 1017–1029 (1991).
5 R. J. Bartlett and J. F. Stanton, *Reviews in Computational Chemistry* **5**, 65 (1994).
6 T. Helgaker, P. Jørgensen and J. Olsen, *Molecular Electronic Structure Theory* (John Wiley & Sons, 2000).
7 C. D. Sherrill and H. F. Schaefer, in *Advances in Quantum Chemistry*, edited by P. O. Lowdin, J. R. Sabin, M. C. Zerners and E. Brandas (Academic Press, 1999), Vol. 34, pp. 143–269.
8 D. Cremer, *Wiley Interdisciplinary Reviews– Computational Molecular Science* **3** (5), 482–503 (2013).
9 J. Olsen, F. Jorgensen, H. Koch, A. Balkova and R. J. Bartlett, *Journal of Chemical Physics* **104** (20), 8007–8015 (1996).
10 J. Olsen, P. Jorgensen and J. Simons, *Chemical Physics Letters* **169** (6), 463–472 (1990).
11 E. R. Davidson, *Journal of Computational Physics* **17** (1), 87–94 (1975).
12 H. G. Hiscock and A. J. W. Thom, *Journal of Chemical Theory and Computation* **10** (11), 4795–4800 (2014).
13 K. Yamaguchi, F. Jensen, A. Dorigo and K. N. Houk, *Chemical Physics Letters* **149** (5–6), 537–542 (1988).
14 P. G. Szalay, T. Mueller, G. Gidofalvi, H. Lischka and R. Shepard, *Chemical Reviews* **112** (1), 108–181 (2012).
15 A. I. Krylov, in *Annual Review of Physical Chemistry* (2008), Vol. 59, pp. 433–462.
16 Y. H. Shao, M. Head-Gordon and A. I. Krylov, *Journal of Chemical Physics* **118** (11), 4807–4818 (2003).
17 J. Olsen, B. O. Roos, P. Jorgensen and H. J. A. Jensen, *Journal of Chemical Physics* **89** (4), 2185–2192 (1988).
18 D. Ma, G. L. Manni and L. Gagliardi, *Journal of Chemical Physics* **135** (4), 044128 (2011).
19 J. Ivanic, *Journal of Chemical Physics* **119** (18), 9364–9376 (2003).
20 W. T. Borden and E. R. Davidson, *Accounts of Chemical Research* **29** (2), 67–75 (1996).
21 C. Moller and M. S. Plesset, *Physical Review* **46** (7), 0618–0622 (1934).
22 D. Cremer, *Wiley Interdisciplinary Reviews – Computational Molecular Science* **1** (4), 509–530 (2011).
23 S. Grimme, L. Goerigk and R. F. Fink, *Wiley Interdisciplinary Reviews– Computational Molecular Science* **2** (6), 886–906 (2012).

24 J. Olsen, O. Christiansen, H. Koch and P. Jorgensen, *Journal of Chemical Physics* **105** (12), 5082–5090 (1996).

25 J. Olsen, P. Jorgensen, T. Helgaker and O. Christiansen, *Journal of Chemical Physics* **112** (22), 9736–9748 (2000).

26 A. V. Sergeev and D. Z. Goodson, *Journal of Chemical Physics* **124** (9), 094111 (2006).

27 N. C. Handy, P. J. Knowles and K. Somasundram, *Theoretica Chimica Acta* **68** (1), 87–100 (1985).

28 P. M. Kozlowski and E. R. Davidson, *Journal of Chemical Physics* **100** (5), 3672–3682 (1994).

29 T. J. Lee, A. P. Rendell, K. G. Dyall and D. Jayatilaka, *Journal of Chemical Physics* **100** (10), 7400–7409 (1994).

30 K. R. Glaesemann and M. W. Schmidt, *Journal of Physical Chemistry A* **114** (33), 8772–8777 (2010).

31 H. B. Schlegel, *Journal of Physical Chemistry* **92** (11), 3075–3078 (1988).

32 P. J. Knowles and N. C. Handy, *Journal of Physical Chemistry* **92** (11), 3097–3100 (1988).

33 J. Finley, P. A. Malmqvist, B. O. Roos and L. Serrano-Andres, *Chemical Physics Letters* **288** (2–4), 299–306 (1998).

34 P. Celani and H. J. Werner, *Journal of Chemical Physics* **112** (13), 5546–5557 (2000).

35 H. J. Werner, *Molecular Physics* **89** (2), 645–661 (1996).

36 R. J. Bartlett and M. Musial, *Reviews of Modern Physics* **79** (1), 291–352 (2007).

37 B. Cooper and P. J. Knowles, *Journal of Chemical Physics* **133** (23), 234102 (2010).

38 D. I. Lyakh, M. Musial, V. F. Lotrich and R. J. Bartlett, *Chemical Reviews* **112** (1), 182–243 (2012).

39 W. Duch and G. H. F. Diercksen, *Journal of Chemical Physics* **101** (4), 3018–3030 (1994).

40 J. D. Watts, J. Gauss and R. J. Bartlett, *Journal of Chemical Physics* **98** (11), 8718–8733 (1993).

41 J. J. Eriksen, D. A. Matthews, P. Jorgensen and J. Gauss, *Journal of Chemical Physics* **143** (4), 041101 (2015).

42 K. A. Brueckner, *Physical Review* **96** (2), 508–516 (1954).

43 J. F. Stanton, J. Gauss and R. J. Bartlett, *Journal of Chemical Physics* **97** (8), 5554–5559 (1992).

44 E. A. Salter, H. Sekino and R. J. Bartlett, *Journal of Chemical Physics* **87** (1), 502–509 (1987).

45 C. Hampel, K. A. Peterson and H. J. Werner, *Chemical Physics Letters* **190** (1–2), 1–12 (1992).

46 T. J. Lee, R. Kobayashi, N. C. Handy and R. D. Amos, *Journal of Chemical Physics* **96** (12), 8931–8937 (1992).

47 A. Balkova and R. J. Bartlett, *Chemical Physics Letters* **193** (5), 364–372 (1992).

48 J. A. Pople, M. Headgordon and K. Raghavachari, *Journal of Chemical Physics* **87** (10), 5968–5975 (1987).

49 G. E. Scuseria and H. F. Schaefer, *Journal of Chemical Physics* **90** (7), 3700–3703 (1989).

50 T. J. Lee, A. P. Rendell and P. R. Taylor, *Journal of Physical Chemistry* **94** (14), 5463–5468 (1990).

51 M. Bohme and G. Frenking, *Chemical Physics Letters* **224** (1–2), 195–199 (1994).

52 O. Christiansen, H. Koch and P. Jorgensen, *Chemical Physics Letters* **243** (5–6), 409–418 (1995).

53 H. Koch, O. Christiansen, P. Jorgensen, A. M. S. deMeras and T. Helgaker, *Journal of Chemical Physics* **106** (5), 1808–1818 (1997).

54 A. I. Krylov, *Journal of Chemical Physics* **113** (15), 6052–6062 (2000).

55 T. J. Lee and P. R. Taylor, *International Journal of Quantum Chemistry* **S23**, 199–207 (1989).

56 M. Musial, A. Perera and R. J. Bartlett, *Journal of Chemical Physics* **134** (11), 114108 (2011).

57 M. K. Sprague and K. K. Irikura, *Theoretical Chemistry Accounts* **133** (9), 1544 (2014).

58 T. Kato, *Communications on Pure and Applied Mathematics* **10** (2), 151–177 (1957).

59 E. A. Hylleraas, *Zeitschrift Fur Physik* **65** (3–4), 209–225 (1930).

60 V. I. Korobov, *Physical Review A* **61** (6), 064503 (2000).

61 W. Kolos and Wolniewi. L, *Journal of Chemical Physics* **49** (1), 404 (1968).

62 W. Kutzelnigg and J. D. Morgan, *Journal of Chemical Physics* **96** (6), 4484–4508 (1992).

63 W. Kutzelnigg and W. Klopper, *Journal of Chemical Physics* **94** (3), 1985–2001 (1991).

64 G. M. J. Barca, P.-F. Loos and P. M. W. Gill, *Journal of Chemical Theory and Computation* **12** (4), 1735–1740 (2016).
65 C. Haettig, W. Klopper, A. Koehn and D. P. Tew, *Chemical Reviews* **112** (1), 4–74 (2012).
66 W. Klopper, *Journal of Chemical Physics* **102** (15), 6168–6179 (1995).
67 W. Klopper, *Journal of Chemical Physics* **120** (23), 10890–10895 (2004).
68 E. F. Valeev, *Chemical Physics Letters* **395** (4–6), 190–195 (2004).
69 D. S. Lambrecht and C. Ochsenfeld, *Journal of Chemical Physics* **123** (18), 184101 (2005).
70 M. Head-Gordon, J. A. Pople and M. J. Frisch, *Chemical Physics Letters* **153** (6), 503–506 (1988).
71 M. Haser, *Theoretica Chimica Acta* **87** (1–2), 147–173 (1993).
72 S. A. Maurer, D. S. Lambrecht, D. Flaig and C. Ochsenfeld, *Journal of Chemical Physics* **136** (14), 144107 (2012).
73 V. M. Anisimov, G. H. Bauer, K. Chadalavada, R. M. Olson, J. W. Glenski, W. T. C. Krarner, E. Apra and K. Kowalski, *Journal of Chemical Theory and Computation* **10** (10), 4307–4316 (2014).
74 H. J. Werner, F. R. Manby and P. J. Knowles, *Journal of Chemical Physics* **118** (18), 8149–8160 (2003).
75 D. G. Liakos, M. Sparta, M. K. Kesharwani, J. M. L. Martin and F. Neese, *Journal of Chemical Theory and Computation* **11** (4), 1525–1539 (2015).
76 J. J. Eriksen, P. Baudin, P. Ettenhuber, K. Kristensen, T. Kjaergaard and P. Jorgensen, *Journal of Chemical Theory and Computation* **11** (7), 2984–2993 (2015).
77 M. S. Gordon, D. G. Fedorov, S. R. Pruitt and L. V. Slipchenko, *Chemical Reviews* **112** (1), 632–672 (2012).
78 M. A. Collins and R. P. A. Bettens, *Chemical Reviews* **115** (12), 5607–5642 (2015).
79 S. Tanaka, Y. Mochizuki, Y. Komeiji, Y. Okiyama and K. Fukuzawa, *Physical Chemistry Chemical Physics* **16** (22), 10310–10344 (2014).
80 M. Kobayashi and H. Nakai, *International Journal of Quantum Chemistry* **109** (10), 2227–2237 (2009).
81 U. Benedikt, K.-H. Boehm and A. A. Auer, *Journal of Chemical Physics* **139** (22), 224101 (2013).
82 E. Epifanovsky, D. Zuev, X. Feng, K. Khistyaev, Y. Shao and A. I. Krylov, *Journal of Chemical Physics* **139** (13), 134105 (2013).
83 T. Helgaker, J. Gauss, P. Jorgensen and J. Olsen, *Journal of Chemical Physics* **106** (15), 6430–6440 (1997).
84 A. T. B. Gilbert, N. A. Besley and P. M. W. Gill, *Journal of Physical Chemistry A* **112** (50), 13164–13171 (2008).
85 M. Head-Gordon, R. J. Rico, M. Oumi and T. J. Lee, *Chemical Physics Letters* **219** (1–2), 21–29 (1994).
86 S. P. A. Sauer, M. Schreiber, M. R. Silva-Junior and W. Thiel, *Journal of Chemical Theory and Computation* **5** (3), 555–564 (2009).
87 K. B. Wiberg, A. E. de Oliveira and G. Trucks, *Journal of Physical Chemistry A* **106** (16), 4192–4199 (2002).
88 A. Dreuw and M. Head-Gordon, *Chemical Reviews* **105** (11), 4009–4037 (2005).
89 M. Headgordon, A. M. Grana, D. Maurice and C. A. White, *Journal of Physical Chemistry* **99** (39), 14261–14270 (1995).
90 R. L. Martin, *Journal of Chemical Physics* **118** (11), 4775–4777 (2003).
91 T. Etienne, X. Assfeld and A. Monari, *Journal of Chemical Theory and Computation* **10** (9), 3896–3905 (2014).
92 M. J. G. Peach, P. Benfield, T. Helgaker and D. J. Tozer, *Journal of Chemical Physics* **128** (4), 044118 (2008).
93 T. Le Bahers, C. Adamo and I. Ciofini, *Journal of Chemical Theory and Computation* **7** (8), 2498–2506 (2011).
94 A. Luechow, *Wiley Interdisciplinary Reviews – Computational Molecular Science* **1** (3), 388–402 (2011).
95 B. M. Austin, D. Y. Zubarev and W. A. Lester, Jr., *Chemical Reviews* **112** (1), 263–288 (2012).
96 N. Nemec, M. D. Towler and R. J. Needs, *Journal of Chemical Physics* **132** (3), 034111 (2010).

5

Basis Sets

Ab initio methods try to derive information by solving the Schrödinger equation without fitting parameters to experimental data. Actually, *ab initio* methods also make use of experimental data, but in a somewhat more subtle fashion. Many different approximate methods exist for solving the Schrödinger equation; which one to use for a specific problem is usually chosen by comparing the performance against known experimental data. Experimental data thus guides the *selection* of the computational model, rather than directly *entering* into the computational procedure.

One of the approximations inherent in essentially all *ab initio* methods is the introduction of a basis set. Expanding an unknown function, such as a molecular orbital (MO), in a set of known functions is not an approximation if the basis set is *complete.* However, a complete basis set means that an infinite number of functions must be used, which is impossible in actual calculations. An unknown MO can be thought of as a function in the infinite coordinate system spanned by the complete basis set. When a finite basis set is used, only the components of the MO along those coordinate axes corresponding to the selected basis functions can be represented. The smaller the basis set, the poorer the representation. The *type* of basis functions used also influence the accuracy. The better a single basis function is able to reproduce the unknown function, the fewer basis functions are necessary for achieving a given level of accuracy. Knowing that the computational effort of *ab initio* methods scales formally as at least M_{basis}^4, it is of course of prime importance to make the basis set as small as possible, without compromising the accuracy.[1,2] The expansion of the molecular orbitals leads to integrals of quantum mechanical operators over basis functions, and the ease by which these integrals can be calculated also depends on the type of basis function. In some cases the *accuracy-per-function* criterion produces a different optimum function type than the *efficiency-per-function* criterion. We will in the following use the terms *basis function* and *basis set,* where a basis function is a specific type of mathematical function, while a basis set is a collection of basis functions containing a specific set of parameters.

The basis set desiderate can be listed as follows:

1. The basis functions should reflect the nature of the problem, such that a good accuracy can be obtained by a relatively small number of functions.
2. The basis functions should be able to generate a complete basis set, such that a well-defined basis set limit can be obtained.

Introduction to Computational Chemistry, Third Edition. Frank Jensen.

Companion Website: http://www.wiley.com/go/jensen/computationalchemistry3

3. Basis sets should be available in several hierachical levels, where each level provides a well-defined accuracy and the hierachy systematically converges the result towards the basis set limit. Ideally the basis set convergence should be monotomic and fast.
4. For a given accuracy, the basis set should be as computationally efficient as possible, that is delivering the target accuracy for as low a computational cost as possible. The computational cost is often related to the number of basis functions, but other factors may also be important.
5. Basis sets should ideally be universal, that is suitable for different methods (HF, DFT, electron correlation methods, relativistic methods) and different properties (energy, molecular structure, vibrational frequences, polarizabilities, NMR spin–spin coupling constants, etc.).
6. Be available for all atoms, or at least for a large fraction of the periodic table.

Unfortunately points 4 and 5 are mutually incompatible, because different methods and different properties have different basis set demands. A universal basis set would need to include basis functions meeting all the requirements for each method/property and thus become very large, and therefore is not computationally efficient. The wealth of different basis sets proposed can thus be seen as various compromises between universability and efficiency. Older basis sets were designed as being of general purpose, but modern basis sets are usually optimized for specific tasks. HF and DFT methods (Chapters 3 and 6), for example, have different basis set requirements than electron correlation methods (Chapter 4), and explicitly correlated methods (Section 4.11) have requirements intermediate between these. Relativistic methods (Chapter 9) are sufficiently different from non-relativistic methods that they require different basis sets. Molecular properties (Chapter 11) likewise often require specialized basis sets in order to achive a smooth and fast convergence towards the basis set limiting value.

5.1 Slater- and Gaussian-Type Orbitals

There are two types of basis functions (also called *Atomic Orbitals* (AO), although they in general are not solutions to an atomic Schrödinger equation) commonly used in electronic structure calculations: *Slater-Type Orbitals* (STOs) and *Gaussian-Type Orbitals* (GTOs). Slater-type orbitals[3] have the functional form shown in

$$\chi_{\zeta,n,l,m}(r,\theta,\varphi) = NY_{l,m}(\theta,\varphi)r^{n-1}\mathrm{e}^{-\zeta r} \tag{5.1}$$

Here N is a normalization constant and $Y_{l,m}$ are spherical harmonic functions. The exponential dependence on the distance between the nucleus and electron mirrors the exact orbitals for hydrogen-like atoms. The STOs, however, do not have radial nodes; nodes in the radial part are introduced by making linear combinations of STOs. The exponential dependence ensures a fairly rapid convergence with an increasing numbers of functions, but, as noted in Section 3.5, the calculation of three- and four-center two-electron integrals cannot be performed analytically. STOs are therefore primarily used for atomic and diatomic systems, where high accuracy is required, and in semi-empirical methods (Chapter 7), where all three- and four-center integrals are neglected. They can also be used with density functional methods (Chapter 6) that do not include exact exchange and where the Coulomb energy is calculated by fitting the density to a set of auxiliary functions (Section 3.8.7).

Gaussian type orbitals[4] can be written in terms of polar (Equation (5.2)) or Cartesian (Equation (5.3)) coordinates:

$$\chi_{\zeta,n,l,m}(r,\theta,\varphi) = NY_{l,m}(\theta,\varphi)r^{2n-2-l}e^{-\zeta r^2} \tag{5.2}$$

$$\chi_{\zeta,l_x,l_y,l_z}(x,y,z) = Nx^{l_x}y^{l_y}z^{l_z}e^{-\zeta r^2} \tag{5.3}$$

The sum of l_x, l_y and l_z in Equation (5.3) determines the type of orbital (e.g. $l_x + l_y + l_z = 1$ is a p-orbital). Although a GTO appears similar in the two set of coordinates, there is a subtle difference for angular momentum of 2 or higher. A *d*-type GTO written in the spherical form has five components ($Y_{2,2}$, $Y_{2,1}$, $Y_{2,0}$, $Y_{2,-1}$, $Y_{2,-2}$), but there appear to be six components in the Cartesian coordinates ($x^2, y^2, z^2, xy, xz, yz$). The latter six functions, however, may be transformed to the five spherical *d*-functions and one additional *s*-function ($x^2 + y^2 + z^2$). Similarly, there are ten Cartesian "*f*-functions" that may be transformed into seven spherical *f*-functions and one set of spherical *p*-functions. Modern programs for evaluating two-electron integrals are geared to Cartesian coordinates and they generate pure spherical *d*-functions by transforming the six Cartesian components into the five spherical functions. When only one *d*-function is present per atom the saving by removing the extra *s*-function is small, but if many *d*-functions and/or higher angular momentum functions (*f*-, *g*-, *h*-, etc., functions) are present, the savings can be substantial. Furthermore, the use of only the spherical components reduces the problems of linear dependence for large basis sets, as discussed in Section 5.8.

The r^2 dependence in the exponential makes the GTOs inferior to the STOs in two respects. At the nucleus a GTO has a zero slope, in contrast to a STO which has a "cusp" (discontinuous derivative), and GTOs consequently have problems representing the proper behavior near the nucleus. The other problem is that the GTO falls off too rapidly far from the nucleus compared with an STO, and the "tail" of the wave function is consequently represented poorly. Both STOs and GTOs can be chosen to form a complete basis, but the above considerations indicate that more GTOs are necessary for achieving a certain accuracy compared with STOs. A rough guideline says that three times as many GTOs as STOs are required for reaching a given level of accuracy. Figure 5.1 shows how a 1s-STO can be modeled by a linear combination of three GTOs.

The increase in the number of GTO basis functions, however, is more than compensated for by the ease with which the required integrals can be calculated. In terms of computational efficiency, GTOs are therefore preferred and are used almost universally as basis functions in molecular electronic structure calculations. Furthermore, essentially all applications take the GTOs to be centered at the nuclei. For certain types of calculations the center of a basis function may be taken not to coincide with a nucleus, for example being placed at the center of a bond or between non-bonded atoms for improving the calculation of van der Waals interactions.

5.2 Classification of Basis Sets

Having decided on the type of basis function (STO/GTO) and the location (nuclei), the most important factor is the number of functions to be used. The smallest basis set employs only enough functions for a minimum description of the occupied orbitals of the neutral atom(s) and is called a *minimum* or *Single Zeta* (*SZ*) basis set. The term zeta stems from the fact that the exponent of STO basis functions is often denoted by the Greek letter ζ. For hydrogen (and helium), a SZ basis set has only a single *s*-function. For the second row in the periodic system it means two *s*-functions (1s and 2s) and one set of *p*-functions ($2p_x$, $2p_y$ and $2p_z$). Lithium and beryllium formally only require two *s*-functions,

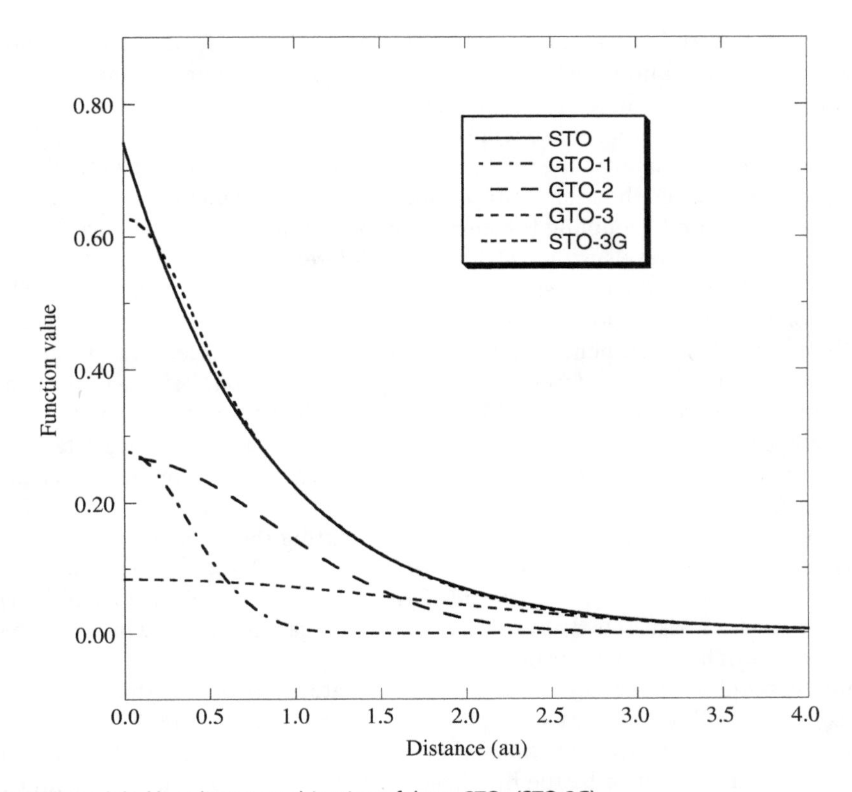

Figure 5.1 A 1s-STO modeled by a linear combination of three GTOs (STO-3G).

but a set of *p*-functions is usually also added. For the third row elements, three *s*-functions (1s, 2s and 3s) and two sets of *p*-functions (2p and 3p) are used.

The next improvement of the basis set is a doubling of all basis functions, producing a *Double Zeta* (*DZ*)-type basis. A DZ basis set thus employs two *s*-functions for hydrogen (1s and 1s′), four *s*-functions (1s, 1s′, 2s and 2s′) and two sets of *p*-functions (2p and 2p′) for second row elements and six *s*-functions and four sets of *p*-functions for third row elements. The importance of a DZ over a minimum basis set can be illustrated by considering the bonding in the HCN molecule, shown in Figure 5.2. The H—C bond will primarily consist of the hydrogen s-orbital and the p_z-orbital on C.

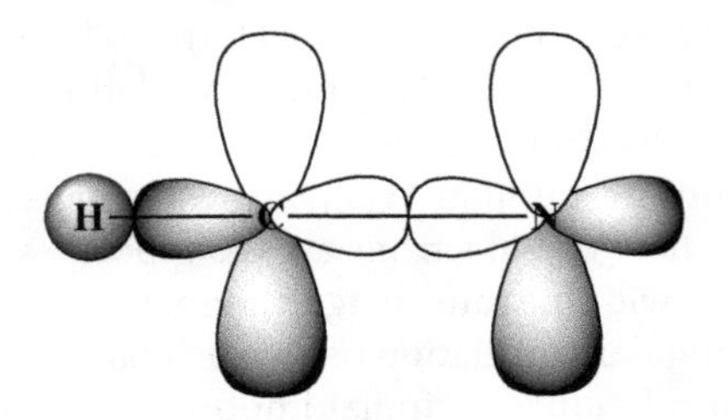

Figure 5.2 A double zeta basis set allows for different bonding in different directions.

The π-bond between C and N will consist of the p_x (and p_y) orbitals of C and N, and will have a more diffuse electron distribution than the H—C σ-bond. The optimum exponent for the carbon p-orbital will thus be smaller for the x-direction than for the z-direction. If only a single set of p-orbitals is available (SZ basis set), a compromise will be necessary. A DZ basis set, however, has two sets of p-orbitals with different exponents. The tighter function (larger exponent) can enter the H—C σ-bond with a large coefficient, while the more diffuse function (small exponent) can be used primarily for describing the C—N π-bond. Doubling the number of basis functions thus allows for a much better description of the fact that the electron distribution is different in different directions.

The chemical bonding occurs between valence orbitals. Doubling the 1s-functions in, for example, carbon allows for a better description of the 1s-electrons. However, the 1s-orbital is essentially independent of the chemical environment, being very close to the atomic case. A variation of the DZ-type basis set only doubles the number of valence orbitals, producing a *split valence basis*. In actual calculations, a doubling of the core orbitals would rarely be considered, and the term DZ basis set is used also for split valence basis sets (or sometimes denoted VDZ, for valence double zeta).

The next step up in basis set size is a *Triple Zeta* (*TZ*). Such a basis set contains three times as many functions as the SZ basis, that is six *s*-functions and three *p*-functions for the second row elements. Some of the core orbitals may again be saved by only splitting the valence, producing a *triple split valence* basis set. Again the term TZ is used to cover both cases. The series continues with the names and acronyms *Quadruple Zeta* (QZ), *Quintuple Zeta* (5Z), *Hextuble Zeta* (6Z) and *Heptuble Zeta* (7Z) for the next levels of basis sets that are also used, but large basis sets are often given explicitly in terms of the number of basis functions of each type.

So far, only the number of *s*- and *p*-functions for each atom (second or third row in the periodic table) has been discussed. In most cases, higher angular momentum functions are also important, and these are denoted *polarization functions*. Consider again the bonding in HCN in Figure 5.2. The H—C bond is primarily described by the hydrogen s-orbital(s) and the carbon s- and p_z-orbitals. It is clear that the electron distribution *along* the bond near the hydrogen atom will be different than *perpendicular* to the bond. If only *s*-functions are present on hydrogen, this cannot be described. However, if a set of p-orbitals is added to hydrogen, the p_z component can be used for improving the description of the H—C bond. The p-orbital introduces a polarization of the s-orbital(s). Similarly, d-orbitals can be used for polarizing p-orbitals, f-orbitals for polarizing d-orbitals, etc. Once a p-orbital has been added to polarize a hydrogen s-orbital, it may be argued that the p-orbital should be polarized by adding a d-orbital, which should be polarized by an f-orbital, etc. For independent-particle wave functions (Chapters 3 and 6), where electron correlation is not considered, the first set of polarization functions (i.e. *p*-functions for hydrogen and *d*-functions for non-hydrogen p-block atoms) is by far the most important, and will in general describe most of the important charge polarization effects.

If methods including electron correlation (Chapter 4) are used, higher angular momentum functions are essential. Electron correlation describes the energy lowering by the electrons "avoiding" each other, beyond the average effect taken into account by Hartree–Fock methods. Two types of atomic correlation can be defined, an "*in–out*" and an "*angular*" correlation. The in–out or *radial correlation* refers to the situation where one electron is close to, and the other far from, the nucleus. To describe this, the basis set needs functions of the same type, but with different exponents. The *angular correlation* refers to the situation where two electrons are on opposite sides of the nucleus. To describe this, the basis set needs functions with the same magnitude exponents, but different angular momentum. For example, to describe angular correlation of an *s*-orbital, *p*-functions (and *d*-, *f*-, *g*-functions, etc.) are needed. The angular correlation is of similar importance to the radial correlation, and higher angular momentum functions are consequently essential for correlated calculations. Although these should properly be labeled correlation functions, they also serve as polarization functions for HF wave functions, and it is common to denote them as polarization functions.

In most cases only the correlation of the valence electrons is considered, and the exponents of the polarization functions should be of the same magnitude as the valence *s*- and *p*-functions (actually slightly larger in order to have the same maximum in the radial distribution function). In contrast to HF methods, the higher angular momentum functions (beyond the first set of polarization functions) are quite important, or, alternatively formulated, the convergence in terms of angular momentum is slower for correlated wave functions than at the HF level. For a basis set that is complete up to angular momentum L, both numerical[5,6] and theoretical[7] analyses suggest that the asymptotic convergence at the HF level is exponential (i.e. $\sim\exp(-\sqrt{L})$), while it is $\sim L^{-3}$ at correlated levels (Section 5.9).

Polarization functions are added to the chosen *sp*-basis set. Adding a single set of polarization functions (*p*-functions on hydrogens and *d*-functions on non-hydrogen p-block atoms) to a DZ basis set forms a *Double Zeta plus Polarization* (*DZP*)-type basis set. There is a variation where polarization functions are only added to non-hydrogen atoms. This does not mean that polarization functions are not important on hydrogen. Hydrogen, however, often has a "passive" role, sitting at the end of bonds that do not take an active part in the property of interest. The error introduced by not including hydrogen polarization functions is often rather constant and, as the interest usually is in energy differences, tends to cancel out. Hydrogen often accounts for a large number of atoms in the system and a saving of three basis functions for each hydrogen is significant. If hydrogen plays an important role in the property of interest, it is of course not a good idea to neglect polarization functions on hydrogen.

Similarly to the *sp*-basis sets, multiple sets of polarization functions with different exponents may be added. If two sets of polarization functions are added to a TZ *sp*-basis, a *Triple Zeta plus Polarization* (TZP)-type basis is obtained. For larger basis sets with many polarization functions the explicit composition in terms of number and types of functions is usually given. At the HF level there is usually little gained by expanding the basis set beyond TZP, and even a DZP-type basis set usually gives "good" results (compared with the HF limit). Correlated methods, however, require more, and higher angular momentum, polarization functions to achieve the same level of convergence.

Before moving on we need to introduce the concept of *basis set balance*. In principle, many sets of polarization functions may be added to a small *sp*-basis set, but this is a poor idea. If an insufficient number of *sp*-functions has been chosen for describing the fundamental electron distribution, the optimization procedure used in obtaining the wave function (and possibly also the geometry) may try to compensate for inadequacies in the *sp*-basis set by using higher angular momentum functions, thereby producing artefacts. A rule of thumb says that the number of functions of a given type should at most be one less than the type with one lower angular momentum. A 3s2p1d basis is balanced, but a 3s2p2d2f1g is too heavily polarized. It may not be necessary to polarize the basis all the way up; thus a 5s4p3d2f1g basis is balanced, but if it is known (e.g. by comparison with experimental data) that *f*- and *g*-functions are unimportant, they may be left out.

The hierachy levels mentioned in the introduction as a basis set desiderate is usually associated with the *n*-Zeta (*n*Z) notation. Since modern basis sets implicitly define the polarization functions that should be added at each zeta level, the *n*Z (DZP, TZP, QZP, ...) notation also indicates the highest angular momentum function included in the basis set, where it is implicitly assumed that the number of lower angular momentum functions has been chosen to ensure a balanced description. It should be noted that the highest angular momentum function included in the basis set for a given *n*Z depends on the atom, where a DZP basis set for hydrogen implies $L_{max} = 1$ (*p*-functions), but $L_{max} = 2$ (*d*-functions) for carbon and silicium, for example, and $L_{max} = 3$ (*f*-functions) for iron, for example.

Another aspect of basis set balance is the occasional use of *mixed* basis sets, for example a DZP quality on the atoms in the "interesting" part of the molecule and an SZ basis for the "spectator" atoms. Another example would be the addition of polarization functions for only a few hydrogens that are located "near" the reactive part of the system. For a large molecule, this may lead to a substantial saving in the number of basis functions, but it may also bias the results and can create artefacts. For example, a calculation on the H_2 molecule with an SZ basis on one of the nuclei and a DZ basis on the other will predict that H_2 has a dipole moment, since the variational principle will preferentially place the electrons near the nucleus with the most basis functions. The majority of calculations are therefore performed with basis sets of the same quality (SZ, DZP, TZ2P, etc.) on all atoms, possibly removing polarization and/or diffuse (small exponent) functions on hydrogen. Even so, it may be argued that small basis sets inherently tend to be unbalanced. Consider, for example, the LiF molecule in a minimum or DZ-type basis. This will have a very ionic structure, Li^+F^-, with nearly all the valence electrons being located at the fluorine. In terms of number of basis functions per electron, the Li basis set is thus of a much higher quality than the F basis set, and thereby unbalanced. This effect of course diminishes as the size of the atomic basis set increases.

Except for very small systems, it is impractical to saturate the basis set such that the absolute error in the energy is reduced below chemical accuracy, say a few kJ/mol. The important point in choosing a balanced basis set is to keep the error as constant as possible. The use of mixed basis sets should therefore only be done after careful consideration. Furthermore, the use of small basis sets for systems containing elements with substantially different numbers of valence electrons (such as LiF) may produce artefacts.

5.3 Construction of Basis Sets

The n-Zeta notation introduced in the previous section is closely connected to the idea of one basis function being able to represent one (atomic) orbital. While this is a reasonable approximation for STO basis functions, this is not the case for GTO functions. A basis function is often composed as a linear combination of several GTOs, as indicated in Figure 5.1. Each individual GTO is called a *primitive* function, while the fixed linear combination is called a *contracted* function. A given basis set thus contains a specific number of (contracted) basis functions, each of which contains basis set parameters in terms of the number of primitive GTOs, and their exponents and contraction coefficients. We will in this section discuss how the basis set parameters can be obtained.

5.3.1 Exponents of Primitive Functions

The exponents of the primitive GTOs determine the radial extent of the functions, with large values providing "tight" functions capable of describing the shape of the orbitals near the nucleus, while small exponents provide "diffuse" functions that can be used for describing the long-range "tail" behavior of the orbitals far from the nucleus. It is clear that a basis set must contain a series of GTOs covering a range of exponents, but the number of functions and the exact values of the exponents clearly will affect the attainable accuracy.

A few basis sets employ a fitting strategy for determining the optimum GTO exponents, for example by fitting to an STO function (Section 5.4.1). A more common procedure, however, is to determine the best set of exponents by minimizing the energy of isolated atoms using the exponents as variational parameters. The variational principle suggests that the exponent values that give the lowest energy are the "best", at least for the atom. The exponents for functions describing occupied orbitals

(s- and p-functions for second and third row elements and also d-functions for fourth and fifth row elements) can thus be determined by minimizing the HF or DFT energy. The exponents of polarization functions cannot be optimized at the HF or DFT level for atoms, since these functions are unoccupied and therefore make no contribution to the energy. Suitable polarization exponents may be chosen by performing variational calculations on molecular systems (where the HF or DFT energy does depend on polarization functions) or on atoms with a correlated wave function, such as CISD. Basis sets designed for HF or DFT typically employ the former, while basis sets designed for electron correlation typically employ the latter. Polarization exponents can alternatively be assigned based on a maximum overlap criterion with the occupied orbitals they are designed to polarize and/or correlate. In some cases, only the optimum exponent is determined for a single polarization function and multiple polarization functions are generated by splitting the exponents symmetrically around the optimum value for a single function. The splitting factor is typically taken in the range 2–4. For example, if a single d-function for carbon has an exponent value of 0.8, two polarization functions may be assigned with exponents of 0.4 and 1.6 (splitting factor of 4). The details of how the exponents are determined for various basis sets are discussed in the following sections.

5.3.2 Parameterized Exponent Basis Sets

The optimization of basis function exponents is an example of a highly non-linear optimization problem (Chapter 13). While the exponent optimization for atoms is relatively straightforward, especially if analytical gradients of the energy with respect to the basis function exponents is available, it becomes progressively more difficult as the size of the basis set increases. The basis functions start to become linearly dependent (the basis set approaches completeness) and the energy becomes a very flat function of the exponents. As an alternative to performing a full optimization of all the exponents in a basis set, the exponents may be generated by a parameterized formula, and only the (fewer) parameters in the generating formula need to be optimized.

Analyses of basis sets that have been fully optimized by variational methods reveal that the ratio between two successive exponents is approximately constant. Taking this ratio to be constant reduces the optimization problem to only two parameters for each type of basis function, independent of the size of the basis. Such basis sets have been labeled *even-tempered* basis sets, with the ith exponent given as $\zeta_i = \alpha\beta^i$, where α and β are fixed constants for a given type of function and nuclear charge. It was later discovered that the *optimum* α and β constants to a good approximation can be written as functions of the size of the basis set, M:[8]

$$\begin{aligned} \zeta_i &= \alpha\beta^i; \quad i = 1, 2, \dots, M \\ \ln(\ln\beta) &= b\ln M + b' \\ \ln\alpha &= a\ln(\beta - 1) + a' \end{aligned} \tag{5.4}$$

The constants a, a', b and b' depend only on the atom and the type of function (s or p). Even-tempered basis sets have the advantage that it is easy to generate a sequence of basis sets that converge towards a complete basis. This is useful if the attempt is to extrapolate a given property to the basis set limit. The disadvantage is that the convergence is somewhat slow, and an explicitly optimized basis set of a given size will usually give a better answer than an even-tempered basis of the same size.

Even-tempered basis sets have the same ratio between exponents over the whole range. From chemical considerations it is usually preferable to cover the valence region better than the core region. This may be achieved by *well-tempered* basis sets.[9] The idea is similar to the even-tempered basis sets, with the exponents being generated by a suitable formula containing only a few parameters to be optimized. The exponents in a well-tempered basis of size M are generated according to

$$\zeta_i = \alpha\beta^{i-1}\left(1+\gamma\left(\frac{i}{M}\right)^{\delta}\right); \quad i = 1,2,\ldots,M \tag{5.5}$$

The α, β, γ and δ parameters are optimized for each atom. The exponents are the same for all types of angular momentum functions, and s-, p- and d-functions (and higher angular momentum) consequently have the same radial part.

A well-tempered basis set has four parameters, compared with two for an even-tempered one, and is consequently capable of giving a better result for the same number of functions. Petersson *et al.*[10] have proposed a more general parameterization based on expanding the logarithmic exponents in *Legendre* polynomials as shown in the equation below. The use of Legendre polynomials is advantageous since they are orthogonal, which significantly improves the optimization of the A_k parameters:

$$\ln \zeta_i = \sum_{k=0}^{K_{\max}} A_k P_k\left(\frac{2i-2}{M-1}-1\right) \quad ; \quad i = 1,2,\ldots,M \tag{5.6}$$

$$P_0(x) = 1 \quad ; \quad P_1(x) = x \quad ; \quad P_2(x) = \tfrac{1}{2}(3x^2-1) \quad ; \quad P_3(x) = \tfrac{1}{2}(5x^3-3x)$$

Setting $K_{\max} = 1$ is equivalent to generating an even-tempered basis set and a fourth-order polynomial ($K_{\max} = 3$) expansion produces much better results than the well-tempered formula, despite having the same number of variables. The difference in energy between a fully optimized basis set and one determined from a fourth-order Legendre parameterization corresponds typically to only a few functions, that is the penalty for reducing the number of optimization variables from M to four is only a few functions. The Legendre parameterization also solves the potential problem of variational collapse, that is two neighboring exponents collapsing to the same value during optimization, and Equation (5.6) provides an efficient way of systematically approaching the basis set limit.

A disadvantage that all parameterized exponent methods share is that the exponents for a fully optimized basis set with relatively few functions display a clear orbital node structure. Orbital radial nodes correspond to distances from the nucleus where there is a reduced probability of finding an electron, and these regions are therefore less important in an energetic sense. A set of basis functions with energy-optimized exponents will therefore display "gaps" in the exponent sequence corresponding to values near orbital nodes. Such irregular exponent sequences are very difficult to obtain by parameterized formulas.

5.3.3 Basis Set Contraction

Optimizing the exponents for a set of primitive functions by minimizing the total energy will distribute the functions in exponent space according to their energetic importance. Since the inner-shell orbitals account for a large fraction of the total energy, this means that many functions will be used for obtaining a good description of the core orbitals, while relatively few will have sufficiently low exponents to be suitable for describing the energetically less important valence orbitals. Chemistry, however, is mainly dependent on the valence orbitals. The fact that many basis functions focus on describing the energetically important, but chemically unimportant, core electrons is the foundation for *contracted* basis sets. Consider, for example, a basis set consisting of ten s-functions (and some p-functions) for carbon. Having optimized these ten exponents by a variational calculation on a carbon atom, maybe six of the ten functions are found primarily to be used for describing the 1s-orbital and two of the four remaining describe the "inner" part of the 2s-orbital. The important chemical region is the outer valence. Out of the ten functions, only two are actually used for describing the chemically interesting phenomena. Considering that the computational cost increases as the fourth power (or higher) of the number of basis functions, this is inefficient. As the core orbitals change very little depending on the chemical bonding situation, the MO expansion coefficients in front of these inner basis functions also change very little. The majority of the computational effort is therefore spent describing the chemically uninteresting part of the wave function, which is furthermore almost constant.

Consider now making the variational coefficients in front of the inner basis functions constant, that is they are no longer parameters to be determined by the variational principle. The 1s-orbital is thus described by a *fixed* linear combination of, say, six basis functions. Similarly, the remaining four basis functions may be contracted into only two functions, for example by fixing the coefficient in front of the inner three functions. In doing this the number of basis functions to be handled by the variational procedure has been reduced from ten to three.

Combining the full set of primitive GTOs (PGTOs) into a smaller set of functions by forming fixed linear combinations is known as basis set contraction, and the resulting functions are called contracted GTOs (CGTOs):

$$\chi(\text{CGTO}) = \sum_{i}^{k} a_i \chi_i(\text{PGTO}) \tag{5.7}$$

The nZ notation introduced in Section 5.2 refers to the number of contracted basis functions. Contraction is especially useful for orbitals describing the inner (core) electrons, since they require a relatively large number of functions for representing the wave function cusp near the nucleus, and furthermore are largely independent of the environment. Contracting a basis set will always increase the energy, since it is a restriction of the number of variational parameters and makes the basis set less flexible, but it will also reduce the computational cost significantly. The decision is thus how much loss in accuracy is acceptable compared with the gain in computational efficiency.

The *degree of contraction* is the number of PGTOs entering the CGTO, typically varying between one and ten. The specification of a basis set in terms of primitive and contracted functions is done by the notation (10s4p1d/4s1p) → [3s2p1d/2s1p]. The basis set in parentheses is the number of primitives functions with non-hydrogen atoms (second row elements) before the slash and hydrogen after. The basis set in the square brackets is the number of contracted functions. Note that this does not indicate how the contraction is done; it only indicates the size of the final basis (and thereby the size of the variational problem in HF calculations).

There are two different ways of contracting a set of primitive GTOs to a set of contracted GTOs: *segmented* and *general* contraction. Segmented contraction is the older method and the one used in the above example. A given set of PGTOs is partitioned into smaller sets of functions that are made into CGTOs by determining suitable contraction coefficients. A 10s basis set may be contracted to 3s by taking the inner six functions as one CGTO, the next three as the second CGTO and the one remaining PGTO as the third "contracted" GTO:

$$\begin{aligned}
\chi_1(\text{CGTO}) &= \sum_{i=1}^{6} a_i \chi_i(\text{PGTO}) \\
\chi_2(\text{CGTO}) &= \sum_{i=7}^{9} a_i \chi_i(\text{PGTO}) \\
\chi_3(\text{CGTO}) &= \chi_{10}(\text{PGTO})
\end{aligned} \tag{5.8}$$

In a segmented contraction each primitive function as a rule is used only in one contracted function, that is the primitive set of functions is partitioned into disjoint sets. This often requires that some

PGTOs in two adjacent CGTOs need to be duplicated, since a PGTO may be important for both describing the outer part of a core orbital and the inner part of a valence orbital. The contraction coefficients in a segmented contraction are usually determined by a variational optimization of the atomic energy, where both the exponents and contraction coefficients are optimized simultaneously. It should be noted that this optimization often contains multiple minima, and selecting a suitable "optimum" solution may be non-trivial.[11]

In a general contraction *all* primitive functions (on a given atom) enter *all* the contracted functions, but with different contraction coefficients:

$$\chi_1(\mathrm{CGTO}) = \sum_{i=1}^{8} a_i \chi_i(\mathrm{PGTO})$$
$$\chi_2(\mathrm{CGTO}) = \sum_{i=1}^{8} b_i \chi_i(\mathrm{PGTO}) \tag{5.9}$$
$$\chi_3(\mathrm{CGTO}) = \sum_{i=1}^{8} c_i \chi_i(\mathrm{PGTO})$$

The contraction coefficient can be taken as the HF or DFT MO coefficients for different orbitals (e.g. 1s, 2s, 3s, ...) or, more generally, as coefficients from *Atomic Natural Orbitals* (ANOs) from a correlated calculation, as discussed in Section 5.4.4. The difference between segmented and general contraction may be illustrated as shown in Figure 5.3. The requirement of duplicating primitive

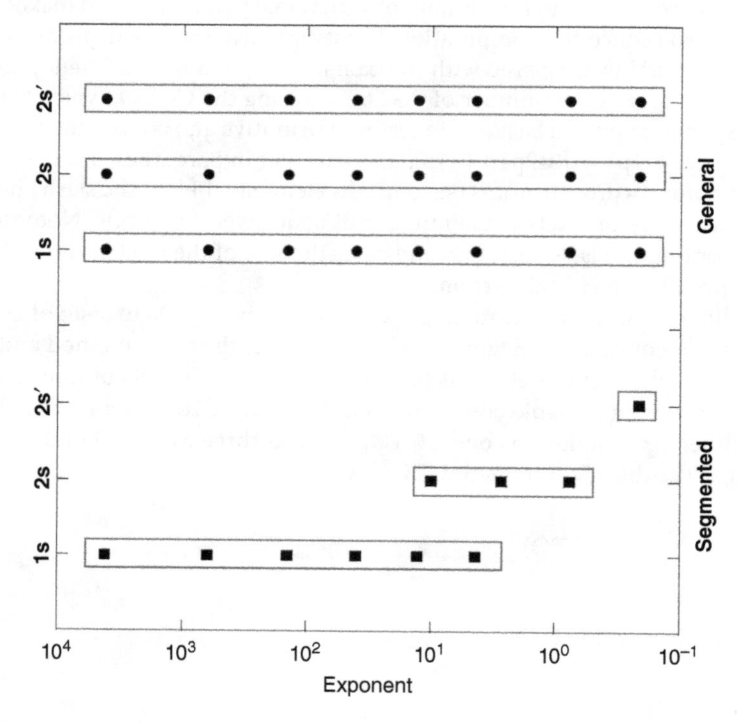

Figure 5.3 Illustrating segmented and general contraction.

functions in a segmented contraction usually means that more primitive functions are required to obtain the same accuracy as a general contraction of a smaller primitive set of functions.

In reality, there are only few truly segmented or general contracted basis sets. General contracted basis sets normally leave the outermost primitive function(s) uncontracted, and the disjoint nature of the primitive set of functions in a segmented contraction often necessitates a duplication of one or more functions, that is effectively a general contraction. The segmented–general classification should thus be seen as limiting cases, with actual basis sets having varying characteristics of both types. In practise it is usually easy to classify a basis set as either segmented and general contracted, based on whether only a few or the majority of primitive functions are shared between many (all) contracted functions.

A general contracted basis set is normally constructed from a set of energy-optimized primitive functions that in a separate step is contracted by a set of MO or ANO coefficients. The outermost primitive functions are then sequentially uncontracted until the contraction error becomes acceptably small. Different sized contracted basis sets can thus be constructed from the same set of primitive functions, which makes it easy to define and control the contraction error. A segmented contracted basis set, on the other hand, requires a decision on how to partition the primitive functions into disjoint sets of contracted functions and whether to duplicate one or more primitive functions. The exponents and contraction coefficients are then optimized simultaneously. This means that different primitive functions are used for each different contraction schemes of the same number of primitive functions, which makes it difficult to define and control the contraction error. The combined exponent and coefficient optimization is furthermore a complicated non-linear multiminima problem, which often makes it difficult to decide whether the "correct" minimum has been located by the optimization procedure.

Segmented and general contraction was originally seen as two unrelated contraction methods, but it has been shown that a general contracted basis set can be converted into a segmented contracted one in a unique fashion, which effectively corresponds to removal of the redundancy between the general contracted functions.[12] Segmented contracted basis sets are computationally more efficient than general contracted ones, since integral screening (Section 3.8.5) works more efficiently for the former. The integral screening is perfomed using contracted functions, and a general contraction implies that if a single contracted function is important beyond a given threshold, then integrals over all primitive functions must be calculated. In a segmented contraction, on the other hand, only integrals over those primitive functions that contribute to a given contracted function need to be calculated. This is especially important for methods like HF and DFT where the computational time for calculating integrals over basis functions dominates the total computational effort. For highly correlated methods, like CCSD(T), the time for calculating integrals over basis functions is only a small fraction of the total computational time and general contracted basis sets are often used in such cases.

5.3.4 Basis Set Augmentation

Molecular properties (Chapter 11) often have basis set requirements different from those obtained by energy minimization of the exponents and perhaps also the contraction coefficients. The easiest way of improving the basis set performance is usually to augment a standard energy-optimized basis set with specific basis functions for the particular property. The basis function optimization in these cases is in terms of the property of interest, and not in terms of energy, that is basis functions are added until the change upon addition of one extra function is less than a given threshold.

Many properties depend mainly on the wave function "tail" (far from the nucleus), which energetically is unimportant. An energy-optimized basis set that gives a good description of the outer part of the wave function therefore needs to be very large, with the majority of the functions being used to describe the 1s-electrons with an accuracy comparable with the outer electrons in an energetic sense. This is not the most efficient way of designing basis sets for describing the outer part of the wave function. Instead energy-optimized basis sets are augmented explicitly with *diffuse* functions (basis functions with small exponents). Diffuse functions are needed whenever loosely bound electrons are present (e.g. anions or excited states) or when the property of interest primarily depends on the most loosely bound electrons (e.g. polarizability). The exponents of the diffuse functions can be determined by energy minimization of atomic anions, where only the exponents of the additional diffuse functions are optimized. For basis sets designed for molecular properties, the exponents of the diffuse function can be determined, for example, by maximizing the atomic polarizability.[13] Alternatively, the exponents of diffuse functions can be assigned by simply dividing the smallest exponent in an energy-optimized basis set by a suitable scale factor, typically with a value close to 3. The scaling approach is particularly useful for generating multiple sets of diffuse function, which is required for calculating molecular properties such as hyperpolarizabilites.

Some properties are sensitive to an accurate representation of the wave function in the core region, and in some cases depend on the electron density *at* the nucleus position (e.g. the Fermi contact contribution to spin–spin coupling constants (Section 11.8.7)). In such cases a standard basis set can be augmented with *tight* basis functions, that is functions with (very) large exponents. For core properties, it is usually also necessary to reduce the contraction level of an energy-optimized basis set, since additional freedom is required to calculate the (small) variations in the core orbitals due to the molecular environment.

5.4 Examples of Standard Basis Sets

Optimization of basis function exponents, deciding a suitable contraction scheme and how to define augmenting functions are not things the average user needs to consider. Optimized basis sets of many different sizes and qualities are available either from websites[14] or stored internally in the computer programs. The user “merely” has to select a suitable quality basis for the calculation. Below is a short description of some basis sets that often are used in routine calculations, and reviews are available for more information.[15,16] The contractions are given for a second row element (such as carbon), while the corresponding ones for other elements can be found in the references.

5.4.1 Pople Style Basis Sets

STO-nG basis sets These are *Slater-type orbitals* consisting of n PGTOs.[17] This is an SZ-type basis set where the exponents of the PGTO are determined by *fitting* to the STO, rather than optimizing them by a variational procedure. Although basis sets with $n = 2$–6 have been derived, it has been found that using more than three PGTOs for representing the STO gives little improvement and the STO-3G basis is a widely used SZ basis set. This type of basis set has been determined for many elements of the periodic table. The designation of the carbon STO-3G basis is (6s3p) $\rightarrow$ [2s1p].

k-nlmG basis sets These basis sets, designed by Pople and coworkers, are of the split valence type, with the k in front of the dash indicating how many PGTOs are used for representing the core orbitals. The *nlm* after the dash indicate both how many functions the valence orbitals are split into and how many PGTOs are used for their representation. Two values (*nl*) indicate a split valence, while three values (*nlm*) indicate a triple split valence. The values *before* the G (for Gaussian) indicate the *s*- and *p*-functions in the basis; the polarization functions are placed *after* the G. These types of basis sets have the further restriction that the same exponent is used for both the *s*- and *p*-functions in the valence. This increases the computational efficiency, but of course decreases the flexibility of the basis set. The exponents and contraction coefficients for the *s*- and *p*-functions have been optimized by variational procedures at the HF level for atoms, while polarization functions have been assigned based on molecular calculations for a few small systems.

3-21G This is a split valence basis set, where the core orbitals are a contraction of three PGTOs, the inner part of the valence orbitals is a contraction of two PGTOs and the outer part of the valence is represented by one PGTO.[18] The designation of the carbon 3-21G basis is (6s3p) → [3s2p]. Note that the 3-21G basis set contains the same number of primitive GTOs as the STO-3G, but it is much more flexible as there are twice as many valence functions that can combine freely to make MOs.

6-31G This is also a split valence basis set, where the core orbitals are a contraction of six PGTOs, the inner part of the valence orbitals is a contraction of three PGTOs and the outer part of the valence is represented by one PGTO.[19] The designation of the carbon 6-31G basis set is (10s4p) → [3s2p]. In terms of contracted basis functions it contains the same number as 3-21G, but the representation of each function is better since more PGTOs are used.

6-311G This is a triple split valence basis set, where the core orbitals are a contraction of six PGTOs and the valence split into three functions, represented by three, one and one PGTOs, respectively, that is (11s5p) → [4s3p].[20]

To each of these basis sets can be added diffuse[21] and/or polarization functions.[22] Diffuse functions are *s*- and *p*-functions (only) and consequently go before the G. They are denoted by + or ++, with the first + indicating one set of diffuse *s*- and *p*-functions on non-hydrogen atoms and the second + indicating that a diffuse *s*-function is added also to hydrogen. The argument for only adding diffuse functions on non-hydrogen atoms is the same as for only adding polarization functions on non-hydrogens (Section 5.2). Polarization functions are indicated after the G, with a separate designation for non-hydrogen atoms and hydrogen. The 6-31+G(d), for example, is a split valence basis set with one set of diffuse *sp*-functions and a single *d*-type polarization function on non-hydrogen atoms. The 6-311++G(2df,2pd) is a triple split valence basis set with additional diffuse *sp*-functions, two *d*-functions and one *f*-function on non-hydrogen atoms, and diffuse *s*- and two *p*- and one *d*-functions on hydrogen. The largest standard Pople style basis set is 6-311++G(3df,3pd). These types of basis set have been derived for hydrogen and the second row elements, and some of the basis sets have also been derived for third and higher row elements. The composition in terms of contracted and primitive functions for the *s*- and *p*-parts is given in Table 5.1.

If only one set of polarization functions is used, an alternative notation in terms of * is also widely used. The 6-31G* basis set is identical to 6-31G(d) and 6-31G** is identical to 6-31G(d,p). A special note should be made for the 3-21G* basis. The 3-21G basis set is fundamentally too small to support polarization functions (it becomes unbalanced). However, the 3-21G basis set by itself performs poorly for hypervalent molecules, such as sulfoxides and sulfones. This can be improved substantially by adding a set of *d*-functions. The 3-21G* basis set has only *d*-functions on third row elements (it is sometimes denoted 3-21G(*) to indicate this) and should not be considered a polarized basis. Rather the addition of a set of *d*-functions is an *ad hoc* repair of a known flaw.

Table 5.1 Composition in terms of contracted and primitive basis functions for some Pople style basis sets.

	Hydrogen		Second row elements		Third row elements	
Basis	**Contracted**	**Primitive**	**Contracted**	**Primitive**	**Contracted**	**Primitive**
STO-3G	1s	3s	2s1p	6s3p	3s2p	9s6p
3-21G	2s	3s	3s2p	6s3p	4s3p	9s6p
6-31G(d,p)	2s1p	4s	3s2p1d	10s4p	4s3p1d	16s10p
6-311G(2df,2pd)	3s2p1d	5s	4s3p2d1f	11s5p	6s4p2d1f[a]	13s9p[a]

[a] McLean–Chandler basis set.

The Pople style basis sets have been used extensively and continue to be used as general-purpose basis sets for a variety of methods and properties, despite the fact that more modern basis sets provide better performance, that is deliver lower basis set errors for a similar computational cost. Part of the success of the Pople style basis sets is that they have been used in so many applications that certain "magic" combinations of methods and basis sets have been identified, where basis set errors partly compensate for method errors for a limited test set of systems and properties, the ubiquous B3LYP/6-31G* combination being a vivid example. These "right-answers-for-the-wrong-reason" combinations often employ small basis sets in order to generate compensating errors and consequently lead to computationally very efficient procedures; this without doubt has fuelled the popularity. Furthermore, the large number of calibration studies reported means that it is possible to get a fairly good idea of the level of accuracy that can be attained with a given method/basis set combination. This is of course a self-sustaining procedure; the more calculations that are reported with a given basis set, the more popular it becomes, since the calibration set becomes larger and larger.

5.4.2 Dunning–Huzinaga Basis Sets

Huzinaga has determined uncontracted energy-optimized basis sets up to (10s6p) for second row elements.[23] This was latter extended to (14s9p) by van Duijneveldt,[24] and up to (18s13p) by Partridge.[25] Dunning has used the Huzinaga primitive GTOs to derive various contraction schemes, and these are known as Dunning–Huzinaga (DH) type basis sets.[26] A DZ-type basis set can be made by a contraction of the (9s5p) PGTO to [4s2p]. The contraction scheme is 6,1,1,1 for *s*-functions and 4,1 for the *p*-functions. A widely used split valence-type basis set is a contraction of the same primitive set to [3s2p] where the *s*-contraction is 7,2,1 (note that one primitive enters twice). A widely used TZ-type basis set (actually only a triple split valence) is a contraction of the (10s6p) to [5s3p], with the contraction scheme 6,2,1,1,1 for *s*-functions and 4,1,1 for *p*-functions. Again, a duplication of one of the *s*- and *p*-primitives has been allowed.

McLean and Chandler have developed a similar set of contracted basis sets from Huzinaga primitive optimized sets for third row elements.[27] A DZ-type basis set is derived by contracting (12s8p) → [5s3p] and a TZ-type is derived by contracting (12s9p) → [6s5p]. The latter contraction is 6,3,1,1,1,1 for the *s*-functions (note a duplication of one function) and 4,2,1,1,1 for the *p*-functions, and is often used in connection with the Pople 6-311G when third row elements are present.

The Dunning–Huzinaga-type basis sets do not have the restriction of the Pople style basis sets of equal exponents for the *s*- and *p*-functions, and they are therefore somewhat more flexible. These basis sets are, with the exception of the McLean–Chandler version for third row atoms in connection with the Pople style basis sets, only used very rarely.

Table 5.2 Composition in terms of contracted and primitive basis functions for the Karlsruhe-type basis sets.

	Hydrogen		Second row elements		Third row elements	
Basis	Contracted	Primitive	Contracted	Primitive	Contracted	Primitive
SVP	2s1p	4s	3s2p1d	7s4p	4s3p1d	10s7p
TZV	3s2p1d	5s	5s3p2d1f	11s6p	5s4p2d1f	14s9p
QZV	4s3p2d1f	7s	7s4p3d2f1g	15s8p	9s6p4d2f1g	20s14p

5.4.3 Karlsruhe-Type Basis Sets

The group centered around R. Ahlrichs has designed basis sets of DZ, TZ and QZ quality for the elements up to Kr (Table 5.2). The *Split Valence Polarized* (SVP) basis set is a [3s2p] contraction of a (7s4p) set of primitive functions (contraction 5,1,1 and 3,1), while the *Triple Zeta Valence* (TZV) basis set is a [5s3p] contraction of an (11s6p) set of primitive functions (contraction 6,2,1,1,1 and 4,1,1).[28,29] The series has been extended by a *Quadruple Zeta Valence* (QZV) basis set, being a [7s4p] contraction of a (15s8p) set of primitive functions with the contraction 8,2,1,1,1,1,1 and 5,1,1,1, and this work also included a general update of the SVP and TVP to produce the so-called Def2-SVP, -TVP, -QZV basis sets.[30] Note that both the TZV and QZV basis sets employ more contracted *s*-functions than indicated by the TZ and QZ acronyms. The *s*- and *p*-exponents and corresponding contraction coefficients are optimized at the HF level, while the polarization functions are taken from the cc-pVXZ basis sets. Two slightly different sets of polarization functions are available for some atoms, depending on whether the basis set is used for HF/DFT or electron correlation methods.

The Karlsruhe basis sets have been used extensively for both HF/DFT and electron correlation methods, since they are computationally quite efficient and furthermore available for a large fraction of atoms in the periodic table.

5.4.4 Atomic Natural Orbital Basis Sets

All of the above basis sets are of the segmented contraction type. Modern contracted basis sets aimed at producing very accurate wave functions often employ a general contraction scheme. The *Atomic Natural Orbitals* (ANOs) and correlation consistent basis sets below are of the general contraction type.

The idea in the ANO-type basis sets is to contract a large PGTO set to a fairly small number of CGTOs by using *natural orbitals* from a correlated calculation on the free atom, typically at the CISD level.[31,32] The natural orbitals are those that diagonalize the density matrix and the eigenvalues are called *orbital occupation numbers* (see Section 10.5). The orbital occupation number is the number of electrons in the orbital. For an RHF wave function, ANOs would be identical to the canonical orbitals with occupation numbers of exactly 0 or 2. When a correlated wave function is used, however, the occupation number may have any value between 0 and 2. The ANO contraction selects the important combinations of the PGTOs from the magnitude of the occupation numbers. A large primitive basis set, often generated as an even-tempered sequence, may generate several different contracted basis sets by gradually lowering the selection threshold for the occupation number. The nice feature of the ANO contraction is that it more or less "automatically" generates balanced basis sets, for example for neon the ANO procedure generates the following basis set: [2s1p], [3s2p1d], [4s3p2d1f] and

[5s4p3d2f1g]. Furthermore, in such a sequence the smaller ANO basis sets are true subsets of the larger, since the same set of primitive functions is used.

Roos and coworkers have constructed ANO-S (Small) and ANO-L (Large) basis sets for atoms up to Kr by a general contraction of a set of primitive functions using coefficients from natural orbitals calculated by CASPT2.[33] For second row atoms the designations are (14s9p4d) → [3s2p1d] and (14s9p4d3f) → [4s3p2d1f], respectively, while the corresponding designations for a third row atom are (17s12p5d) → [4s3p1d] and (17s12p5d4f) → [5s4p2d1f].

5.4.5 Correlation Consistent Basis Sets

The primary disadvantage of ANO basis sets is that a large number of primitive GTOs are employed, which makes them computationally inefficient. Dunning, Peterson and coworkers have proposed a somewhat smaller set of primitives that yields comparable results to the ANO basis sets.[34,35] The *correlation consistent* (cc; the convention is to use lower case letters as the acronym, to distinguish it from coupled cluster (CC)) basis sets are geared towards recovering the correlation energy of the valence electrons. The name correlation consistent refers to the fact that the basis sets are designed such that functions that contribute similar amounts of correlation energy are included at the same stage, independent of the function type. For example, the first *d*-function provides a large energy lowering, but the contribution from a second *d*-function is similar to that from the first *f*-function. The energy lowering from a third *d*-function is similar to that from the second *f*-function and the first *g*-function. The addition of polarization functions should therefore be done in the order: 1d, 2d1f and 3d2f1g. An additional feature of the correlation consistent basis sets is that the energy error from the *sp*-basis should be comparable with (or at least not exceed) the correlation error arising from the incomplete polarization space, and the *sp*-basis therefore also increases as the polarization space is extended. The *s*- and *p*-basis set exponents are optimized at the HF level for the atoms, while the polarization exponents are optimized at the CISD level as an even-tempered sequence and the primitive functions are contracted by a general contraction scheme using natural orbital coefficients.

Several different sizes of correlation consistent basis sets are available in terms of the number of contracted functions. These are known by their acronyms: cc-pVDZ, cc-pVTZ, cc-pVQZ, cc-pV5Z and cc-pV6Z (*correlation consistent polarized Valence Double/Triple/Quadruple/Quintuple/Sextuple Zeta*). The composition in terms of contracted and primitive (for the *s*- and *p*-parts) functions is shown in Table 5.3. Note that each step up in terms of quality increases each type of basis function by one, and adds a new type of higher-order polarization function. For third row systems it has been found that the performance is significantly improved by adding an extra tight *d*-function.[36]

Table 5.3 Composition in terms of contracted and primitive basis functions for the correlation consistent basis sets.

	Hydrogen		Second row elements		Third row elements	
Basis	Contracted	Primitive	Contracted	Primitive	Contracted	Primitive
cc-pVDZ	2s1p	4s	3s2p1d	9s4p	4s3p2d	12s8p
cc-pVTZ	3s2p1d	5s	4s3p2d1f	10s5p	5s4p3d1f	15s9p
cc-pVQZ	4s3p2d1f	6s	5s4p3d2f1g	12s6p	6s5p4d2f1g	16s11p
cc-pV5Z	5s4p3d2f1g	8s	6s5p4d3f2g1h	14s8p	7s6p5d3f2g1h	20s12p
cc-pV6Z	6s5p4d3f2g1h	10s	7s6p5d4f3g2h1i	16s10p	8s7p6d4f3g2h1i	21s14p

The energy-optimized cc-pVXZ basis sets can be augmented with diffuse functions, indicated by adding the prefix aug- to the acronym.[37] The augmentation consists of adding one extra function with a smaller exponent for each angular momentum, that is the aug-cc-pVDZ has additionally one *s*-, one *p*- and one *d*-function, the cc-pVTZ has 1s1p1d1f extra for non-hydrogens and so on. The exponents for the diffuse functions are determined by energy minimization of atomic anions. The aug′-cc-pVXZ notation is sometimes used to indicate that diffuse functions are added only to non-hydrogen atoms.

The cc-pVXZ basis sets may also be augmented with additional *tight* functions (large exponents) if the interest is in recovering core–core and core–valence electron correlation, producing the acronym cc-pCVXZ (X = D, T, Q, 5). The cc-pCVDZ has additionally one tight *s*- and one *p*-function, the cc-pCVTZ has 2s2p1d tight functions, the cc-pCVQZ has 3s3p2d1f and the cc-pCV5Z has 4s4p3d2f1g for non-hydrogens.[38] The exponents for the tight functions are determined by energy minimization of the core–core and core–valence correlation energy. The cc-pwCVXZ basis sets[39] employ a weighting scheme in order to focus on the chemically more important core–valence, rather than the core–core, correlation energy. The main difference is that the exponents of the additional tight functions are somewhat smaller for the cc-pwCVXZ basis sets relative to the cc-pCVXZ basis sets.

The first paper in 1989 describing the correlation consistent basis sets marks the birth of modern basis sets designed to provide systematic convergence towards the basis set limiting result. They have been very succesful in establishing benchmark results for highly correlated wave function methods for a variety of systems and properties. Since they have been specifically optimized for wave function correlation methods, they should be the first choice for such calculations.

5.4.6 Polarization Consistent Basis Sets

The basis set convergence of electron correlation methods is inverse polynomial in the highest angular momentum functions included in the basis set, while the convergence of the independent-particle HF and DFT methods is exponential (Section 5.9).[7,40,41] This difference in convergence properties suggests that the optimum basis sets for the two cases will also be different, especially should low angular momentum functions be more important for HF/DFT methods than for electron correlation methods as the basis set becomes large. Since DFT (Chapter 6) often is the preferred method for routine calculations, it is of interest to have basis sets that are optimized for DFT-type calculations and that are capable of systematically approaching the basis set limit. The *polarization consistent* (pc) basis sets have been developed analogously to the correlation consistent basis sets except that they are optimized for DFT methods.[42,43] The name indicates that they are geared towards describing the polarization of the (atomic) electron density upon formation of a molecule, rather than describing the correlation energy. Since there is little difference between HF and DFT, and even less difference between different DFT functionals, these basis sets are suitable for independent-particle methods in general.

The polarization consistent basis sets also employ an energetic criterion for determining the importance of each type of basis function. The level of polarization beyond the isolated atom is indicated by a value after the acronym, that is a pc-0 basis set is unpolarized, pc-1 contains a single polarization function with one higher angular momentum, pc-2 contains polarization functions up to two beyond that required for the atom, etc. In contrast to the cc-pVXZ basis sets, the importance of the polarization functions must be determined at the molecular level, since the atomic energies only depend on *s*- and *p*-functions (at least for elements in the first three rows in the periodic table). For the DZP- and TZP-type basis sets (pc-1 and pc-2), the consistent polarization is the same as for the cc-pVXZ basis sets (1d and 2d1f), but at the QZP and 5ZP levels (pc-3 and pc-4) there are one and two additional *d*-functions (4d2f1g and 6d3f2g1h), respectively (Table 5.4). The *s*- and *p*-basis set exponents are optimized at the DFT level for the atoms, while the polarization exponents are selected as suitable average

Table 5.4 Composition in terms of contracted and primitive basis functions for the segmented polarization consistent basis sets.

	Hydrogen		Second row elements		Third row elements	
Basis	Contracted	Primitive	Contracted	Primitive	Contracted	Primitive
pc-0	2s	3s	3s2p	6s3p	4s3p	9s6p
pc-1	2s1p	4s	3s2p1d	8s4p	4s3p1d	12s9p
pc-2	3s2p1d	6s	4s3p2d1f	11s6p	5s4p2d1f	15s11p
pc-3	4s4p2d1f	9s	5s4p4d2f1g	15s9p	6s5p4d2f1g	21s14p
pc-4	5s6p3d2f1g	11s	6s5p6d3f2g1h	20s11p	7s6p6d3f2g1h	25s17p

values from optimizations for a selection of molecules. The primitive functions are subsequently contracted by a general contraction scheme by using the atomic orbital coefficients, where the degree of contraction is determined such that the contraction error is smaller than the error relative to the basis set limit. The general contracted pc-n basis sets have subsequently been converted into a segmented contracted version, with the acronym pcseg-n, which has essentially the same accuracy, but is significantly more efficient computationally.[12] The conversion from pc-n to pcseg-n also included a general update, and the original general contracted pc-n basis sets should therefore be considered obsolete.

For properties dependent on the wave function tail, such as electric moments and polarizabilities, the convergence towards the basis set limit can be improved by explicitly adding a set of diffuse functions, producing the acronym aug-pcseg-n. The pcseg-n basis sets are available for atoms up to Kr, and since they have been specifically optimized for DFT methods, should be a first choice for such type of calculations (the author admits to being biased in this statement).

5.4.7 Correlation Consistent F12 Basis Sets

The slow basis set convergence of wave function electron correlation methods is due to difficulties in representing the Coulomb hole by an expansion in nuclear centered Gaussian functions. The description of the short-range electron correlation can be substantially improved by including a correlation function depending explicitly on the interelectronic distance, and this leads to the so-called F12 methods (Section 4.11). Not surprisingly, these methods have different basis set requirements than HF/DFT and standard electron correlation methods, and Table 5.5 shows optimized basis sets for F12 methods developed along the same procedures as the cc-pVXZ basis sets.

Table 5.5 Composition in terms of contracted and primitive basis functions for the cc-pVXZ-F12 basis sets.

	Hydrogen		Second row elements		Third row elements	
Basis	Contracted	Primitive	Contracted	Primitive	Contracted	Primitive
cc-pVDZ-F12	3s2p	5s	5s5p2d	11s6p	6s6p3d	16s12p
cc-pVTZ-F12	4s3p1d	6s	6s6p3d2f	13s7p	7s7p4d2f	17s13p
cc-pVQZ-F12	5s4p2d1f	8s	7s7p4d3f2g	15s9p	8s8p5d3f2g	21s13p

5.4.8 Relativistic Basis Sets

Inclusion of scalar relativistic effects (Chapter 9) has the qualitative consequence that s-orbitals contract while other types of orbitals expand, and these effects become progressively larger as the atomic number increases. Basis sets that are optimized for relativistic methods thus have s-functions with larger exponents than those optimized for non-relativistic methods, and the contraction coefficients also differ to reflect the different radial behavior. The requirement for s-functions with very large exponents is to some extent relaxed if a finite size nucleus is used in place of a point-nucleus. Relativistic methods of the four-component type include spin–orbit interactions and have different spinors, such as $p_{1/2}$ and $p_{3/2}$, that are expanded in the same type of basis function. The small component $p_{1/2}$ spinor acquires substantial s-character and thus also contracts compared to the non-relativistic case, while the $p_{3/2}$ spinor expands. This leads to a requirement of p-type basis functions with both larger and smaller exponents compared to the non-relativistic case, and the contraction coefficients for the $p_{1/2}$ and $p_{3/2}$ spinors in this set of primitive p-functions are also different. Similar considerations apply for the spinors expanded in d- and f-type basis functions, although the differences compared to the non-relativistic case are smaller than for the s- and p-type functions. Basis sets optimized for four-component relativistic methods thus tend to have more primitive functions than corresponding non-relativistic ones.

A common procedure for designing basis sets for scalar relativistic methods is to simply recontract a set of basis functions optimized for a non-relativistic method with contraction coefficients from a relativistic calculation. This ignores the need for primitive s-functions with larger exponents in order to represent the 1s-orbital near the nucleus, but the resulting error is insensitive to the (molecular) environment and therefore tends to cancel out when calculating relative energies. The cc-pVXZ and ANO-L basis sets are available in such relativistic recontracted versions where the contraction coefficients are taken from Douglas–Kroll–Hess calculations, with the labels cc-pVXZ-DK[44] and ANO-RCC.[45]

K. Dyall constructed general contracted basis sets of DZP, TZP and QZP quality that are suitable for four-components methods. These basis sets are constructed along the same principles as used for constructing the non-relativistic cc-pVXZ basis sets and are available for all the atoms in the periodic table. The functions describing the atomic occupied orbitals are optimized at the Dirac–Hartree–Fock level with a finite Gaussian nucleus model while correlating functions are determined at the CISD level.[46]

5.4.9 Property Optimized Basis Sets

The basis set requirements for obtaining a certain accuracy of a given molecular property is often different from that required for a corresponding accuracy in energy. There is no analogy to the variational principle for molecular properties, which makes the design of property optimized basis sets less straightforward. Basis sets for properties are therefore often tailored by adding functions until the desired accuracy is obtained.

Properties that depend directly on the potential energy surface, like equilibrium geometries and vibrational frequencies (first and second derivatives of the energy with respect to nuclear positions), usually do not have basis set requirements beyond those in standard energy-optimized basis sets. The basis set requirements for other types of properties, however, depend on the specific perturbation operator defining the property. Electric polarizabilities, for example, measure how easy the wave function distorts when applying a perturbing electric field. An electric field adds a potential energy operator to the Hamiltonian, and thus influences the wave function most where the potential from the nuclei and other electrons is small, that is in the region far from the nuclei where the electron density is low. The important part of the wave function is thus the "tail", necessitating diffuse function in the basis set. Furthermore, an electric field polarizes the electron cloud and diffuse polarization functions are therefore also important. For perturbation-independent basis functions, there is a "$2n + 1$" rule, that is if the unperturbed system is reasonably described by basis functions up to angular momentum L, then a basis set that includes functions up to angular momentum $L + n$ can predict properties up to order $2n + 1$. A minimum description of molecules containing up to third row atoms requires s- and p-functions, implying that d-functions are necessary for the polarizability and the first hyperpolarizability and f-functions should be included for the second and third hyperpolarizabilities. A more realistic description, however, would include d-functions for the unperturbed system, necessitating f-functions for the polarizability. The standard diffuse functions available for many basis sets (Section 5.3.4) are often sufficient to substantially improve the basis set convergence of electric properties, but for certain properties, like high-order hyperpolarizabilities, addition of a second set of even more diffuse functions is sometimes found to be necessary. The exponents for the diffuse functions have for some types of basis sets been assigned by maximizing the calculated polarizability.[13]

A completely different type of property is the indirect spin–spin coupling constant, which contains information about the interactions of nuclear spins coupled by the electron spins. There are four operators involved in calculating the indirect spin–spin coupling constant (Section 11.8.7): Fermi-contact, spin-dipole, paramagnetic-spin-orbit and diamagnetic-spin-orbit (Table 11.2). The Fermi-contact operator is a δ-function (Equation (11.109)), which measures the electron density of the wave function at a single point, the nuclear position. Since Gaussian functions have an incorrect behavior at the nucleus (zero derivative compared with the "cusp" displayed by an exponential function), this requires the addition of a number of very "tight" (large exponents) s-functions in order to predict this term accurately. The paramagnetic-spin-operator is found to require additional tight p-functions, while the spin-dipole operator requires addition of tight p-, d- and f-functions. In order to allow sufficient flexibility to describe the (small) differences in the wave function near the nuclei due to differences in the valence bonding region, the basis set furthermore needs to be much less contracted than for energetic properties. The pcJ-n and ccJ-pVXZ basis sets have been specifically optimized to reduce the basis set errors using DFT and correlated wave function methods, respectively, and both are available in several zeta-quality levels.[47,48]

Nuclear magnetic shielding constants measure the interaction of nuclear spins with an external magnetic field as shielded by the electrons, and the relevant operators are the diamagnetic shielding, paramagnetic-spin-orbit and angular momentum operators (Equation (11.128)). The paramagnetic-spin-orbit operator, in analogy with the indirect spin–spin constant, requires addition of tight p-functions and a somewhat reduced contraction of the p-space in the basis set. For fourth row elements, which have occupied d-orbitals, the d-contraction must also be slightly reduced compared to basis sets suitable for energetic properties. The pcS-n basis sets have been designed to reduce basis set errors in the calculation of nuclear magnetic shielding constants using DFT methods, but they have been shown also to be suitable for MP2 methods.[49] In analogy with the energy optimized pc-n basis sets, the general contracted pcS-n have been converted to the computationally more efficient segmented contracted pcSseg-n.[50]

5.5 Plane Wave Basis Functions

Rather than starting with basis functions aimed at modeling the atomic orbitals (STOs or GTOs) and forming linear combination of these to describe orbitals for the whole system, one may instead use functions aimed directly at the full system. For modeling extended (infinite) systems, for example a unit cell with periodic boundary conditions, this suggests the use of functions with an "infinite" range. The outer valence electrons in metals behave almost like free electrons, which leads to the idea of using solutions for the free electron as basis functions. The solutions to the Schrödinger equation for a free electron in one dimension can be written either in terms of complex exponentials or cosine and sine functions:

$$\phi(x) = A\,\mathrm{e}^{ikx} + B\,\mathrm{e}^{-ikx} \tag{5.10}$$

$$\phi(x) = A\cos(kx) + B\sin(kx) \tag{5.11}$$

The energy depends quadratically on the k factor, as shown in

$$E = \tfrac{1}{2}k^2 \tag{5.12}$$

For infinite systems, the molecular orbitals coalesce into *bands*, since the energy spacing between distinct levels vanishes. The electrons in a band can be described by orbitals expanded in a basis set of *plane waves*, which in three dimensions can be written as a complex function:

$$\chi_k(\mathbf{r}) = \mathrm{e}^{i\mathbf{k}\cdot\mathbf{r}} \tag{5.13}$$

The wave vector **k** plays the same role as the exponent ζ in a GTO (Equations (5.2) and (5.3)), and is related to the energy by means of Equation (5.12) (conventionally given in units of eV). As seen in Equation (5.11), **k** can also be thought of as a frequency factor, with high **k** values indicating a rapid oscillation. The permissible **k** values are given by the unit cell translational vector **t**, that is $\mathbf{k}\cdot\mathbf{t} = 2\pi m$, with m being a positive integer. The number of plane waves is thus determined by the highest energy **k** vector included, which implicitly defines the maximum kinetic energy that can be represented (Equation (5.12)), and the volume V of the unit cell, as shown in

$$M_{\mathrm{PW}} = \frac{1}{2\pi^2} V E_{\mathrm{max}}^{3/2} \tag{5.14}$$

Basis sets of increasing quality can be defined by increasing the E_{max} parameter. A typical energy cutoff of 200 eV and a modest-sized unit cell of dimension ~15 Å corresponds to a basis set with ~20 000 functions, that is plane wave basis sets tend to be significantly larger than typical Gaussian basis sets. Note also that the size of a plane wave basis set depends only on E_{max} and the size of the periodic cell, not on the actual system described within the cell. This is in contrast to the linear increase with system size for nuclear-centered Gaussian functions, that is plane wave basis sets become more favorable for large systems.

While plane wave basis sets have primarily been used for periodic systems, they can also be used for molecular species by a *supercell* approach, where the molecule is placed in a sufficiently large unit cell.[51] Sufficiently large in this context is defined by the condition that the molecule does not interact with its own images in the neighboring cells. For charged or strongly polar systems, this would require a very large unit cell, which becomes computationally inefficient. Specialized techniques can be used for handling such cases.[52]

Plane wave basis functions are ideal for describing delocalized slowly varying electron densities, such as the valence and conduction bands in a metal. The core electrons, however, are strongly localized around the nuclei, and the valence orbitals have a number of rapid oscillations in the core region to maintain orthogonality. Describing the core region adequately thus requires a large number of rapidly oscillating functions, that is a plane wave basis with very large E_{max}. The singularity of the nucleus–electron potential is furthermore almost impossible to describe in a plane wave basis, and this type of basis set is therefore often used in connection with pseudo-potentials (Section 5.12) for smearing the nuclear charge and modeling the effect of the core electrons. Note that a pseudopotential is also required for smearing the potential near the nucleus in hydrogen, even though hydrogen does not have core electrons. Gaussian functions, on the other hand, are very efficient for describing the core electrons, but poor for describing the wave-like character of the valence electrons. The latter requires linear combinations of Gaussian functions with small exponents, which in practice leads to severe numerical problems with linear dependency.

The most efficient way of explicitly describing both core and valence electrons for periodic systems is a combination of a localized basis set for describing the core electrons and a plane wave basis set for describing the valence electrons. This has the price of an increased computational complexity, since new integrals involving different types of basis functions are required or the physical space must be partitioned into core and valence regions with continuity conditions at the boundary. The localized basis functions can be radial polynomials, numerical solutions to the atomic problem[53] or Gaussian functions,[54,55] and such mixed basis sets are often collectively called *Augmented Plane Wave* (APW) methods.

5.6 Grid and Wavelet Basis Sets

The Hartree–Fock Equations (3.42) are a set of coupled second-order differential equations that can be transformed into a matrix Equation (3.54) by expansion into a basis set. For atoms and diatomic molecules, the differential equations can be solved directly by transformation into a prolate spheroidal coordinate system (ξ, η, θ), where θ is the angle around the molecular axis and the two other coordinates are defined in the following equation by the internuclear distance R_{AB} and electron–nuclear distances r_A and r_B:

$$\xi = \frac{r_A + r_B}{R_{AB}} \quad ; \quad \eta = \frac{r_A - r_B}{R_{AB}} \tag{5.15}$$

The angular dependence can be solved analytically, while the equations in the two other coordinates are solved on a two-dimensional grid; such solutions are often referred to as *numerical Hartree–Fock* results.[56] Even modest-sized grids are capable of achieving milli-Hartree accuracy in terms of total energy, but higher accuracy requires careful control of the grid size and cutoff criteria. Grid-based techniques for solving the Hartree–Fock or Kohn–Sham equations for polyatomic molecules[57,58] and periodic systems[59] have also been reported, and can be considered as solving the problem by expansion into a basis set composed of δ-functions. For independent particle methods (HF/DFT) the number of grid points is typically a few thousands per atom (analogous to the integration of the exchange-correlation contribution in the Kohn–Sham DFT method), but an extension to correlated methods would require an unmangeable many millions of grid points per atom.

Harrison and coworkers have proposed a multiresolution procedure where the molecular orbitals are expanded into a set of *wavelets*.[60] The essence of the method is to place the molecule in a suitably sized box and to repeatedly subdivide the space by a factor of 2, leading to 2^n boxes at level n. The part of the orbitals within each box is expanded into a set of (orthogonal) Legendre polynomials of order k, the first few of which are given in Equation (5.6). By increasing the number of boxes and the number of Legendre polynomials, the accuracy can be tuned to any desired degree, and the procedure can be made self-adaptive, such that a given box is not further subdivided once the target accuracy has been reached. For small systems, this approach is able to provide HF energies accurate to within 10^{-6} au.[61] One significant advantage of the multiresolution method is that it by construction is a low-scaling method (near-linear for HF and DFT and quadratic for MP2), but the associated prefactor is large, and these methods are therefore not competitive for small systems.

The advantage of nucleus positioned Gaussian basis sets is that they allow for a compact representation of the wave function in terms of only tens of functions per atom, while grid-based methods require thousands of δ-functions per atom for an accurate result. Multiresolution methods can be considered as grid-based methods, where the δ-functions are replaced by finite extent polynomials and are intermediate in terms of the number of basis functions required for a given accuracy.

The main problem in multiresolution methods is the need to handle the large number of expansion coefficients associated with the basis functions. For a given molecule and target accuracy, the box containing the molecule is subdivided into many smaller boxes (of different sizes) each containing a number of one-dimensional polynomial functions. The wave function is represented in this basis set of polynomials in boxes by a set of coefficients for products of polynomials. Independent particle methods like HF and DFT only need to represent one-electron quantities (orbital/density) and the number of expansion coefficients thus increases as p^3, where p is a measure of the number of polynomials and p increases with increasing desired accuracy of the final result. Methods including electron-pair correlation, such as MP2, need to represent two-electron quantities, and the corresponding scaling with accuracy is therefore p^6, and methods including three-electron correlation, such as CCSDT, scales as p^9. The increase in the number of expansion coefficients with complexity has been called the *curse of dimensionality*, and an essential part in handling the computational complexity is to use tensor decomposition methods (Section 17.6.3) to reduce the number of coefficients that need to be stored.[62]

5.7 Fitting Basis Sets

Reduced prefactor methods (Section 3.8.7) often employ an auxiliary basis set where products of (orbital) basis functions are replaced by an expansion in a single set of fitting functions:

$$|\chi_\alpha \chi_\beta\rangle \approx \sum_k^{M_{\text{aux}}} c_k \chi_k \tag{5.16}$$

The set of auxiliary basis functions must span the product space of the orbital basis functions, to within a given threshold accuracy. For computational efficiency reasons the fitting functions are usually taken to be Gaussian-type functions. Construction of auxiliary basis sets can be done using similar techniques as for optimizing orbital basis sets, by minimizing a suitable error function in place of minimizing the total energy.[63] The requirement of the auxiliary basis set to span the orbital product function space means that it is somewhat larger and contains higher angular momentum functions than the orbital basis set, and can usually only be contracted to a small extent. The optimum set of fitting functions will, in analogy with the orbital basis set, depend on the property (Coulomb, exchange, MP2, F12 methods) and to some extend also on the quality of the orbital basis set.[64]

5.8 Computational Issues

The computational problem is formally the same whether a Gaussian, plane wave or polynomial basis is used – calculate matrix elements of quantum mechanical (kinetic and potential) operators over basis functions and solve the variational problem by an iterative procedure – but the nature of the functions results in some differences. With a GTO basis set the matrix elements can be calculated by analytical integration and implementation of the corresponding formulas. The electron–electron repulsion operator requires integrals of all combinations of four basis functions, and HF and DFT methods therefore formally scales as M^4_{basis}.

With plane wave basis sets the integrals involving the kinetic energy operator are trivial, since they just correspond to multiplying by half the squared momentum operator, $^1/_2\mathbf{k}^2$. Matrix elements involving local potential energy operators are similarly easy, since they correspond to simple multiplication in the Fourier transformed space, and the (fast) Fourier transform only requires $M_{PW}\log M_{PW}$ computational effort. The Coulomb operator is local, as is the exchange operator in density functional theory, but the exchange energy at the Hartree–Fock level involves a non-local operator. Incorporating HF exchange with plane wave basis sets requires a separate Fourier transform for each orbital pair, and is therefore computationally expensive compared to the pure DFT method.[65,66] This difference at least partly explains why (pure) density functional methods (as opposited to hybrid) have dominated in solid-state physics, while HF traditionally has been preferred for molecular systems in connection with GTO basis sets. Another reason is of course that HF theory cannot describe metallic systems – the large band gap predicted by HF makes all periodic systems insulators or semiconductors.

A GTO basis set will typically have tens of functions per atom, with perhaps a few hundred or thousand functions for the whole system. A plane wave basis set, on the other hand, will often have hundreds of thousands of functions. In a traditional implementation, the variational problem is solved by repeated diagonalization of a Fock-type matrix but this becomes problematic when the number of basis functions exceeds a few thousand owing to the cubic scaling of matrix diagonalization. For large plane wave basis sets, the variational problem is therefore often solved by other methods, such as conjugate gradient optimization, quenched dynamics methods or DIIS-type extrapolations for direct minimization of the energy or by reduced space methods (Section 17.2.5) for selectively extracting a limited number of eigenvalue and eigenvectors of the energy matrix.[67] In Car–Parrinello-type dynamics (Section 15.2.5), the variational problem is solved by propagating the orbital parameters with fictive masses along with the nuclear degrees of freedom. The computational scaling of solving the variations problem with these methods is $N^2{}_{\text{electron}}M_{\text{PW}}$, that is the computational time increases linearly with the number of plane wave functions. Electron correlation methods, like MP2, employ an expansion of the correlation correction in the virtual orbitals and is therefore computationally expensive with plane wave basis sets.[68]

A significant advantage of plane wave basis sets is that they are independent of the nuclear positions and essentially complete up to the maximum energy included. This means that the problem of basis set superposition error (Section 5.13) does not occur, and the calculation of the gradient of the energy is easy, as it is given directly by the Hellman–Feynman force (Section 11.3), that is there are no components associated with the change of basis function position ("Pulay forces").

5.9 Basis Set Extrapolation

The main advantage of basis sets that are available in several well-defined quality levels, such as the ANO, correlation consistent and polarization consistent basis sets, is the ability to generate results that converge toward the basis set limit in a systematic fashion. For example, from a series of calculations with the 3-21G, 6-31G(d,p), 6-311G(2d,2p) and 6-311++G(3df,3pd) basis sets it may not be obvious whether the property of interest is "converged" with respect to further increases in the basis, and it is difficult to estimate what the basis set limit would be. The cc-pVXZ basis sets, on the other hand, consistently reduce errors (both HF and correlation) for each step up in quality. In test cases it has been found that the cc-pVDZ basis can provide ~65% of the total (valence) correlation energy, the cc-pVTZ ~85%, cc-pVQZ ~93%, cc-pV5Z ~96% and cc-pV6Z ~98%, with similar reductions of the HF error. The analysis of basis set convergence is often done in terms of the maximum angular momentum function included in the basis set, but to avoid the ambiguity arising by having atoms from different rows and/or blocks in the periodic table, the basis set quality is in the following quantified in terms of the cardinal number X ($X = 2$ for a DZP-type basis set, $X = 3$ for a TZP-type basis set, etc.).

Theoretical and numerical analyses suggest that the basis set convergence at HF and DFT levels is square-root exponential:[7]

$$E_{\mathrm{HF/DFT}}(X) = E(\infty) + Ae^{-B\sqrt{X}} \tag{5.17}$$

Similar analyses for the *correlation* energy itself (i.e. not the *total* energy, which includes the HF contribution) suggest a convergence with an inverse power dependence, with the leading term for singlet electron pairs being X^{-3} while the leading term for triplet pairs is X^{-5}:[6,69,70]

$$\Delta E_{\mathrm{corr}}(X) = E(\infty) + AX^{-3} + BX^{-4} + CX^{-5} + \cdots \tag{5.18}$$

Electron correlation methods incorporating interelectronic distances, so-called F12 methods (Section 4.11), provide an accurate representation of the electron–electron cusp in the wave function and consequently converge much faster in terms of basis set quality. In analogy with standard electron correlation methods, the convergence is inverse polynomial, but the leading term is X^{-7} rather than X^{-3}. This in effect means that the total correlation energy with F12 methods and a given zeta-quality basis set is usually similar to a non-F12 method and basis sets of one or two higher zeta-quality. This lower basis set requirement of F12 methods is to a large extent offset by the requirement of calculating many additional integrals involving interelectronic coordinates.

Taking only the leading term in Equation (5.18) into account, the basis set convergence in Equations (5.17) and (5.18) can be generalized to the following equation by a function depending on the cardinal number X:

$$E_X = E_\infty + Af_X \tag{5.19}$$

The complete basis set energy can thus be estimated from two calculations with basis sets having cardinal numbers X and Y according to

$$E_\infty = \frac{f_X^{-1}E_X - f_Y^{-1}E_Y}{f_X^{-1} - f_Y^{-1}} \tag{5.20}$$

For an inverse polynomial dependence X^{-n}, as in Equation (5.18), the extrapolation becomes

$$\Delta E_{\mathrm{corr},\infty} = \frac{X^n \Delta E_{\mathrm{corr},X} - Y^n \Delta E_{\mathrm{corr},Y}}{X^n - Y^n} \tag{5.21}$$

A direct use of Equation (5.17) requires three points for extrapolation, but it can be used as a two-point formula by taking the B parameter as a universal constant (typical value ~6):

$$E_{\mathrm{HF/DFT},\infty} = \frac{e^{B\sqrt{X}} \Delta E_{\mathrm{HF/DFT},X} - e^{B\sqrt{Y}} \Delta E_{\mathrm{HF/DFT},Y}}{e^{B\sqrt{X}} - e^{B\sqrt{Y}}} \tag{5.22}$$

The theoretical assumption underlying the formulas in Equations (5.17) and (5.18) is that the basis set is saturated in the radial part (e.g. a TZP- type basis set should be *complete* in the *s*-, *p*-, *d*-

and f-function space). This is rarely the case as basis sets typically are designed to have comparable (residual) errors for each type of angular momentum function. Nevertheless, it has been found that extrapolations using the correlation-consistent basis sets and taking only the leading X^{-3} term in Equation (5.21) into account give good results when compared with accurate results generated by, for example, R12 methods.[71] It has been suggested that a separate extrapolation of the singlet (opposite spin) and triplet (same spin) correlation energies with X^{-3} and X^{-5} function forms, respectively, may provide better results.[72]

The main difficulty in using the cc-pVXZ or pcseg-n basis sets is that each step up in quality roughly doubles the number of basis functions. The simplest two-point extrapolation is using results from X, $Y = 2$, 3 basis sets (e.g. cc-pVDZ and cc-pVTZ), but a DZP-type basis set is usually too small to give good extrapolated correlation energies. An extrapolation based on X, $Y = 3$, 4 basis sets usually provides an improvement, but the requirement of performing calculations with at least a QZP-type basis set places severe constraints on the size of the systems that can be treated.

More empirical extrapolation formulas can be designed by rearranging Equation (5.20) and introducing a generalized fitting function depending on both cardinal numbers:[73]

$$E_\infty = \frac{f_X^{-1}E_X - f_Y^{-1}E_Y}{f_X^{-1} - f_Y^{-1}} = \left(1 - \frac{f_X}{f_Y}\right)^{-1}(E_X - E_Y) + E_Y \tag{5.23}$$

$$E_\infty = F(X,Y)(E_X - E_Y) + E_Y \tag{5.24}$$

The function $F(X,Y)$ can be parameterized to partly compensate for inadequacies in the two basis sets used for the extrapolation, and will then depend explicitly on the X, Y values. Extrapolation may provide only a marginal improvement, especially when used with DZP/TZP-type basis sets, but very rarely causes the results to deteriorate and can in favorable cases improve them corresponding to roughly one higher zeta level. Since extrapolation carries no additional computational effort, it should be considered when the ability to perform calculations with (at least) two different basis sets of well-defined quality levels exist.

Perhaps the most interesting aspect of the analyses that led to the development of the correlation-consistent basis sets is the fact that high angular momentum functions are necessary for achieving high accuracy. While d-polarization functions are sufficient for a DZ-type basis, a TZ-type should also include f-functions. Similarly, it is questionable to use a QZ-type basis for the sp-functions without also including three d-, two f- and one g-function in order to systematically reduce the errors. It can therefore be argued that an extension of, for example, the 6-31G(d,p) basis to 6-311G(d,p) is inconsistent as the second set of d-orbitals (and second set of p-orbitals for hydrogen) and a set of f-functions (d-functions for hydrogen) will give similar contributions as the extra set of sp-functions. Similarly, the extension of the 6-311G(2df,2pd) basis to 6-311G(3df,3pd) may be considered inconsistent, as the third d-function is expected to be as important as the fourth valence set of sp-functions, the second set of f-functions and the first set of g-functions, all of which are neglected.

In the search for a basis set converged value, other approximations should be kept in mind. Basis sets with many high angular momentum functions are normally designed for recovering a large fraction of the correlation energy. In the majority of cases, only the electron correlation of the valence electrons is considered (frozen-core approximation), since the core orbitals usually are insensitive to the molecular environment. As the valence space approaches completeness in terms of basis functions, the error from the frozen-core approximation will at some point become comparable to the remaining valence error. From studies of small molecules, where good experimental data are available, it is suggested that the effect of core electron correlation for unproblematic systems is comparable with the change observed upon enlarging the cc-pV5Z basis, that is of a similar magnitude as the introduction of h-functions.[74] Improvements beyond the cc-pV6Z basis set have been argued to produce changes of similar magnitude to those expected from relativistic corrections for second row elements, and further increases to cc-pV7Z and cc-pV8Z-type basis sets would be comparable with corrections due to breakdown of the Born–Oppenheimer approximation for systems with hydrogen. Within the non-relativistic realm, it would therefore appear that basis sets larger than cc-pV6Z would be of little use, except for extrapolating to the non-relativistic, clamped nuclei limit for testing purposes. In attempts at obtaining results of "spectroscopic accuracy" (~0.01 kJ/mol), a brute force calculation with, for example, the cc-pV7Z quality basis set combined with explicit extrapolation has been shown to become problematic,[72] and such high-quality results require explicit correlated techniques, such as the F12 method discussed in Section 4.11.

There is a practical aspect of using large basis sets, especially those including diffuse functions, that requires special attention, namely the problem of *linear dependence*. Linear dependence means that one (or more) of the basis functions can be written as a linear combination of the other, that is the basis set is overcomplete. A diffuse function has a small exponent and consequently extends far away from the nucleus on which it is located. An equally diffuse function located on a nearby atom will therefore span almost the same space. A measure of the degree of linear dependence in a basis set can be obtained from the eigenvalues of the overlap matrix **S** (Equation (3.54)). A truly linearly dependent basis set will have at least one eigenvalue of exactly zero, and the smallest eigenvalue of the **S** matrix is therefore an indication of how close the actual basis set is to linear dependence. As described in Section 17.2.3, solution of the SCF equations requires orthogonalization of the basis set by means of the $\mathbf{S}^{-1/2}$ matrix (or a related matrix that makes the basis orthogonal). If one of the **S** matrix eigenvalues is close to zero, this means that the $\mathbf{S}^{-1/2}$ matrix is essentially singular, which in turn will cause numerical problems if trying to carry out an actual calculation. In practice, there is therefore an upper limit on how close to completeness a basis set can be chosen to be, and this limit is determined by the finite precision with which the calculations are carried out. If the selected basis set turns out to be too close to linear dependence to be handled numerically for the particular system, the linear combinations of basis functions with low eigenvalues in the **S** matrix may be discarded.

5.10 Composite Extrapolation Procedures

The large majority of systems can, in principle, be calculated with a high accuracy by using a highly correlated method such as CCSD(T) and performing a series of calculations with systematically larger basis sets in order to extrapolate to the basis set limit. In practice, even a single water molecule is demanding to treat in this fashion (Chapter 12). Various approximate procedures have therefore been developed for estimating the "infinite correlation, infinite basis" limit (Figure 4.3) as efficiently as possible. These models rely on the fact that different properties converge with different rates as the level of sophistication increases and that effects from extending the basis set to a certain degree are additive. There are four main steps in these procedures:

1. Selecting the geometry.
2. Estimating the Hartree–Fock energy.
3. Estimating the electron correlation energy.
4. Estimating the energy from translation, rotation and vibrations.

Given a predefined target accuracy, the error from each of these four steps should be reduced below the desired tolerance. The error at a given level may be defined as the change that would occur if the calculation was taken to the "infinite correlation, infinite basis" limit. A typical target accuracy is ~4 kJ/mol (~1 kcal/mol), the so-called "chemical accuracy", although more advanced methods aim for an accuracy of ~1 kJ/mol.

Geometries converge relatively fast: at the HF level with a DZP-type basis the "geometry error" is often already ~4 kJ/mol or less, and an MP2 or DFT geometry optimized with a DZP or TZP basis set is normally sufficient for most applications. The translational and rotational contributions are trivial to calculate, as they depend only on the molecular mass and the geometry (Sections 14.5.1 and 14.5.2) and are very small in absolute values. The error from these can be neglected. The vibrational effect is mainly the zero-point energy, and it requires calculation of the frequencies. An accurate prediction of frequencies is fairly difficult. However, since the absolute value of the zero-point energy is fairly small, a large relative error is tolerable. Furthermore, the errors in calculated frequencies are to a certain extent systematic and can therefore be improved by a uniform scaling.[75]

The HF error depends only on the size of the basis set. The energy, however, behaves asymptotically as $\sim\exp(-\sqrt{X})$, where X is the basis set cardinal number. For example, with a basis set of TZP quality (4s3p2d1f for second row elements) the results are already quite stable. Combined with the fact that an HF calculation is the least expensive *ab initio* method, this means that the HF error is rarely the limiting factor.

The main problem is estimating the correlation energy. All electron correlation methods have a rather steep increase in the computational cost as the size of the basis set is enlarged, and the convergence in terms of the basis set cardinal number is quite slow ($\sim X^{-3}$). The main contribution to the correlation energy is from pairs of electrons in the same spatial MO. This effect is reasonably well described at the MP2 level, but requires a large basis set in order to recover a large fraction of the absolute value. The remaining correlation energy is much harder to calculate: coupled cluster is the preferred method here but, since the absolute value is substantially smaller than the MP2 correlation energy, a smaller basis set can be employed. This means that the relative error is quite large but the absolute error is of the same magnitude as the correlation error from the MP2 calculation with the large basis set.

Several families of composite extrapolation procedures have been proposed, each with a number of variations, and they differ in the exact procedure for estimating the infinite-correlation-infinite-basis-set limit and the resulting accuracy for a chosen reference data set.[76]

5.10.1 Gaussian-n Models

The *Gaussian-n* (G*n*, $n = 1, 2, 3, 4$) sequence of methods represents improvements in the target accuracy, driven by the continuing incease in computational capabilities.[77] Each of the G*n* methods exist in several different versions, depending on the specific combinations of methods and basis sets for each of the steps, and carry acronyms such as G*n*(MP2), G*n*B3, G*n*(MP2)B3 and G*n*X(MP2). The G*n* methods are characteristic by employing mainly the Pople style basis sets and utilizing empirical fitting parameters for improving the performance. The performance and parameterization is done using experimental molecular benchmark data sets carrying acronyms such as G2/97, G3/99 and G3/05, consisting of mainly enthalpies of formation, ionization potentials and electron affinities. As an example, the G4 method involves the following steps:[78]

1. The geometry is optimized at the B3LYP/6-31G(2df,p) level, which is used as the reference geometry.
2. Vibrational frequencies are calculated at the B3LYP/6-31G(2df,p) level and scaled by 0.9854 to produce zero-point energies.
3. The basis set limit HF energy is estimated by an exponential extrapolation from results obtained with slightly reduced versions of the aug-cc-pVQZ and aug-cc-pV5Z basis sets.
4. The energy is calculated at the CCSD(T)/6-31G(d) level, which automatically generates the MP4 value as an intermediate result. Additive basis set effects corresponding to augmenting with diffuse

functions on non-hydrogen atoms (6-31+G(d)) and increasing the polarization space to (2df,p) is estimated at the MP4 level.

5. Non-additive basis set effects are estimated at the MP2 level from calculations with the G3LargeXP, 6-31G(2df,p), 6-31+G(d) and 6-31G(d) basis sets, where G3LargeXP denotes a slightly extended version of the 6-311G(2df,2p) basis set.
6. Spin-orbit corrections are taken from experiments or (other) theoretical estimates.
7. To correct for electron correlation beyond CCSD(T) and basis set limitations, an empirical *Higher-Level Correction* (HLC) is added to the total energy, $\Delta E_{\text{HLC}} = -\, b'N_\beta$ for closed-shell systems and $\Delta E_{\text{HLC}} = -aN_\alpha - bN_\beta$ for open-shell systems, where N_α/N_β denotes the number of α and β electrons and it is assumed that the number of α electrons is larger than or equal to the number of β electrons. The numerical constants a,b,b' are determined by fitting to the reference data and are different for atoms and molecules. It should be noted that this correction makes the G4 method non-size extensive.

The steps involved in a G4 calculation and associated extrapolations/combinations are illustrated in Figure 5.4. In the G4(MP2) version, the effects of extending the basis set beyond 6-31G(d) in steps 4 and 5 are replaced by a single MP2/G3LargeXP calculation. The performance measured in terms of the mean absolute deviation over three different sets of data is shown in Table 5.6.

5.10.2 Complete Basis Set Models

The *Complete Basis Set* (CBS) models also employ the Pople style basis sets but attempt to perform an explicit extrapolation for estimating the correlation energy, rather than assuming additivity as in the G*n* methods. The main part of the correlation energy is due to electron pairs, that is described by doubly excited configurations. In terms of perturbation theory, this may again be divided into

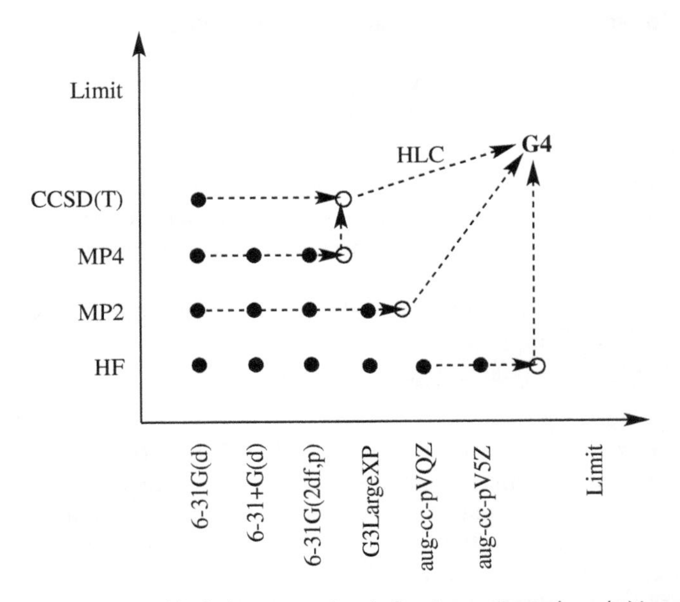

Figure 5.4 Illustrating the calculations (black dots, open dots indicating estimated results) involved in a G4 composite calculation, and how they are combined to the G4 result. Note that the highest level correlated calculation with a given basis set automatically provides lower correlated level results.

Table 5.6 Mean absolute deviations for G*n* models (kJ/mol).[77]

Data	N_{data}	G2	G2(MP2)	G3	G3(MP2)	G4	G4(MP2)
G2/97	298	6.2	7.9	4.1	5.4	3.4	4.4
G3/99	376			4.4	5.5	3.4	4.3
G3/05	454			4.7	5.8	3.5	4.4

contributions from different orders, the most important being from second order (MP2). By using pair natural orbitals (being eigenvectors of the density matrix, see Section 10.5) as the expansion parameter and assuming that enough pairs have been included to reach the asymptotic limit, it may be shown that the MP2 energy calculated by a limited natural orbitals expansion (of the size N) behaves as $1/N$, and can therefore be extrapolated to the complete basis set limit. There are several different CBS methods, each having their own set of prescriptions and resulting computational cost and accuracy, which are known by the acronyms CBS-4, CBS-q, CBS-Q and CBS-APNO,[79] and updated versions labeled CBS-4M and CBS-QB3. As an explicit example, we will take the CBS-QB3 model:[80]

1. The geometry is optimized at the B3LYP/6-311G(d,p) level, which is used as the reference geometry.
2. Vibrational frequencies are calculated at the B3LYP/6-311G(d,p) level and scaled by 0.9854 to produce zero-point energies.
3. An MP2/6-311+G(2df,2p) calculation is carried out, which automatically yields the corresponding HF energy. The MP2 result is extrapolated to the basis set limit by the pair natural orbital method.
4. The energy is calculated at the MP4(SDQ)/6-31+G(d,p) and CCSD(T)/6-31+G(d†) level to estimate the effect from higher-order electron correlation (d† indicates that the exponents for the *d*-functions are taken from the 6-311G(d) basis set).
5. Corrections due to remaining correlation effects are estimated by an empirical expression,

$$\Delta E_{emp} = -0.00579 \sum_i \left(\sum_\mu C_{\mu ii} \right)^2 |S|_{ii}^2 \tag{5.25}$$

   where the sum over $C_{\mu ii}$ is the trace of the first-order wave function coefficients for the natural orbital pair ii, $|S|_{ii}$ is the absolute value of the spatial overlap between the α and β spin components of the ith MO and the factor 0.00579 is determined by fitting to the reference data. This empirical correction is size extensive.
6. For open-shell species the UHF method is used, which in some cases suffers from spin contamination. To correct for this an empirical correction based on the deviation of $\langle S^2 \rangle$ from the theoretical value is added, $\Delta E_{emp} = -0.00954[\langle \mathbf{S}^2 \rangle - S(S-1)]$, where the factor of -0.00954 is derived by fitting.

The mean absolute deviations for CBS-QB3 and CBS-4M over the G2/97 data set are 4.6 and 13.6 kJ/mol, respectively, which can be compared with the results in Table 5.6.

It should be noted that the G*n* data set primarily includes data for molecules containing few non-hydrogen atoms. It is likely that the typical error for a given model to a certain extent depends on the size of the system, that is the error in the calculated heat of formation of, say, C_{60} (if it was computationally feasible) is likely to be substantially larger than the average errors shown in Table 5.6. Furthermore, the properties included (atomization energies, ionization potentials, electron and proton affinities) all correspond to energy differences between well-separated systems: atomization energies are energy differences between a molecule and isolated atoms, and the other three properties correspond to removal or addition of a single electron or proton. As illustrated in Chapter 12, such energy differences are easier to calculate than between systems containing a half broken/formed bond. As with any scheme that has been parameterized on experimental data, it is questionable to assume that the typical accuracy for a selected set of properties will be true in general. A good performance for the G*n* data set does not necessarily indicate that the same level of accuracy can be obtained over a wide variety of geometries, for example including transition structures. It should furthermore be recognized that the values in Table 5.6 are average values, but the maximum error for the reference data set is often 5–10 times larger. Since it is difficult to know in advance whether the particular system of interest behaves as the average or the exceptional case, the predicted value must realistically be assumed to have an uncertainty of perhaps 10–20 kJ/mol. Part of the reason for the relatively large spread in the errors is the assumption of additivity in basis sets effect, which has little theoretical foundation, although the empirical corrections at least partly absorb some of these errors.

5.10.3 Weizmann-n Models

If higher accuracy is desired, for example "subchemical" (~0.5 kJ/mol) or "spectroscopic accuracy" (~1 cm^{-1}, ~0.01 kJ/mol), a number of other factors must also be considered:

1. Including correlation energy between the core and core–valence electrons. This becomes progressively more important as systems with heavier elements are considered.
2. Inclusion of high-order correlation effects, such as connected triple, quadruple and quintuple excitations.
3. Relativistic effects, such as mass–velocity, Darwin and spin–orbit coupling perturbative corrections, or more sophisticated relativistic treatments. Obviously, these corrections become important even at the chemical accuracy level if atoms from the lower part of the periodic table are present in the system.
4. Non-Born–Oppenheimer corrections. These will be most important for systems containing hydrogen.
5. Basis set superposition corrections.
6. Anharmonic vibrations.
7. Vibrational–rotational coupling.

At present, there is no standard procedure for achieving "spectroscopic accuracy", but Martin and coworkers have developed the W*n* ($n = 1, 2, 3, 4$) methods aimed at a target accuracy of ~1 kJ/mol on the average for atomization energies, with worst-case systems having errors below ~5 kJ/mol.[81] The W*n* methods employ the cc-pVXZ basis sets and rely on explicit extrapolation to the infinite basis set limit for the HF and correlation energies without empirical fitting parameters, as well as addition of relativistic and non-Born–Oppenheimer effects. Each of the W*n* methods exist in different versions, with W4 being the most sophisticated and consisting of the following steps:[81]

1. The geometry is optimized at the CCSD(T)/cc-pVQZ level.
2. Anharmonic frequencies are calculated at the CCSD(T)/cc-pVQZ level.
3. The HF limit is estimated by extrapolating the results from the aug-cc-pV5Z and aug-cc-pV6Z basis sets using Equation (5.17) with a fixed B value of 9.
4. The CCSD valence correlation energy is estimated from two-point extrapolation of the results from the aug-cc-pV5Z and aug-cc-pV6Z basis sets using separate extrapolation for the

singlet and triplet coupled electron pairs using Equation (5.18) with exponents of 3 and 5, respectively.
5. The (T) contribution to the valence correlation energy is estimated from two-point extrapolation of the results from the aug-cc-pVQZ and aug-cc-pV5Z basis sets.
6. The CCSDT-CCSD(T) difference in valence correlation energy is estimated from two-point extrapolation of the results from the cc-pVTZ and cc-pVDZ basis sets.
7. The Q-contribution to the valence correlation energy is estimated from an additive correction based on CCSDT(Q) calculations with the cc-pVTZ and cc-pVDZ basis sets and a CCSDTQ calculation with the cc-pVDZ basis sets.
8. The 5-contribution to the valence correlation energy is estimated from a CCSDTQ5 calculations with a DZ basis sets and a CCSDTQ calculation with the cc-pVDZ basis sets.
9. The core and core–valence correlation energy is estimated by extrapolation of the CCSD(T) results using the aug-cc-pwCVQZ and aug-cc-pwCVTZ basis sets.
10. Scalar relativistic corrections are estimated from a second-order Douglas–Kroll–Hess CCSD(T)/DK-aug-cc-pVQZ (relativistic version of the aug-cc-pVQZ basis) calculation.
11. Spin-orbit corrections are taken from experiments.
12. Non-Born–Oppenheimer effects are estimated as the diagonal-BO correction calculated at the HF/aug-cc-pVTZ level.

The steps 3 to 8 involved in a W4 calculation and associated extrapolations/combinations are illustrated in Figure 5.5.

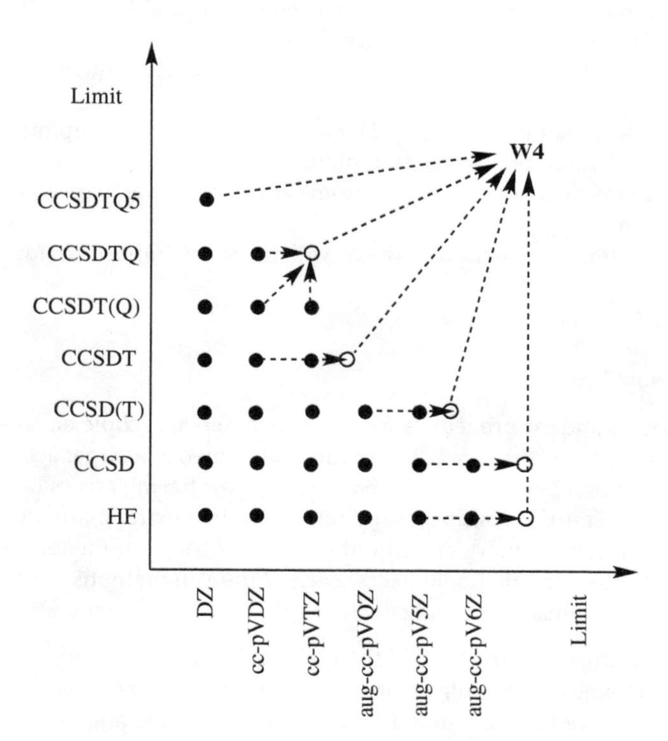

Figure 5.5 Illustrating the calculations (black dots, open dots indicating extrapolated results) involved in estimating the non-relativistic valence correlation W4 composite energy, and how they are combined to the W4 result. Note that the highest level correlated calculation with a given basis set automatically provides lower correlated level results.

The W4 method provides a mean absolute deviation for atomization energies of 26 molecules of 0.38 kJ/mol, with the worst case having an error of 1.3 kJ/mol, and these values can be compared with the average experimental error of 0.22 kJ/mol for the same set of data.[81] It can be noted that the experimental data were carefully selected to have small experimental uncertainties, a more typical experimental error is 5–10 kJ/mol. Such elaborate composite methods are thus capable of yielding results with accuracies comparable to or surpassing experimental methods, but the associated computational cost is so high that only systems containing less than ~20 valence electrons (4–5 atoms) can be handled. The W3, W2 and W1 (and their variants) trade accuracy (W3 errors are roughly twice as large as W4 and W2 errors are roughly twice as large as W3) against computational cost and can be applied for somewhat larger systems.

5.10.4 Other Composite Models

There are a number of similar composite extrapolation procedures, such as the ccCA (correlation consistent Composite Approach)[82] and HEAT (High-accuracy Extrapolated *Ab initio* Thermochemistry)[83] methods, each differing in the exact methods and basis set combinations and how these results are combined. Hybrid methods mixing results from wave function and density functional methods with a selection of basis sets by a set of fitted parameters have also been proposed with the acronym MCCM (MultiCoefficient Correlation Method).[84]

There are in addition some simple correction procedures that may be considered as extrapolation schemes. The *Scaled External Correlation* (SEC) and *Scaled All Correlation* (SAC) methods scale the correlation energy by a factor such that calculated dissociation energy agrees with the experimental value:[85]

$$
\begin{aligned}
E_{\text{SEC/SAC}} &= E_{\text{ref}} + \frac{E_{\text{corr}} - E_{\text{ref}}}{F} \\
F &= \frac{D_{\text{e}}(\text{corr}) - D_{\text{e}}(\text{ref})}{D_{\text{e}}(\text{exp}) - D_{\text{e}}(\text{ref})}
\end{aligned}
\tag{5.26}
$$

The SEC acronym refers to the case where the reference wave function is of the MCSCF type and the correlation energy is calculated by a MR-CISD procedure. When the reference is a single determinant (HF), the SAC nomenclature is used. In the latter case the correlation energy may be calculated, for example, by MP2, MP4 or CCSD, producing acronyms such as MP2-SAC, MP4-SAC and CCSD-SAC. In the SEC/SAC procedure, the scale factor F is assumed to be constant over the whole surface. If more than one dissociation channel is important, a suitable average F may be used.

The *Parameterized Configuration Interaction X* (PCI-X) method[86] simply takes the correlation energy and scales it by a constant factor X (typical value ~1.2), that is it is assumed that the given combination of method and basis set recovers a constant fraction of the correlation energy.

The introduction of various empirical corrections, such as scale factors for frequencies and energy corrections based on the number of electrons and degree of spin contamination, blurs the distinction between whether these methods should be considered *ab initio*, or as belonging to the semi-empirical class of methods described in Chapter 7. Nevertheless, the accuracy that these methods are capable of delivering makes it possible to calculate absolute stabilities (heat of formation) for small- and medium-sized systems that rival (or surpass) experimental data, often at a substantially lower cost than that for actually performing the experiments.

5.11 Isogyric and Isodesmic Reactions

The most difficult part in calculating absolute stabilities (heat of formation) is the correlation energy. For calculating energies relative to isolated atoms, which is the goal of the composite models described in the previous section, essentially all the correlation energy of the bond(s) being broken must be recovered. This in turn necessitates large basis sets and sophisticated correlation methods. This is also the reason why *ab initio* energies are not converted into heat of formation, as is normally done for semi-empirical methods (Equation (7.41)), since the resulting values are poor unless a very high level of theory is employed.

In many cases, however, it is possible to choose less demanding reference systems than the isolated atoms. Consider, for example, calculating the C—H dissociation energy of CH_4 (Figure 5.6). In a direct calculation this is given as the difference in total energy of CH_4 and CH_3 + H.

$$CH_4 \longrightarrow CH_3 + H$$

Figure 5.6 Dissociation of CH_4.

In order to calculate an accurate value for this energy difference, essentially all the electron correlation (and HF) energy for the C—H bond must be recovered. Consider now the reaction in Figure 5.7.

$$CH_4 + H \longrightarrow CH_3 + H_2$$

Figure 5.7 An example of an isogyric reaction.

The difference between the two reactions in Figures 5.6 and 5.7 is that the latter has the same number of electron pairs on both sides; such reactions are called *isogyric*. The task of calculating *all* the correlation energy of a C—H bond is replaced by calculating the *difference* in correlation energy between a C—H and an H—H bond. The latter will benefit from cancellation of errors and therefore stabilize much earlier in terms of theoretical level. Isogyric reactions can thus be used for obtaining *relative* values. In the above example the CH_4 dissociation energy is given relative to that of H_2. By using the experimental value for H_2, the CH_4 dissociation energy may be calculated quite accurately, even at relatively low levels of theory.

The concept may be taken one step further. It is often possible to set up reactions where not only the number of electron pairs is constant but also the formal type of bonds is the same on both sides. Consider, for example, calculating the stability of propene by the reaction in Figure 5.8.

$$H_2C{=}CH{-}CH_3 + CH_4 \longrightarrow H_2C{=}CH_2 + H_3C{-}CH_3$$

Figure 5.8 An example of an isodesmic reaction.

In this case the number of C=C, C—C and C—H bonds is the same on both sides and the “reaction” energy is therefore relatively easy to calculate since the electron correlation is to a large extent the same on both sides. Such reactions that conserve both the number and types of bonds are called *isodesmic* reactions. Combining the calculated energy difference for the left- and right-hand sides with experimental values for $H_2C{=}CH_2$, $H_3C{-}CH_3$ and CH_4, the (absolute) stability of propene can be obtained reasonably accurately at a quite low level of theory. It does, however, require that the experimental values for the chosen reference compounds are available. Furthermore, there are several possible ways of constructing isodesmic or isogyric reactions (e.g. replacing H with Cl in Figure 5.7); that is such methods are not unique.

5.12 Effective Core Potentials

Systems involving atoms from the lower part of the periodic table have a large number of core electrons. These are, as already mentioned, unimportant in a chemical sense, but it is necessary to use a large number of basis functions to expand the corresponding orbitals, otherwise the valence orbitals will not be properly described (due to a poor description of the electron–electron repulsion). In the lower half of the periodic table relativistic effects further complicate matters (see Chapter 9). These two problems may be "solved" simultaneously by modeling the core electrons by a suitable function and treating only the valence electrons explicitly.

The function modeling the core electrons is usually called an *Effective Core Potential* (ECP) in the chemical community,[87,88] while the physics community uses the term *Pseudo-Potential* (PP).[67] The neglect of an explicit treatment of the core electrons, analogous to the semi-empirical methods in Chapter 7, often gives quite good results at a fraction of the cost of a calculation involving all electrons, and part of the relativistic effects (especially the scalar effects) may also be taken care of, without having to perform the full relativistic calculation.

There are four major steps in designing a pseudo-potential:

1. Generate a good-quality all-electron wave function for the atom. This will typically be from a numerical Hartree–Fock, a relativistic Dirac–Hartree–Fock or a density functional calculation.
2. Replace the valence orbitals by a set of nodeless pseudo-orbitals. The regular valence orbitals will have radial nodes in order to make them orthogonal to the core orbitals, and the pseudo-orbitals are designed such that they behave correctly in the outer part, but without the nodal structure in the core region.
3. Replace the core electrons by a potential parameterized by expansion into a suitable set of analytical functions of the nuclear–electron distance, for example a polynomial or a set of spherical Bessel or Gaussian functions. Since relativistic effects are mainly important for the core electrons, this potential can effectively include relativity. The potential may be different for each angular momentum.
4. Fit the parameters of the potential such that the solutions of the Schrödinger (or Dirac) equation produce pseudo-orbitals matching the all-electron valence orbitals.

Molecular systems have traditionally been described by Gaussian-type basis sets, while plane waves have been favored for extended (periodic) systems; this difference has resulted in some differences for the corresponding pseudo-potentials. When using Gaussian functions for describing the valence orbitals, it is natural to also use Gaussian functions to describe the ECP. Since Gaussian functions are continuous, there is no fixed distance to characterize the extent of the core potential and the quality of the ECP is determined by the number of electrons chosen to be represented by the ECP. For transition metals, it is clear that the outer $(n + 1)$s-, $(n + 1)$p- and (n)d-orbitals constitute the valence space. While such "full-core" potentials give reasonable geometries, it has been found that the energetics are not always satisfactory. Better results can be obtained by also including the orbitals in the next lower shell in the valence space, albeit at an increase in the computational cost. For silver

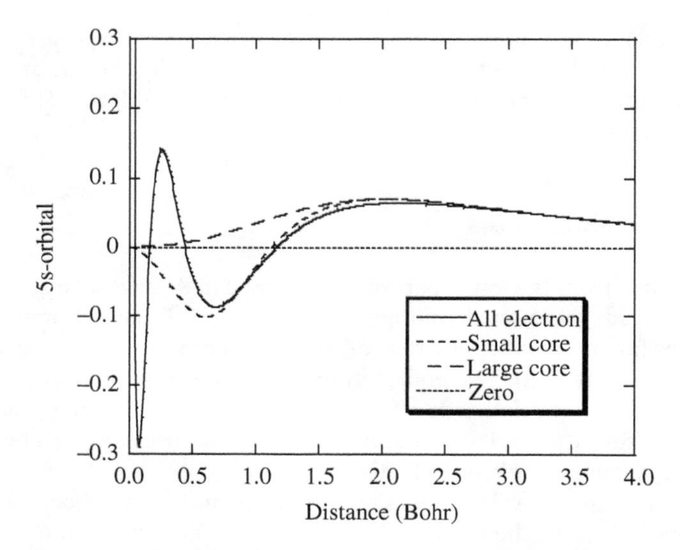

Figure 5.9 The 5s-orbital for Ag with either an all-electron, large- or small-core effective core potential.

with an atomic number of 47, for example, one may consider two different choices of core size, where the electrons replaced by an ECP are indicated in *italic* and the remaining electrons in **bold**:

- "Large-core" ECP: 11 electrons considered explicitly: *(1s)*^{2} *(2s)*^{2} *(2p)*^{6} *(3s)*^{2} *(3p)*^{6} *(4s)*^{2} *(3d)*^{10} *(4p)*^{6} **(4d)**^{10} **(5s)**^{1}
- "Small-core" ECP: 19 electrons considered explicitly: *(1s)*^{2} *(2s)*^{2} *(2p)*^{6} *(3s)*^{2} *(3p)*^{6} **(4s)**^{2} *(3d)*^{10} **(4p)**^{6} **(4d)**^{10} **(5s)**^{1}
- All-electron: 47 electrons considered explicitly: **(1s)**^{2} **(2s)**^{2} **(2p)**^{6} **(3s)**^{2} **(3p)**^{6} **(4s)**^{2} **(3d)**^{10} **(4p)**^{6} **(4d)**^{10} **(5s)**^{1}

The shape of the resulting 5s-(pseudo)-orbital for these choices is shown in Figure 5.9.

The gain by using ECPs is largest for atoms in the lower part of the periodic table, especially those where relativistic effects are important.[89] Since fully relativistic results are scarce, the performance of ECPs is somewhat difficult to evaluate by comparing with other calculations,[90] but they often reproduce known experimental results, thereby justifying the approach. ECPs have also been designed for second row elements (Li—Ne),[91] although the savings in these cases are marginal relative to all-electron calculations.

The size of a plane-wave basis set is given by the maximum kinetic energy, which is inversely related to the smallest variation of the wave function that can be described. The singularity of the nuclear potential ($\mathbf{V}_{ne}$) and the resulting strongly localized core electrons are essentially impossible to describe by any reasonable-sized plane-wave basis set. Pseudo-potentials are therefore used for smearing the nuclear charge and modeling the core electrons. These potentials are typically characterized by a "core radius" r_c (which may depend on the angular momentum of the valence orbitals), that is the pseudo-potentials used in connection with plane waves have a finite physical extent. The potential for distances smaller than r_c is described by a suitable analytical function, typically a polynomial or spherical Bessel function, and the pseudo-wave function and its first and second derivatives are required to match those of the reference wave function at r_c. It is clear that a "hard" (small r_c)

pseudo-potential will require more plane wave basis functions for describing the region beyond r_c than a "soft" (large r_c) pseudo-potential, but a too large r_c will deteriorate the quality of the calculated results and also make the pseudo-potential less transferable.

The *norm-conserving* pseudo-potentials proposed by Hamann, Schlüter and Chiang require in addition to the above matching conditions at r_c that the integral of the square of the reference and pseudo-wave from 0 to r_c agree, that is conservation of the wave function norm.[92] For the late second row elements (C to F) and the 3d transition metals (Sc to Zn), these pseudo-potentials are rather "hard" and therefore require a relatively large energy cutoff for the plane waves. Vanderbilt proposed to relax the norm-conserving requirement to give the so-called *ultrasoft* pseudo-potentials,[93,94] thereby reducing the necessary number of plane waves for expanding the valence orbitals by roughly a factor of two.

While there is essentially no basis set error when using plane waves for expanding the orbitals, the requirement of using a pseudo-potential to describe the core region means that there is a fundamental limitation in how accurate the results can be. For systems composed of atoms from the first three rows in the periodic table, the error in DFT calculations is roughly equivalent to that imposed by using a Gaussian basis set of TZP quality in an all-electron calculation,[95] and the differences between different pseudo-potentials are of the same magnitude.[96] An implicit limitation of pseudo-potential methods is of course difficulty in describing molecular properties that depend directly on the core electrons (as in X-ray photoelectron spectroscopy) or the electron density near the nucleus (as in NMR spin–spin coupling constants). NMR shielding constants are sufficiently valence-like that quite accurate results can be obtained by assuming that the contribution from the electrons neglected by the pseudo-potential is constant, and can be added based on an atomic all-electron calculation.[97]

The *Projector Augmented Wave* (PAW) method is usually also considered a pseudo-potential method, although it formally retains all the core electrons.[98,99] Indeed, the Vanderbilt ultrasoft pseudo-potential can be derived by linearization of two terms in the PAW expression. The PAW wave function is written as a valence term expanded in a plane-wave basis plus a contribution from the region within the core radius of each nucleus, evaluated on a grid. The contribution from a core region is expanded as a difference between two sets of densities, one arising from the (all-electron) atomic orbitals, the other from a set of nodeless pseudo-atomic orbitals, that is this term allows the wave function within the core region to adjust for different environments. In most applications the (all-electron) atomic orbitals have been kept fixed at their form for the isolated atoms, that is a "frozen-core" approach. Allowing for a full self-consistent optimization of the atomic core[100] makes the PAW equivalent to the all-electron mixed basis sets methods discussed in Section 5.5.

A closely related idea arising from the chemical community is the use of the *frozen-core* approximation.[101,102] The core electrons are here included in the treatment, but the corresponding orbitals are fixed at their atomic values and represented by a fixed expansion in a suitable basis set. This preserves the full electron–electron interaction but ignores the change in the core orbitals due to the molecular environment. The savings are marginal for systems with only up to second row atoms, but for systems with heavier elements the computational cost may be significantly reduced. The frozen-core approximation may furthermore be useful for calculations using relativistic wave functions, as it effectively prevents a variational collapse.

A common feature of all pseudo-potential methods is that the parameters depend on the employed method, that is the potential derived for the *Local Spin Density Approximation* (LSDA) functional (Section 6.5.1), for example, is different from that derived from a generalized gradient functional such as *Perdew–Burke–Ernzerhof* (PBE) (Section 6.5.2). In practice, the difference is relatively small and pseudo-potentials optimized for one functional are often used for other functionals without re-optimization.

5.13 Basis Set Superposition and Incompleteness Errors

By far the most common type of basis set for molecular applications is a set of Gaussian functions centered on the nuclei. Such a basis set can be made to be near-complete if a very large number of functions are used, but reduction of the error in the total energy to chemical accuracy (~ few kJ/mol) requires several hundred functions per atom, which severely limits the size of molecules that can be handled. For commonly used basis sets of DZP or TZP quality, the errors in the absolute energy from basis set incompleteness are large, maybe several au (thousands of kJ/mol). The interest is usually in relative energies, however, and the primary goal is therefore to make the basis set error as constant as possible. This is one of the reasons why it is important to choose a "balanced" basis set. The first, perhaps obvious, step is that the same basis set must be used when comparing energies: comparing energies of two isomers where the 6-31G** basis set has been used for one of them and the cc-pVDZ basis set for the other is meaningless, although both basis sets are of polarized double zeta quality.

Fixing the position of the basis functions to the nuclei allows for a compact basis set; otherwise sets of basis functions positioned at many points in the geometrical space would be needed. When comparing energies for different molecular structures, however, a nuclear fixed basis set introduces an error. The quality of the basis set is not the same at all geometries, owing to the fact that the electron density around one nucleus may be described by functions centered at another nucleus. This is especially troublesome when calculating small effects, such as energies of van der Waals complexes and hydrogen bonds. Consider, for example, the hydrogen bond between two water molecules. The simplest approach consists of calculating the energy of the dimer and subtracting two times the energy of an isolated molecule (assuming a size extensive method). The electron distribution within each water molecule in the dimer is very close to that of the monomer. In the dimer, however, basis functions from one molecule can help compensate for the basis set incompleteness on the other molecule, and vice versa. The dimer will therefore be artificially lowered in energy and the strength of the hydrogen bond overestimated. This effect is known as the *Basis Set Superposition Error* (BSSE). In the limit of a complete basis set, the BSSE will be zero, and adding more basis functions will not give any change. The conceptually simplest approach for eliminating BSSE is therefore to add more and more basis functions, until the interaction energy no longer changes. Unfortunately, this requires very large basis sets. Since non-bonded interactions are weak, the desired accuracy is often ~0.5 kJ/mol. Using the correlation consistent basis sets, the water dimer interaction energy stabilizes at this level with the aug-cc-pVTZ basis (184 basis functions for H_2O) at the HF level, but requires (at least) the aug-cc-pV5Z basis (574 basis functions) at the MP2 level.[103] As inclusion of electron correlation is mandatory for calculating the dispersion interaction between molecules, even the water dimer potential is computationally challenging.

An *approximate* way of assessing BSSE is the *CounterPoise* (CP) correction.[104] In this method the BSSE is estimated as the difference between monomer energies with the regular basis and the energies calculated with the full set of basis functions for the whole complex. Consider two molecules A and B, each having regular nuclear-centered basis sets denoted with subscripts a and b, and the complex AB having the combined basis set ab. The geometries of the two isolated molecules and of the complex are first optimized or otherwise assigned. The geometries of the A and B molecules in the complex will usually be slightly different than for the isolated species, and the complex geometry will be denoted with a *. The dimer energy minus the monomer energies is the directly calculated complexation energy:

$$\Delta E_{\text{complexation}} = E(AB)^*_{ab} - E(A)_a - E(B)_b \tag{5.27}$$

The CP estimate of how much of this complexation energy is due to BSSE requires four additional energy calculations. Using the a basis set for A and the b basis set for B, the energies of each of the two fragments are calculated with the geometry they have in the complex. Two additional energy calculations of the fragments at the complex geometry are then carried out with the full ab basis set. This means that the energy of A is calculated in the presence of both the normal a basis functions *and* with the b basis functions of fragment B located at the corresponding nuclear positions, but *without* the B nuclei present, and vice versa. Such basis functions located at fixed points in space are often referred to as *ghost orbitals*. The fragment energy for A will be lowered due to these ghost functions, since the a basis becomes more complete. The CP correction to the interaction energy is defined by

$$\Delta E_{\text{CP}} = E(A)^*_{ab} - E(A)^*_{a} + E(B)^*_{ab} - E(B)^*_{b} \tag{5.28}$$

The counterpoise-corrected complexation energy is then given as $\Delta E_{\text{complexation}} - \Delta E_{\text{CP}}$.

There have been attempts at developing methods where the BSSE is excluded explicitly in the computational expressions, such as the *Chemical Hamiltonian Approach* (CHA).[105] The CHA has the disadvantage that the Hamiltonian becomes non-Hermitian and, since the results are very similar to those from the CP method,[106] it is rarely used. It has also been argued that the full set of ghost orbitals should not be used, since some of the functions in the complex are used for describing the occupied orbitals of the other component, and only the virtual orbitals are available for "artificial" stabilization. This has been put on formal ground by the *Same Number Of Optimized Parameters* (SNOOP) approach, where the number of parameters describing the electronic structure in the two monomers and the dimer are required to be the same in order to provide a balance of errors. The SNOOP approach calculates the monomer energy by taking the wave functions for the isolated monomers, allowing the orbitals for A to relax in the additional space described by (only) the virtual orbitals of B (and vice versa), and using only the combined virtual space for describing electron correlation.[107]

The CP correction for methods including electron correlation is usually larger and more sensitive to the size of the basis set than at HF or DFT levels. This is in line with the fact that HF/DFT methods converge faster with respect to the size of the basis set than correlated wave functions. It is also well recognized that the CP correction for intermolecular interactions only provides an *estimate* of the BSSE effect, but does not provide either an upper or lower limit. It has in several cases been observed that the interaction energy with the CP correction converges more regularly and stabilizes at the basis set limiting value earlier than uncorrected values when regular basis sets are used, but not necessarily if diffuse functions are included in the basis set. Furthermore, the CP corrected and uncorrected interaction energies often converge towards the limiting value from opposite sides, which has led to the suggestion that only half of the CP correction should be added.[108] The SNOOP approach by construction provides results that are intermediate between the CP corrected and uncorrected values, and thus rationalizes this heuristic scheme. Detailed analyses suggest that the situation may be more complicated for very weak interactions, and the CP correction can in some cases even be in the wrong direction.[109] This is due to the fact that a basis set for simple systems (e.g. atoms) is more complete than for the complex systems, and the basis set therefore preferentially favors the isolated fragments over the complex. The complexation energy can consequently be underestimated, even when BSSE is present, and adding the CP correction will further disfavor the complex.

A heuristic method for substantially reducing BSSE is to use *bond functions*, where Gaussian functions are placed at the midpoint between atoms/fragments. This is quite efficient for calculating accurate intermolecular potentials with respect to a single variable (distance), but it is cumbersome to generalize to many-dimensional and intramolecular cases. It furthermore often leads to unpredictable basis set convergence, which prohibits the use of extrapolation methods (Section 5.9).

The CP method is well-defined for non-bonded interactions where the A and B monomers can be uniquely identified, but it should be recognized that the BSSE effect is always present when comparing energies of different molecules (e.g. ethanol and dimethyl ether) or energies of the same molecule with different structures (e.g. staggered and eclipsed ethane).[110] The *Basis Set Incompleteness Error* (BSIE) for an atom can be defined as the difference in the result obtained with a specific basis set and the complete basis set value. The corresponding difference for a molecule, however, is a mixture of BSIE and BSSE, and it is very difficult to disentangle the two effects. In a large molecule there may be conformations where parts of the molecule are spatially close without being close in terms of bonding, and such conformations will be artificially stabilized by the intramolecular analog of BSSE. For small peptides with DZP-type basis sets, this may lead to errors of ~20 kJ/mol in relative conformational energies and completely change rotational energy profiles.[111,112] Even atoms that are directly bonded will have components of intramolecular BSSE in addition to BSIE. Such directly bonded intramolecular BSSE has been blamed for the spurious imaginary frequencies found for planar aromatic compounds at correlated levels with certain basis sets, although it could also be attributed to basis set imbalance, that is an inconsistent number of angular momentum functions of each type.[113,114]

Attempts to extend the CP scheme to account for intramolecular BSSE face the problem of how to partition the system into fragments. The least biased choice is to use individual atoms as fragments, which requires a total of N_{atom} CP calculations, each employing all basis functions for the whole molecular system. While this at face value suggests a prohibitive computational cost, the use of distance cutoffs and integral screening mean that the atomic CP correction can be calculated at roughly the same cost as the uncorrected energy itself.[115]

Kruse and Grimme have proposed a parameterized correction for intramolecular BSSE denoted *geometrical* CP (gCP) which only requires geometrical information and easy to calculate overlap integrals.[116] Local electron correlation methods (Section 4.12.2) tend to have reduced BSSE compared to cannonical orbital methods, since only a (geometrically) limited number of orbitals are used in the correlation treatment.

Calculations with periodic boundary conditions for modeling solids or liquids is strongly affected by BSSE when localized (Gaussian) basis sets are employed,[117] while plane wave basis sets are free of BSSE, since the basis functions do not depend on the nuclear positions.

References

1 D. Feller and E. R. Davidson, *Reviews in Computational Chemistry* **1**, 1 (1990).
2 T. Helgaker, P. Jørgensen and J. Olsen, *Molecular Electronic Structure Theory* (John Wiley & Sons, 2000).
3 J. C. Slater, *Physical Review* **36** (1), 0057–0064 (1930).
4 S. F. Boys, *Proceedings of the Royal Society of London Series A – Mathematical and Physical Sciences* **200** (1063), 542–554 (1950).
5 W. Klopper and W. Kutzelnigg, *Journal of Molecular Structure –Theochem* **28**, 339–356 (1986).
6 W. Kutzelnigg and J. D. Morgan, *Journal of Chemical Physics* **96** (6), 4484–4508 (1992).
7 L. K. McKemmish and P. M. W. Gill, *Journal of Chemical Theory and Computation* **8** (12), 4891–4898 (2012).
8 M. W. Schmidt and K. Ruedenberg, *Journal of Chemical Physics* **71** (10), 3951–3962 (1979).
9 S. Huzinaga, M. Klobukowski and H. Tatewaki, *Canadian Journal of Chemistry – Revue Canadienne De Chimie* **63** (7), 1812–1828 (1985).

10 G. A. Petersson, S. J. Zhong, J. A. Montgomery and M. J. Frisch, *Journal of Chemical Physics* **118** (3), 1101–1109 (2003).
11 F. Jensen, *Journal of Chemical Physics* **122** (7), 074111 (2005).
12 F. Jensen, *Journal of Chemical Theory and Computation* **10** (3), 1074–1085 (2014).
13 D. Rappoport and F. Furche, *Journal of Chemical Physics* **133** (13), 134105 (2010).
14 K. L. Schuchardt, B. T. Didier, T. Elsethagen, L. Sun, V. Gurumoorthi, J. Chase, J. Li and T. L. Windus, *Journal of Chemical Information and Modeling* **47** (3), 1045–1052 (2007).
15 J. G. Hill, *International Journal of Quantum Chemistry* **113** (1), 21–34 (2013).
16 F. Jensen, *Wiley Interdisciplinary Reviews – Computational Molecular Science* **3** (3), 273–295 (2013).
17 W. J. Hehre, R. F. Stewart and J. A. Pople, *Journal of Chemical Physics* **51** (6), 2657 (1969).
18 J. S. Binkley, J. A. Pople and W. J. Hehre, *Journal of the American Chemical Society* **102** (3), 939–947 (1980).
19 W. J. Hehre, Ditchfie.R and J. A. Pople, *Journal of Chemical Physics* **56** (5), 2257 (1972).
20 R. Krishnan, J. S. Binkley, R. Seeger and J. A. Pople, *Journal of Chemical Physics* **72** (1), 650–654 (1980).
21 M. J. Frisch, J. A. Pople and J. S. Binkley, *Journal of Chemical Physics* **80** (7), 3265–3269 (1984).
22 M. M. Francl, W. J. Pietro, W. J. Hehre, J. S. Binkley, M. S. Gordon, D. J. Defrees and J. A. Pople, *Journal of Chemical Physics* **77** (7), 3654–3665 (1982).
23 S. Huzinaga, *Journal of Chemical Physics* **42** (4), 1293 (1965).
24 F. B. van Duijneveldt, *IBM Technical Research Report* RJ945 (1971).
25 H. Partridge, *Journal of Chemical Physics* **90** (2), 1043–1047 (1989).
26 T. H. Dunning, *Journal of Chemical Physics* **55** (2), 716 (1971).
27 A. D. McLean and G. S. Chandler, *Journal of Chemical Physics* **72** (10), 5639–5648 (1980).
28 A. Schafer, H. Horn and R. Ahlrichs, *Journal of Chemical Physics* **97** (4), 2571–2577 (1992).
29 A. Schafer, C. Huber and R. Ahlrichs, *Journal of Chemical Physics* **100** (8), 5829–5835 (1994).
30 F. Weigend and R. Ahlrichs, *Physical Chemistry Chemical Physics* **7** (18), 3297–3305 (2005).
31 J. Almlof and P. R. Taylor, *Journal of Chemical Physics* **92** (1), 551–560 (1990).
32 J. Almlof and P. R. Taylor, *Advances in Quantum Chemistry* **22**, 301–373 (1991).
33 K. Pierloot, B. Dumez, P. O. Widmark and B. O. Roos, *Theoretica Chimica Acta* **90** (2–3), 87–114 (1995).
34 T. H. Dunning, *Journal of Chemical Physics* **90** (2), 1007–1023 (1989).
35 A. K. Wilson, T. van Mourik and T. H. Dunning, *Theochem – Journal of Molecular Structure* **388**, 339–349 (1996).
36 T. H. Dunning, K. A. Peterson and A. K. Wilson, *Journal of Chemical Physics* **114** (21), 9244–9253 (2001).
37 R. A. Kendall, T. H. Dunning and R. J. Harrison, *Journal of Chemical Physics* **96** (9), 6796–6806 (1992).
38 D. E. Woon and T. H. Dunning, *Journal of Chemical Physics* **103** (11), 4572–4585 (1995).
39 K. A. Peterson and T. H. Dunning, *Journal of Chemical Physics* **117** (23), 10548–10560 (2002).
40 F. Jensen, *Theoretical Chemistry Accounts* **104** (6), 484–490 (2000).
41 K. A. Christensen and F. Jensen, *Chemical Physics Letters* **317** (3–5), 400–403 (2000).
42 F. Jensen, *Journal of Chemical Physics* **115** (20), 9113–9125 (2001).
43 F. Jensen, *Journal of Chemical Physics* **116** (8), 3502–3502 (2002).
44 N. B. Balabanov and K. A. Peterson, *Journal of Chemical Physics* **123** (6), 064107 (2005).
45 B. O. Roos, R. Lindh, P. A. Malmqvist, V. Veryazov and P. O. Widmark, *Journal of Physical Chemistry A* **109** (29), 6575–6579 (2005).
46 K. G. Dyall, *Theoretical Chemistry Accounts* **131** (3), 1172 (2012).

47 F. Jensen, *Theoretical Chemistry Accounts* **126** (5–6), 371–382 (2010).
48 U. Benedikt, A. A. Auer and F. Jensen, *Journal of Chemical Physics* **129** (6), 064111 (2008).
49 D. Flaig, M. Maurer, M. Hanni, K. Braunger, L. Kick, M. Thubauville and C. Ochsenfeld, *Journal of Chemical Theory and Computation* **10** (2), 572–578 (2014).
50 F. Jensen, *Journal of Chemical Theory and Computation* **11** (1), 132–138 (2015).
51 M. Preuss, W. G. Schmidt, K. Seino, J. Furthmuller and F. Bechstedt, *Journal of Computational Chemistry* **25** (1), 112–122 (2004).
52 C. Freysoldt, J. Neugebauer and C. G. Van de Walle, *Physical Review Letters* **102** (1), 016402 (2009).
53 G. K. H. Madsen, P. Blaha, K. Schwarz, E. Sjostedt and L. Nordstrom, *Physical Review B* **64** (19), 195134 (2001).
54 F. A. Pahl and N. C. Handy, *Molecular Physics* **100** (20), 3199–3224 (2002).
55 L. Fusti-Molnar and P. Pulay, *Journal of Chemical Physics* **116** (18), 7795–7805 (2002).
56 J. Kobus, *Computer Physics Communications* **184** (3), 799–811 (2013).
57 T. Shiozaki and S. Hirata, *Physical Review A* **76** (4), 040503 (2007).
58 O. Cohen, L. Kronik and A. Brandt, *Journal of Chemical Theory and Computation* **9** (11), 4744–4760 (2013).
59 J. Enkovaara, C. Rostgaard, J. J. Mortensen, J. Chen, M. Dulak, L. Ferrighi, J. Gavnholt, C. Glinsvad, V. Haikola, H. A. Hansen, H. H. Kristoffersen, M. Kuisma, A. H. Larsen, L. Lehtovaara, M. Ljungberg, O. Lopez-Acevedo, P. G. Moses, J. Ojanen, T. Olsen, V. Petzold, N. A. Romero, J. Stausholm-Moller, M. Strange, G. A. Tritsaris, M. Vanin, M. Walter, B. Hammer, H. Hakkinen, G. K. H. Madsen, R. M. Nieminen, J. K. Norskov, M. Puska, T. T. Rantala, J. Schiotz, K. S. Thygesen and K. W. Jacobsen, *Journal of Physics – Condensed Matter* **22** (25), 253202 (2010).
60 R. J. Harrison, G. I. Fann, T. Yanai, Z. Gan and G. Beylkin, *Journal of Chemical Physics* **121** (23), 11587–11598 (2004).
61 T. Yanai, G. I. Fann, Z. T. Gan, R. J. Harrison and G. Beylkin, *Journal of Chemical Physics* **121** (14), 6680–6688 (2004).
62 F. A. Bischoff and E. F. Valeev, *Journal of Chemical Physics* **139** (11), 114106 (2013).
63 F. Weigend, *Physical Chemistry Chemical Physics* **8** (9), 1057–1065 (2006).
64 S. Kritikou and J. G. Hill, *Journal of Chemical Theory and Computation* **11** (11), 5269–5276 (2015).
65 S. Chawla and G. A. Voth, *Journal of Chemical Physics* **108** (12), 4697–4700 (1998).
66 J. Hafner, *Journal of Computational Chemistry* **29** (13), 2044–2078 (2008).
67 M. C. Payne, M. P. Teter, D. C. Allan, T. A. Arias and J. D. Joannopoulos, *Reviews of Modern Physics* **64** (4), 1045–1097 (1992).
68 M. Marsman, A. Grueneis, J. Paier and G. Kresse, *Journal of Chemical Physics* **130** (18), 184103 (2009).
69 W. Kutzelnigg, *Journal of Chemical Physics* **97** (11), 8821–8821 (1992).
70 C. Haettig, W. Klopper, A. Koehn and D. P. Tew, *Chemical Reviews* **112** (1), 4–74 (2012).
71 T. Helgaker, W. Klopper, H. Koch and J. Noga, *Journal of Chemical Physics* **106** (23), 9639–9646 (1997).
72 E. F. Valeev, W. D. Allen, R. Hernandez, C. D. Sherrill and H. F. Schaefer, *Journal of Chemical Physics* **118** (19), 8594–8610 (2003).
73 D. W. Schwenke, *Journal of Chemical Physics* **122** (1), 014107 (2005).
74 K. A. Peterson, A. K. Wilson, D. E. Woon and T. H. Dunning, *Theoretical Chemistry Accounts* **97** (1–4), 251–259 (1997).
75 A. P. Scott and L. Radom, *Journal of Physical Chemistry* **100** (41), 16502–16513 (1996).
76 J. M. Simmie and K. P. Somers, *The Journal of Physical Chemistry A* **119** (28), 7235–7246 (2015).
77 L. A. Curtiss, P. C. Redfern and K. Raghavachari, *Wiley Interdisciplinary Reviews – Computational Molecular Science* **1** (5), 810–825 (2011).
78 L. A. Curtiss, P. C. Redfern and K. Raghavachari, *Journal of Chemical Physics* **126** (8), 084108 (2007).

79 J. A. Montgomery, J. W. Ochterski and G. A. Petersson, *Journal of Chemical Physics* **101** (7), 5900–5909 (1994).

80 J. A. Montgomery, M. J. Frisch, J. W. Ochterski and G. A. Petersson, *Journal of Chemical Physics* **112** (15), 6532–6542 (2000).

81 A. Karton, E. Rabinovich, J. M. L. Martin and B. Ruscic, *Journal of Chemical Physics* **125** (14), 144108 (2006).

82 N. J. De Yonker, T. R. Cundari and A. K. Wilson, *Journal of Chemical Physics* **124** (11), 114104 (2006).

83 M. E. Harding, J. Vazquez, B. Ruscic, A. K. Wilson, J. Gauss and J. F. Stanton, *Journal of Chemical Physics* **128** (11), 114111 (2008).

84 Y. Zhao, B. J. Lynch and D. G. Truhlar, *Physical Chemistry Chemical Physics* **7** (1), 43–52 (2005).

85 I. Rossi and D. G. Truhlar, *Chemical Physics Letters* **234** (1–3), 64–70 (1995).

86 P. E. M. Siegbahn, M. Svensson and P. J. E. Boussard, *Journal of Chemical Physics* **102** (13), 5377–5386 (1995).

87 G. Frenking, I. Antes, M. Böhme, S. Dapprich, A. W. Ehlers, V. Jonas, A. Nauhaus, M. Otto, R. Stegmann, A. Veldkamp and S. F. Vyboishchikov, *Reviews in Computational Chemistry* **8**, 63 (1996).

88 T. R. Cundari, M. T. Benson, M. L. Lutz and S. O. Sommerer, *Reviews in Computational Chemistry* **8**, 145 (1996).

89 M. Dolg and X. Cao, *Chemical Reviews* **112** (1), 403–480 (2012).

90 K. G. Dyall, *Journal of Chemical Physics* **96** (2), 1210–1217 (1992).

91 W. J. Stevens, H. Basch and M. Krauss, *Journal of Chemical Physics* **81** (12), 6026–6033 (1984).

92 D. R. Hamann, M. Schlüter and C. Chiang, *Physical Review Letters* **43** (20), 1494–1497 (1979).

93 D. Vanderbilt, *Physical Review B* **41** (11), 7892-7895 (1990).

94 G. Kresse and J. Hafner, *Journal of Physics – Condensed Matter* **6** (40), 8245–8257 (1994).

95 S. Tosoni, C. Tuma, J. Sauer, B. Civalleri and P. Ugliengo, *Journal of Chemical Physics* **127** (15), 154102 (2007).

96 K. Lejaeghere, G. Bihlmayer, T. Bjorkman, P. Blaha, S. Blugel, V. Blum, D. Caliste, I. E. Castelli, S. J. Clark, A. Dal Corso, S. de Gironcoli, T. Deutsch, J. K. Dewhurst, I. Di Marco, C. Draxl, M. Dulak, O. Eriksson, J. A. Flores-Livas, K. F. Garrity, L. Genovese, P. Giannozzi, M. Giantomassi, S. Goedecker, X. Gonze, O. Granas, E. K. U. Gross, A. Gulans, F. Gygi, D. R. Hamann, P. J. Hasnip, N. A. W. Holzwarth, D. Iusan, D. B. Jochym, F. Jollet, D. Jones, G. Kresse, K. Koepernik, E. Kucukbenli, Y. O. Kvashnin, I. L. M. Locht, S. Lubeck, M. Marsman, N. Marzari, U. Nitzsche, L. Nordstrom, T. Ozaki, L. Paulatto, C. J. Pickard, W. Poelmans, M. I. J. Probert, K. Refson, M. Richter, G.-M. Rignanese, S. Saha, M. Scheffler, M. Schlipf, K. Schwarz, S. Sharma, F. Tavazza, P. Thunstrom, A. Tkatchenko, M. Torrent, D. Vanderbilt, M. J. van Setten, V. Van Speybroeck, J. M. Wills, J. R. Yates, G.-X. Zhang and S. Cottenier, *Science (New York, N.Y.)* **351** (6280), aad3000–aad3000 (2016).

97 F. Vasconcelos, G. A. de Wijs, R. W. A. Havenith, M. Marsman and G. Kresse, *Journal of Chemical Physics* **139** (1), 014109 (2013).

98 P. E. Blochl, *Physical Review B* **50** (24), 17953–17979 (1994).

99 G. Kresse and D. Joubert, *Physical Review B* **59** (3), 1758–1775 (1999).

100 M. Marsman and G. Kresse, *Journal of Chemical Physics* **125** (10), 104101 (2006).

101 E. J. Baerends, D. E. Ellis and P. Ros, *Chemical Physics* **2** (1), 41–51 (1973).

102 G. te Velde, F. M. Bickelhaupt, E. J. Baerends, C. F. Guerra, S. J. A. Van Gisbergen, J. G. Snijders and T. Ziegler, *Journal of Computational Chemistry* **22** (9), 931–967 (2001).

103 A. Halkier, H. Koch, P. Jorgensen, O. Christiansen, I. M. B. Nielsen and T. Helgaker, *Theoretical Chemistry Accounts* **97** (1–4), 150–157 (1997).

104 F. B. van Duijneveldt, J. G. C. M. van Duijneveldt-van de Rijdt and J. H. van Lenthe, *Chemical Reviews* **94** (7), 1873–1885 (1994).

105 I. Mayer and A. Vibok, *Molecular Physics* **92** (3), 503–510 (1997).
106 P. Salvador, B. Paizs, M. Duran and S. Suhai, *Journal of Computational Chemistry* **22** (7), 765–786 (2001).
107 K. Kristensen, P. Ettenhuber, J. J. Eriksen, F. Jensen and P. Jorgensen, *Journal of Chemical Physics* **142** (11), 114116 (2015).
108 L. A. Burns, M. S. Marshall and C. D. Sherrill, *Journal of Chemical Theory and Computation* **10**, 49–57 (2014).
109 Ł. M. Mentel and E. J. Baerends, *Journal of Chemical Theory and Computation* **10**, 252–267 (2014).
110 F. Jensen, *Chemical Physics Letters* **261** (6), 633–636 (1996).
111 H. Valdes, V. Klusak, M. Pitonak, O. Exner, I. Stary, P. Hobza and L. Rulisek, *Journal of Computational Chemistry* **29** (6), 861–870 (2008).
112 L. F. Holroyd and T. van Mourik, *Chemical Physics Letters* **442** (1–3), 42–46 (2007).
113 D. Moran, A. C. Simmonett, F. E. Leach, III, W. D. Allen, P. V. R. Schleyer and H. F. Schaefer, III, *Journal of the American Chemical Society* **128** (29), 9342–9343 (2006).
114 D. Asturiol, M. Duran and P. Salvador, *Journal of Chemical Physics* **128** (14), 144108 (2008).
115 F. Jensen, *Journal of Chemical Theory and Computation* **6** (1), 100–106 (2010).
116 H. Kruse and S. Grimme, *Journal of Chemical Physics* **136** (15), 154101 (2012).
117 F. Jensen, *Theoretical Chemistry Accounts* **132** (8), 1380 (2013).

6

Density Functional Methods

The basis for *Density Functional Theory* (DFT) is the proof by Hohenberg and Kohn[1] that the ground state electronic energy is determined completely by the electron density ρ (see Appendix B for details). In other words, there exists a one-to-one correspondence between the electron density of a system and the energy. The "intuitive" proof of why the density completely defines the system is due to E. B. Wilson,[2] who argued that:

- The integral of the density defines the number of electrons.
- The cusps in the density define the position of the nuclei.
- The heights of the cusps define the corresponding nuclear charges.

The significance of the Hohenberg–Kohn theorem is perhaps best illustrated by comparing it with the wave function approach. A wave function for an N electron system contains $4N$ variables, three spatial and one spin coordinate for each electron. The electron density is the square of the wave function, integrated over $N - 1$ electron coordinates, and each spin density only depends on three spatial coordinates, *independent* of the number of electrons. While the complexity of a wave function increases exponentially with the number of electrons, the electron density has the same number of variables, independent of the system size. The "only" problem is that although it has been proven that each different density yields a different ground state energy, the functional connecting these two quantities is not known. The goal of DFT methods is to design functionals connecting the electron density with the energy.[3–9]

A note on semantics: a *function* is a prescription for producing a number from a set of *variables* (coordinates), while a *functional* is a prescription for producing a number from a *function*, which in turn depends on variables. A wave function and the electron density are thus functions, while the energy depending on a wave function or an electron density is a functional. We will denote a function depending on a set of variables with parenthesis, $f(\mathbf{x})$, while a functional depending on a function is denoted with square brackets, $F[f]$.

Early attempts at designing DFT models (actually predating wave mechanics) tried to express all the energy components as a functional of the electron density, but these methods had poor performance, and wave function-based methods were consequently preferred. The success of modern DFT methods is based on the suggestion by Kohn and Sham in 1965 that the electron kinetic energy should be calculated from an auxiliary set of orbitals used for representing the electron density.[10] The exchange–correlation energy, which is a rather small fraction of the total energy, is then the only

Introduction to Computational Chemistry, Third Edition. Frank Jensen.

Companion Website: http://www.wiley.com/go/jensen/computationalchemistry3

unknown functional, and even relatively crude approximations for this term provide quite accurate computational models. The simplest model is the local density approximation, where the electron density is assumed to be slowly varying, such that the exchange–correlation energy can be calculated using formulas derived for a uniform electron density. A significant improvement in the accuracy can be obtained by making the exchange–correlation functional dependent also on the first derivative of the density, and further refinements also add the second derivative and mix Hartree–Fock exchange into the functional. Density functional theory is conceptually and computationally very similar to Hartree–Fock theory, but provides much better results and has consequently become a very popular method. The main problem in DFT is the inability to systematically improve the results, which has spawned a bewildering zoo of different functionals with (slightly) different performances for different properties.

6.1 Orbital-Free Density Functional Theory

Compared with the wave mechanics approach, it seems clear that the energy functional may be divided into three parts, kinetic energy, $T[\rho]$, attraction between the nuclei and electrons, $E_{ne}[\rho]$, and electron–electron repulsion, $E_{ee}[\rho]$ (the nuclear–nuclear repulsion is a constant within the Born–Oppenheimer approximation). Furthermore, with reference to Hartree–Fock theory (Equation (3.33)), the $E_{ee}[\rho]$ term may be divided into Coulomb and exchange parts, $J[\rho]$ and $K[\rho]$, implicitly including correlation energy in all the terms. The $E_{ne}[\rho]$ and $J[\rho]$ functionals are given by their classical expressions, where the factor of $^1/_2$ in $J[\rho]$ allows the integration to be over all space for both variables:

$$E_{ne}[\rho] = -\sum_{a}^{N_{\text{nuclei}}} \int \frac{Z_a(\mathbf{R}_a)\rho(\mathbf{r})}{|\mathbf{R}_a - \mathbf{r}|} d\mathbf{r} \tag{6.1}$$

$$J[\rho] = \frac{1}{2} \iint \frac{\rho(\mathbf{r})\rho(\mathbf{r}')}{|\mathbf{r} - \mathbf{r}'|} d\mathbf{r}d\mathbf{r}' \tag{6.2}$$

Early attempts of deducing functionals for the kinetic and exchange energies considered a uniform electron gas, where it may be shown that $T[\rho]$ and $K[\rho]$ are given by

$$\begin{aligned} T_{TF}[\rho] &= C_F \int \rho^{5/3}(\mathbf{r}) d\mathbf{r} \\ C_F &= \tfrac{3}{10}(3\pi^2)^{2/3} \end{aligned} \tag{6.3}$$

$$\begin{aligned} K_D[\rho] &= -C_x \int \rho^{4/3}(\mathbf{r}) d\mathbf{r} \\ C_x &= \frac{3}{4}\left(\frac{3}{\pi}\right)^{1/3} \end{aligned} \tag{6.4}$$

The energy functional $E_{TF}[\rho] = T_{TF}[\rho] + E_{ne}[\rho] + J[\rho]$ is known as the *Thomas–Fermi* (TF) theory, while inclusion of the $K_D[\rho]$ exchange part (first derived by Bloch,[11] but commonly associated with the name of Dirac[12]) constitutes the *Thomas–Fermi–Dirac* (TFD) model.

The assumption of a uniform electron gas is fair for the valence electrons in certain metallic (periodic) systems, but is poor for atoms and molecules. The TF kinetic energy furthermore leads to an incorrect asymptotic behavior for the electron density (r^{-6} instead of exponential fall-off) and fails to reproduce the shell structure of atoms. A serious flaw from a chemical point of view is that neither TF nor TFD theories predict bonding: molecules simply do not exist.

The kinetic and exchange functionals can be improved by the addition of terms depending on the derivative(s) of the electron density. This is equivalent to considering a non-uniform electron gas and performing a Taylor-like expansion with the density as a variable.[13,14] The expansion for the kinetic energy is given in Equation (6.5), where odd terms vanish owing to the rotational invariance of the energy with respect to **r**:

$$T[\rho] = T_{\mathrm{TF}}[\rho] + T_2[\rho] + T_4[\rho] + T_6[\rho] + \cdots \tag{6.5}$$

$$T_2[\rho] = \lambda \tau_{\mathrm{w}}[\rho] \quad ; \quad \tau_{\mathrm{w}}[\rho] = \int \frac{|\nabla \rho(\mathbf{r})|^2}{8\rho(\mathbf{r})} \mathrm{d}\mathbf{r} \tag{6.6}$$

$$T_4[\rho] = (540(3\pi)^{2/3})^{-1} \int \rho^{1/3}(\mathbf{r}) \left\{ \left(\frac{\nabla^2 \rho(\mathbf{r})}{\rho(\mathbf{r})} \right)^2 - \frac{9}{8} \left(\frac{\nabla^2 \rho(\mathbf{r})}{\rho(\mathbf{r})} \right) \left(\frac{\nabla \rho(\mathbf{r})}{\rho(\mathbf{r})} \right)^2 + \frac{1}{3} \left(\frac{\nabla \rho(\mathbf{r})}{\rho(\mathbf{r})} \right)^4 \right\} \mathrm{d}\mathbf{r} \tag{6.7}$$

The T_2 correction contains the *von Weizsacker* kinetic energy, τ_{W}, where λ has a value of 1/9. Various empirical values for λ have been used in cases where the expansion is terminated after the T_2 term, motivated by the fact that the von Weizsacker expression is equivalent to the Hartree–Fock kinetic energy for one- and two-electron systems. The kinetic energy at the Thomas–Fermi level is typically underestimated by ~10%, which is reduced to ~1% by addition of the T_2 term, while inclusion of the T_4 correction leads to an overestimation of slightly larger magnitude. In terms of absolute energies, this is comparable to the HF method, but energy differences (e.g. atomization energies) are calculated with much lower accuracy than with the HF model. Unfortunately, the sixth- and higher-order T terms diverge in regions far from the nuclei, preventing further improvements.

The second-order exchange term K_2 is given in Equation (6.9) and the K_4 term has an expression similar to T_4, except that not all the expansion coefficients are known:[15]

$$K[\rho] = K_{\mathrm{D}}[\rho] + K_2[\rho] + K_4[\rho] + \cdots \tag{6.8}$$

$$K_2[\rho] = -\frac{5}{216}(3\pi^5)^{-1/3} \int \frac{|\nabla \rho(\mathbf{r})|^2}{\rho^{4/3}(\mathbf{r})} \mathrm{d}\mathbf{r} \tag{6.9}$$

Addition of gradient correction terms improves the Thomas–Fermi results, for example bonding is now allowed, but the lack of sufficient error cancellation and the divergence of higher-order correction terms mean that this is not a viable approach for constructing DFT models capable of yielding results comparable with those obtained by wave mechanics methods.

Although there have been attempts at constructing *orbital-free* (as opposed to the Kohn–Sham version discussed in the next section) T functionals depending directly on the electron density, the accuracy is still too low to be of general use.[16,17] If such functionals could be derived, however, the full potential of DFT in having only three variables independent of system size could be fully realized.

6.2 Kohn–Sham Theory

The foundation for the use of DFT methods in computational chemistry is the introduction of orbitals, as suggested by Kohn and Sham (KS).[5] The main flaw in orbital-free models is the poor representation of the kinetic energy, and the idea in the KS formalism is to split the kinetic energy functional into two parts, one which can be calculated exactly and a small correction term. The price to be paid is that orbitals are re-introduced, thereby increasing the complexity from 3 to $3N$ variables, and that electron correlation re-emerges as a separate term. The KS model is closely related to the HF method, sharing identical formulas for the kinetic, electron–nuclear and Coulomb electron–electron energies.

The division of the electron kinetic energy into two parts, with the major contribution being equivalent to the HF kinetic energy, can be justified as follows. Assume for the moment a Hamiltonian operator of the form in the following equation with $0 \leq \lambda \leq 1$:

$$\mathbf{H}_\lambda = \mathbf{T} + \mathbf{V}_{\text{ext}}(\lambda) + \lambda \mathbf{V}_{\text{ee}} \tag{6.10}$$

The external potential operator $\mathbf{V}_{\text{ext}}$ is equal to $\mathbf{V}_{\text{ne}}$ for $\lambda = 1$, but for intermediate λ values it is assumed that $\mathbf{V}_{\text{ext}}(\lambda)$ is adjusted such that the *same* density is obtained for $\lambda = 1$ (the real system), for $\lambda = 0$ (a hypothetical system with non-interacting electrons) and for all intermediates λ values. For the $\lambda = 0$ case, the electrons are non-interacting and the *exact* solution to the Schrödinger equation is given as a Slater determinant composed of (molecular) orbitals, ϕ_i. The *exact* kinetic energy functional is given by

$$T_{\text{S}} = \sum_{i=1}^{N_{\text{elec}}} \left\langle \phi_i \left| -\tfrac{1}{2}\nabla^2 \right| \varphi_i \right\rangle \tag{6.11}$$

The subscript $_{\text{S}}$ denotes that it is the kinetic energy calculated from a Slater determinant. The $\lambda = 1$ case corresponds to *interacting* electrons, and Equation (6.11) is therefore only an approximation to the real kinetic energy, but a substantial improvement over the TF formula (Equation (6.3)).

Another way of justifying the use of Equation (6.11) for calculating the kinetic energy is by reference to natural orbitals (eigenvectors of the density matrix, Section 10.5). The *exact* kinetic energy can be calculated from the natural orbitals (NO) arising from the *exact* density matrix:

$$T[\rho_{\text{exact}}] = \sum_{i=1}^{\infty} n_i \left\langle \phi_i^{\text{NO}} \left| -\tfrac{1}{2}\nabla^2 \right| \phi_i^{\text{NO}} \right\rangle \tag{6.12}$$

$$\rho_{\text{exact}} = \sum_{i=1}^{\infty} n_i \left|\phi_j^{\text{NO}}\right|^2 \quad ; \quad N_{\text{elec}} = \sum_{i=1}^{\infty} n_i \tag{6.13}$$

The orbital occupation numbers n_i (eigenvalues of the density matrix) will be between 0 and 1, corresponding to the number of electrons in the (spin) orbital. Representing the exact density will require an infinite number of natural orbitals, with the N_{elec} first having occupation numbers close to 1 and the remaining numbers close to 0. Since the exact density matrix is not known, an (approximate) density can be written in terms of a set of auxiliary one-electron functions, that is orbitals:

$$\rho_{\text{approx}} = \sum_{i=1}^{N_{\text{elec}}} |\phi_i|^2 \tag{6.14}$$

This corresponds to Equation (6.13) with occupation numbers of exactly 1 or 0. The "missing" kinetic energy from Equation (6.11) is thus due to the occupation numbers deviating from being exactly 1 or 0. Since the occupation numbers of an HF (single-determinant) wave function are also exactly 1 or 0, the missing kinetic energy can also be considered as the (kinetic) correlation energy.

The key to Kohn–Sham theory is to calculate the kinetic energy under the assumption of non-interacting electrons (in the same sense that HF orbitals in wave mechanics describe non-interacting electrons) from Equation (6.11). In reality, the electrons *are* interacting and Equation (6.11) does not provide the total kinetic energy. However, just as HF theory provides ~99% of the correct answer, the difference between the exact kinetic energy and that calculated by assuming non-interacting electrons is small. The remaining kinetic energy is absorbed into an exchange–correlation term, and a general DFT energy expression can be written as

$$E_{\text{DFT}}[\rho] = T_{\text{S}}[\rho] + E_{\text{ne}}[\rho] + J[\rho] + E_{\text{xc}}[\rho] \tag{6.15}$$

By equating E_{DFT} to the exact energy, this expression *defines* E_{xc}, that is it is the part that remains after subtraction of the non-interacting kinetic energy, and the E_{ne} and J potential energy terms:

$$E_{\text{xc}}[\rho] = (T[\rho] - T_{\text{S}}[\rho]) + (E_{\text{ee}}[\rho] - J[\rho]) \tag{6.16}$$

The first parenthesis in Equation (6.16) may be considered as the kinetic correlation energy, while the last contains both potential correlation and exchange energy.

The task in developing orbital-free models is to derive approximations to the kinetic and exchange energy functionals (including correlation), while the corresponding task in Kohn–Sham theory is to derive approximations to the exchange–correlation energy functional only. For the neon atom, for example, the kinetic energy is 128.93 au, the exchange energy is −12.10 au and the correlation energy is −0.38 au (as calculated by wave mechanics methods). Since the exchange–correlation energy is roughly a factor of 10 smaller than the kinetic energy, Kohn–Sham theory is much less sensitive to inaccuracies in the functional(s) than orbital-free theory. While orbital-free theory is a true density functional theory (three variables), Kohn–Sham methods are independent-particle models ($3N$ variables), analogous to Hartree–Fock theory, but are still much less complicated than many-particle (correlation) wave function models.

6.3 Reduced Density Matrix and Density Cumulant Methods

Before embarking on a more detailed analysis of how to design exchange–correlation energy functionals in Kohn–Sham theory, it may be instructive to take a slight detour and consider methods using *Reduced Density Matrices* (RDM), rather than the electron density itself.[9] We will start by defining the first- and second-order reduced density matrices γ_1 and γ_2:

$$\gamma_1(\mathbf{r}_1, \mathbf{r}_1') = N_{elec} \int \Psi^*(\mathbf{r}_1', \mathbf{r}_2, \dots, \mathbf{r}_{N_{\text{elec}}}) \Psi(\mathbf{r}_1, \mathbf{r}_2, \dots, \mathbf{r}_{N_{\text{elec}}}) d\mathbf{r}_2 \cdots d\mathbf{r}_{N_{\text{elec}}} \tag{6.17}$$

$$\gamma_2(\mathbf{r}_1, \mathbf{r}_2, \mathbf{r}_1', \mathbf{r}_2') = N_{elec}(N_{elec} - 1) \int \Psi^*(\mathbf{r}_1', \mathbf{r}_2', \mathbf{r}_3, \dots, \mathbf{r}_{N_{\text{elec}}}) \Psi(\mathbf{r}_1, \mathbf{r}_2, \mathbf{r}_3, \dots, \mathbf{r}_{N_{\text{elec}}}) d\mathbf{r}_3 \cdots d\mathbf{r}_{N_{\text{elec}}} \tag{6.18}$$

We will ignore electron spin, except when required. The corresponding reduced spin density matrices are defined completely analogously to Equations (6.17) and (6.18), if the $\mathbf{r}$ variables are taken to represent both spatial and spin coordinates.

The diagonal components of the first-order density matrix (setting $\mathbf{r}_1' = \mathbf{r}_1$) gives the *electron density* function ρ_1, often written without the subscript 1 when higher-order densities are not involved:

$$\rho_1(\mathbf{r}_1) = \gamma_1(\mathbf{r}_1, \mathbf{r}_1) = N_{\text{elec}} \int \Psi^*(\mathbf{r}_1 \cdots \mathbf{r}_{N_{\text{elec}}}) \Psi(\mathbf{r}_1 \cdots \mathbf{r}_{N_{\text{elec}}}) dr_2 \cdots dr_{N_{\text{elec}}} \tag{6.19}$$

The integral is the probability of finding an electron (it does not matter which, since they are indistinguishable) at position $\mathbf{r}_1$, and the N_{elec} prefactor ensures that the density integrates to the number of electrons.

The corresponding second-order density matrix yields the *electron pair-density* upon setting $\mathbf{r}_1' = \mathbf{r}_1$ and $\mathbf{r}_2' = \mathbf{r}_2$:

$$\begin{aligned} \rho_2(\mathbf{r}_1, \mathbf{r}_2) &= \gamma_2(\mathbf{r}_1, \mathbf{r}_2, \mathbf{r}_1, \mathbf{r}_2) \\ &= N_{\text{elec}}(N_{\text{elec}} - 1) \int \Psi^*(\mathbf{r}_1, \mathbf{r}_2 \cdots \mathbf{r}_{N_{\text{elec}}})\Psi(\mathbf{r}_1, \mathbf{r}_2 \cdots \mathbf{r}_{N_{\text{elec}}}) d\mathbf{r}_3 \cdots d\mathbf{r}_{N_{\text{elec}}} \end{aligned} \tag{6.20}$$

The integral is the probability of finding an electron at position $\mathbf{r}_1$ and another electron at position $\mathbf{r}_2$, and the $N_{\text{elec}}(N_{\text{elec}} - 1)$ prefactor ensures that ρ_2 integrates to the number of electron pairs (note that the number of *unique* electron pairs is only $^1/_2 N_{\text{elec}}(N_{\text{elec}} - 1)$).

The *exact* kinetic and potential energies are given by the (exact) first- and second-order density matrices, with an implicit summation over electron spin:

$$T = -\frac{1}{2} \int \nabla^2 \gamma_1(\mathbf{r}_1, \mathbf{r}_1')|_{r=r'} \mathrm{d}\mathbf{r}_1 \tag{6.21}$$

$$V_{\text{ne}} = -\sum_a^{N_{\text{nuclei}}} \int \frac{Z_a \gamma_1(\mathbf{r}_1, \mathbf{r}_1)}{|\mathbf{R}_a - \mathbf{r}_1|} \mathrm{d}\mathbf{r}_1 = -\sum_a^{N_{\text{nuclei}}} \int \frac{Z_a \rho_1(\mathbf{r}_1)}{|\mathbf{R}_a - \mathbf{r}_1|} \mathrm{d}\mathbf{r}_1 \tag{6.22}$$

$$V_{\text{ee}} = \frac{1}{2} \int \frac{\gamma_2(\mathbf{r}_1, \mathbf{r}_2, \mathbf{r}_1, \mathbf{r}_2)}{|\mathbf{r}_1 - \mathbf{r}_2|} \mathrm{d}\mathbf{r}_1 \mathrm{d}\mathbf{r}_2 = \frac{1}{2} \int \frac{\rho_2(\mathbf{r}_1, \mathbf{r}_2)}{|\mathbf{r}_1 - \mathbf{r}_2|} \mathrm{d}\mathbf{r}_1 \mathrm{d}\mathbf{r}_2 \tag{6.23}$$

Note that the potential energy terms in Equations (6.22) and (6.23) can be written in terms of ρ_1 and ρ_2; only the kinetic energy requires γ_1. Note also that the V_{ee} term can be split into a Coulomb term that only depends on ρ_1, while the exchange part depends on ρ_2.

For the special case of a single-determinant wave function, the first- and second-order reduced density matrices are given by Equations (6.25) and (6.26):

$$\mathbf{\Phi}_{\text{SD}} = \frac{1}{\sqrt{N!}} \begin{vmatrix} \phi_1(1) & \phi_2(1) & \cdots & \phi_N(1) \\ \phi_1(2) & \phi_2(2) & \cdots & \phi_N(2) \\ \vdots & \vdots & \ddots & \vdots \\ \phi_1(N) & \phi_2(N) & \cdots & \phi_N(N) \end{vmatrix} \tag{6.24}$$

$$\gamma_1(\mathbf{r}_1, \mathbf{r}_1') = \sum_{i=1}^{N_{\text{occ}}} \phi_i(\mathbf{r}_1')\phi_i(\mathbf{r}_1) \tag{6.25}$$

$$\gamma_2(\mathbf{r}_1, \mathbf{r}_2, \mathbf{r}_1', \mathbf{r}_2') = \sum_{i,j=1}^{N_{\text{occ}}} \{\phi_i(\mathbf{r}_1')\phi_j(\mathbf{r}_2')\phi_i(\mathbf{r}_1)\phi_j(\mathbf{r}_2) - \phi_i(\mathbf{r}_1')\phi_j(\mathbf{r}_2')\phi_j(\mathbf{r}_1)\phi_i(\mathbf{r}_2)\} \tag{6.26}$$

Using Equations (6.25) and (6.26) in Equations (6.21) to (6.23) is readily seen to provide the Hartree–Fock energy expression, Equation (3.32).

Since γ_1 can be obtained by straightforward integration of γ_2, an appealing idea is to use the elements of γ_2 as variables for solving the Schrödinger equation or its equivalent density formulation by a variational procedure. Unfortunately, the γ_2 elements cannot be varied freely, since they must correspond to an antisymmetric wave function.[18] Although a formal solution of this N-*representability* problem exists,[19] it does not lend itself to an efficient computational implementation. Work by

D. Mazziotti has shown that good approximations to the N-representability can be obtained by enforcing positive semi-definiteness (non-negative eigenvalues) of three matrices during the optimization of γ_2.[20] These three matrices are the two-particle density matrix γ_2 itself and the corresponding representations in terms of hole–hole and particle–hole creations, commonly denoted **D**, **Q** and **G**. The semi-definite condition arises since these matrices describe probabilities. Performing such constrained optimizations is a non-trivial computational task, but methods have been proposed that (only) scales as M^6_{basis}, making such calculation tractable for general systems.[21] The accuracy of the results can be improved by imposing semipositivity of higher-order density matrices,[22] albeit at a significantly higher computational cost. The quality of the results are typically comparable to CCSD(T) but somewhat more tolerant toward multireference character.

An alternative approach, called *Density Cumulant functional Theory* (DCT), is to write γ_2 as a product of γ_1 (compare to Equation (6.15)) and a reminder term:[23]

$$\gamma_2(\mathbf{r}_1,\mathbf{r}_2,\mathbf{r}'_1,\mathbf{r}'_2) = \gamma_1(\mathbf{r}_1,\mathbf{r}'_1)\gamma_1(\mathbf{r}_2,\mathbf{r}'_2) - \gamma_1(\mathbf{r}_2,\mathbf{r}'_1)\gamma_1(\mathbf{r}_1,\mathbf{r}'_2) + \boldsymbol{\lambda}_2(\mathbf{r}_1,\mathbf{r}_2,\mathbf{r}'_1,\mathbf{r}'_1) \tag{6.27}$$

The cumulant $\boldsymbol{\lambda}_2$ term accounts for the pair-correlation (correlation is also present in γ_1, see Equation (6.28)) and must implicitly ensure the N-representability. The γ_1 is similarly written as a sum of an idempotent component ($\boldsymbol{\kappa}_1$) and a correlation ($\boldsymbol{\tau}_1$) term:

$$\begin{aligned}\gamma_1(\mathbf{r}_1,\mathbf{r}'_1) &= \boldsymbol{\kappa}_1(\mathbf{r}_1,\mathbf{r}'_1) + \boldsymbol{\tau}_1(\mathbf{r}_1,\mathbf{r}'_1)\\ \boldsymbol{\kappa}_1^2 &= \boldsymbol{\kappa}_1\end{aligned} \tag{6.28}$$

The $\boldsymbol{\kappa}_1$ term corresponds to the orbital product in Equation (6.25) for a single determinant wave function, where the $\boldsymbol{\tau}_1$ term is zero. The $\boldsymbol{\tau}_1$ term can be formulated in terms of $\boldsymbol{\lambda}_2$, which is then the central variable. The N-representability of $\boldsymbol{\lambda}_2$ can be approximately fulfilled to a given order by a perturbation expansion.[24] The advantage of DCT over RDM methods is that the N-representability in the former only affects the correlation energy, not the total energy itself. A practical advantage is also that DCT involves solving non-linear equations similar to those in coupled cluster theory, while RDM methods require solving semi-definite equations involving constraints. At the time of writing, DCT typically provides results of a quality intermediate between CCSD and CCSD(T).[25]

The electron pair-density ρ_2 is sufficient for calculating the exact electron–electron interaction, and *Pair Density Functional Theory* (PDFT) attempts to use this as the fundamental variable. The electron–nucleus interaction requires the density ρ_1, which can be obtained by integration of ρ_2, but the kinetic energy requires the reduced density matrix γ_1. The challenge in PDFT is thus to design kinetic energy functionals depending on ρ_2, in addition to enforcing the N-representability condition.

The electron density matrix ρ_1 can be diagonalized to produce eigenvalues and eigenvectors, called occupation numbers n_i and natural orbitals ϕ_i^{NO}. The γ_1 can be written in terms of these quantities (compare with Equation (6.25), but note that the summation now includes all orbitals, since $n_i \neq 0$ in general):

$$\gamma_1(\mathbf{r}_1,\mathbf{r}'_1) = \sum_i^{N_{\text{orb}}} n_i \phi_i^{\text{NO}}(\mathbf{r}'_1)\phi_i^{\text{NO}}(\mathbf{r}_1) \tag{6.29}$$

As an alternative to using γ_2 as the fundamental variable, the first-order reduced density matrix, or its parameterization in terms of natural orbitals and occupation numbers, can be used. The N-representability problem for γ_1 is easy to fulfill, since the only requirements are that the eigenvalues are between 0 and 1, and that they sum to N_{elec}. The task in this case is to derive a suitable approximation for the V_{ee} energy term in terms of γ_1, rather than ρ_2.[26,27] The Coulomb and exchange parts of V_{ee} are given by the analogous formulas (Equations (6.23) and (6.26)) using natural orbitals and occupation numbers, leaving the correlation energy as the only unknown function of γ_1. The correlation part can be incorporated by multiplying the orbital products with functions of the occupation numbers as shown in Equation (6.31):

$$V_{ee} = \frac{1}{2} \int \frac{\rho_2(\mathbf{r}_1, \mathbf{r}_2)}{|\mathbf{r}_1 - \mathbf{r}_2|} d\mathbf{r}_1 d\mathbf{r}_2 \tag{6.30}$$

$$\rho_2(\mathbf{r}_1, \mathbf{r}_2) = \sum_{ij}^{N_{orb}} \left\{ \begin{array}{l} f(n_i, n_j)\phi_i^{NO}(\mathbf{r}_1)\phi_i^{NO}(\mathbf{r}_1)\phi_j^{NO}(\mathbf{r}_2)\phi_j^{NO}(\mathbf{r}_2) \\ -g(n_i, n_j)\phi_i^{NO}(\mathbf{r}_1)\phi_j^{NO}(\mathbf{r}_1)\phi_i^{NO}(\mathbf{r}_2)\phi_j^{NO}(\mathbf{r}_2) \end{array} \right\} \tag{6.31}$$

The choice of $f(n_i, n_j) = g(n_i, n_j) = n_i n_j$ implies the Hartree–Fock model, and optimization of the occupation numbers and natural orbitals with this choice indeed returns the HF wave function, that is the HF wave function cannot be improved by allowing fractional occupation numbers. The $f(n_i, n_j)$ function is usually set equal to $n_i n_j$, since this is just the Coulomb interaction, and the exchange–correlation part is modeled by $g(n_i, n_j)$.[28] A number of such *Natural Orbital Functionals* (NOFs) has been proposed and tested for performance.[29]

The Hohenberg–Kohn theorem, which states that the energy is uniquely determined by the one-electron density ρ_1, forms the basis for what is commonly called density functional theory. As discussed in Section 6.1, it is difficult to construct a sufficiently accurate total energy functional depending only on ρ_1. The Kohn–Sham version, where only the exchange–correlation part of the energy must be estimated as a functional of ρ_1, provides viable models, and these will be discussed in more detail in Section 6.5.

Perhaps the most surprising result of the Hohenberg–Kohn theorem is that the correlation energy is completely determined by the one-electron density function ρ_1. Electron correlation is inherently a two-electron phenomenon, and it is difficult to envisage how an accurate correlation functional depending on only the one-electron density should be constructed from theoretical arguments, although one certainly can understand that the correlation will affect the electron density. Indeed, the interpretation of electron correlation in terms of correlation holes, as discussed in the next section, requires the two-electron density ρ_2. P.M.W. Gill has suggested that one could consider a quantity similar to the second-order reduced density matrix, except that the arguments are the position and momentum of the two electrons:[30]

$$\begin{aligned} W(\mathbf{r}_1, \ldots, \mathbf{r}_{N_{elec}}, \mathbf{p}_1, \ldots, \mathbf{p}_{N_{elec}}) &= \pi^{-3N_{elec}} \int \Psi^*(\mathbf{r}_1 + \mathbf{q}_1, \ldots, \mathbf{r}_{N_{elec}} + \mathbf{q}_{N_{elec}}) \\ &\times \Psi(\mathbf{r}_1 - \mathbf{q}_1, \ldots, \mathbf{r}_{N_{elec}} - \mathbf{q}_{N_{elec}}) e^{2i(\mathbf{p}_1 \cdot \mathbf{q}_1, \ldots \mathbf{p}_{N_{elec}} \cdot \mathbf{q}_{N_{elec}})} d\mathbf{q}_1 \cdots d\mathbf{q}_{N_{elec}} \\ W_2(\mathbf{r}_1, \mathbf{r}_2, \mathbf{p}_1, \mathbf{p}_2) &= \int W(\mathbf{r}_1, \ldots, \mathbf{r}_{N_{elec}}, \mathbf{p}_1, \ldots, \mathbf{p}_{N_{elec}}) d\mathbf{r}_3 \cdots d\mathbf{r}_{N_{elec}} d\mathbf{p}_3 \cdots d\mathbf{p}_{N_{elec}} \end{aligned} \tag{6.32}$$

The W_2 is called the second-order *Wigner intracule* and represents a quasi-probability function for finding two electrons at positions $\mathbf{r}_1$ and $\mathbf{r}_2$ with momentum $\mathbf{p}_1$ and $\mathbf{p}_2$. It cannot be interpreted as a genuine probability function, as it can achieve negative values. Since one would expect the correlation energy to depend on the distance between the two electrons, the relative momentum between them and the orientation of the distance and momentum difference vectors, Gill *et al.* have suggested that the corresponding Ω intracule depending on the "internal" coordinates may contain information sufficient for determining the correlation energy:

$$\Delta E_{corr} = \int \Omega(u, v, \omega) G(u, v, \omega) \mathrm{d}u\, \mathrm{d}v\, \mathrm{d}\omega \tag{6.33}$$

$$W_2(\mathbf{r}_1, \mathbf{r}_2, \mathbf{p}_1, \mathbf{p}_2) \rightarrow \Omega(u, v, \omega)$$
$$u = |\mathbf{r}_1 - \mathbf{r}_2|; \quad v = |\mathbf{p}_1 - \mathbf{p}_2|; \quad \cos\omega = \frac{(\mathbf{r}_1 - \mathbf{r}_2)'(\mathbf{p}_1 - \mathbf{p}_2)}{|\mathbf{r}_1 - \mathbf{r}_2||\mathbf{p}_1 - \mathbf{p}_2|} \tag{6.34}$$

Analogously to density functional methods, the task is to construct a correlation kernel G that yields the correlation energy upon integration with the Ω intracule, and use this in connection with the exchange energy calculated at the Hartree–Fock level.

The differences between wave mechanics and density-based methods can be summarized as follows:

- Wave mechanics employ the exact Hamiltonian operator, but makes approximations in the form of the wave function.
- Density functional methods make approximations in the energy functional (Hamiltonian), but allow a free variation of the electron density ρ_1. The functional must therefore *implicitly* enforce the N-representability. This is difficult to achieve in orbital-free methods, but the Kohn–Sham approach with a determinantal orbital product takes care of the majority of this problem.[31] Furthermore, in orbital-free methods the kinetic energy functional is unknown, and since this is equal in magnitude to the total energy, even minor inaccuracies cause large absolute errors. In the Kohn–Sham version only the exchange–correlation functional is unknown, and since this is a relatively minor component of the total energy, the results are less sensitive to inaccuracies in the functional.
- Methods using the first-order reduced density matrix as variable can be chosen to strictly enforce the N-representability of γ_1, and employ the exact energy functional for all the terms except the correlation energy. The latter, however, inherently depends on γ_2, which in this approach must be approximated as a function of γ_1. One can argue that Hartree–Fock belongs to this class of methods, with implicit neglect of the electron correlation.
- Methods using the second-order reduced density matrix as variable employ the exact energy functional in terms of γ_2, but must make approximations for enforcing the N-representability of γ_2.
- Methods using the electron-pair density ρ_2 as a variable employ the exact energy functional for the electron–nuclear and electron–electron interactions, but must make approximations for calculating the kinetic energy, in addition to enforcing the N-representability of ρ_2.

Solving the Schrödinger equation by means of reduced density matrices has many appealing features, such as being able to describe the whole potential energy curve with equal accuracy and to account for a very large fraction of the correlation energy. At the time of writing, however, they produce results of roughly the same quality as a conventional wave function-based CCSD.

6.4 Exchange and Correlation Holes

We now return to the problem of expressing the exchange–correlation energy as a functional of ρ ($= \rho_1$). Since the exchange energy is by far the largest contributor to E_{xc} (compare the values for the neon atom in Section 6.2), one may reasonably ask why not calculate this term "exactly" from orbitals (analogous to the kinetic energy), by the formula known from wave mechanics (Equation (3.32)), and only calculate the computationally difficult part, the correlation energy, by DFT. Although this has been tried, it gives poor results. The basic problem is that the DFT definitions of exchange and correlation energies are not completely equivalent to their wave mechanics counterparts.[32] The DFT exchange energy may be defined by the same formula as in HF theory (Equation (3.31)), except that Kohn–Sham orbitals are used. This leads to a non-local potential, that is the exchange potential at a given point is strongly dependent on the density at distant points. The correlation energy in wave mechanics is defined as the difference between the exact energy and the corresponding Hartree–Fock value. Both the exchange and correlation energies have a short- and long-range part (in terms of the distance between two electrons). The long-range correlation is essentially the "static" correlation energy (i.e. the "multireference" part, see Section 4.6) while the short-range part is the "dynamical" correlation. The long-range part of the correlation energy in wave mechanics effectively cancels the delocalized part of the exchange energy. The definitions of exchange and correlation in DFT (at least in current implementations) are local (short range), since they only depend on the density at a given point and the immediate vicinity (via derivatives of the density). The cancellation at long range is (or should be) implicitly built into the exchange–correlation functional. Calculating the exchange energy by wave mechanics and the correlation by DFT thus destroys the cancellation, although research is continuing towards designing correlation functionals to be used in connection with HF exchange.[33]

A more detailed discussion of these features is most easily given in terms of *exchange* and *correlation holes*.[34] Electrons avoid each other owing to their electric charges, and the energy associated with this repulsion is given classically by the Coulomb Equation (6.2). Quantum mechanically, however, this repulsion must be modified to take into account that electrons have spins of $^1/_2$. The Pauli principle states that two fermions (particles with half-integer spin) cannot occupy the same spatial position or, equivalently, that the total wave function must be antisymmetric upon interchange of any two particles. This leads to the exchange energy (see Section 3.3), which can be considered as a quantum correction to the classical Coulomb repulsion and includes the electron self-repulsion. The exchange term is already present in Hartree–Fock theory and must also be incorporated into DFT. There is in addition a dynamical effect where electrons tend to avoid each other more than given by an HF wave function, and this is the correlation energy calculated by wave mechanics methods.

These qualitative considerations can be put into quantitative terms by probability *holes*. If electrons did not have charge or spin, the probability of finding an electron at a given position would be independent of the position of a second electron, and the electron-pair density ρ_2 would be given as a simple product of two one-electron densities ρ_1, with a proper normalization factor:

$$\rho_2^{\mathrm{indep}}(\mathbf{r}_1, \mathbf{r}_2) = \frac{N_{\mathrm{elec}} - 1}{N_{\mathrm{elec}}} \rho_1(\mathbf{r}_1)\rho_1(\mathbf{r}_2) = \left(1 - \frac{1}{N_{\mathrm{elec}}}\right) \rho_1(\mathbf{r}_1)\rho_1(\mathbf{r}_2) \tag{6.35}$$

Since electrons have both charge and spin, however, there is a reduced probability of finding an electron near another electron. We can write this formally in terms of a conditional probability factor $h_{\mathrm{xc}}(\mathbf{r}_1, \mathbf{r}_2)$, which includes the $1/N_{\mathrm{elec}}$ self-interaction factor in Equation (6.35):

$$\rho_2(\mathbf{r}_1, \mathbf{r}_2) = \rho_1(\mathbf{r}_1)\rho_1(\mathbf{r}_2) + \rho_1(\mathbf{r}_1) h_{\mathrm{xc}}(\mathbf{r}_1, \mathbf{r}_2) \tag{6.36}$$

The reduced probability is called the *exchange–correlation hole* and can be written in terms of ρ_2 and ρ_1 by solving Equation (6.36):

$$h_{\mathrm{xc}}(\mathbf{r}_1, \mathbf{r}_2) = \frac{\rho_2(\mathbf{r}_1, \mathbf{r}_2)}{\rho_1(\mathbf{r}_1)} - \rho_1(\mathbf{r}_2) \tag{6.37}$$

The exchange–correlation hole represents the reduced probability of finding electron 2 at a position $\mathbf{r}_2$ given that electron 1 is located at $\mathbf{r}_1$. The exchange part of h_{xc} is called the *Fermi hole*, while the dynamical correlation gives rise to the *Coulomb hole*. Since exchange only occurs between electrons of the same spin, the total hole can also be written in terms of individual spin contributions:

$$\begin{aligned} h_{xc} &= h_x + h_c \\ h_x &= h_x^{\alpha\alpha} + h_x^{\beta\beta} \\ h_c &= h_c^{\alpha\alpha} + h_c^{\beta\beta} + h_c^{\alpha\beta} \end{aligned} \tag{6.38}$$

From the definitions of ρ_2 and ρ_1, it follows that the integral of h_{xc} over $\mathbf{r}_2$ equals -1; this is the electron self-repulsion part:

$$\int h_{xc}(\mathbf{r}_1,\mathbf{r}_2)d\mathbf{r}_2 = \int \frac{\rho_2(\mathbf{r}_1,\mathbf{r}_2)}{\rho_1(\mathbf{r}_1)}d\mathbf{r}_1 d\mathbf{r}_2 - \int \rho_1(\mathbf{r}_2)d\mathbf{r}_2 = \frac{N_{elec}(N_{elec}-1)}{N_{elec}} - N_{elec} = -1 \tag{6.39}$$

A similar argument for the separate spin densities shows that the Fermi hole itself is negative everywhere and integrates to -1, which means that the integral of the Coulomb hole is 0. The Fermi (exchange) hole describes a static reduction in the probability function corresponding to one electron. The Coulomb (correlation) function, on the other hand, reduces the probability of finding an electron near the reference electron, but increases the probability of finding it far from the reference electron.

The exchange energy in Hartree–Fock theory is a non-local function, that is the HF exchange hole is delocalized over the whole system (or at least a large part of it). For a diatomic system, for example, the exchange hole is delocalized over both nuclei. When electron correlation is added explicitly, the left–right correlation to a large extent serves to cancel the delocalized nature of the HF exchange hole. The exchange functional in DFT, on the other hand, is local, that is the cancellation of the delocalized HF exchange hole by the left–right correlation in wave function approaches should be inherent in the functional, and this is the main difference between the definitions of exchange and correlation in wave function and current density functional descriptions.

A closely related phenomenon is the electron self-interaction energy. The Coulomb energy functional given in Equation (6.2) only depends on ρ_1. This means that the density arising from a single electron will interact with itself (e.g. there will be a non-zero electron–electron Coulomb repulsion even for a one-electron system), and this self-repulsion is clearly non-physical. For a multielectron system, there will be such a self-interaction term for the density associated with each electron, although it is difficult to define this rigorously. The HF model takes care of this problem elegantly, since the expression for the exchange energy exactly cancels the Coulomb self-interaction (Equation (3.33)).

A one-electron system should within the DFT model have an exchange energy exactly opposing the Coulomb energy, $E_x = -J$, and the correlation energy should be zero. In a multielectron system, one may thus consider part of the exchange energy as a correction for the self-interaction energy, with the remaining being the "true" Fermi hole. The self-interaction correction is in this view the major part of the exchange energy, with the "true" Fermi hole being comparable to the Coulomb (correlation) hole. It should be noted, however, that such a partitioning is not invariant to a unitary transformation of the occupied orbitals.[35,36] The self-interaction cancellation by the exchange energy is not guaranteed in DFT, and very few of the current exchange–correlation functionals are completely self-interaction-free. It has been proven that a completely self-interaction-free *local* potential does not exist.[37] Perdew and Zunger have suggested an approximate correction scheme, where each

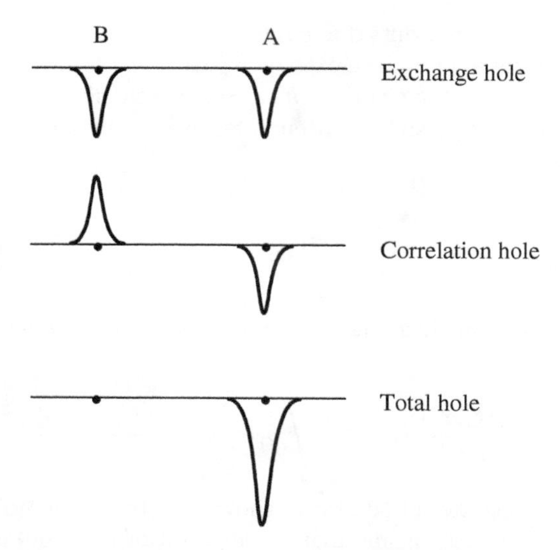

Figure 6.1 Illustrating the exchange and correlation holes for the H_2 molecule at the dissociation limit, with the reference electron located near nucleus *A* and the vertical axis representing probability.

orbital becomes self-interaction-free.[38] The procedure formally changes the underlying functional and destroys the invariance of the energy with respect to mixing of the occupied orbitals.

These concepts can be illustrated for the H_2 molecule for increasing internuclear distances.[39,40] The ground state for H_2 has in wave mechanics two electrons of opposite spin in the same spatial orbital, and the exchange hole is thus entirely the self-interaction correction (no same-spin exchange):

$$\begin{aligned} \phi &= N(\chi_A + \chi_B) \\ \rho_1 &= 2\phi^2 \\ h_x &= -\tfrac{1}{2}\rho_1 \end{aligned} \tag{6.40}$$

For the case of H_2, this is just the negative of the occupied molecular orbital and, by symmetry arguments, the HF exchange hole at each nucleus thus integrates to $-^1/_2$, independent of the internuclear separation. This delocalization is clearly non-physical in the dissociation limit, since the correct wave function must have one electron localized at each nucleus. This means that for a reference electron near nucleus *A*, the total probability hole is localized at nucleus *A*. The correlation hole must therefore exactly cancel the exchange hole at nucleus *B*, while increasing the hole at nucleus *A* in order to integrate to −1 (Figure 6.1); this is the reason why the wave function correlation energy increases as a function of internuclear distance.

Within wave mechanics, the exchange hole is static and delocalized over the whole molecule, while the long-range part of the electron correlation serves to cancel the exchange hole away from the reference electron.

6.5 Exchange–Correlation Functionals

The difference between various DFT methods is the choice of functional form for the exchange–correlation energy. It can be proven that the exchange–correlation potential is a unique functional, valid for all systems, but an explicit functional form of this potential has been elusive, except for special cases such as a uniform electron gas. It is possible, however, to derive a number of properties that the exact functional should have, of which some of the more important ones are:[41]

1. The energy functional should be self-interaction-free, that is the exchange energy for a one-electron system, such as the hydrogen atom, should exactly cancel the Coulomb energy and the correlation energy should be zero. Although these seem like obvious requirements, none of the common functionals have this property, and this leads to a systematic overstabilization of electronic states with delocalized electrons.
2. When the density becomes constant, the uniform electron gas result should be recovered. While this surely is a valid mathematical requirement, and important for applications in solid-state physics, it may not be as important for chemical applications, as molecular densities are relatively poorly described by uniform electron gas models.
3. The coordinate scaling of the exchange energy should be linear, that is multiplying the electron coordinates with a constant factor should result in a similar linear scaling of the exchange energy:[42]

$$\begin{aligned} \rho_\lambda(x, y, z) &= \lambda^3 \rho(\lambda x, \lambda y, \lambda z) \\ E_\mathrm{x}[\rho_\lambda] &= \lambda E_\mathrm{x}[\rho] \end{aligned} \tag{6.41}$$

4. No direct scaling law applies for the correlation energy, but scaling the electron coordinates by a factor larger than 1 should increase the magnitude of the correlation (and vice versa).[43] In the low-density limit, the scaling becomes linear, as for the exchange energy:

$$-E_\mathrm{c}[\rho_\lambda] > -\lambda E_\mathrm{c}[\rho]; \quad \lambda > 1 \tag{6.42}$$

5. As the scaling parameter goes to infinity, the correlation energy for a finite system approaches a negative constant.
6. The *Lieb–Oxford* condition places an upper bound for the exchange–correlation energy relative to the *local density approximation* (LDA) (see Section 6.5.1) exchange energy:[43]

$$E_\mathrm{x}[\rho] \geq E_\mathrm{xc}[\rho] \geq 2.273 E_\mathrm{x}^\mathrm{LDA}[\rho] \tag{6.43}$$

7. The exchange potential should show an asymptotic $-r^{-1}$ behavior as $r \to \infty$.[44] This follows since for a neutral atom it must exactly cancel the Coulomb potential corresponding to a single positive charge. The exchange–correlation potential is furthermore discontinuous as a function of the number of electrons, by an amount corresponding to the difference between the ionization potential and electron affinity.[45] Functionals that depend only on the density will display an exponential ($\mathrm{e}^{-\alpha r}$) fall-off for the potential, which causes problems in the long-range regime where the potential decays too fast to zero. Functionals incorporating a fraction of exact exchange will retain the same fraction of the correct potential.
8. The correlation potential should show an asymptotic $-{}^1/{}_2\alpha r^{-4}$ behavior, with α being the polarizability of the $N_\mathrm{elec} - 1$ system.

The difference in scaling behavior (points 3 and 4) is a strong argument for separating the corresponding exchange and correlation functionals but, on the other hand, it implies a difficult task for getting the correlation component to exactly cancel the long-range exchange component.

Exchange–correlation functionals have, in analogy with other (partly) empirical methods, a mathematical form containing parameters. There are two main philosophies for assigning values to these parameters, either by requiring the functional to fulfill the above criteria (or a suitable selection thereof) or by fitting the parameters to experimental data, although in practice a combination of these approaches is often used. The quality of exchange–correlation functionals will ultimately have to be settled by comparing the performance with experiments or high-level wave mechanics calculations. Such calibration studies, however, only evaluate the quality for the chosen selection of systems and properties. It has indeed been found that the "best" functionals depend on the system and properties, some being good for molecular systems, others for delocalized (periodic) systems, and others again for properties such as excitation energies or NMR chemical shifts. At present, there are no clear "standard" methods, like MP2 and CCSD in traditional *ab initio* theory, although the hybrid methods discussed below usually give good performance. Since DFT is an active area of research, new and improved functionals are likely to emerge. The total number of proposed functionals is several hundreds, but only a relatively small number of these are in common use. The bewildering number of different functionals often has the consequence that "old" and well-tested functionals are used in place of more modern ones, since the choice in many cases is based on a popularity count. Modern functionals often offer only marginal improvement, and often only for selected properties and/or systems. It should be noted that many of the proposed functionals have never made it past the research stage and are not available in commonly available programs. Below we will give a short summary of functionals that have been proposed by different research groups. We will give the explicit forms for some of the more commonly used functionals for illustration, although they do not contain much physical insight by themselves.

It is customary to separate E_{xc} into two parts, a pure exchange E_x and a correlation part E_c, which seems reasonable based on the discussion of the exchange and correlation holes above, and their different scaling properties. It should be noted, however, that only the combined exchange–correlation hole has a physical meaning, and it could be argued that this calls for a combined E_{xc}. Early work tended to focus on only one of the components, and subsequently combined these, while the current trend is to construct the two parts in a combined fashion.

Each of the exchange and correlation energies is often written in terms of the energy per particle (energy density), ε_x and ε_c:

$$E_{xc}[\rho] = E_x[\rho] + E_c[\rho] = \int \rho(\mathbf{r})\varepsilon_x[\rho(\mathbf{r})]d\mathbf{r} + \int \rho(\mathbf{r})\varepsilon_c[\rho(\mathbf{r})]d\mathbf{r} \tag{6.44}$$

As mentioned at the start of Chapter 4, the correlation between electrons with parallel spin is different from that between electrons with opposite spin. The exchange energy is "by definition" given as a sum of contributions from the α and β spin densities, as exchange energy only involves electrons of the same spin. The kinetic energy, the nuclear–electron attraction and Coulomb terms are trivially separable in terms of electron spin:

$$E_x[\rho] = E_x^{\alpha}[\rho_\alpha] + E_x^{\beta}[\rho_\beta] \tag{6.45}$$

$$E_c[\rho] = E_c^{\alpha\alpha}[\rho_\alpha] + E_c^{\beta\beta}[\rho_\beta] + E_c^{\alpha\beta}[\rho_\alpha, \rho_\beta] \tag{6.46}$$

The total density is the sum of the α and β contributions, $\rho = \rho_\alpha + \rho_\beta$, and these are identical ($\rho_\alpha = \rho_\beta$) for a closed-shell singlet. Functionals for the exchange and correlation energies may be formulated in terms of separate spin densities, but they are often given instead as functions of the spin polarization ζ (normalized difference between ρ_α and ρ_β) and the radius of the effective volume containing one electron, r_s:

$$\zeta = \frac{\rho_\alpha - \rho_\beta}{\rho_\alpha + \rho_\beta} \tag{6.47}$$

$$\tfrac{4}{3}\pi r_s^3 = \rho^{-1} \tag{6.48}$$

It is implicitly assumed in the formulas below that the exchange and correlation energies are summed over both α and β densities.

The difference between various wave function-based methods is how the electron correlation is included, and the quality of these methods can be characterized by an ordering parameter, such as the perturbation order or the level of excitations included. There are no similar theoretically founded ordering parameters for DFT methods, as the exchange–correlation functionals to a large extent are empirical. A heuristic characterization can be done by considering the fundamental variables used for defining the exchange–correlation functional. J.P. Perdew has suggested such a "Jacob's ladder" approach, where one can expect or at least hope for an improvement in the accuracy for each step up the ladder,[46,47] and this is the approach taken here to systematize the plethora of functionals that have been proposed.

6.5.1 Local Density Approximation

In the *Local Density Approximation* (LDA) it is assumed that the density locally can be treated as a uniform electron gas, or equivalently that the density is a slowly varying function. The exchange energy for a uniform electron gas is given by the Dirac formula (Equation (6.4)):

$$\begin{aligned} E_\mathrm{x}^\mathrm{LDA}[\rho] &= -C_\mathrm{x} \int \rho^{4/3}(\mathbf{r})d\mathbf{r} \\ \varepsilon_\mathrm{x}^\mathrm{LDA} &= -C_\mathrm{x}\rho^{1/3} \end{aligned} \tag{6.49}$$

In the more general case, where the α and β densities are not equal, LDA (where the *sum* of the α and β densities is raised to the $^4/_3$ power) has been virtually abandoned and replaced by the *Local Spin Density Approximation* (LSDA) (which is given as the sum of the *individual* densities raised to the $^4/_3$ power:

$$E_\mathrm{x}^\mathrm{LSDA}[\rho] = -2^{1/3}C_\mathrm{x} \int \left(\rho_\alpha^{4/3} + \rho_\beta^{4/3}\right) d\mathbf{r} \tag{6.50}$$

LSDA may also be written in terms of the total density and a spin-polarization function:

$$\begin{aligned} \varepsilon_\mathrm{x}^\mathrm{LSDA} &= -C_\mathrm{x} f_1(\zeta)\rho^{1/3} \\ f_1(\zeta) &= \tfrac{1}{2}[(1+\varsigma)^{4/3} + (1-\varsigma)^{4/3}] \end{aligned} \tag{6.51}$$

For closed-shell systems, LSDA is equal to LDA and, since this is the most common case, LDA is often used interchangeably with LSDA, although this is not true in the general case. The X_α method proposed by Slater in 1951[48] can be considered as an LDA method where the correlation energy is neglected and the exchange term is as given in

$$\varepsilon_{X_\alpha} = -\tfrac{3}{2}\alpha C_\mathrm{x}\rho^{1/3} \tag{6.52}$$

With $\alpha = {}^2/_3$ this is identical to the Dirac expression. The original X_α method used $\alpha = 1$, but a value of $^3/_4$ has been shown to give better agreement for atomic and molecular systems. The name Slater is often used as a synonym for the L(S)DA exchange energy involving the electron density raised to the $^4/_3$ power.

The analytical form for the correlation energy of a uniform electron gas, which is purely dynamical correlation, has been derived in the high- and low-density limits.[49,50] For intermediate densities, the correlation energy has been determined to a high precision by quantum Monte Carlo methods (Section 4.15). In order to use these results in DFT calculations, it is desirable to have a suitable analytic interpolation formula, and such formulas have been constructed by *Vosko, Wilk* and *Nusair* (VWN) and by *Perdew* and *Wang* (PW), and are considered to be accurate fits.[51,52] The VWN parameterization is given in Equation (6.53), where a slightly different spin-polarization function has been used:

$$\varepsilon_c^{VWN}(r_s,\varsigma) = \varepsilon_c(r_s,0) + \varepsilon_a(r_s)\left[\frac{f_2(\varsigma)}{f_2''(0)}\right](1-\varsigma^4) + [\varepsilon_c(r_s,1)-\varepsilon_c(r_s,0)]f_2(\varsigma)\varsigma^4$$
$$f_2(\varsigma) = \frac{(f_1(\varsigma)-2)}{(2^{1/3}-1)} \tag{6.53}$$

The $\varepsilon_c(r_s,\zeta)$ and $\varepsilon_a(r_s)$ functions are parameterized as in the following equation, with A, x_0, b and c being suitable fitting constants:

$$\varepsilon_{c/a}(x) = A\left\{\begin{array}{l}\ln\dfrac{x^2}{X(x)} + \dfrac{2b}{Q}\tan^{-1}\left(\dfrac{Q}{2x+b}\right)\\ -\dfrac{bx_0}{X(x_0)}\left[\ln\dfrac{(x-x_0)^2}{X(x)} + \dfrac{2(b+2x_0)}{Q}\tan^{-1}\left(\dfrac{Q}{2x+b}\right)\right]\end{array}\right\} \tag{6.54}$$
$$x = \sqrt{r_s} \quad ; \quad X(x) = x^2 + bx + c \quad ; \quad Q = \sqrt{4c-b^2}$$

Several slightly different parameterizations were proposed in the original paper, which has caused some confusion, since different implementations have used different parameterizations and therefore produce slightly different numerical results.The PW parameterization for $\varepsilon_{c/a}$ is given below, with a, α, β_1, β_2, β_3 and β_4 again being fitting parameters:

$$\varepsilon_{c/a}^{PW}(x) = -2a\rho(1+\alpha x^2)\ln\left(1+\frac{1}{2a(\beta_1 x + \beta_2 x^2 + \beta_3 x^3 + \beta_4 x^4)}\right) \tag{6.55}$$

The LSDA method is an *exact* DFT method for the special case of a uniform electron gas, except for small differences depending on the interpolation formula chosen for the correlation energy. The LSDA approximation underestimates the exchange energy by ~10% for a molecular system, thereby creating errors that are larger than the whole correlation energy. Electron correlation is overestimated, often by a factor close to 2, and bond strengths are as a consequence overestimated, often by ~100 kJ/mol. The overestimation of the exchange energy furthermore leads to an artificial stabilization of high spin states, that is energy differences between states having different spin contain large systematic errors. Despite the simplicity in the fundamental assumptions, LSDA methods are often found to provide results with an accuracy similar to that obtained by wave mechanics Hartree–Fock methods. It has furthermore been used extensively in the physics community for describing extended systems, such as metals, where the approximation of a slowly varying electron density is reasonable.

6.5.2 Generalized Gradient Approximation

Improvements over the LSDA approach must consider a non-uniform electron gas. A step in this direction is to make the exchange and correlation energies dependent not only on the electron density but also on derivatives of the density. The first-order correction for the exchange energy is given in Equation (6.9), and the corresponding quantity for the correlation energy is also known.[53] While inclusion of the first-order exchange term improves the exchange energy, inclusion of the first-order correlation correction often makes the correlation energy *positive*. A straightforward inclusion of these first-order terms leads to a model that performs worse than the simple LSDA model. The main reason for the success of the LSDA approach is that it fulfills the requirements of the Fermi hole integrating to −1 and the Coulomb hole to 0, while the addition of gradient terms destroys these important properties.

In *Generalized Gradient Approximation* (GGA) methods, the first derivative of the density is included as a variable, and in addition it is required that the Fermi and Coulomb holes integrate to the required values of −1 and 0. GGA methods are also sometimes referred to as *non-local* methods, although this is somewhat misleading since the functionals only depend on the density (and derivative) at a given point, not on a space volume as the Hartree–Fock exchange energy.

One of the earliest and most popular GGA exchange functionals was proposed by A.D. Becke (B or B88) as a correction to the LSDA exchange energy:[54]

$$\begin{aligned} \varepsilon_x^{B88} &= \varepsilon_x^{LDA} + \Delta\varepsilon_x^{B88} \\ \Delta\varepsilon_x^{B88} &= -\beta\rho^{1/3}\frac{x^2}{1+6\beta x\sinh^{-1}x} \\ x &= \frac{|\nabla\rho|}{\rho^{4/3}} \end{aligned} \tag{6.56}$$

The β parameter is determined by fitting to known data for the rare gas atoms using the dimensionless gradient variable x. The B88 exchange functional has the correct asymptotic behavior for the energy density (but not for the exchange potential).[55] It reduces the error in the exchange energy by almost two orders of magnitude relative to the LSDA result, and thus represents a substantial improvement for a simple functional form containing only one adjustable parameter.

Handy and Cohen have investigated several forms related to Equation (6.56) where the parameters were optimized with respect to exchange energies calculated at the HF level. The best resulting model had two parameters and was labeled *OPTX* (*OPTimized eXchange*).[56] It was also found that no significant improvement could be made by including higher-order derivatives (discussed in the next section). *Hamprecht, Cohen, Tozer* and *Handy* have further extended the B97 model discussed in the next section, but using only the pure density components (i.e. no exact exchange) to a functional containing 15 parameters that were fitted to experimental and *ab initio* data, giving the acronyms *HCTH93, HCTH147* and *HCTH407*, where the number refers to the number of molecules in the fitting data set.[57]

There have similarly been various GGA functionals proposed for the correlation energy. One popular functional is due to *Lee, Yang* and *Parr* (LYP),[58,59] which has the rather intimidating form shown by

$$\begin{aligned} \varepsilon_c^{LYP} &= -4a\frac{\rho_\alpha\rho_\beta}{\rho^2(1+d\rho^{-1/3})} \\ &\quad - ab\omega\left\{\begin{array}{l} \dfrac{\rho_\alpha\rho_\beta}{18}\left[\begin{array}{l} 144(2^{2/3})C_F\left(\rho_\alpha^{8/3}+\rho_\beta^{8/3}\right)+(47-7\delta)|\nabla\rho|^2 \\ -(45-\delta)\left(|\nabla\rho_\alpha|^2+|\nabla\rho_\beta|^2\right)+2\rho^{-1}(11-\delta)\left(\rho_\alpha|\nabla\rho_\alpha|^2+\rho_\beta|\nabla\rho_\beta|^2\right) \end{array}\right] \\ +\frac{2}{3}\rho^2\left(|\nabla\rho_\alpha|^2+|\nabla\rho_\beta|^2-|\nabla\rho|^2\right)-\left(\rho_\alpha^2|\nabla\rho_\beta|^2+\rho_\beta^2|\nabla\rho_\alpha|^2\right) \end{array}\right\} \\ \omega &= \frac{e^{-c\rho^{-1/3}}}{\rho^{14/3}(1+d\rho^{-1/3})} \quad ; \quad \delta = c\rho^{-1/3}+\frac{d\rho^{-1/3}}{(1+d\rho^{-1/3})} \end{aligned} \tag{6.57}$$

The a, b, c and d parameters are determined by fitting to data for the helium atom. Although not obvious from the form shown in Equation (6.57), the LYP functional does not include parallel spin correlation when all the spins are aligned (e.g. the LYP correlation energy for ^{3}He is zero). The LYP correlation functional is often combined with the B88 or OPTX exchange functional to produce the BLYP and OLYP acronyms.

J. P. Perdew and coworkers have proposed several related exchange–correlation functionals based on removing spurious oscillations in the Taylor-like expansion to first order and ensuring that the exchange and correlation holes integrate to the required values of −1 and 0. The associated acronyms are *PW86* (*Perdew–Wang 1986*),[60] *PW91* (*Perdew–Wang 1991*)[61] and *PBE* (*Perdew–Burke–Ernzerhof*).[62] These three functionals should be considered as refinements of the same underlying model, that is the PBE version should be used in favor of the PW86 and PW91 versions. The exchange part is written as an enhancement factor multiplied on to the LSDA functional, where the dimensionless gradient variable x is defined in Equation (6.56):

$$\begin{aligned} \varepsilon_x^{\text{PBE}} &= \varepsilon_x^{\text{LDA}} F(x) \\ F(x) &= 1 + a - \frac{a}{1 + bx^2} \end{aligned} \tag{6.58}$$

The correlation part is similarly written as an enhancement factor added to the LSDA functional, where the t variable is related to the x variable by means of yet another spin-polarization function:

$$\begin{aligned} \varepsilon_c^{\text{PBE}} &= \varepsilon_c^{\text{LDA}} + H(t) \\ H(t) &= cf_3^3 \ln\left[1 + \mathrm{d}t^2\left(\frac{1 + At^2}{1 + At^2 + A^2t^4}\right)\right] \\ A &= d\left[\exp\left(-\frac{\varepsilon_c^{\text{LDA}}}{cf_3^3}\right) - 1\right]^{-1} \\ f_3(\zeta) &= \tfrac{1}{2}\left[(1+\zeta)^{2/3} + (1-\zeta)^{2/3}\right] \\ t &= \left[2(3\pi^3)^{1/3} f_3\right]^{-1} x \end{aligned} \tag{6.59}$$

The a, b, c and d parameters in these functionals are non-empirical, that is they are not obtained by fitting to experimental data, but derived from some of the conditions in Section 6.5. The PW91 functional has been tuned to improve the performance for weak interactions, producing the *mPW91* acronym.[63] The PBE functional has similarly been slightly modified (*RPBE*) to improve the performance for periodic systems,[64] but this modification actually destroys the hole condition (Equation (6.39)) for the exchange energy. An alternative modification using one additional parameter to give the acronym *mPBE* has also been proposed.[65]

The *KT3* (*Keal–Tozer*) functional has been constructed as a combination of LDA and OPTX exchange combined with the LYP correlation functional, and modified with an additional gradient term, all multiplied with fitting coefficients, as shown by[66]

$$E_{xc}^{\text{KT3}} = aE_x^{\text{LDA}} + bE_x^{\text{OPTX}} + cE_c^{\text{LYP}} + d\int \frac{|\nabla\rho|^2}{\rho^{4/3} + e} \tag{6.60}$$

The a, b and c coefficients have been optimized with respect to experimental quantities such as atomization energies and geometries, while the d and e parameters are fitted to NMR shielding constants. The primary focus of KT3 and earlier versions (KT1 and KT2) is to provide a functional suitable for calculating NMR shielding constants, which other standard functionals have difficulties with. Part of this deficiency may be due to the neglect of current density terms.[67]

6.5.3 Meta-GGA Methods

The logical extension of GGA methods is to allow the exchange and correlation functionals to depend on higher-order derivatives of the electron density, with the Laplacian ($\nabla^2\rho$) being the second-order term. Alternatively, the functional can be taken to depend on the orbital kinetic energy density τ, which for a single orbital is identical to the von Weizsäcker kinetic energy τ_W (Equation (6.6)):

$$\tau(\mathbf{r}) = \frac{1}{2}\sum_i^{\text{occ}} |\nabla\phi_i(\mathbf{r})|^2 \tag{6.61}$$

$$\tau_W(\mathbf{r}) = \frac{|\nabla\rho(\mathbf{r})|^2}{8\rho(\mathbf{r})} \tag{6.62}$$

The orbital kinetic energy density and the Laplacian of the density essentially carry the same information, since they are related via the orbitals and the effective potential (all potential terms in the KS equation):

$$\tau(\mathbf{r}) = \tfrac{1}{2}\sum_i^{\text{occ}} \varepsilon_i|\phi_i(\mathbf{r})|^2 - v_{\text{eff}}(\mathbf{r})\rho(\mathbf{r}) + \tfrac{1}{2}\nabla^2\rho(\mathbf{r}) \tag{6.63}$$

This may also be seen from the gradient expansion of τ for slowly varying densities:[68]

$$\tau(\mathbf{r}) = \frac{3}{10}(6\pi^2)^{2/3}\rho(\mathbf{r})^{5/3} + \frac{1}{72}\frac{|\nabla\rho(\mathbf{r})|^2}{\rho(\mathbf{r})} + \frac{1}{6}\nabla^2\rho(\mathbf{r}) + O(\nabla^4\rho(\mathbf{r})) \tag{6.64}$$

Inclusion of either the Laplacian or the orbital kinetic energy density as a variable leads to the so-called *meta-GGA* functionals, and functionals which in general use orbital information may also be placed in this category. Calculation of the orbital kinetic energy density is numerically more stable than calculation of the Laplacian of the density, and the two τ functions in Equations (6.61) and (6.62) are common components of meta-GGA functionals.

One of the earliest attempts to include kinetic energy functionals was by *Becke* and *Roussel* (BR), who proposed the exchange functional shown by[69]

$$\begin{aligned} \varepsilon_x^{\text{BR}} &= -\frac{2-(2+ab)e^{-ab}}{4b} \\ a^3e^{-ab} &= 8\pi\rho \quad ; \quad a(ab-2) = b\frac{\nabla^2\rho - 4(\tau-\tau_W)}{\rho} \end{aligned} \tag{6.65}$$

A similar correlation functional shown below was proposed somewhat later by A. D. Becke (B95) and is one of the few functionals that does not have the self-interaction problem:[70]

$$\begin{aligned} \varepsilon_c^{\text{B95}} &= \varepsilon_c^{\alpha\beta} + \varepsilon_c^{\alpha\alpha} + \varepsilon_c^{\beta\beta} \\ \varepsilon_c^{\alpha\beta} &= \left[1 + a\left(x_\alpha^2 + x_\beta^2\right)\right]^{-1}\varepsilon_c^{\text{PW},\alpha\beta} \\ \varepsilon_c^{\sigma\sigma} &= \left[1 + bx_\sigma^2\right]^{-2}\frac{(\tau-\tau_W)_\sigma}{2^{5/3}C_F\rho_\sigma^{5/3}}\varepsilon_c^{\text{PW},\sigma\sigma} \end{aligned} \tag{6.66}$$

Here σ runs over α and β spins, x_σ is defined in Equation (6.56) with the implicit spin dependence denoted by the subscript σ, a and b are fitting parameters and $\varepsilon_c^{\text{PW}}$ is the Perdew–Wang parameterization of the LSDA correlation functional (Equation (6.55)).

The HCTH functional has been extended to also include the kinetic energy density as a variable, producing the acronym τ-HCTH.[71] The *VSXC* (*Voorhis–Scuseria exchange–correlation*) functional similarly includes the kinetic energy density and contains 21 parameters that are fitted to experimental data.[72] The *TPSS* (*Tao–Perdew–Staroverov–Scuseria*) exchange–correlation functional, on the other hand, is a non-empirical version that represents a further development of the *PKZB* (*Perdew–Kurth–Zupan–Blaha*) functional,[73] and can be considered as the next improvement over the PBE functional.[74] The inclusion of second-order density information to produce level-3 meta-GGA functionals historically postdates the inclusion of exact HF exchange, leading to level-4 hybrid methods, and there are thus relatively few pure meta-GGA functionals.

6.5.4 Hybrid or Hyper-GGA Methods

From the Hamiltonian in Equation (6.10) and the definition of the exchange–correlation energy in Equation (6.16), an exact connection can be made between the exchange–correlation energy and the corresponding hole potential connecting the non-interacting reference and the actual system (see Appendix B for details). The resulting equation is called the *Adiabatic Connection Formula* (ACF)[75] and involves integration over the parameter λ, which "turns on" the electron–electron interaction:

$$E_{xc} = \int_0^1 \left\langle \Psi_\lambda \left| \mathbf{V}_{xc}^{hole}(\lambda) \right| \Psi_\lambda \right\rangle d\lambda \tag{6.67}$$

In the crudest approximation (taking $\mathbf{V}_{xc}^{hole}$ to be linear in λ), the integral is given as the average of the values at the two end-points:

$$E_{xc} \approx \frac{1}{2}\left(\left\langle \Psi_0 \left| \mathbf{V}_{xc}^{hole}(0) \right| \Psi_0 \right\rangle + \left\langle \Psi_1 \left| \mathbf{V}_{xc}^{hole}(1) \right| \Psi_1 \right\rangle \right) \tag{6.68}$$

In the $\lambda = 0$ limit, the electrons are non-interacting and there is consequently no correlation energy, only exchange energy. Furthermore, since the *exact* wave function in this case is a single Slater determinant composed of KS orbitals, the exchange energy is *exactly* that given by Hartree–Fock theory (Equation (3.34)). If the KS orbitals were identical to the HF orbitals, the exchange energy would be precisely the energy calculated by HF wave mechanics methods; the energy calculated by using the HF formula and KS orbitals is referred to as *exact exchange* or HF exchange. The last term in Equation (6.68) is still unknown. Approximating it by the LSDA result defines the *Half-and-Half* (H + H) method:[76]

$$E_{xc}^{H+H} = \frac{1}{2}E_x^{HF} + \frac{1}{2}\left(E_x^{LSDA} + E_c^{LSDA}\right) \tag{6.69}$$

Since the GGA methods give a substantial improvement over LDA, a generalized version of the H+H method may be defined by writing the exchange energy as a combination of LSDA and HF exchange and a gradient correction term. The correlation energy may similarly be taken as the LSDA formula plus a gradient correction term. Models that include HF exchange are often denoted *hybrid* methods, and a prototypical example is the B3LYP functional, which consists of B88 and HF exchange, and the LYP correlation functional, and contains three parameters:[77,78]

$$E_{xc}^{B3LYP} = (1-a)E_x^{LSDA} + aE_x^{HF} + b\Delta E_x^{B88} + (1-c)E_c^{LSDA} + cE_c^{LYP} \tag{6.70}$$

The a, b and c parameters are determined by fitting to experimental data and are $a = 0.20$, $b = 0.72$ and $c = 0.81$ for B3LYP. The choice of different VWN parameterizations of the LSDA correlation energy (Section 6.5.1) unfortunately means that there are several different definitions of B3LYP, which give slightly different results. The BHLYP model builds upon Equation (6.69) and can be considered as a variation of B3LYP, where the *a*, *b* and *c* parameters are 0.50. Other hybrid models can be constructed by choosing other forms for E_x^{GGA} and E_c^{GGA}, and different fitting parameters, producing acronyms such as B3PW91 and O3LYP. Increasing the number of fitting parameters to ten in the B97[79] and B98[80] models was found to provide rather marginal improvements relative to the three-parameter version, but the B97 model has subsequently been reparameterized to the B97-1, B97-2 and B97-3 models.[66]

The τ-HCTH functional has been augmented with exact exchange to produce the acronym τ-HCTH-hybrid.[71] The PBE functional has also been improved by addition of exact exchange to give the PBE0 functional (also denoted PBE1PBE in the literature),[81] where the mixing coefficient for the exact exchange is argued to have a value of 0.25 from perturbation arguments.[82] Similarly, the third-rung TPSS functional has been augmented with ~10% exact exchange to give the TPSSh method.[83,84] The hybrid classification is defined by inclusion of exact exchange and does not distinguish between whether second-order density information (orbital kinetic energy or density Laplacian) is included or not.

Inclusion of HF exchange is often found to improve the calculated results, although the optimum fraction to include depends on the specific property of interest. The improvement of new functionals by inclusion of a suitable fraction of exact exchange is now a standard feature. At least part of the improvement may arise from reducing the self-interaction error, since HF theory is completely self-interaction-free. A more pragmatic view is that LSDA (part of most higher-rung functionals) and HF often display systematic errors in opposite directions, and by taking a suitable linear combination the average error can be made close to zero. Hybrid methods are used less frequently for solid-state systems, since the calculation of exact exchange is computationally difficult for periodic systems.

6.5.5 Double Hybrid Methods

At the fifth level of the Jacob's ladder classification, the full information of the KS orbitals is employed, that is not only the occupied but also the virtual orbitals are included. An example of such methods is to include part of an MP2-like energy, calculated by the MP2 formula in Equation (4.58) but using the DFT orbitals and orbital energies. In the spirit of the hybrid methods, the resulting MP2-like energy term is included with an empirical parameter, and a general double hybrid DFT energy expression is given by[85]

$$E_{xc}^{DHDFT} = (1-a)E_x^{DFT} + aE_x^{HF} + (1-b)E_c^{DFT} + bE_c^{MP2} \tag{6.71}$$

The MP2-like term describes "real" electron correlation energy, and such double hybrid methods allow a qualitative description of dispersion. One of the first examples of a double hybrid method was B2PLYP, which builds on the B88 exchange and LYP correlation functionals,[86] and PBE0-DH similarly builds upon the PBE0 functional.[87]

Electron correlation can alternatively be calculated within a DFT framework by the *Random Phase Approximation* (RPA), as described in Section 6.9.1, and it can similarly be used to define double hybrid functionals.

The main disadvantage of the double hybrid methods is the computational cost. The MP2-like expression has an N^5 computational dependence, which prevents the application to large systems. The presence of the MP2 expression furthermore means that the basis set convergence for double hydrid methods is slower than for pure DFT methods. An extension to "multiple hybrid" methods by including MP3, MP4 and CCSD(T) terms has been found not to provide any significant improvement over just including an MP2 term.[88]

6.5.6 Range-Separated Methods

DFT methods sometimes are found to display systematic errors for certain properties and, once identified, efforts are made to fix these errors. One such identified error was in the calculation of excitation energies for charge transfer and to some extent also Rydberg states (Section 4.14) by TDDFT methods (Section 6.9.1), where excitation energies often were underestimated by a factor of two. This is related to the general problem of most DFT methods to overdelocalize the electron density, that is delocalized systems are predicted to be artificially lower in energy than localized systems.

A related problem is the presence of self-interaction errors, arising from the incomplete cancellation of the exchange and Coulomb energies calculated from the electron density for a single electron and the incorrect long-range behavior of the exchange–correlation potential failing to cancel the Coulomb potential. This is especially problematic for loosely bound electrons, for example in Rydberg-type excited states and for anions formed from systems with low electron affinities. Since loosely bound electrons by definition have most of the associated density far from the nuclei, the self-interaction error may be larger than the actual binding energy, and for anions erroneously lead to an unbound electron. In actual calculations using a limited basis set, this may not be obvious, since the outer electron is confined by the most diffuse basis function. An anion with a positive HOMO energy, however, is a clear warning sign, and extending the basis set with many diffuse functions in such cases may cause the outer electron to drift away from the atom. This means that only systems with high electron affinities have a well-defined basis set limiting value. Nevertheless, a medium-sized basis set with a single set of diffuse functions will in many cases give a reasonable estimate of the experimental electron affinity.[89] The basis set confines the outer electron to be in the correct physical space and the exchange–correlation functional gives a reasonable estimate of the energy of this density. It should be noted that the relatively good performance is in essence due to a correct physical description, rather than a correct theoretical methodology.

A possible fix to these problems is to partition the electron–electron Coulomb operator for the exchange energy (only) into a short- and a long-range part, usually done by the standard error function as shown below, since this facilitates the calculation of the resulting integrals, where the ω parameter controls the partitioning between the two parts:[90]

$$\begin{aligned} \frac{1}{|\mathbf{r}_1 - \mathbf{r}_2|} &= \frac{1}{r_{12}} = \frac{1 - erf(\omega r_{12})}{r_{12}} + \frac{erf(\omega r_{12})}{r_{12}} \\ erf(x) &= \frac{2}{\sqrt{\pi}} \int_0^x \mathrm{e}^{-t^2} \mathrm{d}t \end{aligned} \tag{6.72}$$

The trick is to use different exchange functionals for the short- and long-range parts. The most common is to employ a density exchange functional for the short-range part and the (exact) HF expression for the long-range part. If the density exchange part by itself contains a fraction of HF exchange, the interpolation is between that fraction and 100%, rather than between 0 and 100%. The long-range HF exchange ensures that the DFT overdelocalization of charge separation is removed and yields much improved excitation energies for charge-transfer states. This is primarily a consequence of the long-range HF part taking care of the proper limiting behavior of the exchange potential (Section 6.5, item 7) and thus ensuring no self-interaction in the large-distance limit. Such *range-separated*, also called *long-range corrected*, functionals also solve the problem of dissociation into fractionally charged systems and the problem of anions capable of binding only a factional number of electrons. They similarly yield much improved electronic structure predictions of zwitterionic structures, that is systems that have formally positively and negatively charged sites within the same molecule.[91]

The ω parameter in Equation (6.72) determines when to switch between the short- and long-range parts, and is a free variable. Range-separated versions can be constructed for all standard DFT functionals (with acronyms like LC-BLYP, LC-PBE, etc.), but may require a retuning of parameters once the new additional ω parameter is introduced. One of the first functionals to employ this idea was the CAM-B3LYP,[92] which builds upon the famous B3LYP method. The ωB97X is similarly a range-separated version of the B97 functional containing a moderate (17) number of empirical parameters.[93] Determining the optimal ω parameter can be done, for example, by fitting to experimental thermodynamic data, and a typical ω value is 0.30–0.50/bohr.

The ω parameter influences the orbital energies, and an interesting idea by Livshits and Baer is to choose ω as non-empirical by requiring that Koopmans theorem is fulfilled, that is the HOMO energy is equal to the ionization potential.[94] This *optimally tuned* approach, however, requires a non-linear search and leads to a different ω value for each system.

Functionals employing the "inverse" separation, that is using the HF exchange as the short-range component and a density exchange expression as the long-range part, like the Heyd–Scuseria–Ernzerhof (HSE) functional, has been advocated for use in solid-state calculations.[95,96] These functionals are sometimes called *screened-exchange* to distinguish them from the range-separated ones. A variation using two ω parameters such that HF exchange is initially increased and then decreased as a function of r_{12} has also been proposed by Henderson, Izmaylov, Scuseria and Savin (HISS).[97]

6.5.7 Dispersion-Corrected Methods

One of the serious shortcomings of standard DFT methods is the inability to describe dispersion forces (part of van der Waals-type interactions).[98] Many functionals provide a purely repulsive interaction between rare gas atoms, while others describe a weak stabilizing interaction, but fail to have the correct R^{-6} long-range distance behavior. The stabilization most likely arises due to charge penetration effects in combination with parameterization against experimental results that include dispersion bound species, and for smaller basis sets may also include basis set superposition errors. Although dispersion is a short-ranged weak interaction, it is cumulative, and therefore becomes increasingly important as the system gets larger. Hydrogen bonding, however, is mainly electrostatic and is reasonably well accounted for by many DFT functionals.

S. Grimme has proposed to include dispersion by additive empirical terms, much like the van der Waals energy term in force field methods (Section 2.2.5).[99] The parameterization in the earliest models simply included an R^{-6} energy term for each atom pair, with an atom-dependent C_6 parameter. The method has been refined by including higher-order terms (R^{-8}, R^{-10}) to better describe the medium-range dispersion and making the parameters depend on the atomic environment in terms of the number of directly bonded atoms. Adding such attractive energy terms has the potential problem of divergence for short interatomic distances and is consequently often used in connection with a damping function:

$$\Delta E_{\mathrm{disp}} = -\sum_{n=6(8,10)} s_n \sum_{\mathrm{AB}}^{\mathrm{atoms}} \frac{C_n^{\mathrm{AB}}}{R_{\mathrm{AB}}^n} f_{\mathrm{damp}}(R_{\mathrm{AB}}) \tag{6.73}$$

A complication is that the parameterization of the dispersion correction depends on the underlying exchange–correlation functional, as different functionals via their parameterization may include some of the short-range interaction, and this can be taken into account by a functional-dependent scaling factor s_n in Equation (6.73). Such dispersion corrected methods are denoted with a D/D2/D3 after the DFT acronym, as, for example, ωB97XD.[93] They can in addition be extended to include also three-body effects.

Becke and Johnson (BJ) have proposed a physics-inspired dispersion model where the C_6 parameter is written as a function of the atomic polarizabilities α and the average dipole moment associated with the exchange hole $\langle \mu^2 \rangle$, which introduces an explicit dependence on the actual electronic structure of the system:[100]

$$C_6^{\mathrm{AB}} = \frac{\alpha_\mathrm{A}\alpha_\mathrm{B}\langle \mu^2 \rangle_\mathrm{A}\langle \mu^2 \rangle_\mathrm{B}}{\alpha_\mathrm{A}\langle \mu^2 \rangle_\mathrm{B} + \alpha_\mathrm{B}\langle \mu^2 \rangle_\mathrm{A}} \tag{6.74}$$

Similar formulas can be derived for the C_8 and C_{10} parameters in terms of averages over higher-order multipole moments.

Another approach is to make the dispersion correction directly dependent on the actual electron density by writing it as a six-dimensional integral over densities with an appropriate dispersion kernel Φ (the factor $1/2$ corrects for double counting):

$$\Delta E_{\mathrm{disp}} = \frac{1}{2}\int \rho(\mathbf{r})\Phi(\mathbf{r},\mathbf{r}')\rho(\mathbf{r}')\mathrm{d}\mathbf{r}\mathrm{d}\mathbf{r}' \tag{6.75}$$

Such dispersion methods are often denoted non-local as they depend on electron densities that can be far apart. The *van Vorhis–Vydrov* (VV10) kernel has the form shown in[101]

$$\Phi(\mathbf{r},\mathbf{r}') = -\frac{3}{2}[g(\mathbf{r})g(\mathbf{r}')(g(\mathbf{r}) + g(\mathbf{r}'))]^{-1} \tag{6.76}$$

Here the $g(\mathbf{r})/g(\mathbf{r}')$ functions are defined below with the C and b parameters chosen to provide the correct asymptotic C_6 coefficients and controlling the short-range damping, respectively:

$$g(\mathbf{r}) = |\mathbf{r} - \mathbf{r}'|^2\sqrt{C\left(\frac{\nabla\rho(\mathbf{r})}{\rho(\mathbf{r})}\right)^4 + \frac{4\pi}{3}\rho(\mathbf{r})} + b\frac{3\pi}{2}\left(\frac{\rho(\mathbf{r})}{9\pi}\right)^{1/6} \tag{6.77}$$

The VV10 dispersion expression has been combined with the B97 exchange functional to give a family of functionals at the GGA, meta-GGA and hybrid-GGA with acronyms B97M-V and ωB97X-V, each employing a moderate (10–12) number of parameters.[102] The development of these functionals employed a search of a parameter space with dimension ~10^{40} and an explicit testing of ~10^{10} combinations in order to determine the optimum number of parameters and their optimum values.

The double hybrid methods described in Section 6.5.5 account explicitly for dispersion, for example by an MP2 energy term scaled by an empirical parameter, but they can also be combined with empirical dispersion terms, producing acronyms like B2PLYP-D3.

The above dispersion corrections can be classified according to whether they only depend on the atom type (and perhaps bonding environment) or on the actual electron density calculated for the specific system, and in the latter case, whether it is included self-consistently or as *a posteriori* correction. The associated computational cost ranges from essentially zero for the additive preparameterized methods of Grimme, to ~50% overhead for the BJ model and to ~100% overhead for the VV10 method. While the BJ and VV10 methods are capable of adjusting the dispersion interaction, for example according to the oxidation state of the atom, all of the above methods have been found to provide similar performance in benchmark investigations,[103] and the computationally inexpensive additive D3 correction is consequently rapidly becoming the method of choice for routine applications.

6.5.8 Functional Overview

The introduction of GGA and hybrid functionals during the early 1990s yielded a major improvement in terms of accuracy for chemical applications, and resulted in the Nobel Prize in Chemistry being awarded to W. Kohn and J. A. Pople in 1998. Progress since this initial exciting development has been slower, and the (in)famous B3LYP functional[77,78] proposed in 1993 still represents one of the most successful in terms of overall performance. Unfortunately, neither the addition of more fitting parameters, the addition of more variables in the functionals, nor imposing more fundamental restrictions for the functional form have (yet) provided models with a significantly better overall performance.[104] Although the performance for a given property can be improved by tailoring the functional form or parameters, such measures often result in the deterioration of the results for other properties.

It should be noted that the implicit cancellation of the long-range part of the exchange and correlation energies implies that the two functional parts should be at the same level of the ladder, and preferably developed in an integrated fashion. A popular topic in the literature is to search for "magic" combinations of exchange and correlation functionals, perhaps with a few adjustable scaling parameters and a choice of basis set, in order to reproduce a selected set of experimental data. This is not a theoretically justified procedure and should be considered merely as data fitting without much physical relevance. Nevertheless, such a procedure can of course be taken as an "experimental" fitting function that can be useful for predicting specific properties for a series of compounds.

Table 6.1 shows an overview of commonly used functionals given by their acronym[101] and placed in the Jacob's ladder classification. One may furthermore differentiate the functionals based on their use (or lack) of experimental data for assigning values to the parameters in the functional forms. The non-empirical functional such as PW86, PW91, PBE and TPSS use the free parameters to fulfill as many of the requirements in Section 6.5 as possible at each level. Empirical functionals such as BLYP, B3LYP, HTCT and VSXC, on the other hand, attempt to improve the performance by fitting a handful of free parameters to give good agreement with experimental data. Some functionals contain a large (30–60) number of fitting parameters, and they tend to perform better (by construction) than the non-empirical ones for systems that resemble those in the parameterization set. Since the parameterization data usually are molecular systems composed of atoms from the first three rows in the periodic table (H to Ar), this means that they are often preferred for chemical purposes, but may give

Table 6.1 Perdew classification of exchange–correlation functionals. *Italic* indicates that these include range-separated XC potentials while **bold** indicates inclusion of dispersion.

Level	Name	Variables	Examples	
			Few (≤12) parameters	Many (>12) parameters
1	Local density	ρ	LDA, LSDA, X_α	
2	GGA	ρ, $\nabla\rho$	PW91, PBE, BLYP, OLYP, HCTH, KT3	SOGGA11, N12
3	Meta-GGA	ρ, $\nabla\rho$, $\nabla^2\rho$ or τ	BR, B95, PKZB, TPSS, τ-HCTH, **B97M-V**	VSXC, M11L, MN12L
4	Hybrid-GGA	ρ, $\nabla\rho$, $\nabla^2\rho$ or τ *HF exchange*	BHLYP, B3LYP, B3PW91, O3LYP, PBE0, TPSSh, τ-HCTH-hyb, HSE, *CAM-B3LYP*, ***ωB97X-V***	B97-3, SOGGA11X, *M11*, *N12SX*, *MN12SX*, *ωB97X*, ***ωB97XD***
5	Double hybrid	ρ, $\nabla\rho$, $\nabla^2\rho$ or τ *HF exchange* *Virtual orbitals*	B2PLYP, PBE0-DH	

inferior performance for systems with transition metals or periodic systems. The "need" for a different "optimum" functional for different systems/properties is not very satisfying from a purist perspective and raises the nagging question as to whether these are just fancy non-linear fitting functions for the reference data.[105] From a pragmatic point of view this is irrelevant, as any method capable of providing results with a useful accuracy is a valuable tool.

Most commonly used functionals belong to levels 2 and 4, as the inclusion of HF exchange historically has preceded the development of functionals using derivatives beyond first order. As one moves along the rungs of the ladder, it is expected (or hoped) that the accuracy will improve, but there is no guarantee that this is the case. The trend in functional development is to design families of functionals belonging to different ladder levels but with the same underlying philosophy. The group centered on D. G. Truhlar has proposed a series of functionals in categories 2, 3 and 4 with acronyms containing a number reflecting the publication year.[106] These families of functionals developed in 2005, 2006, 2008, 2011 and 2012 belong to the heavily empirical ones and contain 25–60 parameters that are fitted to experimental data. The SOGGA11 (Second Order GGA) and N12 are GGA-type functionals containing 24[107] and 30[108] parameters, respectively. The M11L (Minnesota Local) and MN12L (Minnesota Non-separable Local, where non-separable indicates that the exchange and correlation contributions cannot be written as separate terms) are level 3 functionals containing 55 parameters[109] and 58 parameters,[110] respectively. The SOGGA11X, M11, N12SX and MN12SX are all level 4 hybrid models, where the latter three include range separation of the exchange, and contain 25,[111] 46,[112] 28 and 58 parameters,[113] respectively. The N12SX and MN12SX (SX = Screening eXchange) functionals use in analogy with the HSE an inverse range separation, where a finite amount of HF exchange is included at short range, but none at long range. The above versions carrying the year-label 11 or 12 should be considered as further developments of previous versions carrying labels such as 05, 06 or 08 (e.g. M05-2X, M06-2X, M08-HX, M11 form a progression of functionals with the same underlying philosophy).

6.6 Performance of Density Functional Methods

An evaluation of the performance of the plethora of different functionals for a variety of properties is a major undertaking and has resulted in a number of benchmarks probing different aspects of the performance. A few results for a selection of some of the more commonly used functionals have been collected in Table 6.2,[106] and additional results can be found in reference 114. The quality measure is the Mean Absolute Deviation (MAD) from accurate reference values. The MGAE109 benchmark set contains 109 atomization energies for systems composed of main group elements and the MAD is per bond. The SRMBE13 contains 13 bond energies for single-reference systems containing metal atoms and MRBE10 contains 10 bond energies for systems having multireference character. IP21 and EA13 are data sets for ionization potentials and electron affinities (21 and 13 points, respectively), while BH76 contains 76 barrier heights (activation energies) for atom transfer reactions (actually 38 reactions with barrier heights for the reaction in both directions, of which 8 are identity reactions). The basis set is of TZP quality, which should be sufficiently large that residual basis set errors at the DFT and HF levels at most should be a few kJ/mol. The methods that include wave function electron correlation (B2PLYP, MP2, CCSD, CCSD(T)) will display significantly slower basis set convergence, and the MAD values obtained by extrapolation to the complete basis set limit are given in parenthesis. The IP21 and SRMBE13 benchmarks contain systems with atoms from the fifth row in the periodic table (Zr, Mo, Ru, Rh, Pd, Ag) and have been calculated using pseudo-potentials for the core electrons, which limits the attainable accuracy using high-level methods like CCSD(T). Note that

Table 6.2 Comparison of the performance of DFT methods in terms of Mean Absolute Deviations from reference values for different benchmarks (values in kJ/mol).

Functional	Ladder level[a]	MGAE109[b]	SRMBE13	MRBE10	IP21	EA13	BH76
HF	–	129	144	304	92	113	48
MP2	–	9 (10)	22 (21)	161 (141)	27 (21)	13 (8)	20 (19)
CCSD	–	18 (8)	36 (26)	111 (93)	24 (12)	22 (11)	12 (10)
CCSD(T)	–	12 (1)	23 (14)	73 (43)	21 (8)	16 (2)	5 (2)
LSDA	1	70	89	127	41	24	63
BLYP	2	6	28	27	27	11	34
OLYP	2	4	39	27	12	15	23
PBE	2	13	29	60	26	9	37
HCTH407	2	5	27	37	28	15	25
SOGGA11	2	7	52	41	26	22	23
N12	2	5	38	30	14	18	29
τ-HCTH	3	4	40	44	19	9	27
TPSS	3	4	24	28	17	10	35
τ-HCTHhyb	4	3	28	36	17	8	20
TPSSh	4	4	19	68	14	12	27
B3LYP	4	4	25	91	24	10	18
PBE0	4	4	21	117	14	12	16
MN12L	4	3	48	31	15	11	7
CAM-B3LYP	4-RS	3	33	120	21	9	12
ωB97XD	4-RS	2	24	106	13	8	13
MN12-SX	4-RS	2	45	44	22	9	5
B2PLYP	5	4 (2)	18 (15)	74 (64)	10 (10)	10 (6)	8 (8)

[a] Ladder level: 1 = local spin density approximation, 2 = generalized gradient approximation, 3 = meta generalized gradient approximation, 4 = hybrid functional including HF exchange, 4-RS = 4 with range-separated exchange, 5 = double hybrid method. All results refer to calculations using a TZP-type basis set. The values in parenthesis are from extrapolation to the complete basis set limit.
[b] MAD per bond.

the high accuracy of the CCSD(T) method only materializes when a large basis set is used, the performance with the TZP basis set is only at par with the better of the DFT methods. One should as usual be aware that statistical measures may hide specific failures; a ~20 kJ/mol error for an activation energy may be acceptable for high-energy barriers, but for low-barrier reactions it may lead to negative activation energies, and thus a qualitatively incorrect energy surface. Similarly, the error per bond measure for the MGAE109 benchmark implies that the absolute error for a large system with many bonds most likely will be significantly higher than for a small system with few bonds.

The LSDA method performs somewhat better than Hartree–Fock, but all the gradient-corrected methods are clearly far superior. The PBE functional performs somewhat poorer than the other GGA functionals for the MGAE109 data set, which likely reflects that it does not contain fitting parameters that have been adjusted to reproduce atomization energies. There is only a small improvement in the performance upon going from GGA to meta-GGA-type functionals. Hybrid methods including exact exchange tend to perform (slightly) better than the corresponding pure functionals (e.g. BLYP/B3LYP and PBE/PBE0), and inclusion of exact exchange also tends to improve meta-GGA (τ-HCTH/τ-HCTHhyb and TPSS/TPSSh). Exact exchange, however, is clearly a disadvantage for systems with multireference character (MRBE10). The MP2, CCSD, CCSD(T) sequence systematically improves the results, but the convergence is slow for multireference systems. Non-hybrid functionals, on the other hand, handle multireference systems with similar accuracy as the other systems, and often produce results comparable to or better than those obtained by coupled cluster methods (see also Section 12.7.3). Table 6.2 illustrates that functionals high on the Jacob's ladder typically perform better than those at lower rungs, but certain functionals may perform better for specific systems and properties. For comparison, it can be noted that semi-empirical methods (Chapter 7) have typical errors of 30–40 kJ/mol for the MGAE109 benchmark.[115]

6.7 Computational Considerations

The strength of DFT is that only the total density needs to be considered. In order to calculate the kinetic energy with sufficient accuracy, however, orbitals have to be reintroduced. Nevertheless, Kohn–Sham DFT displays a computational cost similar to HF theory, with the possibility of providing more accurate (exact, in principle) results.

Once an exchange–correlation functional has been selected, the computational problem is very similar to that encountered in wave mechanics HF theory: determine a set of orthogonal orbitals that minimizes the energy. Since the $J[\rho]$ (and $E_{xc}[\rho]$) functional depends on the total density, a determination of the orbitals involves an iterative sequence. The orbital orthogonality constraint may be enforced by the Lagrange method (Section 13.5), again in complete analogy with wave mechanics HF methods (Equation (3.35)):

$$L[\rho] = E_{\mathrm{DFT}}[\rho] - \sum_{ij}^{N_{\mathrm{orb}}} \lambda_{ij}(\langle\phi_i|\phi_i\rangle - \delta_{ij}) \tag{6.78}$$

Requiring the variation of L to vanish provides a set of equations involving an effective one-electron operator ($\mathbf{h}_{\mathrm{KS}}$), similar to the Fock operator in wave mechanics (Equation (3.37)):

$$\begin{aligned}
\mathbf{h}_{\mathrm{KS}}\phi_i &= \sum_{j}^{N_{\mathrm{orb}}} \lambda_{ij}\phi_j \\
\mathbf{h}_{\mathrm{KS}} &= \frac{1}{2}\nabla^2 + \mathbf{v}_{\mathrm{eff}} \\
\mathbf{v}_{\mathrm{eff}}(\mathbf{r}) &= \mathbf{V}_{\mathrm{ne}}(\mathbf{r}) + \int \frac{\rho(\mathbf{r}')}{|\mathbf{r}-\mathbf{r}'|}d\mathbf{r}' + \mathbf{V}_{\mathrm{xc}}(\mathbf{r})
\end{aligned} \tag{6.79}$$

The effective potential contains the nuclear contribution, the electronic Coulomb repulsion and the exchange–correlation potential, which is given as the functional derivative of the energy with respect to the density:

$$\mathbf{V}_{\mathrm{xc}}(\mathbf{r}) = \frac{\delta E_{\mathrm{xc}}[\rho]}{\delta\rho(\mathbf{r})} = \varepsilon_{\mathrm{xc}}[\rho(\mathbf{r})] + \int \rho(\mathbf{r}')\frac{\delta\varepsilon_{\mathrm{xc}}(\mathbf{r}')}{\delta\rho(\mathbf{r})}\mathrm{d}\mathbf{r}' \tag{6.80}$$

A unitary transformation that makes the matrix of the Lagrange multiplier diagonal may again be chosen, producing a set of canonical KS orbitals. The resulting pseudo-eigenvalue equations are known as the Kohn–Sham equations:

$$\mathbf{h}_{\mathrm{KS}}\phi_i = \varepsilon_i\phi_i \tag{6.81}$$

The KS orbitals can be determined completely by a numerical procedure, analogously to numerical HF methods. Such procedures are in practice limited to small systems, and essentially all calculations employ an expansion of the KS orbitals in an atomic basis set:

$$\phi_i = \sum_{\alpha}^{M_{\text{basis}}} c_{\alpha i}\chi_\alpha \tag{6.82}$$

The basis functions are often the same as used in wave mechanics for expanding the HF orbitals, although basis sets specifically optimized for DFT have been designed to reduce the basis set error at a given quality level (see Section 5.4.6 for details).

The variational procedure again leads to a matrix equation in the atomic orbital basis that can be written in the following form (compare to Equation (3.54)):

$$\begin{aligned} &\mathbf{h}_{\text{KS}}\mathbf{C} = \mathbf{SC}\boldsymbol{\varepsilon} \\ &h_{\alpha\beta} = \langle \chi_\alpha | \mathbf{h}_{\text{KS}} | \chi_\beta \rangle \quad ; \quad S_{\alpha\beta} = \langle \chi_\alpha | \chi_\beta \rangle \end{aligned} \tag{6.83}$$

The $\mathbf{h}_{\text{KS}}$ matrix is analogous to the Fock matrix in wave mechanics and the one-electron and Coulomb parts are identical to the corresponding Fock matrix elements. The exchange–correlation part, however, is given in terms of the electron density, and possibly also involves derivatives of the density or orbitals:

$$\int \chi_\alpha(\mathbf{r})\mathbf{V}_{\text{xc}}[\rho(\mathbf{r}), \nabla\rho(\mathbf{r})]\chi_\beta(\mathbf{r})\mathrm{d}\mathbf{r} \tag{6.84}$$

Since the $\mathbf{V}_{\text{xc}}$ functional depends on the integration variables implicitly via the electron density, these integrals cannot be evaluated analytically but must be generated by a numerical integration:

$$\int \chi_\alpha(\mathbf{r})\mathbf{V}_{\text{xc}}[\rho(\mathbf{r}), \nabla\rho(\mathbf{r})]\chi_\beta(\mathbf{r})d\mathbf{r} \approx \sum_{k}^{G} \mathbf{V}_{\text{xc}}[\rho(\mathbf{r}_k), \nabla\rho(\mathbf{r}_k)]\chi_a(\mathbf{r}_k)\chi_\beta(\mathbf{r}_k)\Delta\mathbf{v}_k \tag{6.85}$$

As the number of grid points G goes to infinity, the approximation becomes exact. The number of points is in practice selected based on the desired accuracy of the final results; that is if the energy is only required with an accuracy of 10^{-3}, the number of integration points can be smaller than if the energy is required with an accuracy of 10^{-5}.[116] There are also some technical skills involved in selecting the optimum distribution of a given number of points to yield the best accuracy, that is the points should be dense where the function $\mathbf{V}_{\text{xc}}$ varies most. The grid is usually selected as being spherical around each nucleus, making it dense in the radial direction near the nucleus and dense in the angular part in the valence space. For typical applications, 1000–10 000 points are used for each atom.[117] It should be noted that only the larger of such grids approach saturation, that is the energy will depend on the number (and location) of grid points. In order to compare energies for different systems, the same grid must therefore be used. The grid plays the same role for E_{xc} as the basis set for the other terms. Just as it is improper to compare energies calculated with different basis sets, it is not justified to compare DFT energies calculated with different grid sizes. An incomplete grid may furthermore lead to "grid superposition errors" analogous to basis set superposition errors (Section 5.10).[118]

With an expansion of the orbitals in basis functions, the number of integrals necessary for solving the KS equations rises as M_{basis}^4, owing to the Coulomb integrals in the J functional (and possibly also "exact" exchange in the hybrid methods). The number of grid points for the numerical E_{xc} integration (Equation (6.85)) increases linearly with the system size, and the computational effort for the exchange–correlation term rises as GM_{basis}^2, that is a cubic dependence of the system size. When the

Coulomb (and possibly "exact" exchange) term is evaluated directly from integrals over basis functions, DFT methods scale formally as M_{basis}^4. The reduced scaling and reduced prefactor methods discussed in Sections 3.8.6 and 3.8.7 can be used for reducing the computational cost of the Coulomb part. The numerical integration required for the exchange and correlation parts can also be reduced to a computation cost that scales linearly with system size.[119] The use of grid-based techniques for the numerical integration of the exchange–correlation contribution has some disadvantages when derivatives of the energy are desired, and grid-free DFT methods based on resolution-of-identity techniques where the exchange–correlation potential is expressed completely in terms of analytical integrals have also been proposed.[120]

The computational cost of a DFT calculation depends strongly on the implementation strategy. The use of DFT in the chemical community has to a large extent been introduced by modifying existing programs designed for wave function methods, and in these cases the numerical integration of the exchange–correlation energy adds a small overhead relative to an HF calculation. Programs designed for DFT from the outset, on the other hand, can exploit the reductions arising from reduced scaling and prefactor methods and can consequently run significantly faster than a wave function HF calculation.[121]

Finally, *DFT methods are one-dimensional* just like HF methods, and increasing the size of the basis set allows a better and better description of the KS orbitals. Since the DFT energy depends directly on the electron density, it has an exponential convergence with respect to basis set size, analogously to HF methods, and a polarized triple zeta-type basis usually gives results close to the basis set limit.

6.8 Differences between Density Functional Theory and Hartree-Fock

DFT methods based on unrestricted determinants (analogous to UHF, Section 3.7) for open-shell systems have the significant advantage that they are not very prone to "spin contamination", that is $\langle \mathbf{S}^2 \rangle$ is normally close to $S(S+1)$ (see also Sections 4.4 and 12.5.3). This can be considered a consequence of electron correlation being included in the single-determinant wave function (by means of E_{xc}) or as a consequenc of incomplete cancellation of the electron self-interaction energy. It has furthermore been argued that "spin contamination" is not well defined in DFT methods and that $\langle \mathbf{S}^2 \rangle$ should *not* be equal to $S(S+1)$.[122,123] The argument is that real systems display "spin polarization", that is there are points in space where ρ_α is larger than ρ_β (assuming that the number of α electrons is larger than the number of β electrons). This effect cannot be achieved by a restricted open-shell-type determinant (analogous to ROHF), but only by an unrestricted treatment that allows the α and β orbitals to be different. It is somewhat unclear whether this argument holds for cases with $\langle \mathbf{S}^2 \rangle$ values very different from $S(S+1)$, as in, for example, systems with multiple open-shell fragments.[124] Another consequence of the presence of E_{xc} is that restricted-type determinants are much more stable toward symmetry breaking to an unrestricted determinant (Section 3.8.3) than Hartree–Fock wave functions. For ozone (Section 4.4), for example, it is not possible to find a lower energy solution corresponding to UHF for "pure" DFT methods (such as LSDA or BLYP), although those including exact exchange (such as B3LYP) display a triplet instability. This "inverse" symmetry breaking is a consequence of DFT having a tendency of favoring electron delocalization, which may be problematic for some cases. DFT methods for radical cations, for example, usually refuse to localize the spin and charge, and thereby create unrealistic energy surfaces.

The Lagrange multipliers arising in Hartree–Fock theory from the orthogonality constraints of the orbitals are molecular orbital energies, and the occupied orbital energies correspond to ionization potentials in a frozen orbital approximation via Koopmans' theorem. The corresponding Lagrange multipliers in DFT do not have the same formal relationship, since Koopmans' theorem does not hold unless the *exact* exchange–correlation functional is employed. The Lagrange multipliers can for approximate XC functionals be interpreted as the derivative of the total energy with respect to the occupation number of the orbital, often called the *Janak theorem*[125] but discussed first by Slater,[126] and this is of course also closely related to experimentally measured ionization potential:

$$\frac{\partial E}{\partial n_i} = \varepsilon_i \tag{6.86}$$

The Lagrange multipliers may also be considered as approximations to ionization potentials using *relaxed* orbitals, and in practice gives quite accurate results for the valence orbitals.[127,128] The orbital energies resulting from Kohn–Sham calculations were in early work not considered to have any physical relevance, since they often showed poor agreement with ionization potentials and orbital energy differences correlated poorly with excitation energies. It is now clear that part of the poor agreement was due to the self-interaction error embedded in LDA and GGA methods, while more modern functionals yield much improved results. The *optimally tuned* approach can be used to specifically tune a range-separated functional such that the HOMO energy is equal to the ionization potential,[94] in which case the orbital energy obviously has a well-defined interpretation.

Another difference is that the unoccupied orbital energies in Hartree–Fock theory are determined in the field of N electrons and therefore correspond to *adding* an electron, that is the electron affinity. The virtual orbitals in density functional theory, on the other hand, are determined in the field of $N-1$ electrons and therefore correspond to *exciting* an electron, that is unoccupied orbitals in DFT tend to be significantly lower in energy than the corresponding HF ones, and the highest occupied molecular orbital–lowest unoccupied molecular orbital (HOMO–LUMO) gaps are therefore much smaller with DFT methods than for HF. This also means that orbital energy differences in DFT are reasonable estimates of excitation energies, in contrast to HF methods where excitation energies involve additional Coulomb and exchange integrals.[129] The LSDA method usually underestimates the HOMO–LUMO gap, leading to the incorrect prediction of metallic behavior for certain semiconducting materials.

Although it is clear that there are many similarities between wave mechanics HF theory and DFT, there is an important difference. If the *exact* $E_{xc}[\rho]$ was known, DFT would provide the *exact* total energy, *including* electron correlation. DFT methods therefore have the potential of including the computationally difficult part in wave mechanics, the correlation energy, at a computational effort similar to that for determining the uncorrelated HF energy. Although this certainly is the case for approximations to $E_{xc}[\rho]$, this is not necessarily true for the *exact* $E_{xc}[\rho]$. It may well be that the exact $E_{xc}[\rho]$ functional is so complicated that the computational effort for solving the KS equations will be similar to that required for solving the Schrödinger equation (exactly) with a wave mechanics approach. Indeed, unless one believes that the Schrödinger equation contains superfluous information, this is likely to be the case. Since exact solutions are generally not available in either approach, the important question is instead what the computational cost is for generating a solution of a given accuracy. In this respect, DFT methods have very favorable characteristics.

6.9 Time-Dependent Density Functional Theory (TDDFT)

Density functional theory can in analogy to wave function methods be extended to include time-dependent (external) electric potentials,[130] and this allows a description of, for example, excited states and frequency-dependent polarizability (Section 11.10). The Hohenberg–Kohn theorem only holds for electron densities describing the time-independent ground state, but the Runge–Gross theorem[131] states that the unique one-to-one correspondence between an external potential and the electron density holds both in time-dependent and time-independent cases.

The time-dependent Schrödinger equation can be written as given below for the typical situation where the perturbation consists of a time-dependent external electric potential:

$$i\frac{\partial}{\partial t}\tilde{\Psi}(\mathbf{r},t) = \mathbf{H}\tilde{\Psi}(\mathbf{r},t)$$
$$\mathbf{H}(\mathbf{r},t) = \mathbf{H}_0(\mathbf{r}) + \mathbf{V}_{\text{ext}}(\mathbf{r},t) \quad (6.87)$$

We will assume that the time-independent part ($\mathbf{H}_0$) has been solved and the external potential is turned on at time t_0 by means of the Heaviside step function θ ($\theta(t) = 0$ for $t < 0$ and $\theta(t) = 1$ for $t \geq 0$):

$$\mathbf{V}_{\text{ext}}(\mathbf{r},t) = \theta(t - t_0)\mathbf{V}_{\text{ext}}(\mathbf{r},t) \quad (6.88)$$

The time-dependent wave function $\tilde{\Psi}$ can be written as a phase factor ϖ times a *regular* wave function Ψ, where the phase factor depends on time but not on space coordinates:[132]

$$\tilde{\Psi}(\mathbf{r},t) = e^{-i\varpi(t)}\Psi(\mathbf{r},t) \quad (6.89)$$

The formal solution to the time-dependent Schrödinger equation is given by

$$\Psi(\mathbf{r},t) = \mathbf{U}(t,t_0)\Psi(\mathbf{r},t_0)$$
$$\mathbf{U}(t,t_0) = \exp\left[-i\int_{t_0}^{t}\mathbf{H}(\tau)\mathrm{d}\tau\right] \quad (6.90)$$

The exponential time-evolution operator $\mathbf{U}$, however, is very demanding to calculate, and practical methods rely on approximating it as a sequence of finite time-steps, analogous to classical molecular dynamics (Section 15.2.1):

$$\mathbf{U}(t_0 + \Delta t) \simeq \exp(-i\mathbf{H}\Delta t) \quad (6.91)$$

The exponential Hamiltonian term may in actual applications be further approximated, for example based on a Taylor expansion.

The time-independent Schrödinger equation leads to an energy minimum criterion (the variational principle), which can be used to derive, for example, the Hartree–Fock equations. Energy is not a conserved quantity in the time-dependent case, and the analogy to the variational principle is a stationary condition for the action defined by

$$S[\Psi] = \int_{t_0}^{t}\langle\Psi|\left(i\frac{\partial}{\partial t'} - \mathbf{H}\right)|\Psi\rangle\mathrm{d}t' \quad (6.92)$$

Note that the stationary condition ($\nabla S = 0$) is a less useful criterion than the energy minimum condition ($\Delta E > 0$) for generating (approximate) solutions. The action furthermore depends on the initial time t_0, and solutions to the stationary action condition will therefore depend on the initial state Ψ_0.

If the wave function consists of a single Slater determinant composed of molecular orbitals the time-dependent Schrödinger equation results in a set of time-dependent Hartree–Fock equations for the orbitals, which must be solved self-consistently:

$$i\frac{\partial}{\partial t}\phi_i(\mathbf{r},t) = (\mathbf{F} + \mathbf{V}_{\text{ext}}(t))\phi_i(\mathbf{r},t) = (\mathbf{T} + \mathbf{V}_{\text{eff}}(t))\phi_i(\mathbf{r},t) \quad (6.93)$$

The evolution operator is the sum of the (time-dependent) Fock operator **F** and the external potential, which can be written as the kinetic energy **T** and an effective potential operator $\mathbf{V}_{\text{eff}}$. The effective potential can be decomposed into a nuclear-electron term, Coulomb and exchange potentials from the electron–electron interaction and the external potential:

$$\mathbf{V}_{\text{eff}}(\mathbf{r},t) = \mathbf{V}_{\text{ne}}(\mathbf{r},t) + \mathbf{J}(\mathbf{r},t) + \mathbf{K}(\mathbf{r},t) + \mathbf{V}_{\text{ext}}(\mathbf{r},t) \tag{6.94}$$

The Coulomb (**J**) and exchange (**K**) potentials are defined in terms of the orbitals, analogous to Equation (3.34), but are time-dependent since the orbitals are time-dependent:

$$\mathbf{J}_i(\mathbf{r},t)\phi_i(\mathbf{r},t) = \left[\sum_{j=1}^{N_{\text{elec}}} \int \frac{\phi_j(\mathbf{r}',t)\phi_j(\mathbf{r}',t)}{|\mathbf{r}-\mathbf{r}'|} \mathrm{d}\mathbf{r}'\right] \phi_i(\mathbf{r},t) \tag{6.95}$$

$$\mathbf{K}_i(\mathbf{r},t)\phi_i(\mathbf{r},t) = \left[\sum_{j=1}^{N_{\text{elec}}} \int \frac{\phi_j(\mathbf{r}',t)\phi_i(\mathbf{r}',t)}{|\mathbf{r}-\mathbf{r}'|} \mathrm{d}\mathbf{r}'\right] \phi_j(\mathbf{r},t) \tag{6.96}$$

TDDFT is, in analogy with the time-independent case, similar to TDHF. If the initial state is the ground state, the Hohenberg–Kohn theorem ensures that this only depends on the initial density. Under the same assumptions as in the Kohn–Sham theory, the time-dependent density is obtained from a set of orbitals corresponding to a non-interacting reference system where the wave function is a Slater determinant composed of Kohn–Sham orbitals:

$$\rho(\mathbf{r},t) = \sum_{i=1}^{N} \phi_i^2(\mathbf{r},t) \tag{6.97}$$

This leads to a set of equations where the orbitals are determined by solving self-consistently a set of time-dependent Kohn–Sham equations identical in form to Equation (6.93) but with a different effective potential:

$$\mathbf{V}_{\text{eff}}(\mathbf{r},t) = \mathbf{V}_{\text{ne}}(\mathbf{r},t) + \mathbf{J}(\mathbf{r},t) + \mathbf{V}_{\text{xc}}(\mathbf{r},t,\Psi_0) + \mathbf{V}_{\text{ext}}(\mathbf{r},t) \tag{6.98}$$

The nuclear-electron, Coulomb and external potentials are identical to the HF case, but the exchange potential is replaced by the exchange–correlation potential. The Coulomb potential can be written in terms of the instantaneous time-dependent density $\rho(\mathbf{r}',t)$ (compare with Equation (6.95)):

$$\mathbf{J}(\mathbf{r},t) = \int \frac{\rho(\mathbf{r}',t)}{|\mathbf{r}-\mathbf{r}'|} \mathrm{d}\mathbf{r}' \tag{6.99}$$

The exchange–correlation potential is more complicated. In the time-independent case, it is given as the functional derivative of the exchange–correlation energy (Equation (6.80)), but in the time-dependent case, it is given as the functional derivative of the exchange–correlation action:[133]

$$\mathbf{V}_{\text{xc}}(\mathbf{r},t,\rho_0) = \frac{\delta S_{\text{xc}}[\rho]}{\delta \rho(\mathbf{r},t)} \tag{6.100}$$

The important difference relative to the time-independent case is that the exchange–correlation potential now depends on the initial density at time t_0, as well as the whole time-interval t–t_0. This means that solutions to the time-dependent Kohn–Sham Equations (6.93) has a memory and must be solved self-consistently also in the time domain. Essentially all commonly employed TDDFT methods make the adiabatic approximation of neglecting the time–memory effect on the exchange–correlation potential. This is equivalent to assuming that the density is a slowly varying function of time, analogous to the LSDA assumption that the density is a slowly varying function of the position. Formulated alternatively in frequency space, corresponding to a Fourier transformation, the exchange–correlation potential is taken as its zero-frequency limit. This is clearly an approximation, since the response of the electron density will depend on how rapidly the external potential varies. With the adiabatic approximation the exchange–correlation potential in Equation (6.100) reduces to the corresponding time-independent case in Equation (6.80).

6.9.1 Weak Perturbation – Linear Response

TDHF and TDDFT can be used to study time-dependent phenomenon in general, such as laser-induced ionization, but the most common application is in the regime of weak perturbations, where only the linear response is considered. Rather than solving the full time-dependent Equation (6.87), the response of the system is calculated by perturbation theory. The perturbation is usually an oscillating electric field, which in the dipole approximation can be written as

$$\mathbf{V}_{\text{ext}}(t) = \boldsymbol{\mu} F \cos(\omega t) \tag{6.101}$$

Here ω is the frequency, F is the strength of the field and $\boldsymbol{\mu}$ is the dipole operator. The weak perturbation regime corresponds to the situation where the external electric field is significantly smaller than the field generated by the nuclear charge. For an electron in a hydrogen atom at a distance of 1 bohr, the nuclear field is $\sim 5 \times 10^9$ V/m, which corresponds to a laser intensity of $\sim 4 \times 10^{16}$ W/cm^2.

The general problem of determining the wave function or density response to a time-dependent weak perturbation is treated in Section 11.10. Using a $(\mathbf{Y}, \mathbf{Z})$ vector to represent the real and imaginary parts of the first-order response, it can be obtained by solving a set of linear equations, and can be written as given below (which is the same as Equation (11.164), with * denoting complex conjugation):[134]

$$\left(\begin{bmatrix} \mathbf{A} & \mathbf{B} \\ \mathbf{B}^* & \mathbf{A}^* \end{bmatrix} - \omega \begin{bmatrix} 1 & 0 \\ 0 & -1 \end{bmatrix} \right) \begin{bmatrix} \mathbf{Y} \\ \mathbf{Z} \end{bmatrix} = - \begin{bmatrix} \mathbf{P} \\ \mathbf{P}^* \end{bmatrix} \tag{6.102}$$

The **A** matrix elements are for an HF wave function, given as the difference in orbital energies between an occupied and virtual orbital, and a term corresponding to differences in electron–electron interaction upon changing orbital occupancy, while the **B** matrix elements are given as the difference in electron–electron interaction (using the notation in Equation (3.62) with ij being occupied and ab being virtual orbitals) (which is the same as Equations (11.66) to (11.68)):

$$A_{ij}^{ab} = \left\langle \Phi_i^a \middle| \mathbf{H}_0 \middle| \Phi_j^b \right\rangle - E_0 \delta_{ij} \delta_{ab} = \delta_{ij} \delta_{ab} (\varepsilon_a - \varepsilon_i) + \langle ij|ab\rangle - \langle ia|jb\rangle \tag{6.103}$$

$$B_{ij}^{ab} = \left\langle \Phi_0 \middle| \mathbf{H}_0 \middle| \Phi_{ij}^{ab} \right\rangle = \langle ij|ab\rangle - \langle ij|ba\rangle \tag{6.104}$$

$$P_i^a = \langle i|\mathbf{r}|a\rangle \tag{6.105}$$

The last two-electron integrals in Equation (6.103) and (6.104) are often called exchange terms. Solving Equation (6.102) for the wave function or density response allows a calculation of time-dependent molecular properties, as discussed in Section 11.10.

If the right-hand side of Equation (6.102) is set equal to zero, the equation becomes a generalized eigenvalue problem, with ω being the eigenvalues:

$$\begin{bmatrix} \mathbf{A} & \mathbf{B} \\ \mathbf{B}^* & \mathbf{A}^* \end{bmatrix} \begin{bmatrix} \mathbf{Y} \\ \mathbf{Z} \end{bmatrix} = \omega \begin{bmatrix} 1 & 0 \\ 0 & -1 \end{bmatrix} \begin{bmatrix} \mathbf{Y} \\ \mathbf{Z} \end{bmatrix} \tag{6.106}$$

Solution of Equation (6.106) provides excitation and de-excitation energies occurring in pairs $\pm\omega$. In CI language, the **A** matrix contains matrix elements between singly excited Slater determinants while the **B** matrix contains matrix elements between doubly excited Slater determinants and the HF reference determinant. If the **B** matrix is set to zero, known as the *Tamm–Dancoff approximation* (TDA), TDHF therefore becomes identical to CIS (Section 4.14). The TDHF Equations (6.106) within the TDA correspond to a diagonalization of **A**, where the eigenvalues provides the excitation energies and the **Y** vector containing the corresponding eigenvectors provides the excited state wave function expressed in singly excited Slater determinants. The **B** matrix can be considered as a correction term corresponding to also allowing de-excitations, which provide inclusion of some electron correlation for the reference state.

Equation (6.106) can also be derived by the so-called *Random Phase Approximation* (RPA) in response theory. Unfortunately the RPA acronym is used both for the case where the **A** and **B** matrix elements are given in Equations (6.103) and (6.104), and for the case where the last two-electron exchange integrals are neglected in both **A** and **B** matrix elements. The first case is denoted *full-RPA* or *RPAX* (RPA with eXchange) and is thus synonymous with TDHF, while the latter is denoted *direct-RPA* but is often just called RPA. Note that "direct" in this context does not refer to the technical procedure of solving the equation by (re)generating the required integrals as they are needed (Section 3.8.5), rather than calculating them only once and storing them on an external media.

The **B** matrix in Equation (6.106) describes, as noted above, electron correlation to some extent, and the differences in excitation energies ω between solving Equation (6.106) with and without the **B** matrix thus provides a measure of the electron correlation in the reference state. Solving the full Equation (6.106) can either be full-RPA/TDHF or direct-RPA, where the exchange terms are neglected, while solving the equation without the **B** matrix is TDA/CIS. Since the exchange terms can be computationally expensive, and furthermore can lead to problems due to triplet instabilities, the most common option is to use the direct-RPA version. The RPA correlation energy can be defined as a sum over differences in RPA and TDA excitation energies (positive ω only), and can be shown to be an approximation to the CCD energy:[135]

$$\Delta E_{\text{corr}}^{\text{RPA}} = \frac{1}{2}\sum_i \left(\omega_i^{\text{RPA}} - \omega_i^{\text{TDA}}\right) \tag{6.107}$$

The RPA correlation energy is of similar quality to MP2 but is computationally more efficient, and can be used either as a way of including electron correlation from an HF reference wave function or as a component in double hybrid functionals in a DFT framework (Section 6.5.5). An advantage of RPA is that it is insensitive to small orbital energy differences between occupied and virtual orbitals, in contrast to MP2, and therefore can be used, for example, for semiconductor and metallic periodic systems.

The **A** and **B** matrices also arise in the second derivative of the HF energy with respect to the orbitals coefficients, where **A** + **B** is the matrix associated with real variations of the orbitals while **A** − **B** is the matrix associated with imaginary variations.[136] Diagonalization of the **A** + **B** and **A** − **B** matrices therefore allows an evaluation of whether the solution to the HF equations obtained by setting the first derivative equal to zero is a minimum or saddle point in the parameter space. It is in practice sufficient to extract the lowest eigenvalue of these matrices to determine whether the solution is a minimum or not. The **A** and **B** matrices can be separated into their α and β spin components, which for a singlet RHF reference wave function can be combined to probe whether there is a lower energy RHF state (internal or singlet instability) or a lower energy UHF-type state (external or triplet instability).

Both the **A** and **B** matrices have the dimension $N_{occ}N_{vir}$ corresponding to all single excitations. Solving the TDHF equations is roughly twice as expensive as solving the CIS equations, where only the **A** part of Equation (6.106) needs to be solved. TDHF often provides similar quality results as CIS for singlet states, but worse results for triplet states, which is related to the presence of triplet instabilities in the HF wave function; thus TDHF is consequently not a commonly used method.

The TDDFT equations are formally identical to Equation (6.106), but the last exchange-type two-electron terms in the **A** and **B** matrix elements are replaced with corresponding XC terms:

$$A_{ij}^{ab} = \delta_{ij}\delta_{ab}(\varepsilon_a - \varepsilon_i) + \langle ij|ab\rangle + \langle ij|f_{xc}|ab\rangle \tag{6.108}$$

$$B_{ij}^{ab} = \langle ij|ab\rangle + \langle ij|f_{xc}|ab\rangle \tag{6.109}$$

The f_{xc} is called the XC kernel and is given as the functional derivative of the corresponding potential, and thereby also as the second derivative of the XC action:

$$f_{xc}(\mathbf{r}, t, \mathbf{r}', t') = \frac{\delta V_{xc}[\rho](\mathbf{r}, t)}{\delta \rho(\mathbf{r}', t')} = \frac{\delta^2 S_{xc}[\rho]}{\delta \rho(\mathbf{r}, t)\delta \rho(\mathbf{r}', t')} \tag{6.110}$$

The adiabatic approximation corresponds to neglecting the time-dependence, in which case the action becomes the energy:

$$f_{xc}^{adia}(\mathbf{r}, \mathbf{r}') = \frac{\delta V_{xc}[\rho](\mathbf{r})}{\delta \rho(\mathbf{r}')} = \frac{\delta^2 E_{xc}[\rho]}{\delta \rho(\mathbf{r})\delta \rho(\mathbf{r}')} \tag{6.111}$$

The RPA acronym in TDDFT corresponds to the direct RPA in the HF case, that is neglecting the XC contribution:

$$f_{xc}^{RPA}(\mathbf{r}, \mathbf{r}') = 0 \tag{6.112}$$

The TDA (neglect of the **B** matrix) can also be used for TDDFT to improve the computational efficiency, and it is usually a good approximation. TDDFT excitation energies for valence states are usually an improvement relative to TDHF/CIS, which primarily is due to the orbital energy differences in the **A** matrix elements, with Equation (6.108)/Equation (6.109) being a much better approximation to the excitation energies than the HF orbital energy differences.[137] The adiabatic approximation restricts TDDFT to describe only singly excited states since doubly excited states require the XC kernel to be time/frequency-dependent.[138] Inclusion of double excitations is also required for a description of conical intersections.

The TDDFT formalism allows calculation of excitation energies and, by addition of these to the ground state energy, leads to construction of excited state energy surfaces. Combining derivatives of the ground state energy surface with derivatives of the excitation energies allows calculation of nuclear gradients of the excited state surface, and thus geometry optimization and dynamic simulations on excited states.

6.10 Ensemble Density Functional Theory

The Hohenberg–Kohn theorem ensures that the electron density contains all the necessary information, but the Kohn–Sham version expresses the kinetic energy in terms of the more general first-order reduced density matrix in order to obtain sufficient accuracy. While any valid electron density can be obtained from a single Slater determinant, the corresponding first-order reduced density matrix may be inaccurate. There may therefore be systems where it is necessary to write the density matrix as arising from a linear combination of more than one Slater determinant. This is called *ensemble* DFT and is analogous to the extension of HF to MCSCF in wave mechanics. Ensemble DFT is expected to be important for systems with large amounts of static correlation in the wave function language, that is when two (or more) electron configurations are near-degenerate, but the near-degeneracy condition is much stronger in DFT compared to wave mechanics methods. The determination of the optimum ensemble DFT can be reformulated as a simultaneous optimization of a set of orbitals and their fractional electron occupancies.[139]

6.11 Density Functional Theory Problems

Despite the many successes of DFT, there are some areas where the current functionals are known to perform poorly.[140]

The embarrassingly simple problem of dissociating H_2^+ is incorrectly described by essentially all functionals as resulting in a state where a $\frac{1}{2}$ electron is residing at each nucleus.[139] The dissociation energy profile as a result displays an artificial barrier and an incorrect dissociation energy, often in error by ~200 kJ/mol. While H_2^+ is of limited interest by itself, the same problem of dissociation into species with a fractional number of electrons is present for radical cation and anion systems in general, and leads to energetic artefacts for stretched bonds, as, for example, in atom transfer transition structures.[141] Activation barriers are usually underestimated by functionals that do not included exact exchange, but since Hartree–Fock overestimates activation barriers, hybrid methods often give reasonable results. These failures may be viewed as a manifestation of the incomplete cancellation of the electron self-interaction energy by the exchange functional, which leads to a favorization of delocalized electron distributions.

Relative energies of states with different spin multiplicity are often poorly described. The energy difference between a singlet and triplet state with the same orbital occupancy is in HF theory given by an exchange integral. This must in DFT be described by the exchange–correlation functional, which only depends on the electron density. If the two spin states arise from the same electron configuration the two electron densities are very similar, and this makes the results sensitive to the details of the exchange–correlation functional. These problems are especially problematic for transition metal systems, where several low-energy spin states often are possible, and many of these cannot be described by a single determinant. Pure DFT methods favor low spin states while HF favors high spin states, and hybrid methods with a suitable parameterized amount of exact exchange perform better, but the optimum mixing parameter tends to be system-dependent.[142,143] These problems can perhaps be improved by adding current density terms to the DFT formalism, but this is not yet a commonly used procedure since it requires that the orbitals are allowed to become complex.

Individual spatial components of a spin multiplet may have different energies, even in the absence of a magnetic field. The boron atom, for example, has the electron configuration $1s^2 2s^2 2p^1$, and the single p-electron can be in either a p_{-1}, p_0 or p_{+1} orbital. These should all have the same energy, but since the density associated with the p_0 orbital is different from that of a $p_{\pm 1}$ orbital, their energies as a result differ by ~25 kJ/mol. This is clearly non-physical, but can be significantly improved by introducing current density terms.[144,145]

6.12 Final Considerations

Should DFT methods be considered *ab initio* or semi-empirical? If *ab initio* is taken to mean the absence of fitting parameters, LSDA methods are *ab initio* but gradient-corrected methods may or may not be. The LSDA exchange energy contains no parameters and the correlation functional is known accurately as a tabulated function of the density. The use of a parameterized interpolation formula in practical calculations does not represent fitting in order to improve the performance for atomic and molecular systems. Some gradient-corrected methods (e.g. the B88 exchange and the LYP correlation), however, contain parameters that are fitted to give the best agreement with experimental atomic data, but the number of parameters is significantly smaller than for semi-empirical methods. The semi-empirical PM3 method (Section 7.4.5), for example, has 18 parameters for each atom, while the B88 exchange functional only has one fitting constant, valid for the whole periodic table. Functionals such as VSXC or the Minnesota family contain a substantial number of parameters (20–60), while other functionals such as PBE are derived entirely from theory and can consequently be considered "pure" *ab initio*.

If *ab initio* is taken to mean that the method is based on theory, which in principle is able to produce the exact results, DFT methods are *ab initio*. The only caveat is that current methods cannot yield the exact results, even in the limit of a complete basis set, since the functional form of the exact exchange–correlation energy is not known. At present it is easier to systematically improve on a wave function description than adding corrections to the energy functional in DFT. Methods using reduced density matrices are not yet sufficiently mature to allow any definitive conclusions to be drawn.

It is perhaps a little disturbing that seemingly very different functionals give similar-quality results.[146] It is possible to take an electron density from a wave function of near-exact quality (such as CCSD(T)) and "invert" this to produce the corresponding KS exchange–correlation potential.[147] Comparisons of such "exact" $\mathbf{V}_{xc}$ potentials with those discussed in the previous subsections have revealed large deviations and erroneous functional behavior.[148] Since many of these functionals perform well in practical applications, it is clear that the performance is not particularly sensitive to details in the functional and that the good performance to some extent is due to error cancellations.

Although gradient-corrected DFT methods have been shown to give impressive results, even for theoretically difficult problems, the lack of a systematic way of extending a series of calculations to approach the exact result is a major drawback of DFT. The results converge toward a certain value as the basis set is increased, but theory does not allow an evaluation of the errors inherent in this limit (such as the systematic overestimation of vibrational frequencies with wave mechanics HF methods). Furthermore, although a progression of methods such as LSDA, BLYP and B3LYP provides successively lower errors for a suitable set of reference data (such as that used for calibrating the Gaussian-2 model), there is no guarantee that the same progression will provide better and better results for a specific property of a given system. Indeed, LSDA methods may in some cases provide better results, even in the limit of a large basis set, than either of the more "complete" gradient-corrected models. The quality of a given result can therefore only be determined by comparing the performance for similar systems where experimental or high-quality wave mechanics results are available, and DFT in this respect resembles semi-empirical methods. Nevertheless, DFT methods, especially those involving gradient corrections and hybrid methods, are significantly more accurate (and the errors are much more uniform) than those of, for example, the MNDO family, and DFT is consequently a valuable tool for systems where (very) high accuracy is not needed.

References

1 P. Hohenberg and W. Kohn, *Physical Review B* **136** (3B), B864 (1964).
2 P. O. Lowdin, *International Journal of Quantum Chemistry* **S19**, 19–37 (1985).
3 R. G. Parr and W. Yang, *Density Functional Theory* (Oxford University Press, 1989).
4 T. Ziegler, *Chemical Reviews* **91** (5), 651–667 (1991).

5 L. J. Bartolotti and K. Flurchick, *Reviews in Computational Chemistry* 7, 187 (1996).
6 A. St-Amant, *Reviews in Computational Chemistry* 7, 217 (1996).
7 E. J. Baerends and O. V. Gritsenko, *Journal of Physical Chemistry A* **101** (30), 5383–5403 (1997).
8 W. Koch and M. C. Holthausen, *A Chemist's Guide to Density Functional Theory* (Wiley-VCH, 2001).
9 E. S. Kryachko and E. V. Ludena, *Physics Reports – Review Section of Physics Letters* **544** (2), 123–239 (2014).
10 W. Kohn and L. J. Sham, *Physical Review* **140** (4A), 1133 (1965).
11 F. Bloch, *Zeitschrift Fur Physik* **57** (7–8), 545–555 (1929).
12 P. A. M. Dirac, *Proceedings of the Cambridge Philosophical Society* **26**, 376–385 (1930).
13 D. R. Murphy, *Physical Review A* **24** (4), 1682–1688 (1981).
14 E. K. U. Gross and R. M. Dreizler, *Zeitschrift Fur Physik A Hadrons and Nuclei* **302** (2), 103–106 (1981).
15 P. S. Svendsen and U. von Barth, *Physical Review B* **54** (24), 17402–17413 (1996).
16 S. Laricchia, L. A. Constantin, E. Fabiano and F. Della Sala, *Journal of Chemical Theory and Computation* **10**, 164–179 (2014).
17 D. Garcia-Aldea and J. E. Alvarellos, *Journal of Chemical Physics* **127** (14), 144109 (2007).
18 A. J. Coleman and V. I. Yukalov, *Reduced Density Matrices: Coulson's Challenge* (Springer-Verlag, New York, 2000).
19 C. Garrod and J. K. Percus, *Journal of Mathematical Physics* **5** (12), 1756 (1964).
20 D. A. Mazziotti, *Chemical Reviews* **112** (1), 244–262 (2012).
21 D. A. Mazziotti, *Physical Review Letters* **106** (8), 083001 (2011).
22 D. A. Mazziotti, *Physical Review Letters* **108** (26), 263002 (2012).
23 W. Kutzelnigg, *Journal of Chemical Physics* **125** (17), 171101 (2006).
24 A. Y. Sokolov, A. C. Simmonett and H. F. Schaefer, III, *Journal of Chemical Physics* **138** (2), 024107 (2013).
25 J. W. Mullinax, A. Y. Sokolov and H. F. Schaefer, *Journal of Chemical Theory and Computation* **11** (6), 2487–2495 (2015).
26 S. Goedecker and C. J. Umrigar, *Physical Review Letters* **81** (4), 866–869 (1998).
27 O. Gritsenko, K. Pernal and E. J. Baerends, *Journal of Chemical Physics* **122** (20), 204102 (2005).
28 C. Kollmar, *Journal of Chemical Physics* **121** (23), 11581–11586 (2004).
29 M. Piris, *International Journal of Quantum Chemistry* **113** (5), 620–630 (2013).
30 P. M. W. Gill, D. L. Crittenden, D. P. O'Neill and N. A. Besley, *Physical Chemistry Chemical Physics* **8** (1), 15–25 (2006).
31 P. W. Ayers and S. Liu, *Physical Review A* **75** (2), 022514 (2007).
32 O. V. Gritsenko, P. R. T. Schipper and E. J. Baerends, *Journal of Chemical Physics* **107** (13), 5007–5015 (1997).
33 A. D. Becke, *Journal of Chemical Physics* **138** (7), 074109 (2013).
34 C. Haettig, W. Klopper, A. Koehn and D. P. Tew, *Chemical Reviews* **112** (1), 4–74 (2012).
35 S. Lehtola and H. Jonsson, *Journal of Chemical Theory and Computation* **10** (12), 5324–5337 (2014).
36 S. Lehtola and H. Jonsson, *Journal of Chemical Theory and Computation* **11** (10), 5052–5053 (2015).
37 R. K. Nesbet and R. Colle, *Journal of Mathematical Chemistry* **26** (1–3), 233–242 (1999).
38 J. P. Perdew and A. Zunger, *Physical Review B* **23** (10), 5048–5079 (1981).
39 E. J. Baerends, *Physical Review Letters* **87** (13), 133004 (2001).
40 A. D. Becke, *Journal of Chemical Physics* **119** (6), 2972–2977 (2003).
41 J. P. Perdew, J. M. Tao, V. N. Staroverov and G. E. Scuseria, *Journal of Chemical Physics* **120** (15), 6898–6911 (2004).
42 M. Levy and J. P. Perdew, *Physical Review A* **32** (4), 2010–2021 (1985).

43 E. H. Lieb and S. Oxford, *International Journal of Quantum Chemistry* **19** (3), 427-439 (1981).
44 R. van Leeuwen and E. J. Baerends, *Physical Review A* **49** (4), 2421–2431 (1994).
45 J. P. Perdew, R. G. Parr, M. Levy and J. L. Balduz, *Physical Review Letters* **49** (23), 1691–1694 (1982).
46 S. Kurth, J. P. Perdew and P. Blaha, *International Journal of Quantum Chemistry* **75** (4–5), 889–909 (1999).
47 J. P. Perdew, A. Ruzsinszky, J. M. Tao, V. N. Staroverov, G. E. Scuseria and G. I. Csonka, *Journal of Chemical Physics* **123** (6), 062201 (2005).
48 J. C. Slater, *Physical Review* **81** (3), 385–390 (1951).
49 W. J. Carr, *Physical Review* **122** (5), 1437 (1961).
50 W. J. Carr and A. A. Maradudin, *Physical Review A – General Physics* **133** (2A), A371 (1964).
51 S. H. Vosko, L. Wilk and M. Nusair, *Canadian Journal of Physics* **58** (8), 1200–1211 (1980).
52 J. P. Perdew and Y. Wang, *Physical Review B* **45** (23), 13244–13249 (1992).
53 S. K. Ma and Brueckne.Ka, *Physical Review* **165** (1), 18 (1968).
54 A. D. Becke, *Physical Review A* **38** (6), 3098–3100 (1988).
55 G. Ortiz and P. Ballone, *Physical Review B* **43** (8), 6376–6387 (1991).
56 N. C. Handy and A. J. Cohen, *Molecular Physics* **99** (5), 403–412 (2001).
57 A. D. Boese and N. C. Handy, *Journal of Chemical Physics* **114** (13), 5497–5503 (2001).
58 C. T. Lee, W. T. Yang and R. G. Parr, *Physical Review B* **37** (2), 785–789 (1988).
59 B. Miehlich, A. Savin, H. Stoll and H. Preuss, *Chemical Physics Letters* **157** (3), 200–206 (1989).
60 J. P. Perdew and Y. Wang, *Physical Review B* **33** (12), 8800–8802 (1986).
61 J. P. Perdew, J. A. Chevary, S. H. Vosko, K. A. Jackson, M. R. Pederson, D. J. Singh and C. Fiolhais, *Physical Review B* **46** (11), 6671–6687 (1992).
62 J. P. Perdew, K. Burke and M. Ernzerhof, *Physical Review Letters* **77** (18), 3865–3868 (1996).
63 C. Adamo and V. Barone, *Journal of Chemical Physics* **108** (2), 664–675 (1998).
64 B. Hammer, L. B. Hansen and J. K. Norskov, *Physical Review B* **59** (11), 7413–7421 (1999).
65 C. Adamo and V. Barone, *Journal of Chemical Physics* **116** (14), 5933–5940 (2002).
66 T. W. Keal and D. J. Tozer, *Journal of Chemical Physics* **123** (12), 121103 (2005).
67 S. Reimann, U. Ekstrom, S. Stopkowicz, A. M. Teale, A. Borgoo and T. Helgaker, *Physical Chemistry Chemical Physics* **17** (28), 18834–18842 (2015).
68 M. Brack, B. K. Jennings and Y. H. Chu, *Physics Letters B* **65** (1), 1–4 (1976).
69 A. D. Becke and M. R. Roussel, *Physical Review A* **39** (8), 3761–3767 (1989).
70 A. D. Becke, *Journal of Chemical Physics* **104** (3), 1040–1046 (1996).
71 A. D. Boese and N. C. Handy, *Journal of Chemical Physics* **116** (22), 9559–9569 (2002).
72 T. Van Voorhis and G. E. Scuseria, *Journal of Chemical Physics* **109** (2), 400–410 (1998).
73 J. P. Perdew, S. Kurth, A. Zupan and P. Blaha, *Physical Review Letters* **82** (12), 2544–2547 (1999).
74 J. M. Tao, J. P. Perdew, V. N. Staroverov and G. E. Scuseria, *Physical Review Letters* **91** (14), 146401 (2003).
75 J. Harris, *Physical Review A* **29** (4), 1648–1659 (1984).
76 A. D. Becke, *Journal of Chemical Physics* **98** (2), 1372–1377 (1993).
77 A. D. Becke, *Journal of Chemical Physics* **98** (7), 5648–5652 (1993).
78 P. J. Stephens, F. J. Devlin, C. F. Chabalowski and M. J. Frisch, *Journal of Physical Chemistry* **98** (45), 11623–11627 (1994).
79 A. D. Becke, *Journal of Chemical Physics* **107** (20), 8554–8560 (1997).
80 H. L. Schmider and A. D. Becke, *Journal of Chemical Physics* **108** (23), 9624–9631 (1998).
81 M. Ernzerhof and G. E. Scuseria, *Journal of Chemical Physics* **110** (11), 5029–5036 (1999).
82 J. P. Perdew, M. Emzerhof and K. Burke, *Journal of Chemical Physics* **105** (22), 9982–9985 (1996).

83 V. N. Staroverov, G. E. Scuseria, J. M. Tao and J. P. Perdew, *Journal of Chemical Physics* **119** (23), 12129–12137 (2003).
84 V. N. Staroverov, G. E. Scuseria, J. M. Tao and J. P. Perdew, *Journal of Chemical Physics* **121** (22), 11507–11507 (2004).
85 L. Goerigk and S. Grimme, *Wiley Interdisciplinary Reviews – Computational Molecular Science* **4** (6), 576–600 (2014).
86 S. Grimme, *Journal of Chemical Physics* **124** (3), 034108 (2006).
87 D. Bousquet, E. Bremond, J. C. Sancho-Garcia, I. Ciofini and C. Adamo, *Journal of Chemical Theory and Computation* **9** (8), 3444–3452 (2013).
88 B. Chan, L. Goerigk and L. Radom, *Journal of Computational Chemistry* **37** (2), 183–193 (2016).
89 J. C. Rienstra-Kiracofe, G. S. Tschumper, H. F. Schaefer, S. Nandi and G. B. Ellison, *Chemical Reviews* **102** (1), 231–282 (2002).
90 T. Tsuneda and K. Hirao, *Wiley Interdisciplinary Reviews – Computational Molecular Science* **4** (4), 375390 (2014).
91 S. Jakobsen, K. Kristensen and F. Jensen, *Journal of Chemical Theory and Computation* **9** (9), 3978–3985 (2013).
92 T. Yanai, D. P. Tew and N. C. Handy, *Chemical Physics Letters* **393** (1-3), 51-57 (2004).
93 J.-D. Chai and M. Head-Gordon, *Physical Chemistry Chemical Physics* **10** (44), 6615–6620 (2008).
94 E. Livshits and R. Baer, *Physical Chemistry Chemical Physics* **9** (23), 2932–2941 (2007).
95 J. Heyd, G. E. Scuseria and M. Ernzerhof, *Journal of Chemical Physics* **118** (18), 8207–8215 (2003).
96 J. Heyd, J. E. Peralta, G. E. Scuseria and R. L. Martin, *Journal of Chemical Physics* **123** (17), 174101 (2005).
97 T. M. Henderson, A. F. Izmaylov, G. E. Scuseria and A. Savin, *Journal of Chemical Theory and Computation* **4** (8), 1254–1262 (2008).
98 J. Klimes and A. Michaelides, *Journal of Chemical Physics* **137** (12), 120901 (2012).
99 S. Grimme, J. Antony, S. Ehrlich and H. Krieg, *Journal of Chemical Physics* **132** (15), 154104 (2010).
100 A. D. Becke and E. R. Johnson, *Journal of Chemical Physics* **127** (15), 154108 (2007).
101 O. A. Vydrov and T. van Voorhis, *Journal of Chemical Physics* **133** (24), 244103 (2010).
102 N. Mardirossian and M. Head-Gordon, *Journal of Chemical Physics* **142** (7), 074111 (2015).
103 W. Hujo and S. Grimme, *Journal of Chemical Theory and Computation* **9** (1), 308–315 (2013).
104 G. K. L. Chan and N. C. Handy, *Journal of Chemical Physics* **112** (13), 5639–5653 (2000).
105 T. Schwabe, *Physical Chemistry Chemical Physics* **16** (28), 14559–14567 (2014).
106 R. Peverati and D. G. Truhlar, *Philosophical Transactions of the Royal Society A – Mathematical Physical and Engineering Sciences* **372**, 20120476 (2014).
107 R. Peverati, Y. Zhao and D. G. Truhlar, *Journal of Physical Chemistry Letters* **2** (16), 1991–1997 (2011).
108 R. Peverati and D. G. Truhlar, *Journal of Chemical Theory and Computation* **8** (7), 2310–2319 (2012).
109 R. Peverati and D. G. Truhlar, *Journal of Physical Chemistry Letters* **3** (1), 117–124 (2012).
110 R. Peverati and D. G. Truhlar, *Physical Chemistry Chemical Physics* **14** (38), 13171–13174 (2012).
111 R. Peverati and D. G. Truhlar, *Journal of Chemical Physics* **135** (19), 191102 (2011).
112 R. Peverati and D. G. Truhlar, *Journal of Physical Chemistry Letters* **2** (21), 2810–2817 (2011).
113 R. Peverati and D. G. Truhlar, *Physical Chemistry Chemical Physics* **14** (47), 16187–16191 (2012).
114 N. Mardirossian and M. Head-Gordon, *Physical Chemistry Chemical Physics* **16** (21), 9904–9924 (2014).
115 P. O. Dral, X. Wu, L. Spoerkel, A. Koslowski and W. Thiel, *Journal of Chemical Theory and Computation* **12** (3), 1097–1120 (2016).
116 O. Treutler and R. Ahlrichs, *Journal of Chemical Physics* **102** (1), 346–354 (1995).
117 P. M. W. Gill, B. G. Johnson and J. A. Pople, *Chemical Physics Letters* **209** (5–6), 506–512 (1993).

118 S. E. Wheeler and K. N. Houk, *Journal of Chemical Theory and Computation* **6** (2), 395–404 (2010).
119 R. E. Stratmann, G. E. Scuseria and M. J. Frisch, *Chemical Physics Letters* **257** (3–4), 213–223 (1996).
120 K. R. Glaesemann and M. S. Gordon, *Journal of Chemical Physics* **108** (24), 9959–9969 (1998).
121 C. F. Guerra, J. G. Snijders, G. te Velde and E. J. Baerends, *Theoretical Chemistry Accounts* **99** (6), 391–403 (1998).
122 J. A. Pople, P. M. W. Gill and N. C. Handy, *International Journal of Quantum Chemistry* **56** (4), 303–305 (1995).
123 C. R. Jacob and M. Reiher, *International Journal of Quantum Chemistry* **112** (23), 3661–3684 (2012).
124 E. R. Davidson and A. E. Clark, *Physical Chemistry Chemical Physics* **9** (16), 1881–1894 (2007).
125 J. F. Janak, *Physical Review B* **18** (12), 7165–7168 (1978).
126 J. C. Slater, *The Self-Consistent Field for Molecules and Solids: Quantum Theory of Molecules and Solids* (McGraw-Hill, 1974).
127 D. P. Chong, O. V. Gritsenko and E. J. Baerends, *Journal of Chemical Physics* **116** (5), 1760–1772 (2002).
128 O. V. Gritsenko and E. J. Baerends, *Journal of Chemical Physics* **120** (18), 8364–8372 (2004).
129 R. van Meer, O. V. Gritsenko and E. J. Baerends, *Journal of Chemical Theory and Computation* **10** (10), 4432–4441 (2014).
130 C. A. Ullrich, *Time-Dependent Density Functional Theory* (Oxford University Press, 2012).
131 E. Runge and E. K. U. Gross, *Physical Review Letters* **52** (12), 997–1000 (1984).
132 F. Pawlowski, J. Olsen and P. Jorgensen, *Journal of Chemical Physics* **142** (11), 114109 (2015).
133 R. van Leeuwen, *Physical Review Letters* **80** (6), 1280–1283 (1998).
134 A. Dreuw and M. Head-Gordon, *Chemical Reviews* **105** (11), 4009–4037 (2005).
135 G. E. Scuseria, T. M. Henderson and D. C. Sorensen, *Journal of Chemical Physics* **129** (23), 231101 (2008).
136 R. Seeger and J. A. Pople, *Journal of Chemical Physics* **66** (7), 3045–3050 (1977).
137 E. J. Baerends, O. V. Gritsenko and R. van Meer, *Physical Chemistry Chemical Physics* **15** (39), 16408–16425 (2013).
138 N. T. Maitra, F. Zhang, R. J. Cave and K. Burke, *Journal of Chemical Physics* **120** (13), 5932–5937 (2004).
139 C. R. Nygaard and J. Olsen, *Journal of Chemical Physics* **138** (9), 094109 (2013).
140 A. J. Cohen, P. Mori-Sanchez and W. Yang, *Chemical Reviews* **112** (1), 289–320 (2012).
141 O. V. Gritsenko, B. Ensing, P. R. T. Schipper and E. J. Baerends, *Journal of Physical Chemistry A* **104** (37), 8558–8565 (2000).
142 A. Droghetti, D. Alfe and S. Sanvito, *Journal of Chemical Physics* **137** (12), 124–303 (2012).
143 V. E. J. Berryman, R. J. Boyd and E. R. Johnson, *Journal of Chemical Theory and Computation* **11** (7), 3022–3028 (2015).
144 A. D. Becke, *Journal of Chemical Physics* **117** (15), 6935–6938 (2002).
145 S. Pittalis, S. Kurth, S. Sharma and E. K. U. Gross, *Journal of Chemical Physics* **127** (12), 124103 (2007).
146 R. Neumann, R. H. Nobes and N. C. Handy, *Molecular Physics* **87** (1), 1–36 (1996).
147 L. O. Wagner, T. E. Baker, E. M. Stoudenmire, K. Burke and S. R. White, *Physical Review B* **90** (4), 045109 (2014).
148 R. J. Bartlett, *Molecular Physics* **108** (21–23), 3299–3311 (2010).

7

Semi-empirical Methods

Independent particle methods like Hartree–Fock and Kohn–Sham density functional theories (Chapters 3 and 6) employ a single determinant wave function composed of molecular orbitals. When the latter are expanded in a basis set, the variational principle leads to a matrix equation for determining the expansion coefficient and the total energy (Section 3.5):

$$\mathbf{FC} = \mathbf{SC}\varepsilon \tag{7.1}$$

Without any further approximations, the cost of forming the Fock (or Kohn–Sham) matrix scales formally as the fourth power of the number of basis functions, since this is the number of two-electron integrals necessary for constructing **F**. Semi-empirical methods reduce the computational cost by reducing the number of these integrals.[1–4] Although linear scaling methods can reduce the scaling of *ab initio* HF methods to $\sim M_{\text{basis}}$, this is only the limiting behavior in the large basis set limit, and *ab initio* methods will still require a significantly larger computational effort than semi-empirical methods.

The first step in reducing the computational problem is to consider only the valence electrons explicitly; the core electrons are accounted for by reducing the nuclear charge or introducing functions to model the combined repulsion due to the nuclei and core electrons. Furthermore, only a minimum basis set (the minimum number of functions necessary for accommodating the electrons in the neutral atom) is used for the valence electrons. Hydrogen thus has one basis function, and all atoms in the second and third rows of the periodic table have four basis functions (one s- and one set of p-orbitals, p_x, p_y and p_z).

The central assumption of semi-empirical methods is the *Zero Differential Overlap* (ZDO) approximation, which neglects all products of basis functions that depend on the same electron coordinates when located on different atoms. Denoting an atomic orbital on center A as μ_A (it is customary to denote basis functions with μ, ν, λ and σ in semi-empirical theory, while we are using χ_α, χ_β, χ_γ and χ_δ for *ab initio* methods), the ZDO approximation corresponds to $\mu_A \nu_B = 0$. Note that it is the *product* of functions on different atoms that is set equal to zero, not the *integral* over such a product. This has the following consequences (Equations (3.54), (3.59) and (3.60)):

1. The overlap matrix **S** is reduced to a unit matrix.
2. One-electron integrals involving three centers (two from the basis functions and one from the operator) are set to zero.

Introduction to Computational Chemistry, Third Edition. Frank Jensen.

Companion Website: http://www.wiley.com/go/jensen/computationalchemistry3

3. All three- and four-center two-electron integrals, which are by far the most numerous of the two-electron integrals, are neglected.

Since only one- and two-center integrals survive the ZDO approximation, it is possible to employ Slater-type orbitals (Chapter 5) as basis functions. To compensate for the above approximations, the remaining integrals are usually made into parameters, and their values are assigned based on other calculations or experimental data. Exactly how many integrals are neglected, and how the parameterization is done, defines the various semi-empirical methods.

Removal of the large majority of the integrals for constructing the Fock matrix in many cases transfers the computational bottleneck to solving the matrix Equation (7.1), which by standard diagonalization techniques scale as the cube of the matrix dimension, but it is possible to solve the variational problem by other methods that scale less steeply with system size.[5,6] The focus in this chapter is on the computational cost for construction of the Fock matrix **F**, and for simplicity we will assume that Equation (7.1) is solved by standard diagonalization methods.

In order to discuss the differences in semi-empirical methods, Equation (3.55) is rewritten with semi-empirical labels to give the following expression for a Fock matrix element, where a two-electron integral is abbreviated as $\langle\mu\nu|\lambda\sigma\rangle$ (Equation (3.61)):

$$\begin{aligned} F_{\mu\nu} &= h_{\mu\nu} + \sum_{\lambda\sigma}^{M_{\text{basis}}} D_{\lambda\sigma}(\langle\mu\nu|\lambda\sigma\rangle - \langle\mu\lambda|\nu\sigma\rangle) \\ h_{\mu\nu} &= \langle\mu|\mathbf{h}|\nu\rangle \end{aligned} \tag{7.2}$$

Approximations are made for the one- and two-electron parts as follows.

7.1 Neglect of Diatomic Differential Overlap (NDDO) Approximation

In the *Neglect of Diatomic Differential Overlap* (NDDO) approximation there are no further approximations than those mentioned above. Using μ and ν to denote either an s- or p-type (p_x, p_y or p_z) orbital, the NDDO approximation is defined by Equations (7.3) to (7.7).

Overlap integrals (Equation (3.54)):

$$S_{\mu\nu} = \langle\mu|\nu\rangle = \delta_{\mu\nu}\delta_{\text{AB}} \tag{7.3}$$

One-electron operator (Equation (3.25)):

$$\mathbf{h} = -\frac{1}{2}\nabla^2 - \sum_{A}^{N_{\text{nuclei}}} \frac{Z'_A}{|\mathbf{R}_A - \mathbf{r}|} = -\frac{1}{2}\nabla^2 - \sum_{A}^{N_{\text{nuclei}}} \mathbf{V}_A \tag{7.4}$$

where Z' denotes that the nuclear charge has been reduced by the number of core electrons.

One-electron integrals (Equation (3.61)):

$$\langle\mu_\text{A}|\mathbf{h}|\nu_\text{A}\rangle = \left\langle\mu_\text{A}\left|-\frac{1}{2}\nabla^2 - \mathbf{V}_\text{A}\right|\nu_\text{A}\right\rangle - \sum_{a\neq\text{A}}^{N_{\text{nuclei}}} \langle\mu_\text{A}|\mathbf{V}_a|\nu_\text{A}\rangle \tag{7.5}$$

$$\langle\mu_\text{A}|\mathbf{h}|\nu_\text{B}\rangle = \left\langle\mu_\text{A}\left|-\frac{1}{2}\nabla^2 - \mathbf{V}_\text{A} - \mathbf{V}_\text{B}\right|\nu_\text{B}\right\rangle \tag{7.6}$$

$$\langle\mu_\text{A}|\mathbf{V}_\text{C}|\nu_\text{B}\rangle = 0 \tag{7.7}$$

Due to the orthogonality of the atomic orbitals, the first one-center matrix element in Equation (7.5) is zero unless the two functions are identical:

$$\left\langle \mu_A \left| -\frac{1}{2}\nabla^2 - \mathbf{V}_A \right| \nu_A \right\rangle = \delta_{\mu\nu} \left\langle \mu_A \left| -\frac{1}{2}\nabla^2 - \mathbf{V}_A \right| \mu_A \right\rangle \tag{7.8}$$

Two-electron integrals (Equation (3.61)):

$$\langle \mu_A \nu_B | \lambda_C \sigma_D \rangle = \delta_{AC}\delta_{BD} \langle \mu_A \nu_B | \lambda_A \sigma_B \rangle \tag{7.9}$$

7.2 Intermediate Neglect of Differential Overlap (INDO) Approximation

The *Intermediate Neglect of Differential Overlap* (INDO) approximation neglects all two-center two-electron integrals that are not of the Coulomb type, in addition to those neglected by the NDDO approximations. Furthermore, in order to preserve rotational invariance, that is the total energy should be independent of a rotation of the coordinate system, integrals such as $\langle \mu_A | \mathbf{V}_a | \mu_A \rangle$ and $\langle \mu_A \nu_B | \mu_A \nu_B \rangle$ must be made independent of the orbital type (i.e. an integral involving a p-orbital must be the same as with an s-orbital). This has the consequence that one-electron integrals involving two different functions on the same atom and a $\mathbf{V}_a$ operator from another atom disappear. The INDO method involves the following additional approximations, besides those for NDDO.

One-electron integrals (Equation (7.5)):

$$\langle \mu_A | \mathbf{h} | \mu_A \rangle = \left\langle \mu_A \left| -\frac{1}{2}\nabla^2 - \mathbf{V}_A \right| \mu_A \right\rangle - \sum_{a \neq A}^{N_{\text{nuclei}}} \langle \mu_A | \mathbf{V}_a | \mu_A \rangle \tag{7.10}$$

$$\langle \mu_A | \mathbf{h} | \nu_A \rangle = -\delta_{\mu\nu} \sum_{a \neq A}^{N_{\text{nuclei}}} \langle \mu_A | \mathbf{V}_a | \mu_A \rangle \tag{7.11}$$

Two-electron integrals are approximated as in the following equation, except that one-center integrals $\langle \mu_A \lambda_A | \nu_A \sigma_A \rangle$ are preserved:

$$\langle \mu_A \nu_B | \lambda_C \sigma_D \rangle = \delta_{AC}\delta_{BD}\delta_{\mu\lambda}\delta_{\nu\sigma} \langle \mu_A \nu_B | \mu_A \nu_B \rangle \tag{7.12}$$

The surviving integrals are commonly denoted by γ:

$$\langle \mu_A \nu_A | \mu_A \nu_A \rangle = \langle \mu_A \mu_A | \mu_A \mu_A \rangle = \gamma_{AA} \tag{7.13}$$

$$\langle \mu_A \nu_B | \mu_A \nu_B \rangle = \gamma_{AB} \tag{7.14}$$

The INDO method is intermediate between the NDDO and CNDO methods in terms of approximations.

7.3 Complete Neglect of Differential Overlap (CNDO) Approximation

In the *Complete Neglect of Differential Overlap* (CNDO) approximation all the Coulomb two-electron integrals are subjected to the condition in Equation (7.12), including the one-center integrals, and are again parameterized as in Equations (7.13) and (7.14). The approximations for the one-electron integrals in CNDO are the same as for INDO. The *Pariser–Pople–Parr* (PPP) method can be considered as a CNDO approximation where only π-electrons are treated.

The main difference between CNDO, INDO and NDDO is in the treatment of the two-electron integrals. While CNDO and INDO reduce these to just two parameters (γ_{AA} and γ_{AB}), all the one- and two-center integrals are retained in the NDDO approximation. Within an *sp*-basis, however, there are only 27 different types of one- and two-center integrals, while the number rises to over 500 for a basis containing both *s*-, *p*- and *d*-functions.

7.4 Parameterization

An *ab initio* HF calculation with a minimum basis set is rarely able to give more than a qualitative description of the MOs, and it is of very limited value for predicting quantitative features. Introducing the ZDO approximation decreases the quality of the (already poor) wave function, that is a direct employment of the above NDDO/INDO/CNDO schemes is not useful. To "repair" the deficiencies due to the approximations, parameters are introduced in place of some or all of the integrals.

There are three methods that can be used for transforming the NDDO/INDO/CNDO approximations into working computational models;

1. The remaining integrals can be calculated from the functional form of the atomic orbitals.
2. The remaining integrals can be made into parameters, which are assigned values based on a few (usually atomic) experimental data.
3. The remaining integrals can be made into parameters, which are assigned values based on *fitting* to many (usually molecular) experimental data.

Method 2 derives specific atomic properties, such as ionization potentials and excitation energies, in terms of the parameters, and assigns their values accordingly. Method 3 takes the parameters as fitting constants and assigns their values based on a least-squares fit to a large set of experimental data, analogously to the fitting of force field parameters (Section 2.3).

The CNDO, INDO and NDDO *methods* use a combination of methods 1 and 2 for assigning parameters.[7] Some of the non-zero integrals are calculated from the atomic orbitals and others are assigned values based on atomic ionization potentials and electron affinities. Many different versions exist; they differ in the exact way in which the parameters have been derived. Some of the names associated with these methods are CNDO/1, CNDO/2, CNDO/S, CNDO/FK, CNDO/BW, INDO/1, INDO/2, INDO/S and SINDO1. These methods are rarely used in modern computational chemistry, mainly because the "modified" methods described below usually perform better. Exceptions are INDO-based methods, such as SINDO1[8] and INDO/S.[9] SINDO (*Symmetric orthogonalized INDO*) methods employ the INDO approximations described above, but not the ZDO approximation for the overlap matrix. The INDO/S method (INDO parameterized for *Spectroscopy*) is especially designed for calculating electronic spectra of large molecules or systems involving heavy atoms.

M. J. S. Dewar and coworker used a combination of methods 2 and 3 for assigning parameter values, resulting in a class of commonly used methods. The molecular data used for parameterization are geometries, heats of formation, dipole moments and ionization potentials. These methods are denoted "modified" as their parameters have been obtained by fitting.

7.4.1 Modified Intermediate Neglect of Differential Overlap (MINDO)

Three versions of *Modified Intermediate Neglect of Differential Overlap* (MINDO) models exist, MINDO/1, MINDO/2 and MINDO/3. The first two attempts at parameterizing INDO gave quite poor results, but MINDO/3, introduced in 1975,[10] produced the first general-purpose quantum chemical method that could successfully predict molecular properties at a relatively low computational cost. The parameterization of MINDO contains *diatomic* variables in the two-center one-electron term; thus the β_{AB} parameters must be derived for all *pairs* of bonded atoms. The I_μ parameters are ionization potentials:

$$\begin{aligned}\langle \mu_A|\mathbf{h}|\nu_B\rangle &= \left\langle \mu_A \left| -\frac{1}{2}\nabla^2 - \mathbf{V}_A - \mathbf{V}_B \right| \nu_B \right\rangle \\ &= S_{\mu\nu}\beta_{AB}(I_\mu + I_\nu) \\ S_{\mu\nu} &= \langle \mu_A|\nu_B\rangle\end{aligned} \tag{7.15}$$

MINDO/3 has been parameterized for H, B, C, N, O, F, Si, P, S and Cl, although certain combinations of these elements have been omitted. MINDO/3 is rarely used in modern computational chemistry, having been succeeded in accuracy by the NDDO methods below. Since there are parameters in MINDO that depend on two atoms, the number of parameters rises as the square of the number of elements. It is unlikely that MINDO will be parameterized in the future beyond those mentioned above.

7.4.2 Modified NDDO Models

The MNDO, AM1 and PM3 methods[11] are parameterizations of the NDDO model where the parameterization is in terms of *atomic* variables, that is referring only to the nature of a single atom. MNDO, AM1 and PM3 are derived from the same basic approximations (NDDO) and differ only in the way in which the core–core repulsion is treated and in how the parameters are assigned. Each method considers only the valence *s*- and *p*-functions, which are taken as Slater-type orbitals with corresponding exponents ζ_s and ζ_p.

The *one-center one-electron integrals* have a value corresponding to the energy of a single electron experiencing the nuclear charge (U_s or U_p) plus terms from the potential due to all the other nuclei in the system (Equation (7.5)). The latter is parameterized in terms of the (reduced) nuclear charges Z' and a two-electron integral:

$$h_{\mu\nu} = \langle \mu_A|\mathbf{h}|\nu_A\rangle = \delta_{\mu\nu}U_\mu - \sum_{a\neq A}^{N_{\text{nuclei}}} Z'_a\langle \mu_A s_a|\nu_A s_a\rangle \tag{7.16}$$

$$U_\mu = \left\langle \mu_A \left| -\frac{1}{2}\nabla^2 - \mathbf{V}_A \right| \mu_A \right\rangle \tag{7.17}$$

The *two-center one-electron integrals* given by Equation (7.6) are written as a product of the corresponding overlap integral multiplied by the average of two atomic "resonance" parameters, β:

$$\langle \mu_A|\mathbf{h}|\nu_B\rangle = \frac{1}{2}S_{\mu\nu}(\beta_\mu + \beta_\nu) \tag{7.18}$$

The overlap element $S_{\mu\nu}$ is calculated explicitly (note that this is not consistent with the ZDO approximation and the inclusion is the origin of the "modified" label).

There are only five types of *one-center two-electron integrals* surviving the NDDO approximation within an *sp*-basis (Equation (7.19)):

$$\begin{aligned} \langle ss|ss\rangle &= G_{ss} \\ \langle sp|sp\rangle &= G_{sp} \\ \langle ss|pp\rangle &= H_{sp} \\ \langle pp|pp\rangle &= G_{pp} \\ \langle pp'|pp'\rangle &= G_{p2} \end{aligned} \tag{7.19}$$

The G-type parameters are Coulomb terms, while the H parameter is an exchange integral. The G_{p2} integral involves two different types of p-functions (i.e. p_x, p_y or p_z).

There are a total of 22 different *two-center two-electron integrals* arising from an *sp*-basis, and these are modeled as interactions between multipoles. Electron 1 in an $\langle s\mu|s\mu\rangle$-type integral, for example, is modeled as a monopole, in an $\langle s\mu|p\mu\rangle$-type integral as a dipole and in a $\langle p\mu|p\mu\rangle$-type integral as a quadrupole. The dipole and quadrupole moments are generated as fractional charges located at specific points away from the nuclei, where the distance is determined by the orbital exponents ζ_s and ζ_p. The main reason for adapting a multipole expansion of these integrals was the limited computational resources available when these methods were developed. In the limit of the two nuclei being placed on top of each other, a two-center two-electron integral becomes a one-center two-electron integral, which puts boundary conditions on the functional form of the multipole interaction. The bottom line is that all two-center two-electron integrals are written in terms of the orbital exponents and the one-center two-electron parameters given in Equation (7.19).

The *core–core repulsion* is the repulsion between nuclear charges, properly reduced by the number of core electrons. The "exact" expression for this term is simply the product of the charges divided by the distance, $Z'_A Z'_B / R_{AB}$. Due to the inherent approximations in the NDDO method, however, this term is not cancelled by electron–electron terms at long distances, resulting in a net repulsion between uncharged molecules or atoms even when their wave functions do not overlap. The core–core term must consequently be modified to generate the proper limiting behavior, which means that two-electron integrals must be involved. The specific functional form depends on the exact method and is given below.

Each of the MNDO, AM1 and PM3 methods involves at least 12 parameters per atom: orbital exponents, $\zeta_{s/p}$; one-electron terms, $U_{s/p}$ and $\beta_{s/p}$; two-electron terms, G_{ss}, G_{sp}, G_{pp}, G_{p2}, H_{sp}; and parameters used in the core–core repulsion, α, and for the AM1 and PM3 methods also a, b and c constants, as described below.

7.4.3 Modified Neglect of Diatomic Overlap (MNDO)

The core–core repulsion of the *Modified Neglect of Diatomic Overlap* (MNDO) model[12] has the form given in

$$V_{nn}^{MNDO}(A,B) = Z'_A Z'_B \langle s_A s_A | s_B s_B \rangle \left(1 + e^{-\alpha_A R_{AB}} + e^{-\alpha_B R_{AB}}\right) \tag{7.20}$$

The α exponents are taken as fitting parameters.

Interactions involving O—H and N—H bonds are treated differently:

$$V_{nn}^{MNDO}(A,H) = Z'_A Z'_H \langle s_A s_A | s_H s_H \rangle \left(1 + R_{AH} e^{-\alpha_A R_{AH}} + e^{-\alpha_H R_{AH}}\right) \tag{7.21}$$

In addition, MNDO uses the approximation $\zeta_s = \zeta_p$ for some of the lighter elements. The G_{ss}, G_{sp}, G_{pp}, G_{p2} and H_{sp} parameters are taken from atomic spectra, while the others are fitted to molecular data. MNDOC[13,14] (C for correlation) has the same functional form as MNDO, but electron correlation is *explicitly* calculated by second-order perturbation theory. The derivation of the MNDOC parameters is done by fitting the correlated MNDOC results to experimental data. Electron correlation in MNDO is only included implicitly via the parameters, from fitting to experimental results. Since the training set only includes ground-state stable molecules, MNDO has problems treating systems where the importance of electron correlation is substantially different from "normal" molecules. MNDOC consequently performs significantly better for a system where this is not the case, such as transition structures and excited states.

7.4.4 Austin Model 1 (AM1)

After some experience with MNDO, it became clear that there were certain systematic errors. For example, the repulsion between two atoms that are 2–3 Å apart is too high. This has as a consequence that activation energies in general are too large. The source was traced to a too-repulsive interaction in the core–core potential. In order to remedy this, the core–core function was modified by adding Gaussian functions, and the whole model was reparameterized. The result was called *Austin Model 1* (AM1)[15], in honor of Dewar's move to the University of Austin at the time. The core–core repulsion of AM1 has the form given in

$$V_{\mathrm{nn}}^{\mathrm{AM1}}(\mathrm{A},\mathrm{B}) = V_{\mathrm{nn}}^{\mathrm{MNDO}}(\mathrm{A},\mathrm{B}) + \frac{Z'_{\mathrm{A}} Z'_{\mathrm{B}}}{R_{\mathrm{AB}}} \sum_k \left(a_{k\mathrm{A}} \mathrm{e}^{-b_{k\mathrm{A}}(R_{\mathrm{AB}}-c_{k\mathrm{A}})^2} + a_{k\mathrm{B}} \mathrm{e}^{-b_{k\mathrm{B}}(R_{\mathrm{AB}}-c_{k\mathrm{B}})^2} \right) \tag{7.22}$$

Here k is between 2 and 4, depending on the atom. It should be noted that the Gaussian functions were added more or less as patches on to the underlying parameters, which explains why a different number of Gaussians is used for each atom. As for MNDO, the G_{ss}, G_{sp}, G_{pp}, G_{p2} and H_{sp} parameters are taken from atomic spectra, while the rest, including the a_k, b_k and c_k constants, are fitted to molecular data.

7.4.5 Modified Neglect of Diatomic Overlap, Parametric Method Number 3 (PM3)

The parameterization of MNDO and AM1 had been done essentially by hand, taking the G_{ss}, G_{sp}, G_{pp}, G_{p2} and H_{sp} parameters from atomic data and varying the rest until a satisfactory fit had been obtained. Since the optimization was done by hand, only relatively few reference compounds could be included. J.J.P. Stewart made the optimization process automatic by deriving and implementing formulas for the derivative of a suitable error function with respect to the parameters.[16,17] All parameters could then be optimized simultaneously, including the two-electron terms, and a significantly larger training set with several hundred data could be employed. In this reparameterization, the AM1 expression for the core–core repulsion (Equation (7.22)) was kept, except that only two Gaussians were assigned to each atom. These Gaussian parameters were included as an integral part of the model and allowed to vary freely. The resulting method was denoted *Modified Neglect of Diatomic Overlap, Parametric Method Number 3* (MNDO-PM3 or PM3 for short) and is essentially AM1 with all the parameters fully optimized. In a sense, it is *the* best set of parameters (or at least a good local minimum) for the given set of experimental data. The optimization process, however, still requires some human intervention in selecting the experimental data and assigning appropriate weight factors to each set of data.

The MNDO, AM1 and PM3 methods have been parameterized for most of the main group elements,[18] and parameters for many of the transition metals are also being developed under the name

PM3(tm), which includes d-orbitals. The PM3(tm) set of parameters are determined exclusively from geometrical data (X-ray) since there are few reliable energetic data available for transition metal compounds.

7.4.6 The MNDO/d and AM1/d Methods

With only *s*- and *p*-functions included, the MNDO/AM1/PM3 methods are unable to treat a large part of the periodic table. Furthermore, from *ab initio* calculations it is known that d-orbitals significantly improve the results for compounds involving third row elements, especially hypervalent species. The main problem in extending the NDDO formalism to include d-orbitals is the significant increase in distinct two-electron integrals that ultimately must be assigned suitable values. For an *sp*-basis there are only five one-center two-electron integrals, while there are 17 in an *spd*-basis. Similarly, the number of two-center two-electron integrals rises from 22 to 491 when *d*-functions are included.

Thiel and Voityuk have constructed a workable NDDO model that also includes d-orbitals for use in connection with MNDO, called *MNDO/d*.[19,20] With reference to the above description for MNDO/AM1/PM3, it is clear that there are immediately three new parameters: ζ_d, U_d and β_d (Equations (7.17) and (7.18)). Of the 12 new one-center two-electron integrals, only one (G_{dd}) is taken as a freely varied parameter. The other 11 are calculated analytically based on pseudo-orbital exponents, which are assigned such that the analytical formulas regenerate G_{ss}, G_{pp} and G_{dd}.

With only *s*- and *p*-functions present, the two-center two-electron integrals can be modeled by multipoles up to order 4 (quadrupoles); however, with *d*-functions present multipoles up to order 16 must be included. In MNDO/d all multipoles beyond order 4 are neglected. The resulting MNDO/d method typically employs 15 parameters per atom and contains parameters for the following elements (beyond those already present in MNDO): Na, Mg, Al, Si, P, S, Cl, Br, I, Zn, Cd and Hg.

7.4.7 Parametric Method Numbers 6 and 7 (PM6 and PM7)

PM3 is likely to represent close to the limit of accuracy that can be obtained within the NDDO approximation. Building upon the experience with the PM3(tm) and MNDO/d models, J. J. P. Stewart has developed the PM6[21] and PM7[22] methods (PM4 and PM5 being unpublished experimental versions) with parameters covering most of the periodic table (70 elements). The main difference relative to PM3 is a change in the MNDO core–core repulsion term (Equation (7.20)) from atomic to diatomic parameters (x_{AB} and α_{AB}):

$$V_{\mathrm{nn}}^{\mathrm{PM6}}(\mathrm{A,B}) = Z'_{\mathrm{A}} Z'_{\mathrm{B}} \langle s_{\mathrm{A}} s_{\mathrm{A}} | s_{\mathrm{B}} s_{\mathrm{B}} \rangle \left(1 + x_{\mathrm{AB}} \mathrm{e}^{-\alpha_{\mathrm{AB}} R_{\mathrm{AB}}}\right) \tag{7.23}$$

Certain combinations of atoms have slightly modified expressions, analogous to Equation (7.21). The introduction of diatomic parameters was found to provide an increased accuracy, but significantly increases the number of core–core parameters, from ~70 atomic to ~5000 diatomic parameters. This additional flexibility made it possible to reduce the number of Gaussian repulsion terms (Equation (7.22)) to only one for each atom.

The PM7 method can be considered as a reparameterized version of PM6 where a number of errors have been corrected and the reference data has been extended to include periodic systems (crystals). The latter necessitated a modification of the γ_{AB} parameter (Equation (7.14)) to avoid infinite energies.

PM7 in addition employs specific energy terms for hydrogen bonding and dispersion, and this allowed removing the Gaussian core–core terms for all but the H, C, N, O atoms:

$$V^{\mathrm{PM7}}_{\mathrm{H\text{-}bond}}(\mathrm{A},\mathrm{B}) = -2.5(\cos\theta_{\mathrm{AHB}})^4\,\mathrm{e}^{-80(R_{\mathrm{AB}}-2.67)^2} \tag{7.24}$$

The parameterization strategy developed for PM3 employing analytical gradients of the error function with respect to the parameter enabled the optimization of the large number of diatomic parameters. The PM7 model performs relatively poorly for activation energies, with average errors ~45 kJ/mol, but a reparameterized model denoted PM7-TS reduced the average error to ~15 kJ/mol. A common flaw of all the NDDO methods is that rotational barriers for bonds that have partly double bond character are significantly too low. The barrier for rotation around the central bond in butadiene, for example, is calculated to be only 2–8 kJ/mol, in contrast to the experimental value of 25 kJ/mol.[23] Similarly, the rotational barrier around the C—N bond in amides is calculated to be 30–50 kJ/mol, which is roughly a factor of two smaller than the experimental value. A purely *ad hoc* fix has been made by adding a force field rotational term to the C—N bond that raises the value to ~100 kJ/mol and brings it into better agreement with experimental data.

7.4.8 Orthogonalization Models

The above NDDO methods enforce the ZDO approximation on the overlap matrix which reduces it to a unit matrix. Qualitative molecular orbital theory (Section 15.3), however, suggests that the orbital overlap elements play an important role, for example, in differentiating the energy change between bonding and antibonding orbitals upon bond formation. This suggests that the parameterization should not be done on the secular equation $\mathbf{FC} = \mathbf{C}\varepsilon$ (Equation (7.1) with $\mathbf{S} = \mathbf{I}$) but on the symmetrical orthogonalized equation $\mathbf{F}'\mathbf{C}' = \mathbf{C}'\varepsilon$ ($\mathbf{F}' = \mathbf{S}^{-1/2}\mathbf{F}\mathbf{S}^{-1/2}$, $\mathbf{C}' = \mathbf{S}^{1/2}\mathbf{C}$, Section 17.2.3) where overlap elements are included in the $\mathbf{F}'$ elements. By analyzing the effects of the orthogonalization on the one- and two-electron terms in the MNDO family of models, W. Thiel and coworkers proposed the *orthogonalization model* 1/2/3 (OM1/2/3).[24] The effect on the two-electron term was deemed to be small enough to be ignored, OM1 introduced changes in the one-center-one-electron term, while OM2 also changed the two-center one-electron terms. OM3 is a minor updated version of OM2. The OMx models have at the time of writing been parameterized for the elements H, C, N, O, F.[25,26]

7.5 Hückel Theory

7.5.1 Extended Hückel theory

The Hückel methods perform the parameterization on the Fock matrix elements (Equation (7.2)) and not at the integral level, as do NDDO/INDO/CNDO. This means that Hückel methods are non-iterative and they only require a single diagonalization of the Fock (Hückel) matrix. The *extended Hückel theory* (EHT) or *method* (EHM), developed primarily by R. Hoffmann, again only considers the valence electrons.[27] It makes use of Koopmans' theorem (Equation (3.50)) and assigns the diagonal elements in the $\mathbf{F}$ matrix to be atomic ionization potentials. The off-diagonal elements are parameterized as averages of the diagonal elements, weighted by an overlap integral. The overlap integrals are actually calculated, that is the ZDO approximation is not invoked. The basis functions are taken as Slater-type orbitals, with the exponents assigned according to the rules of Slater:[28]

$$F_{\mu\mu} = -I_\mu \tag{7.25}$$

$$F_{\mu\nu} = -\frac{1}{2}K(I_\mu + I_\nu)S_{\mu\nu} \tag{7.26}$$

The K constant is usually taken as 1.75, as this value reproduces the rotational barrier in ethane.

Since the diagonal elements only depend on the nature of the atom (i.e. the nuclear charge), this means, for example, that all carbon atoms have the same ability to attract electrons. After having performed a Hückel calculation, the actual number of electrons associated with atom A, ρ_A, can be calculated by a Mulliken population analysis according to (see Equations (10.5) and (10.6))

$$\rho_A = \sum_i^{MO} n_i \sum_{\alpha \in A}^{AO} \sum_\beta^{AO} c_{\alpha i} c_{\beta i} S_{\alpha\beta} \tag{7.27}$$

The effective (net) atomic charge Q_A is given as the (reduced) nuclear charge minus the electronic contribution:

$$Q_A = Z'_A - \rho_A \tag{7.28}$$

In general, it is unlikely that all carbon atoms have the exact same charge, that is owing to the different environments their ability to attract electrons is no longer equal. This may be argued to be inconsistent with the initial assumption of all carbons having the same diagonal elements in the Hückel matrix. In order to achieve "self-consistency", a diagonal element $F_{\mu\mu}$ belonging to atom A may be modified by the calculated atomic charge:

$$F_{\mu\mu} = -I_\mu + \omega Q_A \tag{7.29}$$

The ω parameter determines the weight of the charge on the diagonal elements. Since Q_A is calculated from the *results* (MO coefficients, Equation (7.27)) but enters the Hückel matrix that *produces* the results (by diagonalization), such schemes become iterative. Methods where the matrix elements are modified by the calculated charge are often called *charge iteration* or *self-consisten*t (Hückel) methods.

The main advantage of extended Hückel theory is that only atomic ionization potentials are required, and it is easily parameterized to the whole periodic table. Extended Hückel theory can be used for large systems involving transition metals, and until the development of DFTB methods (Section 7.6), this was often the only possible computational model. The very approximate method of extended Hückel theory makes it unsuitable for geometry optimizations without additional modifications[29] or for calculations of energetic features at any reasonable level of accuracy. It is primarily used for obtaining qualitatively correct MOs, which can, for example, be used as an initial guess of the density matrix for *ab initio* SCF calculations, or for use in connection with qualitative theories, as discussed in Chapter 15. Orbital energies (and thereby the total energy), however, in many cases show the correct trend for geometry perturbations corresponding to bond bending or torsional changes, and thus qualitative features regarding molecular shapes may often be predicted or rationalized from EHT calculations.

7.5.2 Simple Hückel Theory

In the simple Hückel model the approximations are taken to the limit.[30] Only planar conjugated systems are considered. The σ-orbitals, which are symmetric with respect to a reflection in the molecular plane, are neglected. Only electrons in the π-orbitals (antisymmetric with respect to the molecular mirror plane) are considered. The overlap matrix is taken as a unit matrix and a diagonal element of the **F** matrix is assigned a value of α, which depends on the atom type. Off-diagonal elements are taken either as β (depending on the two atom types) or zero, conditioned on whether the two atoms are "neighbors" (i.e. connected by a σ-bond) or not:

$$\begin{aligned} F_{\mu_A\mu_A} &= \alpha_A \\ F_{\mu_A\mu_B} &= \beta_{AB} \quad \text{(A and B are neighbors)} \\ F_{\mu_A\mu_B} &= 0 \quad \text{(A and B are not neighbors)} \end{aligned} \tag{7.30}$$

Atoms are assigned "types", much as in force field methods, that is the parameters depend on the nuclear charge and the bonding situation. The α_A and β_{AB} parameters for atom types A and B are related to the corresponding parameters for sp^2-hybridized carbon by means of the dimensionless constants h_A and k_{AB}:

$$\begin{aligned} \alpha_A &= \alpha_C + h_A\beta_{CC} \\ \beta_{AB} &= k_{AB}\beta_{CC} \end{aligned} \tag{7.31}$$

The carbon parameters α_C and β_{CC} are normally just denoted α and β, and are rarely assigned numerical values. Simple Hückel theory thus only considers the *connectivity* of the π-atoms: there is no information about the molecular geometry entering the calculations (e.g. whether some bonds are shorter or longer than others, or differences in bond angles).

In analogy to extended Hückel theory, there are also charge iterative methods for simple Hückel theory. The equivalent of Equation (7.27) is given in

$$\rho_A = \sum_i^{MO} n_i c_{Ai}^2 \tag{7.32}$$

Equation (7.29) becomes

$$\alpha'_A = \alpha_A + \omega(n_A - \rho_A)\beta \tag{7.33}$$

Here n_A is the number of π-electrons involved from atom A.

The Hückel method is essentially only used for educational purposes or for very qualitative orbital considerations. It has the ability to produce qualitatively correct MOs, involving a computational effort that is within reach of doing by hand.

7.6 Tight-Binding Density Functional Theory

The semi-empirical methods of the NDDO family can be considered as approximation to Hartree–Fock theory. Hartree–Fock theory has largely been replaced by Density Functional Theory, since DFT provides more accurate results for a similar computational cost. It is therefore not surprising that semi-empirical methods have also been developed as approximations to DFT. A widespread model is called *tight-binding DFT*, with the acronym DFTB,[31] with several different flavors in common use. In analogy with the NDDO methods, DFTB assumes a minimal basis set, neglects all three- and four-center integrals and only considers valence electrons explicitly. The core–core repulsion energy can be considered as the zeroth-order energy without any electronic contribution (i.e. equivalent to the bare nucleus repulsion $\mathbf{V}_{nn}$ in all-electron models), with the factor $^1/_2$ compensating for the summation over all pairs of atoms:

$$E_{\text{DFTB0}} = \frac{1}{2} \sum_{\text{A} \neq \text{B}} E_{\text{core}}^{\text{AB}} \tag{7.34}$$

The core–core repulsion is typically parameterized in terms of spline functions and fitted to corresponding all-electron DFT results. The valence electronic energy is calculated from the Fock (Kohn–Sham) energy matrix, where the surviving one- and two-center terms are parameterized as shown in

$$F_{\mu\nu} = \begin{cases} \varepsilon_{\mu}^{\text{atom}} \quad ; \quad \mu = \nu \\ \left\langle \mu_{\text{A}} \left| -\frac{1}{2}\nabla^2 + \mathbf{V}_{\text{A}} + \mathbf{V}_{\text{B}} \right| \nu_{\text{B}} \right\rangle \quad ; \quad \text{A} \neq \text{B} \\ 0 \quad ; \quad otherwise \end{cases} \tag{7.35}$$

DFTB has primarily been used for solid-state purposes, where the basis functions μ, ν should be more compact than the free atoms orbitals and they are generated by calculating atomic orbitals subject to a confinement potential. The diagonal elements ε^{atom} correspond to orbital energies for the free atoms to ensure proper dissociation, and can be considered as approximations to the ionization potential, in line with the EHT model (Equation (7.25)). The off-diagonal elements contain the effective potential $\mathbf{V}$ for the free atom, which includes the attraction to the nuclei and the electron–electron interaction (Coulomb and exchange–correlation term within the DFT framework). The energy and overlap matrix elements depend on internuclear distances, but can be precomputed in a tabulated form using a suitable exchange–correlation functional and interpolated for the actual geometry in a calculation.

The parameterization in terms of energy matrix elements and explicit consideration of the overlap matrix is analogous to the extended Hückel method, but the energy matrix elements are parameterized independently and are not assumed proportional to the overlap matrix elements as in Equation (7.26). The parameterization at the energy matrix element level means that the eigenvalue problem of Equation (7.1) can be solved by a single diagonalization. This provides the valence electronic energy as a simple sum of eigenvalues multiplied with occupation numbers and is combined with the core–core interaction to form the DFTB1 model:

$$E_{\text{DFTB1}} = \sum_{i=1}^{N} n_i \varepsilon_i + E_{\text{DFTB0}} \tag{7.36}$$

The DFTB1 energy can, in analogy with the extended Hückel method, be improved by including corrections due to charge equilibration (Equation (7.29)), to either second (DFTB2) or third (DFTB3) order. DFTB2 modifies the energy matrix elements by a term corresponding to the Mulliken charge Q on the atoms:

$$F_{\mu\nu} = F_{\mu\nu}^{0} + \frac{1}{2} S_{\mu\nu} \sum_{\text{C}} Q_{\text{C}} (\gamma_{\text{AC}} + \gamma_{\text{BC}}) \quad ; \quad \begin{matrix} \mu \in \text{A} \\ \nu \in \text{B} \end{matrix} \tag{7.37}$$

The γ function interpolates between R^{-1} in the long-distance limit, which is just the Coulomb interaction, and twice the chemical hardness U ($U = 2\eta$ = IP-EA, Section 16.2) in the R = 0 limit. A possible interpolation function is the Klopman–Ohno formula given in

$$\gamma_{\text{AB}} = \left(R_{\text{AB}}^2 + \frac{1}{4} \left(U_{\text{A}}^{-1} + U_{\text{B}}^{-1} \right) \right)^{-1/2} \tag{7.38}$$

The corresponding DFTB2 (alternative acronym is SCC-DFTB, for *Self Consistent Charge* DFTB) energy can be written as

$$E_{\text{DFTB2}} = E_{\text{DFTB1}} + \frac{1}{2}\sum_{A\neq B} Q_A Q_B \gamma_{AB} \tag{7.39}$$

Since the Mulliken charges depend on the LCAO coefficients (Equations (7.27) and (7.28)), this leads to an iterative scheme where the atomic charges Q are calculated by solving Equation (7.1) self-consistently with Fock matrix elements from Equation (7.37). Including charge corrections to third order defines the DFTB3 model:

$$E_{\text{DFTB3}} = E_{\text{DFTB2}} + \frac{1}{3}\sum_{A,B} Q_A^2 Q_B \left.\frac{\partial \gamma_{AB}}{\partial Q_A}\right|_{Q_A=0} \tag{7.40}$$

DFTB, in analogy with DFT, does not account for dispersion, but this can be added by empirical terms, as discussed in Section 6.5.7. Additional empirical energy terms may be added to improve on specific interactions described poorly by DFTB models, such as, for example, "halogen bonding" due to the positive electrostatic potential at the end of an R–X bond.[32]

7.7 Performance of Semi-empirical Methods

The electronic energy (including the core–core repulsion) calculated by semi-empirical methods is, in analogy with *ab initio* methods, the total energy relative to a situation where the nuclei (with their core electrons) and the valence electrons are infinitely separated. It is customary, however, to convert the electronic energy to a heat of formation by subtracting the electronic energy of the isolated atoms that make up the system and adding the experimental atomic heat of formation. It should be noted that thermodynamic corrections (e.g. zero-point energies, see Section 14.5.5) should *not* be added to the ΔH_f values, as these are included implicitly by the parameterization:

$$\Delta H_f(\text{molecule}) = E_{\text{elec}}(\text{molecule}) - \sum^{M_{\text{atoms}}} E_{\text{elec}}(\text{atoms}) - \sum^{M_{\text{atoms}}} \Delta H_f(\text{atoms}) \tag{7.41}$$

Mean absolute deviations for heat of formation for consistent data sets for NDDO methods are given in Table 7.1. Mean absolute deviations for bond distance are given in Table 7.2. Errors in bond angles are typically in the 5–7° range.

The method progression MNDO–AM1–PM3–PM6–PM7 in most cases reflects an increase in the number of parameters and better optimization with respect to reproducing the reference data, and it is thus expected that the error decreases in the same order. This is indeed the general trend in the above tables. The inclusion of solid-state reference data in addition to molecular data in the PM7 model, however, lead to a slight deterioration in the performance for some of the molecular species. The OM2/3 methods for these benchmarks display very similar performance as PM7.

Consistent benchmark data covering a variety of methods are scarce. Table 7.3 shows results for a subset of the GMTKN24/30 benchmarks comparing AM1, PM3/6/7, OM1/2/3, DFTB2 and the all-electron DFT method B3LYP. The GMTKN24/30 benchmark sets contain a selection of molecular properties covering relative energies, reaction and activation energies, ionization potentials,

Table 7.1 Mean absolute deviations for heat of formation (kJ/mol).[26]

Data set	Number	MNDO/d	AM1/d	PM3	PM6	PM7	OM1	OM2	OM3
HCNO	1141	49	39	23	19	16	35	20	20
Core	1572	50	41	26	19	17			
Extended	3163		141	122	88	85			
All	4417				79	86			
S22+S66	88				12	3.3		3.7	3.5

HCNO = compounds composed of H, C, N, O.
Core = HCNO set + elements F, P, S, Cl.
Extended = core + S-block, fourth and fifth period P-block, and selected other elements.
All = full periodic table.
S22+S66 = benchmarks containing intermolecular interactions.
MNDO, AM1, PM3, PM6 and PM7 data by courtesy of J. J. P. Stewart.

Table 7.2 Mean absolute deviations for bond distances (Å).

Data set	Number	MNDO/d	AM1/d	PM3	PM6	PM7
HCNO	313	0.019	0.021	0.018	0.018	0.019
Core	424	0.021	0.031	0.024	0.018	0.019
Extended	6605	0.100	0.090	0.084	0.078	0.073
All	9118	0.095	0.090	0.084	0.084	0.080

MNDO, AM1, PM3, PM6 and PM7 data by courtesy of J. J. P. Stewart.

Table 7.3 Mean absolute deviations for a subset of the GMTKN24 and GMTKN30 benchmarks having H, C, N, O, F elements only (in kJ/mol).[26,33]

	GMTKN24 (370 entries)								
	AM1	PM3	PM6	PM7	OM1	OM2	OM3	DFTB2	B3LYP
Without dispersion	57		61		41	30	28	58	20
With dispersion	51		60		35	26	27	57	19
	GMTKN30 (480 entries)								
Without dispersion	69	60	68	69	50	33	30		
With dispersion						32	30		

electron affinities, proton affinities, atomization energies, radical stabilization energies and water cluster binding energies.

The OM1/2/3 methods display a progression in accuracy and for these benchmarks perform better than the non-orthogonalized NDDO models AM1 and PM6/7; they also outperform the DFTB2 method. The OM3 results for these systems are only slightly inferior to the all-electron DFT B3LYP method.

7.8 Advantages and Limitations of Semi-empirical Methods

The parameterization of the NDDO models is performed by adjusting the constants involved in the different methods such that the results of HF calculations fit experimental data as closely as possible. This is in a sense wrong. We know that the HF method cannot give the correct result, even in the limit of an infinite basis set and without approximations. The HF results lack electron correlation, as discussed in Chapter 4, but the experimental data of course include such effects. This may be viewed as an advantage; the electron correlation effects are implicitly taken into account in the parameterization and we need not perform complicated calculations to improve deficiencies in the HF procedure. However, it becomes problematic when the HF wave function cannot describe the system even qualitatively correctly, as with, for example, biradicals and excited states. In such cases, additional flexibility can be introduced in the trial wave function by adding more Slater determinants, for example by means of a CI procedure (see Chapter 4 for details). However, electron correlation is then taken into account twice, once in the parameterization at the HF level and once explicitly by the CI calculation.

Semi-empirical methods share the advantages and disadvantages of force field methods: they perform best for systems where much experimental information is already available while results for totally unknown compound types are associated with much higher uncertainties. The dependence on experimental data is not as severe as for force field methods, owing to the more complex functional form of the model. The MNDO/AM1/PM3/OMx methods require only atomic parameters, while PM6 and PM7 in addition require some diatomic parameters, but not tri- and tetra-atomic parameters as do force field methods. Once a given atom, or atom pair, has been parameterized, all possible compound types involving this/these element(s) can be calculated. The smaller number of parameters and the more complex functional form has the disadvantage compared with force field methods that it is very difficult to "repair" a specific problem by reparameterization. The lack of a reasonable rotational barrier in amides, for example, cannot be attributed to an "improper" value for a single (or a few) parameter(s). Too low a rotational barrier in a force field model can easily be fixed by increasing the values of the corresponding torsional parameters. The clear advantage of semi-empirical methods over force field techniques is the ability to describe bond-breaking and bond-forming reactions.

DFTB methods bypass the problem of insufficient experimental data, since the parameterization is against all-electron DFT calculations. This of course means that any deficiencies in the chosen all-electron DFT method will be inherent by the DFTB model, but the errors due to the DFTB approximations themselves are in most cases much larger than those in the parent DFT model.

Semi-empirical methods are zero-dimensional, just as force field methods are. There is no way of assessing the reliability of a given result within the method. This is due to the selection of a minimum basis set. The only way of judging results is by calibration, that is by comparing the accuracy of other calculations on similar systems with experimental data.

Semi-empirical models provide a method for calculating the electronic wave function, which may be used for predicting a variety of properties. There is nothing to hinder the calculation of, say, the polarizability of a molecule (the second derivative of the energy with respect to an external electric field), although it is known from *ab initio* calculations that good results require a large polarized basis set including diffuse functions and the inclusion of electron correlation. Semi-empirical methods only employ a minimum basis (lacking polarization and diffuse functions), electron correlation is only included implicitly by the parameters and no polarizability data have been used for deriving the parameters. Whether such calculations can produce reasonable results, as compared with experimental data, is questionable, and careful calibration is certainly required. Again it should be emphasized: *the ability to perform a calculation is no guarantee that the results can be trusted!*

References

1 J. Sadley, *Semi-Empirical Methods of Quantum Chemistry* (John Wiley & Sons, 1985).
2 W. Thiel, *Wiley Interdisciplinary Reviews – Computational Molecular Science* **4** (2), 145–157 (2014).
3 T. Bredow and K. Jug, *Theoretical Chemistry Accounts* **113** (1), 1–14 (2005).
4 M. C. Zerner, *Reviews in Computational Chemistry* **2**, 313 (1991).
5 J. J. P. Stewart, *International Journal of Quantum Chemistry* **58** (2), 133–146 (1996).
6 A. D. Daniels and G. E. Scuseria, *Journal of Chemical Physics* **110** (3), 1321–1328 (1999).
7 W. P. Anderson, T. R. Cundari and M. C. Zerner, *International Journal of Quantum Chemistry* **39** (1), 31–45 (1991).
8 J. Li, P. C. Demello and K. Jug, *Journal of Computational Chemistry* **13** (1), 85–92 (1992).
9 M. Kotzian, N. Rosch and M. C. Zerner, *Theoretica Chimica Acta* **81** (4–5), 201–222 (1992).
10 R. C. Bingham, M. J. S. Dewar and D. H. Lo, *Journal of the American Chemical Society* **97** (6), 1285–1293 (1975).
11 J. J. P. Stewart, *Reviews in Computational Chemistry* **1**, 45 (1990).
12 M. J. S. Dewar and W. Thiel, *Journal of the American Chemical Society* **99** (15), 4899–4907 (1977).
13 W. Thiel, *Journal of the American Chemical Society* **103** (6), 1413–1420 (1981).
14 A. Schweig and W. Thiel, *Journal of the American Chemical Society* **103** (6), 1425–1431 (1981).
15 M. J. S. Dewar, E. G. Zoebisch, E. F. Healy and J. J. P. Stewart, *Journal of the American Chemical Society* **107** (13), 3902–3909 (1985).
16 J. J. P. Stewart, *Journal of Computational Chemistry* **10** (2), 209–220 (1989).
17 J. J. P. Stewart, *Journal of Computational Chemistry* **10** (2), 221–264 (1989).
18 J. J. P. Stewart, *Journal of Molecular Modeling* **10** (2), 155–164 (2004).
19 W. Thiel and A. A. Voityuk, *Journal of Physical Chemistry* **100** (2), 616–626 (1996).
20 W. Thiel, in *Advances in Chemical Physics, Vol.* **Xciii**: *New Methods in Computational Quantum Mechanics*, edited by I. Prigogine and S. A. Rice (John Wiley & Sons, 1996), Vol. 93, pp. 703–757.
21 J. J. P. Stewart, *Journal of Molecular Modeling* **13** (12), 1173–1213 (2007).
22 J. J. P. Stewart, *Journal of Molecular Modeling* **19** (1), 1–32 (2013).
23 R. Engeln, D. Consalvo and J. Reuss, *Chemical Physics* **160** (3), 427–433 (1992).
24 W. Weber and W. Thiel, *Theoretical Chemistry Accounts* **103** (6), 495–506 (2000).
25 P. O. Dral, X. Wu, L. Spoerkel, A. Koslowski, W. Weber, R. Steiger, M. Scholten and W. Thiel, *Journal of Chemical Theory and Computation* **12** (3), 1082–1096 (2016).
26 P. O. Dral, X. Wu, L. Spoerkel, A. Koslowski and W. Thiel, *Journal of Chemical Theory and Computation* **12** (3), 1097–1120 (2016).
27 R. Hoffmann, *Journal of Chemical Physics* **39** (6), 1397 (1963).
28 J. C. Slater, *Physical Review* **36** (1), 0057–0064 (1930).
29 S. L. Dixon and P. C. Jurs, *Journal of Computational Chemistry* **15** (7), 733–746 (1994).
30 K. Yates, *Hückel Molecular Orbital Theory* (Academic Press, 1978).
31 M. Wahiduzzaman, A. F. Oliveira, P. Philipsen, L. Zhechkov, E. van Lenthe, H. A. Witek and T. Heine, *Journal of Chemical Theory and Computation* **9** (9), 4006–4017 (2013).
32 M. Kubillus, T. Kubar, M. Gaus, J. Rezac and M. Elstner, *Journal of Chemical Theory and Computation* **11** (1), 332–342 (2015).
33 M. Korth and W. Thiel, *Journal of Chemical Theory and Computation* **7** (9), 2929–2936 (2011).

8

Valence Bond Methods

Essentially all practical calculations for generating solutions to the electronic Schrödinger equation have been performed with molecular orbital methods. The zeroth-order wave function is constructed as a single Slater determinant and the MOs are expanded in a set of atomic orbitals, the basis set. In a subsequent step the wave function may be improved by adding electron correlation with either CI, MP or CC methods. There are two characteristics of such approaches: (1) the one-electron functions, the MOs, are delocalized over the whole molecule and (2) an accurate treatment of the electron correlation requires many (millions or billions) "excited" Slater determinants. The delocalized nature of the MOs is partly a consequence of choosing the Lagrange multiplier matrix to be diagonal (canonical orbitals, Equation (3.43)); they may in a subsequent step be mixed to form localized orbitals (see Section 10.4) without affecting the total wave function. Such a localization, however, is not unique. Furthermore, delocalized MOs are at variance with the basic concept, especially in organic chemistry, that molecules are composed of structural units (functional groups), which to a very good approximation are constant from molecule to molecule. The MOs for propane and butane, for example, are quite different, although "common" knowledge is that they contain CH_3 and CH_2 units that in terms of structure and reactivity are very similar for the two molecules. A description of the electronic wave function as having electrons in orbitals formed as linear combinations of all (in principle) atomic orbitals is also at variance with the chemical language of molecules being composed of atoms held together by bonds, where the bonds are formed by pairing unpaired electrons contained in atomic orbitals. Finally, when electron correlation is important (as is usually the case), the need to include many Slater determinants obscures the picture of electrons residing in orbitals.

There is an equivalent way of generating solutions to the electronic Schrödinger equation that conceptually is much closer to the experimentalist's language, known as *Valence Bond* (VB) theory.[1,2] We will start by illustrating the concepts for the H_2 molecule and note how it differs from MO methods.

Introduction to Computational Chemistry, Third Edition. Frank Jensen.

Companion Website: http://www.wiley.com/go/jensen/computationalchemistry3

8.1 Classical Valence Bond Theory

A single-determinant MO wave function for the H_2 molecule within a minimum basis consisting of a single *s*-function on each nucleus is given by (see also Section 4.3)

$$\Phi_0 = \begin{vmatrix} \phi_1(1) & \bar{\phi}_1(1) \\ \phi_1(2) & \bar{\phi}_1(2) \end{vmatrix}$$
$$\phi_1 = (\chi_A + \chi_B)\alpha \quad ; \quad \bar{\phi}_1 = (\chi_A + \chi_B)\beta \tag{8.1}$$

We have here ignored the normalization constants. The Slater determinant can be expanded in AOs, as shown below, with the order of the functions reflecting the electron coordinate:

$$\begin{aligned} \Phi_0 &= \phi_1\bar{\phi}_1 - \bar{\phi}_1\phi_1 = (\phi_1\phi_1)\,[\alpha\beta - \beta\alpha] \\ \Phi_0 &= (\chi_A + \chi_B)(\chi_A + \chi_B)[\alpha\beta - \beta\alpha] \\ \Phi_0 &= (\chi_A\chi_A + \chi_B\chi_B + \chi_A\chi_B + \chi_B\chi_A)[\alpha\beta - \beta\alpha] \end{aligned} \tag{8.2}$$

This shows that the HF wave function consists of equal amounts of ionic ($\chi_A\chi_A$ and $\chi_B\chi_B$) and covalent ($\chi_A\chi_B$ and $\chi_B\chi_A$) terms. In the dissociation limit only the covalent terms are correct, but the single-determinant description does not allow the ratio of covalent to ionic terms to vary. In order to provide a correct description, a second determinant is necessary:

$$\begin{aligned} \Phi_1 &= \begin{vmatrix} \phi_2(1) & \bar{\phi}_2(1) \\ \phi_2(2) & \bar{\phi}_2(2) \end{vmatrix} \\ \phi_2 &= (\chi_A - \chi_B)\alpha \quad ; \quad \bar{\phi}_2 = (\chi_A - \chi_B)\beta \\ \Phi_1 &= (\chi_A\chi_A + \chi_B\chi_B - \chi_A\chi_B - \chi_B\chi_A)[\alpha\beta - \beta\alpha] \end{aligned} \tag{8.3}$$

By including the doubly excited determinant Φ_1, built from the antibonding MO, the amounts of covalent and ionic terms may be varied, and this is determined completely by the variational principle (Equation (4.26)):

$$\begin{aligned} \Psi_{CI} &= a_0\Phi_0 + a_1\Phi_1 \\ \Psi_{CI} &= \{(a_0 - a_1)(\chi_A\chi_B + \chi_B\chi_A) + (a_0 + a_1)(\chi_A\chi_A + \chi_B\chi_B)\}[\alpha\beta - \beta\alpha] \end{aligned} \tag{8.4}$$

This two-configurational CI wave function allows a qualitatively correct description of the H_2 molecule at all distances and in the dissociation limit, where the weights of the two configurations become equal.

The classical VB wave function, on the other hand, is built from the atomic fragments by coupling the unpaired electrons to form a bond. In the H_2 case, the two electrons are coupled into a singlet pair, properly antisymmetrized. The simplest VB description, known as a *Heitler–London* (HL) function, includes only the two covalent terms in the HF wave function:

$$\Phi_{HL}^{cov} = (\chi_A\chi_B + \chi_B\chi_A)[\alpha\beta - \beta\alpha] \tag{8.5}$$

Just as the single-determinant MO wave function can be improved by including excited determinants, the simple VB-HL function can be improved by adding terms that correspond to higher energy configurations for the fragments, in this case ionic structures:

$$\Phi_{HL}^{ion} = (\chi_A\chi_A + \chi_B\chi_B)[\alpha\beta - \beta\alpha] \tag{8.6}$$

$$\Psi_{HL} = a_0\Phi_{HL}^{cov} + a_1\Phi_{HL}^{ion} \tag{8.7}$$

The final description, either in terms of a CI wave function written as a linear combination of two determinants built from delocalized MOs (Equation (8.4)) or as a VB wave function written in terms of two VB-HL structures composed of AOs (Equation (8.7)), is identical.

For the H_2 system, the amount of ionic HL structures determined by the variational principle is 44%, close to the MO-HF value of 50%. The need for including large amounts of ionic structures in the VB formalism is due to the fact that pure atomic orbitals are used.

Consider now a covalent VB function built from "atomic" orbitals that are allowed to distort from the pure atomic shape:

$$\begin{aligned}\Phi_{\text{CF}} &= (\phi_{\text{A}}\phi_{\text{B}} + \phi_{\text{B}}\phi_{\text{A}})[\alpha\beta - \beta\alpha] \\ \phi_{\text{A}} &= \chi_{\text{A}} + c\chi_{\text{B}} \quad ; \quad \phi_{\text{B}} = \chi_{\text{B}} + c\chi_{\text{A}}\end{aligned} \tag{8.8}$$

Such a VB function is known as a *Coulson–Fischer* (CF) type. The c constant is fairly small (for H_2, c is ~0.04), but by allowing the VB orbitals to adopt the optimum shape, the need for ionic VB structures is strongly reduced. Note that the two VB orbitals in Equation (8.8) are not orthogonal – the overlap is given by

$$\begin{aligned}\langle\phi_{\text{A}}|\phi_{\text{B}}\rangle &= (1+c^2)\langle\chi_{\text{A}}|\chi_{\text{B}}\rangle + 2c(\langle\chi_{\text{A}}|\chi_{\text{A}}\rangle + \langle\chi_{\text{B}}|\chi_{\text{B}}\rangle) \\ \langle\phi_{\text{A}}|\phi_{\text{B}}\rangle &= (1+c^2)S_{\text{AB}} + 4c\end{aligned} \tag{8.9}$$

Compared with the overlap of the undistorted atomic orbitals used in the HL wave function, which is just S_{AB}, it is seen that the overlap is increased (c is positive), that is the orbitals distort such that they overlap better in order to make a bond. Although the distortion is fairly small (a few percent), this effectively eliminates the need for including ionic VB terms. When c is variationally optimized, the MO-CI, VB-HL and VB-CF wave functions (Equations (8.4), (8.7) and (8.8)) are all completely equivalent. The MO approach incorporates the flexibility in terms of an "excited" determinant, the VB-HL in terms of "ionic" structures and the VB-CF in terms of "distorted" atomic orbitals.

In the MO-CI language, the correct dissociation of a single bond requires addition of a second doubly excited determinant to the wave function. The VB-CF wave function, on the other hand, dissociates smoothly to the correct limit, the VB-orbitals simply reverting to their pure atomic shapes, with the overlap disappearing.

8.2 Spin-Coupled Valence Bond Theory

The generalization of a Coulson–Fischer-type wave function to the molecular case with an arbitrary-size basis set is known as *Spin-Coupled Valence Bond* (SCVB) theory.[2–4]

It is again instructive to compare with the traditional MO approach, taking the CH_4 molecule as an example. The MO single-determinant description (RHF, which is identical to UHF near the equilibrium geometry) of the valence orbitals is in terms of four delocalized orbitals, each occupied by two electrons with opposite spin. The C—H bonding is described by four different, orthogonal molecular orbitals, each expanded in a set of AOs:

$$\begin{aligned}\Phi^{\text{CH}_4}_{\text{valence-MO}} &= \mathbf{A}[\phi_1\bar{\phi}_1\phi_2\bar{\phi}_2\phi_3\bar{\phi}_3\phi_4\bar{\phi}_4] \\ \phi_i &= \sum_{\alpha=1}^{M_{\text{basis}}} c_{\alpha i}\chi_\alpha\end{aligned} \tag{8.10}$$

Here $\mathbf{A}$ is the antisymmetrizing operator (Equation (3.22)) and a bar above an MO indicates that the electron has a β spin function, where no bar indicates an α spin function.

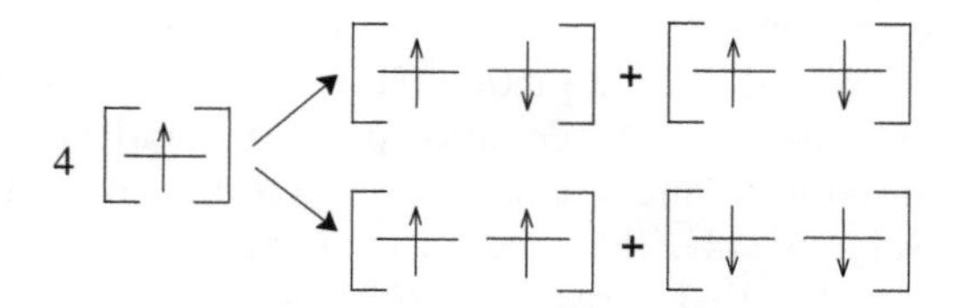

Figure 8.1 Two possible schemes for coupling four electrons to an overall singlet.

The SCVB description, on the other hand, considers the four bonds in CH_4 as arising from coupling of a single electron at each of the four hydrogen atoms with a single unpaired electron at the carbon atom. Since the ground state of the carbon atom is a triplet, corresponding to the electron configuration $1s^22s^22p^2$, the first step is formation of four equivalent "hybrid" orbitals by mixing three parts *p*-function with one part *s*-function, generating four equivalent "sp^3-hybrid" orbitals. Each of these singly occupied hybrid orbitals can then couple with a hydrogen atom to form four equivalent C—H bonds. The electron spins are coupled such that the total spin is a singlet, which can be done in several different ways. The coupling of four electrons to a total singlet state, for example, can be done either by coupling two electrons in a pair to a singlet, and then coupling two singlet pairs, or by first coupling two electrons in a pair to a triplet, and subsequently coupling two triplet pairs to an overall singlet (see Figure 8.1).

The $\Theta_{S,i}^N$ symbol is used to designate the *i*th combination of spin functions coupling *N* electrons to give an overall spin of *S*, and there are f_S^N number of ways of doing this. The value of f_S^N is given by

$$f_S^N = \frac{(2S+1)\,N!}{\left(\frac{1}{2}N+S+1\right)!\left(\frac{1}{2}N-S\right)!} \tag{8.11}$$

For a singlet wave function ($S = 0$), the number of coupling schemes for *N* electrons is given in Table 8.1.

For the eight valence electrons in CH_4 there are 14 possible spin couplings resulting in an overall singlet state. The full SCVB function may be written (again neglecting normalization) as

$$\begin{aligned}\Phi_{\text{valence-SCVB}}^{CH_4} &= \sum_{i=1}^{14} a_i \mathbf{A}\left\{[\phi_1\phi_2\phi_3\phi_4\phi_5\phi_6\phi_7\phi_8]\,\Theta_{0,i}^N\right\} \\ \phi_i &= \sum_{\alpha=1}^{M_{\text{basis}}} c_{\alpha i}\chi_\alpha\end{aligned} \tag{8.12}$$

Table 8.1 Number of possible spin coupling schemes for achieving an overall singlet state.

N	f_0^N
2	1
4	2
6	5
8	14
10	42
12	132
14	429

Figure 8.2 A representation of the dominating spin coupling in CH_4.

Figure 8.3 Molecular orbital energies in benzene.

There are now eight different spatial orbitals, ϕ_i, four of which are essentially carbon sp^3-hybrid orbitals, with the other four being close to atomic hydrogen s-orbitals. The expansion of each of the VB-orbitals in terms of *all* the basis functions located on *all* the nuclei allows the orbitals to distort from the pure atomic shape. The SCVB wave function is variationally optimized, both with respect to the VB-orbital coefficients $c_{\alpha i}$ and the spin coupling coefficients a_i. The result is that a complete set of optimum "distorted" atomic orbitals is determined together with the weight of the different spin couplings. Each spin coupling term (in the so-called *Rumer* basis) is closely related to the concept of a resonance structure used in organic chemistry textbooks. An SCVB calculation of CH_4 gives as a result that one of the spin coupling schemes completely dominates the wave function, namely that corresponding to the electron pair in each of the C—H bonds being singlet-coupled. This is the quantum mechanical analog of the graphical representation of CH_4 shown in Figure 8.2. Each of the lines represents a singlet-coupled electron pair between two orbitals that strongly overlap to form a bond, and the drawing in Figure 8.2 is the only important "resonance" form.

Consider now the π-system in benzene. The MO approach will generate linear combinations of the atomic p-orbitals, producing six π-orbitals delocalized over the whole molecule with four different orbital energies (two sets of degenerate orbitals). The stability of benzene can be attributed to the large gap between the HOMO and LUMO orbitals (see Figure 8.3).

A SCVB calculation considering only the coupling of the six π-electrons gives a somewhat different picture. The VB π-orbitals are strongly localized on each carbon, resembling p-orbitals that are slightly distorted in the direction of the nearest neighbor atoms. It is now found that the five spin coupling combinations shown in Figure 8.4 are important, where a bold line indicates two electrons coupled into a singlet pair.

Each of the two first VB structures contributes ~40% to the wave function and each of the remaining three contributes ~6%.[5] The stability of benzene in the SCVB picture is due to *resonance* between these VB structures. It is furthermore straightforward to calculate the resonance energy by comparing

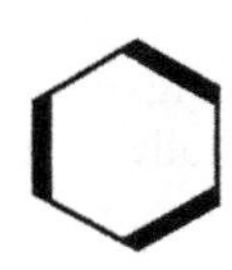

Figure 8.4 Representations of important spin coupling schemes in benzene.

the full SCVB energy with that calculated from a VB wave function omitting certain spin coupling functions.

The MO wave function for CH_4 can be improved by adding configurations corresponding to excited determinants, that is replacing occupied MOs with virtual MOs. Allowing all excitations in the minimal basis valence space and performing a full optimization corresponds to an [8,8]-CASSCF wave function (Section 4.6). Similarly, the SCVB wave function in Equation (8.12) can be improved by adding ionic VB structures such as CH_3^-/H^+ and CH_3^+/H^-, and this corresponds to exciting an electron from one of the singly occupied VB orbitals into another VB orbital, thereby making it doubly occupied. The importance of these excited/ionic terms can again be determined by the variational principle. If all such ionic terms are included, the fully optimized SCVB+CI wave function is for all practical purposes identical to that obtained by the MO-CASSCF approach (the only difference is a possible slight difference in the description of the carbon 1s-core orbital).

Both types of wave function provide essentially the same total energy, and thus include the same amount of electron correlation. The MO-CASSCF wave function attributes the electron correlation to the interaction of 1764 configurations, the Hartree–Fock reference and 1763 excited configurations, with each of the 1763 configurations providing only a small amount of the correlation energy. The SCVB wave function (which includes only one resonance structure), on the other hand, contains 90+% of the correlation energy, and only a few percent is attributed to "excited" structures. The ability of SCVB wave functions to include electron correlation is due to the fact that the VB orbitals are strongly localized and, since they are occupied by only one electron, they have the built-in feature of electrons avoiding each other. In a sense, an SCVB wave function is *the* best wave function that can be constructed in terms of *products* of *spatial* orbitals. By allowing the orbitals to become non-orthogonal, the large majority (80–90%) of what is called electron correlation in an MO approach can be included in a *single*-determinant wave function composed of spatial orbitals, multiplied by proper spin coupling functions.

There are a number of technical complications associated with optimizing the SCVB wave function due to the non-orthogonal orbitals. The MO-CI or MO-CASSCF approaches simplify considerably owing to the orthogonality of the MOs, and thereby also of the Slater determinants. The MO-CI matrix is sparse, since only Slater determinants differing by at most two orbitals can be interacting, while the corresponding SCVB matrix is dense. Computationally, the optimization of an SCVB wave function, where N electrons are coupled in all possible ways, is similar to that required for constructing an $[N,N]$-CASSCF wave function. This effectively limits the size of SCVB wave functions to coupling of 12–16 electrons, although algorithmic improvements have the promise to increase these values.[6] The actual optimization of the wave function is usually done by a second-order expansion of the energy in terms of orbital and spin coupling coefficients, and employing a Newton–Raphson-type scheme, analogously to MCSCF methods (Section 4.6). The non-orthogonal orbitals have the disadvantage that it is difficult to add dynamical correlation on top of an SCVB wave function by perturbation or coupled cluster theory, although (non-orthogonal) CI methods are straightforward. SCVB+CI approaches may also be used to describe excited states, analogously to MO-CI methods.

It should be emphasized that the results obtained from an $[N,N]$-CASSCF and a corresponding N-electron SCVB wave function (or SCVB+CI and MRCI) are virtually identical. The difference is in the way the results can be analyzed. Molecules in the SCVB picture are composed of atoms held together by bonds, where bonds are formed by (singlet) coupling of the electron spins between (two) overlapping orbitals. These orbitals are strongly localized, usually on a single atom, and are basically atomic orbitals slightly distorted by the presence of the other atoms in the molecule. The VB description of a bond as the result of two overlapping orbitals is in contrast to the MO approach, where a bond between two atoms arises as a sum over (small) contributions from many delocalized molecular orbitals. Furthermore, the weights of the different spin couplings in an SCVB wave function carries a direct analogy with chemical concepts such as "resonance" structures.

The SCVB method is a valuable tool for providing *insight* into the problem. This is to a certain extent also possible from an MO-type wave function by localizing the orbitals or by analyzing the natural orbitals (see Sections 10.4 and 10.5 for details). However, there is no unique method for producing localized orbitals and different methods may give different orbitals. Natural orbitals are analogous to canonical orbitals delocalized over the whole molecule. The SCVB orbitals, by contrast, are uniquely determined by the variational procedure, and there is no freedom to further transforming them by making linear combinations without destroying the variational property.

The primary feature of SCVB is the use of non-orthogonal orbitals, which allows a much more compact representation of the wave function. An MO-CI wave function of a certain quality may involve many thousands of Slater determinants, while a similar-quality VB wave function may be written as only a handful of "resonating" VB structures. Furthermore, the VB orbitals, and spin couplings, of a C—H bond in, say, propane and butane are very similar, in contrast to the vastly different MO descriptions of the two systems. The VB picture is thus much closer to the traditional descriptive language used with (organic) molecules composed of functional groups. The widespread availability of programs for performing CASSCF calculations, and the fact that CASSCF calculations are computationally more efficient owing to the orthogonality of the MOs, have prompted developments of schemes for transforming CASSCF wave functions to VB structures, denoted CASVB.[5,7] A corresponding procedure using orthogonal orbitals (which introduce large weights of ionic structures) has also been reported.[8]

8.3 Generalized Valence Bond Theory

The SCVB wave function allows all possible spin couplings to take place and has no restrictions on the form of the orbitals. The *Generalized Valence Bond* (GVB) method can be considered as a reduced version of the full problem where only certain subsets of spin couplings are allowed.[9] For a typical case of a singlet system, the GVB method has two (non-orthogonal) orbitals assigned to each bond, and each pair of electrons in a bond are required to couple to a singlet pair. The coupling of such singlet pairs will then give the overall singlet spin state. This is known as *Perfect Pairing* (PP) and is one of the many possible spin coupling schemes; such two-electron two-orbital pairs are called *geminal pairs*. Just as an orbital is a wave function for one electron, a geminal is a wave function for two electrons. In order to reduce the computational problem, the *Strong Orthogonality* (SO) condition can furthermore be imposed on the GVB wave function. This means that orbitals belonging to different pairs are required to be orthogonal. While the perfect pairing coupling typically is the largest contribution to the full SCVB wave function, the strong orthogonality constraint is often a quite poor approximation, and may lead to artefacts. For diazomethane, for example, the SCVB wave function is dominated (91%) by the PP coupling, leading to the conclusion that the molecule has essentially normal C=N and N=N π-bonds, perpendicular to the plane defined by the CH_2 moiety.[10] Taking into account also the in-plane bonding, this suggest that diazomethane is best described with a triple bond between the two nitrogens, thereby making the central nitrogen "hypervalent", as illustrated in Figure 8.5.

There are strong overlaps between the VB orbitals; the *smallest* overlap (between the carbon and terminal nitrogen) is ~0.4 and that between the two orbitals on the central nitrogen is ~0.9. The GVB-SOPP approach, however, forces these geminal pairs to be orthogonal, leading to the

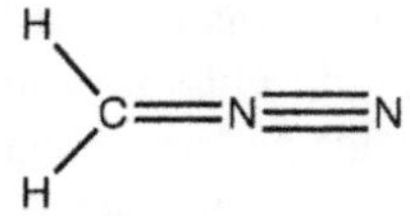

H
C=N≡N
H

Figure 8.5 A representation of the SCVB wave function for diazomethane.

H
C·—N≡N·
H

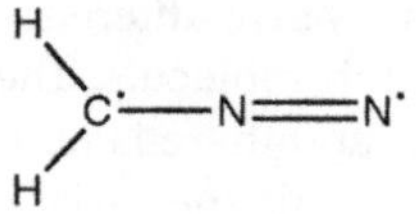

Figure 8.6 A representation of the GVB wave function for diazomethane.

conclusion that the electronic structure of diazomethane has a very strong diradical nature, as illustrated in Figure 8.6.

References

1 S. Shaik and P. C. Hiberty, in *Reviews in Computational Chemistry*, edited by K. B. Lipkowitz, R. Larter and T. R. Cundari (John Wiley &Sons, 2004), Vol. 20, pp. 1–100.
2 W. Wu, P. Su, S. Shaik and P. C. Hiberty, *Chemical Reviews* **111** (11), 7557–7593 (2011).
3 D. L. Cooper, J. Gerratt and M. Raimondi, *Chemical Reviews* **91** (5), 929–964 (1991).
4 J. Gerratt, D. L. Cooper, P. B. Karadakov and M. Raimondi, *Chemical Society Reviews* **26** (2), 87–100 (1997).
5 D. L. Cooper, T. Thorsteinsson and J. Gerratt, *International Journal of Quantum Chemistry* **65** (5), 439–451 (1997).
6 J. Olsen, *Journal of Chemical Physics* **143** (11), 114102 (2015).
7 J. Song, Z. Chen, S. Shaik and W. Wu, *Journal of Computational Chemistry* **34** (1), 38–48 (2013).
8 K. Hirao, H. Nakano, K. Nakayama and M. Dupuis, *Journal of Chemical Physics* **105** (20), 9227–9239 (1996).
9 W. A. Goddard and L. B. Harding, *Annual Review of Physical Chemistry* **29**, 363–396 (1978).
10 D. L. Cooper, J. Gerratt, M. Raimondi and S. C. Wright, *Chemical Physics Letters* **138** (4), 296–302 (1987).

9

Relativistic Methods

The central theme in relativity is that the speed of light, c, is constant in all inertia frames (coordinate systems that move with respect to each other). Augmented with the requirement that physical laws should be identical in such frames, this has as a consequence that time and space coordinates become "equivalent". A relativistic description of a particle thus requires four coordinates, three space and one time coordinate.[1–7] The latter is usually multiplied by c to have units identical to the space variables.

A change between different coordinate systems can be described by a *Lorentz* transformation, which may mix space and time coordinates. The postulate that physical laws should be identical in all coordinate systems is equivalent to the requirement that equations describing the physics must be invariant (unchanged) to a Lorentz transformation. Considering the time-dependent Schrödinger Equation (9.1), it is clear that it is not Lorentz invariant since the derivative with respect to space coordinates is of second order, but the time derivative is only first order. The fundamental structure of the Schrödinger equation is therefore not relativistically correct:

$$\left[-\frac{1}{2m}\left(\frac{\partial^2}{\partial x^2}+\frac{\partial^2}{\partial y^2}+\frac{\partial^2}{\partial z^2}\right)+\mathbf{V}\right]\Psi = i\frac{\partial \Psi}{\partial t} \tag{9.1}$$

For use below, we have elected here to explicitly write the electron mass as m, although it is equal to one in atomic units.

One of the consequences of the constant speed of light is that the mass of a particle, which moves at a substantial fraction of c, increases over the rest mass m_0:

$$m = m_0\left(\sqrt{1-\frac{v^2}{c^2}}\right)^{-1} \tag{9.2}$$

The energy of a 1s-electron in a hydrogen-like system (one nucleus and one electron) is $-{}^1/_2 Z^2$, and classically this is equal to minus the kinetic energy, ${}^1/_2 mv^2$, owing to the virial theorem ($E = -T = {}^1/_2 V$). In atomic units ($m = 1$) the classical velocity of a 1s-electron is thus Z. The speed of light in atomic units is 137.036, and it is clear that relativistic effects cannot be neglected for the core electrons in atoms from the lower part of the periodic table. For atoms with large Z, the 1s-electrons are relativistic and thus heavier, which has the effect that the 1s-orbital shrinks in size, by the same factor as the mass increases (Equation (9.2)). In order to maintain orthogonality, the higher s-orbitals also contract. This provides a more effective screening of the nuclear charge for the higher angular

Introduction to Computational Chemistry, Third Edition. Frank Jensen.

Companion Website: http://www.wiley.com/go/jensen/computationalchemistry3

momentum orbitals, which consequently increase in size. For p-orbitals the spin–orbit interaction, which mixes s- and p-orbitals, counteracts the inflation. The net effect is that p-orbitals are relatively unaffected in size, while d- and f-orbitals become larger and more diffuse.

In terms of total energy, the relativistic correction becomes comparable to the correlation energy already for $Z \sim 10$, but since the majority of the relativistic effects are concentrated in the core orbitals, there is a large error cancellation for relative energies. Relativistic effects for geometries and energetics are normally negligible for the first four rows in the periodic table (up to Kr, $Z = 36$, corresponding to a "mass correction" of 1.04), the fifth row represents an intermediate case, while relativistic corrections are necessary for the sixth and seventh rows, and for lanthanide/actinide metals. For effects involving electron spin (e.g. spin–orbit coupling), which are purely relativistic in origin, there is no non-relativistic counterpart, and the "relativistic correction" is of course everything.

Although an in-depth treatment of relativistic effects is outside the scope of this book, it may be instructive to point out some of the features and problems in a relativistic quantum description of atoms and molecules. Furthermore, we will require some operators derived from a relativistic treatment for calculating molecular properties in Chapter 11.

9.1 The Dirac Equation

For a free electron, Dirac proposed that the (time-dependent) Schrödinger equation should be replaced by

$$[c\boldsymbol{\alpha} \cdot \mathbf{p} + \boldsymbol{\beta} mc^2]\Psi = i\frac{\partial \Psi}{\partial t} \tag{9.3}$$

Here $\boldsymbol{\alpha}$ and $\boldsymbol{\beta}$ are 4×4 matrices, $c\boldsymbol{\alpha}$ is the relativistic velocity operator and $\boldsymbol{\alpha}$ can be written in terms of the three Pauli 2×2 spin matrices $\boldsymbol{\sigma}$ and $\boldsymbol{\beta}$ in terms of a 2×2 unit matrix $\mathbf{I}$:

$$\begin{gathered} \boldsymbol{\alpha}_{x,y,z} = \begin{pmatrix} 0 & \boldsymbol{\sigma}_{x,y,z} \\ \boldsymbol{\sigma}_{x,y,z} & 0 \end{pmatrix}, \quad \boldsymbol{\beta} = \begin{pmatrix} \mathbf{I} & 0 \\ 0 & -\mathbf{I} \end{pmatrix} \\ \boldsymbol{\sigma}_x = \begin{pmatrix} 0 & 1 \\ 1 & 0 \end{pmatrix}, \quad \boldsymbol{\sigma}_y = \begin{pmatrix} 0 & -i \\ i & 0 \end{pmatrix}, \quad \boldsymbol{\sigma}_z = \begin{pmatrix} 1 & 0 \\ 0 & -1 \end{pmatrix}, \quad \mathbf{I} = \begin{pmatrix} 1 & 0 \\ 0 & 1 \end{pmatrix} \end{gathered} \tag{9.4}$$

Except for a factor of $^1/_2$, the $\boldsymbol{\sigma}_{x,y,z}$ matrices can be viewed as representations of the $\mathbf{s}_x$, $\mathbf{s}_y$ and $\mathbf{s}_z$ spin operators, respectively, when the α and β spin functions are taken as (1,0) and (0,1) vectors:

$$\begin{gathered} \mathbf{s}_z = \frac{1}{2}\boldsymbol{\sigma}_z \\ \mathbf{s}_z \begin{pmatrix} 1 \\ 0 \end{pmatrix} = \frac{1}{2}\begin{pmatrix} 1 \\ 0 \end{pmatrix} \quad ; \quad \mathbf{s}_z \begin{pmatrix} 0 \\ 1 \end{pmatrix} = -\frac{1}{2}\begin{pmatrix} 0 \\ 1 \end{pmatrix} \end{gathered} \tag{9.5}$$

The α function is an eigenfunction of the $\mathbf{s}_z$ operator with an eigenvalue of $^1/_2$ and the β function similarly has an eigenvalue of $-^1/_2$.

The Dirac equation is of the same order in all variables (space and time), since the momentum operator $\mathbf{p}$ $(= -i\nabla)$ involves a first-order differentiation with respect to the space variables. It should be noted that the free electron rest energy in Equation (9.3) is mc^2, equal to 0.511 MeV, while this situation is defined as zero in the non-relativistic case. The zero point of the energy scale is therefore shifted by 5.11×10^5 eV, a large amount compared with the binding energy of 13.6 eV for a

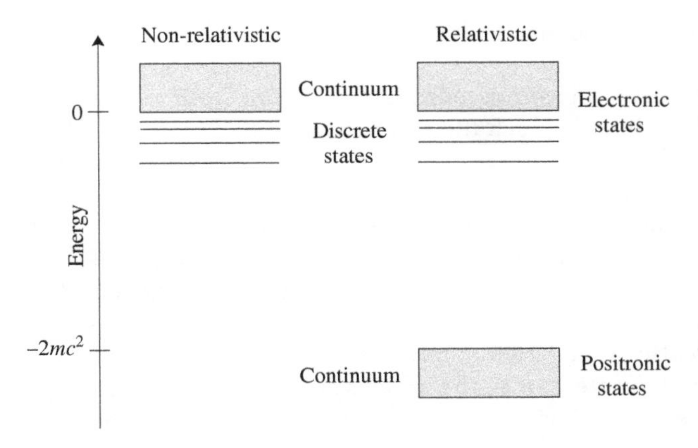

Figure 9.1 Non-relativistic and relativistic solutions. The energy scale of the discrete states is exaggerated relative to the energy separation of the continuum states.

hydrogen atom. The two (relativistic and non-relativistic) energy scales may be aligned by subtracting the electron rest energy, which corresponds to replacing the $\boldsymbol{\beta}$ matrix in Equation (9.3) by $\boldsymbol{\beta}'$:

$$\boldsymbol{\beta}' = \begin{pmatrix} 0 & 0 \\ 0 & -2\mathbf{I} \end{pmatrix} \tag{9.6}$$

The Dirac equation corresponds to satisfying the requirements of *special relativity* in connection with the quantum behavior of the electron. Special relativity considers only systems that move with a constant velocity with respect to each other, which can hardly be considered a good approximation for the movement of an electron around a nucleus. A relativistic treatment of accelerated systems is described by *general relativity*, which is a gravitational theory. For atomic systems, however, the gravitational interaction between electrons and nuclei (or between electrons) is insignificant compared with the electrostatic interaction. Furthermore, a consistent theory describing the quantum aspects of gravitation has not yet been developed.

The Dirac equation is four-dimensional and the relativistic wave function consequently contains *four components*. Two of the degrees of freedom are accounted for by assigning an intrinsic magnetic moment (spin), while the other two are interpreted as two different particles, the electron and positron. The positronic solutions show up as a continuum of "negative" energy states, having energies below $-2mc^2$, as illustrated in Figure 9.1. Note that the spacing between bound states has been exaggerated, as the binding energy is of the order of eV while $2mc^2$ is of the order of MeV.

It is conventional to write the four-component relativistic wave function as two two-component wave functions, as shown in

$$\Psi = \begin{pmatrix} \psi_1 \\ \psi_2 \\ \psi_3 \\ \psi_4 \end{pmatrix} = \begin{pmatrix} \Psi_\mathrm{L} \\ \Psi_\mathrm{S} \end{pmatrix} \tag{9.7}$$

Here Ψ_L and Ψ_S are called the *large* and *small* components of the wave function. For electrons (positive energy solution in Figure 9.1), the large component reduces to the solutions of the Schrödinger equation (the two components corresponding to the α and β spin parts) when $c \to \infty$ (the non-relativistic limit) and the small component disappears, thus giving rise to the names. The large component, one the other hand, disappears for the negative energy solutions as the momentum goes to zero.

Solving the Dirac equation for a hydrogen-like atom (one nucleus and one electron, both taken as point particles) can be done in much the same way as solving the Schrödinger equation, and leads to four quantum numbers, n, l, j and m, with interdependencies on their allowed values. The n and l quantum numbers are very similar to the non-relativistic case ($n = 1, 2, 3, \ldots$, $l = 0, 1, 2, \ldots, (n-1)$), except that l no longer represents the total angular momentum. The total angular momentum is instead quantified by the j value, which can take on two values, $l + {}^1\!/_2$ and $l - {}^1\!/_2$, except for $l = 0$, where only $j = {}^1\!/_2$ is allowed. The j quantum number can be considered as the total angular momentum obtained by coupling the orbital and spin moments, and this is commonly referred to as spin–orbit coupling. The m quantum number can in analogy with the non-relativistic case take on values between $-j$ and j in steps of 1. The orbitals are denoted with the name corresponding to the l quantum number (s, p, d, …) with j given as a subscript, such as $s_{1/2}$, $p_{1/2}$, $p_{3/2}$, etc.

9.2 Connections between the Dirac and Schrödinger Equations

9.2.1 Including Electric Potentials

In the presence of an electric potential **V** (e.g. from nuclei), the time-*independent* Dirac equation may be written as follows, where we have again explicitly indicated the electron mass:

$$[c\boldsymbol{\alpha} \cdot \mathbf{p} + \boldsymbol{\beta}' mc^2 + \mathbf{V}]\Psi = E\Psi \tag{9.8}$$

Since $\boldsymbol{\alpha}$ and $\boldsymbol{\beta}'$ are block matrices in terms of $\boldsymbol{\sigma}$ and $\mathbf{I}$, Equation (9.8) can be factored out in two equations:

$$c(\boldsymbol{\sigma} \cdot \mathbf{p})\Psi_S + \mathbf{V}\Psi_L = E\Psi_L \tag{9.9}$$

$$c(\boldsymbol{\sigma} \cdot \mathbf{p})\Psi_L + (-2mc^2 + \mathbf{V})\Psi_S = E\Psi_S \tag{9.10}$$

Here Ψ_L and Ψ_S are (large and small) two-component wave functions that include the α and β spin functions. The latter equation can be solved for Ψ_S:

$$\Psi_S = (E + 2mc^2 - \mathbf{V})^{-1} c(\boldsymbol{\sigma} \cdot \mathbf{p})\Psi_L \tag{9.11}$$

The inverse quantity can be factorized as in

$$\begin{aligned} (E + 2mc^2 - \mathbf{V})^{-1} &= (2mc^2)^{-1}\left(1 + \tfrac{E-\mathbf{V}}{2mc^2}\right)^{-1} = \left(2mc^2\right)^{-1}\mathbf{K} \\ \mathbf{K} &= \left(1 + \tfrac{E-\mathbf{V}}{2mc^2}\right)^{-1} \end{aligned} \tag{9.12}$$

Equation (9.11) may then be written as

$$\Psi_S = \mathbf{K}\frac{\boldsymbol{\sigma} \cdot \mathbf{p}}{2mc}\Psi_L \tag{9.13}$$

Equation (9.10) then becomes

$$\left[\frac{1}{2m}(\boldsymbol{\sigma} \cdot \mathbf{p})\mathbf{K}(\boldsymbol{\sigma} \cdot \mathbf{p}) + (\mathbf{V} - E)\right]\Psi_L = 0 \tag{9.14}$$

In the non-relativistic limit ($c \to \infty$) the **K** factor is 1, and the first term becomes $(\boldsymbol{\sigma} \cdot \mathbf{p})(\boldsymbol{\sigma} \cdot \mathbf{p})$. Using the vector identity $(\boldsymbol{\sigma} \cdot \mathbf{p})(\boldsymbol{\sigma} \cdot \mathbf{p}) = \mathbf{p} \cdot \mathbf{p} + i\boldsymbol{\sigma}(\mathbf{p} \times \mathbf{p})$, this gives the non-relativistic kinetic energy

$\mathbf{p}^2/2m$, since the vector product of any vector with itself is zero ($\mathbf{p} \times \mathbf{p} = 0$). The equation for the large component therefore reduces to the Schrödinger equation:

$$\left[\frac{\mathbf{p}^2}{2m} + \mathbf{V}\right]\Psi_\mathrm{L} = E\Psi_\mathrm{L} \tag{9.15}$$

The electron spin is still present in Equation (9.15), since Ψ_L is a two-component wave function, but this can trivially be separated out since the operators do not contain any spin dependence.

In the non-relativistic limit the small component of the wave function is given by

$$\Psi_\mathrm{S} = \frac{\sigma \cdot \mathbf{p}}{2mc}\Psi_\mathrm{L} \tag{9.16}$$

For a hydrogenic wave function ($\Psi_\mathrm{L} \approx \mathrm{e}^{-Z\mathbf{r}}$), this gives the following equation in atomic units (setting $m = 1$):

$$\Psi_\mathrm{S} \approx \frac{Z}{2c}\Psi_\mathrm{L} \tag{9.17}$$

For a hydrogen atom the small component accounts for only ~0.4% of the total wave function and 10^{-3}% of the electron density, but for a uranium 1s-electron it is a third of the wave function and ~10% of the density.

We may obtain relativistic corrections by expanding the **K** factor in Equation (9.12):

$$\mathbf{K} = \left(1 + \frac{E - \mathbf{V}}{2mc^2}\right)^{-1} \approx 1 - \frac{E - \mathbf{V}}{2mc^2} + \cdots \tag{9.18}$$

This is only valid when $E - \mathbf{V} \ll 2mc^2$; however, all atoms have a region close to the nucleus where this is not fulfilled (since $\mathbf{V} \to -\infty$ for $\mathbf{r} \to 0$). Inserting (9.18) in (9.14), assuming a Coulomb potential $-Z/\mathbf{r}$ (i.e. **V** is the attraction to a nucleus), gives. after renormalization of the (large component) wave function and some rearrangement. the terms shown in

$$\left[\frac{\mathbf{p}^2}{2m} + \mathbf{V} - \frac{\mathbf{p}^4}{8m^3c^2} + \frac{Z\mathbf{s}\cdot\mathbf{l}}{2m^2c^2r^3} + \frac{Z\pi\delta(\mathbf{r})}{2m^2c^2}\right]\Psi_\mathrm{L} = E\Psi_\mathrm{L} \tag{9.19}$$

Equation (9.19) is called the *Pauli equation*. The first two terms are the usual non-relativistic kinetic and potential energy operators and the $\mathbf{p}^4$ term is called the *mass–velocity* correction, and is due to the dependence of the electron mass on the velocity. The next is the *spin–orbit* term (**s** is the electron spin and **l** is the angular momentum operator $\mathbf{r} \times \mathbf{p}$), which corresponds to an interaction of the electron spin with the magnetic field generated by the movement of the electron. The last term involving the δ function is the *Darwin* correction, which corresponds to a correction that can be interpreted as the electron making a high-frequency oscillation around its mean position, sometimes referred to as *Zwitterbewegung*. The mass–velocity and Darwin corrections are often collectively called the *scalar* relativistic corrections. Since they have opposite signs, they do to a certain extent cancel each other. Methods incorporating (only) scalar relativistic effects can be classified as one-component methods, since the spin-dependence can be separated out analogous to non-relativistic methods. Methods including electron spin, as for example the spin-orbit effect, however, are inherently two-component methods since the spin and angular momentum magnetic moments are mixed. An approximate way of including scalar relativistic effects on the valence orbitals is by using relativistic effective core potentials (Section 5.12).

Owing to the divergence of the **K** expansion near the nuclei, the mass–velocity and Darwin corrections can only be used as first-order corrections. Inclusion of such operators in a variational sense will result in a collapse of the wave function. An alternative method is to partition Equation (9.12) as in Equation (9.20), which avoids the divergence near the nucleus:

$$(E + 2mc^2 - \mathbf{V})^{-1} = (2mc^2 - \mathbf{V})^{-1}\left(1 + \frac{E}{2mc^2 - \mathbf{V}}\right)^{-1} = \left(2mc^2 - \mathbf{V}\right)^{-1}\mathbf{K}'$$
$$\mathbf{K}' = \left(1 + \frac{E}{2mc^2 - \mathbf{V}}\right)^{-1} \tag{9.20}$$

In contrast to Equation (9.18), the factor $E/(2mc^2 - \mathbf{V})$ is always smaller than 1 for energies less than $2mc^2$, and $\mathbf{K}'$ may be expanded in powers of $E/(2mc^2 - \mathbf{V})$, analogously to Equation (9.17). Keeping only the zeroth-order term (i.e. setting $\mathbf{K}' = 1$) gives the *Zeroth-Order Regular Approximation* (ZORA) method:[8]

$$\left[\frac{c^2\mathbf{p}^2}{2mc^2 - \mathbf{V}} + \frac{2c^2}{\left(2mc^2 - \mathbf{V}\right)^2} - \frac{Z\mathbf{s}\cdot\mathbf{l}}{r^3} + \mathbf{V}\right]\Psi_{\mathrm{L}} = E\Psi_{\mathrm{L}} \tag{9.21}$$

Note that in this case the spin–orbit coupling is already included in zeroth order. Including the first-order term from an expansion of $\mathbf{K}'$ defines the *First-Order Regular Approximation* (FORA) method. A disadvantage of these methods is that they are not gauge-invariant, that is changing the zero-point for the potential energy does not lead to an identical change in the zero-point for the energy, and this may be problematic in calculating molecular binding energies.[8]

9.2.2 Including Both Electric and Magnetic Potentials

The presence of a magnetic field can be included in the so-called minimal coupling by addition of a *vector potential* **A** to the momentum operator **p**, forming a *generalized* momentum operator $\boldsymbol{\pi}$, which for an electron (charge of −1) is given by

$$\boldsymbol{\pi} = \mathbf{p} + \mathbf{A} \tag{9.22}$$

The magnetic field is defined as the curl of the vector potential:

$$\mathbf{B} = \nabla \times \mathbf{A} \tag{9.23}$$

For an external magnetic field, it is conventional to write the vector potential as in

$$\mathbf{A}(\mathbf{r}) = \tfrac{1}{2}\mathbf{B} \times (\mathbf{r} - \mathbf{R}_{\mathrm{G}}) \tag{9.24}$$

Here $\mathbf{R}_{\mathrm{G}}$ is the *gauge origin*, that is the "zero" point for the vector potential. The gauge origin is often taken as the center of mass for the system, but this is by no means unique. The results from an exact calculation will be independent of $\mathbf{R}_{\mathrm{G}}$ but, for approximate calculations, this is not guaranteed, and the results may thus depend on where the gauge origin is chosen. Such gauge-dependent properties are clearly undesirable, since different results can be generated by selecting different (arbitrary) gauge origins.

With the generalized momentum operator $\boldsymbol{\pi}$ replacing **p**, the time-independent Dirac equation may be separated analogously to the procedure in Section 8.2.1 to give the equivalent of Equation (9.14):

$$\left[\frac{1}{2m}(\boldsymbol{\sigma}\cdot\boldsymbol{\pi})\mathbf{K}(\boldsymbol{\sigma}\cdot\boldsymbol{\pi}) + (\mathbf{V} - E)\right]\Psi_{\mathrm{L}} = 0 \tag{9.25}$$

Taking the non-relativistic limit corresponding to $\mathbf{K} = 1$ gives $(\boldsymbol{\sigma} \cdot \boldsymbol{\pi})(\boldsymbol{\sigma} \cdot \boldsymbol{\pi})$ for the first term. Using again the vector identity $(\boldsymbol{\sigma} \cdot \boldsymbol{\pi})(\boldsymbol{\sigma} \cdot \boldsymbol{\pi}) = \boldsymbol{\pi} \cdot \boldsymbol{\pi} + i\boldsymbol{\sigma}(\boldsymbol{\pi} \times \boldsymbol{\pi})$, this may be written as in

$$\left[\frac{1}{2m}(\boldsymbol{\pi} \cdot \boldsymbol{\pi} + i\boldsymbol{\sigma} \cdot (\boldsymbol{\pi} \times \boldsymbol{\pi})) + \mathbf{V}\right] \Psi_{\mathrm{L}} = E\Psi_{\mathrm{L}} \tag{9.26}$$

In contrast to the situation without a magnetic field, the latter vector product no longer disappears. The $\boldsymbol{\pi} \times \boldsymbol{\pi}$ term can be expanded by inserting the definition of $\boldsymbol{\pi}$ from Equation (9.22):

$$\begin{aligned} \boldsymbol{\pi} \times \boldsymbol{\pi} &= (\mathbf{p} + \mathbf{A}) \times (\mathbf{p} + \mathbf{A}) \\ &= \mathbf{p} \times \mathbf{p} + \mathbf{p} \times \mathbf{A} + \mathbf{A} \times \mathbf{p} + \mathbf{A} \times \mathbf{A} \end{aligned} \tag{9.27}$$

The first and last terms are zero (since $\mathbf{a} \times \mathbf{a} = 0$). With $\mathbf{p} = -i\nabla$ the other two terms yield

$$\begin{aligned} (\mathbf{p} \times \mathbf{A} + \mathbf{A} \times \mathbf{p})\Psi &= -i\nabla \times (\mathbf{A}\Psi) - i\mathbf{A} \times (\nabla\Psi) \\ &= -i(\nabla \times \mathbf{A})\Psi - i(\nabla\Psi) \times \mathbf{A} - i\mathbf{A} \times (\nabla\Psi) \end{aligned} \tag{9.28}$$

The two last terms cancel (since $\mathbf{a} \times \mathbf{b} = -\mathbf{b} \times \mathbf{a}$) and the curl of the vector potential is the magnetic field (Equation (9.23)). The final result is given in

$$\left[\frac{\boldsymbol{\pi}^2}{2m} + \mathbf{V} + \frac{\boldsymbol{\sigma} \cdot \mathbf{B}}{2m}\right] \Psi_{\mathrm{L}} = E\Psi_{\mathrm{L}} \tag{9.29}$$

The $\boldsymbol{\sigma} \cdot \mathbf{B}$ term is called the (spin) *Zeeman* interaction, and represents the interaction of an (external) magnetic field with the intrinsic magnetic moment associated with the electron. As noted in Equation (9.5), $\boldsymbol{\sigma}$ represents the spin operator (except for a factor of $^1/_2$) and the $\boldsymbol{\sigma} \cdot \mathbf{B}/2m$ interaction can (in atomic units) also be written as $\mathbf{s} \cdot \mathbf{B}$, with $\mathbf{s}$ being the electron spin operator. In a more refined treatment, by including quantum field corrections, it turns out that the electron magnetic moment is not exactly equal to the spin. It is conventional to write the interaction as $g_e\mu_B\mathbf{s} \cdot \mathbf{B}$, where the Bohr magneton μ_B $(= e\hbar/2m)$ has a value of $^1/_2$ in atomic units and the electronic g-factor g_e is approximately equal to 2.0023 (the deviation from the value of 2 (exactly) is due to quantum field fluctuations).

Although electron spin is often said to arise from relativistic effects, the above shows that spin naturally arises in the non-relativistic limit of the Dirac equation. It may also be argued that electron spin is actually present in the non-relativistic case, as the kinetic energy operator $\mathbf{p}^2/2m$ is mathematically equivalent to $(\boldsymbol{\sigma} \cdot \mathbf{p})^2/2m$. If the kinetic energy is written as $(\boldsymbol{\sigma} \cdot \mathbf{p})^2/2m$ in the Schrödinger Hamiltonian, then electron spin is present in the non-relativistic case, although it would only have consequences in the presence of a magnetic field.

The Dirac equation automatically includes effects due to electron spin, while this must be introduced in a more or less *ad hoc* fashion in the Schrödinger equation (the Pauli principle). Furthermore, once the spin–orbit interaction is included, the total electron spin is no longer a "good" quantum number: an orbital no longer contains an integer number of α and β spin functions. The proper quantum number in relativistic theory is the total angular momentum quantified by the j quantum number obtained by vector addition of the orbital and spin moments.

Turning now to the $\boldsymbol{\pi}^2$ term in Equation (9.29), it can, with the use of Equation (9.22), be expanded into

$$\boldsymbol{\pi}^2 = (\mathbf{p} + \mathbf{A})^2 = \mathbf{p}^2 + \mathbf{p} \cdot \mathbf{A} + \mathbf{A} \cdot \mathbf{p} + \mathbf{A}^2 \tag{9.30}$$

The $\mathbf{p}^2$ gives the usual (non-relativistic) kinetic energy operator. Since $\mathbf{p} = -i\nabla$, the $\mathbf{p} \cdot \mathbf{A}$ term gives

$$(\mathbf{p} \cdot \mathbf{A})\Psi = -i(\nabla \cdot \mathbf{A})\Psi = -i\mathbf{A} \cdot (\nabla\Psi) - i\Psi(\nabla \cdot \mathbf{A}) \tag{9.31}$$

The *Coulomb gauge* is defined by $\nabla \cdot \mathbf{A} = 0$, and in this gauge we have $\mathbf{p} \cdot \mathbf{A} = \mathbf{A} \cdot \mathbf{p}$. The two terms involving $\mathbf{A}$ in Equation (9.30) can be evaluated by inserting the expression for the vector potential (9.24):

$$\begin{aligned}\mathbf{A} \cdot \mathbf{p} &= \left(\tfrac{1}{2}\mathbf{B} \times (\mathbf{r} - \mathbf{R}_G)\right) \cdot \mathbf{p} \\ &= \tfrac{1}{2}\mathbf{B} \cdot (\mathbf{r} - \mathbf{R}_G) \times \mathbf{p} \\ &= \tfrac{1}{2}\mathbf{B} \cdot \mathbf{L}_G\end{aligned} \tag{9.32}$$

$$\begin{aligned}\mathbf{A}^2 &= \left(\tfrac{1}{2}\mathbf{B} \times (\mathbf{r} - \mathbf{R}_G)\right) \cdot \left(\tfrac{1}{2}\mathbf{B} \times (\mathbf{r} - \mathbf{R}_G)\right) \\ &= \tfrac{1}{4}\left(\mathbf{B}^2 \cdot (\mathbf{r} - \mathbf{R}_G)\right)^2 - \left(\mathbf{B} \cdot (\mathbf{r} - \mathbf{R}_G)\right)^2\end{aligned} \tag{9.33}$$

Here the vector identities $\mathbf{a} \times \mathbf{b} \cdot \mathbf{c} = \mathbf{a} \cdot \mathbf{b} \times \mathbf{c}$ and $(\mathbf{a} \times \mathbf{b}) \cdot (\mathbf{c} \times \mathbf{d}) = (\mathbf{a} \cdot \mathbf{c})(\mathbf{b} \cdot \mathbf{d}) - (\mathbf{a} \cdot \mathbf{d})(\mathbf{c} \cdot \mathbf{b})$ have been used. In addition to the Zeeman term for electron spin (Equation (9.29)), the presence of a magnetic field introduces two new terms, being linear and quadratic in the field. The linear operator in Equation (9.32) represents an (orbital) Zeeman-type interaction of the magnetic field with the magnetic moment generated by the movement of the electron, as described by the angular momentum operator $\mathbf{L}_G$, while the quadratic term in Equation (9.33) gives rise to a component of the magnetizability in a perturbation treatment, as discussed in Section 11.8.7.

9.3 Many-Particle Systems

A fully relativistic treatment of more than one particle would have to start from a full QED treatment of the system (Chapter 1) and perform a perturbation expansion in terms of the radiation frequency. There is no universally accepted way of doing this, and a full relativistic many-body equation has not yet been developed. For many-particle systems it is assumed that each electron can be described by a Dirac operator ($c\boldsymbol{\alpha} \cdot \boldsymbol{\pi} + \beta' mc^2$) and the many-electron operator is a sum of such terms, in analogy with the kinetic energy in non-relativistic theory. Furthermore, potential energy operators are added to form a total operator equivalent to the Hamiltonian operator in non-relativistic theory. Since this approach gives results that agree with experiments, the assumptions appear justified.

The Dirac operator incorporates relativistic effects for the kinetic energy. In order to describe atomic and molecular systems, the potential energy operator must also be modified. In non-relativistic theory, the potential energy is given by the Coulomb operator:

$$\mathbf{V}(\mathbf{r}_{12}) = \frac{q_1 q_2}{r_{12}} \tag{9.34}$$

According to this equation, the interaction between two charged particles depends only on the distance between them, ***but not on time***. This cannot be correct when relativity is considered, as it implies that the attraction/repulsion between two particles occurs *instantly* over the distance r_{12}, violating the fundamental relativistic principle that nothing can move faster than the speed of light. The interaction between distant particles must be "later" than between particles that are close, and the potential is consequently "retarded" (delayed). The relativistic interaction requires a description, *Quantum ElectroDynamics* (QED), which involves the exchange of photons between charged particles. The photons travel at the speed of light and carry the information equivalent to the classical Coulomb interaction. The relativistic potential energy operator becomes complicated and cannot be written in closed form. For actual calculations, it may be expanded in a Taylor series in $1/c$ and, for chemical purposes, it is normally only necessary to include terms up to $1/c^2$.[9] In this approximation, the potential energy operator for the electron–electron repulsion is given by

$$\mathbf{V}_{ee}(\mathbf{r}_{12}) = \frac{1}{r_{12}} - \frac{1}{r_{12}}\left[\boldsymbol{\alpha}_1 \cdot \boldsymbol{\alpha}_2 - \frac{(\boldsymbol{\alpha}_1 \times \mathbf{r}_{12})(\boldsymbol{\alpha}_2 \times \mathbf{r}_{12})}{2r_{12}^2}\right] \tag{9.35}$$

Note that the subscript on the $\boldsymbol{\alpha}$-matrices refers to the particle, and $\boldsymbol{\alpha}$ here includes all of the $\boldsymbol{\alpha}_x$, $\boldsymbol{\alpha}_y$ and $\boldsymbol{\alpha}_z$ components in Equation (9.4). The first correction term in the square brackets is called the *Gaunt* interaction and the whole term in the square brackets is the *Breit* interaction. The Dirac matrices appear since they represent the velocity operators in a relativistic description. The Gaunt term can be considered as a magnetic interaction (spin), but the whole Breit term represents a retardation effect. Equation (9.35) is more often written in the form shown below, obtained by employing the vector identity $(\mathbf{a} \times \mathbf{b}) \cdot (\mathbf{c} \times \mathbf{d}) = (\mathbf{a} \cdot \mathbf{b})(\mathbf{c} \cdot \mathbf{d}) - (\mathbf{a} \cdot \mathbf{d})(\mathbf{b} \cdot \mathbf{c})$:

$$\mathbf{V}_{ee}^{\text{Coulomb–Breit}}(\mathbf{r}_{12}) = \frac{1}{r_{12}} - \frac{1}{2r_{12}}\left[\boldsymbol{\alpha}_1 \cdot \boldsymbol{\alpha}_2 + \frac{(\boldsymbol{\alpha}_1 \cdot \mathbf{r}_{12})(\boldsymbol{\alpha}_2 \cdot \mathbf{r}_{12})}{r_{12}^2}\right] \tag{9.36}$$

Relativistic corrections to the nuclear–electron attraction ($\mathbf{V}_{ne}$) are of order $1/c^3$ (owing to the much smaller velocity of the nuclei) and are normally neglected.

An expansion in powers of $1/c$ (or, equivalently, in powers of the fine-structure constant $\alpha = 1/c$ in atomic units) is a standard approach for deriving relativistic correction terms. Taking into account electron ($\mathbf{s}$) and nuclear spins ($\mathbf{I}$), and indicating explicitly an external electric potential by means of the field ($\mathbf{F} = -\nabla\phi$ or $-\nabla\phi - \partial\mathbf{A}/\partial t$ if time-dependent), an expansion up to order $1/c^2$ of the Dirac Hamiltonian including the Coulomb–Breit potential gives the following set of operators,[10] where the QED correction to the electron spin has been introduced by means of the $g_e\mu_B$ factor. Note that many of these operators arise from the minimal coupling of the magnetic field via the generalized momentum operator, as discussed in more detail in Section 11.8.

The mass–velocity correction due to the velocity-dependent electron mass, $\mathbf{H}_e^{mv}$, which also is present in Equation (9.19), is

$$\mathbf{H}_e^{mv} = -\frac{1}{8m^3c^2}\sum_{i=1}^{N_{elec}} \mathbf{p}_i^4 \tag{9.37}$$

One-electron operators arising from interaction with external fields are

$$\mathbf{H}_e^{\text{Zeeman}} = g_e\mu_B \sum_{i=1}^{N_{elec}} \left[\mathbf{s}_i \cdot \mathbf{B}_i - \frac{1}{2mc^2}(\mathbf{s}_i \cdot \mathbf{B}_i)\mathbf{p}_i^2\right] \tag{9.38}$$

$$\mathbf{H}_e^{SO} = \frac{g_e\mu_B}{4mc^2}\sum_{i=1}^{N_{elec}} \left[\mathbf{s}_i \cdot (\mathbf{F}_i \times \mathbf{p}_i - \mathbf{p}_i \times \mathbf{F}_i)\right] \tag{9.39}$$

$$\mathbf{H}_e^{\text{Darwin}} = \frac{1}{8m^2c^2}\sum_{i=1}^{N_{elec}} \nabla \cdot \mathbf{F}_i \tag{9.40}$$

Here $\mathbf{F}_i$ and $\mathbf{B}_i$ indicate the (electric and magnetic) fields at the position of particle i. The $\mathbf{H}_e^{\text{Zeeman}}$ has the $\mathbf{s} \cdot \mathbf{B}$ term from Equation (9.29) and a relativistic correction. $\mathbf{H}_e^{SO}$ and $\mathbf{H}_e^{\text{Darwin}}$ are spin–orbit and Darwin-type corrections with respect to an external electric field.

Two electron operators:

$$\mathbf{H}_{\text{ee}}^{\text{SO}} = -\frac{g_e\mu_B}{2mc^2}\sum_{i=1}^{N_{\text{elec}}}\sum_{j\neq i}^{N_{\text{elec}}}\frac{\mathbf{s}_i\cdot(\mathbf{r}_{ij}\times\mathbf{p}_i)}{r_{ij}^3} \tag{9.41}$$

$$\mathbf{H}_{\text{ee}}^{\text{SOO}} = -\frac{g_e\mu_B}{mc^2}\sum_{i=1}^{N_{\text{elec}}}\sum_{j\neq i}^{N_{\text{elec}}}\frac{\mathbf{s}_i\cdot(\mathbf{r}_{ij}\times\mathbf{p}_j)}{r_{ij}^3} \tag{9.42}$$

$$\mathbf{H}_{\text{ee}}^{\text{SS}} = \frac{g_e^2\mu_B^2}{2c^2}\sum_{i=1}^{N_{\text{elec}}}\sum_{j\neq i}^{N_{\text{elec}}}\left(\frac{\mathbf{s}_i\cdot\mathbf{s}_j}{r_{ij}^3} - 3\frac{(\mathbf{s}_i\cdot\mathbf{r}_{ij})(\mathbf{r}_{ij}\cdot\mathbf{s}_j)}{r_{ij}^5} - \frac{8\pi}{3}(\mathbf{s}_i\cdot\mathbf{s}_j)\delta(\mathbf{r}_{ij})\right) \tag{9.43}$$

$$\mathbf{H}_{\text{ee}}^{\text{OO}} = -\frac{1}{4m^2c^2}\sum_{i=1}^{N_{\text{elec}}}\sum_{j\neq i}^{N_{\text{elec}}}\left(\frac{\mathbf{p}_i\cdot\mathbf{p}_j}{r_{ij}} + \frac{(\mathbf{p}_i\cdot\mathbf{r}_{ij})(\mathbf{r}_{ij}\cdot\mathbf{p}_j)}{r_{ij}^3}\right) \tag{9.44}$$

$$\mathbf{H}_{\text{ee}}^{\text{Darwin}} = -\frac{\pi}{2m^2c^2}\sum_{i=1}^{N_{\text{elec}}}\sum_{j\neq 1}^{N_{\text{elec}}}\delta(\mathbf{r}_{ij}) \tag{9.45}$$

The sums run over all values of i and j, excluding the $i = j$ term, and there is consequently a factor of $^1/_2$ included to avoid overcounting. $\mathbf{H}_{\text{ee}}^{\text{SO}}$ is a *spin–orbit* operator, describing the interaction of the electron spin with the magnetic field generated by its own movement, as given by the angular momentum operator $\mathbf{r}_{ij}\times\mathbf{p}_i$. $\mathbf{H}_{\text{ee}}^{\text{SOO}}$ is a *spin–other-orbit* operator, describing the interaction of an electron spin with the magnetic field generated by the movement of the other electrons, as given by the angular momentum operator $\mathbf{r}_{ij}\times\mathbf{p}_j$. $\mathbf{H}_{\text{ee}}^{\text{SS}}$ and $\mathbf{H}_{\text{ee}}^{\text{OO}}$ are *spin–spin* and *orbit–orbit* terms, accounting for additional magnetic interactions, where the orbit–orbit term comes from the Breit correction to $\mathbf{V}_{\text{ee}}$ (Equation (9.36)). The (two-electron) Darwin interaction $\mathbf{H}_{\text{ee}}^{\text{Darwin}}$ contains a δ-function, which arise from the divergence of the field ($\nabla\cdot\mathbf{F}$) from the (electron–electron) potential energy operator, that is $\nabla\cdot(\nabla(1/\mathbf{r})) = -4\pi\delta(\mathbf{r})$. The spin–spin interaction $\mathbf{H}_{\text{ee}}^{\text{SS}}$ also has a δ-function, which comes from taking the curl of the vector potential associated with the magnetic dipole corresponding to the electron spin. A mathematical reformulation leads to a term involving the divergence of the $\mathbf{r}/r^3$ operator, giving $\nabla\cdot(\mathbf{r}/r^3) = (4\pi/3)\delta(\mathbf{r})$. Such terms are often called contact interactions, since they depend on the two particles being at the same position ($\mathbf{r} = 0$). In the spin–spin case, it is normally called the *Fermi-contact* (FC) term.

Operators involving one nucleus and one electron:

$$\mathbf{H}_{\text{ne}}^{\text{SO}} = \frac{g_e\mu_B}{2mc^2}\sum_{i=1}^{N_{\text{elec}}}\sum_{A=1}^{N_{\text{nuclei}}}Z_A\frac{\mathbf{s}_i\cdot(\mathbf{r}_{iA}\times\mathbf{p}_i)}{r_{iA}^3} \tag{9.46}$$

$$\mathbf{H}_{\text{ne}}^{\text{PSO}} = \frac{\mu_N}{mc^2}\sum_{i=1}^{N_{\text{elec}}}\sum_{A=1}^{N_{\text{nuclei}}}g_A\frac{\mathbf{I}_A\cdot(\mathbf{r}_{iA}\times\mathbf{p}_i)}{r_{iA}^3} \tag{9.47}$$

$$\mathbf{H}_{\text{ne}}^{\text{SS}} = -\frac{g_e\mu_B\mu_N}{c^2}\sum_{i=1}^{N_{\text{elec}}}\sum_{A=1}^{N_{\text{nuclei}}}g_A\left(\frac{\mathbf{s}_i\cdot\mathbf{I}_A}{r_{iA}^3} - 3\frac{(\mathbf{s}_i\cdot\mathbf{r}_{iA})(\mathbf{r}_{iA}\cdot\mathbf{I}_A)}{r_{iA}^5} - \frac{8\pi}{3}(\mathbf{s}_i\cdot\mathbf{I}_A)\delta(\mathbf{r}_{iA})\right) \tag{9.48}$$

$$\mathbf{H}_{\text{ne}}^{\text{Darwin}} = \frac{\pi}{2m^2c^2}\sum_{i=1}^{N_{\text{elec}}}\sum_{A=1}^{N_{\text{nuclei}}}Z_A\delta(\mathbf{r}_{iA}) \tag{9.49}$$

The $\mathbf{H}_{ne}^{SO}$ operator is the one-electron part of the spin–orbit interaction, while the $\mathbf{H}_{ee}^{SO}$ and $\mathbf{H}_{ee}^{SOO}$ operators in Equation (9.36) define the two-electron part. The one-electron term dominates and the two-electron contribution is often neglected or accounted for approximately by introducing an effective nuclear charge in $\mathbf{H}_{ne}^{SO}$ (corresponding to a screening of the nucleus by the electrons). The effect of the spin–orbit operators is to mix states having different total spin, as, for example, singlet and triplet states. The one-electron Darwin term in Equation (9.49) similarly dominates the two-electron term in Equation (9.45) by typically two orders of magnitude.

The nucleus–electron equivalent of the electron–electron spin–other-orbit operator in Equation (9.42) splits into two contributions, one involving the interaction of the electron spin with the magnetic field generated by the movement of the nuclei and one describing the interaction of the nuclear spin with the magnetic field generated by the movement of the electrons. Only the latter survives within the Born–Oppenheimer approximation, and it is normally denoted the *Paramagnetic Spin–Orbit* (PSO) operator. The nucleus–electron spin–spin term is analogous to that in Equation (9.43), while the term describing the orbit–orbit interaction disappears owing to the Born–Oppenheimer approximation. The spin–orbit and (one-electron) Darwin terms are the same as given in Equation (9.19), except for the quantum field correction factor of $g_e\mu_B$.

All of the terms in Equations (9.37) to (9.49) may be used as perturbation operators in connection with non-relativistic theory,[11] as discussed in more detail in Chapter 11. It should be noted, however, that some of the operators are inherently divergent and should not be used beyond a first-order perturbation correction.

9.4 Four-Component Calculations

Although relativistic effects can be included by perturbative operators describing corrections to the non-relativistic wave function, this rapidly becomes cumbersome if higher-order corrections are required, and it is then perhaps more satisfying to include relativistic effects by solving the Dirac equation directly. The simplest approximate wave function is a single Slater determinant constructed from four-component one-electron functions, called *spinors*, having large and small components. The spinors are the relativistic equivalents of the spin-orbitals in non-relativistic theory. With such a wave function, the relativistic equation corresponding to the Hartree–Fock equation is the *Dirac–Fock* equation, which in its time-independent form (setting $\boldsymbol{\pi} = \mathbf{p}$ and $m = 1$ in Equation (9.8)) can be written as

$$[c\boldsymbol{\alpha}\cdot\mathbf{p} + \boldsymbol{\beta}'c^2 + \mathbf{V}]\Psi = E\Psi \tag{9.50}$$

The requirement that the wave function energy should be stationary with respect to a variation in the orbitals results in an equation that is formally the same as in non-relativistic theory, $\mathbf{FC} = \mathbf{SC}\varepsilon$ (Equation (3.54)). However, the presence of solutions for the positronic states means that the desired solution is no longer the global minimum (Figure 9.1) and care must be taken that the procedure does not lead to variational collapse. The choice of a proper basis set for expanding the spinors is an essential component in preventing this. Since practical calculations necessarily use basis sets that are far from complete, the large and small component basis sets must be properly balanced. The large component corresponds to the normal non-relativistic wave function and has similar basis set requirements. The small component basis set is chosen to obey the *kinetic balance* condition, which follows from Equation (9.16):

$$\chi^{\text{small}} = \frac{\boldsymbol{\sigma}\cdot\mathbf{p}}{2c}\chi^{\text{large}} \tag{9.51}$$

The use of kinetic balance ensures that the relativistic solution smoothly reduces to the non-relativistic wave function as c is increased. The presence of the momentum operator in Equation (9.51) means that the small component basis set must contain functions that are derivatives of the large component basis set, making the former roughly twice the size of the latter. This means that there are ~8 times as many *large–small* two-electron integrals and ~16 times as many *small–small* integrals than there are *large–large*-type integrals. A relativistic calculation thus requires roughly 25 times as many two-electron integrals compared with a non-relativistic calculation.

When the Dirac operator is invoked, the point charge model of the nucleus also becomes problematic. For a non-relativistic hydrogen atom, the orbitals have a *cusp* (discontinuous derivative) at the nucleus. However, the relativistic solutions have a *singularity*. A singularity is much harder to represent in an approximate treatment (such as an expansion in a Gaussian basis) than a cusp. Consequently, a (more realistic) finite-size nucleus is often used in relativistic methods. A finite nucleus model removes the singularity of the orbitals, which now assume a Gaussian-type behavior within the nucleus. Neither experiments nor theory, however, provide a good model for how the positive charge is distributed *within* the nucleus. The wave function and energy will of course depend on the exact form used for describing the nuclear charge distribution. A popular choice is either a uniformly charged sphere, where the radius is proportional to the nuclear mass to the $^1/_3$ power, or a Gaussian charge distribution (which facilitates the calculation of the additional integrals) with the exponent depending on the nuclear mass. Note that this implies that the energy (and derived properties) depends on the specific isotope, not just the atomic charge, that is the results for, say, ^{81}Br will be (slightly) different from ^{79}Br. The difference between a finite and a point charge nuclear model for an atom in the lower part of the periodic table is large in terms of total energy (~1 au), but the difference due to different finite nucleus models is roughly two orders of magnitude smaller. For valence properties, any "reasonable" model gives essentially the same results.

9.5 Two-Component Calculations

Equations (9.13) and (9.14) show that the large and small components of the four-component wave function can formally be separated by means of the **K** operator, but this is not a practical approach since **K** depends on the energy (Equation (9.12) and is thus different for each energy solution and furthermore depends on its own eigenvalue. Consider the one-electron Dirac Hamiltonian in the following equation, where the $\boldsymbol{\alpha}$ matrix in Equation (9.4) contains off-diagonal $\boldsymbol{\sigma}$ elements that couple the large and small components:

$$\mathbf{H}_D = c\boldsymbol{\alpha}\cdot\mathbf{p} + \beta' mc^2 + \mathbf{V} = \begin{pmatrix} \mathbf{V} & c\boldsymbol{\sigma}\cdot\mathbf{p} \\ c\boldsymbol{\sigma}\cdot\mathbf{p} & \mathbf{V} - 2mc^2 \end{pmatrix} \tag{9.52}$$

The goal of so-called *two-component methods* is to decouple the large and small components by an energy-independent unitary transformation:[12]

$$\mathbf{H}'_D = \mathbf{U}\mathbf{H}_D\mathbf{U}^\dagger = \begin{pmatrix} \mathbf{h}_L & 0 \\ 0 & \mathbf{h}_S \end{pmatrix} \tag{9.53}$$

The desired positive (electronic) energy states in Figure 9.1 are described by (only) the $\mathbf{h}_L$ operator while the negative (positronic) energy states are described by (only) $\mathbf{h}_S$. The wave function is correspondingly transformed and only the transformed large component Ψ'_L needs to be considered:

$$\Psi' = \mathbf{U}\Psi = \mathbf{U}\begin{pmatrix} \Psi_L \\ \Psi_S \end{pmatrix} = \begin{pmatrix} \Psi'_L \\ \Psi'_S \end{pmatrix} \tag{9.54}$$

Although $\mathbf{V}$ in Equation (9.52) contains both one- and two-electron ($\mathbf{V}_{ne}$ and $\mathbf{V}_{ee}$) terms, essentially all commonly used methods consider only the one-electron part corresponding to the nucleus–electron attraction.

Neglecting the transformed small component wave function means that the calculation of all *large–small* and *small–small* integrals is avoided. Since the basis set required for describing the small component wave function typically is twice the size of the basis set for describing the large component wave function, this represents a major computational saving. Furthermore, all matrix operations depending on the size of the basis set become similarly reduced.

The unitary transformation can be parameterized by a generalization of the connection in Equation (9.13) between the small and large components in terms of an energy-independent $\mathbf{X}$ operator:

$$\Psi_S = \mathbf{X}\Psi_L \tag{9.55}$$

The $\mathbf{U}$ transformation can be written as follows, where the square root factors ensure normalization:

$$\mathbf{U} = \begin{pmatrix} \frac{1}{\sqrt{1+\mathbf{X}^\dagger\mathbf{X}}} & \frac{\mathbf{X}^\dagger}{\sqrt{1+\mathbf{X}^\dagger\mathbf{X}}} \\ \frac{-\mathbf{X}}{\sqrt{1+\mathbf{X}\mathbf{X}^\dagger}} & \frac{1}{\sqrt{1+\mathbf{X}\mathbf{X}^\dagger}} \end{pmatrix} \tag{9.56}$$

Using the transformation in Equation (9.56) on the Dirac equation in Equation (9.52) and requiring that the off-diagonal elements vanish, leads to the following equation for the $\mathbf{X}$ operator ($[\mathbf{X},\mathbf{V}]$ being the commutator $= \mathbf{XV} - \mathbf{VX}$):

$$c\mathbf{X}\sigma \cdot \mathbf{pX} - c\sigma \cdot \mathbf{p} + [\mathbf{X}, \mathbf{V}] + 2mc^2\mathbf{X} = 0 \tag{9.57}$$

Unfortunately no closed energy-independent form for the $\mathbf{X}$ operator has been discovered so far. In so-called *eXact 2-Component* (X2C) methods, the $\mathbf{X}$ operator is constructed from Equation (9.55) or Equation (9.57) as a matrix representation in a finite basis set from solutions of the full four-component wave function, and this allows construction of the exact (within the limitations of the basis set) unitary transformation in Equation (9.56). Although solving the full four-component Dirac equation in order to arrive at a two-component equation appears counterintuitive; the key point is that within the approximation of only including the one-electron potential energy operator, the four-component Dirac equation only includes one-electron terms and is relatively undemanding to solve. The X2C transformed Hamiltonian can then be used in subsequent steps involving two-electron terms, for example for solving the Dirac–Fock equation, which usually is computationally more demanding than solving the one-electron four-component Dirac equation.

For a free particle ($\mathbf{V} = 0$) the unitary transformation that diagonalizes the Dirac Hamiltonian is given as follows, which is known as the *Foldy–Wouthuysen* transformation:[13]

$$\mathbf{U}_0 = \mathrm{e}^{\beta\alpha\cdot\frac{\mathbf{p}}{|\mathbf{p}|}\theta} \quad ; \quad \tan 2\theta = \frac{|\mathbf{p}|}{m} \tag{9.58}$$

In terms of the following abbreviations:

$$E_p = \sqrt{\mathbf{p}^2c^2 + m^2c^4} \quad ; \quad A_p = \sqrt{\frac{E_p + mc^2}{2E_p}} \quad ; \quad \mathbf{R}_p = \frac{c\alpha \cdot \mathbf{p}}{E_p + mc^2} \tag{9.59}$$

the Foldy–Wouthuysen transformation can be written as in

$$\mathbf{U}_0 = A_p(1 + \beta\mathbf{R}_p) \tag{9.60}$$

In *Douglas–Kroll–Hess* (DKH) methods, the unitary transformation is written as a sequence of unitary transformations ($\mathbf{U} = \cdots \mathbf{U}_2\mathbf{U}_1\mathbf{U}_0$), which eliminates the off-diagonal elements coupling the large and small components to higher and higher orders in the potential **V**.[14,15] The $\mathbf{U}_0$ transformation is done using Equation (9.60) for a free particle and yields the DKH1 Hamiltonian:

$$\mathbf{H}_{\mathrm{DKH1}} = \mathbf{U}_0\mathbf{H}_D\mathbf{U}_0^{\dagger} = \begin{pmatrix} E_p & 0 \\ 0 & -E_p - 2mc^2 \end{pmatrix} + \begin{pmatrix} A_p(\mathbf{V} + \mathbf{R}_p\mathbf{V}\mathbf{R}_p)A_p & A_p[\mathbf{V},\mathbf{R}_p]A_p \\ -A_p[\mathbf{V},\mathbf{R}_p]A_p & A_p(\mathbf{V} + \mathbf{R}_p\mathbf{V}\mathbf{R}_p)A_p \end{pmatrix} \tag{9.61}$$

The implicit approximation in the DKH approach is to use only the upper-left block element in Equation (9.61) in the actual calculation. The off-diagonal elements in Equation (9.61) define the $\mathbf{U}_1$ operator, which leads to the DKH2 method, etc., where the number indicates the order by which the potential is eliminated in the off-diagonal elements. DKH2 represents a computationally inexpensive method for including relativistic effects and accounts for the large majority of relativistic effects, and including up to fourth-order (DKH4) produces results close to those obtained by solving the full four-component Dirac equation. Higher-order DKH methods are possible but require a number of increasingly complicated unitary transformation operators. Carried out to infinite order, the DKH approach is equivalent to solving the full four-component Dirac equation within a two-component framework.

The key operational step in the DKH scheme is a formulation of all the operators in the basis where the $\mathbf{p}^2$ operator is diagonal. Since the $\mathbf{p}^2$ operator represents twice the non-relativistic kinetic energy, this basis is obtained by diagonalization of the kinetic energy matrix. The $\mathbf{R}_p\mathbf{V}\mathbf{R}_p$ term in Equation (9.61), for example, can in this basis be calculated by scaling the matrix elements by factors involving twice the eigenvalues of the kinetic energy matrix (p^2):

$$R_i V_{ij} R_j = \frac{V_{ij}}{\sqrt{p_i^2c^2 + m^2c^4} + \sqrt{p_j^2c^2 + m^2c^4}} \tag{9.62}$$

If a spin-free DKH calculation (i.e. neglecting spin–orbit effects) is carried out in a standard atom-centered Gaussian basis set, the computational operations consist of decontraction and orthogonalizing the basis functions, diagonalization of the kinetic energy matrix, construction of the new one-electron integrals according to Equation (9.62) and back transformation to the original basis set.[16] This requires only a minor additional computational effort compared to the other steps in a non-relativistic calculation. A corresponding transformation of the two-electron integrals is a significantly larger task, and is usually ignored.

The elimination of the small component wave function leads to a computational more efficient way of accounting for relativistic effects than the full four-component approach, but it has an additional complication when molecular properties are desired. Consider a property P that is calculated as an expectation value of an operator **P** over the four-component wave function Ψ:

$$P = \langle \Psi | \mathbf{P} | \Psi \rangle \tag{9.63}$$

The same quantity can be obtained from the unitary transformed large-component-only wave function Ψ'_{L}, but as seen from the following equation, this involves a transformation of the operator as well:

$$P = \langle \Psi | \mathbf{U}^{\dagger}\mathbf{U}\mathbf{P}\mathbf{U}^{\dagger}\mathbf{U} | \Psi \rangle = \langle \Psi'_{\mathrm{L}} | \mathbf{U}\mathbf{P}\mathbf{U}^{\dagger} | \Psi'_{\mathrm{L}} \rangle = \langle \Psi'_{\mathrm{L}} | \mathbf{P}' | \Psi'_{\mathrm{L}} \rangle \tag{9.64}$$

The difference in the result between using the **P** and **P′** operators in connection with the Ψ'_{L} wave function is called the *picture change effect*.[17] Ignoring the picture change effect and employing the untransformed **U** operator in connection with the Ψ'_{L} wave function can lead to serious errors for properties depending on the wave function close to the nuclei. An advantage of X2C methods is that they readily can avoid picture change effects as the exact unitary transformation is available for transforming the matrix representation of the operator.

9.6 Relativistic Effects

The differences due to relativity can be described as:

1. Differences in the dynamics due to the velocity-dependent mass of the electron. This alters the size of the orbitals: s- and $p_{1/2}$-orbitals contract while $p_{3/2}$-, d- and f-orbitals expand.
2. New (magnetic) interactions in the Hamiltonian operator due to electron spin. The spin–orbit coupling, for example, destroys the picture of an orbital having a definite spin.
3. Introduction of a "small" component of the wave function, leading to a change in the shape of the orbitals: relativistic orbitals, for example, do not have nodes since the small component is finite when the large component is zero.
4. Modification of the potential operator due to the finite speed of light. In the lowest-order approximation, this corresponds to addition of the Breit operator to the Coulomb interaction.

Results from full four-component relativistic calculations are scarce, and the importance of the various effects is still sketchy. The Breit (Gaunt) term is believed to be small and many relativistic calculations neglect this term or include it as a perturbational term evaluated from the converged wave function. For geometries, the relativistic contraction of the s-orbitals normally means that bond lengths become shorter.

Working with a full four-component wave function and the Dirac–Fock operator is significantly more complicated than solving the Roothaan–Hall equations. The spin dependence can no longer be separated out and the basis set for the small component of the wave function must contain derivatives of the corresponding large component basis. This means that the basis set becomes roughly twice as large as in the non-relativistic case for a comparable accuracy. Furthermore, the presence of magnetic terms (spin) in the Hamiltonian operator means that the wave function contains both real and imaginary parts, yielding a factor of two in complexity. In practice, a (single-determinant) Dirac–Fock–Coulomb calculation is about two orders of magnitude more expensive than the corresponding non-relativistic Hartree–Fock case, although integral screening and density fitting techniques can be used to improve the efficiency.[18] Since heavy atom systems by definition contain many electrons, even small systems (in terms of the number of atoms) are demanding. A relativistic calculation for a single radon atom with a DZP quality basis, for example, is computationally equivalent to a non-relativistic calculation of a $C_{13}H_{28}$ alkane for a comparable quality in terms of basis set limitations. To further complicate matters, there are many more systems that cannot be adequately described by a single-determinant wave function in a relativistic treatment owing to the spin–orbit coupling, and therefore require MCSCF-type wave functions.

The two-component X2C and Douglas–Kroll–Hess methods are computationally efficient for including relativistic effects, especially if only scalar relativistic effects are included, but the picture change effect cannot be ignored for many properties. The DKH approach has the technical advantage that it can readily be incorporated into programs designed for non-relativistic methods, since it only requires a few additional integrals and scaling of existing integrals, while the X2C method requires a program capable of performing a full four-component calculation.

Table 9.1 illustrates the magnitude of relativistic effects for dihydrides of the sixth main group in the periodic table, where the relativistic calculations are of the Dirac–Fock–Coulomb type (i.e. a

Table 9.1 Properties of the sixth group dihydrides.

	Non-relativistic				Relativistic correction			
System	Total energy (au)	R_{eq} (Å)	θ_{eq} (°)	ΔE_{atom} (kJ/mol)	Total energy (au)	R_{eq} (Å)	θ_{eq} (°)	ΔE_{atom} (kJ/mol)
H_2O	−76.054	0.9391	107.75	643.8	−0.055	−0.00003	−0.07	−1.6
H_2S	−398.641	1.3429	94.23	514.1	−1.107	−0.00015	−0.09	−4.5
H_2Se	−2400.977	1.4530	93.14	459.4	−28.628	−0.00260	−0.27	−13.3
H_2Te	−6612.797	1.6557	92.57	392.5	−182.072	−0.00720	−0.58	−37.7
H_2Po	−20676.709	1.7539	92.21	350.2	−1555.822	−0.01060	−1.62	−126.8

single-determinant wave function and neglecting the Breit interaction).[19] The relativistic correction to the total energy is significant: even for a second row species such as H_2O the difference is 0.055 au (145 kJ/mol). It increases rapidly down the periodic table and reaches ~7% of the total energy for H_2Po, but the equilibrium distances and angles change only marginally. Similarly, the atomization energy (for breaking both X—H bonds completely) is remarkably insensitive to the large changes in the total energies. This is of course due to a high degree of cancellation of errors; the major relativistic correction is associated with the inner-shell electrons of the heavy atom, with the correction being almost constant for the atom and the molecule. For the lighter elements the effect on the atomization energies is almost solely due to the spin–orbit interaction in the triplet X atom (e.g. $H_2O \rightarrow {}^3O + 2{}^2H$), which is not present in the singlet H_2X molecule.

Similar results have been obtained for the fourth group tetrahydrides, CH_4, SiH_4, SbH_4, GeH_4 and PbH_4, where the Gaunt term has been shown to give corrections typically an order of magnitude less than the other relativistic changes.[20] The general conclusion is that relativistic effects for geometries and energetics can normally be neglected for molecules containing only up to third row elements (i.e. up to argon). This is also true for fourth row elements (up to krypton) unless a high accuracy is required. Although the geometry and atomization energy changes for H_2S and H_2Se in Table 9.1 may be considered significant, it should be noted that the errors due to incomplete basis sets and neglect of electron correlation are much larger than the relativistic corrections. The experimental geometries for H_2S and H_2Se, for example, are 1.3356 Å and 92.12° and 1.4600 Å and 90.57°, respectively. While the relativistic contraction of the H—Se bond is 0.0026 Å, the basis set and electron correlation error is 0.0096 Å. Relativistic effects typically become comparable to those from electron correlation at atomic numbers ~40–50. For molecules involving atoms beyond the fifth row in the periodic table, however, relativistic effects cannot be neglected for quantitative work. It should be noted that an approximate inclusion of the scalar relativistic effects, most notably the change in orbital size, can be modelled by replacing the inner electrons with a relativistic pseudo-potential, as discussed in Section 5.12.

Relativistic methods can be extended to include electron correlation by methods analogous to the non-relativistic cases, for example CI, MCSCF, MP and CC, but such methods are much less developed than for the non-relativistic case.[21,22] Once relativistic effects are considered, one may thus expand the two-dimensional Figure 4.2 with a third axis describing how accurate the relativistic effects are treated, for example measured in terms of one-, two- or four-component wave functions, as illustrated in Figure 9.2.

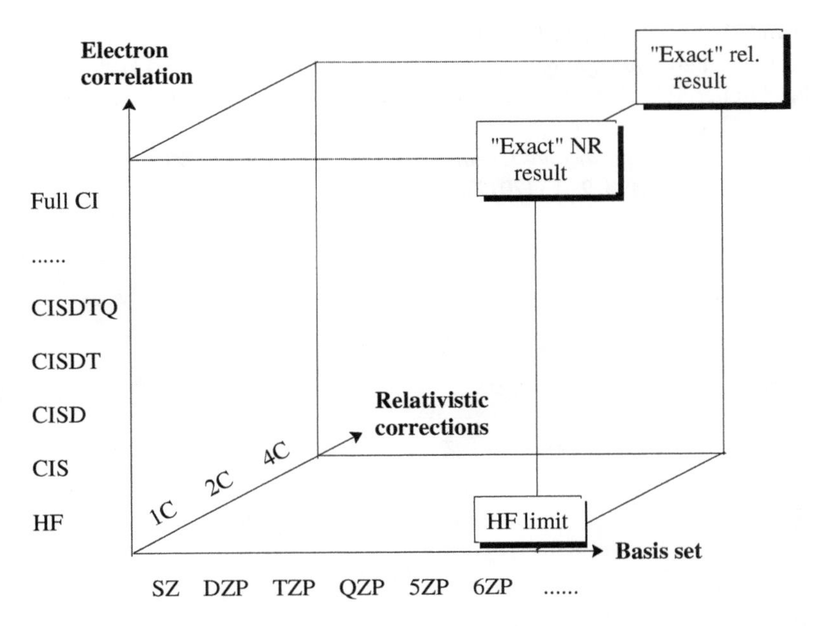

Figure 9.2 Converging the computational results by increasing the basis set, the amount of electron correlation and description of the relativistic effects.

References

1 R. E. Moss, *Advanced Molecular Quantum Mechanics* (Chapman and Hall, 1973).
2 P. Pyykko, *Chemical Reviews* **88** (3), 563–594 (1988).
3 J. Almlöf and O. Gropen, *Reviews in Computational Chemistry* **8**, 203 (1996).
4 K. Balasubramanian, *Relativistic Effects in Chemistry* (John Wiley & Sons, 1997).
5 K. G. Dyall and K. Fægri Jr., *Introduction to Relativistic Quantum Chemistry* (Oxford University Press, 2007).
6 T. Saue, *Chemphyschem* **12** (17), 3077–3094 (2011).
7 J. Autschbach, *Journal of Chemical Physics* **136** (15), 150902 (2012).
8 E. Vanlenthe, E. J. Baerends and J. G. Snijders, *Journal of Chemical Physics* **101** (11), 9783–9792 (1994).
9 S. Bubin, M. Pavanelo, W.-C. Tung, K. L. Sharkey and L. Adamowicz, *Chemical Reviews* **113** (1), 36–79 (2013).
10 R. McWeeny, *Methods of Molecular Quantum Mechanics* (Academic Press, 1992).
11 S. Coriani, T. Helgaker, P. Jorgensen and W. Klopper, *Journal of Chemical Physics* **121** (14), 6591–6598 (2004).
12 D. Peng and M. Reiher, *Theoretical Chemistry Accounts* **131** (1), 1081 (2012).
13 L. L. Foldy and S. A. Wouthuysen, *Physical Review* **78** (1), 29–36 (1950).
14 M. Douglas and N. M. Kroll, *Annals of Physics* **82** (1), 89–155 (1974).
15 B. A. Hess, *Physical Review A* **33** (6), 3742–3748 (1986).
16 A. Wolf, M. Reiher and B. A. Hess, *Journal of Chemical Physics* **117** (20), 9215–9226 (2002).
17 A. Wolf and M. Reiher, *Journal of Chemical Physics* **124** (6), 064102 (2006).

18 M. S. Kelley and T. Shiozaki, *Journal of Chemical Physics* **138** (20), 204113 (2013).
19 L. Pisani and E. Clementi, *Journal of Chemical Physics* **101** (4), 3079–3084 (1994).
20 O. Visser, L. Visscher, P. J. C. Aerts and W. C. Nieuwpoort, *Theoretica Chimica Acta* **81** (6), 405–416 (1992).
21 L. Visscher, *Journal of Computational Chemistry* **23** (8), 759–766 (2002).
22 T. Fleig, *Chemical Physics* **395**, 2–15 (2012).

10

Wave Function Analysis

The previous chapters have focused on various methods for obtaining more or less accurate solutions to the Schrödinger (or Dirac) equation. The natural "byproduct" of determining the electronic wave function is the energy. However, there are many other properties that may be derived. Although the quantum mechanical description of a molecule is in terms of positive nuclei surrounded by a cloud of negative electrons, chemistry is still formulated as "atoms" held together by "bonds". This raises questions such as: given a wave function, how can we define an atom and its associated electron population, or how do we determine whether two atoms are bonded?

Atomic charge is an example of a property often used for discussing/rationalizing structural and reactivity differences.[1] There are three commonly used methods for assigning atomic charges:

1. Partitioning the wave function in terms of the basis functions (Hilbert space partitioning).
2. Derivation from the electrostatic potential.
3. Partitioning the electron density into atomic domains (physical space partitioning).

10.1 Population Analysis Based on Basis Functions

The electron density ρ (probability of finding an electron) at a certain position **r** from a single molecular orbital containing one electron is given as the square of the MO ø:

$$\rho_i(\mathbf{r}) = \phi_i^2(\mathbf{r}) \tag{10.1}$$

Assuming that the MO is expanded in a set of normalized, but non-orthogonal, basis functions χ, this can be written as (see also Equation (3.52)):

$$\phi_i = \sum_{\alpha}^{M_{\text{basis}}} c_{\alpha i}\chi_\alpha \tag{10.2}$$

$$\phi_i^2 = \sum_{\alpha\beta}^{M_{\text{basis}}} c_{\alpha i}c_{\beta i}\chi_\alpha\chi_\beta \tag{10.3}$$

Introduction to Computational Chemistry, Third Edition. Frank Jensen.

Companion Website: http://www.wiley.com/go/jensen/computationalchemistry3

Integrating and summing over all occupied MOs gives the total number of electrons, N_{elec}:

$$\sum_i^{N_{\text{occ}}} \int \phi_i^2 \mathbf{dr} = \sum_i^{N_{\text{occ}}} \sum_{\alpha\beta}^{M_{\text{basis}}} c_{\alpha i} c_{\beta i} \int \chi_\alpha \chi_\beta \mathbf{dr} = \sum_i^{N_{\text{occ}}} \sum_{\alpha\beta}^{M_{\text{basis}}} c_{\alpha i} c_{\beta i} S_{\alpha\beta} = N_{\text{elec}} \tag{10.4}$$

We may generalize this by introducing an occupation number (number of electrons), n, for each MO. For a single-determinant wave function, this will be either 0, 1 or 2, while it may be a fractional number for a correlated wave function (Section 10.5):

$$\sum_i^{N_{\text{orb}}} n_i \int \phi_i^2 \mathbf{dr} = \sum_{\alpha\beta}^{M_{\text{basis}}} \left(\sum_i^{N_{\text{orb}}} n_i c_{\alpha i} c_{\beta i} \right) S_{\alpha\beta} = \sum_{\alpha\beta}^{M_{\text{basis}}} D_{\alpha\beta} S_{\alpha\beta} = N_{\text{elec}} \tag{10.5}$$

The sum of the product of MO coefficients and the occupation numbers is the density matrix defined in Equation (3.55), and the sum over the product of the density and overlap matrices elements is the number of electrons.

The *Mulliken Population Analysis* uses the $\mathbf{D} \cdot \mathbf{S}$ matrix for distributing the electrons into atomic contributions[2] ($\mathbf{D} \cdot \mathbf{S}$ is the entrywise product matrix, Section 17.1, i.e. the products of elements, not elements of the product matrix). A diagonal element $D_{\alpha\alpha} S_{\alpha\alpha}$ in Equation (10.5) is the number of electrons in the α AO and an off-diagonal element $D_{\alpha\beta} S_{\alpha\beta}$ is (half) the number of electrons shared by AOs α and β (there is an equivalent $D_{\beta\alpha} S_{\beta\alpha}$ element). The contributions from all AOs located on a given atom A may be summed up to give the number of electrons associated with atom A. This requires a decision on how a contribution involving basis functions on different atoms should be divided. The simplest, and the one used in the Mulliken scheme, is to partition the contribution equally between the two atoms. The Mulliken electron population is thereby defined as in

$$\rho_{\text{A}} = \sum_{\alpha \in \text{A}}^{M_{\text{basis}}} \sum_{\beta}^{M_{\text{basis}}} D_{\alpha\beta} S_{\alpha\beta} \tag{10.6}$$

The *gross* charge on atom A is the sum of the nuclear and electronic contributions:

$$Q_{\text{A}} = Z_{\text{A}} - \rho_{\text{A}} \tag{10.7}$$

The Mulliken method corresponds to a partitioning of the $\mathbf{D} \cdot \mathbf{S}$ matrix product, while another commonly used method is the *Löwdin* partitioning, which uses the $\mathbf{S}^{1/2} \cdot \mathbf{D} \cdot \mathbf{S}^{1/2}$ matrix for analysis.[3] These are mathematically related as shown in

$$\begin{gathered} \sum \mathbf{D} \cdot \mathbf{S} = N_{\text{elec}} \\ \sum \mathbf{S}^{1/2} \cdot (\mathbf{D} \cdot \mathbf{S}) \cdot \mathbf{S}^{-1/2} = \mathbf{S}^{1/2} N_{\text{elec}} \mathbf{S}^{-1/2} \\ \sum \mathbf{S}^{1/2} \cdot \mathbf{D} \cdot \mathbf{S}^{1/2} = N_{\text{elec}} \end{gathered} \tag{10.8}$$

The Löwdin method is equivalent to a population analysis of the density matrix in the orthogonalized basis set (Section 17.2.3) formed by transforming the original set of functions by $\mathbf{S}^{-1/2}$:

$$\chi' = \mathbf{S}^{-1/2} \chi \tag{10.9}$$

The Mulliken and Löwdin methods are just particular examples of a whole family of population analysis using $\mathbf{S}^n \cdot \mathbf{D} \cdot \mathbf{S}^{1-n}$ matrices.[4] The Mulliken and Löwdin methods give different atomic charges but mathematically there is nothing to indicate which of these partitionings gives the "best" result. It should be noted, however, that the Löwdin method is not rotationally invariant if the basis set contains Cartesian polarization functions rather than spherical functions.[5] The lack of rotational invariance means that symmetry-equivalent atoms may end up having different charges.

There are some common problems with all population analyses based on partitioning the wave function in terms of basis functions:

1. The diagonal elements may be larger than two. This implies more than two electrons in an orbital, violating the Pauli principle.
2. The off-diagonal elements may become negative. This implies a negative number of electrons between two basis functions, which clearly is physically impossible.
3. There is no objective reason for dividing the off-diagonal contributions equally between the two orbitals. It may be argued that the most "electronegative" (which then needs to be defined) atom (orbital) should receive most of the shared electrons.
4. A basis function centered on atom A may have a small exponent, such that it effectively describes the wave function far from atom A. Nevertheless, the electron density is counted as only belonging to A.
5. The dipole, quadrupole, etc., moments are in general not conserved, that is a set of population atomic charges does not reproduce the original multipole moments.

The Mulliken scheme suffers from all of the above, while the Löwdin method solves problems 1, 2 and 3. In the orthogonalized basis, all off-diagonal elements are 0 and the diagonal elements are restricted to values between 0 and 2.

Problem 4 is especially troublesome, as a few examples for the water molecule will demonstrate. A reasonable description of the wave function can be obtained by a Hartree–Fock single determinant with a DZP basis set. An equally good wave function (in terms of energy) may be constructed by having a very large number of basis functions centered on oxygen, and none on the hydrogens (a DZP quality basis set on both oxygen and hydrogen gives an energy similar to having a 5ZP quality basis set on oxygen only). The latter will, according to the above population analysis, have a +1 charge on hydrogen and a −2 charge on oxygen. Worse, another equally good wave function may be constructed by having a large number of basis functions only on the hydrogens. This will give charges of −4 for each of the hydrogens and +8 for the oxygen. Alternatively, the basis functions can be taken to be non-nuclear-centered, in which case the electrons are not associated with any nuclei at all, that is atomic charges of +1 and +8! A less artificial but illustrative example is the WAu_{12} cluster, where a Mulliken population analysis of the B3LYP/QZP electron density yields a charge of −78 on W.[6]

The fundamental problem is that basis functions often describe electron density near a nucleus other than the one they are centered on. An s-type Gaussian function on oxygen with an exponent of 0.15, for example, has a maximum in the radial distribution ($r^2\phi^2$) that peaks at 0.97 Å, that is at the distance where a hydrogen nucleus is located in an OH group. Atomic charges calculated from a Mulliken or Löwdin analysis will therefore not converge to a constant value as the size of the basis set is increased. Enlarging the basis set involves addition of increasingly diffuse basis functions, often leading to unpredictable changes in the atomic charges. This is a case where a "better" theoretical procedure is actually counterproductive. Basis function derived population analyses are therefore most useful for comparing trends in electron distributions, when small- or medium-sized basis sets (which only contain relatively tight functions) are used.

The density matrix can also be used for generating information about bond strengths. A quantitative measure is given by the *Bond Order* ρ. It was originally defined from bond distances, as shown by[7]

$$\rho = e^{-(r-r_0)/a} \tag{10.10}$$

If the bond orders for ethane, ethylene and acetylene are defined to be 1, 2 and 3, respectively, the a constant is found to have a value of approximately 0.3 Å. For bond orders less than 1 (i.e. breaking and forming single bonds), it appears that a value of 0.6 Å is a more appropriate proportionality constant. A "Mulliken" style measure of the bond strength between atoms A and B can be defined from the density matrix as (note that this involves elements of the product of the **D** and **S** matrices)[8]

$$\rho_{\mathrm{AB}} = \sum_{\alpha \in \mathrm{A}}^{M_{\mathrm{basis}}} \sum_{\beta \in \mathrm{B}}^{M_{\mathrm{basis}}} (\mathbf{DS})_{\alpha\beta} (\mathbf{DS})_{\beta\alpha} \tag{10.11}$$

The bond order will, in analogy with Mulliken atomic charges, be basis-set-dependent, with examples shown in Table 10.3 (section 10.7). The concept can be generalized to higher-order quantities, that is three-, four-, five-, etc., center bond indices, which are derived from products of **DS** elements.[9]

Population analysis with semi-empirical methods (Chapter 7) requires a special comment. These methods normally employ the ZDO approximation, that is the overlap **S** is a unit matrix. The population analysis can therefore be performed directly on the density matrix. In some cases, however, a Mulliken population analysis is performed with $\mathbf{D} \cdot \mathbf{S}$, which requires an explicit calculation of the **S** matrix.

10.2 Population Analysis Based on the Electrostatic Potential

One area where the concept of atomic charges is deeply rooted is in force field methods (Chapter 2). A significant part of the non-bonded interaction between polar molecules is described in terms of electrostatic interactions between fragments having an internal asymmetry in the electron distribution. The fundamental interaction is between the *ElectroStatic Potential* (ESP), also called the *Molecular Electrostatic Potential* (MEP), generated by one molecule (or fraction thereof) and the charged particles of another. The ESP at position $\mathbf{r}$ is given as a sum of contributions from the nuclei and the electron density, where the latter is provided by the electronic wave function:

$$\phi_{\mathrm{ESP}}(\mathbf{r}) = \sum_{A}^{\mathrm{nuclei}} \frac{Z_A}{|\mathbf{r} - \mathbf{R}_A|} - \int \frac{\rho(\mathbf{r}')}{|\mathbf{r} - \mathbf{r}'|} \mathrm{d}\mathbf{r}'$$
$$\rho(\mathbf{r}') = |\Psi(\mathbf{r}')|^2 \tag{10.12}$$

The first part of the potential is trivially calculated from the nuclear charges and positions, but the electronic contribution requires knowledge of the wave function. The latter is not available in force field methods, and the simplest way of modeling the electrostatic potential is to assign partial charges to each atom (Section 2.2.6). Atomic charges may be treated as regular force field parameters and assigned values based on fitting to experimental data, such as dipole, quadrupole, octopole, etc., moments, but there are rarely enough data to allow a unique assignment.

A common way of deriving partial atomic charges in force fields is as a set of parameters that in a least-squares sense generates the best fit to the actual electrostatic potential as calculated from an electronic wave function and sampled in a number of points surrounding the molecule.[10] These charges reflect a real physical phenomenon, the ESP, and can be used for analysis purposes as well. The computational problem can be formulated as minimization of an error function (Section 13.1):

$$ErrF(\mathbf{Q}) = N_{\text{points}}^{-1} \sum_{i}^{N_{\text{points}}} (\phi_{\text{ESP}}(\mathbf{r}_i) - \phi_{\text{Approx}}(\mathbf{r}_i))^2 \tag{10.13}$$

$$\phi_{\text{Approx}}(\mathbf{r}_i) = \sum_{\text{A}}^{N_{\text{atoms}}} \frac{Q_{\text{A}}(\mathbf{R}_{\text{A}})}{|\mathbf{r}_i - \mathbf{R}_{\text{A}}|} \tag{10.14}$$

The atomic charges Q_{A} are determined as the parameters that reproduce the electrostatic potential as closely as possible averaged over the sampling points, subject to the constraint that the sum is equal to the total molecular charge. In some cases, the atomic charges may also be constrained to reproduce, for example, the total dipole moment. Additional constraints such as forcing the total charge of a subgroup (such as a methyl group or an amino acid) to be zero are also often employed as this improves the parameter transferability and computational issues related to calculating the electrostatic energy.

The different schemes for deriving ESP atomic charges differ in the number and location of points used for sampling the ESP and whether additional constraints beyond preservation of charge are imposed. A typical selection of sampling points is a few hundred points around each nucleus with distances from just outside the van der Waals radius to about twice that distance. The MSK[11,12] (Merz–Singh–Kollman), CHELP[13] (CHarges from ELectrostatic Potential) and CHELPG[14] (CHarges from ELectrostatic Potential using a Grid-based method) methods only differ by the selection of sampling points, while the HLY (Hu, Lu and Yang) method employs a quite dense grid in a volumen space.[15] Tsiper and Burke, however, have shown that the ESP on an isodensity surface that includes essentially all the electron density *uniquely* determines the ESP at *all* points outside the surface.[16] This implies that the sampling points in Equation (10.13) can be chosen on an isodensity surface with a low density value (e.g. 10^{-4}–10^{-5}) and with a sufficiently dense grid that the sampling becomes a numerical evaluation of a surface integral:

$$\begin{aligned} ErrF(\mathbf{Q}) &= N_{\text{points}}^{-1} \sum_{i}^{N_{\text{points}}} (\phi_{\text{ESP}}(\mathbf{r}_i) - \phi_{\text{Approx}}(\mathbf{r}_i))^2 \\ &\simeq S^{-1} \oint_S (\phi_{\text{ESP}}(\mathbf{r}) - \phi_{\text{Approx}}(\mathbf{r}))^2 \mathrm{d}S \end{aligned} \tag{10.15}$$

Differences in ESP atomic charges due to differences in the selection of sampling points thus primarily reflect an incomplete sampling and associated numerical instabilities in solving the equations arising from minimizing *ErrF*.

The electrostatic potential depends on the total wave function and therefore converges as the size of the basis set and amount of electron correlation is increased. Since the potential depends directly on the electron density ρ, it is fairly insensitive to the level of sophistication, that is an HF or DFT calculation with a DZP-type basis set already gives quite good results. One might thus anticipate that atomic charges based on fitting to the ESP would lead to well-defined values. This, however, is often not the case. The fundamental problem is that the fitting procedure becomes statistically underdetermined even for medium-sized systems, although the severity of this depends on how the fitting is done.[17,18] The difference between the true ESP and that generated by a set of atomic charges on, say, only half the atoms is not significantly reduced by having fitting parameters on all atoms. The ESP experienced outside the molecule is mainly determined by the atoms near the molecular surface, and charges on atoms buried within a molecule can consequently not be assigned with any great confidence. This is rooted in the Coulomb expression for the ESP having an inverse distance dependence (Equation (10.14)), with atoms 2 Å away from a sampling point thus having a five times larger weight than atoms 10 Å away. Furthermore, atoms, say, 9 and 10 Å away from a sampling point have almost the same weight factors, and therefore span nearly the same function space. Having a full set of atomic charges thus forms a near-redundant set of parameters: many different sets of charges may be chosen, all of which are capable of reproducing the true ESP to almost the same accuracy. Although the ESP at a very large number of sampling points (several thousand) is fitted by relatively few (perhaps 20–30) atomic parameters, the fact that the ESP values at the sampling points are highly correlated makes the problem underdetermined. The RESP[19] (Restrained ElectroStatic Potential) method adds a hyperbolic penalty term to *ErrF* to ensure that only those atoms that contribute significantly in a statistical sense acquire non-zero charges.

The larger importance of atoms near the molecular surface relative to buried atoms and the statistically underdetermined fitting procedure at least partly explain that ESP atomic charges depend on the specific molecular conformation[20] and display variability between structurally similar groups in different molecules. The three hydrogens in a freely rotating methyl group, for example, may end up having significantly different charges or two conformations may give two significantly different sets of ESP atomic charges. This is a problem in connection with force field methods that rely on the fundamental assumption that parameters are transferable between similar fragments and that atoms that are easily interchanged (e.g. by bond rotation) should have identical parameters. One way of eliminating the problem with conformationally dependent charges is to add additional constraints, for example forcing the three hydrogens in a methyl group to have identical charges[19] or averaging over different conformations.[21]

Another problem with ESP atomic charges is related to the absolute accuracy of the fitting. Although inclusion of charges on all atoms does not significantly improve the results over that determined from a reduced set of parameters, the *absolute* deviation between the true and fitted electrostatic potentials can be quite large. Interaction energies as calculated by an atom-centered charge model in a force field may be off by several kJ/mol per atom in certain regions of space just outside the molecular surface, an error of one or two orders of magnitude larger than the van der Waals interaction. In order to improve the description of the electrostatic interaction, additional non-nuclear-centered charges may be added,[22] or dipole, quadrupole, etc., moments may be added at nuclear or bond positions.[23] These descriptions are to a certain extent equivalent since a dipole may be generated as two oppositely charged monopoles, a quadrupole as four monopoles, etc (Figure 10.1).

The *Distributed Multipole Analysis* (DMA) developed by A.J. Stone uses the fact that the ESP arising from the charge overlap between two basis functions can be written in terms of a multipole

Figure 10.1 Generation of dipole and quadupole moments by charges.

expansion around a point between the two nuclei.[24] These moments can be calculated *directly* from the density matrix and the basis functions, and are not a result of a fitting procedure. The multipole expansion is furthermore finite; the highest non-vanishing term is given as the sum of the angular momenta for the two basis functions, for example the product of two *p*-functions gives at most rise to a quadrupole moment. For Gaussian orbitals the expansion point is given in the following equation, where $\mathbf{R}_A$ and $\mathbf{R}_B$ are the positions of the two nuclei and α and β are the exponents of the basis functions (this follows since the product of two Gaussians is a single Gaussian located between the two originals, Equation (3.64)):

$$\mathbf{R}_C = \frac{\alpha \mathbf{R}_A + \beta \mathbf{R}_B}{\alpha + \beta} \tag{10.16}$$

If such distributed multipoles are assigned for each pair of basis functions, the electrostatic potential as seen from outside the charge distribution is reproduced exactly. This, however, would mean that $\sim M_{\text{basis}}^2$ different sites are required. In practice, only the nuclei and possibly bond midpoints are selected as multipole points and all the pair expansion points are moved to the nearest multipole point. By moving the origin, the termination after a finite number of terms is destroyed and an infinite sum over all moments must be used for an exact representation. Since most of the pair expansion points are rather close to either a nucleus or the center of a bond, the importance of higher-order moments is usually quite small. Furthermore, since the majority of the electron density can be represented with just *s*- and *p*-functions for atoms belonging to the first three rows of the periodic table, it follows that a representation in terms of charges, dipoles and quadrupoles located on all nuclei centers gives a quite accurate representation of the ESP center.

The original DMA approach has the disadvantage that the calculated multipole moments are quite sensitive to the employed basis set, especially if diffuse basis functions are present, in analogy with other analyses based on the basis functions used for representing the wave function (e.g. Mulliken). A modified DMA version, where the contributions from the diffuse basis functions are evaluated by a real space integration, while the original Hilbert space partitioning is used for the compact basis functions, display a significantly better numerical stability.[25] The DMA decomposition into atomic contributions can exactly reproduce the reference data by going to sufficiently high multipoles, but the representation by a truncated set of multipole moments is necessarily less accurate than a representation in terms of fitted multipoles to the same order. Fitted multipole methods typically reduce the required moments by one or two, that is fitted charges can reproduce DMA results including up to quadrupoles. Another disadvantage of DMA multipole moments is that they are not very transferable between similar molecules or even between different conformations of the same molecule.

10.3 Population Analysis Based on the Electron Density

The examples in Section 10.1 illustrate that it would be desirable to base a population analysis on properties of the wave function or electron density itself, and not on the basis set chosen for representing the wave function. The electron density is the square of the wave function integrated over $N_{\text{elec}} - 1$ coordinates (it does not matter which coordinates since the electrons are indistinguishable):

$$\rho(\mathbf{r}_1) = \int \left|\Psi(\mathbf{r}_1, \mathbf{r}_2, \mathbf{r}_3, \ldots, \mathbf{r}_{N_{\text{elec}}})\right|^2 d\mathbf{r}_2 d\mathbf{r}_3 \cdots d\mathbf{r}_{N_{\text{elec}}} \tag{10.17}$$

The difficulties in partitioning the electron density into atomic contributions is how to define an "atom" within a molecule. An isolated atom is defined as a nucleus and an integer number of electrons and an atom with a molecule can be defined as a nucleus and a suitable fraction of all the electrons. If the total molecular volume is somehow divided into subsections, each belonging to one specific nucleus, then the electron density can be integrated to give the number of electrons N present in each of these atomic basins Ω, and the (net) atomic charge Q is then obtained by adding the nuclear charge Z:

$$N_A = \int_{\Omega_A} \rho(\mathbf{r})d\mathbf{r} \tag{10.18}$$

$$Q_A = Z_A - N_A \tag{10.19}$$

Equation (10.18) can be generalized to Equation (10.20) to allow for soft boundaries between atomic basins, where the weight function $w_A(\mathbf{r})$ determines the fraction of electron density at point $\mathbf{r}$ that belongs to atom A:

$$N_A = \int w_A(\mathbf{r})\rho(\mathbf{r})d\mathbf{r} \tag{10.20}$$

The division into atomic basins by the weight function requires a *choice* to be made and several different schemes have been proposed. We will in the following discuss the different methods with focus on atomic charges, but higher-order atomic multipole (dipole, quadrupole, etc.) moments can be calculated by suitable integrations over atomic basins analogous to Equation (10.20).

10.3.1 Quantum Theory of Atoms in Molecules

Perhaps the most rigorous way of dividing a molecular volume into atomic subspaces is the *Quantum Theory of Atoms In Molecules* (QTAIM) method of R. Bader, also referred to as *Quantum Chemical Topology* (QCT).[26–28] The electron density is a function of three spatial coordinates and may be analyzed in terms of its topology (maxima, minima and saddle points). In the large majority of cases it is found that the only maxima in the electron density occur at the nuclei (or very close to them), which is reasonable since they are the only sources of positive charge. The nuclei thus act as *attractors* of the electron density. At each point in space the gradient of the electron density points in the direction of the strongest (local) attractor. This forms a rigorous way of dividing the physical space into atomic subspaces: starting from a given point in space a series of infinitesimal steps may be taken in the gradient direction until an attractor is encountered. The collection of all such points forms the atomic basin associated with the attractor (nucleus), corresponding to $w = 1$ inside and $w = 0$ outside the basin in Equation (10.20). If the negative of the electron density is considered, the attractors are local minima, and a basin is then defined as points that end up at the local minimum by a steepest descent minimization (Section 13.2.1). In the other direction (away from other nuclei) the gradient goes asymptotically to zero, and the atomic basin stretches into infinity in this direction. The border between two three-dimensional atomic basins is a two-dimensional surface, often called a *zero flux surface*, as illustrated in Figure 10.2.

The carbon and hydrogen atomic basins in cyclopropane are shown in Figure 10.3.

Once the molecular volume has been divided up, the electron density can be integrated within each of the atomic basins to give atomic charges and dipole, quadrupole, etc., moments. As the dividing surface is rigorously defined in terms of the electron density, these quantities will converge to specific values as the quality of the wave function is increased. Furthermore, as only the electron density is involved, the results are fairly insensitive to the theoretical level used for generating the wave function. If the net charges are taken as nuclear centered (analogous to partial charges for force field

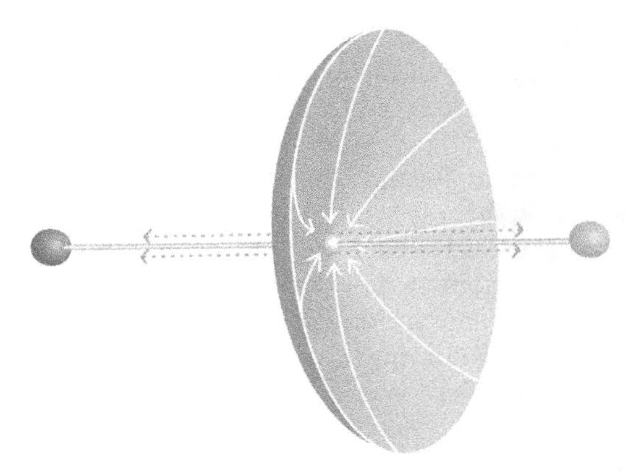

Figure 10.2 Dividing surface between two atomic basins. From N. Singh, R. J. Loader, P. J. O'Malley, P. L. A. Popelier, *J. Phys. Chem. A*, **110** (2006), 6498. Reprinted with permission from The Americal Chemical Society.

methods), they do not reproduce the molecular dipole, quadrupole, etc., moments, and nor do they yield a good representation of the molecular electrostatic potential. They are therefore not suitable for transferring to a force field environment for modeling purposes. Reproducing the molecular electric moment to order N requires that all atomic multipole moments up to order N are included. Including only atomic charges thus only ensures that the total molecular charge is reproduced. If the dipole moments of the atomic basins are also considered, the total molecular dipole moment is reproduced, and similarly for higher-order moments. The dipole moment of CO, for example, is close to zero (0.12 Debye), despite calculated QTAIM charges of ± 1.21. The large dipole moment generated by the two atomic charges positioned at the nuclear positions (−6.54 Debye) is almost exactly cancelled by a compensating contribution from the two atomic dipoles (+6.66 Debye). The QTAIM method is often criticized for generating too large atomic charges for polar bonds, but it should be recognized that this is largely due to the neglect of higher-order moments.

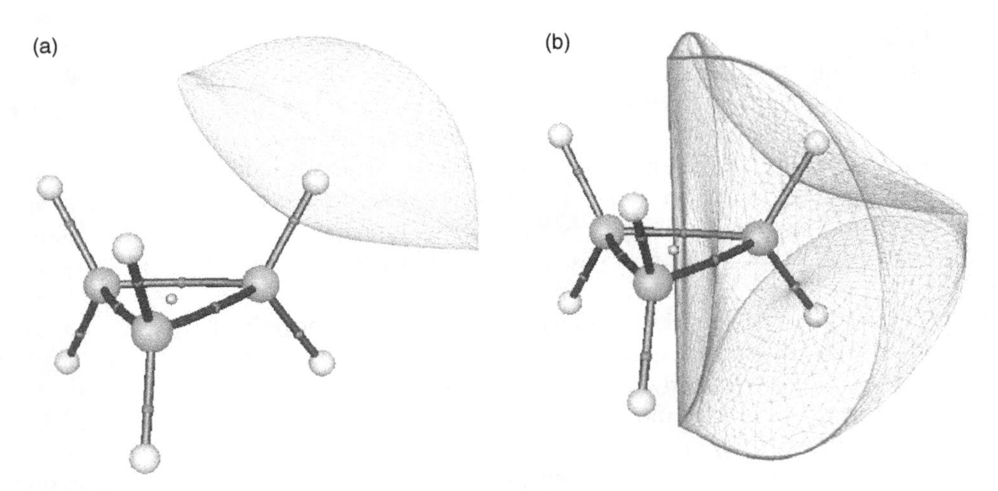

Figure 10.3 Hydrogen and carbon AIM basins for cyclopropane; dots indicate bond and ring critical points. Illustration by courtesy of M. Rafat and P. L. A. Popelier.

A more fundamental problem in the QTAIM approach is the presence of non-nuclear attractors for certain molecular systems, typically involving small clusters of low-valent metals such as Li, Na or Mg, where the topological analysis leads to "atomic" basins that do not contain a nucleus. While non-nuclear attractors are of interest by themselves, they spoil the picture of electrons associated with nuclei, forming atoms within molecules. The presence of non-nuclear attractors depends on both the level of theory, the basis set as well as the molecular geometry, but there are several examples where this feature of the molecular electron density appears genuine.[29,30] For very short internuclear distances an atomic basin may also contain more than one nucleus.

The gradient of the density at a point on a dividing surface between two atomic basins must necessarily be tangential to the surface. Following the gradient path for such a point leads to a stationary point on the surface where the total derivative is zero, marked with a dot in Figure 10.2. The basin attractor is also a stationary point on the electron density surface. The second derivative of the electron density, the Hessian, is a function of the three (Cartesian) coordinates, that is it is a 3×3 matrix. At stationary points, it may be diagonalized and the number of negative eigenvalues determined. The basin attractor is an overall maximum with three negative Hessian eigenvalues. Other stationary points are usually found between nuclei that are "bonded". Such points have a minimum in the electron density in the direction of the nuclei and a maximum in the perpendicular directions, that is the Hessian has one positive and two negative eigenvalues. These are known as *bond critical points*. If the negative of the electron density is considered instead, the attractors are minima (all positive eigenvalues in the second derivative matrix) and the bond critical points are analogous to transition structures (one negative eigenvalue). Comparing with potential energy surfaces (Section 14.1), the (negative) electron density surface may be analyzed in terms of "reaction paths" connecting "transition structures" with minima. Such paths trace the maximum electron density connecting the two nuclei and may be taken as the molecular "bond". It should be noted that bond critical points are not necessarily located on the straight line connecting two nuclei: small strained rings such as cyclopropane, for example, have bond paths that are significantly curved, as illustrated in Figure 10.3. Indeed, the degree of bending tends to correlate with the strain energy.

The value of the electron density at the bond critical point correlates with the strength of the bond, the bond order. As mentioned above, there are certain systems such as metal clusters that have non-nuclear-centered attractors. The corresponding bond critical points have electron densities at least an order of magnitude smaller than "normal" single bonds, and the value of the density at the local maximum is only slightly larger than at the bond critical point. The non-nuclear-centered attractors are thus only weakly defined and may be considered as a special kind of metal bonding, where a "sea" of electrons with weak local maxima surrounds the positive nuclei, which are strong local maxima. In certain cases, bond critical points may also be found between atoms that are not bonded, but experience a strong steric repulsion, corresponding to situations where two atoms are forced to be closer than the sum of their van der Waals radii.[31,32] Such systems usually have values of the electron density at the bond critical point that are at least an order of magnitude smaller than ordinary "bonded" atoms.[33]

There are two other types of critical points, having either one or no negative eigenvalues in the density Hessian. The former are usually found at the center of a ring (illustrated in Figure 10.3 for cyclopropane) and are consequently denoted as a *ring critical point*. The latter are typically found at the center of a cage (e.g. cubane) and are denoted as a *cage critical point*. They correspond to local minima in the electron density in two or three directions.

The second derivative of the electron density, the Laplacian $\nabla^2\rho$, provides information on where electron density is depleted or increased. At a bond critical point the sign of the Laplacian has been used for characterizing the nature of the bond; that is a negative value is taken to indicate a covalent bond, while a positive value is taken to indicate an ionic bond or a van der Waals interaction.

The division of the molecular volume into atomic basins follows from a deeper analysis based on the principle of stationary action. The shapes of the atomic basins and the associated electron densities in a given functional group are very similar in different molecules.[34] The *local* properties of the wave function are therefore transferable to a very good approximation, which rationalizes the basis for organic chemistry, that is functional groups react similarly in different molecules. It may be shown that any observable molecular property may be written as a sum of corresponding atomic contributions:

$$\langle \mathbf{P} \rangle = \sum_{i}^{\text{atomic basins}} \langle \mathbf{P} \rangle_i \tag{10.21}$$

The total energy, for example, may be written as a sum of atomic energies and these atomic energies are again almost constant for the same structural units in different molecules. The QTAIM atomic basins are probably the closest quantum mechanical analogy to the chemical concepts of atoms within a molecule. The good degree of transferability furthermore provides a rationale for defining atom types in force field methods.

10.3.2 Voronoi, Hirshfeld, Stockholder and Stewart Atomic Charges

The QTAIM approach partitions the physical space into atomic basins based on a topological analysis of the electron density itself, but several other methods have been proposed for dividing the molecular space into atomic contributions.

Voronoi charges are based on dividing the physical space according to a distance criterion, that is a given point in space belongs to the nearest nucleus. This is reminiscent of the Mulliken equal partitioning, except that it is the *physical space* between two nuclei that is divided equally to each side, not the *Hilbert space* defined by the basis functions. The atomic basins are bounded by planes perpendicular to the interatomic bonds and are called *Voronoi polyhedra* or *Voronoi cells*. Voronoi charges tend to be rather large. A modified approach where these dividing planes are moved away from the bond midpoint by a distance related to the relative atom sizes, defined by their van der Waals radii, has also been proposed, and this gives significantly smaller charges.[35] In terms of Equation (10.20), the weight function is analogous to the QTAIM approach, a step function with values of either 0 or 1, depending only on the coordinates of the nuclei.

Hirshfeld charges are based on using atomic densities for partitioning the molecular electron density.[36] The *promolecular* density is defined as a sum of atomic densities placed at the nuclear geometries in the molecule. The *actual* molecular electron density at each point in space is then partitioned by Equation (10.20), where the weight function is defined in Equation (10.23) according to the promolecular contributions:

$$\rho_{\text{promolecule}}(\mathbf{r}) = \sum_{A}^{M_{\text{atoms}}} \rho_{\text{A}}^{\text{atomic density}}(\mathbf{r}) \tag{10.22}$$

$$w_{\text{A}}(\mathbf{r}) = \frac{\rho_{\text{A}}^{\text{atomic density}}(\mathbf{r})}{\rho_{\text{promolecule}}(\mathbf{r})} \tag{10.23}$$

Hirshfeld charges may be considered as a soft-boundary version of Voronoi charges. An ambiguity in the original Hirshfeld method is the source of the atomic densities. The normal approach is to use spherically averaged ground state densities for neutral atoms but, in some cases, other valence configurations may be considered, as, for example, the $4s^0 3d^{10}$ electron configuration for Ni rather than $4s^23d^8$.[37] Ambiguities also arise for polar systems like LiF, where either neutral Li and F atoms or charged Li^+ and F^- ions can be used for the promolecular density, and these choices lead to different atomic charges. A similar problem is present in charged systems, where one (or more) atom(s) must be (arbitrarily) selected as charged, rather than neutral, for constructing the promolecular density.

The dependence on the reference atomic densities can be removed by the *iterative Hirshfeld* method.[38] In this approach, the promolecular density and the weight functions are in iteration $N+1$ constructed from atomic densities obtained by linear interpolation between the neutral and cationic/anionic limits based on the Hirshfeld charge at iteration N, and iterated to self-consistency:

$$\begin{aligned} \rho_A^{N+1} &= \left(1 - Q_A^N\right) \rho_{A^0} + Q_A^N \rho_{A^+} \quad ; \quad +1 > Q_A^N > 0 \\ \rho_A^{N+1} &= \left(1 + Q_A^N\right) \rho_{A^0} - Q_A^N \rho_{A^-} \quad ; \quad -1 < Q_A^N < 0 \end{aligned} \tag{10.24}$$

This works well for most organic molecules, but becomes problematic for ionic compounds like metal oxides, where the oxygen atom may end up having a charge larger than −1. In the iterative Hirshfeld method, this requires interpolation between densities for the O^- and O^{2-} ions, but since no isolated atoms are capable of binding two electrons, this leads to an ill-defined interpolation. With an approximate electronic structure method (like HF) and a limited basis set, the O^{2-} density can of course be calculated, but it will depend significantly on the method and basis set, and furthermore extend so far from the nucleus that the weighting function in Equation (10.23) includes regions of space that incorporate other nuclei. The *extended Hirshfeld* method[39] attempts to solve this problem by forming the atomic densities as linear combinations of atomic and monoionic densities only, that is Equation (10.24) without the upper/lower bounds of +1/−1, and the charges are in analogy with the iterative Hirshfeld approach determined self-consistently. This still leaves an open question for atoms where the monoanion is not stable, such as, for example, the nitrogen atom.

The *Charge Model 5* (CM5) is a parameterized method based on calculated Hirshfeld charges that are adjusted such that the total molecular dipole moment is reproduced as well as possible for a selected set of molecules.[40] Earlier CM versions used the same type of parameterization, but build upon Löwdin atomic charges. The electrostatic potential generated by CM5 charges is of similar quality as that from explicitly fitted ESP charges, but do not have the numerical problems associated with the fitting process.

Stockholder charges was originally used as a synonym for Hirshfeld charges, but an *Iterated Stockholder Atom* (ISA) version has been proposed[41] where the weight functions are defined from the atomic spherical average of the (molecular) density as shown below, where the angular brackets $\langle\rangle_A$ denote a spherical average centered on the position of atom A:

$$w_A(\mathbf{r}) = \frac{\langle \rho_A(\mathbf{r}) \rangle_A}{\sum_B \langle \rho_B(\mathbf{r}) \rangle_B} \tag{10.25}$$

The difference relative to Equation (10.23) is that the weight functions are defined without reference to calculated densities for isolated atoms and/or ions. The initial atomic weight functions can be initiated simply as 1 for all atoms and iterated to self-consistency from Equations (10.20) and (10.25), but the convergence is often very slow, requiring thousands of iterations. An improved algorithm for determining the partitioned electron density has been proposed and this also allows defining atomic multipole moments analogous to the distributed multipole analysis.[42]

Stewart atoms are defined as the spherical densities centered at the nuclei that in a least-squares sense fit the molecular density as well as possible, and the resulting densities can be integrated to yield atomic charges and higher-order electric moments.[43] The Stewart atomic densities often have small *negative* contributions far from the nuclei, and the resulting charges are often large and counterintuitive, but give good representations of the molecular electrostatic potential.

10.3.3 Generalized Atomic Polar Tensor Charges

The derivative of the dipole moment with respect to the nuclear coordinates determines intensities of IR absorptions (Section 11.1.6). A central quantity in this respect is the *Atomic Polar Tensor* (APT), which for a given atom is defined by

$$\mathbf{V}^{\mathrm{APT}} = \begin{pmatrix} \frac{\partial \mu_x}{\partial x} & \frac{\partial \mu_x}{\partial y} & \frac{\partial \mu_x}{\partial z} \\ \frac{\partial \mu_y}{\partial x} & \frac{\partial \mu_y}{\partial y} & \frac{\partial \mu_y}{\partial z} \\ \frac{\partial \mu_z}{\partial x} & \frac{\partial \mu_z}{\partial y} & \frac{\partial \mu_z}{\partial z} \end{pmatrix} \tag{10.26}$$

Such a matrix is not independent of the coordinate system, but the trace is. J. Cioslowski has proposed a definition of atomic charges as one-third of the trace over the APT, denoted *Generalized Atomic Polar Tensor* (GAPT) charges.[44] The charge on atom A is defined by

$$Q_{\mathrm{A}}^{\mathrm{GAPT}} = \frac{1}{3}\left(\frac{\partial \mu_x}{\partial x_{\mathrm{A}}} + \frac{\partial \mu_y}{\partial y_{\mathrm{A}}} + \frac{\partial \mu_z}{\partial z_{\mathrm{A}}}\right) \tag{10.27}$$

Since the dipole moment itself is the first derivative of the energy with respect to an external electric field (Section 11.1.1), a calculation of GAPT charges requires the second derivative of the energy. This is a computationally expensive method for generating atomic charges, but if vibrational frequencies are calculated anyway, GAPT charges may be determined with very little additional effort. Dipole derivatives determine intensities of IR absorptions and GAPT charges are therefore directly related to experimentally observable quantities. The GAPT charges are computationally expensive to generate and are sensitive to the amount of electron correlation in the wave function, which has limited the general use of GAPT charges.

10.4 Localized Orbitals

A single Slater determinant (HF or DFT) composed of a set of orthonormal MOs can be written as in

$$\Phi = \frac{1}{\sqrt{N!}}\begin{vmatrix} \phi_1(1) & \phi_2(1) & \cdots & \phi_N(1) \\ \phi_1(2) & \phi_2(2) & \cdots & \phi_N(2) \\ \vdots & \vdots & \ddots & \vdots \\ \phi_1(N) & \phi_2(N) & \cdots & \phi_N(N) \end{vmatrix} \tag{10.28}$$

For computational purposes, it is convenient to work with canonical MOs, that is those that make the matrix of Lagrange multipliers diagonal and are eigenfunctions of the Fock operator at convergence (Equation (3.43)). This corresponds to a specific choice of a unitary (orthogonal) transformation of the occupied MOs. Once the SCF procedure has converged, however, other sets of orbitals may be chosen by forming linear combinations of the canonical MOs. The total wave function, and thus all observable properties, is independent of such a rotation of the MOs:

$$\boldsymbol{\phi}' = \boldsymbol{\phi}\mathbf{U}$$
$$\phi_i' = \sum_{j=1}^{N_{\text{orb}}} u_{ij}\phi_j \tag{10.29}$$

The traditional view of molecular bonds is that they are due to an increased probability of finding electrons between two nuclei, as compared with a sum of the contributions of the pure atomic orbitals. The canonical MOs are delocalized over the whole molecule and do not readily reflect this, since the density between two nuclei is the result of many small contributions from many (all) MOs. There is furthermore little similarity between MOs for systems that by chemical measures should be similar, such as a series of alkanes. The canonical MOs therefore do not reflect the concept of, for example, functional groups in organic chemistry, and nor do they readily allow identification of the bonding properties of the system.

The goal of *Localized Molecular Orbitals* (LMOs) is to define MOs that are spatially confined to a relatively small volume, and therefore clearly display which atoms are bonded and have the property of being approximately constant between structurally similar units in different molecules. LMOs are, furthermore, key elements in linear scaling methods for HF, DFT and electron correlation methods (Section 4.12.2), since they allow a description of the electron density and correlation effects in a given region of the system by only a selected subset of orbitals, rather than all the cannonical orbitals.

A set of LMOs may be defined by optimizing the expectation value of a two-electron operator $\boldsymbol{\Omega}$:[45,46]

$$\langle \boldsymbol{\Omega} \rangle = \sum_{i=1}^{N_{\text{orb}}} \langle \phi_i'\phi_i' | \boldsymbol{\Omega} | \phi_i'\phi_i' \rangle \tag{10.30}$$

The expectation value depends on the u_{ij} parameters in Equation (10.29), which is again a function optimization problem (Chapter 13). It should be noted that the unitary transformation of the orbitals (Equation (10.29)) preserves the orthogonality, that is the resulting LMOs are also orthogonal. Since all observable properties depend only on the total electron density, and not on the individual MOs, there is no unique choice for $\boldsymbol{\Omega}$.

The *Foster–Boys* localization scheme uses the square of the distance between two electrons as the operator and minimizes the expectation value:[47]

$$\langle \boldsymbol{\Omega} \rangle_{\text{FB}} = \sum_{i=1}^{N_{\text{orb}}} \langle \phi_i'\phi_i' | (\mathbf{r}_1 - \mathbf{r}_2)^2 | \phi_i'\phi_i' \rangle \tag{10.31}$$

It can be shown that minimization of $\langle \boldsymbol{\Omega} \rangle_{FB}$ is equivalent to minimizing a sum of orbital contributions, each measuring the spatial extent of the orbital relative to the orbital centroid:

$$\langle \boldsymbol{\Omega}' \rangle_{\text{FB}} = \sum_{i}^{N_{\text{orb}}} \langle \phi_i' | (\mathbf{r} - \langle \phi_i' | \mathbf{r} | \phi_i' \rangle)^2 | \phi_i' \rangle \tag{10.32}$$

Høyvik and Jørgensen have generalized the FB functional to higher-order moments:[48]

$$\langle \boldsymbol{\Omega}' \rangle_{\text{HJ}} = \sum_{i}^{N_{\text{orb}}} \left(\langle \phi_i' | (\mathbf{r} - \langle \phi_i' | \mathbf{r} | \phi_i' \rangle)^n | \phi_i' \rangle \right)^m \tag{10.33}$$

The Forster–Boys functional corresponds to $n = 2$ and $m = 1$, while $n = 4$ serves to decrease the magnitude of the LMO tails and $m = 2$ reduces the spread of the least compact LMO. For extended (periodic) systems described by plane wave basis functions, the equivalent of the Forster–Boys LMOs is called *Wannier* orbitals.[49]

Feng *et al.*[50] have shown that the Forster–Boys LMOs can be made even more compact by 10–25% by allowing the localized orbitals to be non-orthogonal, but this requires a general optimization procedure rather than a simple unitary transformation.[51] Non-orthogonal LMOs achieve the increased compactness because they can remove the orthogonalization tails of orthogonal LMOs.

The *Edmiston–Ruedenberg* localization scheme uses the inverse of the distance between two electrons as the operator and maximizes the expectation value:[52]

$$\langle \boldsymbol{\Omega} \rangle_{\text{ER}} = \sum_{i=1}^{N_{\text{orb}}} \left\langle \phi'_i \phi'_i \left| \frac{1}{|\mathbf{r}_1 - \mathbf{r}_2|} \right| \phi'_i \phi'_i \right\rangle \tag{10.34}$$

This corresponds to determining a set of LMOs that maximizes the self-repulsion energy. It can be shown that this is equivalent to minimizing the interorbital repulsion:

$$\langle \boldsymbol{\Omega}' \rangle_{\text{ER}} = \sum_{\substack{i,j=1 \\ j \neq i}}^{N_{\text{orb}}} \left\langle \phi'_i \phi'_j \left| \frac{1}{|\mathbf{r}_1 - \mathbf{r}_2|} \right| \phi'_i \phi'_j \right\rangle \tag{10.35}$$

The *von Niessen* localization scheme uses the δ-function of the distance between two electrons as the operator and maximizes the expectation value:[53]

$$\langle \boldsymbol{\Omega} \rangle_{\text{vN}} = \sum_{i=1}^{N_{\text{orb}}} \langle \phi'_i \phi'_i | \delta(\mathbf{r}_1 - \mathbf{r}_2) | \phi'_i \phi'_i \rangle \tag{10.36}$$

This corresponds to determining a set of LMOs that maximizes the "self-charge".

The *Pipek–Mezey* localization scheme corresponds to maximizing the sum of Mulliken atomic charges.[45] The contribution to atom A is given in

$$Q_{\text{A}} = Z_{\text{A}} - \sum_{i=1}^{N_{\text{orb}}} \sum_{\alpha \in \text{A}}^{M_{\text{basis}}} \sum_{\beta}^{M_{\text{basis}}} c_{\alpha i} c_{\beta i} S_{\alpha\beta} \tag{10.37}$$

The function to be maximized is given in

$$\langle \boldsymbol{\Omega} \rangle_{\text{PM}} = \sum_{A=1}^{\text{Atoms}} (Q_A)^2 \tag{10.38}$$

The optimization of $\langle \boldsymbol{\Omega} \rangle_{\text{PM}}$ may be problematic for systems where the Mulliken population analysis itself is problematic, for example when using large basis sets with diffuse functions or for extended systems (e.g. clusters with hundreds of atoms), where even a medium-sized basis set may be near-linear-dependent. The Pipek–Mezey approach can be generalized to maximizing other types of atomic charges, but the resulting LMOs are very similar.[54]

There is little experience with the von Niessen method but, for most molecules, the other three schemes tend to give very similar LMOs. The main exception is systems containing both σ- and π-bonds, such as ethylene. The Pipek–Mezey procedure (and its generalizations) will preserve the σ/π-separation, while the Edmiston–Ruedenberg and Foster–Boys schemes produce bent "banana" bonds. Similarly, for planar molecules that contain lone pairs (such as water or formaldehyde), the

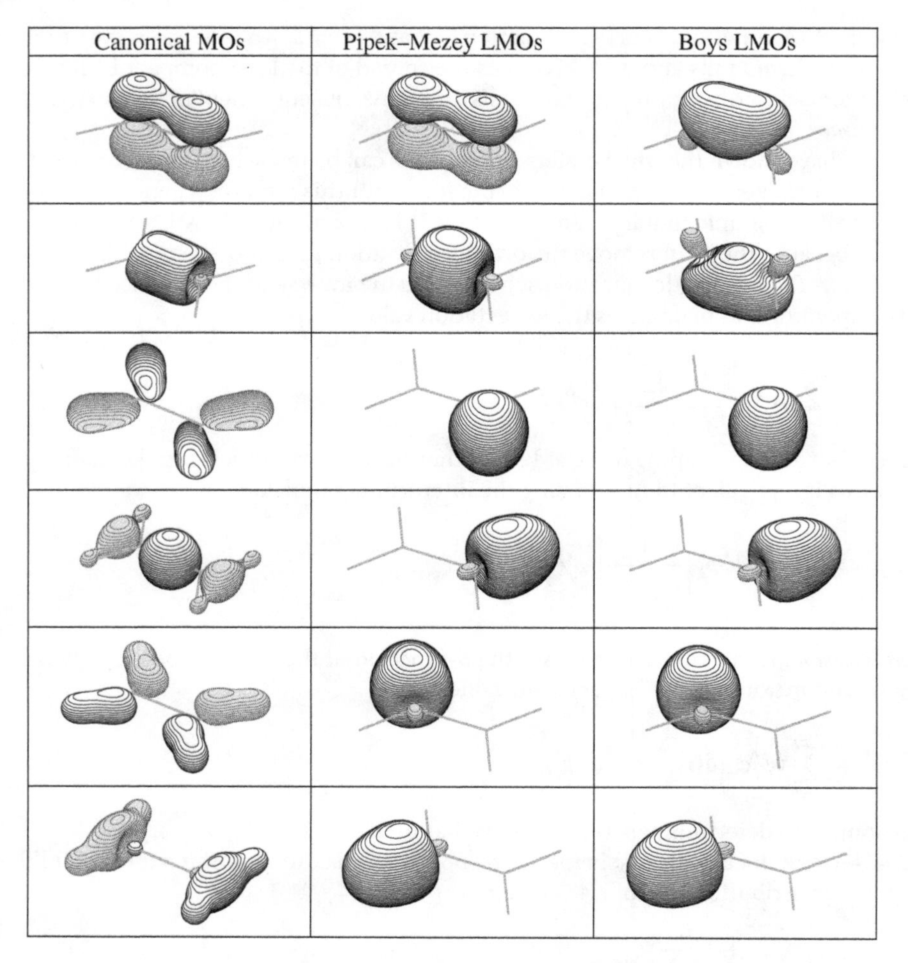

Figure 10.4 Canonical and localized molecular valence orbitals for ethylene.

Pipek–Mezey method will produce one in-plane σ-type lone pair and one out-of-plane π-type lone pair, while the Edmiston–Ruedenberg and Foster–Boys schemes produce two equivalent "rabbit ear" lone pairs. The canonical MOs and the Foster–Boys and Pipek–Mezey LMOs for ethylene are shown in Figure 10.4 for the valence orbitals.

10.4.1 Computational considerations

The dipole integrals in the molecular basis required for the Forster–Boys procedure may be obtained from the corresponding AO integrals:

$$\langle \phi_i | \mathbf{r} | \phi_i \rangle = \sum_{\alpha}^{M_{\text{basis}}} c_{\alpha i} \left(\sum_{\beta}^{M_{\text{basis}}} c_{\beta i} \langle \chi_\alpha | \mathbf{r} | \chi_\beta \rangle \right) \tag{10.39}$$

This is a process that increases as the cube of the basis set size, and the optimization of the $\langle \Omega' \rangle_{FB}$ function is therefore an M^3_{basis} method. The Edmiston–Ruedenberg localization in the above formulation requires standard two-electron integrals over MOs, analogous to those used in electron correlation methods (Equation (4.13)), and it therefore involves a computational effort that increases as M^5_{basis}. If only LMOs for occupied orbitals are desired, only integrals involving occupied MOs are needed and the transformation is not particularly time-consuming for reasonably sized systems,[55] but it will ultimatively require a significant effort for large systems. Head-Gordon and coworkers[56] have shown that the problem can be reformulated as a series of iterative one-index transformations, which reduces the formal scaling to M^3_{basis}. The von Niessen method may be shown to involve a computational effort that increases as M^5_{basis}, while the Pipek–Mezey charge localization only involves overlap integrals between basis functions and consequently has an M^3_{basis} computational dependence. The computational requirements of the PM generalizations to other types of atomic charge depend on the specific type of atomic charge.[54]

The localization of occupied MOs by the FB, ER or PM schemes is a relatively easy optimization problem and has traditionally been done by performing a sequence of 2×2 orbital rotations (Section 17.2). Localization of virtual orbitals and the use of more complicated localization schemes like the HJ functional in Equation (10.33) is computationally more difficult and requires second-order optimization methods (Section 13.2.3).

Although the localization by an energy criterion (Edmiston–Ruedenberg) may be considered more "fundamental" than one based on distance (Forster–Boys) or atomic charge (Pipek–Mezey), the difference in computational effort means that the Forster–Boys or Pipek–Mezey procedures are often used in practice, especially since there is normally little difference in the shape of the final LMOs.

Localized molecular orbitals are generally found to reflect the usual picture of bonding, that is they are localized between two nuclei or in some cases, such as diborane, extended over three nuclei. Although they indicate which atoms are bonded, they do not directly give any information about the strength of the bonds. Furthermore, localizing a set of MOs corresponds to determining orbitals containing electron pairs. In structures with significant open-shell character (e.g. transition structures) it may be difficult to achieve a proper localization of the MOs and molecules with several resonance structures, as, for example, aromatic systems, may have more than one set of LMOs.

10.5 Natural Orbitals

The electron density calculated from a wave function is given as the square of the function, $|\Psi|^2 = \Psi^*\Psi$. The reduced density matrix of order k, γ_k, is defined by[57]

$$\gamma_k(\mathbf{r}_1, \ldots, \mathbf{r}_k, \mathbf{r}'_1, \ldots, \mathbf{r}'_k) = \binom{N_{elec}}{k} \int \Psi^* \left(\mathbf{r}'_1, \ldots, \mathbf{r}'_k, \mathbf{r}_{k+1}, \ldots, \mathbf{r}_{N_{elec}}\right) \times \Psi\left(\mathbf{r}_1, \ldots, \mathbf{r}_k, \mathbf{r}_{k+1}, \ldots, \mathbf{r}_{N_{elec}}\right) d\mathbf{r}_{k+1} \cdots d\mathbf{r}_{N_{elec}} \tag{10.40}$$

Note that some of the coordinates for Ψ^* and Ψ are different. Of special importance in electronic structure theory are the first- and second-order reduced density matrices, $\gamma_1(\mathbf{r}_1, \mathbf{r}'_1)$ and $\gamma_2(\mathbf{r}_1, \mathbf{r}_2, \mathbf{r}'_1, \mathbf{r}'_2)$, since the Hamiltonian operator only contains one- and two-electron operators. Integrating the first-order density matrix over coordinate 1 yields the number of electrons, N_{elec}, while the integral of the second-order density matrix over coordinates 1 and 2 is $N_{elec}(N_{elec} - 1)$, that is the number of electron pairs. The first-order density matrix may be diagonalized and the corresponding eigenvectors and eigenvalues are called *Natural Orbitals* (NO) and *Occupation Numbers.* The corresponding eigenfunctions for the second-order density matrix are called *Natural Geminals.* For a single-determinant RHF wave function, the first-order density matrix is identical to the density matrix used in the formation of the Fock matrix (Equation (3.52)) and the natural orbitals have occupation numbers of either 0 or 2 (exactly). Since there are $^1/_2 N_{\text{elec}}$ orbitals with degenerate eigenvalues of 2, the HF natural orbitals are not uniquely defined and they may be any linear combination of the canonical orbitals. For a multideterminant wave function (MCSCF, CI, MP or CC) the occupation numbers may assume fractional values between 0 and 2. UHF wave functions (when different from RHF) will in general also give fractional occupations.

The original definition of natural orbitals was in terms of the density matrix from a full CI wave function, that is the best possible for a given basis set.[58] These natural orbitals have the significance that they provide the lowest energy for a CI expansion using only a limited set of orbitals, when the orbitals with the largest occupation numbers are selected. The natural orbitals thus provide the fastest convergence towards the full CI result as the size of the orbital space is enlarged. Natural orbitals can also be considered as a principal component analysis of a wave function (Section 18.4.3).

When natural orbitals are determined from a wave function that only includes a limited amount of electron correlation (i.e. not full CI), the fastest convergence property is not rigorously guaranteed but, since most practical methods recover 80–90% of the total electron correlation, the occupation numbers provide a good guideline for how important a given orbital is. This is the reason why natural orbitals are often used for evaluating which orbitals should be included in an MCSCF wave function (Section 4.6).

10.5.1 Natural Atomic Orbital and Natural Bond Orbital Analyses

The concept of natural orbitals may be used for distributing electrons into atomic and molecular orbitals, and thereby derive atomic charges and molecular bonds. The idea in the *Natural Atomic Orbital* (NAO) and *Natural Bond Orbital* (NBO) analysis developed by Weinhold and coworkers[59] is to use the one-electron density matrix for defining the shape of the atomic orbitals in the molecular environment and to derive molecular bonds from electron density between atoms.

Let us assume that the basis functions have been arranged such that all orbitals located on center A are before those on center B, which are before those on center C, etc.:

$$\chi_1^{\text{A}}, \chi_2^{\text{A}}, \chi_3^{\text{A}}, \ldots, \chi_k^{\text{B}}, \chi_{k+1}^{\text{B}}, \chi_{k+2}^{\text{B}}, \ldots, \chi_n^{\text{C}}, \chi_{n+1}^{\text{C}}, \chi_{n+2}^{\text{C}}, \ldots \tag{10.41}$$

The (one-electron) density matrix can be written in terms of blocks of basis functions belonging to a specific center, as shown in

$$\mathbf{D} = \begin{pmatrix} \mathbf{D}^{\text{AA}} & \mathbf{D}^{\text{AB}} & \mathbf{D}^{\text{AC}} & \vdots \\ \mathbf{D}^{\text{AB}} & \mathbf{D}^{\text{BB}} & \mathbf{D}^{\text{BC}} & \vdots \\ \mathbf{D}^{\text{AC}} & \mathbf{D}^{\text{BC}} & \mathbf{D}^{\text{CC}} & \vdots \\ \cdots & \cdots & \cdots & \ddots \end{pmatrix} \tag{10.42}$$

The natural atomic orbitals for atom A in the molecular environment may be defined as those that diagonalize the $\mathbf{D}^{\text{AA}}$ block, NAOs for atom B as those that diagonalize the $\mathbf{D}^{\text{BB}}$ block, etc. These NAOs will in general not be orthogonal and the orbital occupation numbers will therefore not sum to the total number of electrons. To achieve a well-defined partitioning of the electrons, the orbitals must be orthogonalized.

The NAOs will normally resemble the pure atomic orbitals (as calculated for an isolated atom) and may be divided into a "natural minimal basis" (corresponding to the occupied atomic orbitals for the isolated atom) and a remaining set of natural "Rydberg" orbitals based on the magnitude of the occupation numbers. The minimal set of NAOs will normally be strongly occupied (i.e. having occupation numbers significantly different from zero), while the Rydberg NAO usually will be weakly occupied (i.e. having occupation numbers close to zero). There are as many NAOs as the size of the atomic basis set and the number of Rydberg NAOs thus increases as the basis set is enlarged. It is therefore desirable that the orthogonalization procedure preserves the form of the strongly occupied orbitals as much as possible, which is achieved by using an occupancy-weighted orthogonalizing matrix (Equation (17.85)). If all orbital occupancies are exactly 2 or 0, the occupancy-weighted orthogonalization is identical to the Löwdin method (Equation (10.9)). The procedure is as follows (Figure 10.5):

1. Each of the atomic blocks in the density matrix is diagonalized to produce a set of non-orthogonal NAOs, often denoted "pre-NAOs".
2. The strongly occupied pre-NAOs for each center are made orthogonal to all the strongly occupied pre-NAOs on the other centers by an occupancy-weighted procedure. This destroys the orthogonality of the occupied-virtual blocks.
3. The weakly occupied pre-NAOs on each center are made orthogonal to the strongly occupied NAOs on all the centers by a Gram–Schmidt orthogonalization.
4. The (modified) weakly occupied NAOs are made orthogonal to all the weakly occupied NAOs on the other centers by an occupancy-weighted procedure.

The final set of orthogonal orbitals are simply denoted NAOs and the diagonal elements of the density matrix in this basis are the orbital populations. Summing all contributions from orbitals belonging to a specific center produces the atomic charge, often denoted the NPA (Natural Population Analysis) charge. It is usually found that the natural minimal NAOs contribute 99+% of the electron density and they form a very compact representation of the wave function in terms of atomic orbitals. The further advantage of the NAOs is that they are defined from the density matrix, guaranteeing that the electron occupation is between 0 and 2 and that they converge to well-defined values as the size of the basis set is increased. Furthermore, the analysis may also be performed for correlated wave functions. The disadvantage is that the NAOs may still extend quite far from the atom upon which they are derived and, analogously to the Mulliken approach, these NAOs may describe electron density that is near another nucleus but are counted as belonging to the nucleus upon which they are centered.

Once the density matrix has been transformed to the NAO basis, bonds between atoms may be identified from the off-diagonal blocks. The procedure involves the following steps:

1. NAOs for an atomic block in the density matrix that have occupation numbers very close to 2 (say > 1.999) are identified as core orbitals. Their contributions to the density matrix are removed.
2. NAOs for an atomic block in the density matrix that have large occupancy numbers (say > 1.90) are identified as lone pair orbitals. Their contributions to the density matrix are also removed.
3. Each pair of atoms (AB, AC, BC, ...) are now considered and the two-by-two subblocks of the density matrix (with the core and lone pair contributions removed) are diagonalized. Natural bond orbitals (NBO) are identified as eigenvectors that have large eigenvalues (occupation numbers larger than, say, > 1.90).
4. If an insufficient number of NBOs are generated by the above procedure (sum of occupation numbers for core, lone pair and bond orbitals is significantly less than the number of electrons), the criterion for accepting an NBO may be gradually lowered until a sufficiently large fraction of the

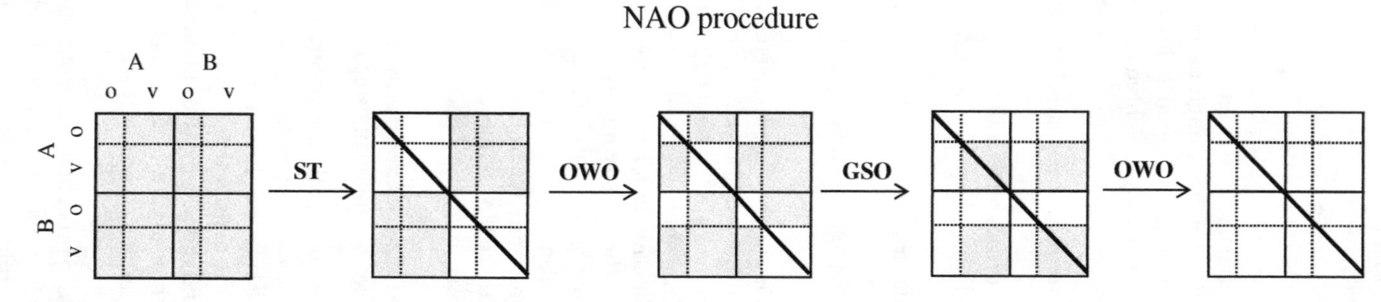

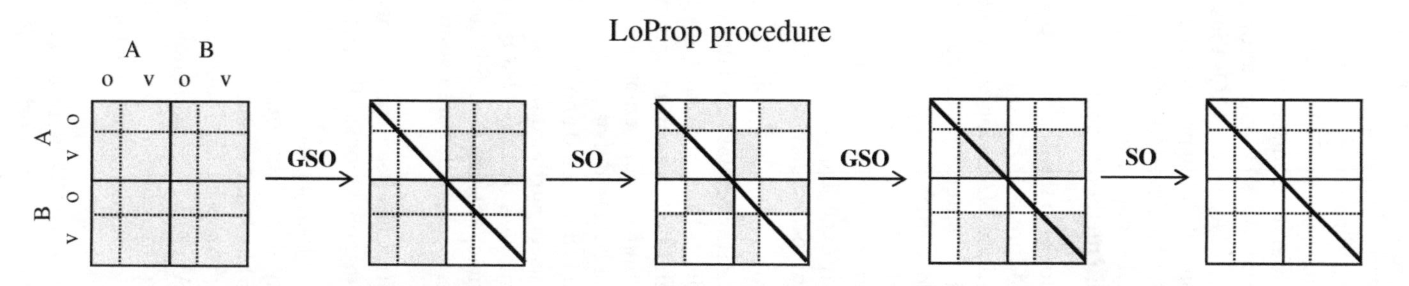

Figure 10.5 Illustration of the orthogonalizations involved in the NAO (using the density matrix) and LoProp (using the overlap matrix) analyses. A/B are two representative atoms, o/v indicate occupied and virtual orbitals for the LoProp approach and strongly/weakly occupied orbitals for NAO. Gray/white indicate non-zero/zero elements in the matrix block, black diagonal indicates non-zero diagonal elements. Notation: ST = Similarity Transform (diagonalization), OWO = Occupancy-Weighted Orthogonalization, GSO = Gram–Schmidt Orthogonalization, SO = Symmetric Orthogonalization (Löwdin orthogonalization).

electrons has been assigned to bonds. Alternatively, a search may be initiated for three-center bonds. The contributions to the density matrix from all diatomic bonds are removed and all three-by-three sub-blocks are diagonalized. Such three-center bonds are quite rare, boron systems being the most notable exception.

Once NBOs have been identified, they may be written as linear combinations of the NAOs, forming a localized picture of which "atomic" orbitals are involved in the bonding.

Gagliardi, Lindh and Karlström have proposed a *LoProp* (*Local Properties*) method,[60] which conceptually is very similar to the natural orbital analysis, with the difference that the analysis is done in terms of the basis function overlap matrix, rather than the density matrix.

The procedure is as follows (Figure 10.5):

1. The basis functions are divided into atomic occupied and virtual orbitals. This is straightforward if the basis set is of the ANO (atomic natural orbital) type (Section 5.4.5), since these are defined by a general contraction of a set of primitive functions with atomic natural orbital coefficients and are therefore already orthogonal and partitioned into occupied and virtual orbitals. For other types of basis sets, where the contracted basis functions are non-orthogonal and all contribute to the atomic occupied orbitals, this requries a recontraction and Gram–Schmidt orthogonalization of the basis functions.
2. All the occupied basis functions on all atoms are made orthogonal to each other by a Löwdin orthogonalization and all virtual orbitals are similarly made orthogonal to each other by a Löwdin orthogonalization. This destroys the orthogonality of the occupied–virtual blocks.
3. All the virtual orbitals are made orthogonal to all the occupied orbitals by a Gram–Schmidt orthogonalization. This destroys the orthogonality of the virtual–virtual blocks.
4. The (modified) virtual orbitals are made orthogonal to each other by a Löwdin orthogonalization.

The LoProp can in analogy with the DMA approach be generalized to generate localized multipole moments and polarizabilitites. A disadvantage of the LoProp scheme is that the convergence by including higher and higher orders of multipole moments often is non-monotomic, for example a representation with atomic charges and dipoles may be worse than with charges alone.[61]

The LoProp partitioning takes place in the Hilbert space defined by the basis set, while the NAO partitioning takes place in a space defined by the electron density represented in the basis set. For a given basis set, the LoProp method only depends on the molecular geometry, while the NAO approach in addition depends on the electronic state. The LoProp method therefore allows analysis of different electronic states with the same atomic partitioning and provides a more smooth connection when the electronic structure changes significantly with geometry than NAO, as, for example, for following bond forming/breaking reactions or near an avoided crossing.

10.6 Computational Considerations

Population analyses based on basis functions (such as Mulliken or Löwdin) require insignificant computational time. The NAO and LoProp analyses involve only matrix diagonalization of small subsets of the density or overlap matrix, and similarly require a negligible amount of computer time, although they are more involved than a Mulliken or Löwdin analysis. The determination of ESP fitted charges requires an evaluation of the potential at many (often several thousand) points in space and a subsequent solution of a matrix equation for minimizing the least-squares error function. For large systems, this is no longer completely trivial in terms of computer time. The QTAIM population

Table 10.1 Atomic charges for carbon in CH_3OH with the B3LYP method and basis sets of increasing quality.

Basis	pcseg-1	pcseg-2	pcseg-3	pcseg-4	aug-pcseg-1	aug-pcseg-2	aug-pcseg-3	aug-pcseg-4
Mulliken	0.04	0.17	−0.01	−0.07	0.16	−0.02	−0.28	−0.75
Löwdin	−0.12	0.01	0.06	0.17	−0.02	0.30	0.41	0.56
NPA	−0.20	−0.26	−0.24	−0.24	−0.21	−0.26	−0.24	−0.25
I-Hirshfeld	−0.02	−0.02	−0.02	−0.02	−0.02	−0.02	−0.02	−0.02
GAPT	0.49	0.49	0.48	0.48	0.48	0.48	0.48	0.48
QTAIM	0.58	0.63	0.59	0.59	0.57	0.63	0.59	0.59
HLY	0.20	0.18	0.18	0.18	0.16	0.17	0.18	0.18

analysis requires a complete topological analysis of the electron density and a subsequent numerical integration of the atomic basins. For medium-sized systems and medium-quality wave functions, such an analysis may be more time-consuming than determining the wave function itself. Voronoi and Hirshfeld charges similarly require a numerical integration of the electron density and the determination of Stewart atoms has proven to be computationally quite difficult. GAPT charges require calculation of the second derivative of the wave function, which is computationally demanding, especially for large molecules and/or correlated wave functions. There is little doubt that these computational considerations partly explain the popularity of especially the Mulliken population analysis, despite its well-known shortcomings. For analysis purposes, the NAO or LoProp procedures are attractive, but for modeling purposes (i.e. force field charges) ESP charges are clearly the logical choice.

10.7 Examples

Table 10.1 shows how the atomic charge on the carbon atom in methanol depends on the method for assigning the charge, using the B3LYP computational level and basis sets of increasing quality. It is evident that the Mulliken and Löwdin values do not converge as the basis set is increased and the values in general behave unpredictably. The charges determined by the NPA, iterative Hirshfeld, GAPT, QTAIM and HLY procedures, on the other hand, attain well-defined values as the basis set is enlarged and are rather insensitive to the presence of diffuse functions. The charges assigned by these methods, however, differ significantly, ranging from the NPA value of −0.25 to the QTAIM value of +0.59. These variations are mirrored by the charges on the other atoms, as shown in Table 10.2. The

Table 10.2 Atomic charges for CH_3OH with the B3LYP method and the aug-pcseg-3 basis set.

	C	O	H (O)	H (C)	H2 (C)
NPA	−0.24	−0.72	0.46	0.19	0.16
I-Hirshfeld	−0.02	−0.25	0.16	0.05	0.03
GAPT	0.48	−0.60	0.24	−0.01	−0.06
QTAIM	0.59	−1.10	0.56	0.00	−0.02
ESP-HLY	0.17	−0.60	0.38	0.06	−0.01

Table 10.3 Bond orders for the C–O and O–H bonds in CH_3OH with the B3LYP method and basis sets of increasing quality.

Basis	pcseg-1	pcseg-2	pcseg-3	pcseg-4	aug-pcseg-1	aug-pcseg-2	aug-pcseg-3	aug-pcseg-4
C–O	0.96	0.90	0.95	0.84	0.79	0.57	0.69	0.59
O–H	0.96	1.03	0.97	0.96	0.87	0.93	0.91	0.87

bond orders calculated by the Mulliken-type approach in Equation (10.11) are shown in Table 10.3, and are likewise sensitive to the basis set quality, especially the presence of diffuse functions.

References

1 S. M. Bachrach, *Reviews in Computational Chemistry* **5**, 171 (1994).
2 R. S. Mulliken, *Journal of Chemical Physics* **23** (10), 1833–1840 (1955).
3 P.-O. Löwdin, *Advancs in Quantitative Chemistry* **5**, 185 (1970).
4 T. Kar, L. Behera and A. B. Sannigrahi, *Chemical Physics Letters* **163** (2–3), 157–164 (1989).
5 I. Mayer, *Chemical Physics Letters* **393** (1–3), 209–212 (2004).
6 J. Autschbach, B. A. Hess, M. P. Johansson, J. Neugebauer, M. Patzschke, P. Pyykko, M. Reiher and D. Sundholm, *Physical Chemistry Chemical Physics* **6** (1), 11–22 (2004).
7 L. Pauling, *Journal of the American Chemical Society* **69** (3), 542–553 (1947).
8 I. Mayer, *Chemical Physics Letters* **97** (3), 270–274 (1983).
9 P. Bultinck, R. Ponec and S. Van Damme, *Journal of Physical Organic Chemistry* **18** (8), 706–718 (2005).
10 D. E. Williams, *Reviews in Computational Chemistry* **2**, 219 (1991).
11 U. C. Singh and P. A. Kollman, *Journal of Computational Chemistry* **5** (2), 129–145 (1984).
12 B. H. Besler, K. M. Merz and P. A. Kollman, *Journal of Computational Chemistry* **11** (4), 431–439 (1990).
13 L. E. Chirlian and M. M. Francl, *Journal of Computational Chemistry* **8** (6), 894–905 (1987).
14 C. M. Breneman and K. B. Wiberg, *Journal of Computational Chemistry* **11** (3), 361–373 (1990).
15 H. Hu, Z. Lu and W. Yang, *Journal of Chemical Theory and Computation* **3** (3), 1004–1013 (2007).
16 E. V. Tsiper and K. Burke, *Journal of Chemical Physics* **120** (3), 1153–1156 (2004).
17 M. M. Francl, C. Carey, L. E. Chirlian and D. M. Gange, *Journal of Computational Chemistry* **17** (3), 367–383 (1996).
18 E. Sigfridsson and U. Ryde, *Journal of Computational Chemistry* **19** (4), 377–395 (1998).
19 C. I. Bayly, P. Cieplak, W. D. Cornell and P. A. Kollman, *Journal of Physical Chemistry* **97** (40), 10269–10280 (1993).
20 M. M. Francl and L. E. Chirlian, *Reviews in Computational Chemistry* **14**, 1–31 (2000).
21 C. A. Reynolds, J. W. Essex and W. G. Richards, *Journal of the American Chemical Society* **114** (23), 9075–9079 (1992).
22 C. Aleman, M. Orozco and F. J. Luque, *Chemical Physics* **189** (3), 573–584 (1994).
23 Y. Yuan, M. J. L. Mills and P. L. A. Popelier, *Journal of Computational Chemistry* **35** (5), 343–359 (2014).
24 A. J. Stone and M. Alderton, *Molecular Physics* **56** (5), 1047–1064 (1985).
25 A. J. Stone, *Journal of Chemical Theory and Computation* **1** (6), 1128–1132 (2005).
26 R. F. W. Bader, *Atoms in Molecules* (Clarendon Press, Oxford, 1990).
27 P. L. A. Popelier, *Atoms in Molecules: An Introduction* (Pearson Education, 1999).
28 R. F. W. Bader, *Chemical Reviews* **91** (5), 893–928 (1991).

29 L. A. Terrabuio, T. Q. Teodoro, M. G. Rachid and R. L. A. Haiduke, *Journal of Physical Chemistry A* **117** (40), 10489–10496 (2013).
30 L.-C. Wu, C. Jones, A. Stasch, J. A. Platts and J. Overgaard, *European Journal of Inorganic Chemistry* (32), 5536–5540 (2014).
31 S. Grimme, C. Mueck-Lichtenfeld, G. Erker, G. Kehr, H. Wang, H. Beckers and H. Willner, *Angewandte Chemie – International Edition* **48** (14), 2592–2595 (2009).
32 J. Dillen, *International Journal of Quantum Chemistry* **113** (18), 2143–2153 (2013).
33 C. F. Matta, N. Castillo and R. J. Boyd, *Journal of Physical Chemistry A* **109** (16), 3669–3681 (2005).
34 P. L. A. Popelier and F. M. Aicken, *Chemphyschem* **4** (8), 824–829 (2003).
35 B. Rousseau, A. Peeters and C. Van Alsenoy, *Journal of Molecular Structure – Theochem* **538**, 235–238 (2001).
36 F. L. Hirshfeld, *Theoretica Chimica Acta* **44** (2), 129–138 (1977).
37 C. F. Guerra, J. W. Handgraaf, E. J. Baerends and F. M. Bickelhaupt, *Journal of Computational Chemistry* **25** (2), 189–210 (2004).
38 P. Bultinck, C. Van Alsenoy, P. W. Ayers and R. Carbo-Dorca, *Journal of Chemical Physics* **126** (14), 144111 (2007).
39 T. Verstraelen, P. W. Ayers, V. Van Speybroeck and M. Waroquier, *Journal of Chemical Theory and Computation* **9** (5), 2221–2225 (2013).
40 A. V. Marenich, S. V. Jerome, C. J. Cramer and D. G. Truhlar, *Journal of Chemical Theory and Computation* **8** (2), 527–541 (2012).
41 T. C. Lillestolen and R. J. Wheatley, *Journal of Chemical Physics* **131** (14), 144101 (2009).
42 A. J. Misquitta, A. J. Stone and F. Fazeli, *Journal of Chemical Theory and Computation* **10** (12), 5405–5418 (2014).
43 A. T. B. Gilbert, P. M. W. Gill and S. W. Taylor, *Journal of Chemical Physics* **120** (17), 7887–7893 (2004).
44 J. Cioslowski, *Journal of the American Chemical Society* **111** (22), 8333–8336 (1989).
45 J. Pipek and P. G. Mezey, *Journal of Chemical Physics* **90** (9), 4916–4926 (1989).
46 D. A. Kleier, T. A. Halgren, J. H. Hall and W. N. Lipscomb, *Journal of Chemical Physics* **61** (10), 3905–3919 (1974).
47 J. M. Foster and S. F. Boys, *Reviews of Modern Physics* **32** (2), 300–302 (1960).
48 I.-M. Hoyvik and P. Jorgensen, *Journal of Chemical Physics* **138** (20), 204104 (2013).
49 N. Marzari, A. A. Mostofi, J. R. Yates, I. Souza and D. Vanderbilt, *Reviews of Modern Physics* **84** (4) (2012).
50 H. S. Feng, J. Bian, L. M. Li and W. T. Yang, *Journal of Chemical Physics* **120** (20), 9458–9466 (2004).
51 G. Cui, W. Fang and W. Yang, *Journal of Physical Chemistry A* **114** (33), 8878–8883 (2010).
52 C. Edmiston and K. Ruedenberg, *Journal of Chemical Physics* **43** (10), S097 (1965).
53 W. von Niessen, *Journal of Chemical Physics* **56** (9), 4290–4297 (1972).
54 S. Lehtola and H. Jónsson, *Journal of Chemical Theory and Computation* **10** (2), 642–649 (2014).
55 R. C. Raffenetti, K. Ruedenberg, C. L. Janssen and H. F. Schaefer, *Theoretica Chimica Acta* **86** (1–2), 149–165 (1993).
56 J. E. Subotnik, Y. H. Shao, W. Z. Liang and M. Head-Gordon, *Journal of Chemical Physics* **121** (19), 9220–9229 (2004).
57 R. G. Parr and W. Yang, *Density Functional Theory* (Oxford University Press, 1989).
58 P. O. Lowdin, *Physical Review* **97** (6), 1474–1489 (1955).
59 A. E. Reed, L. A. Curtiss and F. Weinhold, *Chemical Reviews* **88** (6), 899–926 (1988).
60 L. Gagliardi, R. Lindh and G. Karlström, *Journal of Chemical Physics* **121** (10), 4494–4500 (2004).
61 J. M. H. Olsen, N. H. List, K. Kristensen and J. Kongsted, *Journal of Chemical Theory and Computation* **11** (4), 1832–1842 (2015).

11

Molecular Properties

The focus in Chapters 3 and 4 is on determining the wave function and its energy at a given geometry in the absence of external fields (electric or magnetic). While relative energies are certainly of interest, there are many other molecular properties that can be calculated by electronic structure methods.[1–4] Most properties may be defined as the response of a wave function, an energy or an expectation value of an operator to a perturbation, where the perturbation may be any kind of operator not present in the Hamiltonian used for solving the Schrödinger equation. It may, for example, be terms arising in a relativistic treatment (e.g. spin–orbit interactions), which can be added as perturbations in non-relativistic theory. It may also be external fields (electric or magnetic) or an internal perturbation, such as a nuclear or electron spin. If we furthermore include "perturbations" such as adding or removing an electron, electron affinities and ionization potentials are also included in this definition. There are a few remaining properties that cannot easily be characterized as a response to a perturbation, most notably transition moments, which determine absorption intensities. These depend on matrix elements between two different wave functions, but can be obtained from the same type of calculations that provide, for example, excitation energies.

We will here consider five types of perturbations:

- External electric field (**F**)
- External magnetic field (**B**)
- Nuclear magnetic moment (nuclear spin, **I**)
- Electron magnetic moment (electron spin, **s**)
- Change in the nuclear geometry (**R**).

The external electric and magnetic fields may either be *time-independent*, which lead to *static* properties, or *time-dependent*, leading to *dynamic* properties. Time-dependent fields are usually associated with electromagnetic radiation characterized by a frequency, and static properties may be considered as the limiting case of dynamic properties when the frequency goes to zero. We will focus on properties of a single molecule for a fixed geometry, while a direct comparison with (gas-phase) experimental macroscopic quantities may be done by proper averaging over for example vibrational and rotational states. We will furthermore concentrate on the electronic contribution to properties; the corresponding nuclear contribution (if present) is normally trivial to calculate as it is independent of the wave function.

Introduction to Computational Chemistry, Third Edition. Frank Jensen.

Companion Website: http://www.wiley.com/go/jensen/computationalchemistry3

The nuclear magnetic moment may be considered as an artificial perturbation, since it is an inherent part of a given nucleus (isotope). In most applications, however, the nucleus is modeled as a point particle with an electric charge, and the magnetic moment needs only to be included in the Hamiltonian if the interest is in magnetic interactions involving the nucleus. These interactions are furthermore small and can be treated as a perturbative correction. One may analogously also neglect terms in the Hamiltonian involving electron spins, but one cannot neglect the electron spin in the wave function. The fermion character of the electrons leads to the requirement of wave function antisymmetry, which must be accounted for right from the outset for any theory. With a spin-free Hamiltonian, the spin dependence in the wave function can be integrated out. When properties related to magnetic interactions with the electron spin are desired, the spin-dependent terms in the Hamiltonian can be reintroduced and treated as perturbations.

The theoretical formulation for the calculation of molecular properties is commonly done by one of three closely related approaches:

- As derivatives of the energy.
- From perturbation theory based on the energy.
- As derivatives of the expectation value of an operator, often called propagator methods.

These approaches can be generalized to cases where the perturbation is time-dependent. We will use the electric dipole polarizability as an illustrative example in the following.

The energy derivative formulation for the static case is perhaps the easiest to understand. The energy is here expanded in a Taylor series in the perturbation $\mathbf{P}$:

$$E(\mathbf{P}) = E(0) + \frac{\partial E}{\partial \mathbf{P}}\mathbf{P} + \frac{1}{2}\frac{\partial^2 E}{\partial \mathbf{P}^2}\mathbf{P}^2 + \frac{1}{6}\frac{\partial^3 E}{\partial \mathbf{P}^3}\mathbf{P}^3 + \cdots \tag{11.1}$$

The nth-order property is the nth-order derivative of the energy, $\partial^n E/\partial \mathbf{P}^n$. Note that the perturbation is usually a vector and the first derivative is therefore also a vector, the second derivative a matrix, the third derivative a (third-order) tensor, etc. The polarizability tensor (matrix), for example, is the second derivative with respect to an external electric field $\mathbf{F}$:

$$\boldsymbol{\alpha} = -\frac{\partial^2 E}{\partial \mathbf{F}^2} \tag{11.2}$$

The perturbation approach employs the same methodology as for calculating the electron correlation energy in Section 4.8. The wave function, Hamilton operator and energy are expanded in terms of a perturbation parameter λ and terms are collected according to the order of the perturbation, which then defines the order of the property. The second-order term can be written as a sum of matrix elements over all excited states (Equation (4.49)):

$$W_2 = 2\sum_{i\neq 0} \frac{\langle \Psi_0|\mathbf{P}|\Psi_i\rangle\langle \Psi_i|\mathbf{P}|\Psi_0\rangle}{E_0 - E_i} \tag{11.3}$$

The polarizability corresponds to a perturbing dipolar electric field ($\mathbf{P} = \mathbf{r}$), which yields the formula

$$\boldsymbol{\alpha} = -2\sum_{i\neq 0} \frac{|\langle \Psi_0|\mathbf{r}|\Psi_i\rangle|^2}{E_0 - E_i} \tag{11.4}$$

The additional factor of 2 relative to Equation (11.3) arises from the convention of using an order expansion in perturbation theory, while the derivative formula in Equation (11.1) uses a Taylor expansion which includes a factor of $^1/_2$ for the second-order term.

The propagator formulation expands the expectation value of an operator $\mathbf{O}$ in terms of a perturbation $\mathbf{P}$:

$$\langle\Psi|\mathbf{O}|\Psi\rangle = \langle\Psi_0|\mathbf{O}|\Psi_0\rangle + \frac{\partial\langle\Psi|\mathbf{O}|\Psi\rangle}{\partial\mathbf{P}}\mathbf{P} + \frac{1}{2}\frac{\partial^2\langle\Psi|\mathbf{O}|\Psi\rangle}{\partial\mathbf{P}^2}\mathbf{P}^2 + \cdots \quad (11.5)$$

The polarizability is the first-order change in the dipole moment when subjected to a linear perturbing electric field:

$$\alpha = \frac{\partial\langle\Psi|\boldsymbol{\mu}|\Psi\rangle}{\partial\mathbf{F}} \quad (11.6)$$

Equations for the corresponding dynamic properties must be derived from the time-dependent Schrödinger equation by including a frequency-dependent perturbation, which normally is an external electric or magnetic field; this is discussed in Section 11.10.

11.1 Examples of Molecular Properties

11.1.1 External Electric Field

The interaction of an electronic charge distribution $\rho(\mathbf{r})$ with an electric potential $\phi(\mathbf{r})$ gives an energy contribution:

$$E = \int \rho(\mathbf{r})\phi(\mathbf{r})d\mathbf{r} \quad (11.7)$$

Since the electric field ($\mathbf{F} = -\partial\phi/\partial\mathbf{r}$) is normally fairly uniform at the molecular level, it is useful to write E as a multipole expansion:

$$E = q\phi - \boldsymbol{\mu}\mathbf{F} - \tfrac{1}{2}\mathbf{Q}\mathbf{F}' - \cdots \quad (11.8)$$

Here q is the net charge (monopole), $\boldsymbol{\mu}$ is the (electric) dipole moment, $\mathbf{Q}$ is the quadrupole moment and $\mathbf{F}$ and $\mathbf{F}'$ are the field and field gradient ($\partial\mathbf{F}/\partial\mathbf{r}$), respectively. The dipole moment and electric field are vectors and the $\boldsymbol{\mu}\mathbf{F}$ term should be interpreted as the dot product ($\boldsymbol{\mu}\mathbf{F} = \mu_x F_x + \mu_y F_y + \mu_z F_z$). The quadrupole moment and field gradient are 3×3 matrices and $\mathbf{Q}\mathbf{F}'$ denotes the tensor contraction of these matrices. It is rarely necessary to go beyond the quadrupole term for external fields, but the octupole moment may also be important for molecular interactions (it is, for example, the first non-vanishing moment for spherical molecules such as CH_4).

The unperturbed dipole and quadrupole moments may in the absence of an external field be calculated from the electronic wave function as simple expectation values:

$$\boldsymbol{\mu}_0 = -\langle\Psi|\mathbf{r}|\Psi\rangle \quad (11.9)$$

$$\mathbf{Q}_0 = \langle\Psi|\mathbf{r}\mathbf{r}^t|\Psi\rangle \quad (11.10)$$

The minus sign for the dipole moment arises from the negative charge on the electron. The superscript t denotes a transposition of the $\mathbf{r}$ vector, that is converting it from a column to a row vector. The $\mathbf{r}\mathbf{r}^t$ notation therefore indicates the outer product of $\mathbf{r}$ with itself, and the quadrupole moment is thus a 3×3 matrix, where the Q_{xy} component is calculated as the expectation value of xy. The quadrupole moment is often used in its traceless form where the sum of diagonal elements is zero, and is then usually denoted $\boldsymbol{\Theta}$ to distinuish it from the second-order moment $\mathbf{Q}$ defined in Equation (11.9).

The presence of a field influences the wave function and leads to induced dipole, quadrupole, etc., moments. For the dipole moment this may be written as

$$\boldsymbol{\mu} = \boldsymbol{\mu}_0 + \boldsymbol{\alpha}\mathbf{F} + \tfrac{1}{2}\boldsymbol{\beta}\mathbf{F}^2 + \tfrac{1}{6}\boldsymbol{\gamma}\mathbf{F}^3 + \cdots \tag{11.11}$$

Here $\boldsymbol{\mu}_0$ is the permanent dipole moment, $\boldsymbol{\alpha}$ is the (dipole) *polarizability*, $\boldsymbol{\beta}$ is the (first) *hyperpolarizability*, $\boldsymbol{\gamma}$ is the second hyperpolarizability, etc. The quadrupole moment may similarly be expanded in the field by means of a quadrupole polarizability, hyperpolarizability, etc., and mixed dipole–quadrupole polarizabilities also occur. The energy in an external electric field can thus be written as an expansion in terms of static and induced multipole moments, as shown by

$$E = q\phi - \boldsymbol{\mu}_0\mathbf{F} - \tfrac{1}{2}\boldsymbol{\Theta}_0\mathbf{F}' - \cdots - \tfrac{1}{2}\boldsymbol{\alpha}\mathbf{F}^2 - \tfrac{1}{6}\boldsymbol{\beta}\mathbf{F}^3 - \cdots - \mathbf{A}\mathbf{F}\mathbf{F}' - \mathbf{B}\mathbf{F}^2\mathbf{F}' - \mathbf{C}(\mathbf{F}')^2 - \cdots \tag{11.12}$$

Here $\mathbf{A}$ is the mixed dipole–quadrupole polarizability tensor, $\mathbf{B}$ is the mixed dipole–quadrupole hyperpolarizability tensor and $\mathbf{C}$ is the quadrupole polarizability tensor, with implied tensor contraction of all terms.

For a homogeneous field (i.e. the field gradient and higher derivatives are zero), the total energy of a neutral molecule may be written as a Taylor expansion, where all the derivatives are evaluated at $\mathbf{F} = 0$. There will be a derivative for each individual component of the field, which rapidly leads to a large number of indices and summations in a proper mathematical formulation. In order to avoid this notational cluttering, we will adopt a slightly non-standard notation where the field is indicated by a vector notation, implying that derivatives should be taken along all the individual field components:

$$E(\mathbf{F}) = E(\mathbf{0}) + \left.\frac{\partial E}{\partial \mathbf{F}}\right|_{\mathbf{F}=0}\mathbf{F} + \frac{1}{2}\left.\frac{\partial^2 E}{\partial \mathbf{F}^2}\right|_{\mathbf{F}=0}\mathbf{F}^2 + \frac{1}{6}\left.\frac{\partial^3 E}{\partial \mathbf{F}^3}\right|_{\mathbf{F}=0}\mathbf{F}^3 + \frac{1}{24}\left.\frac{\partial^4 E}{\partial \mathbf{F}^4}\right|_{\mathbf{F}=0}\mathbf{F}^4 + \cdots \tag{11.13}$$

According to Equation (11.8) we also have $\partial E/\partial \mathbf{F} = -\boldsymbol{\mu}$, where $\boldsymbol{\mu}$ is given by the expression in Equation (11.11). Differentiation of Equation (11.13) with respect to $\mathbf{F}$ gives

$$\boldsymbol{\mu} = -\left.\frac{\partial E}{\partial \mathbf{F}}\right|_{\mathbf{F}=0} - \left.\frac{\partial^2 E}{\partial \mathbf{F}^2}\right|_{\mathbf{F}=0}\mathbf{F} - \frac{1}{2}\left.\frac{\partial^3 E}{\partial \mathbf{F}^3}\right|_{\mathbf{F}=0}\mathbf{F}^2 - \frac{1}{6}\left.\frac{\partial^4 E}{\partial \mathbf{F}^4}\right|_{\mathbf{F}=0}\mathbf{F}^3 + \cdots \tag{11.14}$$

Comparing Equations (11.11) and (11.14) shows that the first derivative is the (permanent) dipole moment $\boldsymbol{\mu}_0$, the second derivative is the polarizability $\boldsymbol{\alpha}$, the third derivative is the hyperpolarizability $\boldsymbol{\beta}$, etc.:

$$\boldsymbol{\mu}_0 = -\left.\frac{\partial E}{\partial \mathbf{F}}\right|_{\mathbf{F}=0} \quad ; \quad \boldsymbol{\alpha} = -\left.\frac{\partial^2 E}{\partial \mathbf{F}^2}\right|_{\mathbf{F}=0} \quad ; \quad \boldsymbol{\beta} = -\left.\frac{\partial^3 E}{\partial \mathbf{F}^3}\right|_{\mathbf{F}=0} \quad ; \quad \boldsymbol{\gamma} = -\left.\frac{\partial^4 E}{\partial \mathbf{F}^4}\right|_{\mathbf{F}=0} \tag{11.15}$$

An expansion analogous to Equation (11.13) in the field gradient allows connections to be made with the quadrupole moment and associated polarizabilities as energy derivatives.

11.1.2 External Magnetic Field

The interaction with a magnetic field may similarly be written in terms of magnetic dipole, quadrupole, etc., moments (there is no magnetic monopole, corresponding to electric charge). Since the magnetic interaction is substantially smaller in magnitude than the electric field, only the dipole term is normally considered, with the following equation being the magnetic equivalent of Equation (11.12) where magnetic field gradient terms are neglected:

$$E = -\mathbf{m}_0\mathbf{B} - \tfrac{1}{2}\boldsymbol{\xi}\mathbf{B}^2 - \cdots \tag{11.16}$$

The dipole moment $\mathbf{m}_0$ for an unperturbed system depends on the total electronic angular momentum, which may be written in terms of the orbital angular momentum operator $\mathbf{L}_G$ and the total electron spin $\mathbf{S}$:

$$\begin{aligned} \mathbf{m}_0 &= -\tfrac{1}{2}\langle \Psi | \mathbf{L}_G + g_e \mathbf{S} | \Psi \rangle \\ \mathbf{L}_G &= (\mathbf{r} - \mathbf{R}_G) \times \mathbf{p} \end{aligned} \tag{11.17}$$

Here $\mathbf{R}_G$ is the gauge origin (discussed in Section 11.8.8) and the electronic g_e-factor is a constant approximately equal to 2.0023. The orbital part of the permanent magnetic dipole moment will be zero for all non-degenerate wave functions (i.e. belonging to A, B or Σ representations), since the $\mathbf{L}_G$ operator is purely imaginary ($\mathbf{p} = -i\nabla$) and the wave function in such cases is real. Similarly, only open-shell states (doublet, triplet, etc.) have the spin part of the magnetic dipole moment different from zero. Since the large majority of stable molecules are closed-shell singlets, it follows that permanent magnetic dipole moments are quite rare. The presence of a field, however, may induce a magnetic dipole moment, with the quantity corresponding to the electric polarizability being the *magnetizability* ξ (the corresponding macroscopic quantity is called the magnetic *susceptibility* χ):

$$\mathbf{m} = \mathbf{m}_0 + \xi \mathbf{B} + \cdots \tag{11.18}$$

The energy can again be expanded in a Taylor series:

$$E(\mathbf{B}) = E(\mathbf{0}) + \left.\frac{\partial E}{\partial \mathbf{B}}\right|_{\mathbf{B}=0} \mathbf{B} + \frac{1}{2}\left.\frac{\partial^2 E}{\partial \mathbf{B}^2}\right|_{\mathbf{B}=0} \mathbf{B}^2 + \cdots \tag{11.19}$$

As for the electric field, this leads to the definition of the dipole moment and magnetizability as first and second derivatives of the total energy with respect to the magnetic field:

$$\mathbf{m}_0 = -\left.\frac{\partial E}{\partial \mathbf{B}}\right|_{\mathbf{B}=0} \quad ; \quad \xi = -\left.\frac{\partial^2 E}{\partial \mathbf{B}^2}\right|_{\mathbf{B}=0} \tag{11.20}$$

11.1.3 Nuclear Magnetic Moments

The perturbation can also be a nuclear magnetic moment $\mathbf{I}$, arising from a nuclear spin (the $g_A \mu_N$ factor for converting from spin to magnetic moment has been ignored here):

$$\begin{aligned} E(\mathbf{I}_1, \mathbf{I}_2, \ldots) &= E(\mathbf{0}) + \frac{\partial E}{\partial \mathbf{I}_1}\mathbf{I}_1 + \frac{\partial E}{\partial \mathbf{I}_2}\mathbf{I}_2 + \frac{1}{2}\frac{\partial^2 E}{\partial \mathbf{I}_1 \partial \mathbf{I}_2}\mathbf{I}_1\mathbf{I}_2 + \cdots \\ E(\mathbf{I}_1, \mathbf{I}_2, \ldots) &= E(\mathbf{0}) - h\mathbf{J}_{12}\mathbf{I}_1\mathbf{I}_2 + \cdots \end{aligned} \tag{11.21}$$

There is no energy contribution from the first derivative since there is nothing the magnetic moment can interact with, while the second derivative with respect to two different nuclear spins is the NMR coupling constant $\mathbf{J}$ (Planck's constant appears owing to the convention of reporting coupling constants in hertz, and the factor of $^1/_2$ disappears since we implicitly only consider distinct pairs of nuclei).

11.1.4 Electron Magnetic Moments

Electron spin is implicitly taken into account in non-relativistic theory by the Pauli principle, but without energetic consequences since the Hamilton operator does not depend on electron spin. The actual interaction terms involving electron spin are small and again often accounted for by perturbation theory. Molecules with a singlet state have no net electron spin, and thus only doublet, triplet, etc., molecules can have a non-zero energy correction. There is again no energy contribution from the first derivative, since there is nothing the magnetic moment can interact with, and the equivalent of Equation (11.21) is Equation (11.22), with **s** being the electronic spin:

$$\begin{aligned} E(\mathbf{s}_1, \mathbf{s}_2, \cdots) &= E(\mathbf{0}) + \frac{\partial E}{\partial \mathbf{s}_1}\mathbf{s}_1 + \frac{\partial E}{\partial \mathbf{s}_2}\mathbf{s}_2 + \frac{1}{2}\frac{\partial^2 E}{\partial \mathbf{s}_1 \partial \mathbf{s}_2}\mathbf{s}_1\mathbf{s}_2 + \cdots \\ E(\mathbf{s}_1, \mathbf{s}_2, \cdots) &= E(\mathbf{0}) - \mathbf{D}_{12}\mathbf{s}_1\mathbf{s}_2 + \cdots \end{aligned} \tag{11.22}$$

The first non-zero term corresponds to the interaction of two electron spins, and thus requires at least two unpaired electrons, that is the electronic state must be (at least) a triplet. The **D** tensor is called the *zero field splitting* and lifts the degeneracy of the individual components of a triplet (or higher multiplet) state in the absence of a magnetic field.

11.1.5 Geometry Change

The change in energy for moving a nucleus can similarly be written as a Taylor expansion:

$$\begin{aligned} E(\mathbf{R}) &= E(\mathbf{R}_0) + \frac{\partial E}{\partial \mathbf{R}}(\mathbf{R} - \mathbf{R}_0) + \frac{1}{2}\frac{\partial^2 E}{\partial \mathbf{R}^2}(\mathbf{R} - \mathbf{R}_0)^2 + \frac{1}{6}\frac{\partial^3 E}{\partial \mathbf{R}^3}(\mathbf{R} - \mathbf{R}_0)^3 + \cdots \\ E(\mathbf{R}) &= E(\mathbf{R}_0) + \mathbf{g}(\mathbf{R} - \mathbf{R}_0) + \tfrac{1}{2}\mathbf{H}(\mathbf{R} - \mathbf{R}_0)^2 + \tfrac{1}{6}\mathbf{K}(\mathbf{R} - \mathbf{R}_0)^3 + \cdots \end{aligned} \tag{11.23}$$

The first derivative is the gradient **g** (the negative gradient is the force), the second derivative is the force constant (Hessian) **H**, the third derivative is the anharmonicity **K**, etc. If the $\mathbf{R}_0$ geometry is a stationary point ($\mathbf{g} = \mathbf{0}$) the force constant matrix may be used for evaluating harmonic vibrational frequencies and normal coordinates, **q**, as discussed in Section 17.2.2. If higher-order terms are included in the expansion, it is possible also to determine anharmonic frequencies and phenomena such as Fermi resonance.

11.1.6 Mixed Derivatives

Mixed derivatives refer to cross terms if the energy is expanded in more than one perturbation. There are many such mixed derivatives that translate into molecular properties, a few of which are given below.

The change in the dipole moment with respect to a geometry displacement along a normal coordinate **q** is related to the intensity of an IR absorption. In the so-called double harmonic approximation (terminating the expansion at first order in the electric field and geometry), the intensity is (except for some constants) given by

$$\text{IR intensity} \propto \left(\frac{\partial \boldsymbol{\mu}}{\partial \mathbf{q}}\right)^2 \propto \left(\frac{\partial^2 E}{\partial \mathbf{R} \partial \mathbf{F}}\right)^2 \tag{11.24}$$

Only fundamental bands can have an intensity different from zero in the double harmonic approximation. Including higher-order terms in the expansion allows the calculation of intensities of overtone bands, as well as adding contributions to the fundamental bands.

The intensity of a Raman band in the harmonic approximation is given by the derivative of the polarizability with respect to a normal coordinate:

$$\text{Raman intensity} \propto \left(\frac{\partial \boldsymbol{\alpha}}{\partial \mathbf{q}}\right)^2 \propto \left(\frac{\partial^3 E}{\partial \mathbf{R} \partial \mathbf{F}^2}\right)^2 \tag{11.25}$$

The polarizability derivatives are often calculated in the static limit, but Ramen intensities do in reality depend on the frequency of the incident light.[5]

The mixed derivative of an external and a nuclear magnetic field (nuclear spin) is the NMR shielding tensor $\boldsymbol{\sigma}$:

$$\text{NMR shielding} \propto \left(\frac{\partial^2 E}{\partial \mathbf{B} \partial \mathbf{I}}\right) \tag{11.26}$$

The corresponding quantity related to the electron spin is the ESR **g** tensor:

$$\text{ESR } \mathbf{g} \text{ tensor} \propto \left(\frac{\partial^2 E}{\partial \mathbf{B} \partial \mathbf{s}}\right) \tag{11.27}$$

The mixed derivative of a nuclear and electron magnetic moment is the hyperfine coupling constant:

$$\text{Hyperfine coupling} \propto \left(\frac{\partial^2 E}{\partial \mathbf{I} \partial \mathbf{s}}\right) \tag{11.28}$$

Table 11.1 gives some examples of properties that may be calculated from derivatives of a certain order with respect to the above five perturbations:

$$\text{Property} \propto \frac{\partial^{n_\text{F}+n_\text{B}+n_\text{I}+n_s+n_\text{R}} E}{\partial \mathbf{F}^{n_\text{F}} \partial \mathbf{B}^{n_\text{B}} \partial \mathbf{I}^{n_\text{I}} \partial \mathbf{s}^{n_\text{s}} \partial \mathbf{R}^{n_\text{R}}} \tag{11.29}$$

All of these properties can be calculated at various levels of sophistication (electron correlation and basis sets).

The properties in Table 11.1 cover vibrational effects by the geometry derivatives, but additional properties are related to the overall molecular rotation. The movement of electrically charged particles in a rotating molecule will give rise to a magnetic field, but within the Born–Oppenheimer approximation, the nuclear and electron contributions will cancel exactly.There will in reality be a small lag between the nuclear and electronic motions (a non-Born–Oppenheimer effect) which leads to a magnetic moment that can interact with other magnetic moments. The *rotational* **g** *tensor* can be considered as the second derivative with respect to the molecular rotational angular momentum and an external magnetic field,[6] and the *spin-rotation* constant is the mixed second-order term arising from the coupling of the molecular rotational angular momentum and nuclear magnetic moment.[7] If the molecule has a net electronic spin (non-singlet state), there will be an electronic contribution to the spin-rotation constant from interaction of the electronic spin and molecular rotational angular momentum.[8] Molecules with a non-zero angular momentum electronic wave function (e.g. π or Δ states) have interaction terms with the overall rotation.[9] Magnetic effects arising from the breakdown of the Born–Oppenheimer approximation by vibrational motions and coupling with an external magnetic field provide the *vibrational* **g** *tensor*.[10] The *diagonal Born–Oppenheimer correction* discussed in Section 3.1 can also be considered as a molecular property related to the nuclear movements.

11.2 Perturbation Methods

The presence of perturbations will give rise to extra terms in the Hamiltonian, and we will in the following need to consider operators that are linear and quadratic in the perturbation strength λ:

$$\mathbf{H} = \mathbf{H}_0 + \lambda \mathbf{P}_1 + \lambda^2 \mathbf{P}_2 \tag{11.30}$$

Table 11.1 Examples of properties that may be calculated as derivatives of the energy.

n_F	n_B	n_I	n_s	n_R	Property
0	0	0	0	0	Energy
1	0	0	0	0	Electric dipole moment
0	1	0	0	0	Magnetic dipole moment
0	0	0	0	1	Molecular (nuclear) gradient
2	0	0	0	0	Electric polarizability
0	2	0	0	0	Magnetizability
0	0	2	0	0	Nuclear spin–spin coupling
0	0	0	2	0	Zero-field splitting
0	0	0	0	2	Harmonic vibrational frequencies
1	0	0	0	1	Infrared absorption intensities
1	1	0	0	0	Optical rotation, circular dichroism
0	1	1	0	0	Nuclear magnetic shielding
0	1	0	1	0	ESR **g** tensor
0	0	1	1	0	Hyperfine coupling constant
3	0	0	0	0	Electric hyperpolarizability (first)
0	3	0	0	0	Hypermagnetizability (first)
0	0	0	0	3	Anharmonic corrections to vibrational frequencies (cubic)
2	0	0	0	1	Raman intensities
1	1	0	0	1	Vibrational circular dichroism
1	1	1	0	0	Nuclear magnetic shielding (first) polarizability
1	0	0	0	2	Infrared intensities for overtone and combination bands
4	0	0	0	0	Electric hyperpolarizability (second)
0	4	0	0	0	Hypermagnetizability (second)
0	0	0	0	4	Anharmonic corrections to vibrational frequencies (quartic)
3	0	0	0	1	Hyper-Raman effects
2	1	1	0	0	Nuclear magnetic shielding (second) polarizability
2	0	0	0	2	Raman intensities for overtone and combination bands

$\mathbf{H}_0$ is the normal electronic Hamiltonian operator and the perturbations are described by the operators $\mathbf{P}_1$ and $\mathbf{P}_2$. Based on an expansion in *exact* wave functions, Rayleigh–Schrödinger perturbation theory (Section 4.8) gives the first- and second-order energy corrections upon setting $\lambda = 1$:

$$W_1 = \langle \Psi_0 | \mathbf{P}_1 | \Psi_0 \rangle \tag{11.31}$$

$$W_2 = \langle \Psi_0 | \mathbf{P}_2 | \Psi_0 \rangle + 2 \sum_{i \neq 0} \frac{\langle \Psi_0 | \mathbf{P}_1 | \Psi_i \rangle \langle \Psi_i | \mathbf{P}_1 | \Psi_0 \rangle}{E_0 - E_i} \tag{11.32}$$

The first-order term is identical to Equation (4.46), while the second-order term corresponds to Equation (4.49) including a factor of 2, as mentioned in Equation (11.4), and an additional term involving the expectation value of $\mathbf{P}_2$ over the unperturbed wave function. The first-order energy correction is identified with the first-order property, the second-order correction is the second-order property, etc. Although these expressions only hold for exact wave functions, they may also be used for approximative wave functions. The first-order term is simply the expectation value of the perturbation operator over the unperturbed wave function, and is easy to calculate. The second-order property, however, involves a sum over all excited states. In some cases, mainly associated with semi-empirical methods, second-order properties are evaluated directly from Equation (11.32), known as *Sum Over States* (SOS) methods. Since this involves a determination of all excited states, it is very inefficient for *ab initio* methods.

11.3 Derivative Techniques

Derivative techniques consider the energy in the presence of the perturbation, perform an analytical differentiation of the energy n times to derive a formula for the nth-order property and let the perturbation strength go to zero.

Let us write the energy as

$$E(\lambda) = \langle \Psi(\lambda) | \mathbf{H}_0 + \lambda \mathbf{P}_1 + \lambda^2 \mathbf{P}_2 | \Psi(\lambda) \rangle \tag{11.33}$$

This is strictly true for HF, MCSCF and CI wave functions, and can be generalized to MP and CC methods, as shown in Section 11.5. The perturbation-dependent terms in the operator are written explicitly, while the wave function dependence is implicit, via the parameterization (orbital and state coefficients) and possibly also the basis functions.

The first derivative of the energy can be written as

$$\frac{\partial E}{\partial \lambda} = \left\langle \frac{\partial \Psi}{\partial \lambda} \middle| \mathbf{H}_0 + \lambda \mathbf{P}_1 + \lambda^2 \mathbf{P}_2 \middle| \Psi \right\rangle + \langle \Psi | \mathbf{P}_1 + 2\lambda \mathbf{P}_2 | \Psi \rangle + \left\langle \Psi \middle| \mathbf{H}_0 + \lambda \mathbf{P}_1 + \lambda^2 \mathbf{P}_2 \middle| \frac{\partial \Psi}{\partial \lambda} \right\rangle \tag{11.34}$$

For real wave functions the first and third terms are identical. Letting the perturbation strength go to zero yields

$$\left.\frac{\partial E}{\partial \lambda}\right|_{\lambda=0} = \langle \Psi_0 | \mathbf{P}_1 | \Psi_0 \rangle + 2 \left\langle \frac{\partial \Psi}{\partial \lambda} \middle| \mathbf{H}_0 \middle| \Psi_0 \right\rangle \tag{11.35}$$

The wave function depends on the perturbation indirectly, via parameters in the wave function ($\mathbf{C}$), and possibly also the basis functions (χ). The wave function parameters may be orbital coefficients (HF), state coefficients (CI, MP, CC) or both (MCSCF):

$$\frac{\partial \Psi}{\partial \lambda} = \frac{\partial \Psi}{\partial \mathbf{C}} \frac{\partial \mathbf{C}}{\partial \lambda} + \frac{\partial \Psi}{\partial \chi} \frac{\partial \chi}{\partial \lambda} \tag{11.36}$$

Assuming for the moment that the basis functions are independent of the perturbation ($\partial \chi / \partial \lambda = 0$), the derivative (11.35) may be written as

$$\left.\frac{\partial E}{\partial \lambda}\right|_{\lambda=0} = \langle \Psi_0 | \mathbf{P}_1 | \Psi_0 \rangle + 2 \frac{\partial \mathbf{C}}{\partial \lambda} \left\langle \frac{\partial \Psi}{\partial \mathbf{C}} \middle| \mathbf{H}_0 \middle| \Psi_0 \right\rangle \tag{11.37}$$

If the wave function is variationally optimized with respect to *all* parameters (HF or MCSCF, but *not* CI), the last term disappears since the energy is stationary with respect to a variation of the MO/state coefficients ($\mathbf{H}_0$, $\mathbf{P}_1$ and $\mathbf{P}_2$ do not depend on the parameters $\mathbf{C}$):

$$\frac{\partial E}{\partial \mathbf{C}} = \frac{\partial}{\partial \mathbf{C}} \langle \Psi | \mathbf{H}_0 + \lambda \mathbf{P}_1 + \lambda^2 \mathbf{P}_2 | \Psi \rangle = 2 \left\langle \frac{\partial \Psi}{\partial \mathbf{C}} \middle| \mathbf{H}_0 + \lambda \mathbf{P}_1 + \lambda^2 \mathbf{P}_2 \middle| \Psi \right\rangle \tag{11.38}$$

$$\left.\frac{\partial E}{\partial \mathbf{C}}\right|_{\lambda=0} = 2 \left\langle \frac{\partial \Psi}{\partial \mathbf{C}} \middle| \mathbf{H}_0 \middle| \Psi_0 \right\rangle = 0 \tag{11.39}$$

Variational wave functions thus obey the *Hellmann–Feynman* theorem:

$$\frac{\partial}{\partial \lambda} \langle \Psi | \mathbf{H} | \Psi \rangle = \left\langle \Psi \middle| \frac{\partial \mathbf{H}}{\partial \lambda} \middle| \Psi \right\rangle \tag{11.40}$$

The expression from first-order perturbation theory (11.31) in such cases yields a result identical to the first derivative of the energy with respect to λ. For wave functions that are not completely optimized with respect to all parameters (CI, MP or CC), the Hellmann–Feynman theorem does not hold, and a first-order property calculated as an expectation value will not be identical to that obtained as an energy derivative. Since the Hellmann–Feynman theorem holds for an exact wave function, the difference between the two values becomes smaller as the quality of an approximate wave function increases. However, for practical applications the difference is not negligible. It has been argued that the derivative technique resembles the physical experiment more, and consequently formula (11.35) should be preferred over (11.31).

The second derivative of the energy can for a real wave function be written as

$$\begin{aligned}\frac{1}{2}\frac{\partial^2 E}{\partial \lambda^2} &= \left\langle \frac{\partial^2 \Psi}{\partial \lambda^2} \middle| \mathbf{H}_0 + \lambda \mathbf{P}_1 + \lambda^2 \mathbf{P}_2 \middle| \Psi \right\rangle + 2\left\langle \frac{\partial \Psi}{\partial \lambda} \middle| \mathbf{P}_1 + 2\lambda \mathbf{P}_2 \middle| \Psi \right\rangle \\ &\quad + \left\langle \frac{\partial \Psi}{\partial \lambda} \middle| \mathbf{H}_0 + \lambda \mathbf{P}_1 + \lambda^2 \mathbf{P}_2 \middle| \frac{\partial \Psi}{\partial \lambda} \right\rangle + \langle \Psi | \mathbf{P}_2 | \Psi \rangle\end{aligned} \tag{11.41}$$

In the limit of the perturbation strength going to zero this reduces to

$$\frac{1}{2}\frac{\partial^2 E}{\partial \lambda^2}\bigg|_{\lambda=0} = \left\langle \frac{\partial^2 \Psi}{\partial \lambda^2} \middle| \mathbf{H}_0 \middle| \Psi_0 \right\rangle + 2\left\langle \frac{\partial \Psi}{\partial \lambda} \middle| \mathbf{P}_1 \middle| \Psi_0 \right\rangle + \left\langle \frac{\partial \Psi}{\partial \lambda} \middle| \mathbf{H}_0 \middle| \frac{\partial \Psi}{\partial \lambda} \right\rangle + \langle \Psi_0 | \mathbf{P}_2 | \Psi_0 \rangle \tag{11.42}$$

The implicit wave function dependence on $\mathbf{C}$ allows the derivative to be written as

$$\begin{aligned}\frac{1}{2}\frac{\partial^2 E}{\partial \lambda^2}\bigg|_{\lambda=0} &= \frac{\partial^2 \mathbf{C}}{\partial \lambda^2}\left\langle \frac{\partial \Psi}{\partial \mathbf{C}} \middle| \mathbf{H}_0 \middle| \Psi_0 \right\rangle + \left(\frac{\partial \mathbf{C}}{\partial \lambda}\right)^2 \left\langle \frac{\partial^2 \Psi}{\partial \mathbf{C}^2} \middle| \mathbf{H}_0 \middle| \Psi_0 \right\rangle \\ &\quad + 2\left(\frac{\partial \mathbf{C}}{\partial \lambda}\right)\left\langle \frac{\partial \Psi}{\partial \mathbf{C}} \middle| \mathbf{P}_1 \middle| \Psi_0 \right\rangle + \left(\frac{\partial \mathbf{C}}{\partial \lambda}\right)^2 \left\langle \frac{\partial \Psi}{\partial \mathbf{C}} \middle| \mathbf{H}_0 \middle| \frac{\partial \Psi}{\partial \mathbf{C}} \right\rangle + \langle \Psi_0 | \mathbf{P}_2 | \Psi_{0'} \rangle\end{aligned} \tag{11.43}$$

The first term is again zero for variationally optimized wave functions (Equation (11.39)). Furthermore, the second term, which involves calculation of the second derivative of the wave function with respect to the parameters, can be avoided. This can be seen by differentiating the stationary condition of Equation (11.38) with respect to the perturbation:

$$\begin{aligned}&\frac{\partial}{\partial \lambda}\left\langle \frac{\partial \Psi}{\partial \mathbf{C}} \middle| \mathbf{H}_0 + \lambda \mathbf{P}_1 + \lambda^2 \mathbf{P}_2 \middle| \Psi_0 \right\rangle\bigg|_{\lambda=0} \\ &= \left(\frac{\partial \mathbf{C}}{\partial \lambda}\right)\left\langle \frac{\partial^2 \Psi}{\partial \mathbf{C}^2} \middle| \mathbf{H}_0 \middle| \Psi_0 \right\rangle + \left\langle \frac{\partial \Psi}{\partial \mathbf{C}} \middle| \mathbf{P}_1 \middle| \Psi_0 \right\rangle + \left(\frac{\partial \mathbf{C}}{\partial \lambda}\right)\left\langle \frac{\partial \Psi}{\partial \mathbf{C}} \middle| \mathbf{H}_0 \middle| \frac{\partial \Psi}{\partial \mathbf{C}} \right\rangle = 0\end{aligned} \tag{11.44}$$

The second derivative in Equation (11.43) therefore reduces to

$$\frac{1}{2}\frac{\partial^2 E}{\partial \lambda^2}\bigg|_{\lambda=0} = \langle \Psi_0 | \mathbf{P}_2 | \Psi_0 \rangle + \left(\frac{\partial \mathbf{C}}{\partial \lambda}\right)\left\langle \frac{\partial \Psi}{\partial \mathbf{C}} \middle| \mathbf{P}_1 \middle| \Psi_0 \right\rangle \tag{11.45}$$

In a more compact notation this can be written as

$$\frac{1}{2}\frac{\partial^2 E}{\partial \lambda^2}\bigg|_{\lambda=0} = \langle \Psi_0 | \mathbf{P}_2 | \Psi_0 \rangle + \left\langle \frac{\partial \Psi}{\partial \lambda} \middle| \mathbf{P}_1 \middle| \Psi_0 \right\rangle \tag{11.46}$$

This shows that only the first-order change in the wave function is necessary and quadratic operators only contribute. with an expectation value that is easy to calculate.

For exact wave functions Equation (11.46) becomes identical to the perturbation expression (11.32), since the first derivative of the wave function is the first-order correction to the wave function (Equation (4.45)), which can be expanded in a complete set of eigenfunctions:

$$\frac{\partial \Psi}{\partial \lambda} = \Psi_1 = \sum_{i=1}^{\infty} a_i \Psi_i \quad ; \quad a_i = \frac{\langle \Psi_i | \mathbf{P}_1 | \Psi_0 \rangle}{E_0 - E_i} \tag{11.47}$$

11.4 Response and Propagator Methods

The Taylor expansions in Equations (11.14) and (11.15) suggest that, for example, second-order molecular properties can be defined as the first derivative of a first-order property, and this approach is often called the response or propagator method. It is similar to the derivative technique, but uses the expectation value of an operator instead of the energy for performing a Taylor expansion in the presence of a perturbation:

$$\begin{aligned}\langle \Psi | \mathbf{O} | \Psi \rangle &= \langle \Psi_0 | \mathbf{O} | \Psi_0 \rangle + \frac{\partial \langle \Psi | \mathbf{O} | \Psi \rangle}{\partial \mathbf{P}} \mathbf{P} + \frac{1}{2} \frac{\partial^2 \langle \Psi | \mathbf{O} | \Psi \rangle}{\partial \mathbf{P}^2} \mathbf{P}^2 + \cdots \\ \langle \Psi | \mathbf{O} | \Psi \rangle &= \langle \Psi_0 | \mathbf{O} | \Psi_0 \rangle + \langle\langle \mathbf{O}; \mathbf{P} \rangle\rangle \mathbf{P} + \tfrac{1}{2} \langle\langle \mathbf{O}; \mathbf{P}, \mathbf{P} \rangle\rangle \mathbf{P}^2 + \cdots\end{aligned} \tag{11.48}$$

The *propagator notation* $\langle\langle \mathbf{O}; \mathbf{P} \rangle\rangle$ indicates the first-order change (linear response) in the property subject to perturbation **P**, while $\langle\langle \mathbf{O}; \mathbf{P}, \mathbf{P} \rangle\rangle$ indicates the second-order change (quadratic response) subject to perturbation **P**. Only the wave function depends on the perturbation, and $\langle\langle \mathbf{O}; \mathbf{P} \rangle\rangle$ can for real wave functions be written as

$$\langle\langle \mathbf{O}; \mathbf{P} \rangle\rangle = 2 \left\langle \frac{\partial \Psi}{\partial \mathbf{P}} \middle| \mathbf{O} \middle| \Psi_0 \right\rangle \tag{11.49}$$

This is equivalent to the last term in Equation (11.46). Note that the *linear, quadratic, cubic*, etc., response/propagator properties are associated with the *second-*, *third-*, *fourth-*, etc., derivatives or energy corrections in the perturbation formulation.

11.5 Lagrangian Techniques

For variationally optimized wave functions (HF or MCSCF) there is a $2n + 1$ rule, analogous to the perturbational energy expression (Equation (4.44)): knowledge of the nth derivative (response) of the wave function is sufficient for calculating a property to order $2n + 1$. For non-variational wave functions Equation (11.42) suggests that the nth-order wave function response is required for calculating the nth-order property. This may be avoided, however, by a technique first illustrated for CISD geometry derivatives by Handy and Schaefer, often referred to as the *Z-vector* method.[11] It has later been generalized to cover other types of wave functions and derivatives by formulating it in terms of a Lagrange function.[12]

The idea is to construct a Lagrange function that has the same energy as the non-variational wave function but which is variational in all parameters. Consider, for example, a CI wave function, which is variational in the state coefficients (**a**) but not in the orbital coefficients (**c**), since they are determined by the stationary condition for the HF wave function (note that we employ lower case **c** for the orbital coefficients but capital **C** to denote all wave function parameters, that is **C** contains both **a** and **c**):

$$\frac{\partial E_{\mathrm{CI}}}{\partial \mathbf{a}} = \frac{\partial}{\partial \mathbf{a}}\langle \Psi_{\mathrm{CI}}(\mathbf{a},\mathbf{c})|\mathbf{H}|\Psi_{\mathrm{CI}}(\mathbf{a},\mathbf{c})\rangle = 2\left\langle \frac{\partial \Psi_{\mathrm{CI}}}{\partial \mathbf{a}}\middle|\mathbf{H}\middle|\Psi_{\mathrm{CI}}\right\rangle = 0 \tag{11.50}$$

$$\frac{\partial E_{\mathrm{CI}}}{\partial \mathbf{c}} = \frac{\partial}{\partial \mathbf{c}}\langle \Psi_{\mathrm{CI}}(\mathbf{a},\mathbf{c})|\mathbf{H}|\Psi_{\mathrm{CI}}(\mathbf{a},\mathbf{c})\rangle = 2\left\langle \frac{\partial \Psi_{\mathrm{CI}}}{\partial \mathbf{c}}\middle|\mathbf{H}\middle|\Psi_{\mathrm{CI}}\right\rangle \neq 0 \tag{11.51}$$

$$\frac{\partial E_{\mathrm{HF}}}{\partial \mathbf{c}} = \frac{\partial}{\partial \mathbf{c}}\langle \Psi_{\mathrm{HF}}(\mathbf{a},\mathbf{c})|\mathbf{H}|\Psi_{\mathrm{HF}}(\mathbf{a},\mathbf{c})\rangle = 2\left\langle \frac{\partial \Psi_{\mathrm{HF}}}{\partial \mathbf{c}}\middle|\mathbf{H}\middle|\Psi_{\mathrm{HF}}\right\rangle = 0 \tag{11.52}$$

Consider now the Lagrange function given in

$$L_{\mathrm{CI}} = E_{\mathrm{CI}} + \boldsymbol{\kappa}\frac{\partial E_{\mathrm{HF}}}{\partial \mathbf{c}} \tag{11.53}$$

Here $\boldsymbol{\kappa}$ contains a set of Lagrange multipliers. The derivatives of the Lagrange function with respect to **a**, **c** and $\boldsymbol{\kappa}$ are given in the following equations:

$$\frac{\partial L_{\mathrm{CI}}}{\partial \mathbf{a}} = \frac{\partial E_{\mathrm{CI}}}{\partial \mathbf{a}} = 0 \tag{11.54}$$

$$\frac{\partial L_{\mathrm{CI}}}{\partial \boldsymbol{\kappa}} = \frac{\partial E_{\mathrm{HF}}}{\partial \mathbf{c}} = 0 \tag{11.55}$$

$$\frac{\partial L_{\mathrm{CI}}}{\partial \mathbf{c}} = \frac{\partial E_{\mathrm{CI}}}{\partial \mathbf{c}} + \boldsymbol{\kappa}\frac{\partial^2 E_{\mathrm{HF}}}{\partial \mathbf{c}^2} = 0 \tag{11.56}$$

The first two derivatives are zero due to the properties of the CI and HF wave functions, Equations (11.50) and (11.52). Equation (11.56) is zero by virtue of the Lagrange multipliers, that is we *choose* $\boldsymbol{\kappa}$ such that $\partial L_{\mathrm{CI}}/\partial \mathbf{c} = 0$. It may be written more explicitly as follows:

$$\frac{\partial L_{\mathrm{CI}}}{\partial \mathbf{c}} = 2\left\langle \frac{\partial \Psi_{\mathrm{CI}}}{\partial \mathbf{c}}\middle|\mathbf{H}\middle|\Psi_{\mathrm{CI}}\right\rangle + 2\boldsymbol{\kappa}\left[\left\langle \frac{\partial^2 \Psi_{\mathrm{HF}}}{\partial \mathbf{c}^2}\middle|\mathbf{H}\middle|\Psi_{\mathrm{HF}}\right\rangle + \left\langle \frac{\partial \Psi_{\mathrm{HF}}}{\partial \mathbf{c}}\middle|\mathbf{H}\middle|\frac{\partial \Psi_{\mathrm{HF}}}{\partial \mathbf{c}}\right\rangle\right] = 0 \tag{11.57}$$

Note that no new operators are involved, only derivatives of the CI or HF wave function with respect to the MO coefficients. The matrix elements can thus be calculated from the same integrals as the energy itself, as discussed in Sections 3.3 and 4.2.1.

The derivative with respect to a perturbation can now be written as

$$\frac{\partial L_{\mathrm{CI}}}{\partial \lambda} = \frac{\partial E_{\mathrm{CI}}}{\partial \lambda} + \boldsymbol{\kappa}\frac{\partial}{\partial \lambda}\left(\frac{\partial E_{\mathrm{HF}}}{\partial \mathbf{c}}\right) = 0 \tag{11.58}$$

Expanding out the terms gives

$$\begin{aligned}
\frac{\partial L_{\mathrm{CI}}}{\partial \lambda} &= \frac{\partial}{\partial \lambda}\langle \Psi_{\mathrm{CI}}|\mathbf{H}|\Psi_{\mathrm{CI}}\rangle + 2\boldsymbol{\kappa}\frac{\partial}{\partial \lambda}\left\langle \frac{\partial \Psi_{\mathrm{HF}}}{\partial \mathbf{c}}\middle|\mathbf{H}\middle|\Psi_{\mathrm{HF}}\right\rangle \\
&= \left\langle \Psi_{\mathrm{CI}}\middle|\frac{\partial \mathbf{H}}{\partial \lambda}\middle|\Psi_{\mathrm{CI}}\right\rangle + 2\frac{\partial \mathbf{a}}{\partial \lambda}\left\langle \frac{\partial \Psi_{\mathrm{CI}}}{\partial \mathbf{a}}\middle|\mathbf{H}\middle|\Psi_{\mathrm{CI}}\right\rangle + 2\frac{\partial \mathbf{c}}{\partial \lambda}\left\langle \frac{\partial \Psi_{\mathrm{CI}}}{\partial \mathbf{c}}\middle|\mathbf{H}\middle|\Psi_{\mathrm{CI}}\right\rangle \\
&\quad + \boldsymbol{\kappa}\left[\left\langle \frac{\partial \Psi_{\mathrm{HF}}}{\partial \mathbf{c}}\middle|\frac{\partial \mathbf{H}}{\partial \lambda}\middle|\Psi_{\mathrm{HF}}\right\rangle + \frac{\partial \mathbf{c}}{\partial \lambda}\left(\left\langle \frac{\partial^2 \Psi_{\mathrm{HF}}}{\partial \mathbf{c}^2}\middle|\mathbf{H}\middle|\Psi_{\mathrm{HF}}\right\rangle + \left\langle \frac{\partial \Psi_{\mathrm{HF}}}{\partial \mathbf{c}}\middle|\mathbf{H}\middle|\frac{\partial \Psi_{\mathrm{HF}}}{\partial \mathbf{c}}\right\rangle\right)\right]
\end{aligned} \tag{11.59}$$

The second term disappears since the CI wave function is variational in the state coefficients (Equation (11.50)). The three terms involving the derivative of the MO coefficients ($\partial \mathbf{c}/\partial \lambda$) also disappear owing to our choice of the Lagrange multipliers (Equation (11.50)). If we furthermore adapt the definition that $\partial\mathbf{H}/\partial\lambda = \mathbf{P}_1$ (Equation (11.30)), the final derivative may be written as

$$\frac{\partial L_{\text{CI}}}{\partial\lambda} = \langle\Psi_{\text{CI}}|\mathbf{P}_1|\Psi_{\text{CI}}\rangle + \boldsymbol{\kappa}\left\langle\frac{\partial\Psi_{\text{HF}}}{\partial\mathbf{c}}\middle|\mathbf{P}_1\middle|\Psi_{\text{HF}}\right\rangle \tag{11.60}$$

Here the Lagrange multipliers $\boldsymbol{\kappa}$ are determined from Equation (11.57).

What has been accomplished? The original expression (11.37) contains the derivative of the MO coefficients with respect to the perturbation ($\partial\mathbf{c}/\partial\lambda$), which can be obtained by solving the CPHF equations (Section 11.6.1 below). For geometry derivatives, for example, there will be $3N_{\text{atom}}$ different perturbations, that is we need to solve $3N_{\text{atom}}$ sets of CPHF equations. The Lagrange expression (11.60), on the other hand, contains a set of Lagrange multipliers $\boldsymbol{\kappa}$ that are *independent* of the perturbation, that is we need only to solve one equation for $\boldsymbol{\kappa}$, Equation (11.57). Furthermore, the CPHF equations involve derivatives of the basis functions, while the equation for $\boldsymbol{\kappa}$ only involves integrals of the same type as for calculating the energy itself.

The Lagrange technique may be generalized to other types of non-variational wave functions (MP and CC) and to higher-order derivatives. It is found that the $2n + 1$ rule is recovered; that is if the wave function response is known to order n, the $(2n + 1)$th-order property may be calculated for any type of wave function.

11.6 Wave Function Response

While first-order properties in most cases can be obtained as simple expectation values over the unperturbed wave function, second (and higher-order) properties require the first (and higher)- order change (response) of the wave function in the presense of the perturbation. The first-order change in the wave function $\partial\mathbf{C}/\partial\lambda$ can be obtained from Equation (11.44):

$$\left[\left\langle\frac{\partial^2\Psi}{\partial\mathbf{C}^2}\middle|\mathbf{H}_0\middle|\Psi_0\right\rangle + \left\langle\frac{\partial\Psi}{\partial\mathbf{C}}\middle|\mathbf{H}_0\middle|\frac{\partial\Psi}{\partial\mathbf{C}}\right\rangle\right]\left(\frac{\partial\mathbf{C}}{\partial\lambda}\right) = -\left\langle\frac{\partial\Psi}{\partial\mathbf{C}}\middle|\mathbf{P}_1\middle|\Psi_0\right\rangle \tag{11.61}$$

This is a set of linear equations where the matrix elements on the left-hand side are second derivatives of the energy with respect to the wave function parameters. For the case of a single Slater determinant ($\Psi = \Phi$), the wave function parameters are (only) the molecular orbital coefficients, and these can be parameterized in terms of non-redundant orbital rotation parameters x_i^a (Section 3.6). The wave function derivatives can thus be calculated as matrix elements between the reference and doubly excited determinants and between singly excited determinants (Equation (3.70)):

$$\left\langle\frac{\partial^2\Phi}{\partial x_i^a\partial x_j^b}\middle|\mathbf{H}_0\middle|\Phi_0\right\rangle \propto \left\langle\Phi_{ij}^{ab}\middle|\mathbf{H}_0\middle|\Phi_0\right\rangle - \delta_{ij}\delta_{ab}E_0 \tag{11.62}$$

$$\left\langle\frac{\partial\Phi}{\partial x_i^a}\middle|\mathbf{H}_0\middle|\frac{\partial\Phi}{\partial x_j^b}\right\rangle \propto \left\langle\Phi_i^a\middle|\mathbf{H}_0\middle|\Phi_j^b\right\rangle \tag{11.63}$$

The right-hand side of Equation (11.61) contains matrix elements of the perturbation operator between the reference and singly excited determinants, and is often called a property gradient, in analogy with the energy gradient if the operator is the Hamiltonian (Equation (3.69)):

$$\left\langle\frac{\partial\Phi}{\partial x_i^a}\middle|\mathbf{P}_1\middle|\Phi_0\right\rangle \propto \langle\Phi_i^a|\mathbf{P}_1|\Phi_0\rangle \tag{11.64}$$

The above simplified notation hides a couple of critical details, the most important being that the excitation manifold must include both excitation and de-excitations, that both real and imaginary variations must be allowed and the x_i^a notation implicitly defines both the x_{ai} and x_{ia} elements in the **X** matrix in Equation (3.65). Performing this generalization, the response equations can be written as in the following equation, where (**Y**,**Z**) contains the real and imaginary parts of the wave function response, respectively:

$$\begin{bmatrix} \mathbf{A} & \mathbf{B} \\ \mathbf{B}^* & \mathbf{A}^* \end{bmatrix} \begin{bmatrix} \mathbf{Y} \\ \mathbf{Z} \end{bmatrix} = - \begin{bmatrix} \mathbf{P} \\ \mathbf{P}^* \end{bmatrix} \tag{11.65}$$

The **A** and **B** matrix elements follow from Equations (11.62) and (11.63) and can be obtained from orbital energy differences and two-electron integrals (using the notation in Equation (3.61) with ij being occupied and ab being virtual orbitals), as shown in the following equations:

$$A_{ij}^{ab} = \left\langle \Phi_i^a \middle| \mathbf{H}_0 \middle| \Phi_j^b \right\rangle - E_0 \delta_{ij} \delta_{ab} = \delta_{ij} \delta_{ab} (\varepsilon_a - \varepsilon_i) + \langle ij|ab \rangle - \langle ia|jb \rangle \tag{11.66}$$

$$B_{ij}^{ab} = \left\langle \Phi_0 \middle| \mathbf{H}_0 \middle| \Phi_{ij}^{ab} \right\rangle = \langle ij|ab \rangle - \langle ij|ba \rangle \tag{11.67}$$

$$P_i^a = \langle \Phi_0 | \mathbf{P}_1 | \Phi_i^a \rangle = \langle i | \mathbf{P}_1 | a \rangle \tag{11.68}$$

The response part of the property in Equation (11.46) can with Equation (11.61) be written as

$$\left\langle \frac{\partial \Psi}{\partial \lambda} \middle| \mathbf{P}_1 \middle| \Psi_0 \right\rangle = - \left\langle \frac{\partial \Psi}{\partial \mathbf{C}} \middle| \mathbf{P}_1 \middle| \Psi_0 \right\rangle \left[\left\langle \frac{\partial^2 \Psi}{\partial \mathbf{C}^2} \middle| \mathbf{H}_0 \middle| \Psi_0 \right\rangle + \left\langle \frac{\partial \Psi}{\partial \mathbf{C}} \middle| \mathbf{H}_0 \middle| \frac{\partial \Psi}{\partial \mathbf{C}} \right\rangle \right]^{-1} \left\langle \frac{\partial \Psi}{\partial \mathbf{C}} \middle| \mathbf{P}_1 \middle| \Psi_0 \right\rangle \tag{11.69}$$

Using the notation from Equation (11.65) and generalizing to a propagator notation, the change in the expectation value of operator **O** subject to perturbation **P** can be written as

$$\langle\langle \mathbf{O}; \mathbf{P} \rangle\rangle = -[\mathbf{O} \quad \mathbf{O}^*] \begin{bmatrix} \mathbf{A} & \mathbf{B} \\ \mathbf{B}^* & \mathbf{A}^* \end{bmatrix}^{-1} \begin{bmatrix} \mathbf{P} \\ \mathbf{P}^* \end{bmatrix} \tag{11.70}$$

Both the **A** and **B** matrices have the dimension $N_{\text{occ}} N_{\text{vir}}$ corresponding to all single excitations, which for all but the most trivial cases prohibits a calculation of the inverse matrix in Equation (11.70) directly. The propagator is therefore calculated in two steps, by first solving for the response vector (**Y**,**Z**) in Equation (11.65) by iterative techniques (Section 17.2.5) and subsequently multiplying it with a vector containing the matrix elements of the **O** operator:

$$\langle\langle \mathbf{O}; \mathbf{P} \rangle\rangle = -[\mathbf{O} \quad \mathbf{O}^*] \begin{bmatrix} \mathbf{Y} \\ \mathbf{Z} \end{bmatrix} \tag{11.71}$$

The above expressions for the first-order change in the wave function and the associated second-order property can be generalized to calculating higher-order changes in the wave function required for higher-order properties by repeated differentiations with respect to additional perturbations.

11.6.1 Coupled Perturbed Hartree–Fock

The above response formulation is quite general, except for the matrix elements in Equations (11.66) to (11.68), which assumes an HF-type wave function parameterized in terms of non-redundant orbital rotations. As illustrated in Section 3.6 it is often possible to formulate the same problem in different ways. For a Hartree–Fock wave function, an equation for the change in the MO coefficients may also be formulated from the HF equation in the atomic orbital basis, (Equation (3.54):

$$\mathbf{F}^{(0)}\mathbf{C}^{(0)} = \mathbf{S}^{(0)}\mathbf{C}^{(0)}\boldsymbol{\varepsilon}^{(0)} \tag{11.72}$$

The superscript (0) here denotes the unperturbed system. The orthonormality of the molecular orbitals (Equation (3.20)) can be expressed as

$$\mathbf{C}^{t(0)}\mathbf{S}^{(0)}\mathbf{C}^{(0)} = \mathbf{1} \tag{11.73}$$

Expanding each of the **F**, **C**, **S** and $\boldsymbol{\varepsilon}$ matrices in terms of a perturbation parameter (e.g. $\mathbf{F} = \mathbf{F}^{(0)} + \lambda\mathbf{F}^{(1)} + \lambda^2\mathbf{F}^{(2)} + \cdots$) and collecting all the first-order terms (analogous to the strategy used in Section 4.8) gives

$$\begin{aligned}\mathbf{F}^{(1)}\mathbf{C}^{(0)} + \mathbf{F}^{(0)}\mathbf{C}^{(1)} &= \mathbf{S}^{(1)}\mathbf{C}^{(0)}\boldsymbol{\varepsilon}^{(0)} + \mathbf{S}^{(0)}\mathbf{C}^{(1)}\boldsymbol{\varepsilon}^{(0)} + \mathbf{S}^{(0)}\mathbf{C}^{(0)}\boldsymbol{\varepsilon}^{(1)} \\ [\mathbf{F}^{(0)} - \mathbf{S}^{(0)}\boldsymbol{\varepsilon}^{(0)}]\mathbf{C}^{(1)} &= [-\mathbf{F}^{(1)} + \mathbf{S}^{(1)}\boldsymbol{\varepsilon}^{(0)} + \mathbf{S}^{(0)}\boldsymbol{\varepsilon}^{(1)}]\mathbf{C}^{(0)}\end{aligned} \tag{11.74}$$

The orthonormality condition becomes

$$\mathbf{C}^{t(1)}\mathbf{S}^{(0)}\mathbf{C}^{(0)} + \mathbf{C}^{t(0)}\mathbf{S}^{(1)}\mathbf{C}^{(0)} + \mathbf{C}^{t(0)}\mathbf{S}^{(0)}\mathbf{C}^{(1)} = 0 \tag{11.75}$$

Equation (11.75) is the first-order *Coupled Perturbed Hartree–Fock* (CPHF) equation.[13] The perturbed MO coefficients are given in terms of unperturbed quantities and the first-order Fock, Lagrange ($\boldsymbol{\varepsilon}$) and overlap matrices. The $\mathbf{F}^{(1)}$ term is given as follows, where the (external) perturbation typically is part of $\mathbf{h}^{(1)}$:

$$\mathbf{F}^{(1)} = \mathbf{h}^{(1)} + \mathbf{G}^{(1)}\mathbf{D}^{(0)} + \mathbf{G}^{(0)}\mathbf{D}^{(1)} \tag{11.76}$$

Here **h** is the one-electron (core) matrix, **D** the density matrix and **G** the tensor containing the two-electron integrals. The density matrix is given as a product of MO coefficients (Equation (3.55)):

$$\mathbf{D}^{(0)} = \mathbf{C}^{t(0)}\mathbf{C}^{(0)} \tag{11.77}$$

$$\mathbf{D}^{(1)} = \mathbf{C}^{t(1)}\mathbf{C}^{(0)} + \mathbf{C}^{t(0)}\mathbf{C}^{(1)} \tag{11.78}$$

The $\mathbf{S}^{(1)}$, $\mathbf{h}^{(1)}$ and $\mathbf{g}^{(1)}$ quantities are (first) derivatives of one- and two-electron integrals over basis functions:

$$S_{\alpha\beta}{}^{(1)} = \langle\chi_\alpha|\chi_\beta\rangle^{(1)} = \frac{\partial}{\partial\lambda}\langle\chi_\alpha|\chi_\beta\rangle \tag{11.79}$$

$$h_{\alpha\beta}{}^{(1)} = \langle\chi_\alpha|\mathbf{h}|\chi_\beta\rangle^{(1)} = \frac{\partial}{\partial\lambda}\langle\chi_\alpha|\mathbf{h}|\chi_\beta\rangle \tag{11.80}$$

$$g_{\alpha\beta\gamma\delta}{}^{(1)} = \langle\chi_\alpha\chi_\beta|\mathbf{g}|\chi_\gamma\chi_\delta\rangle^{(1)} = \frac{\partial}{\partial\lambda}\langle\chi_\alpha\chi_\beta|\mathbf{g}|\chi_\gamma\chi_\delta\rangle \tag{11.81}$$

The derivatives of the integrals may involve derivatives of the basis functions or the operator, or both (see Section 11.9). Using Equations (11.76) to (11.78) in Equation (11.74) gives a set of linear equations relating $\mathbf{C}^{(1)}$ to $\mathbf{S}^{(1)}$, $\mathbf{h}^{(1)}$, $\mathbf{g}^{(1)}$ and $\mathbf{C}^{(0)}$.

Just as the variational condition for an HF wave function can be formulated either as a matrix equation or in terms of orbital rotations (Sections 3.5 and 3.6), the CPHF may also be viewed as a rotation of the molecular orbitals, which was the approach taken in the previous section. In the absence of a perturbation, the molecular orbitals make the energy stationary, that is the derivative of the energy with respect to a change in the MOs is zero. This is equivalent to the statement that the off-diagonal elements of the Fock matrix between the occupied and virtual MOs are zero:

$$\langle \phi_i | \mathbf{F} | \phi_a \rangle = 0$$

$$\langle \phi_i | \mathbf{h} | \phi_a \rangle + \sum_{k=1}^{N_{\text{orb}}} [\langle \phi_i \phi_k | \mathbf{g} | \phi_a \phi_k \rangle - \langle \phi_i \phi_k | \mathbf{g} | \phi_k \phi_a \rangle] = 0 \tag{11.82}$$

When a perturbation is introduced, the stationary condition means that the orbitals must change, which can be described as a mixing of the unperturbed MOs. In other words, the stationary orbitals in the presence of a perturbation are given by a unitary transformation of the unperturbed orbitals (see also Section 3.6):

$$\boldsymbol{\phi}' = \boldsymbol{\phi}\mathbf{U} \tag{11.83}$$

$$\phi_i' = \sum_{j=1}^{M_{\text{basis}}} u_{ji}\phi_j \tag{11.84}$$

The $\mathbf{U}$ matrix describes how the MOs change, that is it contains the derivatives of the MO coefficients. In the absence of a perturbation, $\mathbf{U}$ is the identity matrix. Since the energy is independent of a rotation among the occupied or virtual orbitals, only the mixing of occupied and virtual orbitals is determined by requiring that the energy should be stationary. The occupied–occupied and virtual–virtual mixing may be fixed from the orthonormality condition (Equation (11.75)) or, equivalently, by requiring the perturbed Fock matrix to be diagonal also in the occupied–occupied and virtual–virtual blocks. Without these additional requirements, the procedure is called *Coupled Hartree–Fock* (CHF), as opposed to CPHF.

Let us now explicitly make $\mathbf{U}^{(1)}$ the matrix containing the first-order changes in the MO coefficients and being diagonal in the occupied–occupied and virtual–virtual blocks

$$\mathbf{C}^{(1)} = \mathbf{C}^{(0)}\mathbf{U}^{(1)} \tag{11.85}$$

An equation for the $\mathbf{U}^{(1)}$ elements can be obtained from the condition that the Fock matrix is diagonal and by expanding all involved quantities to first order:

$$\langle \phi_i | \mathbf{h} | \phi_a \rangle \to \langle \phi_i | \mathbf{h} | \phi_a \rangle^{(0)} + \langle \phi_i | \mathbf{h} | \phi_a \rangle^{(1)} \tag{11.86}$$

$$\langle \phi_i \phi_k | \mathbf{g} | \phi_a \phi_k \rangle \to \langle \phi_i \phi_k | \mathbf{g} | \phi_a \phi_k \rangle^{(0)} + \langle \phi_i \phi_k | \mathbf{g} | \phi_a \phi_k \rangle^{(1)} \tag{11.87}$$

The $\langle \phi_i | \mathbf{h} | \phi_a \rangle^{(1)}$ and $\langle \phi_i \phi_k | \mathbf{g} | \phi_a \phi_k \rangle^{(1)}$ elements are integral derivatives with respect to the perturbation, analogous to Equations (11.80) and (11.81), but expressed in terms of molecular orbitals. Inserting these expansions into the $\langle \phi_i | \mathbf{F} | \phi_a \rangle = 0$ condition and collecting all terms that are first order in λ gives a set of linear equations that can be written as in Equation (11.65).

The CPHF equations can, as illustrated above, be formulated either in an atomic orbital or molecular orbital basis. Although the latter has computational advantages in certain cases, the former is more suitable for use in connection with direct methods (where the atomic integrals are calculated as required), as discussed in Section 3.8.5.

There is one CPHF equation to be solved for each perturbation. If it is an external electric or magnetic field, there will in general be three components (F_x, F_y, F_z); if it is a geometry perturbation there will be $3N_{\text{atom}}$ (actually only $3N_{\text{atom}} - 6$ independent) components. Since the matrix elements on the left-hand side of Equation (11.65) are independent of the nature of the perturbation, such multiple CPHF equations are often solved simultaneously.

The solution of the CPHF equations corresponds to finding the change in the optimum orbitals when the perturbation is turned on, that is the orbitals are allowed to relax in the presence of the perturbation. For certain properties, it is advantageous to *not* allow the orbitals to relax, because of triplet instabilities in the wave function. The corresponding uncoupled HF corresponds to only including the orbital energy difference in the **A** matrix (Equation (11.66)) and setting the **B** matrix (Equation (11.67)) to zero.

The CPHF procedure may be generalized to higher order. Extending the expansion to second order allows the derivation of an equation for the second-order change in the MO coefficients, by solving a second-order CPHF equation, etc.

For perturbation-dependent basis sets (e.g. geometry derivatives) the (first-order) CPHF equations involve (first) derivatives of the one- and two-electron integrals with respect to the perturbation. For basis functions that are independent of the perturbation (e.g. an electric field), these derivatives are zero. The solution of each CPHF equation (for each perturbation) typically requires approximately half of the time required for solving the HF equations themselves. For basis set-dependent perturbations, the first-order CPHF equations are only needed for calculating second (and higher) derivatives, which have terms involving second (and higher) derivatives of the integrals themselves, and solving the CPHF equations is usually not the computational bottleneck in these cases.

Without the Lagrange technique for non-variational wave functions (CI, MP and CC), the nth-order CPHF is needed for the nth derivative. Consider, for example, the MP2 energy correction:

$$\text{MP2} = \sum_{i<j}^{\text{N}_{\text{occ}}} \sum_{a<b}^{\text{N}_{\text{vir}}} \frac{[\langle \phi_i \phi_j | \phi_a \phi_b \rangle - \langle \phi_i \phi_j | \phi_b \phi_a \rangle]^2}{\varepsilon_i + \varepsilon_j - \varepsilon_a - \varepsilon_b} \tag{11.88}$$

The derivative of a molecular integral is given by Equation (11.57):

$$\frac{\partial}{\partial \lambda} \langle \phi_i \phi_j | \phi_a \phi_b \rangle = \frac{\partial}{\partial \lambda} \sum_{\alpha\beta\gamma\delta}^{M_{\text{basis}}} c_{i\alpha} c_{j\beta} c_{a\gamma} c_{b\delta} \langle \chi_\alpha \chi_\beta | \chi_\gamma \chi_\delta \rangle \tag{11.89}$$

This requires both the derivative of the MO coefficients and the two-electron integrals in the AO basis. The denominator leads to derivatives of the MO energies, which can be obtained by solving the CPHF equations. A straightforward differentiation of Equation (11.88) thus leads to a formula where the first-order response is required.

The following sections exemplify derivative and perturbation expressions for some first- and second-order properties arising from different perturbations when using an HF-type wave function.

11.7 Electric Field Perturbation

11.7.1 External Electric Field

If the perturbation is a homogeneous dipolar electric field $\mathbf{F}$ ($\mathbf{F} = F\mathbf{r}$), the perturbation operator $\mathbf{P}_1$ (Equation (11.31)) is the position vector $\mathbf{r}$ and $\mathbf{P}_2$ is zero. Assuming that the basis functions are independent of the electric field (as is normally the case), the first-order HF property, the dipole moment, is given by the derivative formula (11.35), as shown below (since an HF wave function obeys the Hellmann–Feynman theorem):

$$\boldsymbol{\mu} = -\frac{\partial E_{\mathrm{HF}}}{\partial \mathbf{F}} = -\langle \Phi_{\mathrm{HF}} | \mathbf{r} | \Phi_{\mathrm{HF}} \rangle \tag{11.90}$$

This is equivalent to the expression from first-order perturbation theory, Equation (11.31). For non-variational wave functions the dipole moment calculated by the two approaches will be different, since the derivative of the wave function with respect to the field will not be zero. The second-order property, the dipole polarizability, is given by the derivative formula, Equation (11.46), as shown in

$$\boldsymbol{\alpha} = -\frac{\partial^2 E_{\mathrm{HF}}}{\partial \mathbf{F}^2} = 2\left\langle \frac{\partial \Phi_{\mathrm{HF}}}{\partial \mathbf{F}} \middle| \mathbf{r} \middle| \Phi_{\mathrm{HF}} \right\rangle \tag{11.91}$$

Second-order perturbation theory (Equation (11.32)) yields

$$\boldsymbol{\alpha} = -2\sum_{i \neq 0} \frac{|\langle \Phi_{\mathrm{HF}} | \mathbf{r} | \Phi_i \rangle|^2}{E_0 - E_i} \tag{11.92}$$

11.7.2 Internal Electric Field

Although nuclei are often modeled as point charges in quantum chemistry, they do in fact have a finite size. The internal structure of the nucleus leads to a quadrupole moment for nuclei with spin larger than $^1/_2$ (the dipole and octopole moments vanish by symmetry). This leads to an interaction term that is the product of the nuclear quadrupole moment with the field gradient ($\mathbf{F}' = \nabla \mathbf{F}$) created by the electron distribution:

$$\mathbf{H}_{\Theta} = -\tfrac{1}{2} \sum_{A=1}^{N_{\mathrm{nuclei}}} \boldsymbol{\Theta}_A \mathbf{F}' \tag{11.93}$$

This property is the second derivative of the energy with respect to the nuclear quadrupole moment and the electric field gradient and can thus be considered as the electric analog of the nuclear magnetic shielding constant. The quadrupole moment is a constant for a given nucleus and isotope, and the measured interaction thus provides the electric field gradient at the nuclear position. The electric field gradient is the second derivative of the electric potential with respect to the coordinates and thus contains information regarding the electron distribution surrounding the nucleus. It can be calculated as a simple expectation value over the wave function and terms from the other nuclei:

$$\mathbf{F}'_A = \left\langle \Psi \middle| \frac{3\mathbf{r}_{iA}\mathbf{r}^{\mathrm{t}}_{iA} - \mathbf{r}^{\mathrm{t}}_{iA}\mathbf{r}_{iA}\mathbf{1}}{r^5_{iA}} \middle| \Psi \right\rangle - \sum_{B \neq A} Z_B \frac{3\mathbf{R}_{AB}\mathbf{R}^{\mathrm{t}}_{AB} - \mathbf{R}^{\mathrm{t}}_{AB}\mathbf{R}_{AB}\mathbf{1}}{\mathbf{R}^5_{AB}} \tag{11.94}$$

The quadrupole interaction also determines the peak splitting in Mössbauer spectroscopy, while the electron density at the nuclear position results in an isotope shift.[14]

11.8 Magnetic Field Perturbation

The situation is somewhat more complicated when the perturbation is a magnetic field. An electric field interacts directly with the charged particles (electrons and nuclei) and adds a *potential* energy term to the Hamiltonian operator. A magnetic field, however, interacts with the magnetic moments generated by the *movement* of the charged particles (electrons), that is a magnetic perturbation changes the *kinetic* energy operator. The *generalized* (also called the *canonical*) momentum operator $\boldsymbol{\pi}$ is defined in by

$$\boldsymbol{\pi} = \mathbf{p} - q\mathbf{A} \tag{11.95}$$

Here q is the charge and $\mathbf{A}$ is the *vector potential* associated with the magnetic field $\mathbf{B}$ (more correctly, the *magnetic induction* or *flux* density, being different from the magnetic field by a factor of $4\pi \times 10^{-7}$ H/m), with the latter being given as the curl of the vector potential:

$$\mathbf{B} = \nabla \times \mathbf{A} \tag{11.96}$$

Only the kinetic energy of the electrons is considered within the Born–Oppenheimer approximation and the generalized momentum becomes ($q = -1$):

$$\boldsymbol{\pi} = \mathbf{p} + \mathbf{A} \tag{11.97}$$

The vector potential is not uniquely defined since the gradient of any scalar function may be added (the curl of a gradient is always zero). For an external magnetic field, it is conventional to write it as

$$\mathbf{A}_{\text{ext}}(\mathbf{r}) = \tfrac{1}{2}\mathbf{B}_{\text{ext}} \times (\mathbf{r} - \mathbf{R}_{\text{G}}) \tag{11.98}$$

Here $\mathbf{R}_{\text{G}}$ is referred to as the *gauge origin*, that is the center of the vector potential. One may verify by explicit calculation that the curl of $\mathbf{A}_{\text{ext}}$ in Equation (11.98) indeed gives $\mathbf{B}_{\text{ext}}$.

A nucleus with a non-zero spin acts as a magnetic dipole, giving rise to a vector potential $\mathbf{A}_{\text{A}}$ and producing the associated magnetic field by taking the curl:

$$\mathbf{A}_{\text{A}} = \frac{g_{\text{A}}\mu_{\text{N}}}{c^2}\frac{\mathbf{I}_{\text{A}} \times (\mathbf{r} - \mathbf{R}_{\text{A}})}{|\mathbf{r} - \mathbf{R}_{\text{A}}|^3} \tag{11.99}$$

$$\mathbf{B}_{\text{A}} = -\frac{g_{\text{A}}\mu_{\text{N}}}{c^2}\left[\frac{\mathbf{I}_{\text{A}}}{|\mathbf{r} - \mathbf{R}_{\text{A}}|^3} - 3\frac{(\mathbf{r} - \mathbf{R}_{\text{A}})((\mathbf{r} - \mathbf{R}_{\text{A}}) \cdot \mathbf{I}_{\text{A}})}{|\mathbf{r} - \mathbf{R}_{\text{A}}|^5} - \frac{8\pi}{3}\mathbf{I}_{\text{A}}\delta(\mathbf{r} - \mathbf{R}_{\text{A}})\right] \tag{11.100}$$

Here $g_{\text{A}}\mu_{\text{N}}\mathbf{I}_{\text{A}}$ is the magnetic moment of nucleus A and $\mathbf{R}_{\text{A}}$ is the position (the nucleus is the natural gauge origin). The $\mathbf{B}_{\text{A}}$ expression determines the magnetic field at position $\mathbf{r}$ (not necessarily indicating an electron) due to a magnetic nucleus at position $\mathbf{R}_{\text{A}}$. The δ-function in the last term in the $\mathbf{B}_{\text{A}}$ expression arises from the quantum mechanical possibility of $\mathbf{r} - \mathbf{R}_{\text{A}} = 0$, that is the magnetic field directly at the nuclear position. Note that the presence of c^{-2} emerges from the units of magnetic field ($\mu_0/4\pi = c^{-2}$ in atomic units) and does not indicate a relativistic origin.

The spin associated with an electron also acts as a magnetic dipole ($-g_{\text{e}}\mu_{\text{B}}\mathbf{s}_i$), giving rise to a vector potential $\mathbf{A}_{\text{e}}$ and an associated magnetic field:

$$\mathbf{A}_{\text{e}} = -\frac{g_{\text{e}}\mu_{\text{B}}}{c^2}\frac{\mathbf{s}_i \times (\mathbf{r} - \mathbf{r}_i)}{|\mathbf{r} - \mathbf{r}_i|^3} \tag{11.101}$$

$$\mathbf{B}_{\text{e}} = -\frac{g_{\text{e}}\mu_{\text{B}}}{c^2}\left[\frac{\mathbf{s}_i}{|\mathbf{r} - \mathbf{r}_i|^3} - 3\frac{(\mathbf{r} - \mathbf{r}_i)((\mathbf{r} - \mathbf{r}_i) \cdot \mathbf{s}_i)}{|\mathbf{r} - \mathbf{r}_i|^5} - \frac{8\pi}{3}\mathbf{s}_i\delta(\mathbf{r} - \mathbf{r}_i)\right] \tag{11.102}$$

The $\mathbf{B}_{\text{e}}$ expression similarly determines the magnetic field at position $\mathbf{r}$ due to an electron at position $\mathbf{r}_i$.

The introduction of the generalized momentum operator in the one-electron kinetic energy part of the Dirac equation leads to three new interaction terms, as shown in Equations (9.30) to (9.33). It should be noted that the last two terms will also show up in a non-relativistic treatment when the magnetic vector potential is included, and only the $\mathbf{s} \cdot \mathbf{B}$ term should be considered as a relativistic effect:

$$\begin{array}{c} \mathbf{p}^2 \rightarrow \pi^2 = (\mathbf{p} + \mathbf{A})^2 \\ \Downarrow \\ g_e \mu_B \mathbf{s} \cdot \mathbf{B} \quad ; \quad \mathbf{A} \cdot \mathbf{p} \quad ; \quad \frac{1}{2}\mathbf{A}^2 \end{array} \tag{11.103}$$

If there is more than one type of magnetic field present there will be an additional mixed $\mathbf{AA}'$ term. Depending on the type of magnetic interactions present, these give operators as shown below. To simplify the expressions, we will use the notations $\mathbf{r}_{iA} = \mathbf{r}_i - \mathbf{R}_A$, $\mathbf{r}_{ij} = \mathbf{r}_i - \mathbf{r}_j$ and $\mathbf{R}_{AB} = \mathbf{R}_A - \mathbf{R}_B$.

In addition to the magnetic terms arising from the expansion of the generalized momentum operator, there are also magnetic perturbation terms arising from relativistic corrections, as discussed in Section 9.2. These corrections may be derived by an expansion in the inverse speed of light. For consistency, we will in the following only consider terms up to order c^{-2}, with the exception of the indirect nuclear spin–spin coupling, where the lowest non-vanishing term is of order c^{-4}. Furthermore, in the rest of this section the summation over the number of electrons and nuclei has been omitted for clarity.

11.8.1 External Magnetic Field

For an external magnetic field, the three terms in Equation (11.103) become

$$g_e \mu_B \mathbf{s} \cdot \mathbf{B} = g_e \mu_B \mathbf{s} \cdot \mathbf{B}_{ext} \tag{11.104}$$

$$\begin{aligned} \mathbf{A} \cdot \mathbf{p} &= \left(\tfrac{1}{2}\mathbf{B}_{ext} \times \mathbf{r}_{iG}\right) \cdot \mathbf{p} \\ &= \tfrac{1}{2}\mathbf{B}_{ext} \cdot (\mathbf{r}_{iG} \times \mathbf{p}) \\ &= \tfrac{1}{2}\mathbf{B}_{ext} \cdot \mathbf{L}_G \end{aligned} \tag{11.105}$$

$$\begin{aligned} \tfrac{1}{2}\mathbf{A}^2 &= \tfrac{1}{2}\left(\tfrac{1}{2}\mathbf{B}_{ext} \times \mathbf{r}_{iG}\right) \cdot \left(\tfrac{1}{2}\mathbf{B}_{ext} \times \mathbf{r}_{iG}\right) \\ &= \tfrac{1}{8}\left(\mathbf{B}_{ext}^2 \cdot \mathbf{r}_{iG}^2 - (\mathbf{B}_{ext} \cdot \mathbf{r}_{iG})^2\right) \\ &= \mathbf{B}_{ext} \cdot \mathbf{P}_2^{\xi} \cdot \mathbf{B}_{ext} \end{aligned} \tag{11.106}$$

Here the vector identities $\mathbf{a} \times \mathbf{b} \cdot \mathbf{c} = \mathbf{a} \cdot \mathbf{b} \times \mathbf{c}$ and $(\mathbf{a} \times \mathbf{b}) \cdot (\mathbf{c} \times \mathbf{d}) = (\mathbf{a} \cdot \mathbf{c})(\mathbf{b} \cdot \mathbf{d}) - (\mathbf{a} \cdot \mathbf{d})(\mathbf{c} \cdot \mathbf{b})$ have been used, and the angular momentum operator $\mathbf{L}_G$ is defined implicitly by Equation (11.105). The presence of a magnetic field thus introduces three new terms, two being linear and one being quadratic in the field. The *spin-Zeeman* term $\mathbf{s} \cdot \mathbf{B}_{ext}$ describes the interaction of the electron spin with the magnetic field, while $^1/_2\mathbf{B}_{ext} \cdot \mathbf{L}_G$ is the *orbital-Zeeman* term describing the interaction of the magnetic field with the magnetic moment associated with the movement of the electron. For a many-electron system, the spin-Zeeman term becomes $\mathbf{S} \cdot \mathbf{B}_{ext}$, where $\mathbf{S}$ indicates the total molecular spin. The quadratic $\mathbf{P}_2^{\xi}$ operator arising from $^1/_2\mathbf{A}_{ext}^2$ may be written as

$$\mathbf{P}_2^{\xi} = \tfrac{1}{8}\left(\mathbf{r}_{iG}^t \mathbf{r}_{iG} \mathbf{1} - \mathbf{r}_{iG} \mathbf{r}_{iG}^t\right) \tag{11.107}$$

Here $\mathbf{r}_{iG}^t \mathbf{r}_{iG} \mathbf{1}$ is the inner (dot) product times a unit matrix and $\mathbf{r}_{iG}\mathbf{r}_{iG}^t$ is the outer product, that is a 3×3 matrix containing the products of the x, y, z components, analogous to the quadrupole moment, Equation (11.10). Note that both the $\mathbf{L}_G$ and $\mathbf{P}_2^{\xi}$ operators are gauge-dependent.

11.8.2 Nuclear Spin

The $g_e\mu_B\mathbf{s}\cdot\mathbf{B}$ term in Equation (11.103) in connection with $\mathbf{B}_A$ in Equation (11.99) gives three terms, which conventionally are collected in two operators:

$$\mathbf{H}_{ne}^{SD} = \mathbf{s}_i \cdot \mathbf{P}_{ne}^{SD} \cdot \mathbf{I}_A = -\frac{g_e g_A \mu_B \mu_N}{c^2} \mathbf{s}_i \cdot \left(\frac{\mathbf{r}_{iA}^t \mathbf{r}_{iA} \mathbf{1} - 3\mathbf{r}_{iA}\mathbf{r}_{iA}^t}{r_{iA}^5} \right) \cdot \mathbf{I}_A \tag{11.108}$$

$$\mathbf{H}_{ne}^{FC} = \mathbf{s}_i \cdot \mathbf{P}_{ne}^{FC} \cdot \mathbf{I}_A = \frac{8\pi g_e g_A \mu_B \mu_N}{3c^2} \delta(\mathbf{r}_{iA})(\mathbf{s}_i \cdot \mathbf{I}_A) \tag{11.109}$$

$\mathbf{H}_{ne}^{SD}$ is a (one electron) *Spin-Dipolar* and $\mathbf{H}_{ne}^{FC}$ is a *Fermi Contact* operator, and their sum is the $\mathbf{H}_{ne}^{SS}$ operator in Equation (9.48).

The $\mathbf{A}_A \cdot \mathbf{p}$ term in Equation (11.103) gives the *Paramagnetic Spin–Orbit* operator:

$$\mathbf{H}_{ne}^{PSO} = \mathbf{I}_A \cdot \mathbf{P}_{ne}^{PSO} = \frac{g_A \mu_N}{c^2} \cdot \mathbf{I}_A \cdot \frac{\mathbf{r}_{iA} \times \mathbf{p}_i}{r_{iA}^3} \tag{11.110}$$

$\mathbf{H}_{ne}^{PSO}$ is identical to Equation (9.47).

The $\frac{1}{2}\mathbf{A}_A^2$ term in Equation (11.103) gives a *Diamagnetic Spin–Orbit* operator, which is an operator of order c^{-4}. Although we otherwise only consider terms up to order c^{-2}, the nuclear spin–spin coupling constant only contains terms of order c^{-4}, which is why we need to include $\mathbf{H}_{nn}^{DSO}$:

$$\mathbf{H}_{nn}^{DSO} = \mathbf{I}_A \cdot \mathbf{P}_{nn}^{DSO} \cdot \mathbf{I}_B = \frac{g_A g_B \mu_N^2}{2c^4} \mathbf{I}_A \cdot \frac{(\mathbf{r}_{iA}^t \mathbf{r}_{iB} \mathbf{1} - \mathbf{r}_{iB}\mathbf{r}_{iA}^t)}{r_{iA}^3 r_{iB}^3} \cdot \mathbf{I}_B \tag{11.111}$$

When both nuclear spins and an external magnetic field are present, there is an additional mixed $\mathbf{A}_{ext} \cdot \mathbf{A}_A$ term arising from the expansion of the generalized momentum operator:

$$\mathbf{H}_{ne}^{DS} = \mathbf{B}_{ext} \cdot \mathbf{P}_{ne}^{DS} \cdot \mathbf{I}_A = \frac{g_A \mu_N}{2c^2} \mathbf{B}_{ext} \cdot \frac{\mathbf{r}_{iG}^t \mathbf{r}_{iA} \mathbf{1} - \mathbf{r}_{iA}\mathbf{r}_{iG}^t}{r_{iA}^3} \cdot \mathbf{I}_A \tag{11.112}$$

This nuclear *Diamagnetic Shielding* operator contributes to the NMR shielding tensor.

11.8.3 Electron Spin

The $g_e\mu_B\mathbf{s}\cdot\mathbf{B}$ term in Equation (11.103) in connection with $\mathbf{B}_e$ in Equation (11.101) gives three terms, which again are collected in two operators:

$$\mathbf{H}_{ee}^{SD} = \mathbf{s}_i \cdot \mathbf{P}_{ee}^{SD} \cdot \mathbf{s}_j = \frac{g_e^2 \mu_B^2}{2c^2} \mathbf{s}_i \cdot \left(\frac{\mathbf{r}_{ij}^t \mathbf{r}_{ij} \mathbf{1} - 3\mathbf{r}_{ij}\mathbf{r}_{ij}^t}{r_{ij}^5} \right) \cdot \mathbf{s}_j \tag{11.113}$$

$$\mathbf{H}_{ee}^{FC} = \mathbf{s}_i \cdot \mathbf{P}_{ee}^{FC} \cdot \mathbf{s}_j = -\frac{4\pi g_e^2 \mu_B^2}{3c^2} \delta(\mathbf{r}_{ij})(\mathbf{s}_i \cdot \mathbf{s}_j) \tag{11.114}$$

$\mathbf{P}_{ee}^{SD}$ is a (two-electron) *Spin-Dipolar* and $\mathbf{P}_{ee}^{FC}$ is a *Fermi Contact* operator, with the sum being equal to the $\mathbf{H}_{ee}^{SS}$ operator in Equation (9.43).

The $\mathbf{A}_\mathrm{e} \cdot \mathbf{p}$ term in Equation (11.103) gives the two-electron part of the spin–orbit operator:

$$\mathbf{H}_{\mathrm{ee}}^{\mathrm{SO}} = \mathbf{s}_i \cdot \mathbf{P}_{\mathrm{ee}}^{\mathrm{SO}} = -\frac{g_\mathrm{e}\mu_\mathrm{B}}{2c^2}\mathbf{s}_i \cdot \frac{\mathbf{r}_{ij} \times \mathbf{p}_i + 2\mathbf{r}_{ij} \times \mathbf{p}_j}{r_{ij}^3} \tag{11.115}$$

$\mathbf{H}_{\mathrm{ee}}^{\mathrm{SO}}$ is equivalent to the sum of $\mathbf{H}_{\mathrm{ee}}^{\mathrm{SO}}$ and $\mathbf{H}_{\mathrm{ee}}^{\mathrm{SOO}}$ in Equations (9.41) and (9.42).

The $\frac{1}{2}\mathbf{A}_\mathrm{e}^2$ term in Equation (11.103) gives an operator analogous to Equation (11.111), but that depends on two electron spins instead of two nuclear spins. This, however, is an order c^{-4} operator compared with the order c^{-2} operators $\mathbf{P}_{\mathrm{ee}}^{\mathrm{SD}}$ and $\mathbf{P}_{\mathrm{ee}}^{\mathrm{FC}}$ (Equations (11.113) and (11.114)) describing spin–spin interactions, and is therefore neglected.

The $\mathbf{A}_\mathrm{e} \cdot \mathbf{A}_\mathrm{A}$ term in Equation (11.103) gives a coupling between the electronic and nuclear spins, and is again an operator of order c^{-4}. Compared with the order c^{-2} operators $\mathbf{P}_{\mathrm{ne}}^{\mathrm{SD}}$ and $\mathbf{P}_{\mathrm{ne}}^{\mathrm{FC}}$ (Equations (11.108) and (11.109)), it is again neglected.

When both electron spin and an external magnetic field are considered, there is a mixed $\mathbf{A}_{\mathrm{ext}} \cdot \mathbf{A}_\mathrm{e}$ term:

$$\mathbf{H}_{\mathrm{ee}}^{\mathrm{DS}} = \mathbf{B}_{\mathrm{ext}} \cdot \mathbf{P}_{\mathrm{ee}}^{\mathrm{DS}} \cdot \mathbf{s}_i = -\frac{g_\mathrm{e}\mu_\mathrm{B}}{2c^2}\mathbf{B}_{\mathrm{ext}} \cdot \frac{\left(\mathbf{r}_{iG}^{\mathrm{t}}\mathbf{r}_{ij}\mathbf{1} - \mathbf{r}_{ij}\mathbf{r}_{iG}^{\mathrm{t}}\right)}{r_{ij}^3} \cdot \mathbf{s}_i \tag{11.116}$$

This electronic *Diamagnetic Shielding* operator contributes to the ESR $\mathbf{g}$ tensor.

11.8.4 Electron Angular Momentum

The orbit–orbit magnetic interaction in Equation (9.44) provides a shift in the total energy but does not lead to a splitting of energy levels and thus no observable properties.

11.8.5 Classical Terms

The expansion of the generalized momentum operator only involves the magnetic interactions in the electronic part of the wave function. Since the corresponding nuclear part has been separated out by the Born–Oppenheimer approximations, we need to add a few terms corresponding to the (classical) interaction of the nuclear magnetic moments with an external magnetic field and between nuclei.

The nuclear spin-Zeeman term is analogous to the electronic term in Equation (11.104), except that the electron magnetic moment of $-g_\mathrm{e}\mu_\mathrm{B}\mathbf{s}$ is replaced by the nuclear magnetic moment $g_\mathrm{A}\mu_\mathrm{N}\mathbf{I}$. The nuclear magneton μ_N is defined analogously to the Bohr magneton μ_B, but using the proton mass m_p instead of the electron mass ($\mu_\mathrm{N} = e\hbar/2m_\mathrm{p} = 2.723 \times 10^{-4}$ in atomic units), while the g_A factor depends on the specific nucleus (isotope):

$$\mathbf{H}_{\mathrm{n}}^{\mathrm{Zeeman}} = -\mu_N g_\mathrm{A}\mathbf{I}_\mathrm{A} \cdot \mathbf{B} \tag{11.117}$$

The term involving two nuclei is analogous to the electron–electron spin-dipole term in Equation (11.113). The corresponding Fermi contact term disappears since nuclei cannot occupy the same position at energies relevant for chemistry. Note that the direct spin–spin coupling is independent of the electronic wave function; it only depends on the molecular geometry:

$$\mathbf{H}_{\mathrm{nn}}^{\mathrm{SS}} = \frac{\mu_\mathrm{N}^2 g_\mathrm{A} g_\mathrm{B}}{2c^2}\mathbf{I}_\mathrm{A} \cdot \left(\frac{\mathbf{R}_{\mathrm{AB}}^{\mathrm{t}}\mathbf{R}_{\mathrm{AB}}\mathbf{1} - 3\mathbf{R}_{\mathrm{AB}}\mathbf{R}_{\mathrm{AB}}^{\mathrm{t}}}{R_{\mathrm{AB}}^5}\right) \cdot \mathbf{I}_\mathrm{B} \tag{11.118}$$

11.8.6 Relativistic Terms

The most important relativistic corrections are the one-electron spin–orbit operator and the relativistic correction to the spin-Zeeman operator:

$$\mathbf{H}_{\text{ne}}^{\text{SO}} = \frac{g_e \mu_B Z_A}{2mc^2} \mathbf{s}_i \cdot \frac{\mathbf{r}_{iA} \times \mathbf{p}_i}{r_{iA}^3} \tag{11.119}$$

$$\mathbf{H}_{\text{e}}^{\text{Zeeman-rel}} = \frac{g_e \mu_B}{2mc^2} (\mathbf{s}_i \cdot \mathbf{B}_i) \mathbf{p}_i^2 \tag{11.120}$$

Other relativistic corrections, such as the mass–velocity and Darwin terms, affect the wave function but do not lead to operators associated with molecular properties.

11.8.7 Magnetic Properties

Magnetic properties can be considered as arising from the interaction of magnetic fields generated by either an external magnetic field ($\mathbf{B}_{\text{ext}}$), a nuclear magnetic moment (**I**), an electron magnetic moment (**s**) or the angular moment (**l**) from the electronic movement. Although electron spin and angular momentum are implicitly included in even the simplest wave function by the Pauli principle, their magnetic effects are not included, and these are therefore normally considered as perturbative terms resulting in molecular properties. Table 11.2 shows the perturbation operators arising from

Table 11.2 Magnetic perturbation operators.

Origin	Operator	Equation	Name	Perturbation order				
				$\mathbf{B}_{\text{ext}}$	I	s	l	c^{-n}
$\mathbf{S} \cdot \mathbf{B}_{\text{ext}}$	$\mathbf{S}$	11.104/9.38	Electron spin-Zeeman[a]	1	0	1	0	0
$\mathbf{A}_{\text{ext}} \cdot \mathbf{p}$	$\frac{1}{2}\mathbf{L}_{\text{G}}$	11.105	Orbital-Zeeman	1	0	0	1	0
$\frac{1}{2}\mathbf{A}_{\text{ext}}^2$	$\mathbf{P}_2^{\xi}$	11.107	Diamagnetic magnetizability	2	0	0	0	0
$\mathbf{S} \cdot \mathbf{B}_{\text{A}}$	$\mathbf{P}_{\text{ne}}^{\text{SD}}$	11.108/9.48	Nuclear–electron spin-dipole	0	1	1	0	2
	$\mathbf{P}_{\text{ne}}^{\text{FC}}$	11.109/9.48	Nuclear–electron Fermi contact	0	1	1	0	2
$\mathbf{A}_{\text{A}} \cdot \mathbf{p}$	$\mathbf{P}_{\text{ne}}^{\text{PSO}}$	11.110/9.47	Paramagnetic spin–orbit	0	1	0	1	2
$\frac{1}{2}\mathbf{A}_{\text{A}}^2$	$\mathbf{P}_{\text{nn}}^{\text{DSO}}$	11.111	Diamagnetic nuclear spin–spin	0	2	0	0	4
$\mathbf{A}_{\text{A}} \cdot \mathbf{A}_{\text{ext}}$	$\mathbf{P}_{\text{ne}}^{\text{DS}}$	11.112	Diamagnetic NMR shielding	1	1	0	0	2
$\mathbf{s} \cdot \mathbf{B}_{\text{e}}$	$\mathbf{P}_{\text{ee}}^{\text{SD}}$	11.113/9.43	Electron–electron spin-dipole	0	0	2	0	2
	$\mathbf{P}_{\text{ee}}^{\text{FC}}$	11.114/9.43	Electron–electron Fermi contact	0	0	2	0	2
$\mathbf{A}_{\text{e}} \cdot \mathbf{p}$	$\mathbf{P}_{\text{ee}}^{\text{SO}}$	11.115/9.41/9.42	Two-electron spin–orbit	0	0	1	1	2
$\mathbf{A}_{\text{e}} \cdot \mathbf{A}_{\text{ext}}$	$\mathbf{P}_{\text{ee}}^{\text{DS}}$	11.116	Diamagnetic ESR shielding	1	0	1	0	2
Classic	$\mathbf{I}_{\text{A}}$	11.117	Nuclear spin-Zeeman	1	1	0	0	0
Classic	$\mathbf{P}_{\text{nn}}^{\text{SS}}$	11.118	Nuclear spin–spin coupling	0	2	0	0	0
Relativistic	$\mathbf{P}_{\text{ne}}^{\text{SO}}$	11.119/9.46	One-electron spin–orbit	0	0	1	1	2
Relativistic	$\mathbf{P}_{\text{ee}}^{\text{OO}}$	9.44	Orbit-orbit interaction	0	0	0	2	2
Relativistic	$\mathbf{p}^2\mathbf{s}$	11.120/9.38	Electron spin-Zeeman, relativistic correction	1	0	1	0	2

[a] The spin-Zeeman term involves the total electron spin **S**.

these magnetic fields, where the last five columns indicate the perturbation order with respect to the perturbations and with respect to the inverse speed of light. Only the most important operators up to order c^{-2} have been included, with the exception of the diamagnetic nuclear spin–spin coupling operator, since the leading term for this quantity is of order c^{-4}.

The first-order property is given as an expectation value of operator(s) linear in the perturbation, as seen from Equation (11.31). The second-order property contains two contributions, an expectation value over quadratic (or bilinear) operators and a sum over products of matrix elements involving linear operators connecting the ground and excited states.

The first-order property with respect to an external field is the magnetic dipole moment **m** (Equation (11.17)). When field-independent basis functions are used, the HF magnetic dipole moment is given as the expectation value of the $^1/_2\mathbf{L}_\mathrm{G}$ and **S** (total electron spin) operators over the unperturbed wave function, Equations (11.31) and (11.35). Since the $\mathbf{L}_\mathrm{G}$ operator is imaginary it can only yield a non-zero result for spatially degenerate wave functions and the expectation value of **S** is only non-zero for non-singlet states:

$$\mathbf{m} = -\frac{\partial E_\mathrm{HF}}{\partial \mathbf{B}_\mathrm{ext}} = -\left\langle \Phi_\mathrm{HF} \left| \tfrac{1}{2}\mathbf{L}_\mathrm{G} + g_e \beta_\mathrm{B} \mathbf{S} \right| \Phi_\mathrm{HF} \right\rangle \tag{11.121}$$

Molecules with a permanent magnetic dipole moment are thus quite rare.

The second-order term, the magnetizability ξ, has two components. The derivative expression (11.46) is given by

$$\xi = -\frac{\partial^2 E_\mathrm{HF}}{\partial \mathbf{B}_\mathrm{ext}^2} = -2\left\langle \Phi_\mathrm{HF} \left| \mathbf{P}_2^\xi \right| \Phi_\mathrm{HF} \right\rangle + 2\left\langle \frac{\partial \Phi_\mathrm{HF}}{\partial \mathbf{B}_\mathrm{ext}} \left| \tfrac{1}{2}\mathbf{L}_\mathrm{G} \right| \Phi_\mathrm{HF} \right\rangle \tag{11.122}$$

Second-order perturbation theory (Equation (11.32)) yields

$$\xi = -2\left\langle \Phi_\mathrm{HF} \left| \mathbf{P}_2^\xi \right| \Phi_\mathrm{HF} \right\rangle - 2\sum_{i\neq 0} \frac{\left| \left\langle \Phi_\mathrm{HF} \left| \tfrac{1}{2}\mathbf{L}_\mathrm{G} \right| \Phi_i \right\rangle \right|^2}{E_\mathrm{HF} - E_i} \tag{11.123}$$

The first term is referred to as the *diamagnetic* contribution, while the latter is the *paramagnetic* part of the magnetizability. The total spin operator **S** gives no contribution to the paramagnetic term, since the ground and excited states are orthogonal in the spatial part. Each of the two components depends on the selected gauge origin but these gauge dependencies cancel exactly for an exact wave function. This cancellation is not guaranteed for approximate wave functions, and the total property may consequently depend on where the origin for the vector potential (Equation (11.98)) has been chosen (Section 11.8.8).

The mixed second-order term with respect to a nuclear magnetic moment **I** and electron spin is the *hyperfine coupling* tensor **A**.[15] The leading order term (c^{-2}) can be evaluated as a simple expectation value of the $\mathbf{P}_\mathrm{ne}^\mathrm{FC}$ and $\mathbf{P}_\mathrm{ne}^\mathrm{SD}$ operators. The former provides the isotropic part of the tensor, while the latter gives the anisotropic part:

$$\mathbf{A} = \left\langle \Phi_\mathrm{HF} \left| \mathbf{P}_\mathrm{ne}^\mathrm{FC} + \mathbf{P}_\mathrm{ne}^\mathrm{SD} \right| \Phi_\mathrm{HF} \right\rangle \tag{11.124}$$

The second-order property related to two nuclear spins $\mathbf{I}_\mathrm{A}$ and $\mathbf{I}_\mathrm{B}$ is the nuclear spin–spin coupling tensor. The *direct* interaction is determined entirely by the molecular geometry and is given by

Equation (11.118). For rapidly tumbling molecules (solution or gas phase) this contribution averages out to zero, but it is significant for solid-state NMR.

The *indirect* spin–spin coupling between nuclei A and B, which is the one observed in solution phase NMR, contains several contributions, all being of order c^{-4}:

$$\mathbf{J}_{\mathrm{AB}} = \left\langle \Phi_{\mathrm{HF}} \middle| \mathbf{P}_{\mathrm{nn}}^{\mathrm{DSO}} \middle| \Phi_{\mathrm{HF}} \right\rangle + 2\sum_{i \neq 0} \frac{\langle \Phi_{\mathrm{HF}} | \mathbf{P}_{1,\mathrm{A}} | \Phi_i \rangle \langle \Phi_i | \mathbf{P}'_{1,\mathrm{B}} | \Phi_{\mathrm{HF}} \rangle}{E_{\mathrm{HF}} - E_i} \tag{11.125}$$

The first part can be evaluated as the expectation value of $\mathbf{P}_{\mathrm{nn}}^{\mathrm{DSO}}$ (Equation (11.114)). The second part corresponds to all combinations of operators that are linear in the nuclear spin, that is $\mathbf{P}_{\mathrm{ne}}^{\mathrm{FC}}$, $\mathbf{P}_{\mathrm{ne}}^{\mathrm{SD}}$ and $\mathbf{P}_{\mathrm{ne}}^{\mathrm{PSO}}$ (Equations (11.108) to (11.110)). $\mathbf{P}_{\mathrm{ne}}^{\mathrm{FC}}$ and $\mathbf{P}_{\mathrm{ne}}^{\mathrm{SD}}$ contain the electron spin operators and for a singlet ground state (as is usually the case) this means that the excited state Ψ_i in the summation must be a triplet state. Since $\mathbf{P}_{\mathrm{ne}}^{\mathrm{PSO}}$ does not depend on electron spin, the combination of $\mathbf{P}_{\mathrm{ne}}^{\mathrm{PSO}}$ with either $\mathbf{P}_{\mathrm{ne}}^{\mathrm{FC}}$ or $\mathbf{P}_{\mathrm{ne}}^{\mathrm{SD}}$ gives a zero contribution. For rapidly tumbling molecules, it can be shown that the cross term between $\mathbf{P}_{\mathrm{ne}}^{\mathrm{FC}}$ and $\mathbf{P}_{\mathrm{ne}}^{\mathrm{SD}}$ averages out. For the trace (sum of the diagonal terms) of the 3 × 3 coupling matrix **J**, which is the observed coupling constant, only the three "diagonal" terms ($\mathbf{P}_1 = (\mathbf{P}'_1$) in Equation (11.125) thus survive. The Fermi contact term is the most important for one-bond couplings ($^1\mathbf{J}$) in singly bonded systems, but the other three contributions become important for multiple-bonded systems and for longer-range couplings ($^2\mathbf{J}$, $^3\mathbf{J}$).[16]

The second-order term corresponding to the interaction of the total electron spin **S** with the orbital magnetic moment is the spin–orbit interaction for open-shell molecules with a net angular momentum, as, for example, NO with $a^2\pi$ ground state:

$$\text{SO-interaction} = \left\langle \Phi_{\mathrm{HF}} \middle| \mathbf{P}_{\mathrm{ne}}^{\mathrm{SO}} + \mathbf{P}_{\mathrm{ee}}^{\mathrm{SO}} \middle| \Phi_{\mathrm{HF}} \right\rangle \tag{11.126}$$

The second-order property related to two electron spins is the *Zero Field Splitting* (ZFS) tensor **D**, which is responsible for making the three individual components of a triplet state non-degenerate, with a typical magnitude being of the order of a few per cm.[17,18] It contains contributions from both quadratic and linear operators, the latter being the spin–orbit operator containing both one- and two-electron terms:

$$\mathbf{D} = \left\langle \Phi_{\mathrm{HF}} \middle| \mathbf{P}_{\mathrm{ee}}^{\mathrm{SD}} + \mathbf{P}_{\mathrm{ee}}^{\mathrm{FC}} \middle| \Phi_{\mathrm{HF}} \right\rangle + 2\sum_{i \neq 0} \frac{\left| \left\langle \Phi_{\mathrm{HF}} \middle| \mathbf{P}_{\mathrm{ne}}^{\mathrm{SO}} + \mathbf{P}_{\mathrm{ee}}^{\mathrm{SO}} \middle| \Phi_i \right\rangle \right|^2}{E_{\mathrm{HF}} - E_i} \tag{11.127}$$

The $\mathbf{P}_{\mathrm{ee}}^{\mathrm{FC}}$ operator just introduces a uniform shift of all energy levels, and thus produces no observable effect on the splitting of the energy levels.

The interaction of an external magnetic field with a nuclear spin gives a Zeeman splitting of the energy levels by the $\mathbf{I}_{\mathrm{A}} \cdot \mathbf{B}_{\mathrm{ext}}$ term (Table 11.2). The details of the splitting, however, depend on the molecular environment, since the local magnetic field at a nuclear position is shielded by the electrons relative to the external field, $\mathbf{B}_{\mathrm{local}} = (\mathbf{1} - \boldsymbol{\sigma})\mathbf{B}_{\mathrm{ext}}$. The *NMR shielding tensor* $\boldsymbol{\sigma}$, which is the mixed second derivative with respect to a nuclear spin and an external magnetic field, has, by analogy with the magnetizability, a diamagnetic and paramagnetic part.[16,19] The diamagnetic part arises from $\mathbf{P}_{\mathrm{ne}}^{\mathrm{DS}}$, while the paramagnetic contribution contains products of matrix elements involving operators linear in **B** or **I**. These are given by the angular momentum operator $^1\!/_2\mathbf{L}_{\mathrm{G}}$ (Equation (11.105)) and by the paramagnetic spin–orbit operator $\mathbf{P}_{\mathrm{ne}}^{\mathrm{PSO}}$ (Equation (11.110)). Written in terms of the perturbation formula (11.32), the expression for the nuclear shielding for atom A becomes

$$\sigma_A = \left\langle \Phi_{HF} \left| \mathbf{P}_{ne}^{DS} \right| \Phi_{HF} \right\rangle - \sum_{i \neq 0} \frac{\langle \Phi_{HF} | \mathbf{P}_{ne}^{PSO} | \Phi_i \rangle \left\langle \Phi_i \left| \frac{1}{2}\mathbf{L}_G \right| \Phi_{HF} \right\rangle + \left\langle \Phi_{HF} \left| \frac{1}{2}\mathbf{L}_G \right| \Phi_i \right\rangle \langle \Phi_i | \mathbf{P}_{ne}^{PSO} | \Phi_{HF} \rangle}{E_{HF} - E_i} \tag{11.128}$$

All the operators $\mathbf{P}_{ne}^{DS}$, $\mathbf{P}_{ne}^{PSO}$ and $\mathbf{L}_G$ are gauge-dependent and each of the dia- and paramagnetic terms consequently depends on the chosen gauge. The shielding tensor is a 3 × 3 matrix, which can be diagonalized to give three eigenvalues. These principal components can be observed by solid-state NMR, but for rapidly tumbling molecules, as in the solution phase, only the average can be observed, corresponding to one-third of the trace of the shielding tensor.

The ESR equivalent of the NMR shielding is called the **g** *tensor* and can be considered as the mixed second derivative with respect to the electron spin and an external magnetic field.[20,21] It is a 3 × 3 tensor and can in analogy with the nuclear shielding tensor be written as the diagonal component for the free electron plus a small correction due to the molecular environment:

$$\mathbf{g} = g_e \mathbf{1} + \Delta\mathbf{g} \tag{11.129}$$

The diagonal term $g_e\mathbf{1}$ arises from the spin-Zeeman term $\mathbf{S} \cdot \mathbf{B}_{ext}$, while the anisotropic part has two contributions. The direct term arises from the $\mathbf{P}_{ee}^{DS}$ operator (Equation (11.116)) and the relativistic correction from the spin-Zeeman term (Equation (11.120)), with additional contributions coming from the combination of $^1/_2\mathbf{L}_G$ and the spin–orbit operators $\mathbf{P}_{ne}^{SO}$ and $\mathbf{P}_{ee}^{SO}$:

$$\Delta\mathbf{g} = \left\langle \Phi_{HF} \left| \mathbf{P}_{ee}^{DS} \right| \Phi_{HF} \right\rangle - \frac{g_e \mu_B}{2c^2} \langle \Phi_{HF} | \mathbf{p}^2 | \Phi_{HF} \rangle - \sum_{i \neq 0} \frac{\langle \Phi_{HF} | \mathbf{P}_{ne}^{SO} + \mathbf{P}_{ee}^{SO} | \Phi_i \rangle \left\langle \Phi_i \left| \frac{1}{2}\mathbf{L}_G \right| \Phi_{HF} \right\rangle + \left\langle \Phi_{HF} \left| \frac{1}{2}\mathbf{L}_G \right| \Phi_i \right\rangle \langle \Phi_i | \mathbf{P}_{ne}^{SO} + \mathbf{P}_{ee}^{SO} | \Phi_{HF} \rangle}{E_{HF} - E_i} \tag{11.130}$$

It should be noted that the formulas in Equations (11.121) to (11.130) have been derived by considering only the operators from the leading order in the inverse speed of light. One may obtain relativistic corrections by carrying out the expansion to higher orders, but this rapidly becomes quite involved,[8,9] as many different operators and their combinations can make contributions to a given property.[22] For systems where relativistic effects are important, a full four-component-type calculation (Section 9.4) becomes attractive, at least conceptually, since it automatically includes all effects without the necessity of multiple perturbation operators.

11.8.8 Gauge Dependence of Magnetic Properties

There are two factors that make the calculation of magnetic properties somewhat more complicated than the corresponding electric properties. First, the angular momentum operator $\mathbf{L}_G$ is imaginary (Equation (11.17)), implying that the wave function must be allowed to be complex. Second, the presence of the gauge origin in the operators means that the results may be origin-dependent. An exact wave function will of course give origin-independent results, as will a Hartree–Fock wave function if a complete basis set is employed. In practice, however, a finite basis must be employed, and standard basis sets will yield results that depend on where the user has chosen the origin of the gauge. The center of mass is often used in actual calculations, but this is by no means a unique choice.

The gauge error depends on the distance between the wave function and the gauge origin, and some methods try to minimize the error by selecting separate gauges for each (localized) molecular orbital. Two such methods are known as *individual Gauge for Localized Orbitals* (IGLO)[23] and *Localized Orbital/local oRiGin* (LORG).[24]

A more recent idea, which eliminates the gauge dependence for properties, is to make the basis functions explicitly dependent on the magnetic field by inclusion of a complex phase factor referring to the position of the basis function (usually the nucleus):

$$\begin{aligned} X_{\mathrm{A}}(\mathbf{r}-\mathbf{R}_{\mathrm{A}}) &= \mathrm{e}^{-\frac{i}{c}\mathbf{A}_{\mathrm{A}}\cdot\mathbf{r}}\chi_{\mathrm{A}}(\mathbf{r}-\mathbf{R}_{\mathrm{A}}) \\ \chi_{\mathrm{A}}(\mathbf{r}-\mathbf{R}_{\mathrm{A}}) &= N\mathrm{e}^{-\alpha(\mathbf{r}-\mathbf{R}_{\mathrm{A}})^2} \\ \mathbf{A}_{\mathrm{A}} &= \tfrac{1}{2}\mathbf{B}\times(\mathbf{R}_{\mathrm{A}}-\mathbf{R}_{\mathrm{G}}) \end{aligned} \tag{11.131}$$

Such orbitals are known as *London Atomic Orbitals* (LAOs) or *Gauge Including/Invariant Atomic Orbitals* (GIAOs).[25,26] The effect is that matrix elements involving GIAOs only contain a *difference* in vector potentials, thereby removing the reference to an absolute gauge origin. For the overlap and potential energy, it is straightforward to see that matrix elements become independent of the gauge origin:

$$\begin{aligned} \langle X_{\mathrm{A}}|X_{\mathrm{B}}\rangle &= \langle \chi_{\mathrm{A}}|\mathrm{e}^{\frac{i}{c}(\mathbf{A}_{\mathrm{A}}-\mathbf{A}_{\mathrm{B}})\cdot\mathbf{r}}|\chi_{\mathrm{B}}\rangle \\ \langle X_{\mathrm{A}}|\mathbf{V}|X_{\mathrm{B}}\rangle &= \langle \chi_{\mathrm{A}}|\mathrm{e}^{\frac{i}{c}(\mathbf{A}_{\mathrm{A}}-\mathbf{A}_{\mathrm{B}})\cdot\mathbf{r}}\mathbf{V}|\chi_{\mathrm{B}}\rangle \\ \mathbf{A}_{\mathrm{A}}-\mathbf{A}_{\mathrm{B}} &= \tfrac{1}{2}\mathbf{B}\times(\mathbf{R}_{\mathrm{A}}-\mathbf{R}_{\mathrm{B}}) \end{aligned} \tag{11.132}$$

The kinetic energy is slightly more complicated, but it can be shown that the following relation holds:

$$\begin{aligned} \langle X_{\mathrm{A}}|\pi^2|X_{\mathrm{B}}\rangle &= \left\langle X_{\mathrm{A}}\middle| \left(\mathbf{p}+\tfrac{1}{2}\mathbf{B}\times(\mathbf{r}-\mathbf{R}_{\mathrm{G}})\right)^2 \middle|X_{\mathrm{B}}\right\rangle \\ &= \left\langle \chi_{\mathrm{A}}\middle|\mathrm{e}^{\frac{i}{c}(\mathbf{A}_{\mathrm{A}}-\mathbf{A}_{\mathrm{B}})\cdot\mathbf{r}} \left(\mathbf{p}+\tfrac{1}{2}\mathbf{B}\times(\mathbf{r}-\mathbf{R}_{\mathrm{B}})\right)^2 \middle|\chi_{\mathrm{B}}\right\rangle \end{aligned} \tag{11.133}$$

Note that $\mathbf{R}_{\mathrm{G}}$ has been replaced by $\mathbf{R}_{\mathrm{B}}$ in the last bracket. The use of GIAOs as basis functions makes all matrix elements, and therefore all properties, independent of the gauge origin. The wave function itself, however, is expressed in terms of the basis functions, and therefore becomes gauge-dependent, by means of a complex phase factor. The use of perturbation-dependent basis functions has the further advantage of greatly reducing the need for high angular momentum basis functions; that is the property is typically calculated with an accuracy comparable to that of the unperturbed system.[27]

While LAOs/GIAOs were proposed well before the advent of modern computational chemistry, it was only owing to developments in calculating (geometrical) derivatives of the energy (and wave function) that it became practical to use field-dependent orbitals.[28]

11.9 Geometry Perturbations

The general formula for the first derivative of the energy with respect to a change in geometry, the molecular (nuclear) gradient, is given by Equation (11.35):

$$\mathbf{g} = \frac{\partial E}{\partial \mathbf{R}} = \left\langle \Psi \middle| \frac{\partial \mathbf{H}}{\partial \mathbf{R}} \middle| \Psi \right\rangle + 2\left\langle \frac{\partial \Psi}{\partial \mathbf{R}} \middle| \mathbf{H} \middle| \Psi \right\rangle \tag{11.134}$$

The first term is the Hellmann–Feynman force and the second is the wave function response. The latter contains contributions from a change in the basis functions, the state and the MO coefficients:

$$\frac{\partial \Psi}{\partial \mathbf{R}} = \frac{\partial \Psi}{\partial \boldsymbol{\chi}}\frac{\partial \boldsymbol{\chi}}{\partial \mathbf{R}} + \frac{\partial \Psi}{\partial \mathbf{a}}\frac{\partial \mathbf{a}}{\partial \mathbf{R}} + \frac{\partial \Psi}{\partial \mathbf{c}}\frac{\partial \mathbf{c}}{\partial \mathbf{R}} \tag{11.135}$$

The state and MO dependence disappears for HF, DFT and MCSCF-type wave functions owing to their variational nature ($\partial\Psi/\partial\mathbf{a} = 0$ and $\partial\Psi/\partial\mathbf{c} = 0$). For traditional basis sets consisting of nuclear-centered Gaussian functions, the basis functions are clearly perturbation-dependent since the functions move along with the nuclei, and standard perturbation theory is therefore not suitable for calculating molecular gradients. For a plane wave basis, however, the basis functions are independent of a geometry perturbation and the molecular gradient is just the Hellmann–Feynman term.

Since nuclear derivatives are important for optimizing geometries, it may be useful to look in more detail at the quantities involved in calculating first and second derivatives of a Hartree–Fock wave function with a Gaussian-type basis set, with the expressions for density functional methods being very similar. These formulas are most easily derived directly from the HF energy expressed in terms of the atomic quantities (Equation (3.57)):[29]

$$E_{\text{HF}} = \sum_{\alpha\beta}^{M_{\text{basis}}} D_{\alpha\beta}h_{\alpha\beta} + \frac{1}{2}\sum_{\alpha\beta\gamma\delta}^{M_{\text{basis}}} D_{\alpha\beta}D_{\gamma\delta}(\langle\chi_\alpha\chi_\gamma|\chi_\beta\chi_\delta\rangle - \langle\chi_\alpha\chi_\gamma|\chi_\delta\chi_\beta\rangle) + V_{\text{nn}} \tag{11.136}$$

Differentiation (using λ as a general geometrical displacement of a nucleus) yields

$$\begin{aligned}\frac{\partial E_{\text{HF}}}{\partial\lambda} &= \sum_{\alpha\beta}^{M_{\text{basis}}}\left(\frac{\partial D_{\alpha\beta}}{\partial\lambda}h_{\alpha\beta} + D_{\alpha\beta}\frac{\partial h_{\alpha\beta}}{\partial\lambda}\right)\\ &+\frac{1}{2}\sum_{\alpha\beta\gamma\delta}^{M_{\text{basis}}}\left(\frac{\partial D_{\alpha\beta}}{\partial\lambda}D_{\gamma\delta} + D_{\alpha\beta}\frac{\partial D_{\gamma\delta}}{\partial\lambda}\right)(\langle\chi_\alpha\chi_\gamma|\chi_\beta\chi_\delta\rangle - \langle\chi_\alpha\chi_\gamma|\chi_\delta\chi_\beta\rangle)\\ &+\frac{1}{2}\sum_{\alpha\beta\gamma\delta}^{M_{\text{basis}}} D_{\alpha\beta}D_{\gamma\delta}\frac{\partial}{\partial\lambda}(\langle\chi_\alpha\chi_\gamma|\chi_\beta\chi_\delta\rangle - \langle\chi_\alpha\chi_\gamma|\chi_\delta\chi_\beta\rangle) + \frac{\partial V_{\text{nn}}}{\partial\lambda}\end{aligned} \tag{11.137}$$

The third and fourth terms are identical and may be collected to cancel the factor of $^1/_2$. Rearranging the terms gives

$$\begin{aligned}\frac{\partial E_{\text{HF}}}{\partial\lambda} &= \sum_{\alpha\beta}^{M_{\text{basis}}} D_{\alpha\beta}\frac{\partial h_{\alpha\beta}}{\partial\lambda} + \frac{1}{2}\sum_{\alpha\beta\gamma\delta}^{M_{\text{basis}}} D_{\alpha\beta}D_{\gamma\delta}\frac{\partial}{\partial\lambda}(\langle\chi_\alpha\chi_\gamma|\chi_\beta\chi_\delta\rangle - \langle\chi_\alpha\chi_\gamma|\chi_\delta\chi_\beta\rangle)\\ &+ \sum_{\alpha\beta}^{M_{\text{basis}}}\frac{\partial D_{\alpha\beta}}{\partial\lambda}h_{\alpha\beta} + \sum_{\alpha\beta\gamma\delta}^{M_{\text{basis}}}\frac{\partial D_{\alpha\beta}}{\partial\lambda}D_{\gamma\delta}(\langle\chi_\alpha\chi_\gamma|\chi_\beta\chi_\delta\rangle - \langle\chi_\alpha\chi_\gamma|\chi_\delta\chi_\beta\rangle) + \frac{\partial V_{\text{nn}}}{\partial\lambda}\end{aligned} \tag{11.138}$$

The first two terms involve products of the density matrix with derivatives of the atomic integrals, while the two next terms can be recognized as derivatives of the density matrix times the Fock matrix (Equation (3.55)):

$$\frac{\partial E_{\text{HF}}}{\partial \lambda} = \sum_{\alpha\beta}^{M_{\text{basis}}} D_{\alpha\beta} \frac{\partial h_{\alpha\beta}}{\partial \lambda} + \frac{1}{2} \sum_{\alpha\beta\gamma\delta}^{M_{\text{basis}}} D_{\alpha\beta} D_{\gamma\delta} \frac{\partial}{\partial \lambda} (\langle \chi_\alpha \chi_\gamma | \chi_\beta \chi_\delta \rangle - \langle \chi_\alpha \chi_\gamma | \chi_\delta \chi_\beta \rangle)$$
$$+ \frac{\partial V_{\text{nn}}}{\partial \lambda} + \sum_{\alpha\beta}^{M_{\text{basis}}} \frac{\partial D_{\alpha\beta}}{\partial \lambda} F_{\alpha\beta} \tag{11.139}$$

The derivative in Equation (11.139) of the nuclear repulsion (third term) is trivial since it does not involve electron coordinates. The one-electron derivatives are given in

$$h_{\alpha\beta} = \langle \chi_\alpha | \mathbf{h} | \chi_\beta \rangle$$
$$\frac{\partial h_{\alpha\beta}}{\partial \lambda} = \left\langle \frac{\partial \chi_\alpha}{\partial \lambda} \middle| \mathbf{h} \middle| \chi_\beta \right\rangle + \left\langle \chi_\alpha \middle| \frac{\partial \mathbf{h}}{\partial \lambda} \middle| \chi_\beta \right\rangle + \left\langle \chi_\alpha \middle| \mathbf{h} \middle| \frac{\partial \chi_\beta}{\partial \lambda} \right\rangle \tag{11.140}$$

The central term is recognized as the Hellmann–Feynman force. The two-electron derivatives in Equation (11.139) become

$$\langle \chi_\alpha \chi_\gamma | \chi_\beta \chi_\delta \rangle = \langle \chi_\alpha \chi_\gamma | \mathbf{g} | \chi_\beta \chi_\delta \rangle$$
$$\frac{\partial}{\partial \lambda} \langle \chi_\alpha \chi_\gamma | \chi_\beta \chi_\delta \rangle = \left\langle \frac{\partial \chi_\alpha}{\partial \lambda} \chi_\gamma \middle| \mathbf{g} \middle| \chi_\beta \chi_\delta \right\rangle + \left\langle \chi_\alpha \frac{\partial \chi_\gamma}{\partial \lambda} \middle| \mathbf{g} \middle| \chi_\beta \chi_\delta \right\rangle$$
$$+ \left\langle \chi_\alpha \chi_\gamma \middle| \frac{\partial \mathbf{g}}{\partial \lambda} \middle| \chi_\beta \chi_\delta \right\rangle + \left\langle \chi_\alpha \chi_\gamma \middle| \mathbf{g} \middle| \frac{\partial \chi_\beta}{\partial \lambda} \chi_\delta \right\rangle$$
$$+ \left\langle \chi_\alpha \chi_\gamma \middle| \mathbf{g} \middle| \chi_\beta \frac{\partial \chi_\delta}{\partial \lambda} \right\rangle \tag{11.141}$$

The central term is again the Hellmann–Feynman force, which vanishes since the two-electron operator $\mathbf{g}$ is independent of the nuclear positions.

The last term in Equation (11.139) involves a change in the density matrix, that is the MO coefficients:

$$D_{\alpha\beta} = \sum_{i=1}^{N_{\text{elec}}} n_i c_{\alpha i} c_{\beta i}$$
$$\frac{\partial D_{\alpha\beta}}{\partial \lambda} = \sum_{i=1}^{N_{\text{elec}}} n_i \left(\frac{\partial c_{\alpha i}}{\partial \lambda} c_{\beta i} + c_{\alpha i} \frac{\partial c_{\beta i}}{\partial \lambda} \right) \tag{11.142}$$

Since the HF wave function is variationally optimized, the explicit calculation of the density derivatives can be avoided, as first derived by Pulay.[30] The last term in Equation (11.139) may with Equation (11.142) be written as

$$\sum_{\alpha\beta\gamma\delta}^{M_{\text{basis}}} \frac{\partial D_{\alpha\beta}}{\partial \lambda} F_{\alpha\beta} = \sum_{\alpha\beta\gamma\delta}^{M_{\text{basis}}} \sum_{i=1}^{N_{\text{elec}}} n_i \left(\frac{\partial c_{\alpha i}}{\partial \lambda} F_{\alpha\beta} c_{\beta i} + \frac{\partial c_{\beta i}}{\partial \lambda} F_{\alpha\beta} c_{\alpha i} \right) \tag{11.143}$$

By virtue of the HF condition ($\mathbf{FC} = \mathbf{SC\varepsilon}$), Equation (11.143) may be written in terms of overlap integrals and MO energies:

$$\sum_{\alpha\beta\gamma\delta}^{M_{\text{basis}}} \frac{\partial D_{\alpha\beta}}{\partial\lambda} F_{\alpha\beta} = \sum_{\alpha\beta\gamma\delta}^{M_{\text{basis}}} \sum_{i=1}^{N_{\text{elec}}} n_i \left(\frac{\partial c_{\alpha i}}{\partial\lambda} S_{\alpha\beta}\varepsilon_i c_{\beta i} + \frac{\partial c_{\beta i}}{\partial\lambda} S_{\alpha\beta}\varepsilon_i c_{\alpha i} \right) \tag{11.144}$$

Finally, since the MOs are orthonormal, the derivatives of the coefficients may be replaced by derivatives of the overlap matrix:

$$\begin{aligned}
\langle\phi_i|\phi_j\rangle &= \sum_{\alpha\beta}^{M_{\text{basis}}} c_{\alpha i}c_{\beta i}\langle\chi_\alpha|\chi_\beta\rangle = \sum_{\alpha\beta}^{M_{\text{basis}}} c_{\alpha i}c_{\beta j}S_{\alpha\beta} = \delta_{ij} \\
\frac{\partial}{\partial\lambda}\langle\phi_i|\phi_j\rangle &= \left(\frac{\partial c_{\alpha i}}{\partial\lambda} c_{\beta j}S_{\alpha\beta} + c_{\alpha i}\frac{\partial c_{\beta j}}{\partial\lambda} S_{\alpha\beta} + c_{\alpha i}c_{\beta j}\frac{\partial S_{\alpha\beta}}{\partial\lambda} \right) = 0 \\
2\frac{\partial c_{\alpha i}}{\partial\lambda} c_{\beta j}S_{\alpha\beta} &= -c_{\alpha i}c_{\beta j}\frac{\partial S_{\alpha\beta}}{\partial\lambda}
\end{aligned} \tag{11.145}$$

The final derivative of the energy may thus be written as

$$\begin{aligned}
\frac{\partial E_{\text{HF}}}{\partial\lambda} = &\sum_{\alpha\beta}^{M_{\text{basis}}} D_{\alpha\beta}\frac{\partial h_{\alpha\beta}}{\partial\lambda} + \frac{1}{2}\sum_{\alpha\beta\gamma\delta}^{M_{\text{basis}}} D_{\alpha\beta}D_{\gamma\delta}\frac{\partial}{\partial\lambda}(\langle\chi_\alpha\chi_\gamma|\chi_\beta\chi_\delta\rangle - \langle\chi_\alpha\chi_\gamma|\chi_\delta\chi_\beta\rangle) \\
&+ \frac{\partial V_{\text{nn}}}{\partial\lambda} - \sum_{\alpha\beta}^{M_{\text{basis}}} W_{\alpha\beta}\frac{\partial S_{\alpha\beta}}{\partial\lambda}
\end{aligned} \tag{11.146}$$

Here the energy-weighted density matrix $\mathbf{W}$ has been introduced:

$$W_{\alpha\beta} = \sum_{i=1}^{N_{\text{elec}}} \varepsilon_i c_{\alpha i}c_{\beta i} \tag{11.147}$$

Consider now the case where the perturbation λ is a specific nuclear displacement, $X_k \to X_k + \Delta X_k$. The derivatives of the one- and two-electron integrals are of two types, those involving derivatives of the basis functions and those involving derivatives of the operators. The latter are given by

$$\frac{\partial\mathbf{h}}{\partial X_k} = \frac{\partial}{\partial X_k}\left(-\tfrac{1}{2}\nabla_i^2 - \sum_A^{N_{\text{nuclei}}} \frac{Z_A}{|\mathbf{r}_i - \mathbf{R}_A|} \right) = -Z_k\frac{(x_i - X_k)}{|\mathbf{r}_i - \mathbf{R}_k|^3} \tag{11.148}$$

$$\frac{\partial\mathbf{g}}{\partial X_k} = 0 \tag{11.149}$$

$$\frac{\partial\mathbf{V}_{\text{nn}}}{\partial X_k} = \frac{\partial}{\partial X_k}\left(\sum_{A>B}^{N_{\text{nuclei}}} \frac{Z_A Z_B}{|\mathbf{R}_A - \mathbf{R}_B|} \right) = \sum_{A\neq k}^{N_{\text{nuclei}}} Z_A Z_k \frac{(X_A - X_k)}{|\mathbf{R}_A - \mathbf{R}_k|^3} \tag{11.150}$$

The derivative of the core operator $\mathbf{h}$ is a one-electron operator similar to the nuclear–electron attraction required for the energy itself (Equation (3.59)). The two-electron part yields zero and the $\mathbf{V}_{\text{nn}}$ term is independent of the electronic wave function. The remaining terms in Equations (11.140),

(11.141) and (11.146) all involve derivatives of the basis functions. When these are Gaussian functions (as is usually the case) the derivative can be written in terms of two other Gaussian functions, having one lower and one higher angular momentum:

$$\begin{aligned}
\chi_\alpha(\mathbf{R}_k) &= N(x-X_k)^l(y-Y_k)^m(z-Z_k)^n \mathrm{e}^{-\alpha(\mathbf{r}-\mathbf{R}_k)^2} \\
\frac{\partial \chi_\alpha}{\partial X_k} &= -N(x-X_k)^{l-1}(y-Y_k)^m(z-Z_k)^n \mathrm{e}^{-\alpha(\mathbf{r}-\mathbf{R}_k)^2} \\
&\quad + 2N\alpha(x-X_k)^{l+1}(y-Y_k)^m(z-Z_k)^n \mathrm{e}^{-\alpha(\mathbf{r}-\mathbf{R}_k)^2}
\end{aligned} \tag{11.151}$$

The derivative of a p-function can thus be written in terms of an s- and a d-type Gaussian function. The one- and two-electron integrals involving derivatives of basis functions are therefore of the same type as those used in the energy expression itself, the only difference being the angular momentum and the fact that there are roughly three times as many of these derivative integrals than for the energy itself. Of all the terms in Equation (11.146), the only significant computational effort is the derivatives of the two-electron integrals. Note, however, that the density matrix elements are known at the time when these integrals are calculated, and screening procedures analogous to those used in direct SCF techniques (Section 3.8.5) can be used to avoid calculating integrals that make insignificant contributions to the final result.

The second derivative of the energy with respect to a geometry change can be written as in

$$\begin{aligned}
\mathbf{H} &= \frac{\partial^2 E_{\mathrm{HF}}}{\partial \lambda^2} \\
&= \sum_{\alpha\beta}^{M_{\mathrm{basis}}} D_{\alpha\beta}\frac{\partial^2 h_{\alpha\beta}}{\partial \lambda^2} + \frac{1}{2}\sum_{\alpha\beta\gamma\delta}^{M_{\mathrm{basis}}} D_{\alpha\beta}D_{\gamma\delta}\frac{\partial^2}{\partial \lambda^2}(\langle \chi_\alpha\chi_\gamma|\chi_\beta\chi_\delta\rangle - \langle \chi_\alpha\chi_\gamma|\chi_\delta\chi_\beta\rangle) \\
&\quad + \frac{\partial^2 V_{\mathrm{nn}}}{\partial \lambda^2} - \sum_{\alpha\beta}^{M_{\mathrm{basis}}} W_{\alpha\beta}\frac{\partial^2 S_{\alpha\beta}}{\partial \lambda^2} + \sum_{\alpha\beta}^{M_{\mathrm{basis}}} \frac{\partial D_{\alpha\beta}}{\partial \lambda}\frac{\partial h_{\alpha\beta}}{\partial \lambda} \\
&\quad + \sum_{\alpha\beta\gamma\delta}^{M_{\mathrm{basis}}} \frac{\partial D_{\alpha\beta}}{\partial \lambda} D_{\gamma\delta}\frac{\partial}{\partial \lambda}(\langle \chi_\alpha\chi_\gamma|\chi_\beta\chi_\delta\rangle - \langle \chi_\alpha\chi_\gamma|\chi_\delta\chi_\beta\rangle) \\
&\quad - \sum_{\alpha\beta}^{M_{\mathrm{basis}}} \frac{\partial W_{\alpha\beta}}{\partial \lambda}\frac{\partial S_{\alpha\beta}}{\partial \lambda}
\end{aligned} \tag{11.152}$$

The first four terms only involve derivatives of operators and AO integrals, but the last three terms require the derivative of the density matrix and MO energies. These can be obtained by solving the first-order CPHF equations (Section 11.6.1).

The calculation of the second derivatives is substantially more involved than calculating the first derivative, typically by an order of magnitude, and for large systems a full calculation of the Hessian (force constant) matrix may thus be prohibitively expensive. If the second derivatives are required only for characterizing the nature of a stationary point (minimum or saddle point), the full Hessian is not required: only a few of the lowest eigenvalues are of interest. As shown by Deglmann and Furche, the lowest eigenvalues may be extracted by iterative techniques without explicit construction of the full second derivatives, leading to a substantial saving for large systems.[31]

11.10 Time-Dependent Perturbations

The perturbation, derivative and response/propagator approaches for defining molecular properties can be generalized time-dependent perturbations. The equivalent of Equation (11.30) for a time-dependent perturbation is

$$\mathbf{H}(t) = \mathbf{H}_0 + \mathbf{V}_{\text{ext}}(t) \tag{11.153}$$

The perturbation is usually an oscillating electric field, which in the general case can be written as

$$\mathbf{V}_{\text{ext}}(t) = \sum_k \mathbf{PF}_k e^{-i\omega_k t} \tag{11.154}$$

Here ω_k is the frequency of the field, $\mathbf{F}_k$ is the corresponding field strength and $\mathbf{P}$ is the perturbation operator. The $\mathbf{PF}_k$ term should again be interpreted as a sum over all products of components. If the field is slowly varying over the molecular dimension (wave length of radiation significantly larger than the molecular size), it can be represented by a linear approximation, that is $\mathbf{P}$ is then the dipole operator $\boldsymbol{\mu}$ and $\mathbf{F}_k$ is a vector containing the x, y and z components of the field. Concentrating on a uniform field of strength F with a single frequency ω, Equation (11.154) reduces to

$$\mathbf{V}_{ext}(t) = \boldsymbol{\mu} F \cos(\omega t) \tag{11.155}$$

The time-dependent wave function $\tilde{\Psi}$ can be written as a phase factor ϖ that depends on time but not on space coordinates time a *regular* wave function:

$$\tilde{\Psi}(\mathbf{r}, t) = e^{-i\varpi(t)}\Psi(\mathbf{r}, t) \tag{11.156}$$

Within a perturbation formulation, the time-dependent regular wave function can be Taylor-expanded as in

$$\Psi(\mathbf{r}, t) = \Psi_0(\mathbf{r}) + \Psi_1(\mathbf{r}) \cos \omega t + \Psi_2(\mathbf{r}) \cos 2\omega t + \cdots \tag{11.157}$$

Proceeding along the same lines as in Section 4.8 leads to an expression for the second-order correction term, which for the case of the perturbation in Equation (11.155) becomes the frequency-dependent polarizability given in

$$\alpha_\omega = \sum_{i\neq 0} \frac{|\langle\Psi_0|\mathbf{r}|\Psi_i\rangle|^2}{\omega - E_i + E_0} - \frac{|\langle\Psi_0|\mathbf{r}|\Psi_i\rangle|^2}{\omega + E_i - E_0} \tag{11.158}$$

A time-dependent property should reduce to the time-independent case in the limit of the frequency going to zero, and Equation (11.158) reduces to Equation (11.92) when $\omega = 0$.

The energy derivative approach in Section 11.3 is not suitable for time-dependent properties since energy is not a conserved quantity in the presence of time-dependent perturbations. The analogy to the energy for the time-dependent case is the time-averaged *quasi-energy* W defined below, which is the same as the action integral in Equation (6.92) for the case where the perturbation is periodic:[4,32]

$$W = \frac{1}{T}\int_0^T \left\langle \Psi(t) \middle| \mathbf{H} - i\frac{\partial}{\partial t} \middle| \Psi(t) \right\rangle dt \tag{11.159}$$

The quasi-energy may be differentiated analogously to Equation (11.1), where the perturbation parameter $\mathbf{P}_\omega$ now is associated with a frequency, and each derivative can be equated with a time-dependent molecular property:

$$W(\mathbf{P}_\omega) = E(0) + \frac{\partial W}{\partial \mathbf{P}_\omega}\mathbf{P}_\omega + \frac{1}{2}\frac{\partial^2 W}{\partial \mathbf{P}_\omega^2}\mathbf{P}_\omega^2 + \frac{1}{6}\frac{\partial^3 W}{\partial \mathbf{P}_\omega^3}\mathbf{P}_\omega^3 + \cdots \tag{11.160}$$

Equation (11.160) can be generalized to include multiple perturbations, possibly with different frequencies.

The response/propagator formalism in Equation (11.48) can be generalized to a time/frequency-dependent property written as a Taylor expansion with respect to time-dependent perturbations $\mathbf{P}$, $\mathbf{Q}$ with frequencies ω_1 and ω_2 as in

$$\langle \Psi|\mathbf{O}|\Psi\rangle(t) = \langle \Psi_0|\mathbf{O}|\Psi_0\rangle(0) + \langle\langle \mathbf{O};\mathbf{P}\rangle\rangle_{\omega_1}\mathbf{P}e^{-i\omega_1 t} + \tfrac{1}{2}\langle\langle \mathbf{O};\mathbf{P},\mathbf{Q}\rangle\rangle_{\omega_1,\omega_2}\mathbf{P}\mathbf{Q}e^{-i(\omega_1+\omega_2)t} + \cdots \tag{11.161}$$

A general second-order property can be written in the sum-over-state perturbation expression as

$$\langle\langle \mathbf{O};\mathbf{P}\rangle\rangle_\omega = \sum_{i\neq 0}\frac{\langle \Psi_0|\mathbf{O}|\Psi_i\rangle\langle \Psi_i|\mathbf{P}|\Psi_0\rangle}{\omega - E_i + E_0} - \frac{\langle \Psi_0|\mathbf{P}|\Psi_i\rangle\langle \Psi_i|\mathbf{O}|\Psi_0\rangle}{\omega + E_i - E_0} \tag{11.162}$$

The corresponding third-order expression is given in Equation (11.163), where additional terms (analogous to the second term in Equation (11.162)) involving permutations of the operators and frequencies have been omitted:

$$\langle\langle \mathbf{O};\mathbf{P},\mathbf{Q}\rangle\rangle_{\omega_1,\omega_2} = \sum_{i,j}\frac{\langle \Psi_0|\mathbf{O}|\Psi_i\rangle\langle \Psi_i|\mathbf{P}|\Psi_j\rangle\langle \Psi_j|\mathbf{Q}|\Psi_0\rangle}{(\omega_1 - E_i + E_0)(\omega_2 - E_j + E_0)} + \textit{permutations} \tag{11.163}$$

For the case where $\mathbf{O} = \mathbf{P} = \boldsymbol{\mu}$ (the dipole operator), the $\langle\langle \boldsymbol{\mu};\boldsymbol{\mu}\rangle\rangle_\omega$ propagator describes the linear change in the dipole moment when subjected to an electric field (Equation (11.155)) and is thus the dynamical polarizability $\boldsymbol{\alpha}$ shown in Equation (11.158). The $\langle\langle \boldsymbol{\mu};\boldsymbol{\mu},\boldsymbol{\mu}\rangle\rangle_{\omega 1,\omega 2}$ propagator describes the quadratic change in the dipole moment when subjected to two electric fields with frequencies ω_1 and ω_2, which is the dynamical version of the hyperpolarizability tensor $\boldsymbol{\beta}$ in Table 11.1.

Dynamical properties are typically derived from external electromagnetic perturbations with associated frequencies. Higher-order properties may thus involve one or several different frequencies and associated multiphoton processes,[33] and the frequencies can be indicated either in parenthesis after the tensor or as subscripts on the corresponding propagator. The $\boldsymbol{\beta}(\omega;\omega_1,\omega_2)$ notation thus indicates the hyperpolarizability with observation at the ω frequency as a function of perturbations at frequencies ω_1 and ω_2. If the observation/perturbation is absorption of a photon, the frequency is written as a positive quantity, while emission of a photon is written as the negative quantity $-\omega$. The conservation of energy implies that the sum of all frequencies is zero and the observation frequency is thus given implicity by the (signed) sum of the perturbating frequencies. The propagator notation $\langle\langle \boldsymbol{\mu};\boldsymbol{\mu},\boldsymbol{\mu}\rangle\rangle_{\omega 1,\omega 2}$ only explicitly indicates the frequencies of the two perturbing fields, with the observation frequency given by energy conservation, that is $\langle\langle \boldsymbol{\mu};\boldsymbol{\mu},\boldsymbol{\mu}\rangle\rangle_{\omega 1,\omega 2} = \boldsymbol{\beta}(-\omega;\omega_1,\omega_2)$ with $\omega = \omega_1 + \omega_2$. Perturbations corresponding to static fields are indicated by a 0 for the frequency in these notations.

The sum-over-states expression in Equation (11.158) shows that the poles (where the denominator is zero) correspond to excitation energies and the residues (numerator at the poles) correspond to transition moments between the reference and excited states (excitation or deexcitation). Excitation energies and transition moments can thus be considered as buyproducts of calculating the dynamical polarizability, and higher-order polarizabilities, for example, provide access to energy differences and transition moments between excited states, as seen from Equation (11.163).

Time-dependent electromagnetic fields can be linear, circular or elliptical polarized, and have different orientations with respect to the sample; these combinations lead to a large variety of molecular properties. The molecular response is commonly related to the dipole term, but some properties are related to higher-order moments (e.g. quadrupole) or terms (e.g. field gradients), which further add to the property zoo. The observed effect often involves several individual molecule properties that can be extracted from the residues of suitable propagators, as illustrated in Table 11.3.[4] *Birefringence* is the term used for a difference in the real part of the refractive index of light in different directions, while *Dichroism* refers to a difference in the imaginary part of the refractive index of light in different directions. Each of these can be partitioned into *Linear*, *Circular* and *Axial* cases:

- *Linear Birefringence/Dichroism* (LB/LD) measures the difference between linear polarized light parallel and perpendicular to an external field, and results in eliptical polarized light.
- *Circular Birefringence/Dichroism* (CB/CD) measures the difference between the two circular components of linearly polarized light, and results in a rotation of the plane of the polarized light, commonly associated with the phenomenon of optical activity.
- *Axial Birefringence/Dichroism* (AB/AD) measures the difference between unpolarized light in opposite directions parallel to an external field, and results in a difference of velocities of propagation, thus yielding a phase difference of the two beams.

The practical calculation of dynamical properties proceeds along the same lines as those leading to Equation (11.65), where the only difference is that the diagonal elements are shifted by the energy of the perturbing field:

$$\left[\begin{pmatrix} \mathbf{A} & \mathbf{B} \\ \mathbf{B}^* & \mathbf{A}^* \end{pmatrix} - \omega \begin{pmatrix} 1 & 0 \\ 0 & -1 \end{pmatrix}\right] \begin{bmatrix} \mathbf{Y} \\ \mathbf{Z} \end{bmatrix} = - \begin{bmatrix} \mathbf{P} \\ \mathbf{P}^* \end{bmatrix} \tag{11.164}$$

Solving for the wave function response (**Y**, **Z**) for a given frequency then allows a calculation of the dynamical property from Equation (11.70).

The perturbation, derivative and response formulations can be considered as three different ways of formulating the same properties, and all lead to very similar working equations.[4] The formulation of time-dependent properties as derivatives of a quasi-energy is perhaps the most flexible since it smoothly connects static and dynamical properties, and allows for different parameterizations of the wave function and the time-evolution, as well as providing access to transition moments.[32] A linear parameterization corresponds to a CI-type expansion, while an exponential parameterization corresponds to a CC-type expansion. The combination of a CC expansion for the wave function and a CI expansion for the time-evolution is very closely related to the *Equation-Of-Motion Coupled Cluster* (EOM-CC) method.[35] The response formulation is completely general and can be derived for different types of wave functions, such as DFT (TDDFT, Section 6.9.1), MCSCF or CC. The *Second-Order Polarization Propagator Approximation* (SOPPA) can be considered as employing an MP2-type wave function, while SOPPA(CCSD) is a hybrid method where CCSD amplitudes are used in place of the MP2 coefficients for the excited Slater determinants. Different wave functions lead to different expressions for the matrix elements in Equations (11.66) to (11.68) but otherwise involve solving response equations like Equation (11.164). If the **P** vector is set equal to zero in Equation (11.164), the equation reduces to a generalized eigenvalue problem where the ω eigenvalues are the excitation energies. For an HF reference wave function, this is identical to diagonalizing the CI Hamiltonian matrix (Sections 4.2 and 6.9.1). For other types of reference wave functions (e.g. MCSCF or CC), the formulation allows a generalization of calculating excitation energies, such as the CC2, CCSD, CC3 hierachy discussed in Section 4.14.

Table 11.3 Examples of some experimental effects and the response functions involved in their calculation, where the property in some cases is associated with the residue of the propagator(s). The electric and magnetic dipole operators are denoted $\boldsymbol{\mu}$ and **m**, the electric quadrupole operator is denoted $\boldsymbol{\Theta}$ and the diamagnetic magnetizability operator $\xi = -2\mathbf{P}_2^{\zeta}$ (Equation (11.107)). No distinction is made between the quadrupole moment operator in its traced or untraced form. By courtesy of Professor Antonio Rizzo.

Property (acronym)	Effect	Type	Propagator(s): Highest order	Propagator(s): Lower order
Transition moment	One-photon absorption coefficient		$\langle\langle\boldsymbol{\mu};\boldsymbol{\mu}\rangle\rangle_{\omega}$	
Polarizability			$\langle\langle\boldsymbol{\mu};\boldsymbol{\mu}\rangle\rangle_{\omega}$	
Polarizability at imaginary frequencies	Dispersion coefficients		$\langle\langle\boldsymbol{\mu};\boldsymbol{\mu}\rangle\rangle_{i\omega}$	
Electronic Circular Dichroism (ECD) Optical Rotation (OR)	Optical Activity (OA) Raman Optical Activity (ROA) Vibrational Optical Activity (VOA)[a] Vibrational Raman Optical Activity (VROA)[a]	CD	$\langle\langle\boldsymbol{\mu};\mathbf{m}\rangle\rangle_{\omega}$ $\langle\langle\boldsymbol{\mu};\boldsymbol{\Theta}\rangle\rangle_{\omega}$	
Second-Harmonic Generation (SHG)	Two-photon absorption coefficients Emission of light with a frequency double the incoming frequency		$\langle\langle\boldsymbol{\mu};\boldsymbol{\mu},\boldsymbol{\mu}\rangle\rangle_{\omega,\omega}$	
Electro-Optic Pockels Effect (EOPE) (dc-Pockels effect)	Electric field-dependent polarizability		$\langle\langle\boldsymbol{\mu};\boldsymbol{\mu},\boldsymbol{\mu}\rangle\rangle_{\omega,0}$	
Optical Rectification (OR)			$\langle\langle\boldsymbol{\mu};\boldsymbol{\mu},\boldsymbol{\mu}\rangle\rangle_{\omega,-\omega}$	
Faraday Effect (FE)[b]	Optical activity induced by an external magnetic field	CB	$\langle\langle\boldsymbol{\mu};\boldsymbol{\mu},\mathbf{m}\rangle\rangle_{\omega,0}$	
Magnetic Circular Dichroism (MCD)	Optical activity induced by an external magnetic field	CD	$\langle\langle\boldsymbol{\mu};\boldsymbol{\mu},\mathbf{m}\rangle\rangle_{\omega,0}$	
Magnetochiral Dichroism (MChD)		AD	$\langle\langle\boldsymbol{\mu};\mathbf{m},\mathbf{m}\rangle\rangle_{\omega,0}$ $\langle\langle\boldsymbol{\mu};\boldsymbol{\Theta},\mathbf{m}\rangle\rangle_{\omega,0}$	
Magnetochiral Birefringence		AB	$\langle\langle\boldsymbol{\mu};\mathbf{m},\mathbf{m}\rangle\rangle_{\omega,0}$ $\langle\langle\boldsymbol{\mu};\boldsymbol{\Theta},\mathbf{m}\rangle\rangle_{\omega,0}$	
Buckingham effect	Birefringence induced by an external field gradient	LB	$\langle\langle\boldsymbol{\mu};\boldsymbol{\mu},\boldsymbol{\Theta}\rangle\rangle_{\omega,0}$ $\langle\langle\boldsymbol{\mu};\boldsymbol{\Theta},\boldsymbol{\mu}\rangle\rangle_{\omega,0}$ $\langle\langle\boldsymbol{\mu};\mathbf{m},\boldsymbol{\mu}\rangle\rangle_{\omega,0}$	$\langle\langle\boldsymbol{\mu};\boldsymbol{\mu}\rangle\rangle_{\omega}$ $\langle\langle\boldsymbol{\mu};\mathbf{m}\rangle\rangle_{\omega}$ $\langle\langle\boldsymbol{\mu};\boldsymbol{\Theta}\rangle\rangle_{\omega}$
Third-Harmonic Generation (THG)	Three-photon absorption coefficients Emission of light with a frequency triple the incoming frequency		$\langle\langle\boldsymbol{\mu};\boldsymbol{\mu},\boldsymbol{\mu},\boldsymbol{\mu}\rangle\rangle_{\omega,\omega,\omega}$	

(*continued*)

Table 11.3 *(Continued)*

Property (acronym)	Effect	Type	Propagator(s)	
			Highest order	Lower order
Intensity-Dependent Refractive Index (IDRI) Degenerate Four-Wave Mixing (DFWM)			$\langle\langle \mu; \mu, \mu, \mu \rangle\rangle_{\omega,\omega,-\omega}$	
Optical Kerr Effect (OKE) (ac-Kerr)			$\langle\langle \mu; \mu, \mu, \mu \rangle\rangle_{\omega,\omega',-\omega'}$	
Electric-Field-Induced Second Harmonic Generation (EFISHG)	Second-harmonic generation induced by an external electric field		$\langle\langle \mu; \mu, \mu, \mu \rangle\rangle_{\omega,\omega,0}$	
dc-Optical Rectification (dc-OR)			$\langle\langle \mu; \mu, \mu, \mu \rangle\rangle_{\omega,-\omega,0}$	
Sum Frequency Generation (SFG)	Emission of light with a frequency as the sum of the different incoming frequencies		$\langle\langle \mu; \mu, \mu, \mu \rangle\rangle_{\omega,\omega',0}$	
Difference Frequency Generation (DFG)	Emission of light with a frequency as the difference of the different incoming frequencies		$\langle\langle \mu; \mu, \mu, \mu \rangle\rangle_{\omega,-\omega',0}$	
Magnetic-Field-Induced Second Harmonic Generation (MFISHG)	Second-harmonic generation induced by an external magnetic field		$\langle\langle \mu; \mu, \mu, \mathbf{m} \rangle\rangle_{\omega,\omega,0}$	
EFISHG Circular Intensity Difference (EFISHG-CID)	Optical activity in second-harmonic generation induced by an external electric field	CD	$\langle\langle \mu; \mu, \mathbf{m}, \mu \rangle\rangle_{\omega,\omega,0}$ $\langle\langle \mu; \mu, \mu, \mathbf{m} \rangle\rangle_{\omega,\omega,0}$ $\langle\langle \mu; \mu, \Theta, \mu \rangle\rangle_{\omega,\omega,0}$ $\langle\langle \mu; \mu, \mu, \Theta \rangle\rangle_{\omega,\omega,0}$	
Electro-optic Kerr effect (dc-Kerr)	Birefringence induced by external electric fields	LB	$\langle\langle \mu; \mu, \mu, \mu \rangle\rangle_{\omega,0,0}$	
Jones effect	Birefringence induced by external electric field gradient and external magnetic field	LB	$\langle\langle \mu; \Theta, \mathbf{m}, \mu \rangle\rangle_{\omega,0,0}$ $\langle\langle \mu; \mathbf{m}, \mathbf{m}, \mu \rangle\rangle_{\omega,0,0}$	$\langle\langle \mu; \Theta, \mathbf{m} \rangle\rangle_{\omega,0}$ $\langle\langle \mu; \mathbf{m}, \mathbf{m} \rangle\rangle_{\omega,0}$ $\langle\langle \mu; \xi, \mu \rangle\rangle_{\omega,0}$ $\langle\langle \mu; \xi \rangle\rangle_{\omega}$
Cotton–Mouton effect[b]	Birefringence induced by external magnetic fields	LB	$\langle\langle \mu; \mu, \mathbf{m}, \mathbf{m} \rangle\rangle_{\omega,0,0}$ $\langle\langle \mathbf{m}; \mathbf{m}, \mathbf{m}, \mathbf{m} \rangle\rangle_{\omega,0,0}$	$\langle\langle \mu; \mu, \mathbf{m} \rangle\rangle_{\omega,0}$ $\langle\langle \mu; \mu, \xi \rangle\rangle_{\omega,0}$ $\langle\langle \mathbf{m}; \mathbf{m}, \mathbf{m} \rangle\rangle_{\omega,0}$ $\langle\langle \mathbf{m}; \mathbf{m}, \xi \rangle\rangle_{\omega,0}$ $\langle\langle \mu; \mu \rangle\rangle_{\omega}$ $\langle\langle \mathbf{m}; \mathbf{m} \rangle\rangle_{\omega,0}$

[a] These require derivatives of the propagators with respect to nuclear coordinates; see Table 11.1.
[b] Analogous molecular properties where the magnetic perturbation arises from nuclear spin have also been considered.[34]

11.11 Rotational and Vibrational Corrections

An experimentally derived spectroscopic constant is usually based on the measured response of a macroscopic system and thus corresponds to an average over a large number of molecules in different vibrational and rotational states determined by the Boltzmann distribution at the given temperature. Although the spectroscopic constant calculated at a single geometry corresponding to the optimized molecular geometry may be of sufficient accuracy for comparing to experimental results, higher accuracy requires the ability to average the calculated property over a range of molecular geometries corresponding to the experimental ensemble. The most important correction is normally due to the effect of zero-point vibrations, but finite temperature effects (rotational and vibrational excited states) may also contribute and they can often be included with little additional effort compared to just including the zero-point correction.

Averaging over the complete Boltzmann ensemble of rotational–vibrational states requires the calculation of a semiglobal energy surface in order to obtain detailed rotational–vibrational (nuclear) wave functions and knowledge of the property surface for all geometries that are accessible by the atomic motions at the given temperature. This is impractical for all but the smallest systems, and practical methods for rotational–vibrational averaging employ approximate schemes to capture the main contributions. These can conceptually be divided into two components: (1) a change in geometry due to the anharmonicity of vibrations that lead to a vibrationally averaged geometry (often called an effective geometry) where, for example, bond lengths are slightly elongated and (2) a correction to the non-linear geometry dependence of the property.

Calculation of harmonic vibrational normal coordinates only requires the second derivative of the energy with respect to atomic coordinates at the equilibrium position (Section 17.2.2), while calculation of anharmonic corrections require (at least) third-order derivatives. In the absence of analytical third derivatives, these must be obtained by numerical differentiation of (analytical) second derivatives. Calculation of the geometry dependence of the spectroscopic constant (the property surface) similarly may requires a (large) number of individual calculations at slightly distorted molecular geometries. These numerical differentiations are computationally expensive and the number of required components increases with the size of the molecule. An alternative is to calculate higher-order mixed derivatives at the reference geometry in order to model the geometry dependence of the specific property, which requires a computational efficient method for calculating many different types of derivatives to quite high orders.[36]

A popular method for including rotational and vibrational effects is to first calculate an effective geometry corresponding to averaging over vibrational anhamonicity and rotational centrifugal distortions. At this effective geometry, the harmonic vibrational wave functions are then used to perform a vibrational averaging over the property.[37] It should be noted that the vibrational wave functions depend on the nuclear masses, and thus lead to different vibrational corrections for different isotopes.

The most important vibrational correction for many properties is due to anharmonicity of the bond stretching modes. This effect has as a consequence that the effective (vibrational averaged) bond lengths are slightly longer than the equlibrium bond lengths calculated by minimizing the electronic energy. A heuristic way of modeling this effect is to simply calculate the property with a molecular geometry where the bond length(s) is (are) slightly elongated. This also allows a heuristic way of modeling isotope effects on properties since the effective bond length for the heavier isotope is slightly shorter than for the lighter isotope. By employing an empirically determined standard bond elongation, which is slightly different for different isotopes, it is possible to estimate the isotope effect on the property from only two property calculations at slightly different geometries, and this has, for example, been used for probing isotope effects on nuclear magnetic shielding constants (chemical shifts).

11.12 Environmental Effects

Spectroscopic constants are often obtained from experimental measurements of the compund in a condensed phase, most frequently a solution. Condensed phases can have a significant effect on molecular properties, especially those depending on the outer part of the wave function. The solvent effect can be modeled by a continuum solvation model (Section 15.6.1)[38] or by including explicit solvent molecules in either a quantum or classical description (QM/MM, Section 2.12), where the latter can be in either an electronic or polarization embedding.[39,40]

11.13 Relativistic Corrections

Relativistic effects become increasingly important as the atomic number increases, and for systems containing atoms from the fifth row and below in the periodic table (beyond Xe), two- or four-component relativistic calculations (Chapter 9) are often required to get satisfactory accuracy.[41,42] For systems with atoms from the first four rows in the periodic table, the relativistic effects are often small enough that they can be calculated by perturbational methods employing a non-relativistic wave function as the reference. Spectroscopic parameters derived from perturbation expansions, which may be of relativistic origin, only take into account the lowest-order effects. Higher-order relativistic corrections can be included as (additional) relativistic corrections to the basic spectroscopic parameters.[43,44]

References

1 C. E. Dykstra, J. D. Augspurger, B. Kirtman and D. J. Malik, *Reviews in Computational Chemistry* **1**, 83 (1990).
2 D. B. Chesnut, *Reviews in Computational Chemistry* **8**, 245 (1996).
3 S. P. A. Sauer, *Molecular Electromagnetism* (Oxford University Press, 2011).
4 T. Helgaker, S. Coriani, P. Jorgensen, K. Kristensen, J. Olsen and K. Ruud, *Chemical Reviews* **112** (1), 543–631 (2012).
5 E. J. Heller, Y. Yang and L. Kocia, *Acs Central Science* **1** (1), 40–49 (2015).
6 J. Gauss, K. Ruud and M. Kallay, *Journal of Chemical Physics* **127** (7), 074101 (2007).
7 A. M. Teale, O. B. Lutnaes, T. Helgaker, D. J. Tozer and J. Gauss, *Journal of Chemical Physics* **138** (2), 024111 (2013).
8 B. Minaev, O. Loboda, Z. Rinkevicius, O. Vahtras and H. Agren, *Molecular Physics* **101** (15), 2335–2346 (2003).
9 J. M. Brown and A. J. Merer, *Journal of Molecular Spectroscopy* **74** (3), 488–494 (1979).
10 K. L. Bak, S. P. A. Sauer, J. Oddershede and J. F. Ogilvie, *Physical Chemistry Chemical Physics* **7** (8), 1747–1758 (2005).
11 N. C. Handy and H. F. Schaefer, *Journal of Chemical Physics* **81** (11), 5031–5033 (1984).
12 T. Helgaker and P. Jorgensen, *Theoretica Chimica Acta* **75** (2), 111–127 (1989).
13 J. Gerratt and I. M. Mills, *Journal of Chemical Physics* **49** (4), 1719 (1968).

14 M. Papai and G. Vanko, *Journal of Chemical Theory and Computation* **9** (11), 5004–5020 (2013).
15 R. Improta and V. Barone, *Chemical Reviews* **104** (3), 1231-1253 (2004).
16 T. Helgaker, M. Jaszunski and K. Ruud, *Chemical Reviews* **99** (1), 293–352 (1999).
17 O. Vahtras, O. Loboda, B. Minaev, H. Agren and K. Ruud, *Chemical Physics* **279** (2–3), 133–142 (2002).
18 F. Neese, *Journal of Chemical Physics* **127** (16), 164112 (2007).
19 J. Gauss and J. F. Stanton, in *Advances in Chemical Physics*, edited by I. Prigogine and S. A. Rice (John Wiley & Sons, 2002), Vol. 123, pp. 355–422.
20 D. Jayatilaka, *Journal of Chemical Physics* **108** (18), 7587–7594 (1998).
21 P. Manninen, J. Vaara and K. Ruud, *Journal of Chemical Physics* **121** (3), 1258–1265 (2004).
22 S. Bubin, M. Pavanelo, W.-C. Tung, K. L. Sharkey and L. Adamowicz, *Chemical Reviews* **113** (1), 36–79 (2013).
23 M. Schindler and W. Kutzelnigg, *Journal of Chemical Physics* **76** (4), 1919–1933 (1982).
24 A. E. Hansen and T. D. Bouman, *Journal of Chemical Physics* **82** (11), 5035–5047 (1985).
25 F. London, *Journal de Physique et le Radium* **8**, 397–409 (1937).
26 R. Ditchfie, *Molecular Physics* **27** (4), 789–807 (1974).
27 K. Ruud, T. Helgaker, K. L. Bak, P. Jorgensen and H. J. A. Jensen, *Journal of Chemical Physics* **99** (5), 3847–3859 (1993).
28 K. Wolinski, J. F. Hinton and P. Pulay, *Journal of the American Chemical Society* **112** (23), 8251–8260 (1990).
29 J. A. Pople, R. Krishnan, H. B. Schlegel and J. S. Binkley, *International Journal of Quantum Chemistry* **16**, 225–241 (1979).
30 P. Pulay, *Molecular Physics* **17** (2), 197 (1969).
31 P. Deglmann and F. Furche, *Journal of Chemical Physics* **117** (21), 9535–9538 (2002).
32 F. Pawlowski, J. Olsen and P. Jorgensen, *Journal of Chemical Physics* **142** (11), 114109 (2015).
33 D. H. Friese and K. Ruud, *Physical Chemistry Chemical Physics* **18** (5), 4174–4184 (2016).
34 J. Vaara, A. Rizzo, J. Kauczor, P. Norman and S. Coriani, *Journal of Chemical Physics* **140** (13), 134103 (2014).
35 S. Coriani, F. Pawlowski, J. Olsen and P. Jorgensen, *Journal of Chemical Physics* **144** (2), 024102 (2016).
36 M. Ringholm, D. Jonsson and K. Ruud, *Journal of Computational Chemistry* **35** (8), 622–633 (2014).
37 P. O. Astrand, K. Ruud and P. R. Taylor, *Journal of Chemical Physics* **112** (6), 2655–2667 (2000).
38 K. H. Hopmann, K. Ruud, M. Pecul, A. Kudelski, M. Dracinsky and P. Bour, *Journal of Physical Chemistry B* **115** (14), 4128–4137 (2011).
39 J. M. H. Olsen, C. Steinmann, K. Ruud and J. Kongsted, *Journal of Physical Chemistry A* **119** (21), 5344–5355 (2015).
40 J. M. H. Olsen, C. Steinmann, K. Ruud and J. Kongsted, *Journal of Physical Chemistry A* **119** (26), 6928–6928 (2015).
41 P. Hrobarik, M. Repisky, S. Komorovsky, V. Hrobarikova and M. Kaupp, *Theoretical Chemistry Accounts* **129** (3–5), 715–725 (2011).
42 S. Gohr, P. Hrobarik, M. Repisky, S. Komorovsky, K. Ruud and M. Kaupp, *Journal of Physical Chemistry A* **119** (51), 12892–12905 (2015).
43 P. Manninen, K. Ruud, P. Lantto and J. Vaara, *Journal of Chemical Physics* **122** (11), 114107 (2005).
44 P. Manninen, K. Ruud, P. Lantto and J. Vaara, *Journal of Chemical Physics* **124** (14), 149901 (2006).

12

Illustrating the Concepts

In this chapter we will illustrate some of the methods described in the previous chapters. It is of course impossible to cover all types of bonding and geometries, but for highlighting the features we will look at the H_2O molecule. This is small enough that we can employ the full spectrum of methods and basis sets, and illustrate some general trends.

12.1 Geometry Convergence

The experimental geometry for H_2O has a bond length of 0.9578 Å and an angle of 104.49°.[1,2] Let us investigate how the calculated geometry changes as a function of the theoretical sophistication.

12.1.1 Wave Function Methods

We will look at the convergence as a function of basis set and amount of electron correlation (Figure 4.3). For independent-particle methods (HF and DFT) we will use the segmented polarization consistent basis sets (pcseg-n, $n = 0, 1, 2, 3, 4$), while for correlated methods (MP2 and CCSD(T)) we will use the correlation consistent basis sets (cc-pVXZ, X = D, T, Q, 5, 6). Table 12.1 shows how the optimized geometry changes as a function of basis set at the HF, MP2 and CCSD(T) levels of theory, where the quality of the basis sets is indicated by the cardinal number X that defines the highest angular momentum functions included in the basis set.

The HF results are converged with the pcseg-2 basis set and the HF limit predicts a bond length that is too short, reminiscent of the incorrect dissociation of the single-determinant wave function (Section 4.3). The bond angle as a consequence becomes too large, owing to an overestimation of the repulsion between the two hydrogens. The underestimation of bond lengths at the HF level is quite general for covalent bonds, while the overestimation of bond angles is not. Although the increased repulsion/attraction between atom pairs in general is overestimated owing to too-short bond lengths and too-large charge polarization, these factors may pull in different directions for a larger molecule and bond angles may either be too large or too small. Note that the bond length decreases as the basis set quality is increased, thus an SZ or DZP-type basis may give bond lengths that are longer than the experimental value for some systems. At the HF *limit*, however, covalent bond lengths will normally be too short.

Introduction to Computational Chemistry, Third Edition. Frank Jensen.

Companion Website: http://www.wiley.com/go/jensen/computationalchemistry3

Table 12.1 H_2O geometry as a function of basis set quality at the HF, MP2 and CCSD(T) levels of theory (experimental values are 0.9578 Å and 104.49°).

		HF			MP2		CCSD(T)	
X	Basis set	R_{OH} (Å)	θ_{HOH} (°)	Basis set	R_{OH} (Å)	θ_{HOH} (°)	R_{OH} (Å)	θ_{HOH} (°)
1	pcseg-0	0.9723	109.05					
2	pcseg-1	0.9464	105.46	cc-pVDZ	0.9649	101.90	0.9663	101.91
3	pcseg-2	0.9395	106.39	cc-pVTZ	0.9591	103.59	0.9594	103.58
4	pcseg-3	0.9395	106.38	cc-pVQZ	0.9577	104.02	0.9579	104.12
5	pcseg-4	0.9395	106.36	cc-pV5Z	0.9579	104.29	0.9580	104.38
6				cc-pV6Z	0.9581	104.34	0.9582	104.42

Including electron correlation at the MP2 level increases the bond length by about 0.019 Å, fairly independently of the basis set quality. The bond angle as a consequence decreases, by about 2°. Note that the convergence in terms of basis set quality is much slower than at the HF level. From the observed behavior the MP2 basis set limit may be estimated as 0.9582 ± 0.0001 Å and 104.40° ± 0.04°, which is already in good agreement with the experimental values of 0.9578 Å and 104.49°. H_2O at the equilibrium geometry is a system where HF is a good zeroth-order wave function and perturbation methods should consequently converge fast. Indeed, the MP2 method recovers ~94% of the electron correlation energy, as shown later in Table 12.5. Including additional electron correlation with the CCSD(T) method gives only small changes relative to the MP2 level, and the effect of higher-order correlation diminishes as the basis set is enlarged. For H_2O the CCSD(T) method is virtually indistinguishable from CCSDT and is presumably very close to the full CI limit.[3,4]

The HF wave function contains equal amounts of ionic and covalent contributions (Section 4.3). For covalently bonded systems, such as H_2O, the HF wave function is too ionic and the effect of electron correlation is to increase the covalent contribution. Since the ionic dissociation limit is higher in energy than the covalent, the effect is that the equilibrium bond length increases when correlation methods are used. For dative bonds, for example in metal–ligand compounds, the situation is reversed. In this case the HF wave function dissociates correctly and bond lengths are normally too long. Inclusion of electron correlation adds attraction between ligands (dispersion interaction), which causes the metal–ligand bond lengths to contract.

The MP2 and CCSD(T) values in Table 12.1 are for correlation of the valence electrons only, that is the frozen-core approximation. In order to assess the effect of core electron correlation, the basis sets need to be augmented with tight polarization functions; the corresponding MP2 and CCSD(T) results are shown in Table 12.2.

The effect of core electron correlation is small: a small decrease of the bond length and a corresponding small increase in bond angle. The CCSD(T)/cc-pwCV5Z result with all electrons correlated gives a bond length of 0.9571 Å and an angle of 104.50°. Further basis set increases will presumably slightly increase the angle. Relativistic effects at the Dirac–Fock–Breit level of theory have been reported to give changes of +0.00016 Å and −0.07°.[2] Including these corrections led to a final predicted structure that is in good agreement with the experimental values of 0.9578 Å and 104.49°.

These results show that *ab initio* methods can give results of very high accuracy, provided that sufficiently large basis sets are used. Unfortunately, the combination of highly correlated methods, such as CCSD(T), and large basis sets means that such calculations are computationally expensive. A CCSD(T) calculation with the cc-pV5Z basis is already quite demanding for the H_2O system. The

Table 12.2 H_2O geometry as a function of basis set at the MP2 and CCSD(T) levels of theory including all electrons in the correlation (experimental values are 0.9578 Å and 104.49°).

		MP2		CCSD(T)	
X	**Basis set**	R_{OH} **(Å)**	θ_{HOH} **(°)**	R_{OH} **(Å)**	θ_{HOH} **(°)**
2	cc-pwCVDZ	0.9643	101.94	0.9656	101.97
3	cc-pwCVTZ	0.9578	103.66	0.9582	103.71
4	cc-pwCVQZ	0.9569	104.14	0.9571	104.23
5	cc-pwCV5Z	0.9570	104.41	0.9571	104.50

results also show, however, that a quite respectable level of accuracy is reached at the MP2/cc-pVTZ level, which is applicable to a much larger variety of molecules. Furthermore, the errors at a given level are quite systematic and relative values (comparing, for example, changes in geometries upon introduction of substitutents) will be predicted with a substantially higher accuracy.

12.1.2 Density Functional Methods

The two variables in DFT methods are the basis set and the choice of the exchange–correlation potential. The performance of seven popular functionals for the geometry with the pcseg-n basis sets is given in Tables 12.3 and 12.4. The LSDA functional employs the uniform electron gas approximation, the BLYP and PBE functionals are of the gradient-corrected type, while the B3LYP and PBE0 are hybrid types that contain a fraction of Hartree–Fock exchange. The ωB97XD is a hybrid functional with range-separation while B2PLYP is a double hybrid incorporating an MP2-like term.

Table 12.3 H_2O bond distances (Å) as a function of basis set with various DFT functionals (experimental value is 0.9578 Å).

Basis	**LSDA**	**BLYP**	**PBE**	**B3LYP**	**PBE0**	**ωB97XD**	**B2PLYP**
pcseg-0	1.0076	1.0152	1.0116	1.0002	0.9954	0.9902	0.9968
pcseg-1	0.9765	0.9791	0.9763	0.9683	0.9644	0.9634	0.9657
pcseg-2	0.9698	0.9705	0.9689	0.9604	0.9575	0.9554	0.9586
pcseg-3	0.9700	0.9704	0.9689	0.9603	0.9575	0.9558	0.9587
pcseg-4	0.9700	0.9704	0.9689	0.9603	0.9575	0.9558	0.9586

Table 12.4 H_2O bond angles (°) as a function of basis set with various DFT functionals (experimental value is 104.49°).

Basis	**LSDA**	**BLYP**	**PBE**	**B3LYP**	**PBE0**	**ωB97DX**	**B2PLYP**
pcseg-0	105.33	103.55	103.94	105.31	105.91	106.61	105.88
pcseg-1	103.95	103.04	102.91	103.88	103.83	104.05	103.84
pcseg-2	105.05	104.52	104.23	105.16	104.95	105.21	104.84
pcseg-3	104.99	104.53	104.21	105.14	104.91	105.15	104.87
pcseg-4	104.98	104.52	104.21	105.13	104.90	105.15	104.87

The geometry displays a convergence characteristic similar to the wave mechanics HF method (Table 12.1). A TZP-type basis (pcseg-2) gives good results and a QZP-type basis (pcseg-3) is essentially converged to the basis set limiting value. The deviations between the basis set limiting values and the experimental values are the inherent errors associated with the functionals. Incorporation of exact exchange led to a shorter bond length and larger angle, as expected from the HF values in Table 12.1.

12.2 Total Energy Convergence

The total energy in *ab initio* theory is relative to the separated particles, that is bare nuclei and electrons. The corresponding experimental value for an atom is therefore the sum of all the ionization potentials; for a molecule there are in addition contributions from the molecular bonds and associated zero-point energies. The experimental value for the total energy of H_2O is −76.480 au and the estimated contribution from relativistic effects is −0.045 au. Including also a mass correction of 0.0028 au (a non-Born–Oppenheimer effect that accounts for the difference between finite and infinite nuclear masses) allows the "experimental" non-relativistic Born–Oppenheimer energy to be estimated as -76.438 ± 0.003 au.[5]

The full CI result is available with the cc-pVDZ basis set,[6] which allows an assessment of the performance of various approximate methods. The percent of the electron correlation recovered by different methods is shown in Table 12.5.

The H_2O molecule is, as already mentioned, an easy system, where the HF wave function provides a good reference. Furthermore, since there are only ten electrons in H_2O the effect of higher-order electron correlation is small. The intraorbital correlation between electron pairs dominates the correlation energy for such a small system and the doubly excited configurations, which mainly describe the pair correlation, accounts for a large fraction of the total correlation energy. Consequently even the simple MP2 method performs exceedingly well and the CCSD(T) result is for practical purposes identical to the full CI result. For such simple systems, the MP2 and MP3 percent correlations are probably significantly higher than would be expected for a larger system.

The calculated total energy as a function of the basis set and electron correlation (correlating all electrons) at the experimental geometry is given in Table 12.6. As the cc-pCVXZ basis sets represent a systematic convergence toward the complete basis set limit, there is justification for extrapolating the results to the "infinite" basis set limit (Section 5.4.6). The HF energy is expected to have an exponential behavior and a functional form of the type $A + B\exp(-C\sqrt{X})$ with $X = 4$, 5 and 6

Table 12.5 Percent electron correlation recovered by various methods in the cc-pVDZ basis arranged according to the formal computational scaling.

Scaling	MP		CC		CI	
N^5	MP2	94.0				
N^6	MP3	97.0	CCSD	98.3	CISD	94.5
N^7	MP4	99.5	CCSD(T)	99.7	CISDT	95.8
N^8	MP5	99.8	CCSDT	99.8	CISDTQ	99.8
N^9	MP6	99.9			CISDTQ5	99.9
N^{10}	MP7	100.0	CCSDTQ	100.0	CISDTQ56	100.0

Table 12.6 Total non-relativistic energy (+76 au) as a function of basis set and electron correlation (all electrons) estimated exact value from experimental data is −76.438 ± 0.003 au.[5]

Method	cc-pCVDZ	cc-pCVTZ	cc-pCVQZ	cc-pCV5Z	cc-pCV6Z	cc-pCV∞Z	%EC
HF	−0.027	−0.057	−0.065	−0.067	−0.067	−0.067	0.0
MP2	−0.269	−0.375	−0.408	−0.419	−0.424	−0.430	97.7
MP3	−0.276	−0.380	−0.410	−0.420	−0.423	−0.427	97.0
MP4	−0.282	−0.391	−0.422	−0.433	−0.436	−0.441	100.6
MP5	−0.282	−0.389	−0.420	−0.430	−0.433	−0.438	100.0
CCSD	−0.279	−0.382	−0.411	−0.421	−0.424	−0.429	97.4
CCSD(T)	−0.282	−0.390	−0.421	−0.431	−0.434	−0.439	100.2
CISD	−0.269	−0.368	−0.397	−0.406	−0.426	−0.413	93.2

yields an infinite basis set limit of −76.0674 au, in complete agreement with the estimated HF limit of −76.0674 au.[7] The correlation energy is expected to have an inverse power dependence on the highest angular momentum once the basis set reaches a sufficient (large) size. Extrapolating the correlation contribution for $X = 5$ and 6 with a function of the type $A + BX^{-3}$ yields the cc-pCV∞Z values in Table 12.6. The extrapolated CCSD(T) energy is −76.439 au, which is well within the error limits of the estimated experimental value of −76.438 ± 0.003 au. The total correlation energy is thus −0.373 au, while a corresponding analysis using only the valence electrons yields a value of −0.308 au. The core (and core–valence) electron correlation is thus 0.065 au, which is comparable to the value for the valence electrons (i.e. 0.308 divided between four electron pairs is 0.077 au).

The percent of the total correlation energy at the basis set limit is given in the last column in Table 12.6. The MP2 method recovers 97.7%, which can be compared with the value of 94% in the cc-pVDZ basis (Table 12.5). Notice, however, that while the perturbation series is smoothly convergent with the cc-pCVDZ basis, it becomes oscillating with the larger basis sets. With the cc-pCVTZ basis, the MP5 result is higher in energy than MP4, and with the cc-pCV6Z the MP3 result is higher than the MP2 value. This may be an indication that the perturbation series is actually divergent in a sufficiently large basis set. The CISD method performs rather poorly, yielding results that are worse than MP2 but at a cost similar to an MP4 calculation.

Since the CCSD(T) result is essentially equivalent to a full CI (Table 12.5), the data show that the cc-pCVDZ basis is able to provide 69% of the *total* correlation energy. The corresponding values for the cc-pCVTZ, cc-pCVQZ and cc-pCV5Z basis sets are 90%, 96% and 98%, respectively. Slightly lower percentages have been found in other systems.[8] This illustrates the slow convergence of the correlation energy as a function of the basis set. Each step up in basis set quality roughly doubles the number of functions. The cc-pCVDZ basis is capable of recovering 69% of the correlation energy, and improving the basis from cc-pCVDZ to cc-pCVTZ allows an additional 21% to be calculated. The next step up gives only 6% and the expansion from cc-pCVQZ to cc-pCV5Z gives only 2%. The last 5–10% of the correlation energy is therefore hard to get, requiring very large basis sets. This slow convergence is the principal limitations of traditional *ab initio* methods. The CCSD(T)/cc-pCV5Z total energy is still ~18 kJ/mol off the experimentally derived non-relativistic value, with the remaining error being distributed roughly equally between incomplete basis set and incomplete electron correlation effects. These errors are comparable to the Born–Oppenheimer correction of 7 kJ/mol, and substantially smaller than the relativistic correction of 118 kJ/mol. Calculating the total energy with an accuracy of a few kJ/mol is thus only borderline possible for this simple system.

Although the total energy calculated by DFT methods should in principle converge to the "experimental" value (−76.438 au), there are no upper or lower bounds for the currently employed methods with approximate exchange–correlation functionals, and this can be considered a consequence of not enforcing N-representability of the functionals.[9] The PBE and B3LYP functionals, for example, give total energies of −76.388 and −76.474 at the basis set limit.

12.3 Dipole Moment Convergence

As examples of molecular properties we will look at how the dipole moment and harmonic vibrational frequencies converge as a function of level of theory.

12.3.1 Wave Function Methods

The experimental value for the dipole moment is 1.847 Debye[10] and the calculated value at various levels of theory is shown in Table 12.7.

The dipole moment may be considered as the response of the wave function (energy) to the presence of an external electric field, in the limit where the field strength is vanishingly small (Section 11.1.1). It is consequently sensitive to the representation of the wave function "tail", that is far from the nuclei, and diffuse functions are therefore expected to be important. The results with the regular cc-pVXZ basis sets converge only slowly, as compared with the results for the basis sets augmented with diffuse functions. This illustrates that care must be taken when calculating properties other than the total energy, as standard basis sets may not be able to describe important aspects of the wave function.

The HF dipole moment is too large, which is quite general, as the HF wave function overestimates the ionic contribution. The MP2 procedure recovers the large majority of the correlation effect, but the convergence with the aug-cc-pVXZ basis sets is not smooth and does not readily allow an extrapolation. The CCSD(T) result with the aug-cc-pV5Z basis is very close to the experimental value, although remaining basis set effects and further correlation may change the value slightly. As expected for this property, the effects of core correlation and relativistic corrections are small, as shown by the results in Table 12.8.

12.3.2 Density Functional Methods

Table 12.7 establishes that diffuse functions are mandatory for calculating dipole moments and only the aug-pcseg-n basis set have been used with DFT methods. The calculated results are given in Table 12.9.

Table 12.7 H_2O dipole moment (Debye) as a function of theory (valence correlation only); the experimental value is 1.847 Debye.

Basis	HF	MP2	CCSD(T)	Basis	HF	MP2	CCSD(T)
cc-pVDZ	2.057	1.964	1.936	aug-cc-pVDZ	2.000	1.867	1.848
cc-pVTZ	2.026	1.922	1.903	aug-cc-pVTZ	1.983	1.852	1.839
cc-pVQZ	2.008	1.904	1.890	aug-cc-pVQZ	1.981	1.858	1.848
cc-pV5Z	2.003	1.895	1.884	aug-cc-pV5Z	1.981	1.861	1.851

Table 12.8 H_2O dipole moment (Debye) as a function of theory (all electrons); the experimental value is 1.847 Debye.

Basis	MP2	CCSD(T)	CCSD(T)-DKH2
aug-cc-pwCVDZ	1.868	1.850	1.846
aug-cc-pwCVTZ	1.856	1.843	1.839
aug-cc-pwCVQZ	1.863	1.853	1.849

Table 12.9 H_2O dipole moment (Debye) as a function of DFT functional and basis set; the experimental value is 1.847 Debye.

Basis	LSDA	BLYP	PBE	B3LYP	PBE0	ωB97XD	B2PLYP
aug-pcseg-0	2.744	2.610	2.630	2.661	2.683	2.711	2.704
aug-pcseg-1	1.849	1.788	1.786	1.846	1.851	1.869	1.909
aug-pcseg-2	1.860	1.804	1.802	1.860	1.865	1.881	1.921
aug-pcseg-3	1.856	1.800	1.796	1.856	1.859	1.874	1.916
aug-pcseg-4	1.856	1.799	1.796	1.855	1.858	1.873	1.916

The calculated dipole moment is remarkably insensitive to the size of the basis set, once polarization functions have been included (i.e. at least aug-pcseg-1). Note that the LSDA value in this case is substantially better than the GGA functionals (BLYP and PBE), that is this is a case where the theoretically "poorer" method provides better results than the more advanced gradient methods. Inclusion of "exact" exchange (B3LYP and PBE0) again improves the performance, and provides results very close to the experimental value, even with relatively small basis sets, while the B2PLYP double hybrid method overestimates the value.

12.4 Vibrational Frequency Convergence

The experimental values for the fundamental vibrational frequencies are 1595, 3657 and 3756 cm^{-1}, while the corresponding *harmonic* values are 1649, 3832 and 3943 cm^{-1}.[11] The differences due to anharmonicity are thus 54, 175 and 187 cm^{-1}, that is 3–5% of the harmonic values.

12.4.1 Wave Function Methods

The calculated harmonic frequencies at the HF, MP2 and CCSD(T) levels are given in Table 12.10.

Vibrational frequencies are examples of a slightly more complicated property. The frequencies are obtained from the force constant matrix (second derivative of the energy), *evaluated at the equilibrium geometry* (Section 17.2.2). Both the equilibrium geometry and the shape of the energy surface depend on the theoretical level. Part of the change in frequencies is due to changes in the geometry since the force constant in general decreases with increasing bond length.

The HF vibrational frequencies are too high by about 7% relative to the experimental harmonic values and by 10–13% relative to the anharmonic values. This overestimation is due to the incorrect dissociation and the corresponding bond lengths being too short (Table 12.1) and is consequently

Table 12.10 H_2O harmonic frequencies (cm^{-1}) as a function of method and basis set (only valence electrons are correlated).

	ν_1			ν_2			ν_3		
Basis	**HF**	**MP2**	**CCSD(T)**	**HF**	**MP2**	**CCSD(T)**	**HF**	**MP2**	**CCSD(T)**
cc-pVDZ	1776	1678	1690	4114	3852	3822	4212	3971	3928
cc-pVTZ	1753	1652	1669	4127	3855	3841	4227	3976	3946
cc-pVQZ	1751	1643	1659	4130	3855	3845	4229	3978	3952
cc-pV5Z	1748	1636	1653	4131	3849	3840	4231	3974	3950
cc-pV6Z	1748	1634	1651	4130	3845	3837	4231	3971	3947
Exp		1649			3832			3943	

quite general. Vibrational frequencies at the HF level are therefore often scaled by ~0.9 to partly compensate for these systematic errors.[12]

The inclusion of electron correlation normally lowers the force constants, since the correlation energy increases as a function of bond length. This usually means that vibrational frequencies decrease, although there are exceptions (vibrational frequencies also depend on off-diagonal force constants). The MP2 treatment recovers the majority of the correlation effect, and the CCSD(T) results with the cc-pV6Z basis set are in good agreement with the experimental values (Table 12.10). The remaining discrepancies of 2, 5 and 4 cm^{-1} are mainly due to basis set inadequacies (!). The MP2 values are in respectable agreement with the experimental harmonic frequencies, but of course still overestimate the experimental fundamental ones by the anharmonicity. For this reason, calculated MP2 harmonic frequencies are often scaled by ~0.97 for comparing with experimental results.[12]

The effects of core electron correlation and relativistic corrections are small, as shown in Table 12.11. It should be noted that the valence and core correlation energy per electron pair is of the same magnitude. The core correlation, however, is almost constant over the whole energy surface and consequently contributes very little to properties depending on relative energies, such as vibrational frequencies. The bending frequency is almost unchanged, while the stretching frequencies increase by 7 cm^{-1}, which slightly deteriorates the agreement with the experimental values. The relativistic corrections at the DKH2 level are a few cm^{-1}.

For comparing with experimental frequencies (which necessarily are anharmonic), there is normally little point in improving the theoretical level beyond MP2 with a TZP-type basis set unless anharmonicity constants are calculated explicitly. Although anharmonicity can be approximately

Table 12.11 H_2O MP2 and CCSD(T) harmonic frequencies (cm^{-1}) as a function of basis set (all electrons are correlated).

	ν_1			ν_2			ν_3		
Basis	**MP2**	**CCSD(T)**	**CCSD(T)-DKH2**	**MP2**	**CCSD(T)**	**CCSD(T)-DKH2**	**MP2**	**CCSD(T)**	**CCSD(T)-DKH2**
cc-pwVDZ	1678	1690	1691	3854	3824	3821	3973	3930	3926
cc-pwVTZ	1651	1667	1669	3859	3844	3841	3978	3947	3945
cc-pwVQZ	1641	1657	1659	3862	3851	3849	3985	3959	3956
cc-pwV5Z	1635	1652	1653	3856	3847	3845	3982	3957	3954
Exp		1649			3832			3943	

accounted for by scaling the harmonic frequencies by ~0.97, the remaining errors in the harmonic force constants at this level are normally smaller than the corresponding errors due to variations in anharmonicity.

12.4.2 Density Functional Methods

The harmonic frequencies calculated with various DFT methods as a function of basis set are shown in Tables 12.12 to 12.14.

The convergence as a function of basis set is similar to that observed for the HF method. The Jacob's ladder qualification in this case works very well, with the B2PLYP double hybrid providing amazing agreement with the experimental values. It is also clear from Tables 12.12 to 12.14 that inclusion of "exact" exchange (B3LYP and PBE0) substantially improves the performance. The "pure" DFT gradient methods, BLYP and PBE, have errors of ~150 cm^{-1} for the stretching frequencies and ~50 cm^{-1} for the angle bending.

Table 12.12 H_2O lowest harmonic frequency (cm^{-1}) as a function of basis set with various DFT functionals; the experimental value is 1649 cm^{-1}.

Basis	LSDA	BLYP	PBE	B3LYP	PBE0	ωB97XD	B2PLYP
pcseg-0	1556	1600	1597	1625	1629	1634	1640
pcseg-1	1549	1598	1596	1628	1635	1637	1644
pcseg-2	1543	1594	1590	1624	1630	1632	1636
pcseg-3	1549	1597	1594	1628	1634	1634	1638
pcseg-4	1550	1598	1594	1629	1635	1636	1639

Table 12.13 H_2O second lowest harmonic frequency (cm^{-1}) as a function of basis set with various DFT functionals; the experimental value is 3832 cm^{-1}.

Basis	LSDA	BLYP	PBE	B3LYP	PBE0	ωB97XD	B2PLYP
pcseg-0	3327	3217	3276	3390	3470	3527	3426
pcseg-1	3689	3610	3664	3765	3835	3859	3812
pcseg-2	3727	3667	3707	3809	3868	3902	3835
pcseg-3	3719	3665	3705	3805	3863	3889	3819
pcseg-4	3718	3666	3706	3807	3865	3893	3829

Table 12.14 H_2O highest harmonic frequency (cm^{-1}) as a function of basis set with various DFT functionals; the experimental value is 3943 cm^{-1}.

Basis	LSDA	BLYP	PBE	B3LYP	PBE0	ωB97XD	B2PLYP
pcseg-0	3457	3329	3394	3504	3588	3645	3549
pcseg-1	3812	3722	3780	3877	3951	3975	3932
pcseg-2	3839	3771	3815	3914	3975	4010	3946
pcseg-3	3828	3767	3810	3907	3968	3994	3930
pcseg-4	3827	3768	3811	3909	3970	3998	3939

12.5 Bond Dissociation Curves

As seen in Table 12.6, it is very difficult to converge the total energy to an accuracy of a few kJ/mol. The total energy, however, is in almost all cases irrelevant; the important quantity is the relative energy. Let us now examine how the *shape* of a potential energy surface depends on the theoretical level. We will look at two cases: stretching one of the O—H bonds in H_2O and the HOH bending potential. The O—H dissociation curve is a case where the main change is associated with the difference in electron correlation between the two electrons in the bond being stretched. It should be noted that transition structures typically have bonds that are elongated by 0.5–0.8 Å, and the performance for the dissociation curve in this range will model the behavior for describing bond breaking/forming reactions. The HOH bending energy, on the other hand, does not involve any bond breaking and should therefore be less sensitive to the level of theory.

12.5.1 Wave Function Methods

We will now look at how different types of wave function behave when the O—H bond is stretched. The basis set used in all cases is the aug-cc-pVTZ and the reference curve is taken as the [8, 8]-CASMP2 result, which is slightly larger than a full valence CI followed by a second-order perturbation correction. As mentioned in Section 4.6, this allows a correct dissociation, and since all the valence electrons are correlated, it will generate a curve close to the full CI limit. The bond dissociation energy calculated at this level is 503 kJ/mol, which is comparable to the experimental value of 525 kJ/mol.[13]

H_2O is a closed-shell singlet and the HF wave function near the equilibrium geometry is of the RHF type. As one of the bonds is stretched, however, a UHF type will become lower in energy at some point (Section 4.4). Beyond this instability point, electron correlation methods may be based either on the RHF or UHF reference. The UHF wave function will be spin-contaminated, which has some consequences, as shown below. It should be noted that for open-shell species there is similarly the option of using either an ROHF or UHF reference wave function, but in such cases they will be different at all geometries, and also near the equilibrium. The difference will be small if the UHF wave function is only slightly spin-contaminated.

Figure 12.1 illustrates the behavior of the single-determinant wave functions, RHF, UHF and PUHF (projected UHF, Section 4.4). The RHF energy continues to increase as the bond is stretched since it has the wrong dissociation limit, while the UHF converges to a value of 366 kJ/mol. At the equilibrium geometry the two electrons in the O—H bonding orbital are correlated, but this correlation energy disappears once the bond is broken. The UHF wave function correctly describes the dissociation limit in terms of energy, but does not recover any of the electron correlation at equilibrium (by definition, since UHF = RHF here). The difference between the UHF dissociation energy and the CASMP2 value is therefore a measure of the amount of electron correlation in the O—H bond. With the present basis set this is 137 kJ/mol, which is a typical value for the correlation energy between two electrons in the same spatial MO.

Figure 12.2 shows how the spin contamination as measured by the $\langle \mathbf{S}^2 \rangle$ value of the UHF wave function increases as the bond is stretched. At the dissociation limit the UHF wave function is essentially an equal mixture of a singlet and triplet state, as discussed in Section 4.4. Removal of the triplet state by projection (PUHF) lowers the energy in the intermediate range, but has no effect when the bond is completely broken since the singlet and triplet states are degenerate here.

The RHF/UHF instability point with this basis set occurs when the bond is stretched to 1.36 Å. Figure 12.3 shows the behavior of the energy curves in more detail in this region. It is seen that the PUHF

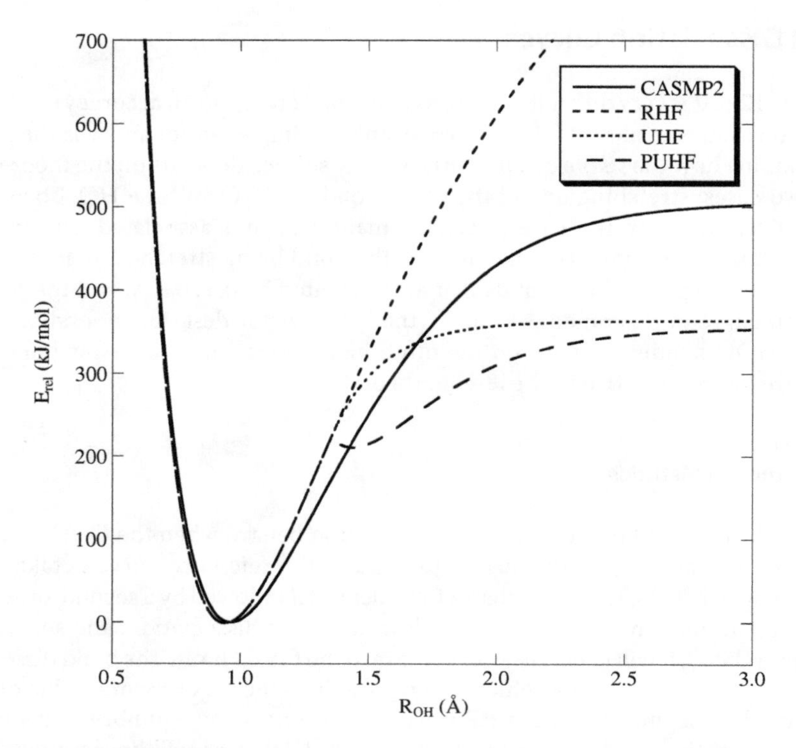

Figure 12.1 RHF, UHF and PUHF dissociation curves for H_2O.

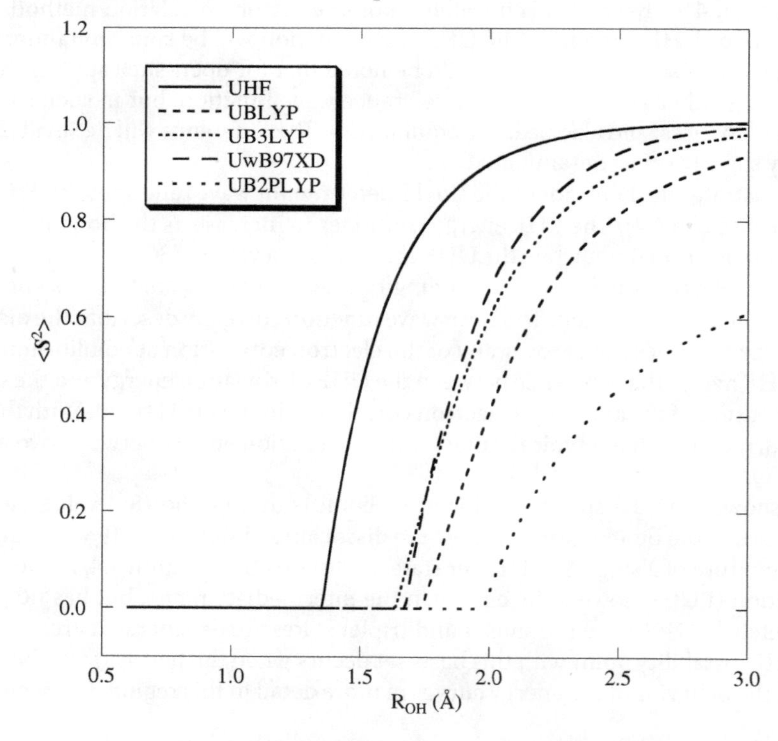

Figure 12.2 Spin contamination, as measured by the $\langle \mathbf{S}^2 \rangle$ value, as a function of the bond length for H_2O.

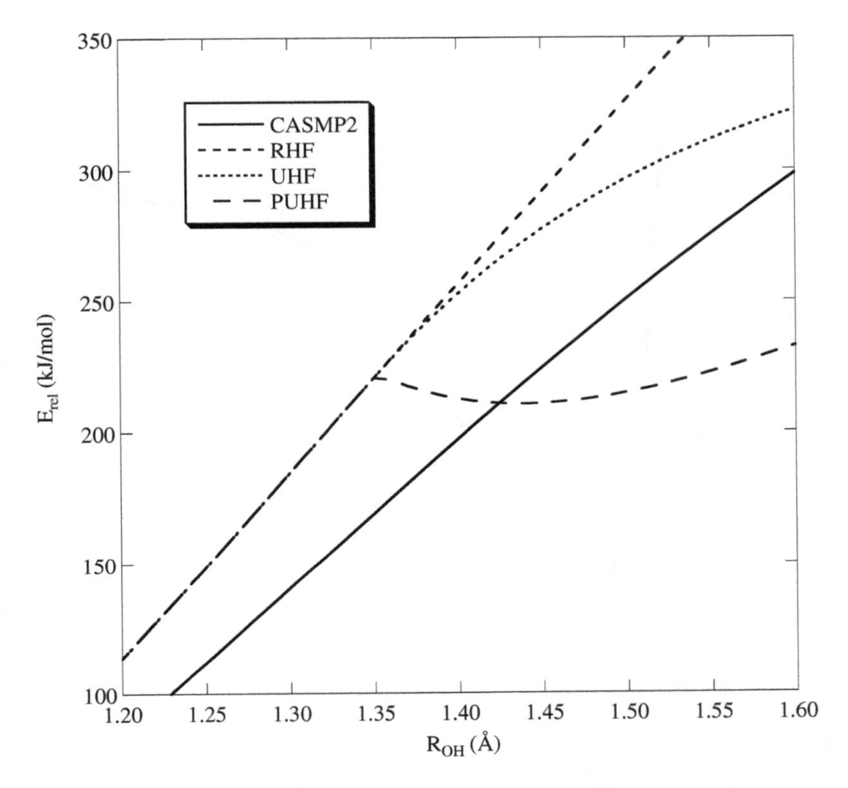

Figure 12.3 RHF, UHF and PUHF dissociation curves for H_2O near the instability point.

has a discontinuous derivative at the instability point, and there is furthermore a shallow minimum right after the instability point, at a bond length of ~1.40 Å.

Since the RHF curve is too high in the transition structure region ($\Delta R \sim 0.5$–0.8 Å), it is clear that RHF activation energies in general will be too large. UHF activation energies may either be too high or too low, but the PUHF value will essentially always be too low. Furthermore, the shape of a spin-contaminated UHF energy surface will be too flat and PUHF surfaces will be qualitatively wrong in the TS region. Spin-contaminated UHF wave functions should consequently not be used for geometry optimizations.

The corresponding difference between restricted, unrestricted and projected unrestricted wave functions at the MP2 level is shown in Figure 12.4.

The RMP2 rises too high, owing to the wrong dissociation limit of the underlying RHF. Both the UMP2 and PUMP2 dissociation energies are in reasonable agreement with the CASMP2 value, but it is clear that the UMP2 energy is too high in the "intermediate" range owing to spin contamination. The PUMP2 curve, on the other hand, traces the reference CASMP2 values more closely. Figure 12.5 shows the curves in more detail near the RHF/UHF instability point.

The UMP2 energy is *higher* than the RMP2, although the UHF energy is *lower* than the RHF. At the HF level, the UHF energy is lowest owing to a combination of spin contamination and inclusion of electron correlation (Section 4.8.2). Since the MP2 procedure recovers most of the electron correlation, only the energy rising effect due to spin contamination remains and the UMP2 energy becomes higher than RMP2. Removing the unwanted spin components makes the PUMP2 energy very similar

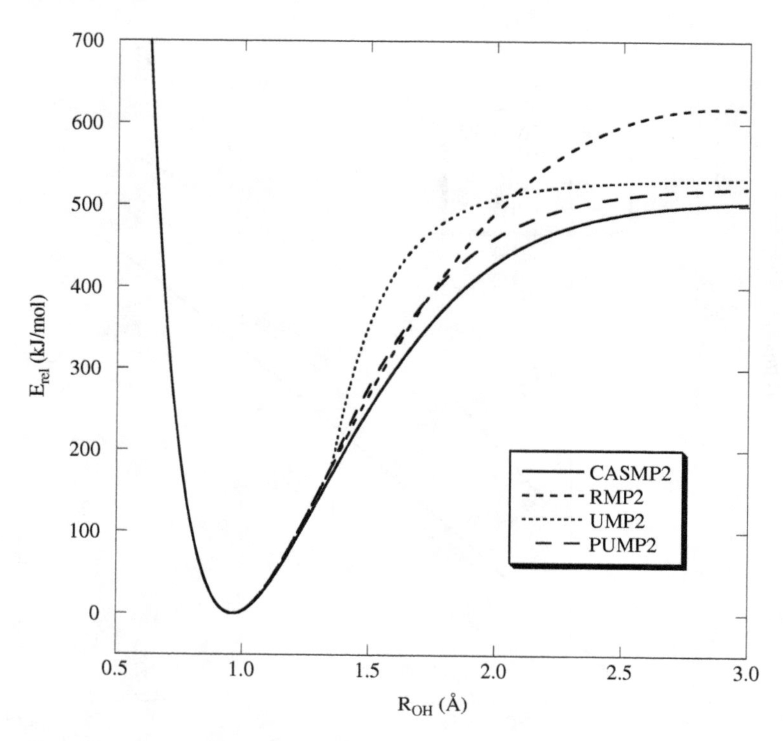

Figure 12.4 RMP2, UMP2 and PUMP2 dissociation curves for H_2O.

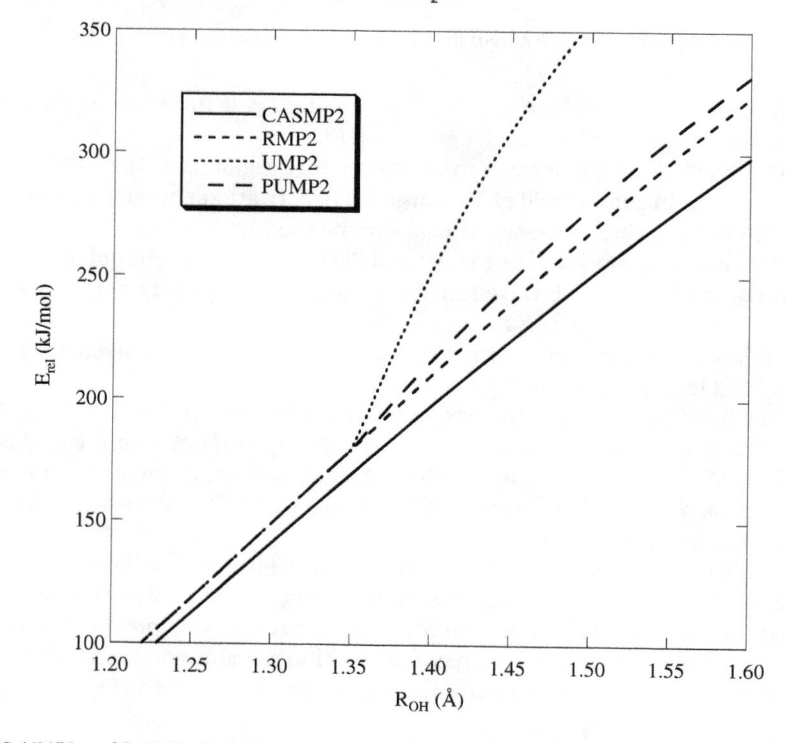

Figure 12.5 RMP2, UMP2 and PUMP2 dissociation curves for H_2O near the instability point.

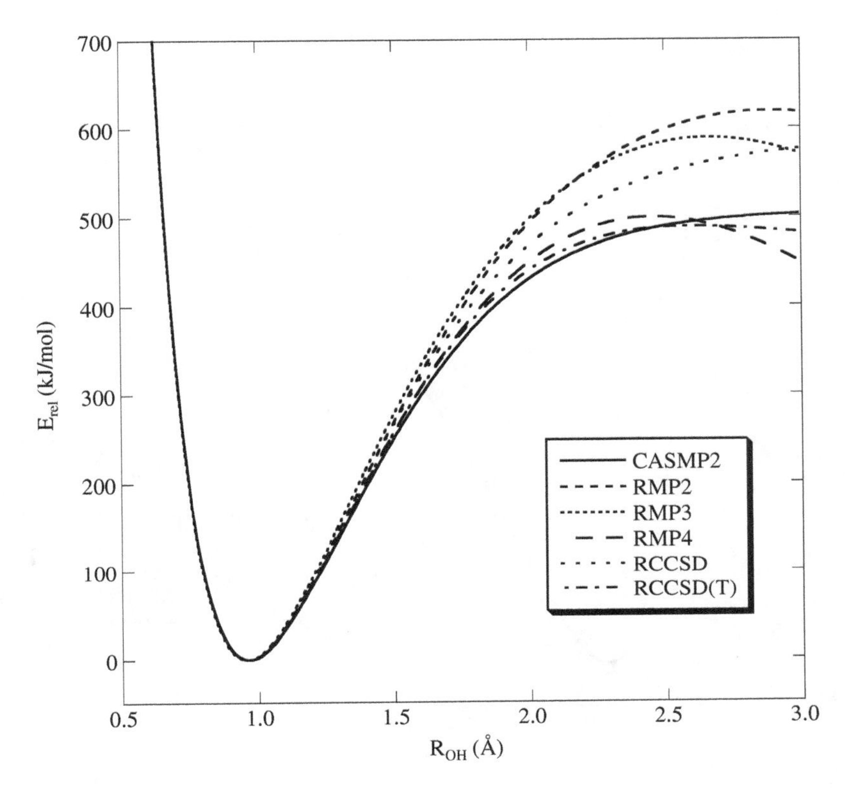

Figure 12.6 RMP2, RMP3, RMP4, RCCSD and RCCSD(T) dissociation curves for H_2O.

to RMP2 for elongations less than ~1 Å, but is significantly better at longer bond lengths owing to the correct dissociation of the UHF wave function. The RMP2 energy follows the "exact" curve closely out to a bond length of ~1.3 Å and is in qualitative agreement out to ~1.8 Å. RMP2 activation energies are therefore often in quite reasonable agreement with experimental or higher level theoretical values. It should also be noted that the discontinuity at the PUHF level essentially disappears when the projection is carried out on the MP2 wave function.

Figures 12.6 and 12.7 show the effect of extending the perturbation series at the RMP and UMP levels and the corresponding CCSD and CCSD(T) results. Addition of more terms in the perturbation series improves the results, although the effect of MP3 compared with MP2 is minute. As the bond is stretched beyond ~2.5 Å, the perturbation series breaks down owing to the RHF wave function becoming too poor a reference and the energies start to decrease. The infinite-order CCSD and CCSD(T) methods outperform the MPn results and the effect of the triples become increasingly important as the molecule dissociates.

The improvement by extending the perturbation series beyond second order is small when a UHF wave function is used as the reference, that is the higher-order terms do very little to reduce the spin contamination. In the dissociation limit the spin contamination is inconsequential and the MP2, MP3 and MP4 results are all in reasonable agreement with the CASMP2 result, but spin contamination leads to significant errors in the intermediate region. The UCCSD and UCCSD(T) energy curves are also somewhat too high in the intermediate region, but the infinite nature of coupled cluster methods is significantly better at removing unwanted spin states as compared with UMP*n* methods.

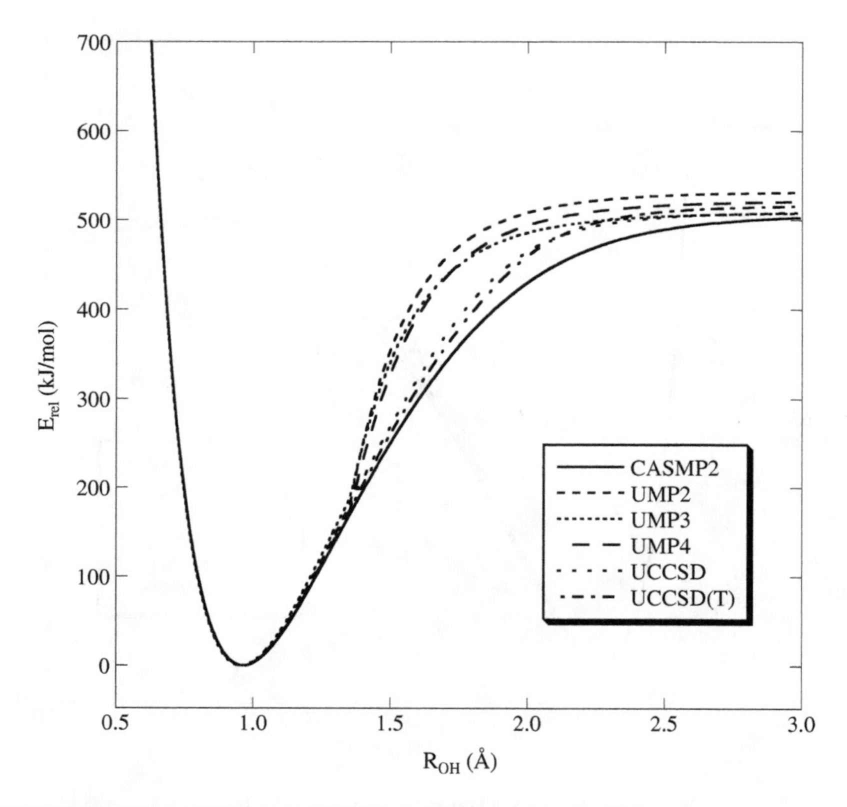

Figure 12.7 UMP2, UMP3, UMP4, UCCSD and UCCSD(T) dissociation curves for H_2O.

12.5.2 Density Functional Methods

The performance of various DFT methods resembles the HF results. A restricted type of determinant leads to an incorrect dissociation, while an unrestricted determinant has the energetically correct dissociation limit. Figures 12.8 and 12.9 show the performance of restricted and unrestricted types of determinants with the BLYP, B3LYP, wB97XD and B2PLYP functionals.

Figure 12.9 shows that DFT methods are much less sensitive to the "spin contamination" problem in the intermediate region, as also shown in Figure 12.2; indeed, spin contamination is not well defined in DFT.[14] Furthermore, as electron correlation is implicitly included, DFT methods are closer in shape to the CASSCF curve. Removing the "spin contamination" by projection methods results in discontinuous derivatives and artificial minima, analogously to the PUHF case in Figures 12.1 and 12.3, and should consequently not be employed.[15] The B2PLYP double hybrid functional inherits the poor performance of the MP2 method at the dissociation limit.

12.6 Angle Bending Curves

The angle bending in H_2O occurs without breaking any bonds and the electron correlation energy is therefore relatively constant over the whole curve. The HF, MP2 and CCSD(T) bending potentials

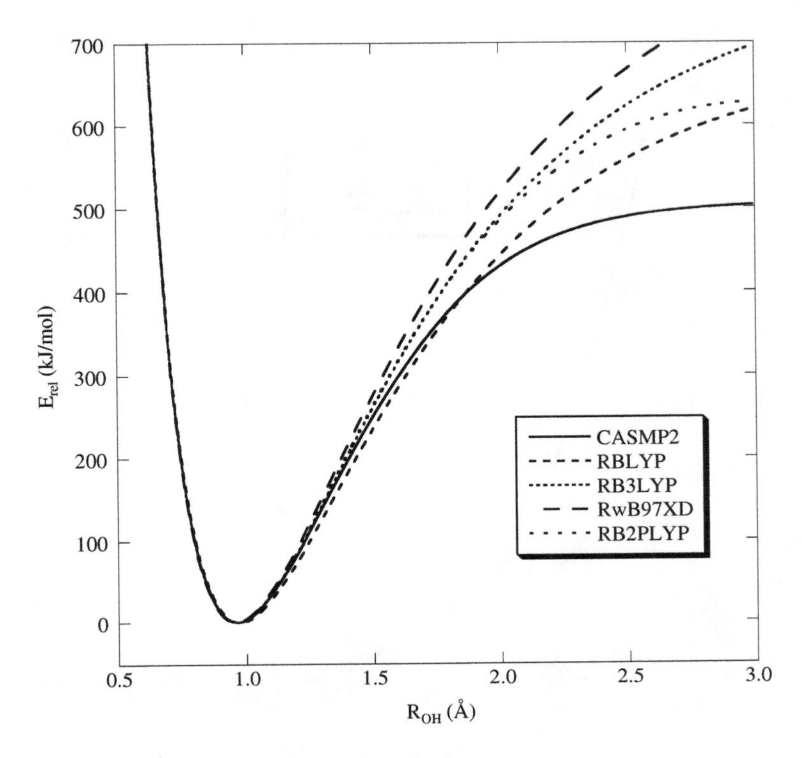

Figure 12.8 Bond dissociation curves for restricted DFT methods.

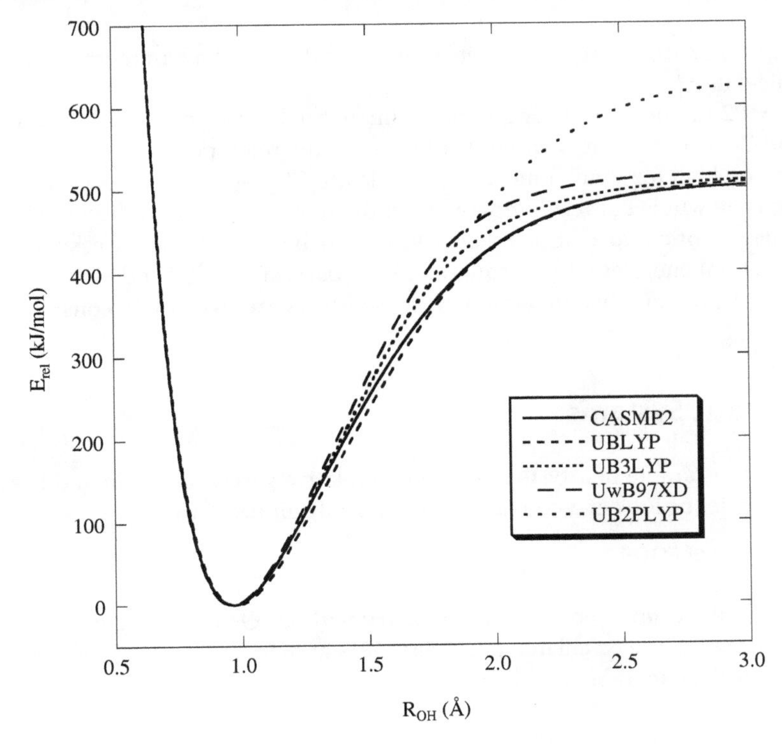

Figure 12.9 Bond dissociation curves for unrestricted DFT methods.

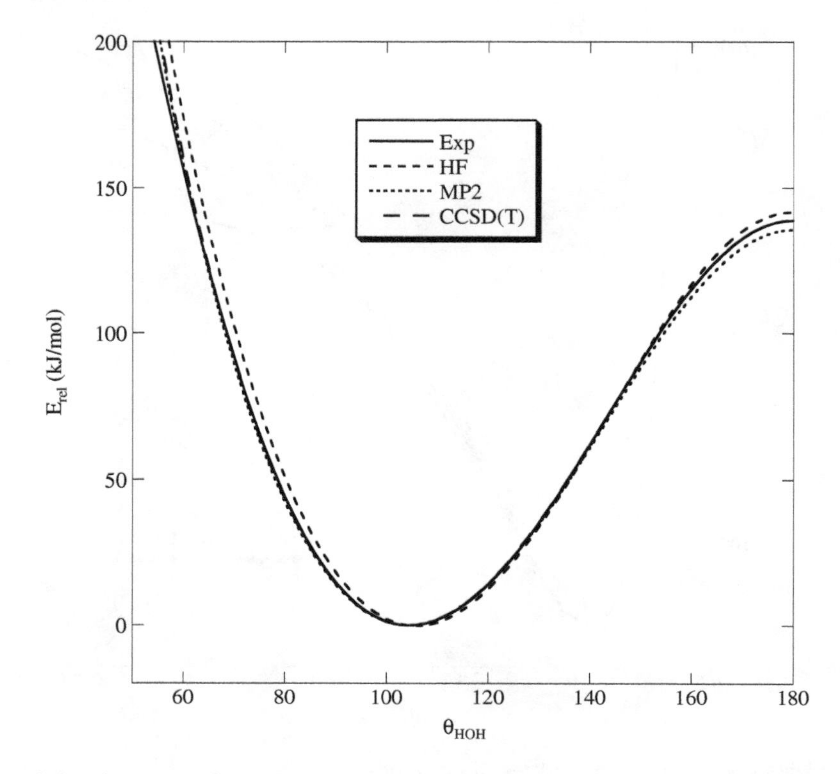

Figure 12.10 Angle bending curves for H_2O.

are shown in Figure 12.10, where the reference curve is taken from a parametric fit to a large number of spectroscopic data.[16]

The HF and MP2 methods over- and underestimate the barrier for linearity by 3 kJ/mol, respectively, while the CCSD(T) result is within 0.1 kJ/mol of the reference value of 138.7 kJ/mol. The HF curve is slightly too high for small bond angles, while the CCSD(T) results are within 1 kJ/mol of the exact result over the whole curve. Compared with the bond dissociation discussed above, it is clear that relative energies of conformations that have similar bonding are fairly easy to calculate. While the HF and MP2 total energies with the aug-cc-pVTZ basis are ~1000 and ~300 kJ/mol higher than the exact values at the equilibrium geometry, these errors are essentially constant over the whole surface.

12.7 Problematic Systems

The H_2O case is an example of a system where it is relatively easy to obtain good results. Nature is not always so kind; let us look at a couple of "theoretically difficult" cases.

12.7.1 The Geometry of FOOF

The FOOF molecule has an experimental geometry with an O—O bond length of 1.217 Å and an F—O bond of 1.575 Å.[17] The calculated bond distances at different levels of theory with the aug-cc-pVTZ basis set are given in Table 12.15.

Table 12.15 Bond distance (Å) in FOOF with the aug-cc-pVTZ basis set.

Method	R_{OO}	R_{FO}
HF	1.300	1.356
MP2	1.166	1.619
MP3	1.300	1.427
MP4(SDQ)	1.291	1.453
CCSD	1.288	1.449
CCSD(T)[a]	1.234	1.545
CISD	1.295	1.389
LSDA	1.188	1.559
BLYP	1.206	1.632
PBE	1.198	1.606
B3LYP	1.227	1.523
PBE0	1.233	1.479
ωB97XD	1.246	1.468
B2PLYP	1.201	1.571
Experimental	1.217	1.575

[a] CCSD(T) basis set limiting values are 1.230 and 1.532 Å.[19]

The results in Table 12.15 clearly show that the results are very sensitive to the inclusion of electron correlation. The MP4(SDQ) geometry is very similar to the CCSD one, but inclusion of the triply excited configurations in the full MP4(SDTQ) method has a huge effect. The F—O bonds are elongated to the point (>2.5 Å) where perturbation theory breaks down since the underlying RHF wave function becomes extremely poor. The MP4(SDTQ) model basically does not predict a stable FOOF molecule. The triples also have a large effect at the CCSD(T) level, but it is clear that the effect is wildly overestimated with the MP4 method. Although the results are not converged with respect to a basis set (aug-cc-pVTZ), the remaining changes are of the order of a few thousandths of an angstrom.[18] Even with the sophisticated CCSD(T) model, the geometry errors are thus ~0.03 Å.

The DFT methods are all well-behaved and perform surprisingly well for such a difficult system, with the B2PLYP results providing better agreement with the experimental values than CCSD(T). The main problem is of course that there is no way of systematically improving the structure, or knowing beforehand whether DFT will be able to give a good description for the specific problem, or which functional will be most suitable.

12.7.2 The Dipole Moment of CO

The experimental value for the dipole moment of CO is 0.122 Debye, with the polarity C^-O^+, for a bond length of 1.1281 Å.[20] Calculated values with the aug-cc-pVXZ basis sets[21] are given in Table 12.16.

The HF level (as usual) overestimates the polarity, in this case leading to an incorrect direction of the dipole moment. The MP perturbation series oscillates and it is clear that the MP4 result is far from converged. The CCSD(T) method apparently recovers the most important part of the electron correlation and is very close to the full CCSDT result in an augmented DZP basis.[22,23] However, even with the aug-cc-pV5Z basis sets, there is still a discrepancy of ~0.01 Debye relative to the experimental

Table 12.16 Dipole moment (Debye) for CO; the experimental value is 0.122 Debye.

Method	aug-cc-pVDZ	aug-cc-pVTZ	aug-cc-pVQZ	aug-cc-pV5Z
HF	−0.259	−0.266	−0.265	−0.264
MP2	0.296	0.280	0.275	0.273
MP3	0.076	0.047	0.036	0.032
MP4	0.220	0.222	0.216	0.214
CCSD	0.097	0.070	0.059	0.055
CCSD(T)	0.141	0.127	0.118	0.115
CISD	0.050	0.023	0.011	0.008
LSDA	0.232	0.226	0.229	0.229
BLYP	0.187	0.184	0.185	0.185
PBE	0.229	0.224	0.224	0.224
B3LYP	0.091	0.086	0.087	0.088
PBE0	0.107	0.101	0.102	0.102
ωB97XD	0.086	0.097	0.099	0.097
B2PLYP	−0.060	−0.067	−0.066	−0.065

value. The DFT methods are not particularly accurate, although for this specific problem the PBE0 method gives a reasonably good result, while the double hybrid B2PLYP provides an incorrect sign.

12.7.3 The Vibrational Frequencies of O_3

Ozone is an example of a molecule where the single-reference RHF is quite poor, since there is considerable biradical character in the wave function (as illustrated in Figure 4.9). The harmonic vibrational frequencies derived from experiments are 716, 1089 and 1135 cm^{-1}, where the band at 1089 cm^{-1} corresponds to an asymmetric stretch.[24] As this nuclear motion changes the relative weights of the ionic and biradical structures, this frequency is very sensitive to the quality of the wave function. Although the wave function is equally poor for all the frequencies, the two other vibrations (symmetric stretch and angle bending) conserve the C_{2v} symmetry, and thus benefit from a significant cancellation of errors. The calculated frequencies at different levels of theory with the cc-pVTZ basis are given in Table 12.17 together with the mean absolute deviation (MAD).

The simple picture with ozone as a resonance structure between ionic and biradical forms suggests that a two-configuration wave function should be able to give a qualitatively correct description. The [2, 2]-CASSCF and [2, 2]-CASPT2 results, however, show that dynamical correlation is also very important. The poor RHF reference wave function is clearly seen by the MPn results, with the MP2 value being in error by a factor of two for the asymmetric stretch and the MP4 result being in error by ~500 cm^{-1} for ν_2, despite reproducing ν_1 and ν_3 to within 30 cm^{-1}. The coupled cluster methods are less sensitive to the quality of the HF wave function and are in somewhat better agreement with the experimental values. The CCSD(T) results are within ~20 cm^{-1} of the experimental values, but part of this agreement is accidental as seen by the CCSDT and CCSDT(Q) results, and even the CCSDT(Q) model has errors of ~25 cm^{-1}. Part of this discrepancy may be due to basis set errors, although the results for the CASPT2 method indicate that larger basis sets will further increase the value of the vibrational frequencies.[26] The DFT methods display varying performance, yielding results comparable to those at the CCSD or CCSD(T) levels, at a fraction of the computational cost. The pure functionals BLYP and PBE perform better than the hybrid DFT methods (B3LYP

Table 12.17 Harmonic frequencies (cm^{-1}) for O_3 with the cc-pVTZ basis.

Method	ν_1	ν_2	ν_3	MAD
HF	867	1418	1537	294
MP2	743	2241	1166	403
MP3	798	1713	1364	312
MP4	695	1592	1107	184
CCSD	762	1266	1278	122
CCSD(T)	716	1054	1153	18
CCSDT[a]	717	1117	1163	19
CCSDT(Q)[a]	709	1112	1133	6
CISD	815	1535	1407	272
[2,2]-CASSCF	799	1497	1189	182
[2, 2]-CASPT2[b]	737	1268	1318	128
[12, 9]-CASPT2[b]	692	1003	1092	51
LSDA	744	1147	1248	66
BLYP	683	980	1129	49
PBE	710	1057	1184	29
B3LYP	746	1193	1251	83
PBEO	777	1295	1322	151
ωB97XD	781	1320	1335	165
B2PLYP	712	1190	1135	35
Experimental	716	1089	1135	

[a] Data from Kucharski and Bartlett.[25]
[b] Data from Ljubic and Sabljic.[26]

and PBEO), while the double hybrid B2PLYP is close to the CCSD(T) results. It can be noted that the cc-pVTZ basis set is sufficiently large that the DFT results are essentially converged, and the results in Table 12.17 thus reflect the intrinsic accuracy of the different DFT methods.

12.8 Relative Energies of C_4H_6 Isomers

The elaborate treatment for the H_2O system is only possible because of its small size. For larger systems, less rigorous methods must be employed. Let us as a more realistic example consider a determination of the relative stability of the C_4H_6 isomers shown in Figure 12.11.

There are experimental heats of formation values for the first eight structures,[27] which allow an evaluation of the performance of different methods. This in turn enables an estimate of how much trust should be put in the predicted values for structures **9**, **10** and **11**. The quoted errors in the experimental values range from 0.6 to 2 kJ/mol, with an average value of 1.1 kJ/mol.

The experimental values are derived from combustion experiments and converted to ΔH_f values, which provide energies relative to the elements in their most stable state. The calculated energies, however, are relative to the isolated nuclei + electron situation. Converting such energies to a theoretical ΔH_f value would require taking a difference between total energies for the molecule and carbon and hydrogen in their most stable states, that is graphite and H_2. These widely different bonding situations would require a very high computational level to achieve a sufficient error cancellation to produce results of a useful accuracy (Section 5.2). It is therefore convenient to convert both sets

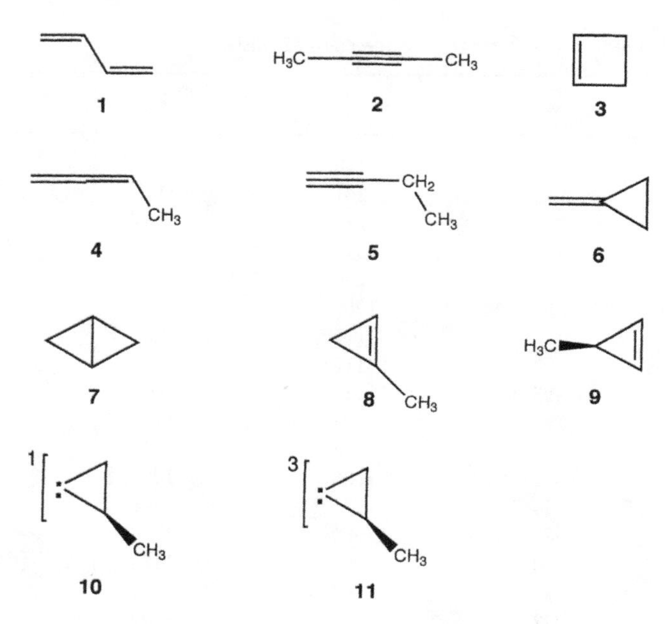

Figure 12.11 C_4H_6 isomers.

of data to relative energies, but this requires a selection of a common reference value for both sets. One could arbitrarily select one of the compounds, for example the most stable, as having an energy of zero, and evaluate the performance of different computational methods by how well the stability of the remaining compounds are predicted. This procedure, however, is sensitive to the choice of reference compound. If a given method, for example, performs poorly for the reference compound, but well for all the remaining compounds, then the method will appear as performing poorly for all the compounds. An unbiased choice is to use the mean of the data set as the reference and evaluate the performance, for example, in terms of the Mean Absolute Deviation (MAD), and the slope and correlation coefficient for the least square line.

The geometry is normally relatively insensitive to the level of theory, and a typical choice for structure optimization is a DFT method with a DZP or TZP-type basis set; this can also be used for evaluating zero-point energies and finite temperature corrections (Section 14.5). The results below have been obtained at the B2PLYP/pcseg-2 level. Improved estimates of the relative energies can then be performed by various combinations of methods and basis sets, and in the limit of a sufficiently large basis set, the error is solely due to the employed method. For the relative energies of the C_4H_6 isomers in Figure 12.11, the basis set error is reduced to below 1 kJ/mol with an extrapolation of the results from TZP and QZP-type basis sets (cc-pVTZ and cc-pVQZ for HF, MP2, CISD, CCSD and CCSD(T) methods and pcseg-1, -2, -3 for DFT).

Table 12.18 shows the performance for four semi-empirical methods: MINDO/3, AM1, PM3 and PM6 for calculating ΔH_f (kJ/mol) for the compounds in Figure 12.11. The MAD values show that typical errors are in the 20–30 kJ/mol range and the R values indicate considerable random errors.

Table 12.19 shows the performance for the wave function methods: HF, MP2, CISD, CCSD, CCSD(T) extrapolated to the basis set limit, as well as the composite G4MP2 and W1 methods for calculating ΔH_f for the compounds in Figure 12.11. The MAD, slope and R values indicate a consistent improvement along the HF, MP2, CISD, CCSD, CCSD(T) series, and the composite G4MP2

Table 12.18 ΔH_f (kJ/mol) values for the compounds in Figure 12.11 using semi-empirical methods.

Structure	MINDO/3	AM1	PM3	PM6	Exp.
1	151.7	129.4	132.2	122.6	108.8 ± 0.79
2	59.4	150.0	129.4	101.2	145.1 ± 1
3	148.0	196.9	159.6	144.3	157 ± 2
4	132.6	163.9	161.5	129.0	162.2 ± 0.59
5	130.8	168.2	153.4	126.4	165.2 ± 0.88
6	158.8	206.2	190.9	154.5	201 ± 2
7	237.6	347.6	306.8	271.3	217 ± 0.8
8	191.4	282.8	246.3	215.6	244 ± 1
9	232.2	293.2	256.8	228.6	
10	387.5	496.3	461.1	401.1	
11	379.1	515.0	476.3	461.9	
MAD	39.2	29.4	23.3	26.5	
Slope	0.75	1.46	1.20	1.04	
R	0.634	0.853	0.842	0.791	

and W1 methods display comparable performances. Residual basis set errors, corrections from core and core-valence correlation and anharmonic vibrations are all below 1 kJ/mol. The MAD values of ~3 kJ/mol should be compared with the quoted experimental errors in the 1–2 kJ/mol range. Roughly half of the MAD error arises from compound **6**, where all the sophisticated methods predict values that are consistently ~9 kJ/mol below the experimental value, and this could point to a possible experimental error.

Table 12.20 shows the performance for the DFT methods: LSDA, BLYP, PBE, B3LYP, PBE0, ωB97XD and B2PLYP extrapolated to the basis set limit for calculating ΔH_f for the compounds in Figure 12.11. The DFT methods display rather similar performances, although there is a weak

Table 12.19 ΔH_f (kJ/mol) values for the compounds in Figure 12.11 using wave function methods.

Structure	HF	MP2	CISD	CCSD	CCSD(T)	G4MP2	W1	Exp.
1	101.0	120.7	114.6	112.3	110.6	108.4	111.1	108.8 ± 0.79
2	136.4	143.1	144.9	145.7	147.2	148.5	147.2	145.1 ± 1
3	167.7	160.6	159.4	159.4	158.9	160.9	159.3	157 ± 2
4	155.4	171.8	165.1	162.3	161.8	160.5	162.0	162.2 ± 0.59
5	159.3	163.1	166.7	165.1	166.0	166.9	166.3	165.2 ± 0.88
6	194.1	194.2	190.9	192.3	193.0	192.1	192.3	201 ± 2
7	241.6	210.6	219.8	223.6	223.3	222.9	222.9	217 ± 0.8
8	244.7	236.1	238.8	239.6	239.4	240.2	239.3	244 ± 1
9	261.3	249.7	253.8	252.7	252.0	253.1	252.3	
10	405.6	462.6	429.9	432.3	438.2	436.8	440.3	
11	407.4	517.7	472.3	492.5	500.1	499.5	501.9	
MAD	9.0	6.3	3.9	3.3	3.2	3.7	3.4	
Slope	1.12	0.86	0.93	0.96	0.96	0.97	0.96	
R	0.977	0.993	0.995	0.994	0.995	0.994	0.995	

Table 12.20 ΔH_f (kJ/mol) values for the compounds in Figure 12.11 using density functional methods.

Structure	LSDA	BLYP	PBE	B3LYP	PBE0	ωB97XD	B2PLYP	Exp.
1	124.7	98.9	116.2	103.9	120.3	116.8	103.4	108.8 ± 0.79
2	162.8	140.8	154.8	143.3	155.9	151.2	141.4	145.1 ± 1
3	152.5	174.6	161.6	169.8	157.3	156.9	168.3	157 ± 2
4	164.2	142.5	157.0	149.3	163.6	160.4	153.0	162.2 ± 0.59
5	192.3	167.5	183.0	169.7	183.5	174.4	166.0	165.2 ± 0.88
6	178.6	190.6	183.1	189.1	182.5	186.5	190.7	201 ± 2
7	195.0	245.7	213.0	236.0	205.6	218.3	236.3	217 ± 0.8
8	230.2	239.7	231.6	239.1	231.6	235.8	241.4	244 ± 1
9	246.9	257.2	249.2	256.4	249.0	251.2	258.2	
10	454.9	444.1	442.9	441.0	438.0	433.1	428.4	
11	505.4	495.1	483.3	492.4	475.7	486.5	467.4	
MAD	15.7	12.1	9.9	9.1	10.6	6.1	7.8	
Slope	0.67	1.10	0.81	1.03	0.76	0.87	1.06	
R	0.917	0.952	0.970	0.969	0.970	0.987	0.977	

tendency that methods high on the Jacob's ladder classification perform better than those at the lower ranks.

The compounds **1** to **8** may be considered as a calibration set of data for evaluating how much faith that should be put on the predicted values for structures **9** to **11**. Structure **9** is very similar to the structures **1** to **8** and should be unproblematic, and indeed all the wave function and DFT methods agree to within a few kJ/mol. Structures **10** and **11**, however, are hypovalence carbenes and **11** has a different spin state than **10**. If we accept the basis set extrapolated CCSD(T) result as a reference, then it is clear that HF overestimates the stability of **10** and **11**, while MP2 underestimates the stability. CISD and CCSD have a smaller systematic error but predict the carbenes to be too stable. The DFT methods in general do not systematically over- or underestimate the carbene stabilities.

The singlet-triplet energy difference between structures **10** and **11** is a more difficult property to predict, since **11** has one fewer electron pair than **10** and the two structures therefore have significantly different amounts of electron correlation. Taking again the CCSD(T) value of 62 kJ/mol as the reference, it is seen (as expected) that the HF method severely underestimates this value. The CISD method also predicts the triplet state to be too stable, while MP2 and CCSD are in good agreement with the reference value. All of the DFT methods underestimate the singlet–triplet energy difference by 10–20 kJ/mol.

References

1 A. R. Hoy and P. R. Bunker, *Journal of Molecular Spectroscopy* **74** (1), 1–8 (1979).
2 A. G. Csaszar, G. Czako, T. Furtenbacher, J. Tennyson, V. Szalay, S. V. Shirin, N. F. Zobov and O. L. Polyansky, *Journal of Chemical Physics* **122** (21), 214305 (2005).
3 A. Halkier, P. Jorgensen, J. Gauss and T. Helgaker, *Chemical Physics Letters* **274** (1–3), 235–241 (1997).
4 G. K. L. Chan and M. Head-Gordon, *Journal of Chemical Physics* **118** (19), 8551–8554 (2003).
5 A. Luchow, J. B. Anderson and D. Feller, *Journal of Chemical Physics* **106** (18), 7706–7708 (1997).

6 J. Olsen, F. Jorgensen, H. Koch, A. Balkova and R. J. Bartlett, *Journal of Chemical Physics* **104** (20), 8007–8015 (1996).
7 T. Helgaker, W. Klopper, H. Koch and J. Noga, *Journal of Chemical Physics* **106** (23), 9639–9646 (1997).
8 A. K. Wilson and T. H. Dunning, *Journal of Chemical Physics* **106** (21), 8718–8726 (1997).
9 P. W. Ayers and S. Liu, *Physical Review A* **75** (2), 022514 (2007).
10 S. A. Clough, Y. Beers, G. P. Klein and L. S. Rothman, *Journal of Chemical Physics* **59** (5), 2254–2259 (1973).
11 N. Gailar and E. K. Plyler, *Journal of Chemical Physics* **24** (6), 1139–1165 (1956).
12 A. P. Scott and L. Radom, *Journal of Physical Chemistry* **100** (41), 16502–16513 (1996).
13 B. Ruscic, A. F. Wagner, L. B. Harding, R. L. Asher, D. Feller, D. A. Dixon, K. A. Peterson, Y. Song, X. M. Qian, C. Y. Ng, J. B. Liu and W. W. Chen, *Journal of Physical Chemistry A* **106** (11), 2727–2747 (2002).
14 J. A. Pople, P. M. W. Gill and N. C. Handy, *International Journal of Quantum Chemistry* **56** (4), 303–305 (1995).
15 J. M. Wittbrodt and H. B. Schlegel, *Journal of Chemical Physics* **105** (15), 6574–6577 (1996).
16 P. Jensen, *Journal of Molecular Spectroscopy* **133** (2), 438–460 (1989).
17 R. H. Jackson, *Journal of the Chemical Society*, 4585 (1962).
18 E. Kraka, Y. He and D. Cremer, *Journal of Physical Chemistry A* **105** (13), 3269–3276 (2001).
19 D. Feller, K. A. Peterson and D. A. Dixon, *Journal of Physical Chemistry A* **114** (1), 613–623 (2010).
20 J. S. Muenter, *Journal of Molecular Spectroscopy* **55** (1–3), 490–491 (1975).
21 K. A. Peterson and T. H. Dunning, *Theochem – Journal of Molecular Structure* **400**, 93–117 (1997).
22 G. E. Scuseria, M. D. Miller, F. Jensen and J. Geertsen, *Journal of Chemical Physics* **94** (10), 6660–6663 (1991).
23 L. A. Barnes, B. W. Liu and R. Lindh, *Journal of Chemical Physics* **98** (5), 3972–3977 (1993).
24 A. Barbe, C. Secroun and P. Jouve, *Journal of Molecular Spectroscopy* **49** (2), 171–182 (1974).
25 S. A. Kucharski and R. J. Bartlett, *Journal of Chemical Physics* **110** (17), 8233–8235 (1999).
26 I. Ljubic and A. Sabljic, *Chemical Physics Letters* **385** (3–4), 214–219 (2004).
27 S. W. Benson, Fr. Cruicksh, D. M. Golden, G. R. Haugen, H. E. Oneal, A. S. Rodgers, R. Shaw and R. Walsh, *Chemical Reviews* **69** (3), 279 (1969).

13

Optimization Techniques

Many problems in computational chemistry can be formulated as an optimization of a multidimensional function (see Figure 13.1).[1–3] Optimization is a general term for finding *stationary* points of a function, that is points where the first derivative is zero. The desired stationary point is in the majority of cases a *minimum* where all the second derivatives are positive. In some cases, the desired point is a first-order *saddle point*, that is the second derivative is negative in one direction and positive in all other directions. In a few special cases, a higher-order saddle point is desired.

Most optimization methods determine the *nearest* stationary point, but a multidimensional function may contain many (in some cases very many!) different stationary points of the same type. The minimum with the lowest value is called the *global* minimum, while all the others are *local* minima. Some examples:

- The energy as a function of nuclear coordinates. Both minima and first-order saddle points (transition structures) are of interest. The energy function may be of the force field type or from solving the electronic Schrödinger equation.
- An error function depending on parameters. Only minima are of interest, and the global minimum is usually (but not always) desired. This may, for example, be determination of parameters in a force field, a set of atomic charges or a set of localized molecular orbitals.
- The energy of a wave function containing variational parameters, such as a Hartree–Fock (one Slater determinant) or multiconfigurational (many Slater determinants) wave function. Parameters are typically the molecular orbital and configurational state coefficients, but may also be, for example, basis function exponents. Usually only minima are desired, although in some cases saddle points may also be of interest (excited states).

When the parameters enter the function to be optimized in a *quadratic* fashion, the stationary point(s) can be obtained by solving a set of *linear* equations by standard matrix techniques. In most cases, however, the parameters enter the function in a *non-linear* fashion, which require *iterative* methods for locating the stationary points. The latter can be further divided into methods for locating minima and methods for locating saddle points. The problem of optimizing quadratic functions

Introduction to Computational Chemistry, Third Edition. Frank Jensen.

Companion Website: http://www.wiley.com/go/jensen/computationalchemistry3

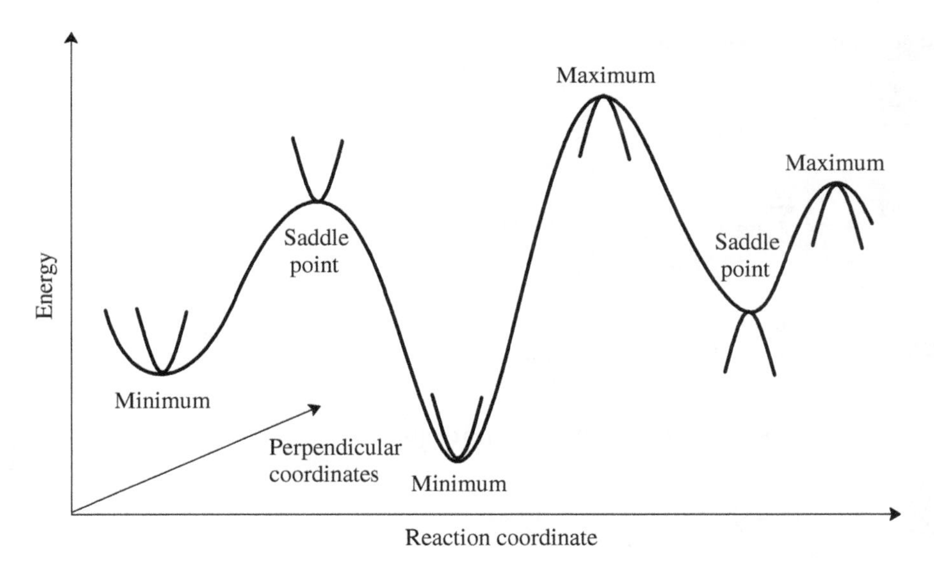

Figure 13.1 Illustrating a multidimensional energy surface.

is treated in Section 13.1, minima and saddle points of general functions in Sections 13.2 and 13.4, respectively, while methods for global optimization are covered in Section 13.6. Optimization of functions subject to external constraints is dealt with in Section 13.5. Finally, Section 13.8 describes methods for following reactions paths, which may be considered either as solving partial differential equations or as a constrained optimization problem.

13.1 Optimizing Quadratic Functions

Data fitting is a typical example of an optimization problem where the parameters enter the function in a quadratic fashion. Consider, for example, the problem of determining a set of force field partial charges Q_a by minimizing an error function measuring the fit to the electrostatic potential sampled at a number of points surrounding the molecule (Sections 2.2.6 and 10.2):

$$ErrF(\mathbf{Q}) = N_{\text{points}}^{-1} \sum_{j}^{N_{\text{points}}} \left(\phi_{\text{ESP}}(\mathbf{r}_j) - \sum_{a}^{N_{\text{atoms}}} \frac{Q_a(\mathbf{R}_a)}{|\mathbf{r}_j - \mathbf{R}_a|} \right)^2 \tag{13.1}$$

We will at present ignore the constraint that the sum of the charges must be equal to the total molecular charge; this can be dealt with by the techniques in Section 13.5. The *ErrF* in Equation (13.1) can be generalized as in the following equation, with b_j being the reference values, x_i the fitting parameters (Q_i) and a_{ij} the coefficients corresponding to the inverse distances $|\mathbf{R}_i - \mathbf{r}_j|^{-1}$:

$$ErrF(\mathbf{x}) = \sum_{j}^{M} \left(b_j - \sum_{i}^{N} x_i a_{ij} \right)^2 \tag{13.2}$$

The best set of parameters is determined by setting all the first derivatives of *ErrF* to zero:

$$\begin{aligned}
\frac{\partial ErrF}{\partial x_1} &= -2\sum_j^M a_{1j}\left(b_j - \sum_i^N x_i a_{ij}\right) = 0 \\
\frac{\partial ErrF}{\partial x_2} &= -2\sum_j^M a_{2j}\left(b_j - \sum_i^N x_i a_{ij}\right) = 0 \\
&\vdots \\
\frac{\partial ErrF}{\partial x_N} &= -2\sum_j^M a_{Nj}\left(b_j - \sum_i^N x_i a_{ij}\right) = 0
\end{aligned} \tag{13.3}$$

By rearrangement, this gives a set of coupled linear equations:

$$\begin{aligned}
\sum_i^N x_i \sum_j^M a_{1j}a_{ij} &= \sum_j^M a_{1j}b_j \\
\sum_i^N x_i \sum_j^M a_{2j}a_{ij} &= \sum_j^M a_{2j}b_j \\
&\vdots \\
\sum_i^N x_i \sum_j^M a_{Nj}a_{ij} &= \sum_j^M a_{Nj}b_j
\end{aligned} \tag{13.4}$$

These N equations with N unknowns can be written in a matrix–vector notation:

$$\mathbf{Ax} = \mathbf{b} \tag{13.5}$$

The formal solution is obtained by multiplying with the inverse **A** matrix:

$$\mathbf{x} = \mathbf{A}^{-1}\mathbf{b} \tag{13.6}$$

In actual applications, the **A** matrix may be singular, or nearly so, and the inverse matrix either does not exist or is prone to numerical errors. A singular matrix indicates that at least one of the linear equations can be written as a combination of the other equations, and such cases can be handled by singular value decomposition methods, as discussed in Section 17.6.3. Indeed, for the example of determining partial charges by fitting to the electrostatic potential, the equations determining the charges on the atoms far from the molecular surface are often poorly conditioned, that is the external electrostatic potential is only weakly dependent on the charges on the buried atoms.

Solving linear equations by matrix inversion as in Equation (13.6) (or by, for example, Gaussian elimination) has a computational requirement that scales as the cube of the problem size, which is problematic if the number of variables is large. The problem can in such cases be reformulated as a minimization of a residual **r** calculated from a given trial solution $\mathbf{x}'$:

$$\mathbf{r} = \mathbf{b} - \mathbf{Ax}' \tag{13.7}$$

$$\frac{\partial(\mathbf{r}^t\mathbf{r})}{\partial \mathbf{x}'} = 0 \quad \Leftrightarrow \quad \mathbf{Ax}' = \mathbf{b} \tag{13.8}$$

Minimization of the residual norm can be done, for example, by conjugate gradient methods described in the next section, where $\mathbf{x}'$ is iteratively converged to the solution $\mathbf{x}$, and this will for a large number of variables have a lower computational cost.

Other examples of optimizing functions that depend quadratically of the parameters include the energy of Hartree–Fock (HF) and configuration interaction (CI) wave functions. Minimization of the energy with respect to the MO or CI coefficients leads to a set of linear equations. In the HF case, the a_{ij} coefficients depend on the parameters x_i and must therefore be solved iteratively. In the CI case, the number of parameters is typically 10^6–10^9, which prohibits a direct solution of the linear equations, and iterative methods are used instead (Section 17.2.5). The use of iterative techniques for solving the CI equations is not due to the mathematical nature of the problem, but due to computational efficiency considerations.

13.2 Optimizing General Functions: Finding Minima

The simple-minded approach for minimizing a function is to step one variable at a time until the function has reached a minimum and then switch to another variable. This requires only the ability to calculate the function value for a given set of variables. As the variables are not independent, however, several cycles through the whole set are necessary for finding a minimum. This is impractical for more than five to ten variables, and may not work anyway.

The *Simplex* method represents a more efficient approach using only function values for constructing an irregular polyhedron in parameters space and moving this polyhedron towards the minimum, while allowing the size to contract or expand to improve the convergence.[4] It is better than the simple-minded "one-variable-at-a-time" approach, but becomes too slow for many-dimensional functions.

Since optimization problems in computational chemistry tend to have many variables, essentially all commonly used methods assume that at least the first derivative of the function with respect to all variables, the gradient $\mathbf{g}$, can be calculated analytically (i.e. directly, and not as a numerical differentiation by stepping the variables). Some methods also assume that the second derivative matrix, the Hessian $\mathbf{H}$, can be calculated.

It should be noted that the target function and its derivative(s) are calculated with a finite precision, which depends on the computational implementation. A stationary point can therefore not be located exactly, the gradient can only be reduced to a certain value. Below this value, the numerical inaccuracies due to the finite precision will swamp the "true" functional behavior. An optimization is in practise considered converged if the gradient is reduced below a suitable threshold value or if the function change between two iterations becomes sufficiently small. Both these criteria may in some cases lead to problems, as a function with a very flat surface in a certain region may meet the criteria without containing a stationary point.

There are three classes of commonly used optimization methods for finding minima, each having their advantages and disadvantages.

13.2.1 Steepest Descent

The gradient vector $\mathbf{g}$ points in the direction where the function increases most, that is the function value can always be lowered by stepping in the opposite direction. In the *Steepest Descent* (SD) method, a series of function evaluations are performed in the negative gradient direction, that is along a search direction defined as $\mathbf{d} = -\mathbf{g}$. Once the function starts to increase, an approximate minimum

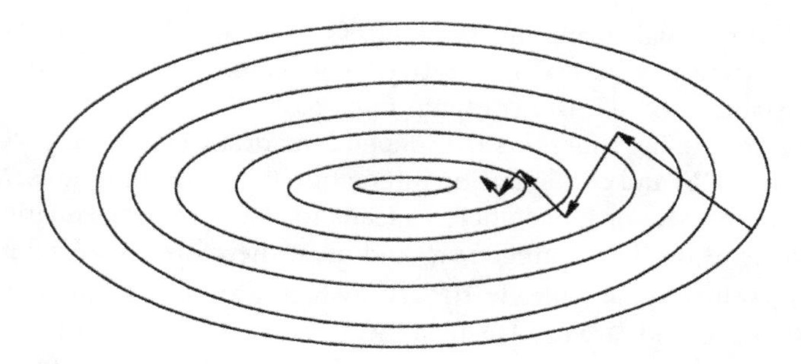

Figure 13.2 Steepest descent minimization.

may be determined by interpolation between the calculated points. At this interpolated point, a new gradient is calculated and used for the next line search.

The steepest descent algorithm is sure-fire. If the line minimization is carried out sufficiently accurately, it will *always* lower the function value and is therefore *guaranteed* to approach a minimum. It has, however, two main problems. Two subsequent line searches are necessarily perpendicular to each other; if there was a gradient component along the previous search direction, the energy could be further lowered in this direction. The steepest descent algorithm therefore has a tendency for each line search to partly spoil the function lowering obtained by the previous search. The steepest descent path oscillates around the minimum path, as illustrated in Figure 13.2, and this is particularly problematic for surfaces having long narrow valleys.

Furthermore, as the minimum is approached, the rate of convergence slows down. The steepest descent will actually never reach the minimum; it will crawl toward it at an ever decreasing speed.

An accurate line search requires several function evaluations along each search direction. Often the minimization along the line is only carried out fairly crudely or a single step is simply taken along the negative gradient direction. In the latter case, the step size is varied dynamically during the optimization; if the previous step reduced the function value, the next step is taken with a slightly longer step size, but if the function values increased, the step size is reduced. Without an accurate line search, the guarantee for lowering the function value is lost and the optimization may potentially end up in an oscillatory state.

By its nature, the steepest descent method can only locate function minima. The advantage is that the algorithm is very simple and requires only storage of a gradient vector. It is furthermore one of the few methods that is guaranteed to lower the function value. Its main use is to quickly relax a poor starting point, before some of the more advanced algorithms take over, or as a "backup" algorithm if the more sophisticated methods are unable to lower the function value.

13.2.2 Conjugate Gradient Methods

The main problem with the steepest descent method is the partial "undoing" of the previous step. The *Conjugate Gradient* (CG) method tries to improve on this by performing each line search not along the current gradient but along a line that is constructed such that it is "conjugate" to the previous search direction(s). If the surface is purely quadratic, the conjugate direction criterion guarantees that each successive minimization will not generate gradient components along any of the previous directions and the minimum is reached after at most N_{var} steps. The first step is equivalent to a steepest descent step, but subsequent searches are performed along a line formed as a mixture of the current negative gradient and the previous search direction:

$$\mathbf{d}_i = -\mathbf{g}_i + \beta_i \mathbf{d}_{i-1} \tag{13.9}$$

The word conjugate denotes that search directions obey the orthogonality condition $\mathbf{d}_i^t \mathbf{A} \mathbf{d}_j = 0$ for $i \neq j$ with a suitable $\mathbf{A}$ matrix (second derivative matrix for a quadratic surface). There are several ways of choosing the β value. Some of the names associated with these methods are *Fletcher–Reeves* (FR), *Polak–Ribiere* (PR) and *Hestenes–Stiefel* (HS). Their definitions of β are given in the following equations:

$$\beta_i^{\mathrm{FR}} = \frac{\mathbf{g}_i^{\mathrm{t}} \mathbf{g}_i}{\mathbf{g}_{i-1}^{\mathrm{t}} \mathbf{g}_{i-1}} \tag{13.10}$$

$$\beta_i^{\mathrm{PR}} = \frac{\mathbf{g}_i^{\mathrm{t}} (\mathbf{g}_i - \mathbf{g}_{i-1})}{\mathbf{g}_{i-1}^{\mathrm{t}} \mathbf{g}_{i-1}} \tag{13.11}$$

$$\beta_i^{\mathrm{HS}} = \frac{\mathbf{g}_i^{\mathrm{t}} (\mathbf{g}_i - \mathbf{g}_{i-1})}{\mathbf{d}_{i-1}^{\mathrm{t}} (\mathbf{g}_i - \mathbf{g}_{i-1})} \tag{13.12}$$

For non-quadratic surfaces, the conjugate property does not hold rigorously and, for real problems, the CG algorithm must often be restarted (i.e. setting $\beta = 0$) during the optimization. The PR formula for β has a tendency to restart the procedure more gracefully than the other two, and is usually preferred in practice. The conjugate property holds best for near-quadratic surfaces, and the convergence properties of CG methods can be improved by scaling the variables by a suitable pre-conditioner matrix, for example containing (approximate) inverse second derivatives, such that the curvatures in different directions are of similar magnitude. Conjugate gradient methods have much better convergence characteristics than the steepest descent, but they are again only able to locate minima. They require slightly more storage than the steepest descent, since two (current gradient and previous search direction) vectors must be stored, but this is rarely a problem.

13.2.3 Newton–Raphson Methods

The *Newton–Raphson* (NR) method expands the true function to second order around the current point $\mathbf{x}_0$:

$$f(\mathbf{x}) \approx f(\mathbf{x}_0) + \mathbf{g}^{\mathrm{t}}(\mathbf{x} - \mathbf{x}_0) + \tfrac{1}{2}(\mathbf{x} - \mathbf{x}_0)^{\mathrm{t}} \mathbf{H} (\mathbf{x} - \mathbf{x}_0) \tag{13.13}$$

Requiring the gradient of the second-order approximation in Equation (13.13) to be zero produces the step

$$(\mathbf{x} - \mathbf{x}_0) = -\mathbf{H}^{-1} \mathbf{g} \tag{13.14}$$

In the coordinate system $(\mathbf{x}')$ where the Hessian is diagonal (i.e. performing a unitary transformation, see Section 17.2), the NR step may be written as

$$\begin{aligned} \Delta \mathbf{x}' &= \left(\Delta x_1', \Delta x_2', \Delta x_3', \ldots, \Delta x_N'\right)^{\mathrm{t}} \\ \Delta x_i' &= -\frac{g'_i}{\varepsilon_i} \end{aligned} \tag{13.15}$$

Here g'_i is the projection of the gradient along the Hessian eigenvector with eigenvalue ε_i (the gradient component pointing in the direction of the ith eigenvector).

As the real function contains terms beyond second order, the NR formula can be used iteratively for stepping towards a stationary point. Near a minimum, all the Hessian eigenvalues are positive (by definition) and the step direction is opposite to the gradient direction, as it should be. If, however, one of the Hessian eigenvalues is negative, the step in this direction will be *along* the gradient component and thus *increase* the function value. In this case, the optimization may end up at a stationary point with one negative Hessian eigenvalue, a first-order saddle point. The NR method thus attempts to converge on the "nearest" stationary point, regardless of whether this is a minimum, saddle point or maximum.

Another problem is the use of the inverse Hessian for determining the step size. If one of the Hessian eigenvalues becomes close to zero, the step size goes toward infinity (except if the corresponding gradient component g'_i is exactly zero). The NR step is thus without bound, and it may take the variables far outside the region where the second-order Taylor expansion is valid. The latter region is often described by a "*Trust Radius*". In some cases, the NR step is taken as a search direction along which the function is minimized, analogously with the steepest descent and conjugate gradient methods. The augmented Hessian methods described below are normally more efficient.

The advantage of the NR method is that the convergence is second order near a stationary point. If the function only contains terms up to second order, the NR step will go to the stationary point in a single step. The function normally contains higher-order terms, but the second-order approximation becomes better and better as the stationary point is approached. Sufficiently close to the stationary point, the gradient is reduced quadratically, that is if the gradient norm is reduced by a factor of 2 between two iterations, it will go down by a factor of 4 in the next iteration and a factor of 16 in the next. The quadratic convergence, however, is often only observed very close to the stationary point and the NR method typically only displays linear convergence, that is the gradient norm is reduced by a roughly constant factor for each iteration.

Besides the above-mentioned problems with step control, there are also other computational aspects that tend to make the straightforward NR problematic for many problem types. The true NR method requires calculation of the full second derivative matrix, which must be stored and inverted (diagonalized). For some types of function, a calculation of the Hessian is computationally demanding. For others, the number of variables is so large that manipulating a matrix the size of the number of variables squared is impossible. The following two sections address some solutions to these problems.

13.2.4 Augmented Hessian Methods

There are two aspects in step control, one is controlling the total *length* of the step, such that it does not exceed the region in which the second-order Taylor expansion is valid, and the second is making sure that the step *direction* is correct. If the optimization is towards a minimum, the Hessian should have all positive eigenvalues in order for the step to be in the correct direction. If, however, the starting point is in a region where the Hessian has negative eigenvalues, the NR step is toward a saddle point or maximum. Both of these problems can be solved by introducing a shift parameter λ (compare to Equations (13.14) and (13.15)):

$$(\mathbf{x} - \mathbf{x}_0) = -(\mathbf{H} - \lambda\mathbf{I})^{-1}\mathbf{g}$$

$$\Delta x'_i = -\frac{g'_i}{\varepsilon_i - \lambda} \tag{13.16}$$

If λ is chosen to be below the lowest Hessian eigenvalue, the denominator is always positive, and the step direction will thus be correct. Furthermore, if λ goes towards $-\infty$, the step size goes towards zero, that is the step size can be made arbitrarily small. Methods that modify the nature of the Hessian matrix by a shift parameter are known by names such as "*augmented Hessian* ", "*level-shifted Newton–Raphson*", "*norm-extended Hessian*" or "*Eigenvector Following*" (EF), depending on how λ is chosen. We will here mention two popular methods for choosing λ.

The *Rational Function Optimization* (RFO) expands the function in terms of a rational approximation instead of a straight second-order Taylor series (Equation (13.13)):[5]

$$f(\mathbf{x}) \approx \frac{f(\mathbf{x}_0) + \mathbf{g}^{\mathrm{t}}(\mathbf{x}-\mathbf{x}_0) + \frac{1}{2}(\mathbf{x}-\mathbf{x}_0)^{\mathrm{t}}\mathbf{H}(\mathbf{x}-\mathbf{x}_0)}{1 + \frac{1}{2}(\mathbf{x}-\mathbf{x}_0)^{\mathrm{t}}\mathbf{S}(\mathbf{x}-\mathbf{x}_0)} \tag{13.17}$$

The $\mathbf{S}$ matrix is eventually set equal to a unit matrix, which leads to the following equation for λ:

$$\sum_i \frac{g_i'^2}{\varepsilon_i - \lambda} = \lambda \tag{13.18}$$

This is a one-dimensional equation in λ which can be solved by standard (iterative) methods. There will in general be one more solution than the number of degrees of freedom, but by choosing the lowest λ solution, it is ensured that the resulting step will be toward a minimum. The RFO step calculated from Equation (13.16) will always be shorter than the pure NR step (Equation (13.15)), but there is no guarantee that it will be within the trust radius. If the RFO step is too long, it may be scaled down by a simple multiplicative factor, but if the factor is much smaller than 1, it follows that the resulting step may not be the optimum for the given trust radius.

Another way of choosing λ is to require that the step length be equal to the trust radius R, which is in essence the best step on a hypersphere with radius R. This is known as the *Quadratic Approximation* (QA) method:[6]

$$|\Delta\mathbf{x}'|^2 = \sum_i \left(\frac{g_i'}{\varepsilon_i - \lambda}\right)^2 = R^2 \tag{13.19}$$

This may again have multiple solutions, but by choosing the lowest λ value, the minimization step is selected. The maximum step size R may be taken as a fixed value, or allowed to change dynamically during the optimization. If the actual energy change between two steps agrees well with that predicted from the second-order Taylor expansion, the trust radius for the next step may be increased, and vice versa.

13.2.5 Hessian Update Methods

The second problem, the computational aspect of calculating the Hessian, is often encountered in electronic structure calculations. Here the calculation of the second derivative matrix can be an order of magnitude more demanding than calculating the gradient. In such cases, an *updating* scheme may be used instead. The idea is to start off with an approximation to the Hessian, maybe just a unit matrix. The initial step will thus resemble a steepest descent step. As the optimization proceeds, the gradients at the previous and current points are used for making the Hessian a better approximation for the actual system. After two steps, the updated Hessian is a rather good approximation to the exact Hessian in the direction defined by these two points (but not in the other directions). There are many such updating schemes, some of the commonly used ones are associated with the names *Davidon–Fletcher–Powell* (DFP), *Broyden–Fletcher–Goldfarb–Shanno* (BFGS) and *Powell.* For minimizations, the BFGS update shown below is usually preferred, as it tends to keep the Hessian positive definite:

$$\mathbf{H}_n = \mathbf{H}_{n-1} + \Delta\mathbf{H}$$
$$\Delta\mathbf{H}_{\mathrm{BFGS}} = \frac{\Delta\mathbf{g}\Delta\mathbf{g}^{\mathrm{t}}}{\Delta\mathbf{g}^{\mathrm{t}}\Delta\mathbf{x}} - \frac{\mathbf{H}\Delta\mathbf{x}\Delta\mathbf{x}^{\mathrm{t}}\mathbf{H}}{\Delta\mathbf{x}^{\mathrm{t}}\mathbf{H}\Delta\mathbf{x}} \tag{13.20}$$

For saddle point searches, the updating must allow the Hessian to develop negative eigenvalues, and the Powell or updates based on combining several methods are usually employed.[7]

The use of approximate Hessians within the NR method is known as *quasi-Newton–Raphson, pseudo-Newton–Raphson* or *variable metric* methods. It is clear that they do not converge as fast as true NR methods, where the exact Hessian is calculated in each step, but if, for example, five steps can be taken for the same computational cost as one true NR step, the overall computational effort may be less. True NR methods converge quadratically near a stationary point, while pseudo-NR methods display a linear convergence. Far from a stationary point, however, the true NR method will typically also only display linear convergence.

The Hessian updating scheme in Equation (13.20) uses only gradient information from the current and previous points to sequentially build a Hessian matrix containing information from all previous points, and this is subsequently used for predicting the next step (Equation (13.16)). Intuitively it may seem that the construction and handling of the Hessian matrix is an unnecessary complication and that the step can be formulated entirely in terms of previous points and associated gradients. This is indeed possible and is known as *limited memory BFGS* (L-BFGS),[8] where the "limited memory" refers to the fact that storage of an N_{var}^2 matrix (Hessian) is replaced by storing a number of vectors (variables and gradients) of (only) dimension N_{var}. As the optimization progresses, the information from early points may be of low value for characterizing the surface curvature at the actual point, and this implies that it may be sufficient to store information from only a fixed number of previous points, which further reduces the storage requirement. The number of previous vectors kept is a free variable and can be considered as a parameter for continously switching from a quasi-NR BFGS (keep all points) to a conjugate gradient-like (keep only the previous point) optimization algorithm. L-BFGS methods are often a good choice for optimizing functions containing a large number of variables.

Quasi-NR methods are usually the best choice in geometry optimizations using an energy function calculated by electronic structure methods. The quality of the initial Hessian of course affects the convergence when an updating scheme is used. The use of an exact Hessian at the first point often gives a good convergence, but this may not be the most cost-efficient strategy. A quite reasonable Hessian for a minimum search may in many cases be generated by simple rules connecting, for example, bond lengths and force constants.[9] Alternatively, the initial Hessian may be taken from a calculation at a lower level of theory. As an initial exploration of an energy surface is often carried out at a low level of theory, followed by frequency calculations to establishing the nature of the stationary points, the resulting force constants can be used for starting an optimization at higher levels. This is especially useful for transition structure searches that require a quite accurate Hessian. The success of this strategy relies on the fact that the qualitative structure of an energy surface is often fairly insensitive to the level of theory, although there certainly are many examples where this is not the case.

13.2.6 Truncated Hessian Methods

A potential problem of all NR-based methods is the storage and handling of the Hessian matrix. For methods where the calculation of the Hessian is easy but the number of variables is large, this may be a problem. A prime example here is geometry optimization using a force field energy function. The computational effort for calculating the Hessian goes up roughly as the *square* of the number of atoms. Diagonalization of the Hessian matrix required for the NR optimization, however, depends on the cube of the matrix size, that is it goes up as the *cube* of the number of atoms. Since matrix diagonalization becomes a significant factor for dimensions ~1000, it is clear that NR methods should not be used for force field optimizations beyond a few hundred atoms. For large systems the computational effort for predicting the geometry step will completely overwhelm the calculation of the energy, gradient and Hessian. The conjugate gradient method avoids handling of the Hessian and only requires storage of two vectors, and it is therefore usually the method of choice for force field optimizations.

For large molecular systems many of the off-diagonal elements in the Hessian are very small, essentially zero (the coupling between distant atoms is very small), and the Hessian for large systems is therefore a *sparse matrix*. NR methods that take advantage of this fact by neglecting off-diagonal blocks are denoted *truncated* NR. Some force field programs use an extreme example of this where only the 3×3 submatrices along the diagonal are retained. These 3×3 matrices contain the coupling elements between the x, y and z coordinates for a single atom. The task of inverting, say, a 3000×3000 matrix is thus replaced by inverting 1000 3×3 matrices, reducing the computational cost for the diagonalization by a factor of 10^6. Such truncated NR methods are often used in connection with an explicit search along the NR step direction.

13.2.7 Extrapolation: The DIIS Method

During an iterative sequence for solving a set of (linear or non-linear) equations, information is generated at a number of points in the parameter space. The idea in extrapolation methods is to use this information for inter- or extrapolation in order to arrive at the desired solution faster (fewer iterations). The best known of these methods is the *Direct Inversion in the Iterative Subspace* (DIIS) proposed by P. Pulay.[10] Assume that the iterative procedure has generated n points in the parameter space $\mathbf{x}_i$ and that an error estimate $\mathbf{e}_i$ is available for measuring how far each point is from the desired solution. Instead of letting the algorithm proceed to point $\mathbf{x}_{n+1}$ from point $\mathbf{x}_n$, the step is instead generated from an interpolated point $\mathbf{x}_n^*$ formed as a linear combination of all previous points:

$$\mathbf{x}_n^* = \sum_i^n c_i \mathbf{x}_i \tag{13.21}$$

The error at the interpolated point is given by

$$\mathbf{e}_n^* = \sum_i^n c_i \mathbf{e}_i \tag{13.22}$$

The coefficients c_i are determined by minimizing the norm of the interpolated error, subject to a normalization condition:

$$ErrF(\mathbf{c}) = (\mathbf{e}_n^* \cdot \mathbf{e}_n^*) \quad ; \quad \sum_i^n c_i = 1 \tag{13.23}$$

The constrained minimization is handled by the Lagrange method (Section 13.5) and leads to the following set of linear equations, where λ is the multiplier associated with the normalization:

$$\begin{pmatrix} a_{11} & a_{12} & \cdots & a_{1n} & -1 \\ a_{21} & a_{22} & \cdots & a_{2n} & -1 \\ \vdots & \vdots & \ddots & \vdots & \vdots \\ a_{n1} & a_{n2} & \cdots & a_{nn} & -1 \\ -1 & -1 & \cdots & -1 & 0 \end{pmatrix} \begin{pmatrix} c_1 \\ c_2 \\ \vdots \\ c_n \\ -\lambda \end{pmatrix} = \begin{pmatrix} 0 \\ 0 \\ \vdots \\ 0 \\ -1 \end{pmatrix} \tag{13.24}$$

$$a_{ij} = (\mathbf{e}_i \cdot \mathbf{e}_j)$$

The set of linear equations can be written in a matrix notation:

$$\mathbf{Ac} = \mathbf{b} \tag{13.25}$$

The **A** matrix in iteration n has dimension $(n+1) \times (n+1)$, where n usually is less than 50. The coefficients **c** can thus be obtained by directly inverting the **A** matrix and multiplying it onto the **b** vector, that is in the "subspace" of the "iterations" the linear equations are solved by "direct inversion", thus the name DIIS:

$$\mathbf{c} = \mathbf{A}^{-1}\mathbf{b} \tag{13.26}$$

If the **A** matrix is close to singular, Equation (13.26) can instead be solved by singular value decomposition methods (Section 17.6.3) or by discarding some of the previous points. The only remaining question is the choice of the error estimate **e**. DIIS was originally proposed for accelerating and stabilizing the SCF convergence (Section 3.8.1) where the **x** parameters are the Fock matrices **F** and the error **e** is taken as the **FDS** − **SDF** difference (**S** and **D** are the overlap and density matrices, respectively), which is related to the gradient of the SCF energy with respect to the MO coefficients; this has been found to work well in practice. A closely related method uses the energy as the error indicator and has the acronym EDIIS.[11]

DIIS has also been used for geometry optimizations in connection with Newton–Raphson methods, and the best known of these is the *Geometry Direct Inversion in the Iterative Subspace* (GDIIS).[12] The **x** parameters are in this case the atomic coordinates, and the NR step in GDIIS is not taken from the last geometry but from the interpolated point $\mathbf{x}_n^*$ with a corresponding interpolated gradient:

$$\mathbf{g}_n^* = \sum_i^n c_i \mathbf{g}_i \tag{13.27}$$

There are two common choices for the error vector, either a "geometry" or "gradient" vector, with the latter usually being preferred:[13]

$$\mathbf{e}_i = \mathbf{H}_n^{-1}\mathbf{g}_i \quad \text{or} \quad \mathbf{e}_i = \mathbf{g}_i \tag{13.28}$$

DIIS requires storage of information (**x**, **e** and perhaps **g**) from all previous iterations, which may be problematic if the number of variables is large, as, for example, when using the method for solving non-linear coupled cluster equations (Section 4.9). It has been shown, however, that by a minor change in the DIIS algorithm, only storage of three sets of information is required, and this is called the *Conjugate Residual with OPtimal trial vectors* (CROP) method.[14]

The DIIS approach attempts to find a low-error point within the subspace already searched. For optimizing electronic wave functions, there is usually only one minimum and the DIIS extrapolation significantly improves both the convergence rate and stability. For geometry optimizations, however, the target function is usually complicated and contains many minima and saddle points, making DIIS extrapolations much less useful or even disadvantageous. It is not uncommon for an optimization to move across a flat part of the surface before entering the local minimum region. This will result in the gradient being small for several steps and then increase as the minimum is approached. The DIIS procedure will in such cases attempt to pull the structure back to the flat energy region, since this is where the gradient is small, and DIIS will in such cases be counterproductive.

13.3 Choice of Coordinates

Naively one may think that any set of coordinates that uniquely describes the function is equally good for optimization. This is not the case! A "good" set of coordinates may transform a divergent optimization into a convergent one, or increase the rate of convergence. We will look specifically at the problem of optimizing a geometry given an energy function depending on nuclear coordinates, but the same considerations hold equally well for other types of optimization. We will furthermore use the straight Newton–Raphson formula (13.14) to illustrate the concepts.

Given the first and second derivatives, the NR formula calculates the geometry step as the inverse of the Hessian times the gradient:

$$(\mathbf{x} - \mathbf{x}_0) = -\mathbf{H}^{-1}\mathbf{g} \tag{13.29}$$

In the coordinate system ($\mathbf{x}'$) where the Hessian is diagonal, the step may be written as

$$\begin{aligned} \Delta\mathbf{x}' &= (\Delta x'_1, \Delta x'_2, \Delta x'_3, \dots, \Delta x'_N)^{\mathrm{t}} \\ \Delta x'_i &= -\frac{g'_i}{\varepsilon_i} \end{aligned} \tag{13.30}$$

Essentially all computational programs calculate the fundamental properties, the energy and derivatives, in Cartesian coordinates. The Cartesian Hessian matrix has the dimension $3N_{\text{atom}} \times 3N_{\text{atom}}$. Of these, three describe the overall translation of the molecule and three describe the overall rotation. In the molecular coordinate system, there are only $3N_{\text{atom}} - 6$ coordinates needed for uniquely describing the nuclear positions. Moving all the atoms in, say, the x-direction by the same amount does not change the energy and the corresponding gradient component (and all higher derivatives) is zero. The Hessian matrix should therefore have six eigenvalues identical to zero and the corresponding gradient components, g'_i, should also be identical zero. In actual calculations, however, these values are certainly small, but not exactly zero. Numerical inaccuracies may introduce errors of perhaps 10^{-14}–10^{-16}, and this can have rather drastic consequences. Consider, for example, a case where the gradient in the x-translation direction is calculated to be 10^{-14}, while the corresponding Hessian eigenvalue is 10^{-16}, leading to an NR step in this direction of 100! This illustrates that care should be taken if redundant coordinates (i.e. more than are necessary for uniquely describing the system) are used in the optimization. In the case of Cartesian geometry optimization, the six translational and rotational degrees of freedom can be removed by projecting these components out of the Hessian prior to formation of the NR step (Section 17.4). The calculated "steps" in the zero eigenvalue directions are then simply neglected.

Another way of removing the six translational and rotational degrees of freedom is to use a set of internal coordinates. For a simple acyclic system, these may be chosen as $N_{\text{atom}} - 1$ distances, $N_{\text{atom}} - 2$ angles and $N_{\text{atom}} - 3$ torsional angles, as illustrated in the construction of Z-matrices in Appendix D. In internal coordinates the six translational and rotational modes are automatically removed (since only $3N_{\text{atom}} - 6$ coordinates are defined), and the NR step can be formed straightforwardly. For cyclic systems, a choice of $3N_{\text{atom}} - 6$ internal variables that span the whole optimization space may be somewhat more problematic to define, especially if symmetry is present.

Diagonalization of the Hessian is an example of a linear transformation; the eigenvectors are just linear combinations of the original coordinates. A linear transformation does not change the convergence/divergence properties or the rate of convergence. We can form the NR step directly in Cartesian coordinates by inverting the Hessian and multiplying it with the gradient vector (Equation (13.29)) or we can transform the coordinates to a system where the Hessian is diagonal, form the ratios $-g_i'/\varepsilon_i$ (Equation (13.30)) and back-transform to the original system. Both methods generate the exact same NR step (except for rounding-off errors). Since we need to give consideration to the six translational and rotational modes, however, the diagonal representation is advantageous.

The transformation from a set of Cartesian coordinates to a set of internal coordinates, which may, for example, be distances, angles and torsional angles, is an example of a non-linear transformation. The internal coordinates are connected with the Cartesian coordinates by means of square root and trigonometric functions, not simple linear combinations. A non-linear transformation will affect the convergence properties. This can be illustrated by considering a minimization of a Morse-type function (Equation (2.5)) with $D = \alpha = 1$ and $x = \Delta R$:

$$E_{\text{Morse}}(x) = [1 - \mathrm{e}^{-x}]^2 \tag{13.31}$$

We will consider two other variables obtained by a non-linear transformation: $y = \mathrm{e}^{-x}$ and $z = \mathrm{e}^{-x}$. The minimum energy is at $x = 0$, corresponding to $y = z = 1$. Consider an NR optimization starting at $x = -0.50$, corresponding to $y = 1.6587$ and $z = 0.6065$. Table 13.1 shows that the NR procedure in the x-variable requires four iterations before x is less than 10^{-4}. In the y-variable the optimization only requires one step to reach the $y = 1$ minimum exactly! The optimization in the z-variable takes six iterations before the value is within 10^{-4} of the minimum.

Consider now the same system starting from $x = 0.30$ ($y = 0.7408$ and $z = 1.3499$) and $x = 1.00$ ($y = 0.3679$ and $z = 2.7183$). The first optimization step in the x-variable for the first case overshoots the minimum but then converges in three additional steps. With the z-variable the first step results in an "non-physical" negative value and subsequent steps do not recover. With the second set of starting conditions, both the x- and z-variable optimizations diverge toward the $x = \infty$ limit. In both cases the y-variable optimization converges (exactly) in one step.

The reason for this behavior is seen when plotting the three functional forms as shown in Figure 13.3.

Table 13.1 Convergence for different choices of variables and starting values.

	$x_{\text{start}} = -0.50$			$x_{\text{start}} = 0.30$			$x_{\text{start}} = 1.00$		
Iteration	x	y	z	x	y	z	x	y	z
0	−0.5000	1.6487	0.6065	0.3000	0.7408	1.3499	1.0000	0.3679	2.7183
1	−0.2176	1.0000	0.7401	−0.2381	1.0000	−0.2229	3.3922	1.0000	4.6352
2	−0.0541		0.8667	−0.0633		−0.3020	4.4283		7.3225
3	−0.0041		0.9570	−0.0055		−0.4110	5.4405		11.2981
4	0.0000		0.9951	0.0000		−0.5628	6.4449		17.2354
5			0.9999						
6			1.0000						

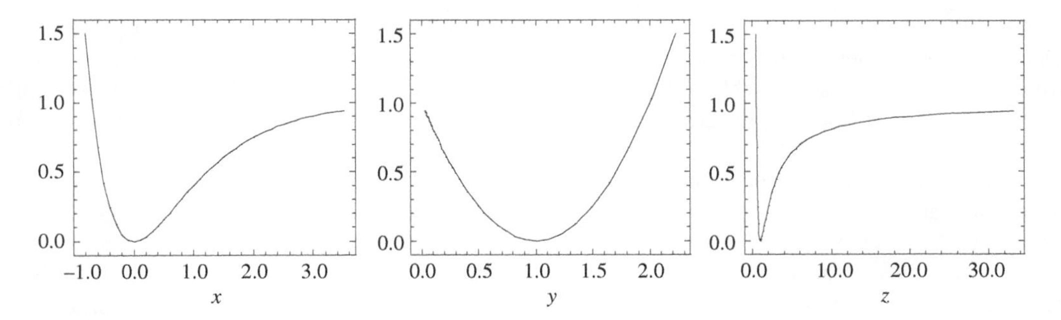

Figure 13.3 Morse curves as a function of *x*, *y* and *z*.

The horizontal axis covers the same range of x-variables for all three figures. In the x-variable space the second derivative is negative beyond $x = \ln 2$ (= 0.69), and if the optimization is started at larger x-values, the optimization is no longer a minimization, but a maximization toward the $x = \infty$ asymptote. The function in the y-variable is a parabola and the second-order expansion of the NR method is *exact*. *All* starting points consequently converge to the minimum in a single step. The transformation to the z-variable introduces a singularity at $z = 0$, and it can be seen from Figure 13.3 that the curve shape is much less quadratic than the original function. Using y as a variable is an example of a "good" non-linear transformation, while z is an example of a "poor" non-linear transformation.

These examples show that non-linear transformations may strongly affect the convergence properties of an optimization. The more "harmonic" the energy function is, the faster the convergence. One should therefore try to choose a set of coordinates where the third- and higher-order derivatives are as small as possible. Cartesian coordinates are not particularly good in this respect but have the advantage that convergence properties are fairly uniform for different systems. A "good" set of internal coordinates may speed up the convergence but a "poor" set of coordinates may slow it down or cause divergence. For acyclic systems the above-mentioned internal coordinates consisting of $N_{atom} - 1$ distances, $N_{atom} - 2$ angles and $N_{atom} - 3$ torsional angles are normally better than Cartesian coordinates. Cyclic systems, however, are notoriously difficult for choosing a good set of internal coordinates. Cyclopropane, for example, has three C–C bonds and three CCC angles, but only three independent variables (not counting the hydrogens). Choosing two distances and one angle introduces a strong coupling between the angle and distances due to the "remote" C—C bond, which is described indirectly by the other three variables. Cartesian coordinates may display better convergence characteristics in such systems.

Another problem is when very soft modes are present. A prototypical example is rotation of a methyl or hydroxy group. Near the minimum the energy changes very little as a function of the torsional angle, that is the corresponding Hessian eigenvalue is small. Consequently, even a small gradient may produce a large change in geometry. The potential is not very harmonic and the result is that the optimization may spend many iterations flopping from side to side. A similar problem is encountered in optimization of molecular clusters where the optimum structure is governed by weak van der Waals-type interactions.

The problem of an "optimum" choice of coordinates has been addressed by Pulay and coworkers, who suggested using *Natural Internal Coordinates*.[15] These are defined by first classifying the atoms into three types: "terminal" (having only one bond), "ring" (part of a ring) or "internal". All distances between bonded atoms are used as variables, using suitable distance criteria to decide which atom pairs are bonded. The ring and internal atoms are assumed to have a local symmetry depending on the number of terminal atoms attached to them, that is an internal atom with three terminal bonds has local C_{3v} symmetry, one with two terminal bonds has C_{2v}, a ring has local D_{nh} symmetry, etc. Suitable linear combinations of bending and torsional angles are then formed such that the coupling between these coordinates is exactly zero if the local symmetry is the exact symmetry. This will usually not be the case, but the local symmetry coordinates tend to minimize the coupling, and thus the magnitude of third and higher derivatives, thereby improving the NR performance. Natural internal coordinates appear to be a good choice for optimization to minima on an energy surface, since the bonding pattern is usually well defined for stable molecules. For locating transition structures, however, it is much less clear whether natural internal coordinates offer any special advantage. The bonding pattern is not as well defined for TSs, and a "good" set of coordinates at the starting geometry may become ill behaved during the optimization. For loosely bound complexes, it has been suggested that coordinates depending on the inverse distance or scaled by an inverse reference distance can improve the convergence.[16,17]

In the original formulation, a set of $3N_{atom} - 6$ independent natural internal coordinates was chosen. It was later discovered that the same optimization characteristics could be obtained by using all distances and bending and torsional angles between atoms within bonding distance as variables.[18] Such a set of coordinates will in general be redundant (i.e. the number of coordinates is larger than $3N_{atom} - 6$) and special care must be taken to handle this. One solution is to extract a set of non-redundant coordinates from a large set of (redundant) internal coordinates by selecting the eigenvectors corresponding to non-zero eigenvalues of the square of the matrix defining the transformation from Cartesian to internal coordinates. These linear combinations have been denoted *delocalized internal coordinates* and can be viewed as either a generalization of the natural internal coordinates or as the principal components of the Cartesian/internal transformation matrix.[19] The original method forms linear combinations of all types of internal coordinates (stretch, bend, torsion) but it can also be formulated as making linear combinations only of interal coordinates of the same type.[20] A major advantage is that delocalized internal coordinates can be generated automatically without any user involvement.

In summary, the efficiency of Newton–Raphson-based optimizations depends on the following factors:

1. Hessian quality (exact or updated).
2. Step control (augmented Hessian, choice of shift parameter(s)).
3. Coordinates (Cartesian, internal).
4. Trust radius update (maximum step size allowed).

A comparison of various combinations of these can be found in the references.[13,21,22]

13.4 Optimizing General Functions: Finding Saddle Points (Transition Structures)

Locating minima for functions is fairly easy. If everything else fails, the steepest descent method is guaranteed to lower the function value. Finding first-order saddle points, *Transition Structures* (TS), is much more difficult. There are *no* general methods that are guaranteed to work! Many different strategies have been proposed, the majority of which can be divided into two general categories, those based on interpolation between two minima and those using only local information.[23,24] Interpolation methods assume that the reactant and product geometries are known and that a TS is located somewhere "between" these two end-points. It should be noted that many of the methods in this group do not actually locate the TS; they only locate a point close to it. Local methods propagate the geometry using only information about the function and its first and possibly also second derivatives at the current point, that is they require no knowledge of the reactant and/or product geometries. Local methods usually require a good estimate of the TS in order to converge. Once the TS has been found, the whole reaction path may be located by tracing the intrinsic reaction coordinate (Section 13.8), which corresponds to a steepest descent path in mass-weighted coordinates, from the TS to the reactant and product.

13.4.1 One-Structure Interpolation Methods

The intuitively simple approach for locating a TS is to select one or a few internal "reaction" coordinates, that is those that describe the main difference between the reactant and product structures. A typical example is a torsional angle for describing a conformational TS or two bond distances for a bond breaking/forming reaction. The selected coordinate(s) is (are) fixed at certain values, while the remaining variables are optimized, thereby adiabatically mapping the energy as a function of the reaction variable(s); such methods are often called "*coordinate driving*". The goal is to find a geometry where the residual gradients for the fixed variables are "sufficiently" small. The success of this method depends on the ability to choose a good set of reaction variables, with a good choice being equated with large coefficients for the selected variables in the actual reaction coordinate vector at the TS (as given by the Hessian eigenvector with a negative eigenvalue). The reaction coordinate at the TS, however, is only known after the TS has actually been found, making the choice strongly user-biased and impossible to verify *a priori*.

If only one or two variables change significantly between the reactant and product, the coordinate driving usually works well and the constrained optimized geometry with the smallest residual gradient is a good approximation to the TS. Some typical examples are rotation of a methyl group (reaction variable is the torsional angle), the HNC to HCN rearrangement (reaction variable is the HCN angle) and S_N2 reactions of the type $X + CH_3Y \rightarrow XCH_3 + Y$ (reaction variables are the XC and CY distances). Good approximations to many conformational TSs can be generated by "driving" a selected torsional angle, and this is often the basis for conformational analysis using force field energy functions. It should be stressed that the highest energy structure located in this fashion is *not* exactly the TS, but it is usually a very good approximation to it. A mapping with more than two reaction variables becomes cumbersome and rarely leads anywhere.

If a bad choice of reaction variables has been made, "hysteresis " is often observed. This is the term used when a series of optimizations made by *increasing* the fixed variable(s) to a given value may produce a different result than when *decreasing* the fixed variable(s) to the same point. If the reaction variable scan is only run in one direction, hysteresis will usually be visible because the optimization suddenly changes the geometry drastically for a small change in the fixed variable(s). This indicates that the chosen reaction variable(s) do not contribute strongly to the actual reaction coordinate at the TS. Some TSs have reaction vectors that are not dominated by a few internal variables and such TSs are difficult to find by constrained optimization methods. Another set of (internal) coordinates may in some cases alleviate the problem, but finding these is part of the "black magic" involved in locating TSs.

The *Linear Synchronous Transit* (LST) method may be considered as a coordinate driving method where all (Cartesian or internal) coordinates are varied linearly between the reactant and product and no optimization is performed.[25] The assumption is that all variables change at the same rate along the reaction path and the TS estimate is simply the highest energy structure along the interpolation line. The assumed synchronous change for all variables is rarely a good approximation and only for simple systems does LST lead to a reasonable estimate of the TS. The *Quadratic Synchronous Transit*

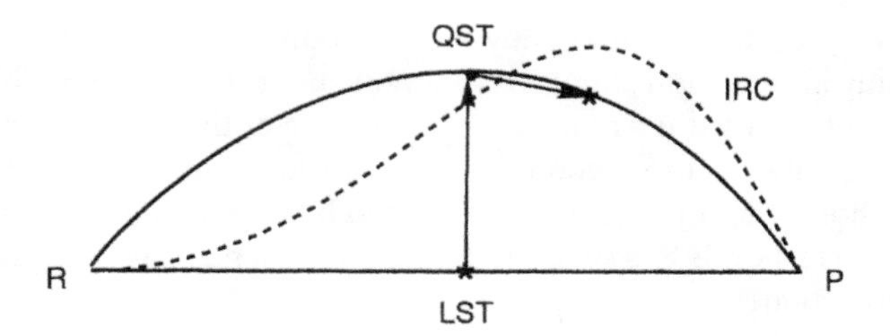

Figure 13.4 Illustration of the linear and quadratic synchronous transit methods; energy maxima and minima are denoted by * and •, respectively.

(QST) approximates the reaction path by a parabola instead of a straight line. After the maximum on the LST is found, the QST is generated by minimizing the energy in the directions perpendicular to the LST path and the QST path may then be searched for an energy maximum. These methods are illustrated in Figure 13.4, where the *Intrinsic Reaction Coordinate* (IRC) represents the "true" reaction coordinate.

Bell and Crighton refined the method by performing the minimization from the LST maximum in the directions *conjugate* to the LST instead of the *orthogonal* directions as in the original formulation.[26] A variation of QST, called *Synchronous Transit-guided Quasi-Newton* (STQN), uses a circle arc instead of a parabola for the interpolation, and uses the tangent to the circle for guiding the search towards the TS region.[27] Once the TS region is located, the optimization is switched to a quasi-Newton–Raphson (Section 13.4.6).

The *Sphere* optimization technique involves a sequence of constrained optimizations on hyperspheres with increasingly larger radii, using the reactant (or product) geometry as a constant expansion point.[28] The lowest energy point on each successive hypersphere thus traces out a low-energy path on the energy surface, as illustrated in Figure 13.5. The sphere method may be considered as a coordinate driving algorithm where the driving coordinate is the distance to the minimum. Ohno and Maeda have suggested a variation where the optimization is done in vibrational normal coordinates scaled by the square root of the corresponding Hessian eigenvalues.[29,30] This makes all directions equivalent in an energetic sense and potentially allows more saddle points to be found, but at the expense of searching the full variable space rather than just the low-energy region. They have suggested that an exhaustive search along all the normal mode directions can potentially find all the

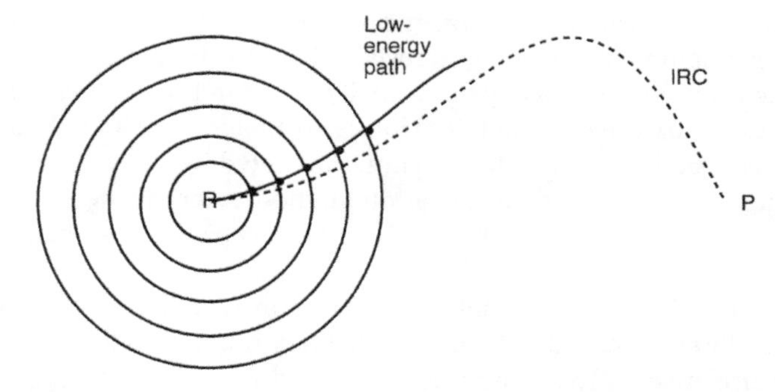

Figure 13.5 Illustration of the sphere method; energy minima on the hyperspheres are denoted by •, while R indicates a (local) minimum in the full variable space.

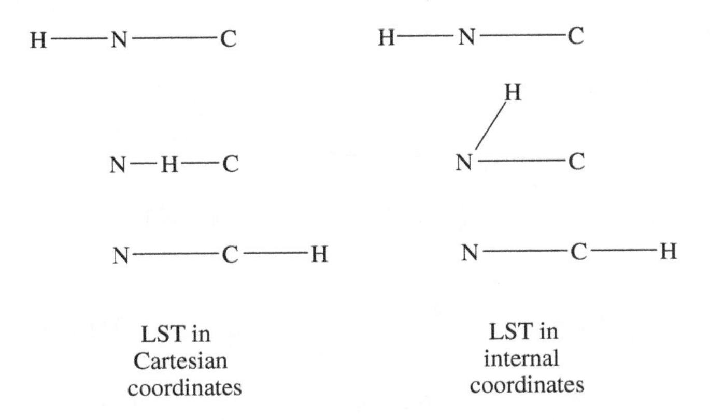

Figure 13.6 LST path in Cartesian and internal coordinates.

TSs connected with a given minimum. Tracing the IRC from all these TSs will lead to other minima, which then can be subjected to a TS search, thereby potentially tracing out all possible reaction paths for a given system.

Barkema and Mousseau have suggested a closely related dynamical version where the gradient at a given point is split into two components parallel and perpendicular to the vector from the minimum to the current point.[31] The gradient component in the perpendicular direction is followed in the downhill direction, while the structure is advanced in the uphill direction along the parallel component.

It should be noted that the success or failure of LST/QST and related interpolation methods, as with all optimizations, depends on the coordinates used in the interpolation. Consider, for example, the HNC to HCN rearrangement. In Cartesian coordinates, the LST path preserves the linearity of the reactant and product, and thus predicts that the hydrogen moves *through* the nitrogen and carbon atoms. In internal coordinates, however, the angle changes from 0° to 180°, and the LST will in this case locate a much more reasonable point with the hydrogen moving around the C–N moiety (see Figure 13.6).

For large complex systems, the LST path, even in internal coordinates, may involve geometries where two or more atoms clash and it may be difficult or impossible to obtain a function value, for example due to an iterative (SCF) procedure failing to converge.

13.4.2 Two-Structure Interpolation Methods

The methods in Section 13.4.1 all optimize one geometrical structure, and differ primarily in how they parameterize the reaction path. The methods in this section operate with two geometrical structures, which attempt to bracket the saddle point and gradually converge on the TS from the reactant and product sides.

In the *Saddle* algorithm,[32] the lowest of the reactant and product minima is first identified. A trial structure is generated by displacing the geometry of the lower energy species a fraction (e.g. 0.05) towards the high-energy minimum. The trial structure is then optimized, subject to the constraint that the distance to the high-energy minimum is constant. The lowest energy structure on the hypersphere becomes the new interpolation end-point and the procedure is repeated. The two geometries will (hopefully) gradually converge on a low-energy structure intermediate between the original two minima, as illustrated in Figure 13.7.

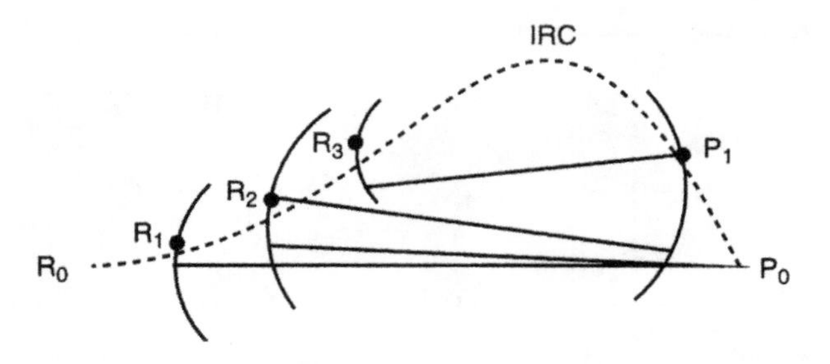

Figure 13.7 Illustration of the saddle method; energy minima on the hyperspheres are denoted by •.

A related idea is used in the *Line-Then-Plane* (LTP) algorithm,[33] where the constrained optimization is done in the hyperplane perpendicular to the interpolation line between the two end-points, rather than on a hypersphere.

The *Ridge* method initially locates the energy maximum along the LST path connecting the reactant and product, and defines two points on either side of the energy maximum.[34] These points are allowed to relax in the downhill direction a given distance and a new energy maximum is located along the interpolation line connecting the two relaxed points, and the cycle is repeated. As the saddle point is approached, the two ridge points gradually contract on the actual TS. This method requires a careful adjustment of the magnitude of the "side" and "downhill" steps as the optimization proceeds.

The *Step-and-Slide* algorithm[35] is a variation where the reactant and product structures are stepped along the LST line until they have energies equal to a preset value. Both structures are then optimized with respect to minimizing the distance between them, subject to being on an isoenergetic contour surface. The energy is increased, followed by another *step-and-slide* optimization, and this sequence is continued until the distance between the two structures decreases to zero, that is converging on the saddle point.

13.4.3 Multistructure Interpolation Methods

The methods in this section operate with multiple (more than two) structures or *images* connecting the reactant and product and are often called *chain-of-state* methods. Relaxation of the images will in favorable cases not only lead to the saddle point but also to an approximation of the whole reaction path. The initial distribution of structures will typically be along a straight line connecting the reactant and product (LST), but may also involve one or more intermediate geometries to guide the search in a certain direction.

The *Self-Penalty Walk* (SPW) method (Figure 13.8) approximates the reaction path by minimizing the average energy along the path, given as a line integral between the reactant and product geometries ($\mathbf{R}$ and $\mathbf{P}$):[36]

$$S(\mathbf{R},\mathbf{P}) = \frac{1}{L}\int_{\mathbf{R}}^{\mathbf{P}} E(\mathbf{x})\mathrm{d}l(\mathbf{x}) \tag{13.32}$$

The line element $\mathrm{d}l(\mathbf{x})$ belongs to the reaction path, which has a total length of L. In practice, the line integral is approximated as a finite sum of M images, where M typically is of the order of 10–20:

$$S(\mathbf{R},\mathbf{x}_1,\mathbf{x}_2,\dots,\mathbf{x}_M,\mathbf{P}) \approx \frac{1}{L}\sum_{i=1}^{M} E(\mathbf{x}_i)\Delta l_i \tag{13.33}$$

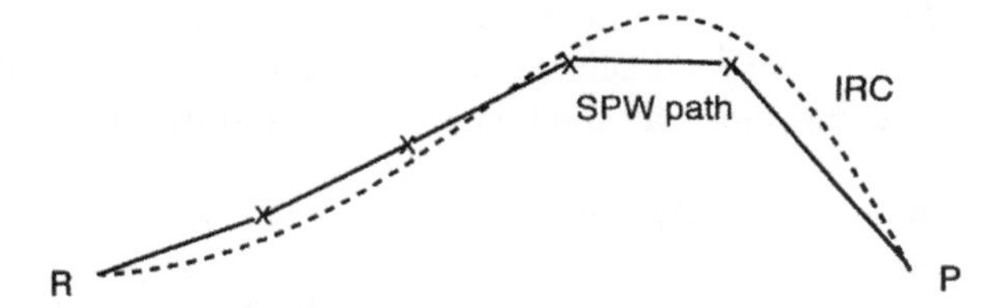

Figure 13.8 Illustration of the SPW method; optimized path points are denoted by x.

In order to avoid all images aggregating near the minima (reactant and product), constraints are imposed for keeping the distance between two neighboring images close to the average distance. Furthermore, repulsion terms between all images are also added to keep the reaction path from forming loops. The resulting target function $T_{SPW}(\mathbf{R}, \mathbf{P})$ may then be minimized by using, for example, a conjugate gradient method:

$$T_{SPW}(\mathbf{R}, \mathbf{x}_1, \mathbf{x}_2, \ldots, \mathbf{x}_M, \mathbf{P}) = \frac{1}{L}\sum_{i=1}^{M} E(\mathbf{x}_i)\Delta l_i + \gamma \sum_{i=0}^{M} (d_{i,i+1} - \bar{d})^2 + \rho \sum_{i>j+1}^{M+1} \exp\left(-\frac{d_{ij}}{\lambda \bar{d}}\right)$$
$$d_{ij} = |\mathbf{x}_i - \mathbf{x}_j| \quad ; \quad \bar{d} = \sqrt{\frac{1}{M+1}\sum_{i=0}^{M} d_{i,i+1}^2} \tag{13.34}$$

The γ, λ and ρ parameters are suitable constants for weighting the distance and repulsion constraints relative to the average path energy. In the original version of SPW, the TS is estimated as the image with the highest energy after minimization of the target function, but Ayala and Schlegel have implemented a version where one of the images is optimized directly to the TS and the remaining images form an approximation to the IRC path.[37]

The *Chain* method (Figure 13.9) initially calculates the energy at a series of images placed at regular intervals (spacing of d_{max}) along a suitable reaction coordinate.[38] The highest energy image is allowed to relax by a maximum step size of d_{max} along a direction defined by the gradient component orthogonal to the line between by the two neighboring images. This process is repeated with the new highest energy image until the gradient becomes tangential to the path (within a specified threshold). When this happens, the current highest energy image cannot be further relaxed, and is instead moved to a maximum *along* the path.

During the relaxation the chain may form loops, in which case intermediate image(s) is (are) discarded. Similarly, it may be necessary to add images to keep the distance between neighbours below d_{max}.

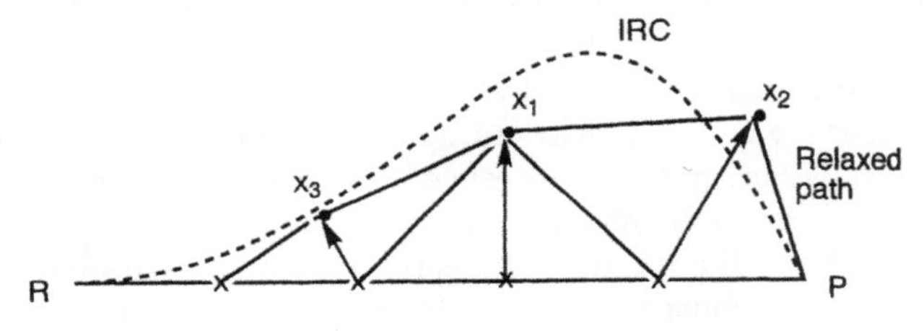

Figure 13.9 Illustration of the chain method; initial points along the path are denoted by x and relaxed points are denoted by •.

The *Locally Updated Planes* (LUP) minimization is related to the chain method, where the relaxation is done in the hyperplane perpendicular to the reaction coordinate, rather than along a line defined by the gradient.[39] Furthermore, all the images are moved in each iteration, rather than one at a time.

The *Conjugate Peak Refinement* (CPR) method may be considered as a dynamical version of the chain method, where images are added or removed based on a sequence of maximizations along line segments and minimizations along the conjugate directions.[40] The first cycle is analogous to the Bell and Crighton version of the QST: location of an energy maximum along a line between the reactant and product, followed by a sequential minimization in the conjugate directions. The corresponding image becomes a new path image and an attempt is made to locate an LST maximum between the reactant and midpoint, and between the midpoint and product. If such a maximum is found, it is followed by a new conjugate minimization, which then defines a new intermediate image, etc. The advantage over the chain and LUP methods is that images tend to be distributed in the important region near the TS, rather than uniformly over the whole reaction path.

In practice, it may not be possible to minimize the energy in *all* the conjugate directions, since the energy surface in general is not quadratic. Once the gradient component along the LST path between two neighboring images exceeds a suitable tolerance during the sequential line minimizations, the optimization is terminated and the geometry becomes a new interpolation point. It may also happen that one of the interpolation images has the highest energy along the path without being sufficiently close to a TS (as measured by the magnitude of the gradient), in which case the image is removed and a new interpolation is performed.

The above multistucture methods are mainly of historical interest, but have played an important role in developing the widely used methods described below.

The *Nudged Elastic Band* (NEB) method defines a target function ("elastic band") as the sum of energies of all images and adds a penalty term having the purpose of distributing the images along the path.[41] A single spring constant k will attempt to distribute the images evenly along the path, but it may also be taken to depend on the energy in order to provide a better sampling near the saddle point:

$$T_{\mathrm{NEB}}(\mathbf{R}, \mathbf{x}_1, \mathbf{x}_2, \ldots, \mathbf{x}_M, \mathbf{P}) = \sum_{i=1}^{M} E(\mathbf{x}_i) + \sum_{i=1}^{M-1} \tfrac{1}{2} k(\mathbf{x}_{i+1} - \mathbf{x}_i)^2 \tag{13.35}$$

A straightforward minimization of T_{NEB} gives a reaction path that has a tendency to cut corners if the spring constant k is too large and a problem of images sliding down towards the minima if the spring constant is too small. These problems can of course be solved by employing a large number of images, but that would render the optimization inefficient. The "corner-cutting" and "down-sliding" problems for a manageable number of images can be alleviated by "nudging" the elastic band, that is using only the component of the spring force *parallel* to the tangent of the path and only the *perpendicular* component of the energy force in the optimization of T_{NEB}. The magnitude of the spring constant influences the optimization efficiency; a small value causes an erratic coverage of the reaction path, while a large value focuses the effort on distributing the images rather than on finding the reaction path, and consequently slows down the convergence. The parallel forces for each image are obtained by projecting the total force on to the reaction path tangent, and the perpendicular components by an orthogonal projection. Since the reaction path is represented by a discrete set of images, the tangent to the path at a given image must be estimated from the neighboring images. The NEB algorithm defines the tangent as the difference vector of the neighboring images, while *String Methods* employ a cubic spline interpolation. String methods furthermore redistribute the images after each optimization cycle, thereby omitting the last term in Equation (13.35) and dispensing with the requirement of using the projected spring force in the optimization.[42] In the *Climbing Image* (CI-NEB) version, one of the images is allowed to move along the elastic band to become the exact saddle point.[43] An adaptive version of NEB has also been proposed, which gradually increases the number of images and concentrates the images near the important saddle point region.[44] The problem of generating an initial path when the LST path is unsuitable has been addressed by gradually adding images from the reactant and product sides, a procedure usually called *Growing String*.[45]

The main computational drawback of multistructure methods is that they require optimization of a target function that contains $\sim 3M_{\text{image}}N_{\text{atom}}$ variables, rather than $\sim 3N_{\text{atom}}$ variables for the single-structure methods. The optimization furthermore consists of two separate tasks, minimizing the gradient perpendicular to the reaction path tangent (which is different for each image) and ensuring a proper distribution of the images along the reaction path. Since the tangent to the reaction path at a given image depends on the neighboring images, this leads to a coupling between all the image structures. If this coupling is ignored the optimization can be split into M_{image} independent optimizations, each involving only $\sim 3N_{\text{atom}}$ variables. This, however, requires a two-step iterative procedure where the reaction path tangent at each image is calculated from the current path and held fixed during the optimization of all images, and then recalculated based on the new image coordinates. The image distribution condition is handled by a penalty approach in NEB, where the gradient from the last term in Equation (13.35) is subjected to a different projection than the gradient from the first term, such that only the component parallel to the reaction path is retained. This different projection of the two terms means that there is not a well-defined target function to minimize and implementation of, for example, conjugate gradient or Newton–Raphson optimization schemes is not straightforward. The minimization can instead be done using a Newtonian dynamics method (e.g. velocity Verlet, Section 15.2.1) where the velocity is quenched regularly, which effectively corresponds to a steepest descent algorithm with a dynamical step size. As discussed in Section 13.2.1, this is a rather inefficient optimization method and minimization of T_{NEB} therefore often requires a large number of iterations.

If the coupling of the NEB images is taken into account, then the tangent direction is allowed to change during the optimization. This has the advantage that the target function has a well-defined gradient and can be optimized with quasi-NR methods.[46] The disadvantage is that the optimization contains $\sim 3M_{\text{image}}N_{\text{atom}}$ variables, which may be borderline possible with NR-based methods, but can be handled with, for example, L-BFGS methods.

The *Artificial Force Induced Reaction* (AFIR) method can be considered as a two-point interpolation method where the end-point structures are defined only by a partitioning of the atoms into two fragments A and B.[47] An approximate reaction path is obtained by minimization of a target function composed of the energy and an artificial force term depending on the distance r_{ij} between all pairs of atoms belonging to each fragment:

$$T_{\text{AFIR}}(\mathbf{x}) = E(\mathbf{x}) + \alpha \frac{\sum\limits_{i \in \text{A}} \sum\limits_{j \in \text{B}} w_{ij} r_{ij}}{\sum\limits_{i \in \text{A}} \sum\limits_{j \in \text{B}} w_{ij}} \tag{13.36}$$

The weighting function can be parameterized as follows, where R_i and R_j are covalent atomic radii:

$$w_{ij} = \left[\frac{(R_i + R_j)}{r_{ij}} \right]^6 \tag{13.37}$$

The strength of the artificial force term is given in the following equation, where the sign can be chosen to make the additional term either attractive or repulsive:

$$\alpha = \pm\gamma R_0^{-1}\left[2^{-\frac{1}{6}} - \left(1 + \sqrt{1 + \frac{\gamma}{\varepsilon}}\right)^{-\frac{1}{6}}\right]^{-1} \tag{13.38}$$

R_0 and ε are parameters typical for van der Waals interactions (i.e. ~3.8 Å and 1 kJ/mol, respectively), while γ is a reaction-specific user-defined parameter that controls the highest energy paths that can be explored. The coordinates of two fragments, or of the atoms in the two fragments, may be randomly chosen or perturbed, which allows an automated search of reaction paths within the user-defined energy window. The energy minima and maxima of the T_{AFIR} function may be refined to genuine minima and TS of the energy function by a local optimization method, such as, for example, a quasi-Newton–Raphson method.

13.4.4 Characteristics of Interpolation Methods

Interpolation methods have the following characteristics:

1. There may not be a TS connecting two minima directly. The algorithm may then find an intermediate geometry having a gradient substantially different from zero, that is no nearby stationary point. This is primarily a problem for the one-structure methods in Section 13.4.1.
2. The TS found is not necessarily one that connects the two minima used in the interpolation. A calculation of the reaction path may reveal that it is a TS for a different reaction. This is primarily a problem for the one- and two-structure methods in Sections 13.4.1 and 13.4.2.
3. There may be several TSs (and therefore at least one minimum) between the two selected end-points. Some algorithms may find one of these, and the two connecting minima can then be found by tracing the reaction path, or all the TSs and intermediate minima may be located. Multistructure methods (e.g. NEB) are examples of the latter behavior.
4. The reaction path formed by a sequence of points generated by constrained optimizations may be discontinuous. For methods where two points are gradually moved from the reactant and product sides (e.g. saddle and LTP), this means that the distance between end-points does not converge towards zero.
5. There may be more than one TS connecting two minima. As many of the interpolation methods start off by assuming a linear reaction coordinate between the reactant and product, the user needs to guide the initial search (e.g. by adding intermediate structures) to find more than one TS.
6. A significant advantage is that the constrained optimization can usually be carried out using only the first derivative of the energy. This avoids an explicit, and computationally expensive, calculation of the second derivative matrix.
7. For the one- and two-structure methods, each successive refinement of the TS estimate requires either location of an energy maximum or minimum along a one-dimensional path (typically a line) or a constrained optimization in an $(N-1)$-dimensional hyperspace. A path minimization or maximization will normally involve several function evaluations, while a multidimensional minimization requires several gradient calculations. Geometry changes are often quite small near convergence, but each step may still require a significant computational effort involving many function and/or gradient calculations. In such cases, it is often advantageous to switch to one of

the Newton–Raphson methods described in Section 13.4.6, but the dimensionality of the problem may prevent this.

8. The multistructure methods in Section 13.4.3 involve an optimization of a target function with M images each having $3N_{atom}$ coordinates, that is optimization of a function with $\sim 3M_{image}N_{atom}$ variables. Since the number of iterations typically increases with the number of variables, the optimization of the target function may require a large number (several hundred or thousand) of gradients.
9. The multistructure methods are quite tolerant toward the presence of many soft degrees of freedom, which often causes problems with the local optimization methods described in Sections 13.4.5 to 13.4.7. Multistructure methods such as NEB are therefore well suited for systems with many degrees of freedom, for example extended (periodic) systems.
10. If the structures corresponding to the two end-points are flexible and contain many conformations or loose van der Waals complexes, it may be difficult to select the correct combination of end-points such that the reaction path only describes the chemical reaction, and not conformational transitions along the two exit channels.[48] Care should also be taken when connecting formally equivalent, but distinct, atoms between the two end-point structures, such as, for example, the three hydrogen atoms in methyl groups.
11. The "automated" structure and reaction path methods, such as AFIR, often require a very large number (10^3–10^5) of gradient calculations.

13.4.5 Local Methods: Gradient Norm Minimization

Since transition structures are points where the gradient is zero, they may in principle be located by minimizing the gradient norm. This is in general not a good approach for two reasons:

1. There are typically many points where the gradient norm has a minimum without being zero.
2. Any stationary point has a gradient norm of zero; thus all types of saddle points and minima/maxima may be found, not just TSs.

Figure 13.10 shows an example of a one-dimensional function and its associated gradient norm. It is clear that a gradient norm minimization will only locate one of the two stationary points if started near $x = 1$ or $x = 9$. Most other starting points will converge on the shallow part of the function near $x = 5$. The often very small convergence radius makes gradient norm minimizations impractical for routine use.

13.4.6 Local Methods: Newton–Raphson

By far the most common local methods are based on the augmented Hessian Newton–Raphson approach (Section 13.2.4). The standard NR formula will locate the TS rapidly when started sufficiently close to the TS. Sufficiently close means that the Hessian should have exactly one negative eigenvalue and the associated eigenvector should be in the correct direction, along the "reaction coordinate". The NR step should furthermore be inside the trust radius. By using augmented Hessian techniques, the convergence radius may be enlarged over the straight NR approach, and first-order saddle points may be located even when started in a region where the Hessian does not have the correct structure, as long as the lowest eigenvector is in the "correct" direction.

The NR step near a first-order saddle point maximizes the energy in one direction (along the Hessian TS eigenvector) and minimizes the energy along all other directions. Such a step may be enforced

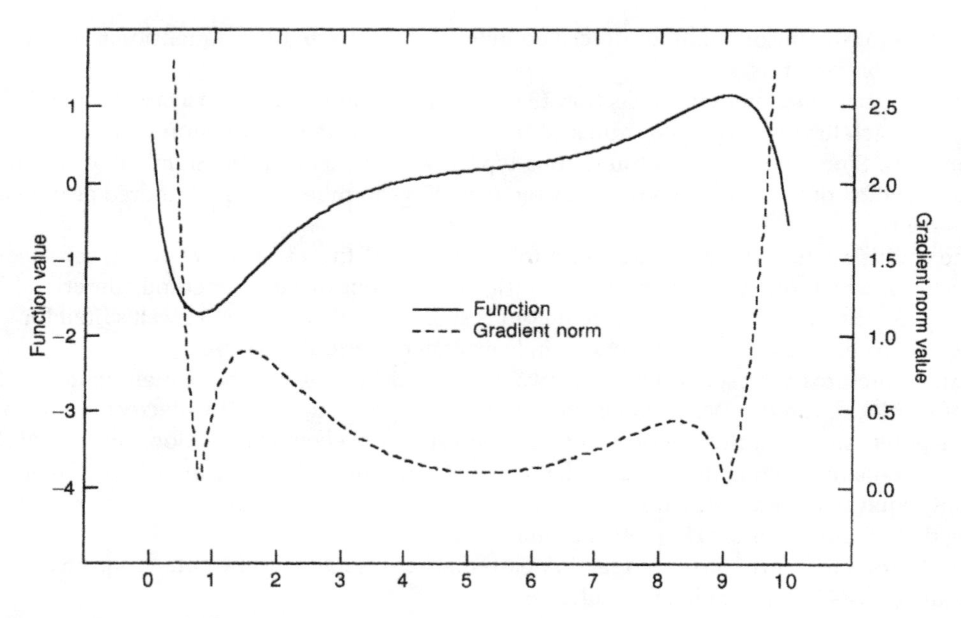

Figure 13.10 An example of a function and the associated gradient norm.

by choosing suitable shift parameters in the augmented Hessian method, that is the step is parameterized as in Equation (13.16). The minimization step is similar to that described in Section 13.2.4 for locating minima; the only difference is for the unique TS direction.

Two shift parameters are employed in the *Partitioned Rational Function Optimization* (P-RFO) method:[5]

$$\sum_{i \neq \mathrm{TS}} \frac{g_i'^2}{\varepsilon_i - \lambda} = \lambda \tag{13.39}$$

$$\frac{g_{\mathrm{TS}}'^2}{\varepsilon_{\mathrm{TS}} - \lambda_{\mathrm{TS}}} = \lambda_{\mathrm{TS}} \tag{13.40}$$

The λ for the minimization modes is determined as for the RFO method (Equation (13.18)). The equation for λ_{TS} is quadratic and by choosing the solution that is larger than $\varepsilon_{\mathrm{TS}}$ it is guaranteed that the step component in this direction is along the gradient, that is a maximization. As for the RFO step, there is no guarantee that the total step length will be within the trust radius.

The *Quadratic Approximation* (QA) method uses only one shift parameter, requiring that $\lambda_{\mathrm{TS}} = -\lambda$, and restricts the total step length to the trust radius (compare with Equation (13.19)):[6]

$$|\Delta \mathbf{x}'|^2 = \sum_{i \neq \mathrm{TS}} \left(\frac{g_i'}{\varepsilon_i - \lambda} \right)^2 + \left(\frac{g_{\mathrm{TS}}'}{\varepsilon_{\mathrm{TS}} + \lambda} \right)^2 = R^2 \tag{13.41}$$

The exact same formula may be derived using the concept of an "image potential" (obtained by inverting the sign of g_{TS}' and λ_{TS}), and the QA name is often used together with the TRIM (*Trust Radius Image Minimization*) acronym.[49]

The ability of augmented Hessian methods for generating a search toward a first-order saddle point, even when started in a region where the Hessian has all positive eigenvalues, suggests that it may be possible to start directly from a minimum and "walk" to the TS by following a selected Hessian eigenvector uphill. Such mode followings, however, are only possible if the eigenvector being followed is only weakly coupled to the other eigenvectors (i.e. third and higher derivatives are small). All NR-based methods assume that one of the Hessian eigenvectors points in the general direction of the TS, but this is only strictly true when the higher-order derivatives are small. If this is not the case, NR-based methods may fail to converge even when started from a "good" geometry, where the Hessian has one negative eigenvalue. Note also that the magnitude of the higher derivatives depends on the choice of coordinates; that is a "good" choice of coordinates may transform a divergent optimization into a convergent one.

All NR methods assume that a "sufficiently good" guess of the TS geometry is available. Generating this guess is part of the magic, but some of the interpolating schemes described in Sections 13.4.1 to 13.4.3 may be useful in this respect.

There are two main problems with all NR-based methods. One is the already mentioned need for a good starting geometry. The other is the requirement of a Hessian, which is quite expensive in terms of computer time for electronic structure methods. Contrary to minimizations, TS optimizations cannot start with a diagonal matrix and update it as the optimization proceeds. An NR TS search requires the definition of a direction along which to maximize the energy, the reaction vector; that is the start Hessian should preferably have one negative eigenvalue. Normally the Hessian needs to be calculated explicitly at the first step; at subsequent steps the Hessian may be updated. An alternative is to use a force field Hessian for starting the optimization, since this effectively removes one of the more expensive steps in a TS optimization.[50] If the geometry changes substantially during the optimization, however, it may be necessary to recalculate the Hessian at certain intervals. Owing to the relatively high cost of calculating the energy, gradient and especially the Hessian, quasi-NR methods have traditionally been the preferred algorithm with *ab initio* wave functions.

13.4.7 Local Methods: The Dimer Method

The main problem with NR methods is the need for generating (calculating or updating) and manipulating (storing and diagonalizing) the Hessian matrix. The main function of the Hessian for saddle point optimizations is to provide the direction along which the energy should be maximized. Sufficiently close to the TS, this direction is along the eigenvector corresponding to the lowest eigenvalue. Determination of this direction, however, can be done *without* calculating the Hessian by placing two symmetrically displaced images, a *dimer*, and minimizing the sum of their energies, subject to a constant distance between them.[51,52] After minimization the lowest mode direction is given by the line connecting the two images, and it can be used for displacing the central structure, followed by a new dimer optimization. Since the dimer optimization can be done using only first derivatives, this alleviates the need for the Hessian matrix. There is relatively little experience with this method, but one would expect it to have the same requirements as NR-based methods, that is a good starting geometry is required for a stable convergence to the TS. Whether the added computational cost of optimizing each dimer configuration outweighs the savings by not having an explicit Hessian is unclear, and will in any case depend on the size of the system.

13.4.8 Coordinates for TS Searches

The choice of a "good" set of coordinates is even more critical in TS optimizations than for minimizations. A good set of coordinates enlarges the convergence region and relaxes the requirement of a good starting geometry. A poor set of coordinates, on the other hand, decreases the convergence

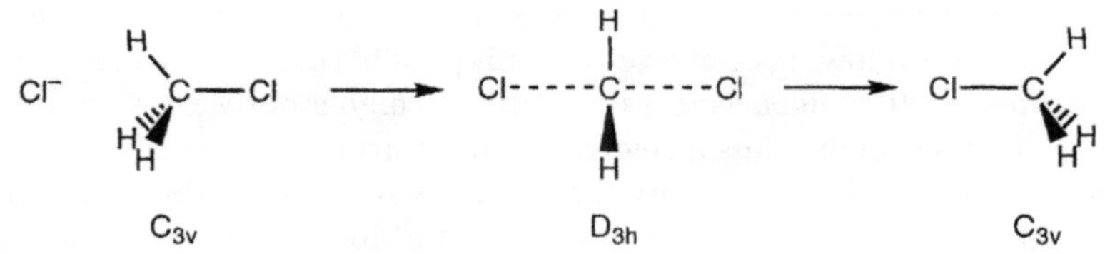

Figure 13.11 The TS for an identity S_N2 reaction has a higher symmetry than the reactant/product.

radius, forcing the user to generate a starting point very close to the actual TS in order for NR methods to work. Furthermore, NR methods are best suited for relatively "stiff" systems; large flexible systems with many small eigenvalues in the Hessian are better handled by some of the interpolations methods, such as NEB.

Mapping out whole reaction pathways by locating minima and connecting TSs is often computationally demanding. The (approximate) geometries of many of the important minima are often known in advance and, as mentioned above, energy minimizations are fairly uncomplicated. Locating TSs is much more involved. On a multidimensional energy surface, there will in general not be TSs connecting all pairs of minima. It is, however, essentially impossible to prove that a TS does not exist.

Symmetry can sometimes be used to facilitate the location of TSs. For some reactions, especially those where the reactant and product are identical, the TS will have a symmetry different from the reactant/product. The reaction vector will belong to one of the non-totally symmetric representations in the point group. The TS can therefore be located by constraining the geometry to a certain symmetry and *minimizing* the energy. Consider, for example, the S_N2 reaction of Cl^- with CH_3Cl (Figure 13.11). The reactant and product have C_{3v} symmetry, but the TS has D_{3h} symmetry. Minimizing the energy under the constraint that the geometry should have D_{3h} symmetry will produce the lowest energy structure within this symmetry, which is the TS.

For non-identity reactions, it is often useful to start a search for stationary points by minimizing high-symmetry geometries. A subsequent frequency calculation on the symmetry-constrained (and minimized) structure will reveal the nature of the stationary point. If it is a minimum or TS we have already obtained useful information. If it turns out to be a higher-order saddle point, the normal coordinates associated with the imaginary frequencies show how the symmetry should be lowered to produce lower energy species, which may be either minima or TSs. As calculations on highly symmetric geometries are computationally less expensive than on non-symmetric structures, it is often a quite efficient strategy to start the investigation by concentrating on structures with symmetry.

13.4.9 Characteristics of Local Methods

Local methods have the following characteristics:

1. A starting geometry close to the saddle point is needed. Especially for reactions that are not dominated by a few (internal) reaction variables, it may be difficult to generate such a guess. The convergence radius is in many cases small, that is the starting geometry must be (very) close to the saddle point in order to converge.
2. Hessian-based methods (Section 13.4.6) require explicit calculation of the second derivative matrix, which may be computationally expensive, and handling of the Hessian matrix furthermore becomes problematic for large systems.

3. Systems with many soft vibrational modes are often problematic, as the resulting low Hessian eigenvalues interfere with the negative curvature along the reaction vector.
4. If a good starting geometry and Hessian is available, the convergence is rapid, often requiring only a few tens of gradient calculations.

13.4.10 Dynamic Methods

The methods in Sections 13.4.1 to 13.4.8 focus on finding a TS connecting a reactant and product, and the resulting activation energy can provide reaction rates via the Arrhenius or Eyring formula (Equations (14.53) and (14.54)). For large complex systems, however, the concept of a single "structure" becomes blurred. In cycloheptadecane, for example, there are hundreds of conformations within 10 kJ/mol of the global minimum, and any experimentally observed property at room temperature will be a Boltzmann average over many individual conformations. The reaction rate for a large system will similarly be a Boltzmann average over perhaps hundreds of TSs, and a single reaction path connecting two minima via a saddle point no longer dominates the reaction rate.[53] A systematic location of all minima (conformations) and corresponding TSs followed by a Boltzmann averaging is a possibility, but this rapidly becomes unmanageable even for medium-sized systems. For large systems, one is therefore forced to perform a sampling of the TSs in order to estimate the reaction rate, in analogy with the ensemble averaging discussed in Section 14.6 for minima. Standard molecular dynamics methods (Section 15.2.1) only sample the low-energy part of the surface and are therefore unsuitable for the (high-energy) saddle point region. A specific part of the surface can be sampled by a biasing potential, such as, for example, in the *umbrella sampling* technique (Section 15.2.9). Such an approach, however, requires *a priori* knowledge of the reaction path or at least the saddle point region. Alternatively, a dynamics simulation may be initiated in the transition state region and the trajectory followed in both directions.[54] Other methods are also available for performing such transition path sampling.[55]

13.5 Constrained Optimizations

In some cases, there are restrictions on the variables used to describe the function, such as, for example:

1. Certain geometrical constraints may be imposed. Experimental data, for example, may indicate that some atom pairs are within a certain distance of each other, or one may for analysis reasons want to impose certain geometrical restrictions on a molecular structure.
2. Fitting atomic charges to give a best match to a calculated electrostatic potential. The constraint is that the sum of atomic charges should equal the net charge of the molecule.
3. A variation of wave function coefficients is subject to constraints such as maintaining orthogonality of the MOs and normalization of the MOs and the total wave function.
4. Finding conical intersections between different energy surfaces. The constraint is that two different energy functions should have the same energy for the same set of nuclear coordinates.

There are three main methods for enforcing constraints during function optimization:

1. Penalty functions
2. Lagrange method of undetermined multipliers
3. Projection methods.

The *penalty function* approach adds a term of the type $k(r - r_0)^2$ to the function to be optimized. The variable r is constrained to be near the target value r_0 and the "force constant" k describes how important the constraint is compared with the unconstrained optimization. By making k arbitrarily large, the constraint may be fulfilled to any given accuracy. It cannot, however, make the constraint variable *exactly* equal to r_0. This would require the constant k to go towards infinity and in practice cause numerical problems when it becomes sufficiently large compared with the other terms. The penalty function approach is often used for restricting geometrical variables, such as distances or angles, during geometry optimizations with force field methods. It may also be used for "driving" a selected variable (Section 13.4.1), such as a torsional angle. In certain cases, the constraint is not to limit a variable to a single value, but rather to keep it between lower and upper limits. This is typically the situation for refining a force field structure subject to constraints imposed by experimental nuclear Overhauser effect (NOE) data, and in such cases the penalty function may be taken as a "flat bottom" potential, that is the penalty term is zero within the limits and rises harmonically outside the limits. The gradient for the penalty function simply has one additional term from each constraint and the penalty function may be optimized using the methods described in Sections 13.2.1 to 13.2.3.

A more elegant method of enforcing constraints is the *Lagrange method*. The function to be optimized depends on a number of variables, $f(x_1, x_2, \ldots, x_N)$, and the constraint condition can always be written as another function, $g(x_1, x_2, \ldots, x_N) = c$. Define now a Lagrange function as the original function minus (or plus) a constant times the constraint condition:

$$L(x_1, x_2, \ldots, x_N, \lambda) = f(x_1, x_2, \ldots, x_N) - \lambda[g(x_1, x_2, \ldots, x_N) - c] \tag{13.42}$$

If there is more than one constraint, one additional multiplier term is added for each constraint. The optimization is then performed on the Lagrange function by requiring that the gradient components with respect to the x- and λ-variable(s) are equal to zero. If the desired stationary point for f is a minimum, then the desired stationary point for the L function is an Nth order saddle point for a Lagrange function having N Lagrange multipliers. The multiplier(s) λ can in many cases be given a physical interpretation at the end. In the variational treatment of an HF wave function (Section 3.3), the MO orthogonality constraints turn out to be MO energies and the multiplier associated with normalization of the total CI wave function (Section 4.2) becomes the total energy.

The Lagrange method increases the number of variables by one for each constraint, which is counterintuitive since introduction of a constraint should decrease the number of variables by one. For simple objective and constraint functions, the reduction can be obtained by solving the constraint condition for one of the variables and substituting it into the object function:

$$\begin{aligned} g(x_1, x_2, \ldots, x_{N-1}, x_N) &= c \quad \Leftrightarrow \quad x_N = h(x_1, x_2, \ldots, x_{N-1}, c) \\ f(x_1, x_2, \ldots, x_{N-1}, x_N) &\Rightarrow f(x_1, x_2, \ldots, x_{N-1}, h(x_1, x_2, \ldots, x_{N-1}, c)) \end{aligned} \tag{13.43}$$

In the large majority of cases, however, the object and constraint functions are so complicated that an analytical elimination of one of the variables is intractable, and this is especially true when there is more than one constraint. The main exception is when the constraint equation is *linear*, in which case it can be considered as a vector in the coordinate space. Instead of eliminating one of the variables explicitly, the constraint condition can be fulfilled by removing the corresponding component of the object function gradient by *projection* (Section 17.4) and performing the optimization using the projected gradient:

$$\begin{aligned} \nabla \mathbf{f}_p &= \nabla \mathbf{f} - \langle \nabla \mathbf{f} | \nabla \mathbf{g}_n \rangle \nabla \mathbf{g}_n \\ \nabla \mathbf{g}_n &= \frac{\nabla \mathbf{g}}{|\nabla \mathbf{g}|} \end{aligned} \tag{13.44}$$

A general (non-linear) constraint condition can be approximately fulfilled by projecting out the first-order (linear) Taylor approximation to the function. Since the optimization normally proceeds by iterative methods, the linear approximation may be sufficient in each step. Alternatively, a microiterate based on successive linear approximations may be performed in each optimization of the objective function.

13.6 Global Minimizations and Sampling

The methods described in Section 13.2 can only locate the "nearest" minimum, which is normally a *local* minimum, when starting from a given set of variables. In some cases, the interest is in the lowest of all such minima, the *global* minimum; in other cases it is important to sample a large (preferably representative) set of local minima. Considering that the number of minima typically grows exponentially with the number of variables, the global optimization problem is an extremely difficult task for a multidimensional function.[56–59] It is often referred to as the *multiple minima* or *combinatorial explosion* problem in the literature.

Consider, for example, the problem of determining the lowest energy conformations of linear alkanes, $CH_3(CH_2)_{n+1}CH_3$, by a force field method, with three possible energy minima for rotation around each C–C bond. For butane, there are thus three conformations, one *anti* and two *gauche* (which are symmetry equivalent). These minima may be generated by starting optimizations from three torsional angles separated by 120°. In the $CH_3(CH_2)_{n+1}CH_3$ case there are n such rotatable bonds, giving a possible 3^n different conformations, and in order to find the global minimum, the energy must be calculated for all of them. Assume for the sake of argument that each conformation optimization takes one second of computer time. Table 13.2 gives the number of possible conformations and the time required for optimizing them all.

The exponential increase in the number of conformations means that it is essentially impossible to perform a complete sampling of systems with more than ~20 degrees of freedom. For the linear alkanes, it is known in advance that *anti* conformations in general are favored over *gauche*; thus we may put some restrictions on the search, such as having a maximum of three *gauche* interactions in total. For most systems, however, there are no good guidelines for such *a priori* selections. Furthermore, for some cases the sampling interval must be less than 120°; in ring systems it may be more like 60°, increasing the potential number of conformations to 6^n. Cycloheptadecane is a frequently used test case for conformational searching and various methods have established that there are 262 different conformations within 12 kJ/mol of the global minimum with the MM2 force field.[60] In the early 1990s, this system was close to the limit for being able to establish the global minimum, but with the increase in computer hardware performance such systems can now be treated within a few hours of computer time.

Table 13.2 Possible conformations for linear alkanes, $CH_3(CH_2)_{n+1}CH_3$.

n	Number of possible conformations (3^n)	Time (1 conformation = 1 second)
1	3	3 seconds
5	243	4 minutes
10	59 049	16 hours
15	14 348 907	166 days

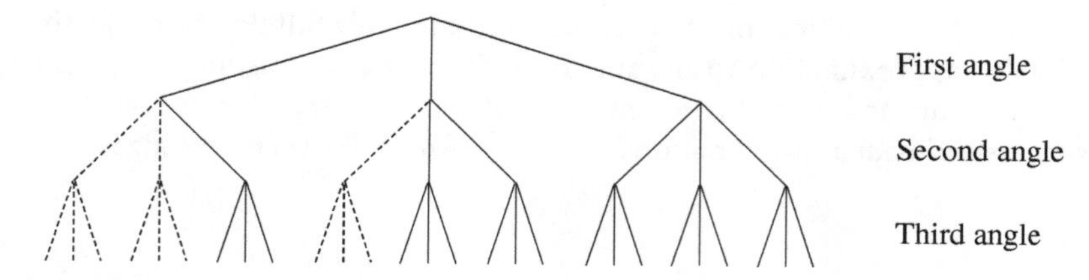

Figure 13.12 Visualizing conformations as a combinatorial tree.

The total number of conformations for a given resolution of each variable (e.g. 120° steps) can be thought of as branches in a combinatorial tree, as illustrated in Figure 13.12.

For a reasonable-sized system, there may be certain combinations of torsional angle that always lead to high-energy structures, for example by atoms clashing. These combinations correspond to specific branches in the combinatorial tree (illustrated by dashed lines in Figure 13.12), and these may consequently be pruned from the search at an early stage. This allows somewhat larger systems to be treated compared with a brute force combinatorial search, but the number of possible conformations still increases rapidly with the size of the system.[61]

Finding "reasonable" minima for large biomolecular systems is heavily dependent on selecting a "good" starting geometry. One way of attempting this is by "building up" the structure. A protein, for example, may be built from amino acid fragments, which have been optimized to their global minimum or to a collection of low-energy minima, and/or smaller fragments of the whole structure may be subjected to a global minimum search. By combining such pre-optimized fragments, it is hoped that the starting geometry for the whole protein will also be "near" the global minimum for the full system.[62]

The systematic, or grid, search is only possible for small systems, while global search methods must be used for larger systems in order to provide estimates of the global minimum. The general problem is to identify a point in a many-dimensional variable space that has the lowest possible object function. Algorithms for global optimizations are numerous and can be used for many purposes, but we will only describe some of the most commonly used:

1. Stochastic and Monte Carlo methods
2. Molecular dynamics
3. Simulated annealing
4. Genetic algorithms
5. Particle swarm optimization methods
6. Diffusion methods
7. Distance geometry methods.

None of these are guaranteed to find the global minimum, but they may in many cases generate a local minimum that is close in energy to the global minimum (but not necessarily close in terms of structure). A brief description of the ideas in these methods is given below. For simplicity, we assume that the optimization is of an energy as a function of atomic coordinates, but it is of course equally valid for any function depending on a set of variables.

13.6.1 Stochastic and Monte Carlo Methods

These methods start from a given geometry, which typically is a (local) minimum, and new configurations are generated by adding a random "kick" to one or more atoms. In *Monte Carlo* (MC) methods, the new geometry is accepted as a starting point for the next perturbing step if it is lower in energy than the current. Otherwise, the Boltzmann factor $e^{-\Delta E/kT}$ is calculated and compared with a random number between 0 and 1. If $e^{-\Delta E/kT}$ is less than this number, the new geometry is accepted, otherwise the next step is taken from the old geometry. This generates a sequence of configurations from which geometries may be selected for subsequent minimization. In order to have a reasonable acceptance ratio, however, the step size must be fairly small, and it is often chosen to give an acceptance ratio of $\sim$0.5.

In *stochastic* methods, the random kick is somewhat larger and is usually performed on all the atoms, and a standard minimization is carried out starting at the perturbed geometry.[63] The optimization may or may not produce a new minimum and a database of all unique structures is gradually built up. A new perturbed geometry is then generated from one of the structures in the database and minimized, etc. There are several variations on how this is done:

- The length of the perturbing step is important; a small kick essentially always returns the geometry to the starting minimum, while a large kick may produce high-energy structures, which minimize to high-energy local minima.
- The perturbing step may be done directly in Cartesian coordinates or in a selected set of internal coordinates, such as torsional angles. The Cartesian procedure has the disadvantage that many of the perturbed geometries are high in energy as two (or more) atoms are moved close together by the kick, although this can be partly alleviated by readjusting all bond lengths to values close to their starting values prior to the optimization. The use of torsional angles as variables is highly efficient for acyclic systems but is problematic for cyclic and confined structures. Cyclic structures can be treated by opening the ring, performing a random perturbation of the torsional angles and attempting to re-close the ring. In the majority of cases this is not possible, and this results in many trial structures being discarded.
- The perturbing step may be taken either from the last minimum found or from all the previous found minima, perhaps weighted by a probability factor such that low-energy minima are used more often than high-energy structures.

Kolossvary and Guida have proposed a method to generate the perturbing step along the eigenvectors with small eigenvalues obtained by diagonalizing the Hessian matrix at each minimum, a method called *low-mode* search.[64] The premise is that the soft deformation modes for a given structure are likely to lead to low-energy transition structures, and consequently to other low-energy minima. The strategy is thus similar to the eigenvector-following tactic discussed in Section 13.4.6 for locating transition structures, except that no attempt is made to find the actual TS. The interest is only in perturbing the geometry sufficiently to get "past" the TS, such that a minimization will locate a new minimum. The advantage of the low-mode search is that the search is concentrated on the low-energy part of the energy surface, and the method furthermore essentially solves the problem of generating trial structures for ring systems. The number of acceptable trial structures generated by the open–perturb–reclose method is often very low, only a few percent, resulting in an inefficient search. Since the Hessian eigenvalues contain information about the coupling of the internal (torsional) coordinates, the low-mode technique can generate trial structures without opening and re-closing the ring.

The disadvantage of the low-mode search is that it requires calculating and diagonalizing the Hessian matrix for each minimum found, which becomes problematic for systems with more than a few hundred atoms. In order to use the method for large systems, the soft Hessian modes can be calculated by an iterative procedure requiring only the gradient.[65] Although this solves the problem of calculating and diagonalizing the Hessian, the computational effort for determining the low-mode directions is still substantial. It has been suggested that for proteins, for example, the low-mode directions for one minimum can be reused for other minima as well, thereby avoiding the expensive low-mode calculations.

The main problem with stochastic methods is generating trial structures. In small flexible molecules, the torsional angles form a good set of coordinates for randomly perturbing the geometry. For cyclic and confined structures, however, a perturbation of a single torsional angle will usually lead to a high-energy structure, either because the remaining (cyclic) structure becomes strained or because of atoms clashing into each other. The low-mode technique solves this by determining a proper combination of internal coordinates to avoid this, but the Hessian diagonalization prevents its use for systems with thousands of atoms. Stochastic methods are therefore primarily useful for searching the conformational space for flexible extended systems, but not for confined molecules such as proteins and DNA. Stochastic methods, however, have the big advantage that they can generate conformations separated by large energy barriers, since the random kick is performed without calculating any energies along the perturbing step, that is the conformations can "tunnel" through large energy barriers.

13.6.2 Molecular Dynamics Methods

Molecular Dynamics (MD) methods solve Newton's equation of motion for atoms on an energy surface (see Section 15.2.1). The available energy for the molecule is distributed between potential and kinetic energy, and molecules are thus able to overcome barriers separating minima if the energy of the barrier is below the total energy. Given a high-enough energy, which is closely related to the simulation temperature, the dynamics will sample the whole surface but will also require an impractically long simulation time. Since quite small time steps must be used for integrating Newton's equation, the simulation time is short (pico- or nanoseconds). Combined with the use of "reasonable" temperatures (a few hundreds of degrees), this means that only the local area around the starting point is sampled and that only relatively small barriers (typically a few tens of a kJ/mol) can be overcome. Different (local) minima may be generated by selecting configurations at suitable intervals during the simulation and subsequently minimize these structures. MD methods use the inherent dynamics of the system to search out the low-energy deformation modes and they can be used for sampling the conformational space for large confined systems. MD methods are typically used for sampling the conformational space when the starting geometry is derived from experimental information, such as an X-ray or NMR structure. The main disadvantage of MD is the inability to overcome barriers larger than the internal energy determined by the simulation temperature. Since this is one of the advantages of MC methods, it is no surprise that mixed MC/MD methods have been developed, of which the *Replica Exchange MD* (REMD) is one of the most commonly used.[66–68]

13.6.3 Simulated Annealing

Both MD and MC methods employ a temperature as a parameter for generating or accepting new geometries. At sufficiently high temperatures and long run times, all the conformational space is sampled. In *Simulated Annealing* (SA) techniques, the initial temperature is chosen to be high, maybe 2000–3000 K.[69–71] An MD or MC run is then initiated, during which the temperature is slowly reduced. Initially the molecule is allowed to move over a large area of the energy surface, but as the temperature is decreased, it becomes trapped in a minimum. If the cooling is done infinitely slowly (implying an infinite run time), the resulting minimum is the global minimum. In practice, however, an MD or MC run is so short that only the local area is sampled. The name, simulated annealing, comes from the analogy of growing crystals. If a melt is cooled slowly, large single crystals can be formed. Such a single crystal represents the global energy minimum for a solid state. A rapid cooling produces a glass (local minimum), that is a disordered solid.

13.6.4 Genetic Algorithms

Genetic or *Evolutionary Algorithms* take their concepts and terminology from biology.[72,73] The idea is to have a "population" of structures, each characterized by a set of "genes". The "parent" structures are allowed to generate "children" having a mixture of the parent genes, allowing for a small amount of "mutations" to occur in the process. The best species from a population are selected based on Darwin's principle, survival of the fittest, and carried on to the next "generation", while the less-fit structures are discarded.

Consider, for example, a molecule having 20 torsional angles, which may have $\sim 10^9$ possible conformations. The species in an initial population of, say, 100 different conformations are characterized by their fitness, for example a low-energy structure is equivalent to one of high fitness. These 100 structures are allowed to produce offsprings with a probability depending on their fitness, that is low-energy structures are more likely to contribute to the next generation than high-energy conformations. Two child conformations can be generated by taking the first n torsional angles from one of the parents and the remaining $20 - n$ from the other ("single-point cross-over"), with the second child being the complementary one. A small amount of mutation is usually allowed in the process, that is randomly changing angles to produce conformations outside the range contained in the current population. Having generated, say, 100 such children, their (minimized) energies are determined and a suitable portion of the best parent and children structures are carried over to the next generation. The population is allowed to evolve for perhaps a few hundred generations or until no change in the best structure has occurred for a suitable number of generations.

There are many variations on genetic algorithms: varying the size of the population, the mutation rate, the breeding selection, the ratio of children to parents surviving to the next generation, single- or multipoint cross-over, etc. Genetic algorithms have become popular as they are easy to implement and have proven to be robust for locating a point in parameter space close to the global minimum. If the parameters are coded into genes, the sampling is pointwise, and the final structures should therefore be refined using a standard gradient optimization. Alternatively, the trial structures may be subjected to a local optimization, making the parameter space continuous.[74] Genetic algorithms may have difficulties if some of the variables are strongly coupled such that the direction along which the target function is lowered corresponds to a simultaneous and coordinated change of two or more variables. In such cases the optimization may end up in a situation where a very large number of small steps in the parameter space is required to reach the (global) minimum, and in practise cause premature (apparent) convergence. Such a situation can be countered by increasing the population or changing the variables, but it is usually difficult to predict or detect whether a particular problem has strongly coupled variables.

13.6.5 Particle Swarm and Gravitational Search Methods

Particle swarm optimization (PSO) algorithms use in analogy with genetic algorithms a population (a swarm) of structures (denoted particles), where each particle moves in the variable space according to certain rules. The biological model in this case is a flock of birds searching for food, where a single bird spotting a potential food source may cause the whole flock to turn toward that direction. The particles in the swarm are initial assigned (random) values of the positions and (random) velocities in the variable space, often limiting the positions to a subsection of the available space and employing low velocities. The algorithm moves the swarm of particles similar to the way a molecular dynamics simulation moves the atoms, by updating the positions and velocities in a sequence of pseudo time steps. Each iteration increments the "time" variable t and updates each particle position $\mathbf{x}_i$ according to the velocity vector $\mathbf{v}_i$ (Equation (13.45)) and updates the velocity vector by a three-parameter formula (Equation (13.46)):[75]

$$\mathbf{x}_i^{t+1} = \mathbf{x}_i^t + \mathbf{v}_i^{t+1} \tag{13.45}$$

$$\mathbf{v}_i^{t+1} = w\mathbf{v}_i^t + a_1 R1_i^t \left(\mathbf{b}_i^t - \mathbf{x}_i^t\right) + a_2 R2_i^t \left(\mathbf{B}^t - \mathbf{x}_i^t\right) \tag{13.46}$$

Equation (13.46) exemplifies the general idea of PSO, but other velocity updating schemes have also been used. In Equation (13.46), w represents an inertia parameter, such that velocities can only change gradually between each step, while a_1 and a_2 are acceleration coefficients controlling the directional change. $R1$ and $R2$ are random numbers in the interval [0,1] that are reassigned for each particle in each iteration, $\mathbf{b}_i$ is the best solution found so far for particle i, while $\mathbf{B}$ is the best solution found so far for all particles, that is the current estimate of the global minimum. The a_1 term thus serves as the particle's own memory of a (local) minimum, while the a_2 term provides communication between the particles regarding a possible global minimum. The $\mathbf{B}$ vector is in some PSO versions not the global best minimum, but only the best within a suitable list of neighbors of particle i, thereby allowing a small pool of global minimum estimates to act as attractors. The two acceleration terms modify the velocities such that they are drawn toward their own best and the collectively best solutions, modified with random factors to provide a spread in the variable space.

The PSO method only requires the ability to calculate the function value, not gradients or higher derivatives, and typical swarm sizes are in the range 20–100. The w, a_1 and a_2 control parameters determine the performance of the optimization, and it has been shown that the velocities remain finite if $a_1 + a_2 > 4w$ (a popular choice is $a_1 = a_2 = 2.05$) and w is in the interval [0,1].[76] A w value close to 1 tends to favor a global optimization, while smaller w values reduce the velocities and make the PSO perform a local optimization. The w parameter is in some cases reduced dynamically during the iterative sequence in order to make a local refinement of an initial wider global search.

A related method called *Gravitational Search Algorithm* (GSA) parameterizes the fitness by a "mass" and allows a collection of "particles" to explore the parameter space by propagating their positions in "time" while interacting by "forces" calculated as the product of their masses and inversely proportional to their distances.[77,78] GSA allows all particles to influence all other particles, modulated by their fitness in terms of mass and distance in parameter space, while PSO has only a few attractors influencing the movement of all particles and emploing a memory effect to prevent particles from collapsing on the current best estimate of the global minimum.

13.6.6 Diffusion Methods

In *Diffusion Methods* the energy function is changed such that it will eventually contain only one minimum (see Figure 13.13).[79] The function may be changed, for example, by adding a contribution proportional to the local curvature of the function (second derivative). This means that minima are raised in energy and saddle points (and maxima) are reduced in energy (negative curvature). Eventually only one minimum remains. Using the single minimum geometry of the modified potential, the process can be reversed, ending up with a minimum on the original surface that often (but not necessarily) is the global minimum. The mathematical description of this process turns out to be identical to the differential equation describing diffusion.

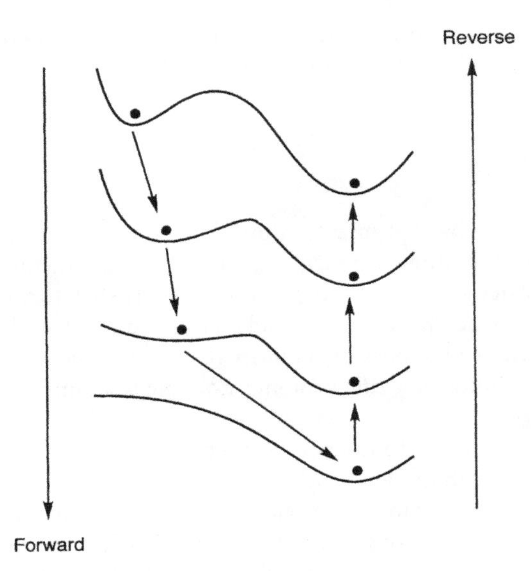

Figure 13.13 Illustration of the diffusion method.

13.6.7 Distance Geometry Methods

The idea in *Distance Geometry* methods is that trial geometries can be generated from a set of lower and upper bounds on distances between all pairs of atoms.[80] The method was originally developed for generating possible geometries based on experimental information such as NMR NOE effects, which place certain constraints on the distance between protons that may be far from each other in terms of bonding. The bonding information itself, however, also places restrictions on distances between all pairs of atoms. Once a set of upper and lower bounds for all pair distances has been assigned, many different trial sets of distances may be generated by selecting random numbers between these limits. Such a random distance matrix can then be translated into a three-dimensional structure, a procedure known as *embedding*. Distance geometry can thus be used for generating trial conformations that can be optimized using conventional methods. The main advantage of the distance geometry method is the ease with which distant constraints between atoms far apart in terms of bonding can be translated into valid trial structures. Without such constraints, some of the other methods in this section are usually more efficient in searching the conformational space.

13.6.8 Characteristics of Global Optimization Methods

As described in the previous sections, it may be clear that MD, MC and stochastic methods tend to primarily sample the local area, generating a relatively large number of local minima in the process. The use of a larger step size in stochastic methods normally means that they are more efficient than MC or MD, at least for systems that are not cyclic or confined. Simulated annealing and diffusion methods, on the other hand, are primarily geared to locating the global minimum, and will in general only produce one final structure, this being the best estimate of the global minimum. Genetic algorithms and swarm optimization methods also focus on the global minimum, but the final population contains a distribution of low-energy structures. Distance geometry methods are more or less random searches, where "impossible" structures are excluded. Simulated annealing normally explores a significantly smaller space than genetic or swarm algorithms. Which method that is best for locating the global minimum depends on the problem at hand.

13.7 Molecular Docking

An important example of a global optimization problem is determining the best alignment of two molecules with respect to each other, a typical example being the fitting of a small molecule into a large protein structure, a process called *docking*. Given an X-ray structure of a protein, preferably with a bound ligand to identify the active site, the ligand can be removed, and other (virtual) compounds may be docked into the active site to possibly identify new molecules with a stronger binding affinity. Since many drugs act by inhibiting specific proteins, docking is an important element in drug design and lead optimization.

The process of docking a ligand into the active site of a rigid protein has six degrees of freedom, three translational and three rotational, besides those arising from the ligand conformations. The three translational degrees of freedom can be sampled on a grid, for example by placing the ligand center of mass within a central box with grid points every 1 Å, which for even a rather small $10 \times 10 \times 10$ Å box generates ~1000 possible points. For each of these points, the overall rotational orientation of the ligand must be sampled, for example by the Euler angles, typically generating a few hundred possibilities. A specific set of intermolecular translational and rotational variables is called a *pose*, and each ligand conformation may thus have $\sim 10^5$ possible poses, even on a rather coarse grid. Even though the majority of these can be rejected based on, for example, atom pair distances between the ligand and receptor, the combinatorial space is still large for even a relatively small ligand. A systematic sampling is often too demanding, so global optimization schemes such as genetic algorithms are often employed for solving the optimization problem. Since the interest is typically to dock perhaps thousands of ligands for (virtual) screening a library of compounds, this furthermore calls for a fast method for estimating the binding energy. Given that the binding energy is the Gibbs free energy, this is clearly a challenging task.

A force field attempt to calculate the enthalpic interaction and an estimate of the free energy by simulation methods (Section 15.5) is much too expensive computationally. Instead, the non-bonded part of a force field function can be augmented with empirical terms, hopefully capturing some of the entropy and solvent effects, and the resulting *scoring function* can be parameterized against experimental binding data. The entropy terms are typically structural descriptors, such as the number of torsional degrees of freedom and the number of hydrogen acceptors and donors, the argument being that fixing of torsional angles by binding to the protein causes a rather constant loss of entropy for each entity:[81–83]

$$\Delta G_{\text{scoring}} = a_1 \Delta E_{\text{vdw}} + a_2 \Delta E_{\text{el}} + a_3 \Delta G_{\text{rot}} + a_4 \Delta G_{\text{H-bond}} + a_5 \Delta G_{\text{solv}} + \cdots \tag{13.47}$$

The a_i weighting factors can be fitted to actual binding data for specific protein–ligand systems.

Developing scoring functions capable of accurately ranking binding energies is an active area of research but it is probably fair to say that no generally accurate scoring function has yet been developed. Some scoring functions employ only force field terms, such as the first two in Equation (13.47), others parameterize it entirely from descriptive terms, such as the last three in Equation (13.47), and some employ a mixture of these. It should be noted that the interaction of the protein atoms with potential ligand atoms at the grid points in the active site is the same for all ligands and their poses, and can therefore be pre-computed to save computational resources. The main purpose of the scoring function is to rank a large number of poses rapidly, from which a smaller number of promising candidates may be subjected to more refined methods for estimating the binding energies.

The main problem of docking ligands into an active site generated by removing an existing ligand from an X-ray structure is that the hole left behind naturally bears a strong resemblance to the compound removed. This will provide a bias for finding compounds differing only slightly from the already-known inhibitor. The fundamental problem is that the flexibility of the protein is neglected, that is the protein is able to some extent to assume different shapes of the active site for different ligands. Taking the enzyme conformational degrees of freedom into consideration during the docking increases the computational problem to essentially unmanageable proportions. A heuristic proposal is to reduce the van der Waals parameters of the protein atoms at the surface of the active site, thereby allowing larger ligands to be docked. Subsequently, the original van der Waals parameters can be reintroduced followed by relaxation of the protein structure, an approach called *induced fit docking*.[84]

13.8 Intrinsic Reaction Coordinate Methods

The optimization methods described in Sections 13.2 to 13.4 concentrate on locating stationary points on an energy surface. The important points for discussing chemical reactions are minima, corresponding to reactant(s) and product(s), and saddle points, corresponding to transition structures. Once a TS has been located, it should be verified that it indeed connects the desired minima. The vibrational normal coordinate associated with the imaginary frequency at the TS is the reaction coordinate (Section 17.2.2), and an inspection of the corresponding atomic motions may be a strong indication that it is the "correct" TS. A rigorous proof, however, requires a determination of the *Minimum Energy Path* (MEP) from the TS to the connecting minima. If the MEP is located in mass-weighted coordinates, it is called the *Intrinsic Reaction Coordinate* (IRC).[85] The IRC path is of special importance in connection with studies of reaction dynamics, since the nuclei will usually stay close to the IRC, and a model for the reaction surface may be constructed by expanding the energy to, for example, second order around points on the IRC (Section 15.2.7). The IRC is formally the reaction path taken in the limit of a zero temperature, and for a modest temperature the deviation from this path is usually small. For high temperatures, however, the favored dynamical path will tend to be the shortest path, regardless of the fact that this may be significantly higher in energy than along the (longer) IRC path.

The IRC path is defined by the differential equation

$$\frac{d\mathbf{x}(s)}{ds} = -\frac{\mathbf{g}}{|\mathbf{g}|} = \mathbf{t} \tag{13.48}$$

Here $\mathbf{x}$ are the (mass-weighted) coordinates, s is the path length and $\mathbf{t}$ is the tangent to the reaction path, defined by the normalized gradient. Determining the IRC requires solving Equation (13.48) starting from a geometry slightly displaced from the TS along the normal coordinate for the imaginary frequency.

The simplest method for integrating Equation (13.48) is the *Euler* method. A series of steps are taken along the tangent direction at the current geometry $\mathbf{x}_n$:

$$\mathbf{x}_{n+1} = \mathbf{x}_n + \Delta s\mathbf{t}(\mathbf{x}_n) \tag{13.49}$$

This corresponds to a steepest descent minimization with a fixed step size Δs. As discussed in Section 13.2.1, such an approach tends to oscillate around the true path and consequently requires a small step size to follow the IRC accurately. The Euler algorithm may be stabilized by using Equation (13.49) as a predictor step followed by a correction step to (partly) account for reaction path curvature.[86] The stabilized Euler integration provides a more accurate IRC following, but requires one additional gradient calculation for each step.

A more advanced method is the *Runge–Kutta* (RK) algorithm. The idea here is to generate some intermediate steps that allow a better and more stable estimate of the next geometry for a given step size. The *second-order* Runge–Kutta (RK2) method first calculates the gradient at a point corresponding to a Euler step with half the step size. The gradient at the halfway point is then used for taking the full step:

$$\begin{gathered}\mathbf{k}_1 = \Delta s\mathbf{t}(\mathbf{x}_n) \quad ; \quad \mathbf{k}_2 = \Delta s\mathbf{t}\left(\mathbf{x}_n + \tfrac{1}{2}\mathbf{k}_1\right) \\ \mathbf{x}_{n+1} = \mathbf{x}_n + \mathbf{k}_2\end{gathered} \tag{13.50}$$

The *fourth-order* Runge–Kutta (RK4) method generates four intermediate gradients, and combines the steps as follows:

$$\begin{gathered}\mathbf{k}_1 = \Delta s\mathbf{t}(\mathbf{x}_n) \quad ; \quad \mathbf{k}_2 = \Delta s\mathbf{t}\left(\mathbf{x}_n + \tfrac{1}{2}\mathbf{k}_1\right) \\ \mathbf{k}_3 = \Delta s\mathbf{t}\left(\mathbf{x}_n + \tfrac{1}{2}\mathbf{k}_2\right) \quad ; \quad \mathbf{k}_4 = \Delta s\mathbf{t}(\mathbf{x}_n + \mathbf{k}_3) \\ \mathbf{x}_{n+1} = \mathbf{x}_n + \tfrac{1}{6}\mathbf{k}_1 + \tfrac{1}{3}\mathbf{k}_2 + \tfrac{1}{3}\mathbf{k}_3 + \tfrac{1}{6}\mathbf{k}_4\end{gathered} \tag{13.51}$$

The Euler and RK methods use only gradient information, but it is also possible to integrate Equation (13.48) using a second-order approximation to the energy surface, analogous to the Newton–Raphson method for locating stationary points. The *Hessian-based Predictor–Corrector* (HPC) algorithm[87] employs the gradient of the second-order Taylor approximation (Equation (13.13)) in Equation (13.48), where $\mathbf{g}$ and $\mathbf{H}$ are the gradient and Hessian at point $\mathbf{x}_0$ and $\Delta\mathbf{x} = \mathbf{x} - \mathbf{x}_0$:

$$\frac{\mathrm{d}\mathbf{x}(s)}{\mathrm{d}s} = -\frac{\mathbf{g} + \mathbf{H}\Delta\mathbf{x}}{|\mathbf{g} + \mathbf{H}\Delta\mathbf{x}|} \tag{13.52}$$

The right-hand side of the differential equation now depends explicitly on $\mathbf{x}$ (Section 16.5.2), in contrast to Equation (13.48). The composite function on the right-hand side of Equation (13.52) can be separated by introducing a parameter p:

$$\frac{\mathrm{d}\mathbf{x}}{\mathrm{d}s} = \frac{\mathrm{d}\mathbf{x}}{\mathrm{d}p}\frac{\mathrm{d}p}{\mathrm{d}s} \tag{13.53}$$

Here p is defined by

$$\frac{\mathrm{d}s}{\mathrm{d}p} = |\mathbf{g} + \mathbf{H}\Delta\mathbf{x}| \tag{13.54}$$

and consequently

$$-\frac{\mathrm{d}\mathbf{x}}{\mathrm{d}p} = (\mathbf{g} + \mathbf{H}\Delta\mathbf{x}) \tag{13.55}$$

The formal solution to Equation (13.55) is given in

$$\mathbf{x}(p) = \mathbf{x}_0 + \mathbf{A}(p)\mathbf{g} \tag{13.56}$$

The **A** matrix depends parametrically on p and the Hessian matrix **H**, as shown below, where λ_i are the Hessian eigenvalues and **U** contains the eigenvectors (see Section 17.2 for how to calculate functions of matrices):

$$\begin{aligned} \mathbf{A}(p) &= \mathbf{U}\alpha\mathbf{U}^t \\ \alpha_i(p) &= \left(\mathrm{e}^{-\lambda_i p} - 1\right)\lambda_i^{-1} \end{aligned} \tag{13.57}$$

The second task is to determine the value of p from Equation (13.54) that produces the desired arc length. Substituting Equation (13.56) into Equation (13.54) gives the following equation expressed in the Hessian eigenvector space, where $\mathbf{g}_i'$ is the projection of the gradient along the Hessian eigenvector i:

$$\frac{\mathrm{d}s}{\mathrm{d}p} = \left(\sum_i (\mathbf{g}_i')^2 \mathrm{e}^{-2\lambda_i p}\right)^{1/2} \tag{13.58}$$

Equation (13.58) is a simple first-order differential equation (Section 17.5.1) that can be solved by numerical integration using a (large) number of (small) steps in p. The predictor step generated from Equations (13.58) and (13.56) is followed by a series of corrector steps on the second-order model surface. Although the HPC algorithm formally requires both first and second derivatives, the exact Hessian matrix can in practise be replaced by an updated version, analogous to the quasi-NR methods in Section 13.2.5. The advantage of the HPC over the Euler or RK methods is that the Hessian information permits larger steps to be taken for a similar accuracy, thus reducing the overall computational cost.

Another method for following the IRC that does not rely on integration of the differential Equation (13.48) has been developed by Gonzales and Schlegel (GS).[88] The idea is to generate points on the IRC by means of a series of constrained optimizations. The algorithm is illustrated in Figure 13.14.

An expansion point is generated by taking a step along the current direction with a step size of $\frac{1}{2}\Delta s$. The energy is then minimized on a hypersphere with radius $\frac{1}{2}\Delta s$, located at the expansion point. This is an example of a constrained optimization that can be handled by means of a Lagrange multiplier (Section 13.5). The GS procedure ensures that the tangent to the IRC path is correct at each point.

Although it is clear that RK4 is more stable and accurate than the Euler method for a given step size, this does not necessarily mean that it is the most efficient method. Since the RK4 method requires four gradient calculations for each step, the simple Euler can employ a step size four times as small for the same computational cost. The HPC method employs Hessian-based information to allow for larger step sizes than the simple Euler scheme. The Gonzales–Schlegel method is also quite tolerant for large

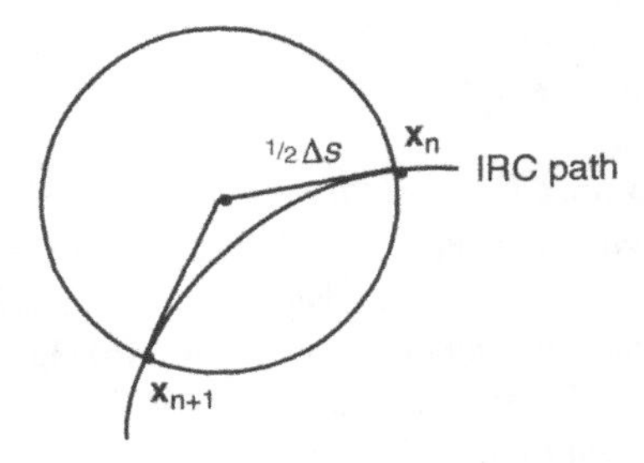

Figure 13.14 Illustration of the Gonzales–Schlegel constrained optimization method for following an IRC.

step sizes, but each constrained optimization may take a number of gradient calculations to converge, which could also be used for advancing the Euler algorithm at a slower pace. Which algorithm is the optimum will depend on the system at hand and the required accuracy of the IRC path. If only the nature of the two minima on each side of the TS is required, a crude IRC is sufficient and a simple Euler algorithm may be the most cost efficient. Note that chain-of-state methods (Section 13.4.3) implicitly define the (approximate) reaction path and by construction ensure that the TS connects the reactant and product. For use in connection with reaction path methods (Section 15.2.7), the IRC needs to be located very accurately, and a sophisticated method and a small step size may be required.

References

1 R. Fletcher, *Practical Methods of Optimization* (John Wiley & Sons, 1980).
2 T. Schlick, *Reviews in Computational Chemistry* **3**, 1 (1992).
3 M. L. McKee and M. Page, *Reviews in Computational Chemistry* **4**, 35 (1993).
4 J. A. Nelder and R. Mead, *Computer Journal* **7** (4), 308–313 (1965).
5 A. Banerjee, N. Adams, J. Simons and R. Shepard, *Journal of Physical Chemistry* **89** (1), 52–57 (1985).
6 P. Culot, G. Dive, V. H. Nguyen and J. M. Ghuysen, *Theoretica Chimica Acta* **82** (3-4), 189-205 (1992).
7 J. M. Anglada and J. M. Bofill, *Journal of Computational Chemistry* **19** (3), 349–362 (1998).
8 D. C. Liu and J. Nocedal, *Mathematical Programming* **45** (3), 503–528 (1989).
9 H. B. Schlegel, *Theoretica Chimica Acta* **66** (5), 333–340 (1984).
10 P. Pulay, *Journal of Computational Chemistry* **3** (4), 556–560 (1982).
11 K. N. Kudin, G. E. Scuseria and E. Cances, *Journal of Chemical Physics* **116** (19), 8255–8261 (2002).
12 P. Csaszar and P. Pulay, *Journal of Molecular Structure* **114** (Mar), 31–34 (1984).
13 F. Eckert, P. Pulay and H. J. Werner, *Journal of Computational Chemistry* **18** (12), 1473–1483 (1997).
14 P. Ettenhuber and P. Jørgensen, *Journal of Chemical Theory and Computation* **11** (4), 1518–1524 (2015).
15 G. Fogarasi, X. F. Zhou, P. W. Taylor and P. Pulay, *Journal of the American Chemical Society* **114** (21), 8191–8201 (1992).
16 J. Baker and P. Pulay, *Journal of Computational Chemistry* **21** (1), 69–76 (2000).
17 P. E. Maslen, *Journal of Chemical Physics* **122** (1), 014104 (2005).
18 C. Y. Peng, P. Y. Ayala, H. B. Schlegel and M. J. Frisch, *Journal of Computational Chemistry* **17** (1), 49–56 (1996).
19 J. Baker, A. Kessi and B. Delley, *Journal of Chemical Physics* **105** (1), 192–212 (1996).
20 F. Jensen and D. S. Palmer, *Journal of Chemical Theory and Computation* **7** (1), 223–230 (2011).
21 V. Bakken and T. Helgaker, *Journal of Chemical Physics* **117** (20), 9160–9174 (2002).
22 E. F. Koslover and D. J. Wales, *Journal of Chemical Physics* **127** (23), 234105 (2007).
23 H. B. Schlegel, *Journal of Computational Chemistry* **24** (12), 1514–1527 (2003).
24 H. B. Schlegel, *Wiley Interdisciplinary Reviews – Computational Molecular Science* **1** (5), 790–809 (2011).
25 T. A. Halgren and W. N. Lipscomb, *Chemical Physics Letters* **49** (2), 225–232 (1977).
26 S. Bell and J. S. Crighton, *Journal of Chemical Physics* **80** (6), 2464–2475 (1984).
27 C. Y. Peng and H. B. Schlegel, *Israel Journal of Chemistry* **33** (4), 449–454 (1993).
28 Y. Abashkin and N. Russo, *Journal of Chemical Physics* **100** (6), 4477–4483 (1994).
29 K. Ohno and S. Maeda, *Chemical Physics Letters* **384** (4–6), 277–282 (2004).
30 S. Maeda and K. Ohno, *Journal of Physical Chemistry A* **109** (25), 5742–5753 (2005).
31 G. T. Barkema and N. Mousseau, *Physical Review Letters* **77** (21), 4358–4361 (1996).

32 M. J. S. Dewar, E. F. Healy and J. J. P. Stewart, *Journal of the Chemical Society – Faraday Transactions II* **80**, 227–233 (1984).
33 C. Cardenaslailhacar and M. C. Zerner, *International Journal of Quantum Chemistry* **55** (6), 429–439 (1995).
34 I. V. Ionova and E. A. Carter, *Journal of Chemical Physics* **98** (8), 6377–6386 (1993).
35 R. A. Miron and K. A. Fichthorn, *Journal of Chemical Physics* **115** (19), 8742–8747 (2001).
36 R. Czerminski and R. Elber, *International Journal of Quantum Chemistry* **S24**, 167–186 (1990).
37 P. Y. Ayala and H. B. Schlegel, *Journal of Chemical Physics* **107** (2), 375–384 (1997).
38 D. A. Liotard, *International Journal of Quantum Chemistry* **44** (5), 723–741 (1992).
39 C. Choi and R. Elber, *Journal of Chemical Physics* **94** (1), 751–760 (1991).
40 S. Fischer and M. Karplus, *Chemical Physics Letters* **194** (3), 252–261 (1992).
41 G. Henkelman and H. Jonsson, *Journal of Chemical Physics* **113** (22), 9978–9985 (2000).
42 H. Chaffey-Millar, A. Nikodem, A. V. Matveev, S. Krüger and N. Rösch, *Journal of Chemical Theory and Computation* **8** (2), 777–786 (2012).
43 G. Henkelman, B. P. Uberuaga and H. Jonsson, *Journal of Chemical Physics* **113** (22), 9901–9904 (2000).
44 P. Maragakis, S. A. Andreev, Y. Brumer, D. R. Reichman and E. Kaxiras, *Journal of Chemical Physics* **117** (10), 4651–4658 (2002).
45 B. Peters, A. Heyden, A. T. Bell and A. Chakraborty, *Journal of Chemical Physics* **120** (17), 7877–7886 (2004).
46 P. Plessow, *Journal of Chemical Theory and Computation* **9** (3), 1305–1310 (2013).
47 S. Maeda, T. Taketsugu and K. Morokuma, *Journal of Computational Chemistry* **35** (2), 166–173 (2014).
48 A. B. Birkholz and H. B. Schlegel, *Journal of Computational Chemistry* **36** (15), 1157–1166 (2015).
49 T. Helgaker, *Chemical Physics Letters* **182** (5), 503–510 (1991).
50 F. Jensen, *Journal of Chemical Physics* **119** (17), 8804–8808 (2003).
51 G. Henkelman and H. Jonsson, *Journal of Chemical Physics* **111** (15), 7010–7022 (1999).
52 A. Heyden, A. T. Bell and F. J. Keil, *Journal of Chemical Physics* **123** (22), 224101 (2005).
53 C. Dellago, P. G. Bolhuis and P. L. Geissler, *Advances in Chemical Physics* **123**, 1–78 (2002).
54 J. E. Basner and S. D. Schwartz, *Journal of the American Chemical Society* **127** (40), 13822–13831 (2005).
55 R. G. Mullen, J.-E. Shea and B. Peters, *Journal of Chemical Theory and Computation* **11** (6), 2421–2428 (2015).
56 A. R. Leach, *Reviews in Computational Chemistry* **2**, 1 (1991).
57 A. E. Howard and P. A. Kollman, *Journal of Medicinal Chemistry* **31** (9), 1669–1675 (1988).
58 C. D. Maranas and C. A. Floudas, *Journal of Chemical Physics* **100** (2), 1247–1261 (1994).
59 C. A. Floudas and C. E. Gounaris, *Journal of Global Optimization* **45** (1), 3–38 (2009).
60 M. Saunders, K. N. Houk, Y. D. Wu, W. C. Still, M. Lipton, G. Chang and W. C. Guida, *Journal of the American Chemical Society* **112** (4), 1419–1427 (1990).
61 F. Villamagna and M. A. Whitehead, *Journal of the Chemical Society – Faraday Transactions* **90** (1), 47–54 (1994).
62 A. Smellie, R. Stanton, R. Henne and S. Teig, *Journal of Computational Chemistry* **24** (1), 10–20 (2003).
63 G. Chang, W. C. Guida and W. C. Still, *Journal of the American Chemical Society* **111** (12), 4379–4386 (1989).
64 I. Kolossvary and W. C. Guida, *Journal of the American Chemical Society* **118** (21), 5011–5019 (1996).
65 I. Kolossvary and G. M. Keseru, *Journal of Computational Chemistry* **22** (1), 21–30 (2001).
66 S. Duane, A. D. Kennedy, B. J. Pendleton and D. Roweth, *Physics Letters B* **195** (2), 216–222 (1987).
67 Y. Sugita and Y. Okamoto, *Chemical Physics Letters* **314** (1–2), 141–151 (1999).
68 D. J. Earl and M. W. Deem, *Physical Chemistry Chemical Physics* **7** (23), 3910–3916 (2005).

69 S. Kirkpatrick, C. D. Gelatt and M. P. Vecchi, *Science* **220** (4598), 671–680 (1983).
70 S. R. Wilson and W. L. Cui, *Biopolymers* **29** (1), 225–235 (1990).
71 L. Ingber, *Mathematical and Computer Modelling* **18** (11), 29–57 (1993).
72 R. S. Judson, *Reviews in Computational Chemistry* **10**, 1 (1997).
73 M. N. Ab Wahab, S. Nefti-Meziani and A. Atyabi, *PloS One* **10** (5), 0122827 (2015).
74 G. M. Morris, D. S. Goodsell, R. S. Halliday, R. Huey, W. E. Hart, R. K. Belew and A. J. Olson, *Journal of Computational Chemistry* **19** (14), 1639–1662 (1998).
75 R. Poli, J. Kennedy and T. Blackwell, *Swarm Intelligence* **1** (1), 33–57 (2007).
76 M. Clerc and J. Kennedy, *IEEE Transactions on Evolutionary Computation* **6** (1), 58–73 (2002).
77 E. Rashedi, H. Nezamabadi-Pour and S. Saryazdi, *Information Sciences* **179** (13), 2232–2248 (2009).
78 T. Dash and P. K. Sahu, *Journal of Computational Chemistry* **36** (14), 1060–1068 (2015).
79 J. Kostrowicki and H. A. Scheraga, *Journal of Physical Chemistry* **96** (18), 7442–7449 (1992).
80 J. M. Blaney and J. S. Dixon, in *Reviews in Computational Chemistry*, edited by K. B. Lipkowitz and D. B. Boyd (John Wiley & Sons, 1994), Vol. 5, pp. 299–335.
81 I. Muegge and M. Rarey, in *Reviews in Computational Chemistry*, edited by K. B. Lipkowitz and D. B. Boyd (John Wiley & Sons, 2001), Vol. 17, pp. 1–60.
82 R. X. Wang, Y. P. Lu and S. M. Wang, *Journal of Medicinal Chemistry* **46** (12), 2287–2303 (2003).
83 T. Cheng, X. Li, Y. Li, Z. Liu and R. Wang, *Journal of Chemical Information and Modeling* **49** (4), 1079–1093 (2009).
84 W. Sherman, T. Day, M. P. Jacobson, R. A. Friesner and R. Farid, *Journal of Medicinal Chemistry* **49** (2), 534–553 (2006).
85 K. Fukui, *Accounts of Chemical Research* **14** (12), 363–368 (1981).
86 K. Ishida, K. Morokuma and A. Komornicki, *Journal of Chemical Physics* **66** (5), 2153–2156 (1977).
87 H. P. Hratchian and H. B. Schlegel, *Journal of Chemical Theory and Computation* **1** (1), 61–69 (2005).
88 C. Gonzalez and H. B. Schlegel, *Journal of Chemical Physics* **90** (4), 2154–2161 (1989).

14

Statistical Mechanics and Transition State Theory

The separation of the nuclear and electronic degrees of freedom by the Born–Oppenheimer approximation leads to a mental picture of a chemical reaction as nuclei moving on a potential energy surface. The easiest path from one minimum to another, that is for transforming one chemical species to another, is along the reaction path having the lowest energy. The highest energy point along this path is the transition structure and the energy relative to the reactant completely determines the reaction rate within *Transition State Theory* (TST). Transition state theory is a semi-classical theory where the quantum nature is taken into account by means of the quantization of vibrational and rotational energy states. The connection between the properties of a single molecule and the experimental conditions employing a very large number of species is given by statistical mechanics, which provides a framework for performing the statistical averaging over a very large number of possible energy distributions. The averaging can for an ideal gas be performed in a closed analytical form within the rigid-rotor harmonic-oscillator approximation. For systems in condensed states, that is liquid or solid states, the averaging must be done by explicitly sampling the phase space.

14.1 Transition State Theory

Consider a chemical reaction of the type A + B → C + D. The rate of reaction may be written as in Equation (14.1), with k_{rate} being the rate constant:

$$\frac{\mathrm{d}\,[\mathrm{C}]}{\mathrm{d}t} = \frac{\mathrm{d}\,[\mathrm{D}]}{\mathrm{d}t} = -\frac{\mathrm{d}\,[\mathrm{A}]}{\mathrm{d}t} = -\frac{\mathrm{d}\,[\mathrm{B}]}{\mathrm{d}t} = k_{\text{rate}}\,[\mathrm{A}]\,[\mathrm{B}] \tag{14.1}$$

If k_{rate} is known, the concentration of the various species can be calculated at any given time from the initial concentrations. At the *microscopic* level, the rate constant is a function of the quantum states of A, B, C and D, that is the electronic, translational, rotational and vibrational quantum numbers. The *macroscopic* rate constant is an average over such "microscopic" rate constants, weighted by the probability of finding a molecule with a given set of quantum numbers. For systems in equilibrium, the probability of finding a molecule in a certain state depends on its energy by means of the Boltzmann distribution and the macroscopic rate constant thereby becomes a function of temperature.

Stable molecules correspond to minima on the potential energy surface within the Born–Oppenheimer approximation and a chemical reaction can be described as nuclei moving from one

Introduction to Computational Chemistry, Third Edition. Frank Jensen.

Companion Website: http://www.wiley.com/go/jensen/computationalchemistry3

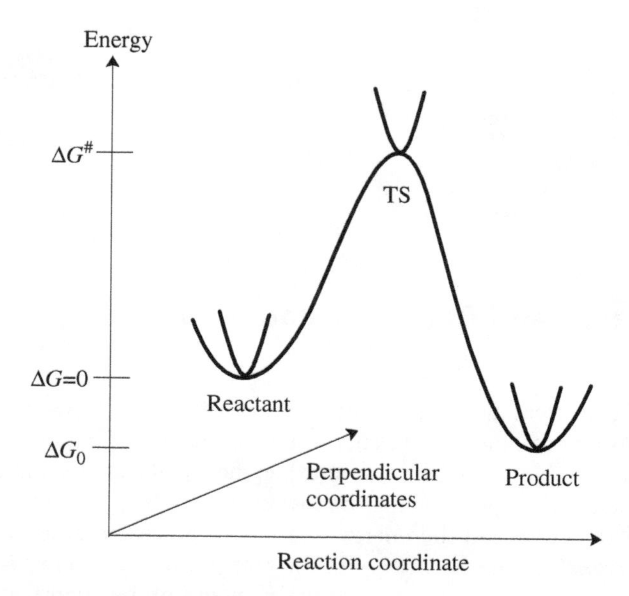

Figure 14.1 Schematic illustration of a reaction path.

minimum to another. In the lowest level of approximation, the motion is assumed to occur along the path of least energy and this path forms the basis for transition state theory.[1] The *Transition State* is the configuration that divides the reactant and product parts of the surface (i.e. a molecule that has reached the transition state will continue on to the product), while the geometrical configuration of the energy maximum along the reaction path (Figure 14.1) is called the *Transition Structure*. The transition state is thus a macroscopic ensemble with a Boltzmann energy distribution, while the transition structure refers to the microscopic system. The two terms are often used interchangeably and share the same acronym, TS. In the multidimensional case, the TS is a first-order saddle point on the potential energy surface, a maximum in the reaction coordinate direction and a minimum along all other coordinates.

TST is a semi-classical theory where the motion along the reaction coordinate is treated classically, while the perpendicular directions take into account the quantization of, for example, the vibrational energy. It furthermore assumes an equilibrium energy distribution among all possible quantum states at all points along the reaction coordinate. The probability of finding a molecule in a given quantum state is proportional to $e^{-\Delta E/kT}$, which is a Boltzmann distribution. Assuming that the molecules at the TS are in equilibrium with the reactant, the macroscopic rate constant can be expressed as

$$\begin{aligned} k_{\text{rate}} &= \frac{kT}{h} e^{-\Delta G^{\neq}/RT} \\ \Delta G^{\neq} &= \Delta G_{\text{TS}} - \Delta G_{\text{reactant}} \end{aligned} \tag{14.2}$$

$\Delta G^{\neq}$ is the Gibbs free energy difference between the TS and reactant, and k is Boltzmann's constant. The TST expression only holds if all molecules that pass from the reactant over the TS go on to the product. The dividing surface separating the reactant from the product is a hyperplane perpendicular to the reaction coordinate at the TS. The TST assumption is that no re-crossings occur, that is all molecules passing through the dividing surface will go on to form the product. These assumptions mean that the rate constant calculated from Equation (14.2) will be an upper limit to the true rate constant. In more refined models, the dividing surface may be located by minimizing the flux through the surface, that is forming the dynamical bottleneck for the reaction, for example by taking dynamics effects into account. To allow for "re-crossings", where a molecule passes over the TS but is reflected back to the reactant side, a *transmission coefficient* κ is sometimes introduced. This factor also allows for the quantum mechanical phenomenon of tunnelling, that is molecules that have insufficient energy to pass *over* the TS may tunnel *through* the barrier and appear on the product side. The transmission coefficient is difficult to calculate but is usually close to 1 and for many systems are in the range 0.5–2. At low temperatures the tunneling contribution dominates, leading to $\kappa > 1$, while the re-crossing effect is the most important at high temperatures, giving $\kappa < 1$. For the majority of reactions the calculated accuracy in $\Delta G^{\neq}$ introduces errors much larger than a factor of 2 and the transmission coefficient is usually ignored.

From the TST expression in Equation (14.2) it is clear that if the free energy of the reactant and TS can be calculated, the reaction rate follows trivially. Similarly, the equilibrium constant for a reaction can be calculated from the free energy difference between the reactant(s) and product(s):

$$K_{\text{eq}} = e^{-\Delta G_0/RT} \tag{14.3}$$

The Gibbs free energy is given in terms of the enthalpy and entropy, $G = H - TS$, and the enthalpy and entropy for a macroscopic ensemble of particles may be calculated from properties of a relatively few molecules by means of statistical mechanics, as discussed in Section 14.4.

The picture in Figure 14.1 relates to chemical reactions occurring on a single energy surface, as is typical for a thermal reaction. Photochemical reactions, on the other hand, occur on at least two and possibly more surfaces. The reaction is initiated by absorption of a photon to produce an excited state with the same nuclear coordinates as the ground state. This geometry will rarely be a stationary point on the excited surface and the resulting nuclear movements may be explored by minimization or dynamical methods analogous to those on the ground state (Chapters 13 and 15). At some point, however, the system must return to the ground electronic surface. While this can occur by a radiative transition (fluorescence or phosphorescence), it may also occur by a non-radiative process where the excess energy is transferred to vibrational energy on the ground state surface. The probability for the latter process depends on the energy difference between the two surfaces, and therefore has a tendency of occurring at nuclear geometries where the two surfaces "touch" each other, and these points are known as *conical intersections*.[2] Two energy surfaces of the same symmetry cannot cross for a diatomic system and instead make an avoided crossing, as illustrated in Figure 3.2. In a multidimensional system, however, there is no such restriction, and two energy surfaces may have the same energy for the same set of nuclear coordinates. Locating conical intersections is a constrained optimization problem, involving finding a set of nuclear coordinates where two different energy functions have the same value, and also involves the non-adiabatic coupling elements between different wave functions, as discussed in Section 3.1.

A conical intersection may be thought of as a "funnel" in the $N - 2$-dimensional subspace that serves as the dynamical bottleneck for a photochemical reaction, analogously to the TS for a thermal reaction. It should be noted, however, that the transition between the excited and ground state surfaces may occur over quite a wide range of nuclear configurations, making the concept of a single "transition structure" somewhat blurred. Furthermore, for many systems it is the movement on the excited state surface to achieve the geometry of the conical intersection that limits the reaction rate, and not the actual transition between the two surfaces at the conical intersection.

14.2 Rice–Ramsperger–Kassel–Marcus Theory

The canonical TST theory in Section 14.1 assumes fast energy exchange with the surroundings, that is that the reacting molecule is in thermal equilibrium with the environment. For unimolecular reactions in the gas phase this assumption may not hold, especially not if the pressure is low (e.g. fragmentations in a mass spectrometer). TST may alternatively be formulated in terms of the total energy, also known as *microcanonical* TST. When applied to unimolecular reactions, this is usually known as *Rice–Ramsperger–Kassel–Marcus* (RRKM) theory. The fundamental assumption here is that no re-crossing occurs for a given total energy of the molecule.

Consider a reaction where a molecule A acquires energy by collision with a molecule M (which may be the same as A) to form an energized molecule A*, with the energy being distributed between the translation, rotation and vibrational degrees of freedom. The vibrational energy can be transferred between the different modes owing to vibrational anharmonicity and if it is higher than the activation energy $E^{\neq}$, it may at some point accumulate in a specific mode to reach an activated state $A^{\#}$ (transition state), leading to a chemical product P:

$$\begin{aligned} &\mathrm{A} + \mathrm{M} \underset{k_{-1}}{\overset{k_1}{\rightleftharpoons}} \mathrm{A}^* + \mathrm{M} \\ &\mathrm{A}^* \xrightarrow{k_2} \mathrm{A}^{\#} \xrightarrow{k^{\#}} \mathrm{P} \end{aligned} \tag{14.4}$$

Assuming that the decay rate $k^{\#}$ for the activated $A^{\#}$ is much faster than k_2, the rate for production of P can be written in terms of the k_1, k_{-1} and k_2 constants by making a steady-state approximation for A^*, as shown by

$$\frac{\mathrm{d[P]}}{\mathrm{d}t} = k_2[\mathrm{A}^*] = \frac{k_1 k_2[\mathrm{M}][\mathrm{A}]}{k_{-1}[\mathrm{M}] + k_2} = k_{\mathrm{eff}}[\mathrm{A}] \tag{14.5}$$

The effective rate constant k_{eff} is thus a function of the concentration of M, that is the pressure of the gas. The amount of energy transferred to A^* by M will be a variable, and the rate constants for the activation and reaction (but not the deactivation) will depend on the energy, that is $k_1(E)$ and $k_2(E)$. The effective rate constant in a small energy interval around E is obtained by rearranging Equation (14.5):

$$k_{\mathrm{eff}}(E + \mathrm{d}E) = \frac{(\mathrm{d}k_1(E)/k_{-1})\, k_2(E)}{1 + k_2(E)/k_{-1}[\mathrm{M}]} \tag{14.6}$$

The ratio k_1/k_{-1} is the equilibrium constant for the first step in Equation (14.4) and $\mathrm{d}k_1(E)/k_{-1}$ is the probability of A^* being in a state with energy E, $P(E)$. The $k_{-1}[\mathrm{M}]$ factor is the collision frequency for deactivation that is usually denoted by ω. The unimolecular rate constant can be obtained by integrating the effective rate constant over all energies higher than the activation energy:

$$k_{\mathrm{uni}} = \int_{E^{\neq}}^{\infty} \frac{k_2(E)}{1 + k_2(E)/\omega} P(E)\, \mathrm{d}E \tag{14.7}$$

The probability factor $P(E)$ is given by a Boltzmann distribution for the reactant, while $k_2(E)$ is determined by the number of vibrational quantum states for the activated state $A^{\#}$. The details are sufficiently complex that the reader is referred to more specialized textbooks,[3,4] but the essence of Equation (14.7) is that the rate constant can be evaluated from the geometries and vibrational frequencies of the reactant and activated complex. In the fast energy exchange limit (i.e. $\omega \rightarrow \infty$) the

RRKM expression becomes equivalent to the TST expression (Equation (14.2)). RRKM calculations typically assume harmonic vibrations, which may be poor for high-barrier reactions where the vibrational anharmonicity significantly increases the state count. An exact calculation of all the anharmonic vibrational states, however, is a significant computational undertaking.

14.3 Dynamical Effects

The inherent assumption of both TST and RRKM is that the internal (vibrational) energy redistribution is significantly faster than the timescale for breaking/forming a bond. This means that the reaction rate only depends on the total *amount* of internal energy, not on *how* the energy is acquired. In other words, the reaction is independent of whether the energy is supplied by excitation of bending or stretching vibrations. In the large majority of chemical reactions, this is probably a valid assumption. For certain reactions where the reaction path involves an intermediate, however, the product distribution indicates that the energy is not completely randomized for the intermediate, that is the timescale for internal redistribution is comparable to that for the progression along the reaction coordinate.[5] A specific example is shown in Figure 14.2.

The reaction in Figure 14.2 involves a biradical intermediate and if it has a sufficiently long lifetime, the thermal randomization of the energy should lead to a symmetric product distribution. Experimentally, however, the *exo* product is found to be favored over the *endo* isomer by 4 : 1. Given that the potential energy surface is symmetric, this has been interpreted as a non-statistical distribution of the internal kinetic energy in the bond-forming step.[6] The bond-breaking reaction occurs from a sample of molecules with a Boltzmann energy distribution, but the small fraction of molecules that in a given timeframe actually reacts must necessarily have the energy localized in the C—N bonds. The molecules passing over the first TS (breaking the C—N bonds) therefore enter the biradical minimum on the potential energy surface with the nuclear kinetic energy in a non-random fashion. In the example shown in Figure 14.2, a direct continuation of the nuclear movement arising from breaking both the C—N bonds leads to the *exo* product. The favoring of the *exo* product can thus be explained by the molecules "surfing" over the intermediate minimum on the potential energy surface, rather than

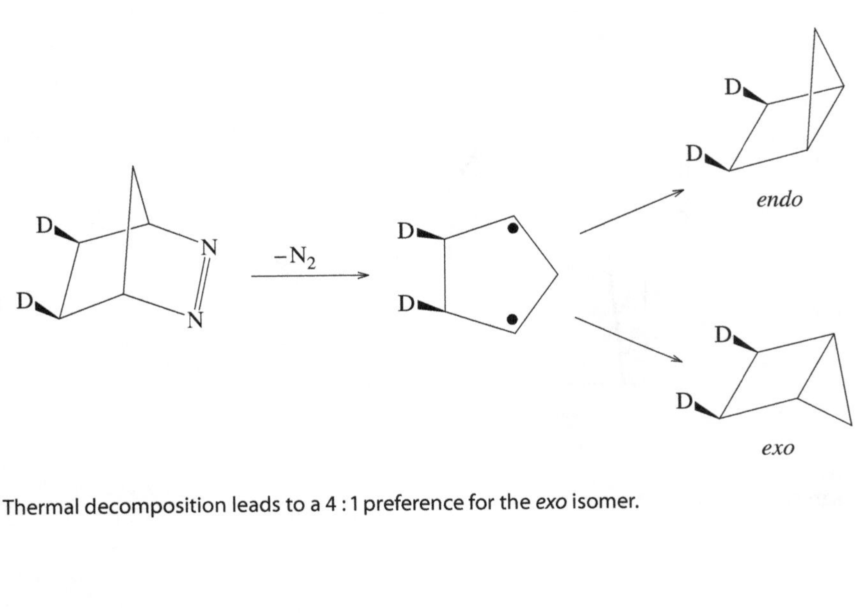

Figure 14.2 Thermal decomposition leads to a 4 : 1 preference for the *exo* isomer.

being trapped and thermally randomized. Describing such effects requires an explicit simulation of the dynamics, as discussed in Section 15.2.

14.4 Statistical Mechanics

Most experiments are performed on macroscopic samples, containing perhaps $\sim 10^{20}$ particles. Calculations, on the other hand, are performed on relatively few particles, typically $1–10^3$, or up to 10^6 in special cases. The (macroscopic) result of an experimental measurement can be connected with properties of the microscopic system. The temperature, for example, is related to the average kinetic energy of the particles:

$$\langle E_{\text{kin}} \rangle = \tfrac{3}{2} RT \tag{14.8}$$

The connection between properties of a microscopic system and a macroscopic sample is provided by *statistical mechanics*.

All molecules are in their energetic ground state at a temperature of 0 K but they have a distribution of all possible (quantum) energy states at a finite temperature. The relative probability P of a molecule being in a state with an energy ε at a temperature T is given by a Boltzmann factor:

$$P \propto e^{-\varepsilon/kT} \tag{14.9}$$

The exponential dependence on the energy means that there is a low (but non-zero) probability for finding a molecule in a high-energy state. This decreased probability for high-energy states is partly offset by the fact that there are many more states with high energy than with low energy. The most probable energy of a molecule in a macroscopic ensemble is therefore not necessarily the one with lowest energy, and a typical distribution is shown in Figure 14.3.

The key feature in statistical mechanics is the *partition function*.[7,8] Just as the wave function is the cornerstone in quantum mechanics (from which everything else can be calculated by applying proper operators), the partition function allows calculation of all macroscopic functions in statistical mechanics.

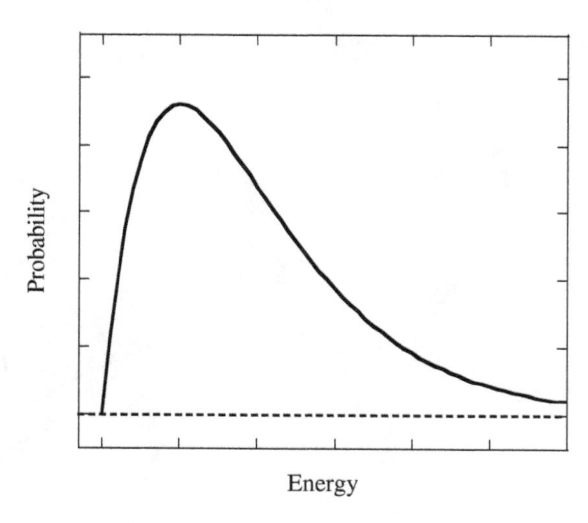

Figure 14.3 Boltzmann energy distribution.

The partition function for a single molecule is usually denoted q and is defined as a sum of exponential terms involving all possible quantum energy *states*:

$$q = \sum_{i=\text{states}}^{\infty} \mathrm{e}^{-\varepsilon_i/kT} \tag{14.10}$$

The partition function can also be written as a sum over all distinct energy *levels*, multiplied by a degeneracy factor g_i that indicates how many states there are with the same energy ε_i:

$$q = \sum_{i=\text{leves}}^{\infty} g_i \mathrm{e}^{-\varepsilon_i/kT} \tag{14.11}$$

The partition function can be considered as an average excited state number-operator, since it is the probability-weighted sum of energy states, each counted with a factor of 1. It may also be viewed as the normalization factor for the Boltzmann probability distribution:

$$P(\varepsilon_i) = q^{-1}\mathrm{e}^{-\varepsilon_i/kT} \tag{14.12}$$

The partition function q is for a single particle; the corresponding quantity Q for a collection of N identical *non-interacting* particles (ideal gas) is given in Equation (14.14):

$$Q = q^N \quad \text{(different particles, non-interacting)} \tag{14.13}$$

$$Q = \frac{q^N}{N!} \quad \text{(identical particles, non-interacting)} \tag{14.14}$$

If the particles are *interacting* (liquid or solid state), the partition function Q must be calculated by summing over all energy states E_i for the whole system. Note that Q here describes the whole system consisting of N interacting particles and the energy states E_i are consequently for all the particles:

$$Q = \sum_{i}^{\infty} \mathrm{e}^{-E_i/kT} \tag{14.15}$$

Owing to the closely spaced energy levels, quantum effects can often be neglected and the state distribution treated as continuous. This corresponds to replacing the discrete sum over energies by an integral over all coordinates (**r**) and momentum (**p**), called the *phase space*:

$$Q = \int \mathrm{e}^{-E(\mathbf{r},\mathbf{p})/kT} \mathrm{d}\mathbf{r}\, \mathrm{d}\mathbf{p} \tag{14.16}$$

More correctly, the partition function in Equation (14.16) should be written in terms of the Hamiltonian for the system, that is replacing E with $\mathbf{H}$. The kinetic and potential energy components, however, can be separated ($\mathbf{H} = \mathbf{T} + \mathbf{V}$). In the absence of a potential energy, the Hamiltonian is purely kinetic energy and the system is an ideal gas. The interesting component is therefore the potential energy part in the partition function, which we denote by E. In the large majority of cases, the energy E is of the force field type described in Chapter 2.

The significance of the partition function Q is that thermodynamic functions, such as the internal energy U and Helmholtz free energy A ($A = U - TS$) can be calculated from it:

$$U = kT^2 \left(\frac{\partial \ln Q}{\partial T}\right)_V \tag{14.17}$$

$$A = -kT\ln Q \tag{14.18}$$

Macroscopic observables, such as pressure P and heat capacity at constant volume C_V, may be calculated as derivatives of thermodynamic functions:

$$P = -\left(\frac{\partial A}{\partial V}\right)_T = kT\left(\frac{\partial \ln Q}{\partial V}\right)_T \tag{14.19}$$

$$C_V = \left(\frac{\partial U}{\partial V}\right)_V = 2kT\left(\frac{\partial \ln Q}{\partial T}\right)_V + kT^2\left(\frac{\partial \ln^2 Q}{\partial T^2}\right)_V \tag{14.20}$$

Other thermodynamic functions, such as the enthalpy H, the entropy S and Gibbs free energy G, can be constructed from these relations:

$$H = U + PV = kT^2\left(\frac{\partial \ln Q}{\partial T}\right)_V + kTV\left(\frac{\partial \ln Q}{\partial V}\right)_T \tag{14.21}$$

$$S = \frac{U - A}{T} = kT\left(\frac{\partial \ln Q}{\partial T}\right)_V + k \ln Q \tag{14.22}$$

$$G = H - TS = kTV\left(\frac{\partial \ln Q}{\partial V}\right)_T - kT \ln Q \tag{14.23}$$

Note the difference between energetic properties such as U, P and H, which all depend on derivatives of Q, and entropic properties such as A, S and G, which depend directly on Q. For simplicity, we will use U and A for illustrations in the following, but other quantities such as H and S can be treated completely analogously.

In order to calculate the partition function q (Q), one needs to know *all* possible quantum states for the system. These can in principle be calculated by solving the nuclear Schrödinger equation, once a suitable potential energy surface is available, for example from solving the electronic Schrödinger equation. Such a rigorous approach is only possible for di- and triatomic systems. The energy levels for a single conformation of an isolated polyatomic molecule can be calculated within the *Rigid-Rotor Harmonic-Oscillator* (RRHO) approximation, where the electronic, vibrational and rotational degrees of freedom are assumed to be separable. Additional conformations can be included straightforwardly by simply offsetting the energy scale relative to the most stable conformation. An isolated molecule corresponds to an ideal gas state and the partition function can be calculated *exactly* for such a system within the RRHO approximation, as discussed in Section 14.5.

The intermolecular interaction for a condensed phase (liquid, solution, solid) is comparable to or larger than a typical kinetic energy, and no separation of degrees of freedom is possible. Calculating the partition function by summing over all energy levels, or integrating over all phase space, is therefore impossible. It is, however, possible by sampling to estimate differences in Q and derivatives such as $\partial \ln Q/\partial T$ from a representative sample of the phase space, as discussed in Section 14.6.

14.5 The Ideal Gas, Rigid-Rotor Harmonic-Oscillator Approximation

The total energy for an isolated molecule can be approximated as a sum of terms involving translational, rotational, vibrational and electronic states, and this is a good approximation for the large majority of systems. For linear, "floppy" (soft bending potential) molecules the separation of the rotational and vibrational modes may be problematic. If two energy surfaces come close together (avoided crossing), the separability of the electronic and vibrational modes may be a poor approximation (breakdown of the Born–Oppenheimer approximation, Section 3.1).

There are in principle also energy levels associated with nuclear spins. In the absence of an external magnetic field, these are degenerate and consequently contribute a constant term to the partition function. As nuclear spins do not change during chemical reactions, we will ignore this contribution.

The assumption that the energy can be written as a sum of terms implies that the partition function can be written as a product of terms. As the enthalpy and entropy contributions involve taking the logarithm of q, the product of q's thus transforms into sums of enthalpy and entropy contributions:

$$\begin{aligned}
\varepsilon_{\text{tot}} &= \varepsilon_{\text{trans}} + \varepsilon_{\text{rot}} + \varepsilon_{\text{vib}} + \varepsilon_{\text{elec}} \\
q_{\text{tot}} &= q_{\text{trans}} q_{\text{rot}} q_{\text{vib}} q_{\text{elec}} \\
H_{\text{tot}} &= H_{\text{trans}} + H_{\text{rot}} + H_{\text{vib}} + H_{\text{elec}} \\
S_{\text{tot}} &= S_{\text{trans}} + S_{\text{rot}} + S_{\text{vib}} + S_{\text{elec}}
\end{aligned} \tag{14.24}$$

The sum over allowed quantum states runs to infinity for each of the partition functions. However, since the energies become larger, the partition functions are finite. Let us examine each of the q factors in a little more detail.

14.5.1 Translational Degrees of Freedom

The translational degrees of freedom can be exactly separated from the other $3N - 3$ coordinates. The allowed quantum states for the translational energy are determined by placing the molecule in a "box", that is the potential is zero inside the box but infinite outside. The only purpose of the box is to allow normalization of the translational wave function, that is the exact size is not important. The solutions to the Schrödinger equation for such a "particle in a box" are standing waves, cosine and sine functions. The energy levels for a one-dimensional box of length L are associated with a quantum number n, and depend only on the total molecular mass M:

$$\varepsilon_n = \frac{n^2 h^2}{8ML^2} \tag{14.25}$$

Although the energy levels are quantized, the energy difference between levels is so small that the distribution can be treated as continuous. The summation involved in the partition function can therefore be replaced by an integral (an integral is just a sum in the limit of infinitely small contributions):

$$q_{\text{trans}} = \sum_{n=0}^{\infty} e^{-\varepsilon_n/kT} \approx \int_0^{\infty} e^{-\varepsilon_n/kT} dn \tag{14.26}$$

Inserting the energy expression and performing the integration gives

$$q_{\text{trans}} = \left(\frac{2\pi MkT}{h^2} \right)^{3/2} V \tag{14.27}$$

The only molecular parameter that enters is the total molecular mass M. The volume depends on the number of particles. It is customary to work on a molar scale, in which case V is the volume of 1 mol of (ideal) gas.

14.5.2 Rotational Degrees of Freedom

The rotation of a molecule is in the lowest approximation assumed to occur with a geometry that is independent of the rotational and vibrational quantum numbers. A more refined treatment allows the geometry to "stretch" with rotational energy, which may be described by adding a "centrifugal" correction, and such corrections are typically of the order of a few percent. The presence of vibrational anharmonicity will furthermore cause the effective geometry to depend on the vibrational quantum state. Within the *rigid-rotor* approximation these effects are neglected, that is the rotation of the molecule is assumed to occur with a fixed geometry.

The energy levels calculated from the Schrödinger equation for a diatomic "rigid rotor" are given in terms of a quantum number J running from zero to infinity and the moment of inertia I:

$$\varepsilon_J = J\,(J+1)\,\frac{h^2}{8\pi^2 I} \tag{14.28}$$

The moment of inertia is calculated from the atomic masses m_1 and m_2 and the distances r_1 and r_2 of the nuclei relative to the center of mass:

$$I = m_1 r_1^2 + m_2 r_2^2 \tag{14.29}$$

The moment of inertia is for all molecules, except very light species such as H_2 and LiH, so large that the spacing between the rotational energy levels is much smaller than kT at ambient temperatures. As for q_{trans}, this means that the summation in Equation (14.10) can be replaced by an integral:

$$q_{\mathrm{rot}} = \sum_{J=0}^{\infty} \mathrm{e}^{-\varepsilon_J/kT} \approx \int_0^{\infty} \mathrm{e}^{-\varepsilon_J/kT}\,\mathrm{d}J \tag{14.30}$$

Performing the integration yields

$$q_{\mathrm{rot}} = \frac{8\pi^2 IkT}{h^2\sigma} \tag{14.31}$$

The symmetry index σ is 2 for a homonuclear system and 1 for a heteronuclear diatomic molecule.

For a polyatomic molecule, the equivalent of Equation (14.29) is a 3×3 matrix:

$$\mathbf{I} = \begin{pmatrix} \sum_i m_i\left(y_i^2+z_i^2\right) & -\sum_i m_i x_i y_i & -\sum_i m_i x_i z_i \\ -\sum_i m_i x_i y_i & \sum_i m_i\left(x_i^2+z_i^2\right) & -\sum_i m_i y_i z_i \\ -\sum_i m_i x_i z_i & -\sum_i m_i y_i z_i & \sum_i m_i\left(x_i^2+y_i^2\right) \end{pmatrix} \tag{14.32}$$

Here the coordinates are again relative to the center of mass. By choosing a suitable coordinate transformation, this matrix may be diagonalized (Section 17.2), with the eigenvalues being the *moments of inertia* and the eigenvectors called *principal axes of inertia*.

The rotational energy levels for a general polyatomic molecule cannot be written in a simple form. A good approximation, however, can be obtained from classical mechanics, resulting in the following partition function:

$$q_{\mathrm{rot}} = \frac{\sqrt{\pi}}{\sigma}\left(\frac{8\pi^2 kT}{h^2}\right)^{3/2}\sqrt{I_1 I_2 I_3} \tag{14.33}$$

Here I_i are the three moments of inertia. The symmetry index σ is the order of the rotational subgroup in the molecular point group (i.e. the number of proper symmetry operations); for H_2O it is 2, for

NH_3 it is 3, for benzene it is 12, etc. The rotational partition function requires only information about the atomic masses and positions (Equation (14.32)), that is the molecular geometry.

14.5.3 Vibrational Degrees of Freedom

The molecular vibrations may in the lowest approximation be described as those of a harmonic oscillator. Harmonic vibrations can be derived by expanding the energy as a function of the nuclear coordinates in a Taylor series around the equilibrium geometry. For a diatomic molecule, the only relevant coordinate is the internuclear distance R:

$$E(R) = E(R_0) + \frac{\mathrm{d}E}{\mathrm{d}R}(R - R_0) + \frac{1}{2}\frac{\mathrm{d}^2E}{\mathrm{d}R^2}(R - R_0)^2 + \frac{1}{6}\frac{\mathrm{d}^3E}{\mathrm{d}R^3}(R - R_0)^3 + \cdots \tag{14.34}$$

The first term may be taken as zero, since this is just the zero point for the energy. The second term (the gradient) vanishes since the expansion is around the equilibrium geometry. Keeping only the lowest non-zero term results in the harmonic approximation, where k is the force constant:

$$E(\Delta R) \cong \frac{1}{2}\frac{\mathrm{d}^2E}{\mathrm{d}R^2}\Delta R^2 = \tfrac{1}{2}k\Delta R^2 \tag{14.35}$$

Including higher-order terms leads to anharmonic corrections to the vibration, and such effects are typically of the order of a few percent. The energy levels obtained from the Schrödinger equation for a one-dimensional harmonic oscillator (diatomic system) are given in

$$\begin{aligned} \varepsilon_n &= \left(n + \tfrac{1}{2}\right) h\nu \\ \nu &= \frac{1}{2\pi}\sqrt{\frac{k}{\mu}} \quad ; \quad \mu = \frac{m_1 m_2}{m_1 + m_2} \end{aligned} \tag{14.36}$$

Here n is a quantum number running from zero to infinity and ν is the vibrational frequency given in terms of the force constant k ($\partial^2E/\partial R^2$) and the reduced mass μ.

In contrast to the translational and rotational energy levels, the spacing between vibrational energy levels is comparable to kT for temperatures around 300 K, and the summation for q_{vib} (Equation (14.10)) cannot be replaced by an integral. Due to the regular spacing, however, the infinite summation can be written in a closed form:

$$\begin{aligned} q_{\mathrm{vib}} &= \sum_{n=0}^{\infty} \mathrm{e}^{-\varepsilon_n/kT} = \mathrm{e}^{-h\nu/2kT} + \mathrm{e}^{-3h\nu/2kT} + \mathrm{e}^{-5h\nu/2kT} + \cdots \\ q_{\mathrm{vib}} &= \mathrm{e}^{-h\nu/2kT}(1 + \mathrm{e}^{-h\nu/kT} + \mathrm{e}^{-2h\nu/kT} + \cdots) \\ q_{\mathrm{vib}} &= \frac{\mathrm{e}^{-h\nu/2kT}}{1-\mathrm{e}^{-h\nu/kT}} \end{aligned} \tag{14.37}$$

Each successive term in the infinite sum is smaller than the previous by a constant factor ($\mathrm{e}^{-h\nu/kT}$, which is <1), and can therefore be expressed in a closed form. Calculating the vibrational partition function for a harmonic oscillator thus requires the second derivative of the energy and the atomic masses.

The force constant k is for a polynuclear molecule replaced by a $3N_{\mathrm{atom}} \times 3N_{\mathrm{atom}}$ matrix containing all the second derivatives of the energy with respect to the coordinates. By mass-weighting and transforming to a new coordinate system called the vibrational normal coordinates, this may be brought to a diagonal form (see Section 17.2.2 for details). In the vibrational normal coordinates, the $3N$-dimensional Schrödinger equation can be separated into $3N$ one-dimensional equations, each having the form of a harmonic oscillator. Of these, three describe the overall translation and three (two for a linear molecule) describe the overall rotation, leaving $3N - 6(5)$ vibrations.

If the stationary point is a minimum on the energy surface, the eigenvalues of the force constant matrix are all positive. If, however, the stationary point is a TS, one (and only one) of the eigenvalues is negative. This corresponds to the energy being a maximum in one direction and a minimum in all other directions. The "frequency" for the "vibration" along the eigenvector with a negative force constant will formally be imaginary, as it is the square root of a negative number (Equation (14.36)), and for a TS there are thus only $3N - 7$ vibrations.

The vibrational degrees of freedom are decoupled in the vibrational normal coordinate system within the harmonic approximation. Since the energy of the $3N - 6$ vibrations can be written as a sum, the partition function can be written as a product over $3N - 6$ vibrational partition functions:

$$E_{\mathrm{vib}} = \sum_{i=1}^{3N_{\mathrm{atom}}-6(7)} \left(n_i + \frac{1}{2}\right) h\nu_i \tag{14.38}$$

$$q_{\mathrm{vib}} = \prod_{i=1}^{3N_{\mathrm{atom}}-6(7)} \frac{\mathrm{e}^{-h\nu_i/2kT}}{1 - \mathrm{e}^{-h\nu_i/kT}} \tag{14.39}$$

The vibrational frequencies are needed for calculating q_{vib}, and can be obtained from the force constant matrix and atomic masses.

14.5.4 Electronic Degrees of Freedom

The electronic partition function involves a sum over electronic quantum states. These are the solutions to the electronic Schrödinger equation, that is the lowest (ground) state and all possible excited states. The energy difference between the ground and excited states is in almost all molecules large compared with kT, which means that only the first term (the ground state energy) in the partition function summation (Equation (14.11)) is important:

$$q_{\mathrm{elec}} = \sum_{i=0}^{\infty} g_i \mathrm{e}^{-\varepsilon_i/kT} \approx g_0 \mathrm{e}^{-\varepsilon_0/kT} \tag{14.40}$$

Defining the zero point for the energy as the electronic energy of the reactant, the electronic partition functions for the reactant and TS are given below:

$$q_{\mathrm{elec}}^{\mathrm{reactant}} = g_0 \tag{14.41}$$

$$q_{\mathrm{elec}}^{\mathrm{TS}} = g_0 \mathrm{e}^{-\Delta E^{\neq}/kT} \tag{14.42}$$

The $\Delta E^{\neq}$ term is the difference in electronic energy between the reactant and TS, and g_0 is the electronic degeneracy of the (ground state) wave function. The degeneracy may be either in the spin part ($g_0 = 1$ for a singlet, 2 for a doublet, 3 for a triplet, etc.) or in the spatial part ($g_0 = 1$ for wave functions belonging to an A, B or Σ representation in the point group, 2 for an E, Δ or Φ representation, 3 for a T representation, etc.). The large majority of stable molecules have non-degenerate ground state wave functions and consequently $g_0 = 1$.

14.5.5 Enthalpy and Entropy Contributions

Given the partition function, the enthalpy and entropy terms may be calculated by carrying out the required differentiations in Equations (14.21) to (14.23). For one mole of molecules, the results for a non-linear system are (R being the gas constant) given in the following equations:

$$H_{\text{trans}} = \tfrac{5}{2}RT \tag{14.43}$$

$$H_{\text{rot}} = \tfrac{3}{2}RT \tag{14.44}$$

$$H_{\text{vib}} = R\sum_{i=1}^{3N-6(7)}\left(\frac{h\nu_i}{2k} + \frac{h\nu_i}{k}\frac{1}{e^{h\nu_i/kT}-1}\right) \tag{14.45}$$

$$H_{\text{elec}}^{\text{reactant}} = 0 \quad ; \quad H_{\text{elec}}^{\text{TS}} = \Delta E^{\neq} \tag{14.46}$$

$$S_{\text{trans}} = \tfrac{5}{2}R + R\ln\left(\frac{V}{N_{\text{A}}}\left(\frac{2\pi MkT}{h^2}\right)^{3/2}\right) \tag{14.47}$$

$$S_{\text{rot}} = R\left(\frac{3}{2} + \ln\left(\frac{\sqrt{\pi}}{\sigma}\left(\frac{8\pi^2kT}{h^2}\right)^{3/2}\sqrt{I_1I_2I_3}\right)\right) \tag{14.48}$$

$$S_{\text{vib}} = R\sum_{i=1}^{3N-6(7)}\left(\frac{h\nu_i}{kT}\frac{1}{e^{h\nu_i/kT}-1} - \ln(1-e^{-h\nu_i/kT})\right) \tag{14.49}$$

$$S_{\text{elec}}^{\text{reactant}} = S_{\text{elec}}^{\text{TS}} = R\ln g_0 \tag{14.50}$$

The rotational terms are slightly different for a linear molecule, and the vibrational terms will contain one more vibrational contribution:

$$H_{\text{rot}}(\text{linear}) = RT \tag{14.51}$$

$$S_{\text{rot}}(\text{linear}) = R\left[1 + \ln\left(\frac{8\pi^2IkT}{\sigma h^2}\right)\right] \tag{14.52}$$

The vibrational enthalpy consists of two parts, the first being a sum of ${}^1/_2h\nu$ contributions giving the zero-point energies. The second part depends on temperature and is a contribution from molecules that are not in the vibrational ground state. This contribution goes toward zero as the temperature goes to zero where all molecules are in the ground state. Note also that the sum over vibrational frequencies runs over $3N-6$ for the reactant(s), but only $3N-7$ for the TS. One of the normal vibrations has been transformed into the reaction coordinate at the TS, which formally has an imaginary frequency.

In order to calculate $\Delta G^{\neq} = G_{\text{TS}} - G_{\text{reactant}}$, we need $\Delta H^{\neq}$ and $\Delta S^{\neq}$. $\Delta H^{\neq}_{\text{elec}}$ is directly the difference in electronic energy between the TS and reactant. Except for complicated reactions involving several electronic states of different degeneracy (e.g. singlet molecules reacting via a triplet TS), $\Delta S^{\neq}_{\text{elec}}$ is zero.

For *unimolecular* reactions $\Delta H^{\neq}_{\text{trans}}$, $\Delta H^{\neq}_{\text{rot}}$ and $\Delta S^{\neq}_{\text{trans}}$ are zero, while $\Delta S^{\neq}_{\text{rot}}$ may be slightly different from zero owing to a change in geometry (thereby changing the moments of inertia). The $\Delta H^{\neq}_{\text{vib}}$ contribution is usually a few kJ/mol negative, as there is one less vibration at the TS (lack of zero-point energy). The TS is normally somewhat more ordered than the reactant, typically giving a slightly negative $\Delta S^{\neq}_{\text{vib}}$.

For *bimolecular* reactions (i.e. where the reactant is two separate molecules) $\Delta H^{\neq}_{trans}$ and $\Delta H^{\neq}_{rot}$ contribute a constant −4RT. The translational and rotational entropy changes are substantially negative, −30 to −50 J/mol K, due to the fact that there are six translational and six rotational modes in the reactants but only three of each at the TS. The six remaining degrees of freedom are transformed into the reaction coordinate and five new vibrations at the TS. These additional vibrations usually make $\Delta H^{\neq}_{vib}$ a few kJ/mol positive and $\Delta S^{\neq}_{vib}$ positive by 5–10 J/mol K. For bimolecular reactions, the entropy typically raises the free energy barrier by 40–60 kJ/mol relative to the electronic energy alone.

Similarly, in order to calculate $\Delta G_0 = G_{product} - G_{reactant}$ we need ΔH_0 and ΔS_0. The generalization for the electronic, translational and rotational contributions to $\Delta H^{\neq}$ and $\Delta S^{\neq}$ given above also holds for ΔH_0 and ΔS_0. The considerations for a unimolecular reaction hold for reactions where the number of reactant and product molecules is the same, while the generalizations for a bimolecular reaction correspond to an addition where two reactants form a single product molecule (the reverse process being a fragmentation). The vibrational contribution to ΔH_0 and ΔS_0 for a "number-conserving" reaction is usually small, since there is the same number of vibrational modes in the reactant and product. For an addition reaction, the number of vibrational modes increases by six and the contributions to ΔH_0 and ΔS_0 are again slightly positive, typically by a few kJ/mol and 5–10 J/mol K.

Tables 14.1 to 14.3 give some examples of the magnitude of each term for two bimolecular reactions (Diels–Alder and S_N2 reactions, forming either one or two molecules as the product) and a unimolecular rearrangement (Claisen reaction) (Figure 14.4).

All values have been calculated at the ωB97XD level with the pcseg-2 basis set for the Diels–Alder and Claisen reactions, and the aug-pcseg-2 basis set for the S_N2 reaction. ΔH and $T\Delta S$ values are given in kJ/mol at a temperature of 298 K ($RT = 2.5$ kJ/mol) and ΔS values are in cal/mol K.

The ωB97XD/pcseg-2 activation enthalpies for the Diels–Alder and Claisen reactions are in good agreement with the experimental values of 105 ± 8 kJ/mol[9] and 125 kJ/mol,[10] respectively, and the

Table 14.1 Diels–Alder reaction of butadiene and ethylene to form cyclohexene.

	$\Delta H^{\ddagger}$	$\Delta S^{\ddagger}$	$-T\Delta S^{\ddagger}$	ΔH_0	ΔS_0	$-T\Delta S_0$
Electronic	95	0	0	−200	0	0
Vibrational	15	5	−6	29	2	−2
Rotational	−4	−11	14	−4	−13	17
Translational	−4	−35	43	−4	−35	43
Total	**103**	**−41**	**52**	**−178**	**−46**	**58**
Experimental	**105**	**−41**	**52**	**−166**	**−45**	**56**

Table 14.2 S_N2 reaction of OH^- with CH_3F to form CH_3OH and F^-.

	$\Delta H^{\ddagger}$	$\Delta S^{\ddagger}$	$-T\Delta S^{\ddagger}$	ΔH_0	ΔS_0	$-T\Delta S_0$
Electronic	−5	0	0	−87	0	0
Vibrational	9	7	−9	11	1	−2
Rotational	−2	0	0	−2	−4	5
Translational	−4	−33	41	0	0	0
Total	**−2**	**−27**	**33**	**−78**	**−3**	**3**

Table 14.3 Claisen rearrangement of allyl vinyl ether to form 5-hexenal.

	$\Delta H^{\ddagger}$	$\Delta S^{\ddagger}$	$-T\Delta S^{\ddagger}$	ΔH_0	ΔS_0	$-T\Delta S_0$
Electronic	140	0	0	−76	0	0
Vibrational	−7	−6	7	−1	2	−2
Rotational	0	0	0	0	0	0
Translational	0	0	0	0	0	0
Total	**134**	**−6**	7	−77	**2**	−2
Experimental	**125**	**−8**	**10**			

corresponding activation entropies are almost perfectly matched. The S_N2 reaction refers to the situation in the gas phase where the reactants initially form an ion–dipole complex, pass over the TS and form another ion–dipole complex. The energies given above are relative to the isolated reactants, which is the reason for the slightly negative activation enthalpy. Note also that the rotational contribution to the reaction enthalpy is not zero; this is due to the fact that one of the reactants is a diatomic molecule, while one of the products is an atom (which has no rotational term).

In summary, to calculate rate and equilibrium constants we need to calculate $\Delta G^{\neq}$ and ΔG_0. This can be done within the RRHO approximation if the geometry, energy and force constants are known for the reactant, TS and product. The translational and rotational contributions are trivial to calculate, while the vibrational frequencies require the full force constant matrix (i.e. all energy second derivatives), which may be a significant computational effort.

The above treatment has made some assumptions, such as harmonic frequencies and "sufficiently small" energy spacing between the rotational levels. If a more elaborate treatment is required, the summation for the partition functions must be carried out explicitly. An approximate account for vibrational anharmonicity can be obtained by using the harmonic form for the partition function (and resulting enthalpy and entropy terms, Equations (14.43) to (14.50)), but using calculated anharmonic frequencies. The latter can be obtained from the third derivative and partial (diagonal components only) fourth-order derivatives of the energy with respect to the nuclear geometry.[11] A computationally cheap alternative is simply to scale the harmonic vibrational frequencies by a constant factor to account for anharmonicity in an average fashion.[12]

Many molecules have internal rotations around bonds with quite small barriers. In the above treatment, these are assumed to be described by simple harmonic vibrations, which may be a poor approximation. The calculated "vibrational frequency" for a low-barrier rotation is often close to zero, and

$$CH_3F + {}^{-}OH \longrightarrow CH_3OH + F^{-}$$

Figure 14.4 The Diels–Alder, S_N2 and Claisen reactions.

inspection of Equation (14.45) shows that the enthalpy term in such cases approaches a constant factor of RT. The entropy term (Equation (14.49)), however, goes towards infinity as the frequency approaches zero. Calculating the energy levels and partition function for a hindered rotor is somewhat complicated,[13] and is rarely done. If the barrier is very low, the motion may be treated as a free rotor, in which case it contributes a constant factor of RT to the enthalpy and $^1/_2R$ to the entropy. The enthalpy contribution is thus asymptotically correct when a low-barrier internal rotation is treated as a harmonic frequency, but the entropy term is not. Even minor inaccuracies in the calculated frequency may thus lead to large errors in the entropy contribution for small frequencies and care must be taken in such cases. A specific problem arises in bimolecular addition reactions (or the reverse fragmentation reaction), where six translational and six rotational degrees of freedom in the reactants are transformed into three translational and three rotational degrees of freedom in the product, that is creating six new internal degrees of freedom. At the TS, several of these often correspond to low-barrier internal rotations, which may be problematic to treat as harmonic vibrations.[14] Although the vibrational entropy reaches a value of $^1/_2R$ already for a harmonic frequency of around 400 cm^{-1}, the difference relative to a hindered internal rotor only becomes significant for frequencies below 100 cm^{-1} and rotational barriers comparable to RT (~3 kJ/mol at room temperature).

It should also be noted that the thermodynamic contributions in Equations (14.45) and (14.47) to (14.49) are calculated using the most common atomic isotopes, while the experimental quantities of course represent an ensemble of molecules containing a statistical mixture of isotopomers. It is straightforward but tedious to construct the thermodynamic contributions corresponding to a mixture of molecules with different atomic isotopes.[15] Since the resulting changes are substantially smaller than the error due to neglect of vibrational anharmonicity, such improvements are usually not considered.

The electronic energy difference between the reactant/TS and reactant/product is the most important contribution to $\Delta G^{\neq}$ and ΔG_0, as can be seen from Tables 14.1 to 14.3. The electronic energy is furthermore the most difficult to calculate accurately. Let us consider three cases.

1. The error in $\Delta E^{\neq}$ or ΔE_0 is ~50 kJ/mol. It is clear that spending significant amounts of computer time in order to include vibrational, rotational and translational corrections has little value.
2. The error in $\Delta E^{\neq}$ or ΔE_0 is ~5 kJ/mol. The corrections from vibrations, rotations and translation now become important and should be included. However, sophisticated treatments such as anharmonic vibrations are unimportant.
3. The error in $\Delta E^{\neq}$ or ΔE_0 is ~0.5 kJ/mol. Corrections from vibrations, rotations and translation are clearly necessary. Explicit calculation of the partition functions for anharmonic vibrations and internal rotations may be considered. However, at this point other factors also become important for the activation energy. These include, for example:
   (a) The position of the TS has been assumed to be at the maximum on the electronic energy surface, whereas in reality it should be at the maximum on the ΔG surface. This would include entropy effects and thus allow the position of the TS to depend on temperature. Such treatments are referred to as *Variational Transition State Theory* (VTST)[16,17] and are important for reactions with small (or zero) enthalpy barriers, such as recombination of radicals or carbene additions.[18]
   (b) The possibility of re-crossings and tunneling (which requires a quantum description of the nuclear motion) should be included in order to produce a transmission coefficient. The effect of tunneling can be estimated from the imaginary frequency at the TS, but an accurate estimate requires elaborate calculations. The re-crossing effect requires simulation of the dynamics of the reaction, again a substantial computational problem.

Calculating electronic activation energies with an accuracy of ~0.5 kJ/mol is only possible for very simple systems. An accuracy of ~5 kJ/mol is usually considered a good, but hard to get, level of accuracy. The situation is slightly better for relative energies of stable species, but a ~5 kJ/mol accuracy still requires significant computational effort. Thermodynamic corrections beyond the rigid-rotor/harmonic vibrations approximation are therefore rarely performed.

A prediction of $\Delta E^{\neq}/\Delta E_0$ to within ~0.5 kJ/mol may produce a $\Delta G^{\neq}/\Delta G_0$ accurate to may be ~1 kJ/mol. This corresponds to an error of a factor of ~1.5 (at $T = 300$ K) in the rate/equilibrium constant, which is poor compared with what is routinely obtained by experimental techniques. Calculating $\Delta G^{\neq}/\Delta G_0$ to within ~5 kJ/mol is still only possible for fairly small systems. This corresponds to predicting the absolute rate constant, or the equilibrium distribution, to within a factor of 10. Theoretical calculations are therefore not very useful for predicting absolute rate or equilibrium constants. Relative rates, however, are somewhat easier. Often the interest is not in how fast a certain product is formed, but on the rate difference between two reactions. The absolute rate (only) influences how long the total reaction time will be or how high the temperature should be. Rate differences, on the other hand, determine what the ratio between products is. When comparing calculated activation parameters for similar reactions, one can always hope for some "cancellation of errors". Theoretical methods are most useful for predicting and rationalizing different reaction pathways, not in predicting absolute rates.

The activation enthalpies and entropies in principle depend on temperature (Equations (14.43) to (14.45) and (14.47) to (14.49)), but only weakly so, and for a limited temperature range they may be treated as constants. Obtaining these quantities experimentally is possible by measuring the reaction rate as a function of temperature and plotting ln (k_{rate}/T) against T^{-1}:

$$\begin{aligned} k_{rate} &= \frac{kT}{h} e^{-\Delta G^{\neq}/RT} \\ \ln\left(\frac{k_{rate}}{T}\right) &= \ln\left(\frac{k}{h}\right) + \frac{\Delta S^{\neq}}{R} - \frac{\Delta H^{\neq}}{RT} \end{aligned} \tag{14.53}$$

Such plots should produce a straight line with the slope being equal to $-\Delta H^{\neq}/R$ and the intercept equal to ln $(k/h) + \Delta S^{\neq}/R$. As the available temperature range often is ~100 °C, the error in $\Delta H^{\neq}$ will typically be 0.5–2 kJ/mol. The activation entropy is determined by extrapolating outside the data points to $T = \infty$ $(1/T = 0)$, and is usually somewhat less well defined; a typical error may be 5 J/mol K.

Experimentalists often analyze their data in terms of an Arrhenius expression instead of the TST expression of Equation (14.53) by plotting ln (k_{rate}) against T^{-1}:

$$\begin{aligned} k_{rate} &= A e^{-\Delta E^{\neq}/RT} \\ \ln(k_{rate}) &= \ln(A) - \frac{\Delta E^{\neq}}{RT} \end{aligned} \tag{14.54}$$

The connection with the TST expression (14.53) may be established from the definition of the activation energy:

$$\Delta E^{\neq} = RT^2 \left(\frac{d \ln k_{rate}}{dT}\right)_V \tag{14.55}$$

This produces the relationship shown below:

$$\Delta H^{\neq} = \Delta E^{\neq} - (1 - \Delta n)\, RT \tag{14.56}$$

$$\Delta S^{\neq} = R\left[\ln A - \ln\left(\frac{kT}{h}\right) - (1 - \Delta n) + \Delta n\,(RT)\right] \tag{14.57}$$

Here Δn is the change in the number of molecules from the reactant to the TS, that is $\Delta n = 0$ for a unimolecular reaction, −1 for a bimolecular reaction, etc. For a solution phase reaction Δn is approximately 0.

For a reaction taking place by multiple reaction paths (e.g. conformational TSs), the *observed* activation energy is obtained from the observed rate constant, which is a sum over individual rate constants:

$$\begin{aligned} k_i &= \frac{kT}{h}\mathrm{e}^{-\Delta G_i^{\neq}/RT} \\ k_{\text{observed}} &= \sum_i k_i \\ \mathrm{e}^{-\Delta G_{\text{observed}}^{\neq}/RT} &= \sum_i \mathrm{e}^{-\Delta G_i^{\neq}/RT} \end{aligned} \tag{14.58}$$

The presence of multiple reaction paths with similar activation energies will thus result in an effective activation energy that is *lower* than the activation energy of the lowest TS.

14.6 Condensed Phases

The (quantum) energy states for a single molecule in the rigid-rotor harmonic-oscillator approximation are sufficiently regular to allow an explicit construction of the partition function. For a collection of many interacting particles (condensed phase), the relevant energy states are those describing the vibrations and translation and rotation of molecules relative to each other. The energy states are in this case not only numerous but also so irregularly spaced that it is impossible to derive them directly from molecular quantities. It is consequently not possible to construct the partition function explicitly. It is, however, possible to estimate *derivatives* of Q and *differences* in Q by a representative sample of the system. Condensed phases can be modeled by periodic boundary conditions and configurations generated by either molecular dynamics or Monte Carlo procedures, as discussed in Sections 15.1 and 15.2.

We can derive formal expressions for U and A from Equations (14.15) and (14.17) by using the fact that $\partial \ln Q/\partial T = Q^{-1}\partial Q/\partial T$:

$$U = \frac{kT^2}{Q}\frac{\partial Q}{\partial T} = \frac{kT^2}{Q}\frac{\partial}{\partial T}\left(\sum_i^{\infty} \mathrm{e}^{-E_i/kT}\right) = \sum_i^{\infty} E_i(Q^{-1}\mathrm{e}^{-E_i/kT}) \tag{14.59}$$

$$A = -kT\ln Q = -kT\ln\left(\sum_i^{\infty} \mathrm{e}^{-E_i/kT}\right) \tag{14.60}$$

The Boltzmann probability function P can be written either in a discrete energy representation or in a continuous phase space formulation:

$$P(E_i) = Q^{-1}\mathrm{e}^{-E_i/kT} \tag{14.61}$$

$$P(\mathbf{r},\mathbf{p}) = Q^{-1}\mathrm{e}^{-E_{(\mathbf{r},\mathbf{p})}/kT} \tag{14.62}$$

Here Q^{-1} is a normalization factor. The internal energy U in Equation (14.59) can thus be written as

$$U = \sum_i^{\infty} E_i P(E_i) \tag{14.63}$$

$$U = \int E(\mathbf{r},\mathbf{p})P(\mathbf{r},\mathbf{p})\mathrm{d}\mathbf{r}\ \mathrm{d}\mathbf{p} \tag{14.64}$$

Equation (14.63) shows that U is simply a sum of energies weighted by the probability of being in that state, that is U is the average (potential) energy of the system. Since high-energy states occur with a low probability, only the low-energy region of the phase space is important for the internal energy.

A similar expression may be derived for A by substituting 1 with $\mathrm{e}^{E/kT}\mathrm{e}^{-E/kT}$ in Equation (14.59) and summing over all N_{states}:

$$\begin{aligned} A &= -kT\ln Q = kT\ln\left(\frac{1}{Q}\right) = kT\ln\left(\frac{\sum_i^{\infty}\mathrm{e}^{-E_i/kT}\mathrm{e}^{E_i/kT}}{N_{\mathrm{states}}Q}\right) \\ A &= -kT\ln(N_{\mathrm{states}}) + kT\ln\left(\sum_i^{\infty}\mathrm{e}^{E_i/kT}P(E_i)\right) \end{aligned} \tag{14.65}$$

The ln (N_{states}) term is constant and corresponds to a change of the zero point, and can consequently be neglected. Alternatively, A may be written as an integral over phase space:

$$A = kT\ln\int \mathrm{e}^{E(\mathbf{r},\mathbf{p})/kT}P(\mathbf{r},\mathbf{p})\mathrm{d}\mathbf{r}\ \mathrm{d}\mathbf{p} \tag{14.66}$$

In contrast to U (Equation (14.63)), the Helmholtz free energy A depends exponentially on the energy; that is although high-energy states occur infrequently, they contribute significantly owing to the exponential weighting factor. Alternatively stated, U depends only on the *derivative* of Q, while A depends *directly* on Q.

It is not possible to carry out the summation over *all* states, or equivalently integrate over *all* phase space in Equations (14.63) to (14.65). The U and A values could in principle be calculated by sampling the phase space in a random fashion (Monte Carlo-type integration), but such an approach will suffer from an extremely slow convergence as the large majority of points will have high energies and consequently contribute with a very small probability. If, however, a *representative* collection of configurations can be generated, the sum over all states can be approximated by an average over a finite set of configurations. Representative here means that the number of configurations with a given energy is proportional to that given by the Boltzman distribution, and that all "important" parts of the phase space are sampled. For a finite number of points M, it is possible to calculate the average value of a given property X according to Equation (14.67), where the points can be denoted either by their energies or by their positions and momentum:

$$\langle X\rangle_M = \frac{1}{M}\sum_{i=1}^{M} X(E_i) = \frac{1}{M}\sum_{i=1}^{M} X(\mathbf{r}_i,\mathbf{p}_i) \tag{14.67}$$

The number of sampling points in a typical simulation is perhaps $\sim 10^6$, which represents only an infinitesimal fraction of the $6N_{\text{atom}}$-dimensional phase space (a rough 10-point sampling in each dimension would give 10^{6N} points). As already mentioned, however, the vast majority of the huge phase space is high in energy and is not accessible at normal temperatures. Consider, for example, placing 1000 water molecules at random in a box with a dimension corresponding to a density of 1 g/cm^3. If any two water molecules have a significant overlap, there will be a large repulsive interaction, and therefore a vanishing probability of such a configuration occurring. Placing all 1000 water molecules in the box without any two molecules having an overlap is difficult, and will essentially never occur by a random placement. Starting from an energy-minimized structure and allowing the system to evolve by a molecular dynamics algorithm, however, will *only* sample those configurations where no serious molecular overlaps occur, that is the important low-energy region.

The "magic" in simulations is generating an ensemble that yields a good representation of the "important" phase space for the given property. A collection of configurations is called an *ensemble* and Equation (14.67) is called an ensemble average, with the subscript indicating what is being averaged. There are two main techniques for generating an ensemble, Monte Carlo and molecular dynamics, which are discussed in Sections 15.1 and 15.2. These methods are based on the *ergodic hypothesis* (which can be proven rigorously only for a hard-sphere gas), which makes the assumption that the average obtained by following a small number of particles over a long time is equivalent to averaging over a large number of particles for a short time. Taken to the limit, this implies that a time average over a single particle is equivalent to an average of a large number of particles at any given time snapshot, that is time-averaging is equivalent to ensemble-averaging:

$$\langle X \rangle = \lim_{\tau \to \infty} \frac{1}{\tau} \int_0^{\tau} X(t)\,\mathrm{d}t = \lim_{M \to \infty} \frac{1}{M} \sum_{i=1}^{M} X_i \tag{14.68}$$

Alternatively stated, the ergodic hypothesis implies that no matter where a system is started, it is possible to get to any other point in phase space. For U and A, this leads to the following expressions:

$$\langle U \rangle_M = \frac{1}{M} \sum_{i=1}^{M} E_i = \langle E \rangle_M \tag{14.69}$$

$$\langle A \rangle_M = kT \ln \left(\frac{1}{M} \sum_{i=1}^{M} \mathrm{e}^{E_i/kT} \right) = kT \ln \langle \mathrm{e}^{E/kT} \rangle_M \tag{14.70}$$

A macroscopic observable can in general be calculated as an average over a corresponding microscopic quantity. The average value of, for example, U calculated from Equation (14.51) has a statistical uncertainty $\sigma(U)$, which is the square root of the variance σ^2 (Equation (18.2), making the approximation $M - 1 \approx M$ for large samples):

$$\sigma^2(U) = \frac{1}{M} \sum_{i=1}^{M} (E_i - \langle E \rangle)^2 \tag{14.71}$$

The statistical uncertainty is therefore inversely proportional to the square root of the number of sampling points M:

$$\sigma(X) \propto \frac{1}{\sqrt{M}} \tag{14.72}$$

Increasing the sample size from 1000 to 4000 thus reduces the standard deviation by a factor of 2. How well the calculated average (from Equation (14.67)) resembles the "true" value, however, depends on whether the ensemble is representative. If a large number of points are collected from a small part of the phase space, the property may be calculated with a small *statistical* error but a large *systematic* error (i.e. the value may be precise, but inaccurate). As it is difficult to establish that the phase space is adequately sampled, this can be a very misleading situation, that is the property *appears* to have been calculated accurately but may in fact be significantly in error. Different parts of the phase space may furthermore be important for different properties. An ensemble that gives an accurate value for one property may not necessarily be suitable for another property. Energy properties, such as U, H and C_V, depend on the derivative of Q for which the low-energy region of the phase space is important, while entropic properties, such as A, S and G, depend directly on Q, where the whole phase space is important. Since Monte Carlo and molecular dynamics techniques preferentially sample the low-energy region, it is computationally difficult to achieve a reasonable statistical error for entropic quantities.

With standard MC or MD simulations, the ensemble reflects the temperature and only configurations that are accessible at the given temperature are represented to any significant extent. This makes it impossible to calculate the absolute value of the entropy, but it is possible to calculate *differences* in entropy properties. Using the Helmholtz free energy for illustration, we can consider two systems A and B described by two different energy functions E^{A} and E^{B}. The energy difference is given as follows and involves a ratio of the corresponding partition functions:

$$A_{\mathrm{A}} - A_{\mathrm{B}} = -kT\,(\ln Q_{\mathrm{A}} - \ln Q_{\mathrm{B}}) = -kT \ln \frac{Q_{\mathrm{A}}}{Q_{\mathrm{B}}} \tag{14.73}$$

The energy difference can analogously to Equation (14.69) be evaluated as an ensemble average:

$$\langle \Delta A_{\mathrm{AB}} \rangle_M = kT \ln \left(\frac{1}{M} \sum_{i=1}^{M} \mathrm{e}^{\left(E_i^{\mathrm{B}} - E_i^{\mathrm{A}}\right)/kT} \right) = kT \ln \langle \mathrm{e}^{\Delta E_{\mathrm{BA}}/kT} \rangle_M \tag{14.74}$$

The important difference is that the exponential now involves an energy difference ΔE_{BA}. Provided that this is sufficiently small compared with kT, the ensemble average will show much better convergence than the absolute entropy of either system. If the energy difference is large compared with kT, we may introduce intermediate states between A and B that can be described in terms of a coupling parameter λ ($0 \le \lambda \le 1$). The simplest approach involves a linear interpolation, but more complicated connections can also be used:[19]

$$E_\lambda = \lambda E_{\mathrm{A}} + (1 - \lambda) E_{\mathrm{B}} \tag{14.75}$$

The sampling can then be performed for each value of λ and all the intermediate results added together to provide the difference ΔA_{AB}. It should be noted that a system corresponding to an intermediate value of λ does not necessarily represent an actual physically realizable system. If, for example, the objective is to calculate the entropy difference between acetone and propane in a solvent, a value of $\lambda = 0.5$ corresponds to a "molecule" with "half" a carbonyl oxygen and two "half" hydrogen atoms on the central carbon. Such artificial intermediate systems do not represent special problems in terms of calculation.

The preference of standard MC or MD methods for sampling the low-energy region of an energy surface is a result of the way one configuration is propagated to the next. In MC methods the probability for accepting a trial move depends on the ratio of the change in energy relative to the temperature, while MD methods have a velocity (direction and magnitude) depending on the temperature. Alternative MC methods have also been proposed where the transition probability instead depends on the inverse density of states, rather than the energy.[20] This in principle makes it possible to simulate an ensemble that provides a uniform coverage of the whole phase space, and therefore allows calculation of absolute values of entropies and free energies. Since the density of states is unknown *a priori*, the method requires a sequence of simulations where the density of the state diagram is gradually constructed and refined.

The main problem in estimating thermodynamic quantities from simulations is the assumption that the generated set of configurations forms a representative set. In practice, this is impossible to guarantee or verify, making simulations somewhat of a "black art". For configurations generated by molecular dynamics, a typical simulation time is of the order of nanoseconds, and it is clear that this is much too short a timespan to adequately sample all the phase space. There is thus a real risk of a simulation being trapped in a small volume of the phase space during the whole simulation, and thereby providing a misleading sampling. In order to evaluate the sensitivity of the results, several simulations are often performed with different starting conditions.

References

1 H. Eyring, *Journal of Chemical Physics* **3** (2), 107–115 (1935).
2 M. A. Robb, M. Garavelli, M. Olivucci and F. Bernardi, *Reviews in Computational Chemistry* **15**, 87–146 (2000).
3 K. J. Laider, *Chemical Kinetics* (Harper and Row, 1987).
4 J. I. Steinfeld, J. S. Francisco and W. L. Hase, *Chemical Kinetics and Dynamics* (Prentice-Hall, 1989).
5 B. K. Carpenter, in *Annual Review of Physical Chemistry* (Annual Reviews, 2005), Vol. 56, pp. 57–89.
6 M. B. Reyes and B. K. Carpenter, *Journal of the American Chemical Society* **122** (41), 10163–10176 (2000).
7 I. N. Levine, *Physical Chemistry* (McGraw-Hill, 1983).
8 K. Lucas, *Applied Statistical Thermodynamics* (Springer-Verlag, 1991).
9 K. N. Houk, R. J. Loncharich, J. F. Blake and W. L. Jorgensen, *Journal of the American Chemical Society* **111** (26), 9172–9176 (1989).
10 F. W. Schuler and G. W. Murphy, *Journal of the American Chemical Society* **72** (7), 3155–3159 (1950).
11 V. Barone, *Journal of Chemical Physics* **120** (7), 3059–3065 (2004).
12 J. P. Merrick, D. Moran and L. Radom, *Journal of Physical Chemistry A* **111** (45), 11683–11700 (2007).
13 Y. Y. Chuang and D. G. Truhlar, *Journal of Chemical Physics* **112** (3), 1221–1228 (2000).
14 L. Masgrau, A. Gonzalez-Lafont and J. M. Lluch, *Journal of Computational Chemistry* **24** (6), 701–706 (2003).
15 F. Jensen, *Molecular Physics* **101** (15), 2315–2318 (2003).
16 D. G. Truhlar and M. S. Gordon, *Science* **249** (4968), 491–498 (1990).
17 S. Y. Yang, I. Hristov, P. Fleurat-Lessard and T. Ziegler, *Journal of Physical Chemistry A* **109** (1), 197–204 (2005).
18 A. E. Keating, S. R. Merrigan, D. A. Singleton and K. N. Houk, *Journal of the American Chemical Society* **121** (16), 3933–3938 (1999).
19 D. Frenkel and B. Smith, *Understanding Molecular Simulations* (Academic Press, 1996).
20 F. G. Wang and D. P. Landau, *Physical Review Letters* **86** (10), 2050–2053 (2001).

15

Simulation Techniques

The analysis of a potential energy surface by locating the minima and saddle points (Chapter 13) corresponds to modeling the system at a temperature of 0 K, where all molecules are in their ground electronic, vibrational and rotational states. The effects of a finite temperature can be incorporated by means of the statistical mechanics methods discussed in Chapter 14. For a system of non-interacting molecules (ideal gas), the partition function can be evaluated quite accurately by the rigid-rotor harmonic-oscillator approximation from relatively simple quantities for the isolated molecule (geometry and vibrational frequencies). Similar approaches are possible for crystalline solid states, where the translational symmetry implies that only properties for the unit cell are required for describing the whole system. For other systems, most notably liquids and solutions, the macroscopic quantities derived from the partition function must be estimated from a representative sampling of the phase space. *Simulations* refer to methods aimed at generating a representative sampling of a system at a finite temperature.[1–5]

Electronic structure methods are typically used for solving the Schrödinger equation for a single or a few molecules, infinitely removed from all other molecules. This physically corresponds to the situation occurring in the gas phase under low pressure (vacuum). Experimentally, however, the majority of chemical reactions are carried out in solution. Biologically relevant processes also occur in solution, aqueous systems with rather specific pH and ionic conditions. Most reactions are both qualitatively and quantitatively different under gas- and solution-phase conditions, especially those involving ions or polar species. Molecular properties are also sensitive to the environment. Simulations are therefore intimately related with describing solute–solvent interactions, but these effects can also be modeled with less rigorous methods.

There are two major techniques for generating an ensemble: Monte Carlo and molecular dynamics.

In *Monte Carlo* (MC) methods,[6] a sequence of points in phase space is generated from an initial geometry by adding a random perturbation to the coordinates of a randomly chosen particle (atom or molecule). In the Markov chain version, the new configuration is accepted if the energy decreases and with a probability of $e^{-\Delta E/kT}$ if the energy increases. This *Metropolis* procedure[7] ensures that the configurations in the ensemble obey a Boltzmann distribution, and the possibility of accepting higher energy configurations allows MC methods to climb uphill and escape from a local minimum. In order to have a reasonable acceptance ratio, however, the step size must be fairly small. This effectively means that even some millions of MC steps (a typical computational limit) only explore the local phase space around the starting geometry.

Introduction to Computational Chemistry, Third Edition. Frank Jensen.

Companion Website: http://www.wiley.com/go/jensen/computationalchemistry3

Monte Carlo methods generate configurations in a random fashion and the Metropolis selection procedure ensures that a proper ensemble is generated. The geometry perturbation in each step may be "non-physical", which is actually an advantage since two consecutive geometries may be separated by a high-energy barrier. MC methods thus have the possibility of "tunneling" between energetically separated regions of phase space, thereby giving a better coverage. The perturbations may be carried out in either internal or Cartesian coordinates, and it is quite easy to freeze out certain degrees of freedom, such as, for example, sampling only the torsional angle space. MC methods are inherently non-deterministic, as each configuration only depends on the previous point and a few random numbers, and two simulations starting from the same geometry will not generate the same sampling since the random numbers will be different. MC simulations require only the ability to evaluate the energy of the system, which may be advantageous if calculating the first derivatives is difficult or time-consuming. Furthermore, since only a single particle is moved in each step, only the energy changes associated with this move must be calculated, not the total energy for the whole system. A disadvantage of MC methods is the lack of the time dimension and atomic velocities, and they are therefore not suitable for studying time-dependent phenomena or properties depending on momentum.

Molecular Dynamics (MD) methods generate a series of time-correlated points in phase space (a *trajectory*) by propagating a starting set of coordinates and velocities according to Newton's second equation by a series of finite time steps. A typical time step is $\sim 10^{-15}$ s and a simulation involving 10^6 steps thus "only" covers $\sim 10^{-9}$ s. This is substantially shorter than many important phenomena, and MD methods, in analogy with MC, tend to only sample the region in phase space close to the starting condition. Furthermore, MD methods simulate the physical evolution of configurations and can easily become trapped in energy wells.

MD simulations are cumbersome to run in anything but Cartesian coordinates and it is somewhat difficult to enforce constraints on the system. MD simulations require small time steps and tend to spend a significant effort describing relatively unimportant bond stretching and angle bending motions. The ability to climb over energy barriers is furthermore limited, as any uphill motion will generate a force trying to pull the system back towards the minimum. MD is in principle deterministic, and starting two simulations with the exact same initial coordinates and velocities should give the same trajectory. Even slight differences ($\sim 10^{-8}$) in the starting conditions, however, rapidly lead to uncorrelated trajectories within a few thousand time steps owing to an exponential divergence. The numerical errors generated in each time step will in addition gradually add up to become significant. As different computers (and compilers) produce different round-off errors, this means that MD simulations in practice are non-deterministic and exhibit chaotic behavior on timescales longer than ~ 50 ps. MD simulations implicitly have both atomic velocities and time dependence, and are thus suitable for modeling, for example, transport phenomena and diffusion. Running an MD simulation requires the ability to calculate the force (first derivatives of the energy) on all particles in the system in addition to the energy. For parameterized energy functions, such as those used in force field methods, this is not a limitation as forces can be calculated almost as easily as the energy. Since all particles are moved in each step, the whole energy function (and gradient) must be recomputed at each step.

The inherently non-deterministic nature of MC methods and the non-deterministic behavior of actual MD simulations might be considered as potential problems. In reality the only concern is generating a representative sample of the phase space, and chaotic behavior may actually help in obtaining a more complete sampling. The random and chaotic elements in simulations, however, make troubleshooting and identification of programming bugs somewhat more problematic than for many other types of computer programs. Verifying that a new simulation program is valid cannot easily be done by comparing exact numbers with another program, as even running the same program on different types of machines may produce different results. Programming bugs that produce small (systematic) errors may thus be swamped by the statistical errors inherent in all simulations and so escape detection. Development of simulation packages must therefore be done with care, involving monitoring many different quantities to ensure that the implementation is valid.[8]

The result of a simulation or an experiment involves averaging over both the number of molecules and time, but usually with significantly different averaging lengths, and it is not completely obvious that the calculated quantities are directly comparable with the experiments. An IR spectrum, for example, records averages over a sample containing perhaps 10^{18} molecules over the timeframe of perhaps 10^{-14} s (the interaction time of radiation with molecules), which is essentially a snapshot of the quantum states for a large selection of molecules. A simulation, on the other hand, may follow the molecular motions of perhaps 10^3 molecules for 10^{-9} s. The *ergodic hypothesis* makes the assumption that the average obtained by following a small number of particles over a long time is equivalent to averaging over a large number of particles for a short time. Taken to the limit, this implies that a time average over a single particle is equivalent to an average of a large number of particles at any given time snapshot, that is time-averaging is equivalent to ensemble-averaging:

$$\langle X \rangle = \lim_{\tau \to \infty} \frac{1}{\tau} \int_0^{\tau} X(t)\,\mathrm{d}t = \lim_{M \to \infty} \frac{1}{M} \sum_{i=1}^{M} X_i \tag{15.1}$$

Alternatively stated, the ergodic hypothesis implies that no matter where a system is started, it is possible to get to any other point in phase space. MC techniques perform an ensemble average, while MD performs a time average.

A simulation can be characterized by quantities such as volume (V), pressure (P), total energy (E), temperature (T), number of particles (N), chemical potential (μ), etc., but not all of these are independent. For a constant number of particles, either the volume or the pressure can be fixed, but not both. Similarly, either the total energy or the temperature can be fixed, but not both, and a constant chemical potential is incommensurable with a constant number of particles. The ensemble is labeled according to the fixed quantities, as shown in Table 15.1, with the remainder being derived from the simulation data, and thus displaying statistical fluctuations.

An MC simulation employs the temperature as the parameter for deciding acceptance or rejection of trial moves and MC simulations are therefore naturally of the *NVT* type. An MD simulation, on the other hand, preserves energy and is therefore naturally of the *NVE* type, but other ensembles for both MC and MD can be generated by the techniques described in Section 15.2.2. Table 15.2 summarizes some of the differences between MC and MD.

Table 15.1 Constants in different ensembles and corresponding equilibrium states.

N	P	V	T	E	μ	Acronym	Equilibrium	Name
×			×	×		NVT	A has minimum	Canonical
×		×		×		NVE	S has maximum	Microcanonical
×	×		×			NPT	G has minimum	Isothermal–isobaric
		×	×		×	$VT\mu$	(PV) has maximum	Grand canonical

N = number of particles; P = pressure; V = volume; T = temperature; E = energy; μ = chemical potential; A = Helmholtz free energy; S = entropy; G = Gibbs free energy.

Table 15.2 Differences between Monte Carlo and molecular dynamics methods.

Property	MC	MD
Basic information needed	Energy	Gradient
Particles moved in each step	One (few)	All
Coordinates	Any	Cartesian
Constraints	Easy	Difficult
Atomic velocities	No	Yes
Time dimension	No	Yes
Deterministic	No	(Yes)
Sampling	Non-physical	Physical
Natural ensemble	*NVT*	*NVE*

It is probably no surprise that hybrid MC/MD methods have been devised, trying to capture the best of both methods.[9] These combined methods typically perform an MD simulation with an occasional MC step thrown in, in order to give a better coverage of the phase space. Alternatively, a trial step may be generated by an MD recipe, using a somewhat larger time step than for pure MD, and this trial step is then accepted or rejected based on an MC criterion.

A popular hybrid method called *Replica Exchange MD* (REMD) or *parallel tempering* consists of running several independent MD simulations of the same system at different temperatures (or other independent variable), and at suitable time intervals use a Metropolis criterion to possibly exchange configurations between these simulations.[10,11] The simulations at high temperatures will sample larger fractions of the phase space, but the Metropolis exchange probability will be essentially zero unless the energy distributions between two neighboring temperatures have a reasonable overlap. This means that a number of replica simulations with sufficiently small temperature differences combined with pair-wise exchanges are required in order to achieve a high temperature that allows an efficient phase space sampling. It should be noted that REMD does not provide direct information on the actual kinetics as the configurations are frequently interchanged between different temperatures.

15.1 Monte Carlo Methods

One of the advantages of Monte Carlo methods is the ease with which they can be implemented in computer programs. The heart of the algorithm is a random number generator and the ability to calculate the energy of the system for a given set of coordinates. Although truly random numbers are difficult to come by, several implementations of pseudo-random number generators are available. A pseudo-random number generator indicates a computer implementation of an algorithm that produces a sequence of seemingly random numbers, but the sequence is repeated (exactly) after some period. A good pseudo-random number generator is characterized by having a long periodicity, and within this periodicity the numbers do not show any systematic correlation with each other. As long as the simulation only uses numbers from within one period, the random aspect is fulfilled and the simulation data should be valid.[12]

An MC simulation starts from a suitable set of coordinates for all the particles. The set of coordinates is perturbed in a random fashion and the new geometry is accepted as a starting point for the next perturbing step if it is lower in energy than the current. If the new geometry is higher in energy, the Boltzmann factor $e^{-\Delta E/kT}$ is calculated and compared with a random number between 0 and 1. If

$e^{-\Delta E/kT}$ is larger than this number the new geometry is accepted; otherwise the old configuration is added to the sampling (again) and a new perturbing step is attempted.

The main variation of MC methods is how the perturbing step is done. For a system composed of spherical particles (atoms), the only variables are the center of mass of each particle and the trial moves are simple translations of particles. For rigid non-spherical particles, the three rotational degrees of freedom must also be sampled, while for flexible molecules, it is usually also of interest to sample the internal degrees of freedom (conformations, vibrations). The latter can be done in Cartesian coordinates or selectively in, for example, only the torsional variables.

A key point in MC methods is to ensure that the chain of configurations arises from a *symmetric* probability decision. Symmetric in this context means that each step is reversible, that is the probability of undoing a step by the next move is equal to the probability of generating the step, sometimes also called the *detailed balance* condition. If this is not fulfilled, the properties derived from the resulting ensemble can (but do not necessarily) display systematic errors, which are usually hard to detect. Generating random moves corresponding only to translations in the positive direction, rather than both positive and negative directions, will almost be sure to lead to artefacts, although one might think that this would be acceptable in a system subjected to periodic boundary conditions. The detailed balance condition is obeyed when a single random particle is subjected to a single random perturbation in each step, but this is not the case if a random perturbing step is applied sequentially to all the particles. It has nevertheless been shown that the sequential update obeys a weaker balance condition and does in fact generate a proper Boltzmann distribution.[13] A procedure where only one particle is moved in each trial step is computationally more efficient than a trial step consisting of moving all particles. When only a single particle is moved, only the change in the energy related to this particle is required, not the whole energy function. This makes the evaluation of each MC trial move somewhat faster than a single time step in an MD simulation. Furthermore, if many (all) particles are allowed to move in each trial step, the acceptance ratio usually becomes prohibitively small, unless very small perturbations are selected.

The size of the perturbing step is an important control parameter. A small step will give a high acceptance ratio but only a slow change of configurations. A large step, on the other hand, gives a low acceptance ratio and therefore a sampling consisting of only a few relatively widely distributed points in the configuration space. The optimum step size can formally be defined as the one that gives the fastest convergence of a given property for a given amount of computer time, but this is difficult to translate into an optimum acceptance ratio. Lacking a more objective criterion, a heuristic acceptance ratio around 0.5 is usually selected, although slightly smaller ratios in many cases may give a better sampling.

The ability to generate non-physical moves means that attention must be paid to the molecular stereochemistry. A random move of atoms makes it possible to invert the configuration of a chiral atom, a process that in reality may require large amounts of energy, but one that can easily be generated by moving a single atom. A Monte Carlo procedure must therefore be able to detect such chirality changes and reject such moves.

The procedure of making random moves of a single or several (all) particles gives MC methods a drawback for describing correlated motions. Exploring the conformational space of a larger molecule such as a protein in a solvent is inefficient, since several simultaneous perturbations of torsional angles are required for generating acceptable conformational changes. Such correlated movements are difficult to generate by random perturbations in either Cartesian or internal coordinates, and almost impossible if only single particle/variable movements are employed in each trial step. Even a small random perturbation of a single torsional angle in a large protein will almost invariably produce large geometrical changes in the overall structure and lead to atomic clashes, with resulting high energies and low acceptance ratios. The necessary correlated movement of several degrees of freedom in each MC step can be introduced by augmenting each random torsional angle perturbation with a sequence of derived torsional angle changes, as has been implemented in the *Concerted Rotations Involving Self-consistent Proposal* (*CRISP*) algorithm.[14] The minimum local perturbation by rotating a single torsional angle is achieved by also rotating the subsequent six torsional angles in order to match the position of the end-atom. Nevertheless, MC methods tend to be best for exploring the translational and rotational space for relatively small molecules, such as a solvent or solution, and internal degrees of freedom for small molecules.

15.1.1 Generating Non-natural Ensembles

A standard MC simulation generates an *NVT* ensemble, that is the pressure and energy will fluctuate. It is quite easy to generate other types of ensembles by MC methods, the most important being the *NPT* ensemble, since this is directly related to most experimental conditions. A constant pressure necessarily means that the volume must be able to change. For simulating an *NPT* ensemble, the total volume of the system is treated as an additional variable and subjected to random perturbations.[15] The acceptance criterion for a volume change is the same as for particle moves, except that the energy change is augmented with two additional terms, that is $\Delta E \rightarrow \Delta E + P\Delta V - NkT \ln(1 + \Delta V/V)$.

15.2 Time-Dependent Methods

The average kinetic energy is directly related to the temperature and at a finite temperature the molecule(s) explores a part of the surface with energies lower than the typical kinetic energy. One possible way of simulating the behavior at a finite temperature is by allowing the system to evolve according to the relevant dynamical equation (Section 1.4). For nuclei this is normally Newton's second law, although the (nuclear) Schrödinger equation must be used for including quantum effects, such as zero-point vibrational energy and tunneling. A dynamics simulation is also required if the interest is in studying time-dependent phenomena, such as transport, and the results of a simulation can yield information about the spectral properties, such as the IR spectrum.

A dynamics simulation requires a set of initial coordinates and velocities, and an interaction potential (energy function). The interaction may for a short time step be considered constant, allowing a set of updated positions and velocities to be estimated, at which point the new interaction can be calculated. By taking a (large) number of (small) time steps, the time behavior of the system can be obtained. Since the phase space is huge and the fundamental time step is short, the simulation will only explore the region close to the starting point, and several different simulations with different starting conditions are required for estimating the stability of the results.

15.2.1 Molecular Dynamics Methods

Nuclei are heavy enough that they, to a good approximation, behave as classical particles and the dynamics can thus be simulated by solving Newton's second equation, $\mathbf{F} = m\mathbf{a}$, which in differential form can be written as

$$-\frac{dV}{d\mathbf{r}} = m\frac{d^2\mathbf{r}}{dt^2} \tag{15.2}$$

Here V is the potential energy at position $\mathbf{r}$. The vector $\mathbf{r}$ contains the coordinates for all the particles, that is in Cartesian coordinates it is a vector of length $3N_{\text{atom}}$. The left-hand side is the negative of the energy gradient, also called the force ($\mathbf{F}$) on the particle(s).

Given a set of particles with positions $\mathbf{r}_i$, the positions a small time step Δt later are given by a Taylor expansion:

$$\begin{aligned} \mathbf{r}_{i+1} &= \mathbf{r}_i + \frac{\partial \mathbf{r}}{\partial t}(\Delta t) + \frac{1}{2}\frac{\partial^2 \mathbf{r}}{\partial t^2}(\Delta t)^2 + \frac{1}{6}\frac{\partial^3 \mathbf{r}}{\partial t^3}(\Delta t)^3 + \cdots \\ \mathbf{r}_{i+1} &= \mathbf{r}_i + \mathbf{v}_i(\Delta t) + \tfrac{1}{2}\mathbf{a}_i(\Delta t)^2 + \tfrac{1}{6}\mathbf{b}_i(\Delta t)^3 + \cdots \end{aligned} \tag{15.3}$$

The velocities $\mathbf{v}_i$ are the first derivatives of the positions with respect to time ($\partial \mathbf{r}/\partial t$) at time t_i, the accelerations $\mathbf{a}_i$ are the second derivatives ($\partial^2 \mathbf{r}/\partial t^2$) at time t_i, the hyperaccelerations $\mathbf{b}_i$ are the third derivatives, etc. The positions a small time step Δt *earlier* are derived from Equation (15.3) by substituting Δt with $-\Delta t$:

$$\mathbf{r}_{i-1} = \mathbf{r}_i - \mathbf{v}_i(\Delta t) + \tfrac{1}{2}\mathbf{a}_i(\Delta t)^2 - \tfrac{1}{6}\mathbf{b}_i(\Delta t)^3 + \cdots \tag{15.4}$$

Addition of Equations (15.3) and (15.4) gives a recipe for predicting the position a time step Δt later from the current and previous positions, and the current acceleration. The latter can be calculated from the force, or equivalently, the potential:

$$\mathbf{r}_{i+1} = (2\mathbf{r}_i - \mathbf{r}_{i-1}) + \mathbf{a}_i(\Delta t)^2 + \cdots \tag{15.5}$$

$$\mathbf{a}_i = \frac{\mathbf{F}_i}{m_i} = -\frac{1}{m_i}\frac{dV}{d\mathbf{r}_i} \tag{15.6}$$

This is the *Verlet* algorithm[16] for solving Newton's equation numerically. Note that the term involving the change in acceleration ($\mathbf{b}$) disappears, that is the equation is correct to third order in Δt. At the initial point, the previous positions are not available, but can be estimated from a first-order approximation of Equation (15.3):

$$\mathbf{r}_{-1} = \mathbf{r}_0 - \mathbf{v}_0\Delta t \tag{15.7}$$

The acceleration must be evaluated from the forces (Equation (15.5)) at each time step, which then allows the atomic positions to be propagated in time and thus generate a trajectory. As the step size Δt is decreased, the trajectory becomes a better and better approximation to the "true" trajectory, until the practical problems of finite numerical accuracy arise (e.g. the forces cannot be calculated with infinite precision). A small time step, however, means that more steps are necessary for propagating the system a given total time, that is the computational effort increases inversely with the size of the time step.

The Verlet algorithm has the numerical disadvantage that the new positions are obtained by adding a term proportional to Δt^2 to a difference in positions ($2\mathbf{r}_i - \mathbf{r}_{i-1}$). Since Δt is a small number and ($2\mathbf{r}_i - \mathbf{r}_{i-1}$) is a difference between two large numbers, this may lead to truncation errors due to finite precision. The Verlet algorithm furthermore has the disadvantage that velocities do not appear explicitly, which is a problem in connection with generating ensembles with constant temperature, as discussed in section 15.2.2.

The numerical aspect and the lack of explicit velocities in the Verlet algorithm can be remedied by the *leap-frog* algorithm.[1] Performing expansions analogous to Equations (15.3) and (15.4) with half a time step followed by subtraction gives

$$\mathbf{r}_{i+1} = \mathbf{r}_i + \mathbf{v}_{i+\frac{1}{2}}\Delta t \tag{15.8}$$

The velocity is obtained by analogous expansions to give

$$\mathbf{v}_{i+\frac{1}{2}} = \mathbf{v}_{i-\frac{1}{2}} + \mathbf{a}_i \Delta t \tag{15.9}$$

Equations (15.8) and (15.9) define the leap-frog algorithm, and it is seen that the position and velocity updates are out of phase by half a time step. In terms of theoretical accuracy it is also of third order, as the Verlet algorithm, but the numerical accuracy is better. The velocities furthermore appear directly, which facilitates a coupling to an external heat bath (Section 15.2.2). The disadvantage is that the positions and velocities are not known at the same time; they are always out of phase by half a time step. The latter abnormality can be removed by the *velocity Verlet* algorithm, where the equations used to propagate the atoms are given as[17]

$$\mathbf{r}_{i+1} = \mathbf{r}_i + \mathbf{v}_i \Delta t + \tfrac{1}{2}\mathbf{a}_i \Delta t^2 \tag{15.10}$$

$$\mathbf{v}_{i+1} = \mathbf{v}_i + \tfrac{1}{2}(\mathbf{a}_i + \mathbf{a}_{i+1})\Delta t \tag{15.11}$$

The preference of Verlet and leap-frog-type algorithms over, for example, Runge–Kutta methods (Section 13.8) for solving the differential Equation (15.2) in MD simulations is that they are time-reversible, which in general tend to improve the energy conservation over long simulation times.[18]

The above solves the dynamical equation by a numerical integration of Newton's second equation, but it is in some cases useful to rewrite the equations in a more general form. Denoting a generalized coordinate with $\mathbf{q}$ and its conjugate moment by $\mathbf{p}$ ($\mathbf{p} = \partial\mathbf{q}/\partial t$), Equation (15.2) becomes

$$-\frac{\partial V}{\partial \mathbf{q}} = \frac{\partial \mathbf{p}}{\partial t} \tag{15.12}$$

This can also be formulated in terms of a Lagrange function $L = T - V$:

$$\frac{\partial L}{\partial \mathbf{q}} - \frac{\mathrm{d}}{\mathrm{d}t}\frac{\partial L}{\partial \mathbf{p}} = 0 \tag{15.13}$$

Since $T = \mathbf{p}^2/2$ and the potential only depend on $\mathbf{q}$, it is easily seen that the Lagrange equation is completely equivalent to the Newton formulation.

The Lagrange formulation can be used to recast the differential form for the equation of motion into an integral form, by defining the action S as the time integral of the Lagrange function:

$$S[q] = \int_{t_0}^{t} L(\mathbf{q}(\tau), \mathbf{p}(\tau))\mathrm{d}\tau \tag{15.14}$$

The action depends on the trajectory q that the system follows as described by the time-dependent generalized variables $\mathbf{q}$. The *principle of stationary action* states that the actual path traversed by the system is determined by the condition that the first-order change δS must vanish. The equivalence of this condition to Equation (15.13) can be seen by requiring the stationary condition for a small perturbation $\boldsymbol{\varepsilon}$:

$$\begin{aligned}\delta S &= S(q+\varepsilon) - S(q) \\ &= \int_{t_0}^{t}\left[L\left(\mathbf{q}+\varepsilon, \mathbf{p}+\frac{d\varepsilon}{dt}\right) - L(\mathbf{q},\mathbf{p})\right]\mathrm{d}\tau \\ &= \int_{t_0}^{t}\left[\varepsilon\frac{\partial L}{\partial \mathbf{q}} + \frac{d\varepsilon}{dt}\frac{\partial L}{\partial \mathbf{p}}\right]\mathrm{d}\tau = 0\end{aligned} \tag{15.15}$$

Integration by parts of the last term and using the boundary condition $\varepsilon(t) = \varepsilon(t_0) = 0$ yields

$$\delta S = \int_{t_0}^{t} \left[\varepsilon \frac{\partial L}{\partial \mathbf{q}} - \varepsilon \frac{d}{dt} \frac{\partial L}{\partial \mathbf{p}} \right] d\tau = \int_{t_0}^{t} \varepsilon \left[\frac{\partial L}{\partial \mathbf{q}} - \frac{d}{dt} \frac{\partial L}{\partial \mathbf{p}} \right] d\tau = 0 \tag{15.16}$$

The stationary action principle means that the integral must be zero for any perturbation ε, which only can be true if Equation (15.13) holds.

Yet another formulation is given by the Hamilton function $H = T + V$ and Equation (15.17), which again may be verified to be completely equivalent to Equation (15.2):

$$\begin{aligned} \frac{\partial H}{\partial \mathbf{q}} + \frac{d\mathbf{p}}{dt} = 0 \\ \frac{\partial H}{\partial \mathbf{p}} - \frac{d\mathbf{q}}{dt} = 0 \end{aligned} \tag{15.17}$$

The main advantage of the Lagrange and Hamilton formulations is that any set of non-redundant variables can be used, while the Newton formulation focuses on spatial coordinates and corresponding velocities. The main difference between the Lagrange and Hamilton formulations is that the former is a single second-order differential equation, while the latter is a coupled set of first-order differential equations. The stationary action principle allows a formulation in terms of a vanishing gradient of a parameterized integral. Depending on the system, one of these formulations may be easier to solve than the other.

The time step employed is an important control parameter for a simulation. The maximum time step that can be taken is determined by the rate of the *fastest* process in the system, that is typically an order of magnitude smaller than the fastest process. Molecular motions (rotations and vibrations) typically occur with frequencies in the range 10^{11}–10^{14} s^{-1} (corresponding to wavenumbers of 3–3300 cm^{-1}), and time steps of the order of femtoseconds (10^{-15} s) or less are required to model such motions with sufficient accuracy. This means that a total simulation time of 1 nanosecond (10^{-9} s) requires $\sim 10^6$ time steps and 1 microsecond (10^{-6}) requires $\sim 10^9$ time steps. A million time steps is already a significant computational effort and typical simulation times are in the pico- or nanosecond range, depending on the complexity of the energy function. Many interesting phenomena unfortunately occur on a substantially longer timescale; protein folding and chemical reactions, for example, occur on the order of milliseconds or seconds. Furthermore, a single trajectory may not be adequate for representing the dynamics, thus requiring that many simulations must be carried out with different starting conditions (positions and velocities) and be properly averaged.

The fastest processes for molecules are the stretching vibrations, especially those involving hydrogen. These degrees of freedom, however, have relatively little influence on many properties. It is therefore advantageous to freeze all bond lengths involving hydrogen atoms, which allow longer time steps to be taken, and consequently longer simulation times to be obtained for the same computational cost. As all atoms move individually according to Newton's equation, constraints must be applied for keeping bond lengths fixed. This is normally done by either the *SHAKE*[19] (Verlet), *LINCS*[20] (leap-frog) or *RATTLE*[21] (velocity Verlet) algorithms, where the distance constraints are incorporated by the method of Lagrange undetermined multipliers (Section 13.5). The atoms are first allowed to move under the influence of the forces, and subsequently forced to obey the constraints by making a few sequential passes through all the variables.

Enforcement of bond length constraints typically allows the time step to be increased by a factor of 2 or 3. Angles may also be frozen by adding a distance constraint on atoms that are 1, 3 relative to each other. Angle bending, however, affects calculated properties more than bond stretching and fixing them may often introduce unacceptable errors. Angle constraints are therefore used less frequently. A simulation can also be performed using fixed molecular geometries, that is only the positions and relative orientations of individual molecules are allowed to change. The natural variables to propagate in time in such cases are the center of mass position and the three Euler angles of each molecule.

15.2.2 Generating Non-natural Ensembles

A standard MD simulation generates an *NVE* ensemble, that is the temperature and pressure will fluctuate. The total energy is a sum of the kinetic and potential energies, and can be calculated from the positions and velocities:

$$E_{\text{tot}} = \sum_{i=1}^{N} \frac{1}{2} m_i \mathbf{v}_i^2 + V(\mathbf{r}) \tag{15.18}$$

Owing to the finite precision with which the atomic forces are evaluated and the finite time step used, the total energy is *not* exactly constant, but this error can be controlled by the magnitude of the time step. Indeed, preservation of the energy to within a given threshold may be used to define the maximum permissible time step.

The temperature of the system is proportional to the average kinetic energy:

$$\langle E_{\text{kin}} \rangle = \frac{1}{2}(3N_{\text{atoms}} - N_{\text{constraint}})kT \tag{15.19}$$

The number of constraints is typically three, corresponding to conservation of linear momentum. Note that for 1 mole of particles, Equation (15.19) reduces to the familiar expression $\langle E_{\text{kin}} \rangle = 3/2RT$. Since the kinetic energy is the difference between the total energy (almost constant) and the potential energy (depends on the positions), the kinetic energy will vary significantly, that is the temperature will be calculated as an average value with an associated fluctuation. Similarly, if the volume of the system is fixed, the pressure will fluctuate.

Although the *NVE* is the natural ensemble generated by an MD simulation, it is possible also to generate *NVT* or *NPT* ensembles by MD techniques by modifying the velocities or positions in each time step. The instant value of the temperature is given by the average of the kinetic energy, as indicated by Equation (15.19). If this is different from the desired temperature, all velocities may be scaled by a factor of $(T_{\text{desired}}/T_{\text{actual}})^{1/2}$ in each time step to achieve the desired temperature. Such an "instant" correction procedure actually alters the dynamics, such that the simulation no longer corresponds to a canonical (*NVT*) ensemble. Performing the scaling at larger intervals introduces some periodicity into the simulation, which is also undesirable. The system may alternatively be coupled to a "*heat bath*", which gradually adds or removes energy to/from the system with a suitable time constant, a procedure often called a *thermostat*.[22] The kinetic energy of the system is again modified by scaling the velocities, but the rate of heat transfer is controlled by a coupling parameter τ:

$$\begin{gathered} \frac{\mathrm{d}T}{\mathrm{d}t} = \frac{1}{\tau}(T_{\text{desired}} - T_{\text{actual}}) \\ \Updownarrow \\ \text{Velocity scale factor} = \sqrt{1 + \frac{\Delta t}{\tau}\left(\frac{T_{\text{desired}}}{T_{\text{actual}}} - 1\right)} \end{gathered} \tag{15.20}$$

Thermostat methods such as Equation (15.20) are widely used but again do not produce a canonical ensemble. They do produce correct averages but give incorrect fluctuations of properties. In *Nosé–Hoover* methods[23,24] the heat bath is considered an integral part of the system and assigned fictive dynamic variables, which are evolved on an equal footing with the other variables. These methods are analogous to the extended Lagrange methods described in Section 15.2.5, and can be shown to produce true canonical ensembles.

The pressure can similarly be held (approximately) constant by coupling to a "pressure bath". Instead of changing the velocities of the particles, the volume of the system is changed by scaling all coordinates according to

$$\begin{gathered}\frac{\mathrm{d}P}{\mathrm{d}t} = \frac{1}{\tau}(P_{\text{desired}} - P_{\text{actual}}) \\ \Updownarrow \\ \text{Coordinate scale factor} = \sqrt[3]{1 + \kappa\frac{\Delta t}{\tau}(P_{\text{actual}} - P_{\text{desired}})}\end{gathered} \tag{15.21}$$

Here the constant κ is the compressibility of the system. Such *barostat* methods are again widely used, both in MC and MD simulations, but do not produce strictly correct ensembles. The pressure may alternatively be maintained by a Nosé–Hoover approach in order to produce a correct ensemble.

15.2.3 Langevin Methods

Molecular dynamics methods generate detailed information about all the particles in the system and are therefore well suited for calculating collective properties. In other cases, the major interest is in the dynamics of a single molecule, in which case the surrounding molecules can be modeled by only including the average interactions. This average interaction is assumed to have a friction term (with a friction coefficient ζ) proportional to the atomic velocity and a random component ($\mathbf{F}_{\text{random}}$) that averages to zero. These terms are in addition to the normal intramolecular forces ($\mathbf{F}_{\text{intra}}$) and possibly also external forces, for example from an electric field. The random force is associated with a temperature and adds energy to the system, while the friction term removes energy. The random force is typically taken to have a Gaussian distribution with a mean value of zero. Newton's second equation then takes the form

$$m\frac{\mathrm{d}^2\mathbf{r}}{\mathrm{d}t^2} = -\zeta\frac{\mathrm{d}\mathbf{r}}{\mathrm{d}t} + \mathbf{F}_{\text{intra}} + \mathbf{F}_{\text{random}} \tag{15.22}$$

Equation (15.22) is called the *Langevin* equation of motion and gives rise to *stochastic* or *Brownian dynamics*.[1,3] The magnitude of the friction coefficient determines the importance of the intramolecular forces compared with the friction term and large values of ζ lead to the Brownian dynamics limit.

15.2.4 Direct Methods

The major computational effort in a molecular or Langevin dynamic simulation is the calculation of the forces on all particles at each time step. Any type of energy function can in principle be used: force field, semi-empirical, *ab initio* electronic structure or DFT methods. Owing to the small time step required, and the resulting many force evaluations necessary, the large majority of simulations are performed with parameterized energy functions of the force field type. For studying macromolecules and solvation, general force fields of the type discussed in Chapter 2 are normally used. While these may be of sufficient accuracy for simulating structural properties, they are unable to describe chemical reactions or to achieve high accuracy. A "global" energy surface may in such cases be constructed by fitting high-level *ab initio* results and experimental data to a suitable functional form.[25,26] For sufficiently small time steps and long simulations times, the result of a simulation is determined entirely by the quality of the energy surface. In order to obtain "converged" results for the dynamics, the energy surface must be accurate to better than 1 kJ/mol, over the *whole* surface that is accessible at the given energy (temperature). Constructing such high-quality "global" energy surfaces is very demanding and has only been done for a few systems. As mentioned in Chapter 1, the sheer dimensionality prevents an adequate sampling of a surface by point calculations for more than three or four atoms, and high-level dynamics have thus been limited to systems of this size.

Even for low-dimensional surfaces (three to six atoms), it is often difficult to design a well-behaved fitting function capable of yielding a balanced description of all reaction channels. A simulation of the reaction between an oxygen atom and methane, for example, requires a balanced energy description of the following (stable) species, as well as the reaction paths connecting these:

- $CH_4 + {}^{1,3}O$
- CH_3OH
- $H_2CO + H_2$
- ${}^{1,3}HCOH$ (*cis* and *trans*) $+ H_2$
- $CH_3O + H$
- ${}^{1,3}CH_2 + H_2O$
- $CH_3 + OH$

Note that both singlet and triplet energy surfaces are important for this reaction. Achieving a high-quality surface will require a large number of MR-CI-type calculations, and designing a suitable interpolation function to reproduce the experimental energy differences between all the above exit channels is a non-trivial exercise.

The surface design and fitting process can be bypassed by performing the dynamics "directly", that is by calculating the required energies and forces in each time step of a simulation. The advantage is that a fitting function is not required, there is no parameterization step and only the part of the surface actually visited by the dynamics has to be calculated. The disadvantage is that the same (or almost the same) points may be calculated many times, and if many trajectories are required, the total amount of points calculated may be larger than required for performing a global fit. It is furthermore difficult to add empirical corrections to the calculated surface. In a global fit approach, deficiencies in the employed computational method can be partly alleviated by enforcing energy differences between experimentally known species in the parameterization step. In some cases, the employed electronic structure method is pre-modified in direct approaches in order to give better agreement with experimental quantities. The latter has especially been used in connection with semi-empirical methods (Chapter 7), where the atomic parameters can be re-tuned to model a specific reaction surface better than the defaults parameters, a procedure called *Specific Reaction Parameterization* (SRP).[27–29]

15.2.5 *Ab Initio* Molecular Dynamics

Molecular dynamics propagate the nuclear positions and velocities by solving Newton's second equation in a time-discretized form, and in the direct dynamics version the atomic forces are calculated using an electronic structure method. When the latter is of the *ab initio* wave function or DFT type, the approach can collectively be denoted *Ab Initio Molecular Dynamics* (AIMD).[30]

The intuitively simplest approach is to determine a converged wave function at each time step and use the corresponding nuclear gradients for propagating the time-evolution. In order to fulfill energy conservation over the whole simulation length, however, such *Born–Oppenheimer Molecular Dynamics* (BOMD) require a very tight convergence of the wave function in each time step, otherwise the electrons will create an artificial frictional term on the nuclei, which makes the procedure computationally expensive.[31] In an elegant breakthrough by *Car and Parrinello* ,[32,33] it was shown that it is not necessary to fully converge the wave function in each time step. After having determined a converged wave function at the first point, the essence of *Car–Parrinello Molecular Dynamics* (CPMD) is to let the wave function parameters (orbitals) evolve *simultaneously* with the changes in nuclear positions. This can be achieved by including the wave function parameters as variables with fictive "masses" in the dynamics, analogous to the nuclear positions and masses. Since this involves generalized variables, the Lagrange formulation (Equation (15.13)) for the dynamical equation is convenient. The use of such *extended Lagrange functions* for describing the evolution of a system with both "real" (nuclear/electronic) and "fictive" (method) parameters is quite general, and is, for example, also used in force field methods incorporating fluctuating charges and/or polarization (Section 2.2.8).

The nuclear contributions in the CPMD method are given by

$$\begin{gathered} L = T_{\text{nuc}} - V_{\text{nuc}} \\ T_{\text{nuc}} = \frac{1}{2}\sum_{a}^{N_{\text{nuc}}} M_a \left(\frac{\mathrm{d}\mathbf{R}_a}{\mathrm{d}t}\right)^2 \quad ; \quad V_{\text{nuc}} = V(\mathbf{R}_{\text{nuc}}) \end{gathered} \tag{15.23}$$

We now add contributions corresponding to treating the orbital expansion coefficients as variables with fictive masses μ_i:

$$\phi_i = \sum_{\alpha}^{M_{\text{basis}}} c_{i\alpha}\chi_\alpha \tag{15.24}$$

$$\begin{gathered} L = T_{\text{nuc}} - V_{\text{nuc}} + T_{\text{orb}} - V_{\text{orb}} \\ T_{\text{orb}} = \frac{1}{2}\sum_{i}^{N_{\text{orb}}} \mu_i \left(\frac{\mathrm{d}c_{i\alpha}}{\mathrm{d}t}\right)^2 \quad ; \quad V_{\text{orb}} = E(\mathbf{c}_{\text{orb}}) \end{gathered} \tag{15.25}$$

The two potential energies can be combined to a single term depending on both the nuclear positions and the orbital coefficients:

$$\begin{gathered} L = T_{\text{nuc}} + T_{\text{orb}} - V_{\text{tot}} \\ V_{\text{tot}} = E(\mathbf{R}_{\text{nuc}}, \mathbf{c}_{\text{orb}}) \end{gathered} \tag{15.26}$$

The orbital orthogonality constraints can be included by addition of terms involving Lagrange multipliers:

$$\langle \phi_i | \phi_j \rangle = \delta_{ij} \tag{15.27}$$

$$\begin{gathered} L = T_{\text{nuc}} + T_{\text{orb}} - V_{\text{tot}} - \sum_{ij}^{N_{\text{orb}}} \lambda_{ij}\sigma_{ij} \\ \sigma_{ij} = \sum_{\alpha\beta}^{M_{\text{basis}}} c_{i\alpha}c_{j\beta}\langle \chi_\alpha | \chi_\beta \rangle - \delta_{ij} = 0 \end{gathered} \tag{15.28}$$

The resulting dynamical equations then become

$$M_a \frac{\partial^2 \mathbf{R}_a}{\partial t^2} = -\frac{\partial E}{\partial \mathbf{R}_a} + \sum_{ij}^{N_{\mathrm{orb}}} \lambda_{ij} \frac{\partial \sigma_{ij}}{\partial \mathbf{R}_a} \tag{15.29}$$

$$\mu_i \frac{\partial^2 c_{i\alpha}}{\partial t^2} = -\frac{\partial E}{\partial c_{i\alpha}} + \sum_{ij}^{N_{\mathrm{orb}}} \lambda_{ij} \frac{\partial \sigma_{ij}}{\partial c_{i\alpha}} \tag{15.30}$$

The constraint forces are handled iteratively, analogously to the constraint of fixed bond lengths in the SHAKE algorithm.

If the nuclear positions **R** are kept constant and the fictive orbital kinetic energy T_{orb} is quenched, the resulting algorithm is essentially a steepest descent minimization of the electronic energy with respect to the orbital coefficients. This is done at the initial point, but at subsequent points the orbital parameters are allowed to evolve along with the nuclear position according to the dynamical equation. This means that the nuclear forces are not strictly correct since the electronic wave function is not converged in the orbital parameter space. This error, however, can be controlled by suitable choices of the fictive masses associated with the orbital parameters, that is small values provide results close to the "true" Born–Oppenheimer results, but also require the use of small time steps since the resulting "orbital parameter frequency" is high.[34, 35] The fictive masses are typically taken to be a few hundred atomic units, giving time steps of ~0.1 femtoseconds, that is roughly an order of magnitude smaller than for classical molecular dynamics. It should be noted that the optimum value for the fictive masses depends on the system, and for metals and semiconductors, for example, it is difficult to choose suitable values. Systems containing hydrogen are especially problematic, since the proton mass (1836 au) is only a factor of 5–10 higher than the fictive orbital parameter mass. This in some cases leads to a coupling of these degrees of freedom, but it can be partly countered by using deuterium instead of hydrogen.

In the CPMD approach, the total energy is conserved, and this now includes the fictive kinetic energy of the orbital parameters. The "real" system of course has no orbital kinetic energy, and this must therefore be kept small compared with the other terms in order for the CPMD method to provide realistic simulation results. The magnitude of the fictive masses for the orbital parameters serves as a coupling parameter between the nuclear and parameter kinetic energies. The temperature associated with the nuclei and orbital parameters are the same in an equilibrium condition, but it is usually desirable to have the parameter kinetic energy to be significantly lower (a value of zero corresponds to the Born–Oppenheimer case). This can be obtained by continuously removing the fictive kinetic energy associated with the orbital parameters, while compensating for the energy loss by adding energy to the nuclei. This can in practice be obtained by allowing the orbital parameters to interact with a "heat bath" of a low temperature, while the nuclei interact with a heat bath of the desired simulation temperature.

The CPMD technique can be used both in a "static" sense, for simultaneously optimizing the wave function and the nuclear positions by periodically quenching the kinetic energies, but it can also be used in a "dynamical" sense for sampling the (nuclear) phase space. The main advantage of the CPMD technique is the much better error cancellation compared with a Born–Oppenheimer dynamics, that is even with non-converged wave function parameters, the long-term energy conservation is fulfilled to a quite high accuracy. The coupling of the real and fictive parameters builds a self-correction into the CPMD method; that is if the nuclei at some point get slightly "ahead" of the electron cloud, they will be slowed down, thus allowing the electrons to "catch up" with the nuclei. Similarly, if the electronic parameters get ahead of the nuclei, the nuclei will be accelerated owing to the Coulomb attraction. The primary disadvantage of CPMD is that the inclusion of the electronic degrees of freedom in the dynamics requires the use of significantly smaller time steps compared to a BOMD approach. The tradeoff is thus that CPMD allows a computationally inexpensive force in each time step, but the time step must be small, while BOMD has a higher computational cost for generating the force, but allows larger time steps.

The main disadvantage of the original CPMD algorithm is the "fast" dynamics of the electronic degree of freedom, which necessitates short time steps. A modified version has been proposed where the electronic parameters are evolved by a predictor–corrector extrapolation from previous points, rather than by their time-dependent dynamics. Using typically four previous points for the predictor step and a single corrector step generate results close to those obtained from the true Born–Oppenheimer surface, and allows time steps of the same magnitude as in the BOMD method.[36] This modified CPMD algorithm thus avoids the multiple SCF iterations of the BOMD method at each step and avoids introducing fictitious variables with high frequencies, as in the original CPMD approach.

AIMD simulations can in principle be carried out with any type of wave function, but they are still significantly more expensive computationally than traditional parameterized energy functions. The CPMD method was originally implemented with DFT methods using plane waves as the basis set, but the technique has also been used with other types of methods (e.g. HF or MP2) and Gaussian-type basis functions, where the density matrix elements are used as variables instead of the molecular orbital coefficients.[37,38] The great advantage over force field-type functions is that electronic structure methods are able to describe bond breaking/formation, that is AIMD methods allow a direct simulation of chemical reactions and processes such as hydrogen exchange in water. Even with the CPMD technique, however, the use of *ab initio* electronic structure calculations is so expensive that only picosecond simulation can be carried out, compared with nano- or microsecond simulations with parameterized energy functions.

AIMD may be considered as a semi-classical dynamics approach where the electrons are treated quantum mechanically while the nuclear motion is treated classically. The latter implies that, for example, zero-point vibrational effects are not included and nor can nuclear tunneling effects be described; this requires fully quantum methods, as described in the next section.

15.2.6 Quantum Dynamical Methods Using Potential Energy Surfaces

In order to incorporate quantum effects into the nuclear motions (vibrational effects and tunneling), the time-dependent (nuclear) Schrödinger equation must be used in place of Newton's equation:

$$\mathbf{H}\Psi = (\mathbf{T} + \mathbf{V})\,\Psi = i\frac{\partial \Psi}{\partial t} \tag{15.31}$$

Here $\mathbf{T}$ is the kinetic energy operator and $\mathbf{V}$ is the potential energy. The square of the wave function is the *probability* of finding a particle at a given position. Heisenberg's uncertainty principle means that a quantum description of a nucleus must be a continuous function, not a single specific position as in classical mechanics.[39,40] Such a continuous function is often denoted a *wave package* and may be modeled by Gaussian functions (semi-classical methods) or numerically (quantum methods). The wave function may, analogously to classical dynamics, be propagated through a series of small, but finite, time steps:

$$\Psi_{i+1} = -i\mathbf{H}\Psi_i\Delta t = -i\,(\mathbf{T} + \mathbf{V})\,\Psi_i\Delta t \tag{15.32}$$

Each time step thus involves a calculation of the effect of the Hamiltonian operator acting on the wave function. In fully quantum methods, the wave function is often represented on a *grid* of points, these being the equivalent of basis functions for an electronic wave function. The effect of the potential energy operator is easy to evaluate, as it just involves a multiplication of the potential at each point with the value of the wave function. The kinetic energy operator, however, involves the derivative of the wave function and a direct evaluation would require a very dense set of grid points for an accurate representation.

The kinetic energy operator is proportional to the square of the momentum, $\mathbf{T} = \mathbf{p}^2/2m$. In a *momentum* representation (i.e. using the particle momentum instead of position as variables), $\mathbf{T}$ is a simple multiplication operator, analogous to $\mathbf{V}$ in position space. The transformation from position to momentum space can be achieved by a Fourier transformation. A numerical solution of the time-dependent Schrödinger equation can thus be done by switching back and forth between a position and momentum representation of the wave function, evaluating the effect of $\mathbf{V}$ in position space and the effect of $\mathbf{T}$ in momentum space. Analogously to the leap-frog algorithm for the classical case (Equations (15.8) and (15.9)), the update of the wave function by the potential and kinetic energy operators may be chosen to be out of phase by half a time step to improve the accuracy. The key to the popularity of this approach is the presence of highly efficient computer routines for performing Fourier transformations.

The requirement of an accurate global energy surface is even more important for a quantum mechanical treatment than for the classical case, since the wave function depends on a finite part of the surface, not just a single point. The updating of the positions and velocities are computationally inexpensive in the classical case, once the forces are available, but the requirement of two Fourier transformations in each time step makes the quantum propagation a significant computational issue. The representation of the wave function on a grid effectively limits the dimensionality to a maximum of three, that is di- and triatomic systems (one and three internal coordinates, respectively). Larger systems necessitate freezing some of the coordinates or treating them classically.[41]

15.2.7 Reaction Path Methods

The main problem in dynamical studies is the requirement of a continuous energy surface over a wide range of geometries. A simulation will normally be done with specification of an energy (or a temperature) and a surface must thus be available for all nuclear configurations that have an energy lower than the chosen simulation value. For quantum methods, the surface must also be available at higher energies as the wave function has a tail that penetrates into classically "forbidden" areas.

Such "global" energy surfaces have traditionally been constructed by fitting a suitable functional form to energies (and possibly also first and second derivatives) calculated by *ab initio* methods at a large number (perhaps a few hundreds or thousands) of geometries.[25] The function may be further refined by including experimental data (such as vibrational frequencies and geometries) in the fitting. For "large" systems (i.e. more than three or four atoms) the generation of an adequate number of fitting points is prohibitively expensive. In order to treat large systems, it is necessary to concentrate the computational effort on the "chemically important" part of the potential energy surface.

A chemical reaction in the simplest description takes place along the lowest energy path connecting the reactant and product, passing over the transition structure (Section 14.1) as the highest point. This is the *Minimum Energy Path* (MEP), which in mass-weighted coordinates is called the *Intrinsic Reaction Coordinate* (IRC) (Section 13.8). The idea in *Reaction Path* (RP) methods[42] is to only consider the energy surface in the immediate vicinity of a suitable one-dimensional reaction path, which usually (but not necessarily) is taken as the IRC. The potential is typically expanded to second order along the reaction path, corresponding to modeling the perpendicular degrees of freedom as harmonic vibrational frequencies. The reaction path potential may be generated by a series of frequency calculations at points along the IRC and the point-wise potential made continuous by interpolation. The potential may be generated prior to the reaction path calculation, or generated "on the fly" in a "direct" fashion.[43] Moyano and Collins have proposed a hybrid method where all the points calculated are stored and used for interpolation if the required point is sufficiently close to prior points.[44] This approach thus starts out as a direct-type dynamics but ends up with an implicitly parameterized surface for sufficiently long simulations times. For long simulation times or for running many trajectories, the savings by interpolation can be substantial.

The reaction path method may be generalized by having two "reaction coordinates" (a *reaction surface*) treated explicitly and the remaining degrees of freedom treated approximately, or by having three "reaction coordinates" (a *reaction volume*).[45] These generalizations are useful for performing mixed classical–quantum dynamics, where the dynamics with the reaction coordinate(s) are treated quantum mechanically while the remaining degrees of freedom are treated classically.

The inclusion of dynamical effects allows the calculation of corrections to simple transition state theory, often described by a transmission coefficient κ to be multiplied with the TST rate constant (Section 14.1) or used in connection with variational TST (Section 14.5.5). Classical dynamics allow corrections due to re-crossing to be calculated, while a quantum treatment is necessary for including tunneling effects. Owing to the stringent requirement of a highly accurate global energy surface, there are only a few systems that have been subjected to a rigorous analysis.

The tunneling effect can be approximated by inclusion of a semi-classical correction based on tunneling through the barrier along the minimum energy path (i.e. the IRC). The *Bell* correction is based on the assumption that the (one-dimensional) energy curve near the transition state can be approximated by a parabola.[46,47] This yields a correction factor that only depends on the activation energy $\Delta E^{\neq}$ and the magnitude of the imaginary frequency ν_i, that is the curvature of the potential energy surface at the TS:

$$\kappa_{\text{Bell}} = \frac{\frac{1}{2}u^{\neq}}{\sin\frac{1}{2}u^{\neq}} - u^{\neq}\mathrm{e}^{-\Delta E^{\neq}/kT}\left(\frac{\mathrm{e}^{-\beta}}{2\pi - u^{\neq}} - \frac{\mathrm{e}^{-2\beta}}{4\pi - u^{\neq}} + \frac{\mathrm{e}^{-3\beta}}{6\pi - u^{\neq}} - \cdots\right)$$
$$u^{\neq} = \frac{h\nu_i}{kT} \quad ; \quad \beta = \frac{2\pi\Delta E^{\neq}}{h\nu_i} \tag{15.33}$$

Except for reactions with low barriers (i.e. <40 kJ/mol at $T = 300$ K) or at high temperatures, the quantity $\Delta E^{\neq}/kT$ is large and the last series can be neglected. The tunneling correction is then given completely in terms of the magnitude of the imaginary frequency. The temperature where the sin $^1/_2u^{\neq}$ term diverges is called the *crossover temperature* T_c and can be considered as the temperature below which the reaction rate is dominated by tunneling. When the imaginary frequency ν_i is given in units of cm^{-1}, the connection is $T_c = 0.23\ \nu_i$, which indicates that tunneling at $T = 300$ K is only important for reactions having ν_i larger than ~1300 cm^{-1}. For small values of $u^{\neq}$ the first term may be Taylor-expanded to give

$$\kappa_{\text{Wigner}} = 1 + \frac{1}{24}(u^{\neq})^2 + \cdots \tag{15.34}$$

The first-order term is known as the *Wigner* correction.[48]

It is possible to derive tunneling corrections for functional forms of the energy barrier other than an inverted parabola, but these cannot be expressed in analytical form. Since any barrier can be approximated by a parabola near the TS, and since tunneling is most important for energies just below the top, they tend to give results in qualitative agreement with the Bell formula.

The main approximation of such one-dimensional corrections is that the tunneling is assumed to occur *along* the MEP. This may be a reasonable assumption for reactions having either early or

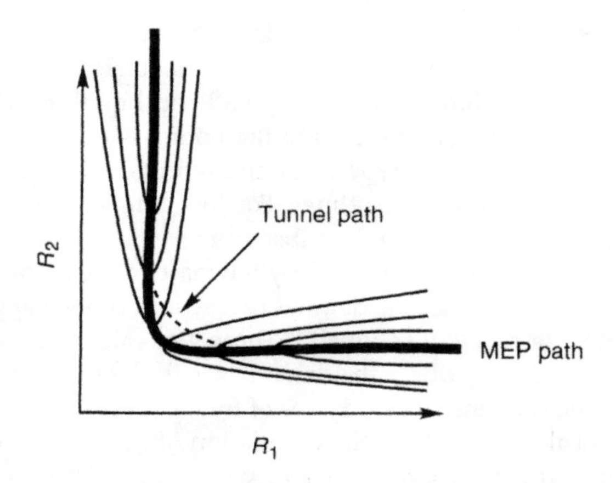

Figure 15.1 A contour plot illustration of the "corner cutting" tunneling path.

late (close to either reactant or product) transition states. For reactions where bond breaking and formation are both significant at the TS (as is usually the case), the dominant tunneling effect is "corner cutting" (Figure 15.1), that is the favored tunnel path is not along the MEP. Although the energy increases away from the MEP, the barrier also becomes narrower on the concave side of the reaction path, which favors the tunneling probability.

Truhlar and coworkers have developed various approximate schemes for including tunneling in multidimensional systems based on the Wentzel–Kramer–Brillouin approximation, where the tunneling correction along a one-dimensional path is given by[43,49]

$$\kappa_{\mathrm{WKB}} = (1 + \mathrm{e}^{2\theta(E)/\hbar})^{-1} \tag{15.35}$$

$$\theta(E) = \int_{r_a}^{r_b} \sqrt{2m(V(r) - E)}\mathrm{d}r \tag{15.36}$$

The barrier penetration integral $\theta(E)$ depends on the mass of the tunneling particle m, the height of the barrier and the length of the path between the two classical turning points r_a and r_b, where the potential V is equal to the energy E. The path corresponding to the maximum tunneling probability is thus a compromise between a long(er) path with a low(er) barrier and a short(er) path with a high(er) barrier. In the *Minimum Energy Path Semi-classical Adiabatic Ground state* (MEPSAG) approximation the tunneling is assumed to occur along the MEP, analogous to the Bell approach, but for an arbitrary shape of the energy surface. The *Small Curvature Semi-classical Adiabatic Ground state* (SCSAG) approximation allows tunneling to occur within one vibrational half-amplitude perpendicular to the reaction path and is expected to be a good approximation at temperatures close to T_c. The *Large Curvature Ground state* (LCG) method approximates the tunneling path by a straight line between the classical turning points on the IRC where the energy is equal to the potential energy, and is expected to be a valid approximation for low temperatures relative to T_c.[50] A linear interpolation between the two paths has been proposed for intermediate temperatures, that is maximizing the tunneling rate as a function of an interpolation parameter.[51] The SCSAG method requires knowledge of the (generalized) frequencies along the IRC (Section 13.8), which can be obtained by calculating the force constant matrix at suitable intervals and interpolating the results. The LCG methods require additional calculations away from the IRC.

Instanton theory can be considered as a generalization of harmonic transition state theory to include tunneling, and allows a determination of the optimal tunneling path without any *a priori* parameterizations.[52] The delocalized nature of quantum particles (nuclei) is modeled by considering multiple paths connecting the reactant and product states, and the quantum statistical partition function is an integral over all such paths, each of which can be characterized by the action S defined in Equation (15.14). The tunneling process can be modeled by considering the action with an imaginary time evolution, called the Euclidian action S_E, which in (non-generalized) coordinates $\mathbf{x}$ can be written as follows, with τ being a path parameterization variable:

$$S_E[\mathbf{x}] = \int_0^{\hbar/kT} \left(\frac{m}{2} \left(\frac{\partial \mathbf{x}(\tau)}{\partial \tau} \right)^2 + V(\mathbf{x}(\tau)) \right) \mathrm{d}\tau \quad (15.37)$$

The stationary path for S_E is the most probable tunneling path and can be considered as the semi-classical generalization of the IRC in classical transition state theory. A manageable approximation to integrating over all paths is constructed by only considering a second-order expansion around the stationary path corresponding to $\delta S_E = 0$. Switching to mass-weighted coordinates ($\mathbf{y} = \sqrt{m}\mathbf{x}$) and approximating the integral by a sum over P images along the path provides the following equation, where the pre-factor for the first summation includes a factor of four owing to the summation being only over unique images, while the full path is a closed loop:

$$S_E[\mathbf{y}] = \frac{2PkT}{\hbar} \sum_{k=1}^{P-1} |\mathbf{y}_{k+1} - \mathbf{y}_k|^2 + \frac{\hbar}{PkT} \sum_{k=1}^{P} V(\mathbf{y}_k) \quad (15.38)$$

This is now a function of P images, each containing $3N_{atom}$ coordinates, and the problem of locating the most probable tunneling path corresponds to finding a first-order saddle point of the $3PN_{atom}$ multidimensional function. The computational problem is thus very similar to the multistructure interpolation methods for locating transition structures described in Section 13.4.3, and essentially the same optimization techniques can be employed. Once the optimized instanton path has been located, the tunneling rate can be obtained from the second derivative of the action S_E, which requires the molecular Hessian for all P images along the path.

15.2.8 Non-Born–Oppenheimer Methods

The methods in Sections 15.2.5 and 15.2.6 attempt to include nuclear quantum corrections based on the Born–Oppenheimer separation of the nuclear and electronic degrees of freedom, that is solving the nuclear dynamics on a potential energy surface obtained by solving the electronic Schrödinger equation. When quantum corrections such as tunneling are large, however, it is an implicit warning that the Born–Oppenheimer approximation may also be problematic. Rather than trying to improve on the underlying model by adding correction terms, it may be both easier and better to treat the nuclei within a quantum framework from the start.

Methods that treat all of the electron and nuclear degrees of freedom within a combined quantum framework are starting to appear; so far they are mostly based on a mean-field (i.e. Hartree–Fock) approximation where the coupling of the nuclear and electron motions is included in an average fashion. Both conceptual and computational developments are required before such methods can be considered to be mature.[53–55] One clear advantage of these methods is the ability to implicitly include both tunneling and vibrational effects, and to selectively treat some nuclei as classical, thereby allowing a simplification for large systems.

In the spirit of the Car–Parrinello approach, the whole set of variables (nuclei and wave function parameters) may also be allowed to evolve simultaneously by solving the time-dependent Schrödinger equation. Öhrn and coworkers have developed an *Electron–Nuclear Dynamics* (END) method,[56,57] where both the orbitals describing the electronic wave function and the nuclear degrees of freedom are described by expansion into a Gaussian basis set, which moves along with the nuclei. Such an approach in principle allows a complete solution of the combined nuclear–electron Schrödinger equation without having to invoke approximations beyond those imposed by the basis set. Inclusion of the electronic parameters in the dynamics, however, means that the fundamental time step is short, and this results in a high computational cost for even quite short simulations and simple wave functions.

15.2.9 Constrained and Biased Sampling Methods

The methods described in Section 15.2.7 focus on the lowest energy reaction path on the electronic energy surface. The activation energy related to the experimental reaction rate, however, depends on the free energy, that is one would optimally like to locate the reaction path on the free energy surface. For small systems, this can be done by adding finite temperature corrections to the enthalpy and entropy in a rigid-rotor harmonic-oscillator approximation based on a second-order expansion of the energy around each point (Equations (14.43) to (14.50)). For large systems, however, the harmonic approximation is less suitable and a more complete sampling by dynamical methods is usually desired. This will yield information about the dynamics in the perpendicular directions and potentially be able to provide a free energy profile along the reaction path.

A straightforward sampling of the reaction path is not possible since the dynamics at ordinary temperatures only very rarely visit the high-energy region near the TS (unless the activation energy is close to zero). In order to achieve a sampling of a specific region of the energy surface with molecular dynamics or Monte Carlo methods, the sampling must be biased toward a specific volume of phase space. A central component in biased sampling methods is the selection of one or a few *Collective Variables* (CV) that can be used to describe the reaction path.[58] A CV can be any function of atomic coordinates, but it must be selected and defined by the user based on the given application. A CV can in the simplest case be the position of a point, a distance between two points, an angle between three points or a torsional angle between four points, where the points can either be single atoms or centers of mass for a larger group of atoms. CVs defined as inverse distances between atom pairs raised to the sixth power may be appropriate variables for biasing MD simulations toward reproducing experimental NMR NOE measurements. CVs involving many (all) atoms can, for example, be normal modes from a harmonic vibrational analysis or principal components from a dynamic simulation or a correlation matrix. More complex CVs can measure the content of α-helix in a protein structure or be defined as the root mean square deviation between all atoms in two structures. Once a set of CVs have been selected, the biasing can be done by two different methods, a penalty and a Lagrange type approach, analogously to the optimization of functions with constraints (Section 13.5). A number of closely related methods have been proposed for performing simulations with bias potentials, typically with the aim of calculating free energy profiles along a suitable reaction coordinate, and the following describes some of these approaches.The penalty approach corresponds to augmenting the energy surface with a biasing potential U, for example a harmonic function centered at position ξ_0 of the CV with a suitable width k_U:

$$\begin{aligned} V_{\text{umbrella}}(\xi) &= V(\xi) + U(\xi) \\ U(\xi) &= k_U(\xi - \xi_0)^2 \end{aligned} \tag{15.39}$$

By making the biasing potential sufficiently steep (large k_U), the energy of the augmented energy surface far from ξ_0 will become so high in energy that only the region near ξ_0 will be sampled at ambient temperatures; this technique is called *umbrella sampling*.[59] The ensemble calculated with the augmented potential V_{umbrella} will of course be non-Boltzmann, but this can be deconvoluted as shown for a property P:

$$\langle P \rangle = \frac{\langle P(\xi) e^{U(\xi)/kT} \rangle_{V_{\text{umbrella}}}}{\langle e^{U(\xi)/kT} \rangle_{V_{\text{umbrella}}}} \tag{15.40}$$

Here $\langle \, \rangle_{V\text{umbrella}}$ indicates an average over the ensemble generated by the augmented potential. By performing a series of simulations with biasing potentials located at different positions along the reaction path, the free energy along the reaction path, often called the *Potential of Mean Force* (PMF), can be simulated.

The Lagrange approach constrains the sampling to the $(N - 1)$-dimensional subspace corresponding to a specific value of the CV, where the constraint is fulfilled by means of an additional term in the Hamiltonian involving a Lagrange multiplier. This is related to the extended Lagrange techniques discussed in Section 15.2.5 and is usually referred to as *Blue Moon* sampling in the literature (the term "Blue Moon" denotes the (relatively) rare phenomenon of having four full moons in a season (three months), and denotes in general a rare phenomenon) and the Lagrange multiplier is called the holonomic constraint.[60, 61]

The main disadvantage of the umbrella or Blue Moon sampling techniques is that the location of the biasing potential must be selected manually and an *a priori* knowledge of an approximate reaction coordinate is therefore required. Once this has been selected, the free energy along this path can be calculated. Since the sampling explores a (small) region around the selected path, the calculated PMF may deviate slightly from the initial selection. If desired, this updated PMF can then be used for a new series of simulations with biasing potentials located along the previously calculated PMF.[62] Such *adaptive* umbrella sampling methods should in principle converge on the true PMF but, in practice, the convergence is sensitive to the selection of a suitable initial reaction path.

Umbrella sampling employs a rather small perturbation potential in order to keep the dynamics close to the unbiased one. The same technique can also be used to force a large conformational change occurring on a long timescale to be observable during a relatively short simulation. This requires a significantly larger bias force, and the technique is then called *Steered Molecular Dynamics* (SMD).[63] It can, for example, be used to model the unfolding of proteins by anchoring one end and attaching the other end to a point in space by a harmonic potential, and then move the point at a constant velocity in a specific direction. SMD can similarly be used to "drag" a ligand from a binding site in a protein, but this requires a careful choice of the direction where the bias force is applied. SMD can also be used to generate a reaction path between two different protein structures by selecting a suitable set of CVs to describe the transition pathway. If the root mean square deviation between the two end structures is used as a CV, the method is called *Targeted MD*.

The binding or unbinding of a ligand to a buried active site in a protein is a key example of a process that occurs on a timescale significantly longer than is achievable by regular MD simulations, and is thus a prime target for enhanced sampling methods. *The Random Expulsion MD* (*REMD*) method attempts to find a pathway for unbinding a ligand by adding an artificial force to the ligand atoms in a random direction and letting the MD simulation run for a relatively short time (~few hundred picoseconds) while monitoring how much the ligand moves in the applied force direction.[64] The underlying idea is to probe the protein to find a low energy deformation mode that will allow the ligand to escape and, by reversibility, to identify the binding pathway. If the randomly chosen direction is towards a "stiff" part of the protein, the ligand displacement will be small and the MD can be aborted after only a short simulation time. If, on the other hand, the chosen direction is towards a "soft" part of the protein, the protein can distort and allow the ligand to move and eventually escape. The magnitude of the applied force is a user-defined parameter and is a compromise between a low value that allows a "natural" but slow protein distortion and a high value that may give a fast but "unnatural" protein distortion.

A common feature of unbiased MD is the tendency of the simulation to revisit the same region of phase space by random thermal motion due to the presence of energy barriers substantially larger than the available thermal energy. *Metadynamics* attempts to prevent this situation by gradually adding repulsive Gaussian-shaped potentials based on the sampling history and can be thought of as gradually filling up the energy minimum currently being sampled until the thermal energy is sufficient for escaping to another minimum.[65] The bias potential is defined in terms of CV and is for a single CV given by

$$U_{\text{metadyn}}(\xi) = \sum_{t_0=0,\Delta,2\Delta,\ldots}^{t-\Delta} W \exp\left(-\frac{(\xi(t)-\xi(t_0))^2}{2\sigma^2}\right) \tag{15.41}$$

Here W and σ are user-defined parameters controlling the high and width of the Gaussian and Δ is the time interval between points on the trajectory where the bias potentials are placed. Equation (15.28) is readily extended to include more than one CV by simply including an additional term in the exponential function for each additional CV. Metadynamics can be considered as a generalization of the *conformational flooding, hyperdynamics* and *local elevation* methods and includes these as special cases. An appealing aspect of metadynamics is that the original free energy surface along the trajectory can be reconstructed *a posterior* if the history of the applied bias potentials is collected during the simulation. The speed of exploring phase space and the accuracy of the reconstructed free energy surface is (not surprisingly) inversely related. Large values of W and σ ensure that the dynamics is rapidly forced to leave the current energy minimum, but also means that each minimum is only sampled sparsely. The accuracy also rapidly deteriorates as the number of CVs is increased due to inefficient sampling.

The *Adaptive Bias Force* (ABF) method[66] can be considered as a dynamical variation of metadynamics where the bias force (potential) is constructed as a running average during the simulation such that it cancels the (current) estimate of the free energy force along the reaction path. The simulation is initiated as an unconstrained MD that samples (only) a small fraction of the phase space along the CV chosen as the reaction variable. As the average force calculated from this sampling gradually stabilizes, a bias force is applied that cancels the restoring force along the reaction path and allows free diffusion to the next part of the path that eventually leads to a flat energy surface along the whole reaction coordinate. When the sampling indicates a flat energy surface along the whole reaction coordinates, the original free energy profile can be constructed from the negative of the bias force. The sampling along the reaction path can be hindered by energy barriers in the orthogonal coordinates, often referred to as hidden barriers, which implies that the selected CV is not a good description of the actual reaction path.

Accelerated Molecular Dynamics (aMD) is similar to metadynamics but employs a time-independent bias potential which is applied only for the part of the energy surface that is below a pre-chosen boost energy value:[67]

$$V_{\text{aMD}}(\mathbf{r}) = \begin{cases} V(\mathbf{r}) & ; \quad V(\mathbf{r}) \geq E_{\text{boost}} \\ V(\mathbf{r}) + U(\mathbf{r}) & ; \quad V(\mathbf{r}) < E_{\text{boost}} \end{cases} \tag{15.42}$$

The bias potential destabilizes the (local) energy minima below E_{boost} and thereby decreases the energy barriers and allows the system to escape and sample other regions of the phase space. The bias potential is given below, where α is a parameter that determines how shallow the modified energy surface is for values below E_{boost}:

$$U(\mathbf{r}) = \frac{(E_{\text{boost}} - V(\mathbf{r}))^2}{\alpha + (E_{\text{boost}} - V(\mathbf{r}))} \tag{15.43}$$

aMD has the advantage over metadynamics that no CVs are required, although the boost energy value can be considered as a generalized CV. The aMD method can be employed in selected internal coordinates, such as, for example, only in the torsional degrees of freedom to enhance conformational transitions.

Temperature accelerated MD is a method where the sampling along one or more CVs is enhanced by artificially increasing the temperature in these degrees of freedom.[68]

A key point in obtaining accurate free energy differences is a proper choice of CVs for describing the reaction path. An internal check can be performed by running the dynamics in both directions, where observed hysteresis is a clear sign that one or more important degrees of freedom are missing from the selected CVs. For a process involving several intermediate metastable states, it may be difficult to find a few CVs describing both end-points as well as the whole reaction path involving metastable structures. As mentioned above, the sampling becomes inefficient if more than 2–3 CVs are selected.

15.3 Periodic Boundary Conditions

A realistic model of a solution requires at least several hundred solvent molecules. To prevent the outer solvent molecules from boiling off into space and minimizing surface effects, *Periodic Boundary Conditions* (PBCs) are normally employed (Figure 15.2). The solvent molecules are placed in a suitable box, often (but not necessarily) having a cubic geometry (it has been shown that simulation results using any of the five types of space-filling polyhedra are equivalent[69]). This box is then duplicated in all directions, that is the central box is surrounded by 26 identical cubes, which are again surrounded by 98 boxes, etc.

If a solvent molecule leaves the central box through the right wall, its image will enter the box through the left wall from the neighboring box. This means that the resulting solvent model becomes quasi-periodic, with a periodicity equal to the dimensions of the box, in contrast to a real solvent where there is no long-range order.

The electrostatic interaction is long-ranged (Section 2.2.6) and will extend beyond the boundary of a box. Truncating the interaction by using a cutoff distance of, say, 10 Å gives discontinuous energies

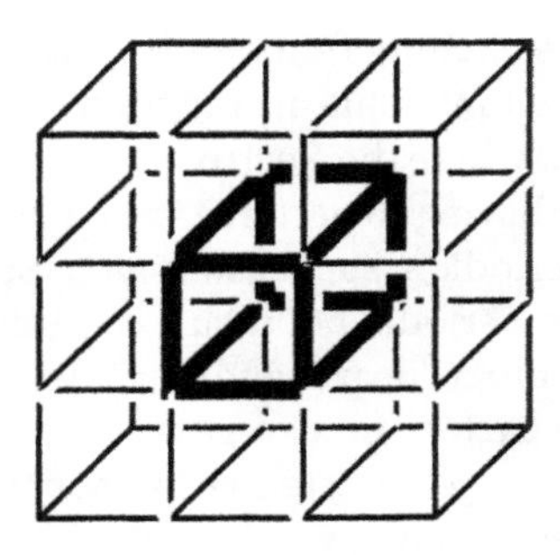

Figure 15.2 Periodic boundary condition.

and forces, and has some rather unfortunate consequences in giving non-physical distributions of the solvent molecules near the cutoff distance and producing "hot" and "cold" spots. A *switching* function approach, where the interaction is gradually reduced to zero over a range of a few angstroms, performs significantly better.[70,71] The switching function is multiplied on to the real potential and has the effect of smoothly reducing the potential from its real value to zero over a distance range from R_1 to R_2. An example of a third-order switching function that has zero first derivatives at both limits[72] is shown by

$$S(r) = \begin{cases} 1 & r \le R_1 \\ \dfrac{(R_2^2 - r^2)^2 (R_2^2 + 2r^2 - 3R_1^2)}{(R_2^2 - R_1^2)^3} & R_1 \le r \le R_2 \\ 0 & r \ge R_2 \end{cases} \tag{15.44}$$

An alternative form for the central part that also has vanishing second derivatives at both limits is given by

$$S(r) = 1 - 10\left(\frac{r - R_1}{R_2 - R_1}\right)^3 + 15\left(\frac{r - R_1}{R_2 - R_1}\right)^4 - 6\left(\frac{r - R_1}{R_2 - R_1}\right)^5 \tag{15.45}$$

A variation of this is to use a *shifting* function, which corresponds to a switching function with $R_1 = 0$. Such functions modify the potential for all r values less than R_2 and an example is given by

$$S(r) = \begin{cases} \left(1 - \dfrac{r^2}{R_2^2}\right)^2 & r \le R_2 \\ 0 & r \ge R_2 \end{cases} \tag{15.46}$$

The use of switching or shifting functions modifies the model, since the potential and forces are changed, and therefore affects the results of a simulation. Whether these changes are significant relative to the other approximations in the model depends on the specific system and properties. It should be noted that some of the errors introduced by using truncation and switching/shifting functions can be absorbed by the force field parameters; that is if the parameters are determined by fitting against experimental data, they will to some extent adjust their value to compensate for the errors in the underlying model. This implies that the error cancellation may be partly destroyed if the values of the switching/shifting distances are changed or the non-bonded interactions are calculated by other methods, such as the Ewald sum methods described below.

Figure 15.3 shows the energy function of two unit charges interacting with a Coulomb potential, one that has been subjected to the switching function (Equation (15.44)) with $R_1 = 10$ Å and $R_2 = 12$ Å, and one that has been subjected to the shifting function (Equation (15.46)) with the same limits.

Methods have also been developed where the electrostatic interaction is treated "exactly" (to within a numerical threshold), but without having to perform the N^2 summation over all atoms. *Ewald sum* methods have been developed for periodic systems (such as crystals) but can also be applied to quasi-periodic models arising by applying periodic boundary conditions (Figure 15.4). The idea in these methods is to split the interaction into a "near"- and "far"-field contribution.[73] The near-field contribution is obtained by embedding each point charge in a screening potential, taken as a Gaussian function with an exactly opposing charge centered at the position of the point charge. Outside the range of the screening function, essentially given by the width of the Gaussian, the net charge is thus zero, and the interaction between these screened point charges is therefore short-ranged and can be evaluated directly. In order to recover the original point charge interaction, the effect of the

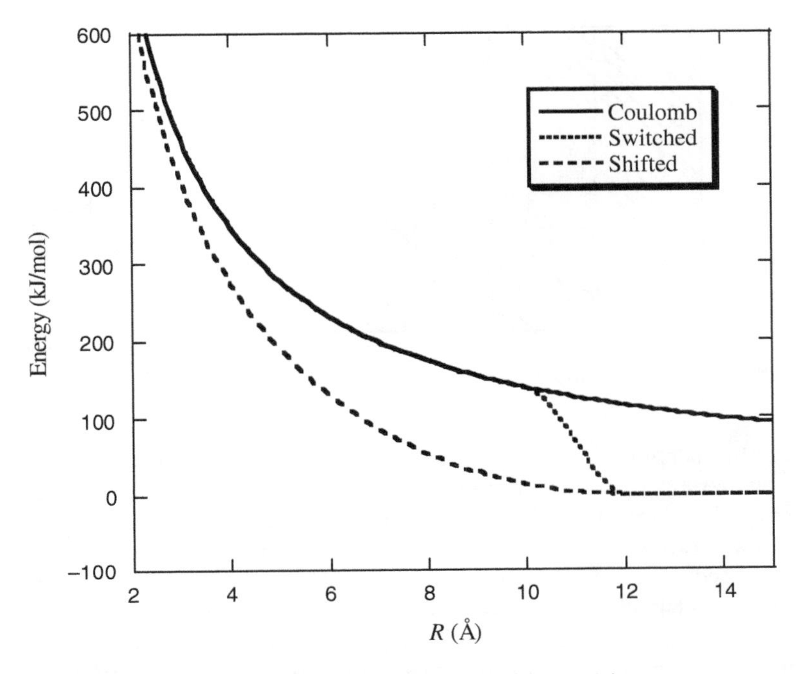

Figure 15.3 Difference between original, switched and shifted Coulomb potentials.

screening potentials must be subtracted again. This compensating term is an interaction between Gaussian charge distributions, which is long-ranged. Since it is a smooth charge distribution, however, it can be evaluated efficiently in reciprocal space by Fourier transform methods. The only free parameter is the width of the Gaussian potential. A narrow Gaussian function makes the direct-space part converge rapidly, but the reciprocal-space part converges slowly, and vice versa for a wide Gaussian function. The optimum width is given by the condition that the computational effort is distributed equally between the direct and reciprocal sums.

A key point in these methods is the existence of computationally efficient methods for performing Fourier transformations, which reduces the scaling from N^2 to $N^{3/2}$. A related method is the *Particle Mesh Ewald* (PME) method, which scales (only) as $N \ln (N)$.[74] PME can also be used for calculating the Lennard–Jones form of the van der Waals energy exactly.[75]

The *Fast Multipole Moment* (FMM) method similarly splits the contribution into a near-field and far-field, and calculates the near-field exactly (Figure 15.5).[76] The far-field energy is calculated by dividing the physical space into boxes and the interaction between all molecules in one box with all

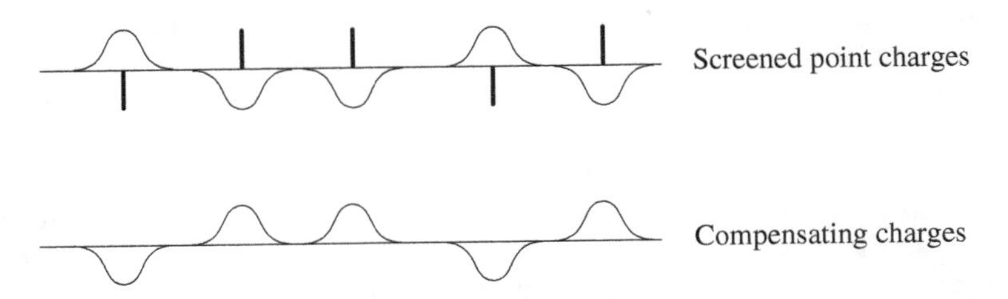

Figure 15.4 Illustration of the Ewald method.

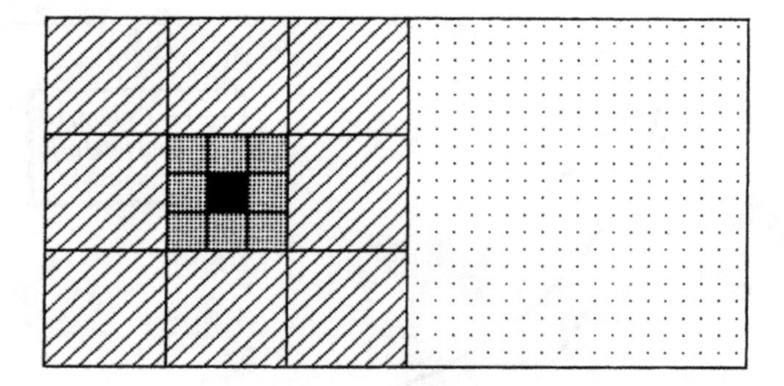

Figure 15.5 Illustration of the fast multipole moment method.

molecules in another is approximated as interactions between multipoles located at the centers of the boxes. The further away from each other two boxes are, the larger the boxes can be for a given accuracy, thereby reducing the formal N^2 scaling into something that approaches linear scaling. The prefactor, however, is rather large and when properly implemented it appears that the cross-over point, where FMM becomes faster than PME, is around 10^5 particles. FMM furthermore works best when the particles are relatively uniformly distributed; for a non-uniform distribution of particles, the multipole order must be significantly increased in order to achieve a given accuracy. A disadvantage of FMM is that the maximum error (relative to an exact calculation) is significantly larger than for Ewald-type methods, that is there are certain particle pairs for which the error is larger than the average error by perhaps a factor of 10. FMM, in contrast to Ewald-based methods, does not have the requirement of periodicity, and is capable of modeling large non-periodic systems.

The original FMM has been refined by also adjusting the accuracy of the multipole expansion as a function of the distance between boxes, producing the *very Fast Multipole Moment* (vFMM) method.[77]

The exact calculation of the electrostatic interaction, albeit by treating the system as being pseudo-crystalline,[78,79] has been shown to give significantly different results than a simple truncation scheme[71] and also different from a switching function approach.[80] Given the existence of computationally efficient methods for performing, for example, PME, there seems to be little reason for employing a non-physical truncation of the electrostatic interaction.

15.4 Extracting Information from Simulations

A necessary (but not sufficient) requirement for producing a representative sampling is that the system is in equilibrium. The starting configuration may be generated by completely random positions (and velocities for MD), but is more often taken either from a previous simulation or by placing the particles at or near the lattice points of a suitable crystal. The system is then equilibrated by running perhaps 10^4–10^5 MC or MD steps, followed by perhaps 10^5–10^7 production steps. Various quantities, such as the average potential energy or correlation functions, can be monitored to validate whether equilibrium has been achieved. The question of how to partition data from a simulation into an equilibration and a sample set is non-trivial. A partition into a small equilibrium set and a large sample set may bias the result towards the starting configuration, while a partition into a large equilibrium set and a small sample set will give a larger statistical error in the result.[81]

The averaging in Equation (15.1) should be over configurations that are uncorrelated, and this is not the case for nearby points in an MD trajectory or sequence of MC steps. The whole set of points should therefore be divided into blocks with a length that is sufficiently long to make equivalent points in two neighboring blocks uncorrelated, but preferably also with a length that is sufficiently short so that no information is lost. Flyvbjerg and Petersen have shown how to determine the optimum block length by a sequence of statistical analyses.[82] For the original data set the mean and variance are calculated according to

$$\bar{x} = \frac{1}{N}\sum_{i=1}^{N} x_i \tag{15.47}$$

$$\sigma^2 = \frac{1}{N-1}\sum_{i=1}^{N} (x_i - \bar{x})^2 \tag{15.48}$$

The variance calculated from Equation (15.48) is only valid for uncorrelated data, which is not the case for the original data. In order to get a realistic estimate of the true variance, we must perform a data compression to filter out the dependence, that is find the block size for producing uncorrelated data and calculate the variance using this blocking. The method of Flyvbjerg and Petersen consists of performing a sequence of data compressions by averaging two neighboring points, thereby reducing the data size by a factor of two and calculating the corresponding variance (the mean is unchanged). The variance divided by the number of data points at a given level, $\sigma^2/(N' - 1)$, will initially increase and then level off to a constant value as the data within two consecutive blocks become uncorrelated. The point where the value becomes constant is the optimum block size for the given property and the $\sigma^2/(N' - 1)$ quantity can be taken as the estimate of the true variance of the property.

The distance between (uncorrelated) data in MD methods has the dimension of time and is called the correlation time. It is important to recognize that different properties may have different correlation times, and for some properties it may be comparable to or exceed the total length of the simulation. A clear advantage of the above procedure for determining the optimum block size is that the statistical error bars associated with the variance can also be calculated, that is the standard deviation of the variance:

$$\sigma^2(x) \approx \frac{\sigma^2(x')}{N'-1}\left(1 \pm \sqrt{\frac{2}{N'-1}}\right) \tag{15.49}$$

Here the prime notation indicates the data set at a given compression level. Equation (15.49) clearly illustrates that the estimate of the variance becomes increasingly uncertain as the number of data blocks decreases, that is when the data have been compressed into only two blocks, the (relative) standard deviation is $\sqrt{2}$. In order to determine whether it is actually possible to obtain uncorrelated data from the simulation, a plot of $\sigma^2/(N' - 1)$ against the compression level should therefore include the associated statistical error from Equation (15.49). If the statistical errors impinge on a conclusion as to whether a constant plateau has been reached, this is an indication that the simulation length is insufficient for obtaining a valid estimate of the given quantity.

Ensembles generated by MC techniques are naturally of the constant *NVT* type, while MD methods naturally generate a constant *NVE* ensemble. Both MC and MD methods, however, may be modified to simulate other ensembles, as described in Sections 15.1.1 and 15.2.2. Of special importance is the constant *NPT* condition, which directly relates to most experimental conditions. The primary advantage of MD methods is that time appears explicitly, that is such methods are natural for simulating time-dependent properties, such as correlation functions, and for calculating properties that depend on particle velocities. Furthermore, if the relaxation time for a given process is (approximately) known, the required simulation time can be estimated beforehand (i.e. it must be at least several multiples of the relaxation time).

In order to reduce the statistical error, the averaging in Equation (15.1) is typically performed on 10^3–10^5 points in phase space. The requirement to calculate this many points and associated energies for a model consisting of several hundred particles means that the use of *ab initio* methods is extremely demanding, even for small systems and simple wave functions. Semi-empirical electronic structure methods may be used for small systems, implicitly accepting the low accuracy of these methods, but the large number of calculations necessary still makes this computationally intensive. The large majority of simulations are therefore carried out with an energy surface generated by a parameterized function of the force field type.

The expressions derived from statistical mechanics (Section 14.4) are often rewritten into computationally more suitable forms that may be evaluated from the basic descriptors: positions **r**, velocities **v** or momenta **p** and energies E. The temperature is related to the average kinetic energy:

$$\frac{1}{2}(3N_{\text{atom}} - N_{\text{constraint}})kT = \left\langle \sum_{i}^{N_{\text{atom}}} \frac{1}{2} m_i \mathbf{v}_i^2 \right\rangle_M \tag{15.50}$$

The temperature is fixed in a standard MC simulation (NVT conditions), while it is a derived quantity in a standard MD simulation (NVE conditions).

The pressure is related to the product of positions and forces (for pairwise potentials):

$$PV = N_{\text{atom}}kT + \frac{1}{3}\left\langle \sum_{i<j}^{N_{\text{atom}}} r_{ij} f_{ij} \right\rangle_M \tag{15.51}$$

Here the first part is for an ideal gas.

The internal (potential) energy is directly a sum of energies, which is normally given as a sum over pair-wise interactions (i.e. van der Waals and electrostatic contribution in a force field description):

$$U = \sum_{i}^{N_{\text{atom}}} E_i = \sum_{i<j}^{N_{\text{atom}}} E_{ij} \tag{15.52}$$

The internal energy will fluctuate around a mean value that may be calculated by averaging over the number of configurations, $\langle U \rangle_M$.

The heat capacity at constant volume is the derivative of the energy with respect to temperature at constant volume (Equation (14.20). There are several ways of calculating such response properties. The most accurate is to perform a series of simulations under NVT conditions and thereby determine the behavior of $\langle U \rangle_M$ as a function of T (e.g. by fitting to a suitable function). Subsequently this function may be differentiated to give the heat capacity. This approach has the disadvantage that several simulations at different temperatures are required. The heat capacity can alternatively be calculated from the *fluctuation* of the energy around its mean value:

$$C_V = \frac{1}{kT^2}\langle (U - \langle U \rangle_M)^2 \rangle_M = \frac{1}{kT^2}\left(\langle U^2 \rangle_M - \langle U \rangle_M^2\right) \tag{15.53}$$

This approach requires only a single simulation. Since the fluctuation has a longer relaxation time than the energy itself, the ensemble average in Equation (15.53) must be over a larger number of points than for $\langle U \rangle_M$ to achieve a similar statistical error, that is the efficiency obtained by avoiding multiple simulations is partly lost owing to a longer simulation time required. Another disadvantage is that Equation (15.53) involves taking differences between large numbers, which is susceptible to round-off errors.

Distribution functions measure the (average) value of a property as a function of an independent variable. A typical example is the radial distribution function $g(r)$, which measures the probability of finding a particle as a function of distance from a "typical" particle relative to that expected from a completely uniform distribution (i.e. with a density of N/V). The radial distribution function is defined by

$$g(r, \Delta r) = \frac{V}{N} \frac{\langle N(r, \Delta r) \rangle_M}{4\pi r^2 \Delta r} \tag{15.54}$$

Here $N(r,\Delta r)$ is the number of molecules between r and $r + \Delta r$ from another particle and $4\pi r^2\ \Delta r$ is the volume of a spherical shell with thickness Δr. The radial distribution function for a solution will typically have a structure as shown in Figure 15.6 for the simulation of a benzene radical anion in water.[83] Figure 15.6 displays the radial distribution function of hydrogen relative to the center of mass of the benzene radical anion. At short distances, the probability is zero due to van der Waals repulsion. The distribution function then rises sharply to a value of ~1.7 for a distance of ~1.8 Å, indicating that it is 1.7 times more likely to find particles with this separation than expected from a uniform distribution. This corresponds to water molecules that are located above or below the molecular plane. A second peak occurs at ~3.2 Å, which corresponds to water molecules located around the edge of the benzene molecule. The integral under a peak gives the number of solvent molecules of a given type. At long range the distribution function levels off to a value of 1, that is the particles no longer sense each other and behave as a uniform distribution.

The radial distribution function can for molecules be extended with orientation degrees of freedom to characterize the angular distribution.

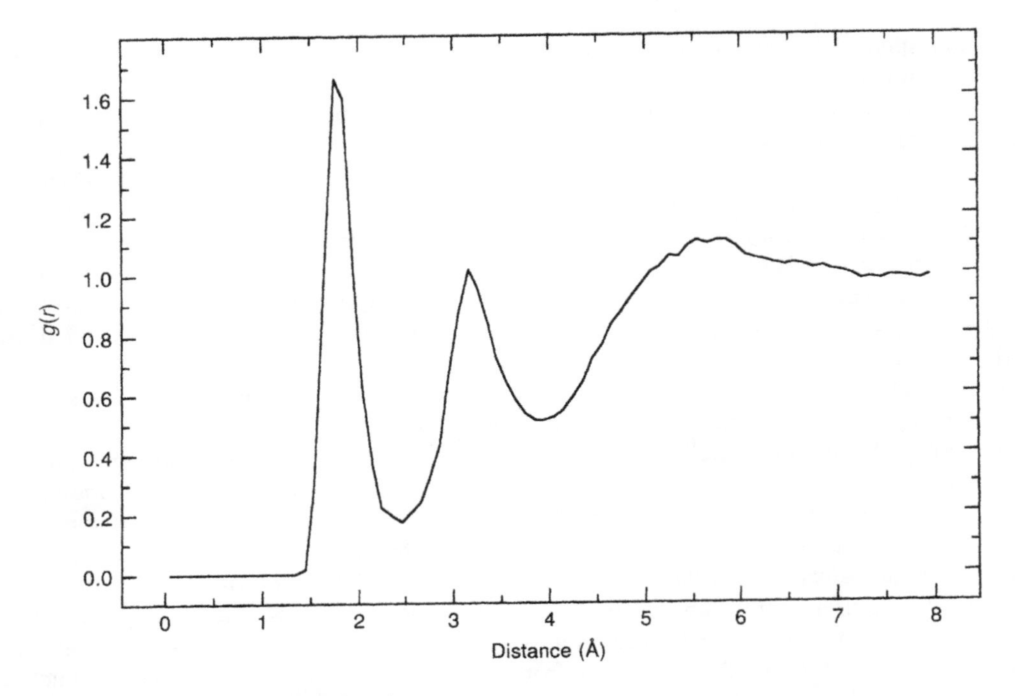

Figure 15.6 A typical radial distribution function.

Correlation functions measure the relationship between two variables, x and y. A common definition is given by

$$C_{xy} = \frac{\langle (x - \langle x \rangle_M)(y - \langle y \rangle_M) \rangle_M}{\sqrt{\langle (x - \langle x \rangle_M)^2 \rangle_M \langle (y - \langle y \rangle_M)^2 \rangle_M}} \tag{15.55}$$

The correlation function is a number between −1 and 1, where 1 indicates that the two quantities are completely correlated, −1 that they are (completely) anticorrelated and 0 means that they are independent (uncorrelated). Such correlation functions are often time-dependent and measure how the correlation between two quantities changes over time. They may be normalized by the corresponding static (i.e. $t = t_0$) limit:

$$C_{xy}(t) = \frac{\langle x(t_0)y(t) \rangle_{N,t_0}}{\langle x(t_0)y(t_0) \rangle_{N,t_0}} \tag{15.56}$$

Notice that the averaging is done over the number of particles N and t_0, but not the number of configurations M. Since an MD simulation produces a set of time-connected configurations, the number of a given configuration is directly related to the simulation time.

In the case where x and y are the same, $C_{xx}(t)$ is called an *autocorrelation* function; if they are different, it is called a *cross-correlation* function. For an autocorrelation function, the initial value at $t = t_0$ is 1 and approaches 0 as $t \to \infty$. How fast it approaches 0 is measured by the *relaxation* time. The Fourier transforms of such correlation functions are often related to experimentally observed spectra; the far IR spectrum of a solvent, for example, is the Fourier transform of the dipole autocorrelation function:[84]

$$I(\omega) \propto \int_{-\infty}^{+\infty} \langle \boldsymbol{\mu}(t) \cdot \boldsymbol{\mu}(0) \rangle e^{i\omega t} dt \tag{15.57}$$

The Raman spectrum can similarly be obtained from the polarizability autocorrelation function.

The averaging in the above focuses on a single long trajectory, but it may alternatively be done over many short trajectories. The latter is trivially parallelizable in terms of computation time and thus has some practical advantages; furthermore, it has been shown that it may give a better performance.[85] This claim, however, is tightly connected with the quality of the force field, as illustrated in Figure 15.7. With a good force field, the results obtained from a single long simulation should be the same as that obtained from several shorter simulations with the same total simulation time (within the usual statistical uncertainty). With a poor force field, however, the single long simulation has time to move sufficiently far away from the starting point in phase space that the resulting average will be "poorer" than averaging over several short simulations, which do not have time to move very far from the starting point.

The key point is that short simulations necessarily can only sample regions of the phase space close to the starting point, while long simulations have the possibility of sampling regions further from the starting point. It should be noted, however, that "long" simulations in practice are only rarely long enough that adequate sampling of the phase space can be expected. Although MD simulations in principle are deterministic (the same trajectory should be obtained from the identical same starting conditions), the behavior is in practice chaotic, and even tiny perturbations in the starting structure lead to different trajectories within a few picoseconds.[85] Simulations on nanosecond timescales at ambient temperatures are only able to explore regions of phase space with low-energy barriers, and different starting conditions may lead to (average) results that differ significantly more than the statistical errors for each simulation. The result from a single simulation thus appears to be much more accurate than it is in reality. It is therefore highly advisable to run multiple simulations with different

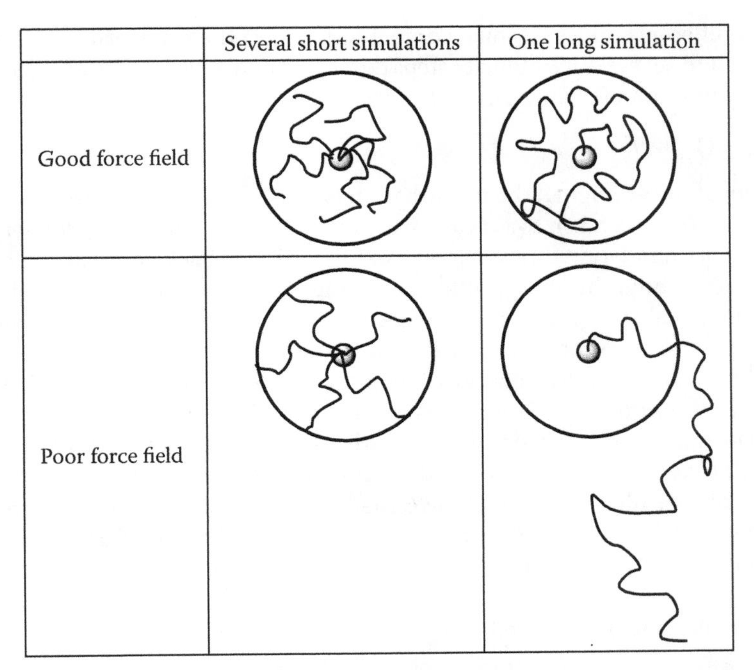

Figure 15.7 Illustrating that differences between averaging over one long versus several short simulations may reflect the quality of the force field. The central dot indicates the starting point in phase space, while the circle indicates the region over which the averaging produces the 'correct' result.

starting conditions, but the combination of many long simulations is of course not desirable from a computational resource point of view. Force field and sampling errors are with present computational resources therefore intertwined and difficult to separate.

15.5 Free Energy Methods

As noted in Section 14.6, it is difficult to calculate entropic quantities with any reasonable accuracy within a finite simulation time. It is, however, possible to calculate *differences* in such quantities.[86,87] Of special importance is the Gibbs free energy, since it is the natural thermodynamic quantity under normal experimental conditions (constant temperature and pressure, Table 15.1), but we will illustrate the principle with Helmholtz free energy instead (constant temperature and volume). As indicated in Equation (14.18), the fundamental problem is the same. There are two commonly used methods for calculating differences in free energy: *Thermodynamic Perturbation* and *Thermodynamic Integration*.[88] Some of the bias MD methods described in Section 15.2.9, such as the adaptive bias force method, can also be used for calculating free energy profiles along a suitable reaction coordinate.

15.5.1 Thermodynamic Perturbation Methods

The difference in entropy properties between two systems A and B can be calculated by an ensemble average, as discussed in Section 14.6:

$$\langle \Delta A_{\mathrm{AB}} \rangle_M = kT \ln \langle \mathrm{e}^{(E_{\mathrm{B}}-E_{\mathrm{A}})/kT} \rangle_M \tag{15.58}$$

Since the energy difference must be small compared with kT, the transformation from A to B must usually be broken into several intermediate steps described by a λ parameter, and the total free energy change is given as the sum of changes in each step:

$$E_\lambda = \lambda E_A + (1 - \lambda)E_B \tag{15.59}$$

To test the quality of the averaging, the perturbation is usually run in both directions (i.e. A → B and B → A), and the difference is taken as a measure of how well ΔA is statistically converged. It should be noted that (too) short simulation times may lead to forward- and backward-calculated values that are in good agreement, without the energy difference being calculated accurately. Establishing a reliable estimate of the statistical error requires running several independent simulations and carefully analyzing the size of the perturbation steps and the correlation times for the various processes occurring in the system.[89] Calculation of free energy differences by means of Equation (15.58) is often called *Thermodynamic Perturbation*[90] or *Free Energy Perturbation* (*FEP*).

Instead of performing a series of simulations with a fixed energy function as in Equation (15.58), it may also be allowed to change *continuously* during a single simulation by changing λ slightly in each time step. This is called the *Slow Growth* method and requires that the increase in λ is slow enough that the system essentially remains at equilibrium at all times. This is difficult to ensure in practice,[91,92] and the slow growth method is therefore used less commonly.

15.5.2 Thermodynamic Integration Methods

Given an energy function as in Equation (15.59), the partition function, and thereby also the free energy, is a function of λ:

$$A(\lambda) = -kT \ln Q(\lambda) \tag{15.60}$$

Differentiating this expressions yields

$$\frac{\partial A}{\partial \lambda} = -\frac{kT}{Q}\frac{\partial Q}{\partial \lambda} = \frac{\partial E}{\partial \lambda} \tag{15.61}$$

Here the definition of Q (Equation (14.15)) has been used. Replacing the right-hand side by an ensemble average and integrating over λ gives

$$A(1) - A(0) = \int_0^1 \left\langle \frac{\partial E(\lambda)}{\partial \lambda} \right\rangle_M \mathrm{d}\lambda \tag{15.62}$$

The left-hand side is the desired free energy difference and the right-hand side may be approximated by a discrete sum:

$$\Delta A = \sum_i \left\langle \frac{\partial E(\lambda)}{\partial \lambda} \right\rangle_M \Delta\lambda_i \tag{15.63}$$

The use of Equation (15.63) for calculating ΔA is normally called *Thermodynamic Integration* (TI).[93] The difference between Equations (15.58) and (15.63) is that the former averages over finite differences in energy functions, while the latter averages over a *differentiated* energy function. For parameterized energy functions, it is fairly easy to form the energy derivative with respect to the coupling parameter analytically, and the averaging in Equation (15.63) is therefore no more complicated than averaging over energy differences as in Equation (15.58). Furthermore, it should be noted that the computational cost of performing the averaging is negligible compared with the cost of generating

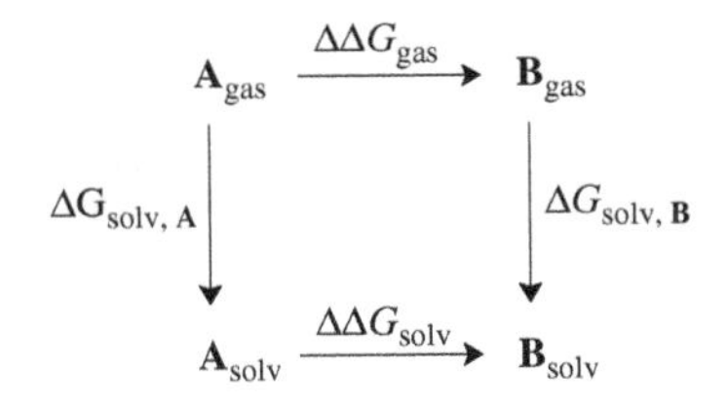

Figure 15.8 An example of a thermodynamic cycle for calculating differences in solvation energies.

the ensemble, and the same ensemble can therefore be used to calculate the free energy difference by either Equation (15.58) or (15.63). This allows a measure of the reliability of the calculated value to be obtained.

Free energy calculations are often combined with *thermodynamic cycles* to calculate properties that would otherwise require impossible long simulation times.[94,95] A direct calculation of, for example, solvating acetone in water would require simulating the transfer of an acetone molecule from the gas phase (vacuum) to an aqueous phase, followed by solvent reorganization. If we wish to calculate the solvation energy of acetone relative to propane, this would require a second (impossibly long) simulation of transferring a propane molecule into the aqueous phase. Alternatively, the *difference* in solvation may be calculated by means of the thermodynamic cycle shown in Figure 15.8.

Since G is a state function, the difference in solvation energy, $\Delta G_{solv,\mathbf{A}} - \Delta G_{solv,\mathbf{B}}$, which is difficult to calculate, may instead be obtained as $\Delta\Delta G_{solv} - \Delta\Delta G_{gas}$. If **A** and **B** are different molecules, such as acetone and propane, the $\Delta\Delta G$ values correspond to non-physical transformations. Theoretically, however, it is quite easy to transform an oxygen atom into two hydrogens. The $\Delta\Delta G_{gas}$ value corresponds to differences in the internal (translational, rotational and vibrational) degrees of freedom, which can be calculated as discussed in Section 14.5. This difference also is part of $\Delta\Delta G_{solv}$, but if the internal energy levels are assumed to be independent of solvent, the solvent part of $\Delta\Delta G_{solv}$ is directly the difference in solvation.

In the acetone/propane example, the **A** to **B** change means that the oxygen atom gradually disappears and two hydrogens gradually appear at the appropriate positions. In a force field energy expression, this corresponds to reducing or increasing van der Waals parameters and atomic charges, as well as changing all other parameters that are affected by the change in atom types. For $\lambda = 0.5$, the **A**/**B** "molecule" thus consists of a CH_3–C–CH_3 framework, with the central carbon having "half" a carbonyl oxygen and two "half" hydrogens attached. Absolute values of solvation energies may be calculated by transforming a solvent molecule to the solute, but if they are structurally very different it may require long simulation times to ensure that equilibrium is attained.

The technique of thermodynamic cycles may be used for calculating relative free energies for a variety of other cases. Differential binding of two ligands to a protein, for example, requires transforming one ligand into the other in a pure solution, and when bound to the protein. The strength of free energy methods is that differences in free energies may be obtained with a statistical accuracy of a few kJ/mol, at quite reasonable computational costs. Whether the calculated values agree with experimental results depends on the quality of the force field, but there are models for many solvents that are capable of providing an accuracy of better than a few kJ/mol in terms of absolute values.

The basic requirement of free energy perturbation or thermodynamic integration methods is that the non-physical transformation is carried out in sufficiently small steps that the sampling of the phase space at two successive points overlaps. Even for quite similar systems, this often means that the transformation must be broken into 10–20 steps, with each step requiring extensive sampling, and this makes such methods computationally intensive. In the *Linear Interaction Energy* (LIE) method, only the physical end-points for the thermodynamic cycle in Figure 15.8 are subjected to a simulation. The difference in binding free energy is then parameterized as a linear combination of the difference in the non-polar (van der Waals) and polar (electrostatic) interactions between the ligands and surroundings (enzyme or solvent):[96]

$$\Delta G_{\mathrm{bind}} = \alpha\Delta\langle E_{\mathrm{vdw}}\rangle + \beta\Delta\langle E_{\mathrm{el}}\rangle + \gamma \tag{15.64}$$

The β constant is expected to have a value of 0.5 from theoretical arguments. Optimization of the three parameters against experimental binding energies for the P450cam system confirms that the optimum β value is close to 0.5, while α and γ have values of ~0.18 and ~ −4.5.[97] It is likely that the optimum parameter values will depend on the specific system[98] but the LIE offers a computational saving of an order of magnitude or more compared with FEP or TI methods.

15.6 Solvation Models

An important aspect of computational chemistry is to evaluate the effect of the environment, such as a solvent. Methods for evaluating the solvent effect may broadly be divided into two types: those describing the individual solvent molecules and those that treat the solvent as a continuous medium.[99] Combinations are also possible, for example by explicitly considering the first solvation shell and treating the rest by a continuum model. Each of these may be subdivided according to whether they use a classical or quantum mechanical description. By far the most important solvent is water, and since it is also one of the most difficult systems to model, the majority of methods have been focused on water, and we will use this for exemplification in the following.

The effects of solvation can be partitioned into two main groups:

- Non-specific (long-range) solvation
  - Polarization
  - Dipole orientation
- Specific (short-range) solvation
  - Hydrogen bonds
  - van der Waals interaction
  - Solvent shell structure
  - Solvent–solute dynamics
  - Charge transfer effects
  - Hydrophobic effects (entropy effects).

The non-specific effects are primarily solvent polarization and orientation of the solvent electric multipole moments by the solute, where the dipole interaction is usually the most important. These effects cause a screening of charge interactions, leading to the (macroscopic) dielectric constant being larger than 1. The microscopic interactions are primarily located in the first solvation shell, although the second solvation shell may also be important for multiple-charged ions. The microscopic interactions depend on the specific nature of the solvent molecule, such as the shape and the ability to form hydrogen bonds.

A molecular description involves periodic boundary conditions and sampling the phase space by simulation methods. Such methods are in principle capable of accounting for all of the above solvent effects but the quality of the results will of course depend on how realistically the solvent–solute and solvent–solvent interactions are described. The requirement of many (hundreds or thousands of) solvent molecules to form a realistic model means that force field methods are often the primary (or only) choice based on computational considerations. Since polarizable force fields are not yet in common use, this means that a major part of the non-specific solvation is lacking. Car–Parrinello molecular dynamics methods using density functional theory for describing the interaction are significantly more expensive and can therefore only give a limited sampling of the phase space. They can account for the polarization but many commonly employed functionals have a poor description of the van der Waals interaction. Semi-empirical electronic structure methods (Chapter 7) are in general not sufficiently accurate for calculating intermolecular potentials unless their parameters are specifically tuned to reproduce a selected solvent. Mixed QM/MM methods, where the solute is described by a (quantum) electronic structure method and the solvent by a (classical) force field, can account for the polarization of the solute, but the backpolarization again requires a polarizable force field.

Explicit solvation models describe the solvent in terms of individual solvent molecules, and given that a realistic solvation model usually requires many thousands of solvent molecules, this implies parameterized energy functions of the force field type. These models can be classified by three criteria:

- Rigid or flexible geometry
- Number of interaction points
- Static or polarizable electrostatic interactions

The number of interaction points, often called sites, refers to how many expansion points that are used for representing the electrostatic interaction. These often coincide with the nucleus positions, but not necessarily, and additional off-nucleus points are often required when fixed point charges are used for representing the electrostatic interaction. A more detailed description of different water models is given in Section 2.5.

Methods involving an explicit description of the solvent molecules require, analogously with other many-body methods, a sampling of the phase space. Since this is computationally expensive, there is a strong interest in developing methods where the solvent is modeled in a less rigorous fashion. The solvent–solute dynamics can be taken into account in an average fashion by the Langevin dynamics method (Section 15.2.3). The non-specific effects of solvation can be modeled by considering the solvent as a homogeneous medium with a dielectric constant, as will be discussed in more detail in the next section.

15.6.1 Continuum Solvation Models

Continuum models consider the solvent as a uniform polarizable medium with a dielectric constant of ε and with the solute **M** placed in a suitably shaped hole in the medium (Figure 15.9).[100]

Creation of a hole in the medium costs energy, that is this is a destabilization, while dispersion interactions between the solvent and solute add a stabilization (this is roughly the van der Waals energy between solvent and solute). The electric charge distribution of **M** will polarize the medium

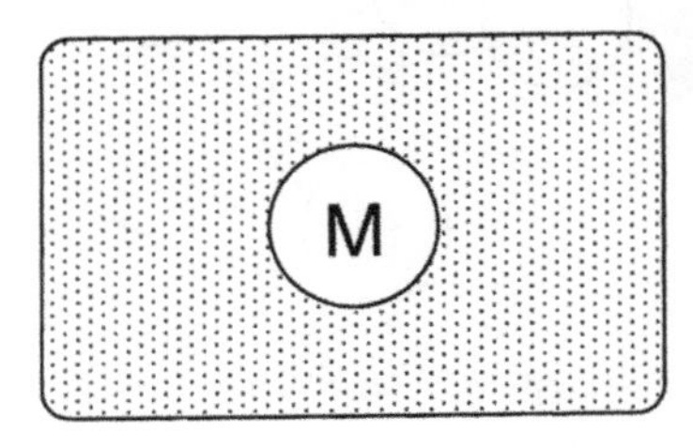

Figure 15.9 Reaction field model.

(induce charge moments), which in turn acts back on the molecule, thereby producing an electrostatic stabilization. One may also consider an exchange-repulsive component as well as a separate term accounting for hydrogen bonding. The solvation (free) energy may thus be written as a sum of components as shown below:

$$\Delta G_{\text{solvation}} = \Delta G_{\text{cavity}} + \Delta G_{\text{elec}} + \Delta G_{\text{dispersion}} + \Delta G_{\text{exchange}} + \Delta G_{\text{H-Bond}} + \cdots \tag{15.65}$$

Reaction field models differ in five aspects:

1. How are the size and shape of the hole defined?
2. How is the charge distribution of **M** represented?
3. How is the solute **M** described, either classical (force field) or quantum (semi-empirical or *ab initio*)?
4. How is the dielectric medium described?
5. How are the non-electrostatic energy terms calculated?

The dielectric medium is normally taken to have a constant value of ε, but may for some purposes also be taken to depend, for example, on the distance from **M**. For dynamical phenomena it can also be allowed to be frequency dependent,[101] that is the response of the solvent is different for a "fast" reaction, such as an electronic transition, and a "slow" reaction, such as a molecular reorientation. It should be noted that ε is the *only* parameter characterizing the electrostatic component, and solvents having the same ε value (such as acetone, $\varepsilon = 20.7$, and 1-propanol, $\varepsilon = 20.1$, or benzene, $\varepsilon = 2.28$, and carbon tetrachloride, $\varepsilon = 2.24$) thus have the same electrostatic energy. The hydrogen bonding capability of 1-propanol compared with acetone will in reality most likely make a difference, and the solvent dynamics of an almost spherical CCl_4 will be different from the planar benzene molecule. These differences must be accounted for by parameterizations of the non-electrostatic energy terms.

The simplest shape for the hole is a sphere or an ellipsoid. This has the advantage that the electrostatic interaction between **M** and the dielectric medium may be calculated analytically. More realistic models employ molecular-shaped holes, generated, for example, by interlocking spheres located on each nucleus. Taking the atomic radius as a suitable factor (a typical value is 1.2) times a van der Waals radius defines a *van der Waals surface*. Such as surface may have small "pockets" where no solvent molecules can enter and a more appropriate descriptor may be defined as the surface traced out by a spherical particle of a given radius (a typical radius of 1.4 Å to model a water molecule) rolling on the van der Waals surface. This is denoted the *Solvent Accessible Surface* (SAS) and is illustrated in Figure 15.10.

Since an SAS is computationally more expensive to generate than a van der Waals surface, and since the difference is often small, a van der Waals surface is often used in practice. An additional

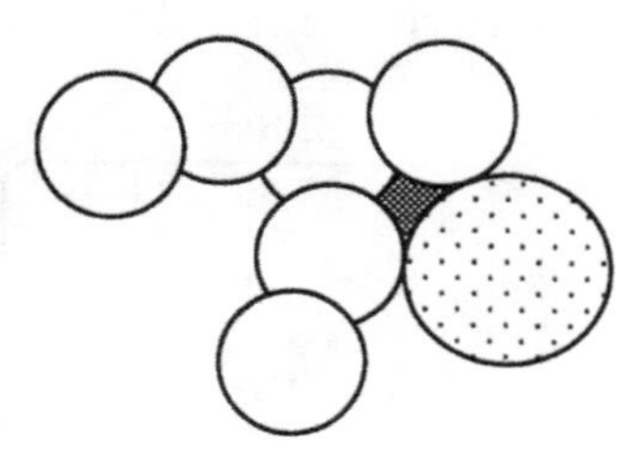

Figure 15.10 On a surface generated by overlapping van der Waals spheres there will be areas (hatched) that are inaccessible to a solvent molecule (dotted sphere).

disadvantage is that a very small displacement of an atom may alter the SAS in a discontinuous fashion, as a "pocket" in the van der Waals surface suddenly becomes too small to allow the solvent probe to enter. The cavity may alternatively be calculated directly from the wave function, for example by taking an isodensity surface corresponding to a value of 10^{-3}–10^{-4}.[102] It is generally found that the shape of the hole is important and that molecular-shaped cavities are necessary to be able to obtain good agreement with experimental data (such as solvation energies).

The energy required in creating the cavity (entropy factors and loss of solvent–solvent interactions), the stabilization due to dispersion between the solute and solvent, as well as the exchange-repulsion energy is usually assumed to be proportional to the surface area. The corresponding energy terms are often parameterized as being proportional to the total cavity area (a single proportionality constant) or parameterized by having a constant ξ specific for each atom type (analogous to van der Waals parameters in force field methods), with the ξ parameters being determined by fitting to experimental solvation data. Explicit calculations have indicated that a linear parameterization in the SAS may be quite poor:[103]

$$G_{\text{cavity}} + \Delta G_{\text{dispersion}} + \Delta G_{\text{exchange}} = \gamma\,\text{SAS} + \beta \tag{15.66}$$

$$G_{\text{cavity}} + \Delta G_{\text{dispersion}} + \Delta G_{\text{exchange}} = \sum_i^{\text{atoms}} \xi_i S_i \tag{15.67}$$

For solvent models where the cavity/dispersion/exchange interaction is parameterized by fitting to experimental solvation energies, the use of a few explicit solvent molecules for the first solvation sphere is *not* recommended, as the parameterization represents a best fit to experimental data *without* any explicit solvent present. Short-range specific solvation effects, of which hydrogen bonding between the solute and solvent is the most important, can be parameterized based on features in the electrostatic component, as described below. The electrostatic component of Equation (15.65) can be described at several different levels of approximation, as discussed in the following sections.

15.6.2 Poisson–Boltzmann Methods

The *Poisson* equation is a second-order differential equation describing the connection between the electrostatic potential ϕ, the charge distribution ρ and the dielectric constant ε[104, 105]

$$\nabla \cdot (\varepsilon(\mathbf{r})\,\nabla\phi(\mathbf{r})) = -4\pi\rho(\mathbf{r}) \tag{15.68}$$

Note that the dielectric "constant" may depend on the position. When it is independent of the position (i.e. truly a constant), Equation (15.68) becomes

$$\nabla^2\phi(\mathbf{r}) = -\frac{4\pi}{\varepsilon}\rho(\mathbf{r}) \tag{15.69}$$

If the charge distribution is a point charge, the solution of Equation (15.69) reduces to the Coulomb interaction. Equation (15.68) can be used to describe, for example, the solvation of a protein in water, where the protein region is taken to have a low dielectric constant ($2 < \varepsilon < 5$) while the solvent has a high dielectric constant ($\varepsilon = 78$). The boundary between the two regions is typically taken as the SAS.

The Poisson equation can be modified by taking into account a (thermal) Boltzmann distribution of ions in the solvent. The negative ions will accumulate where the potential is positive, and vice versa, subject to a thermal fluctuation. The charge densities from a collection of ions with charges q and $-q$ and concentration c are given by

$$\begin{aligned} \rho_+ &= qce^{-q\phi/kT} \\ \rho_- &= -qce^{-q\phi/kT} \end{aligned} \tag{15.70}$$

Addition of these contributions to Equation (15.68) leads to the *Poisson–Boltzmann Equation* (PBE):

$$\begin{aligned} &\nabla \cdot (\varepsilon(\mathbf{r})\nabla\phi(\mathbf{r})) - \kappa^2 \left(\frac{kT}{q}\right) \sinh\left(\frac{q\phi(\mathbf{r})}{kT}\right) = -4\pi\rho(\mathbf{r}) \\ &\kappa^2 = \frac{8\pi q^2 I}{kT} \end{aligned} \tag{15.71}$$

Here I is the ion strength of the solution and the κ^2 factor is inversely related to the Debye–Hückel length, measuring how far the electrostatic effects extend into the solution. The $\sinh(q\phi(\mathbf{r})/kT)$ term only applies for the region corresponding to the solvent, that is for $\mathbf{r}$ outside the cavity. Since $q\phi/kT$ is dimensionless, the PBE is often written in terms of a reduced potential u instead:

$$\nabla \cdot (\varepsilon(\mathbf{r})\nabla u(\mathbf{r})) - \kappa^2 \sinh(u(\mathbf{r})) = -4\pi\rho(\mathbf{r}) \tag{15.72}$$

If the potential is sufficiently small (i.e. the solute is not strongly charged), the $\sinh(x)$ function can be expanded in a Taylor series, $\sinh(x) \approx x + x^3/6 + \cdots$. Keeping only the first term gives the *Linearized Poisson–Boltzmann Equation* (LPBE):

$$\nabla \cdot (\varepsilon(\mathbf{r})\nabla u(\mathbf{r})) - \kappa^2 u(\mathbf{r}) = -4\pi\rho(\mathbf{r}) \tag{15.73}$$

All of these Equations ((15.68) to (15.73)) are differential equations that must be solved numerically, typically by a grid representation, and the results give information about the electrostatic potential at any point in space. It can be mapped on to the surface of the solute where it may suggest regions for interaction with other polar molecules. It can also be used for generating the reaction field, defined as the difference between the potential in the presence of a solvent ($\varepsilon = 78$) and in vacuum ($\varepsilon = 1$), that is $\phi_{\text{reac}} = \phi_{\text{solv}} - \phi_{\text{vac}}$. Multiplication of the reaction field with the solute charges in either a continuous (ρ) or partial charge (Q) description gives the electrostatic component of the free energy:

$$\Delta G_{\text{elec}} = \frac{1}{2}\int \rho(\mathbf{r})\,\phi_{\text{reac}}(\mathbf{r})\,d\mathbf{r} \tag{15.74}$$

$$\Delta G_{\text{elec}} = \frac{1}{2}\sum_i Q(\mathbf{r}_i)\phi_{\text{reac}}(\mathbf{r}_i) \tag{15.75}$$

15.6.3 Born/Onsager/Kirkwood Models

The numerical aspects of solving the Poisson or Poisson–Boltzmann equations make them too demanding for use in connections with, for example, geometry optimizations or simulations of macromolecules. For certain special cases, however, the Poisson Equation (15.68) can be solved analytically, and this forms the basis for many approximate models for estimating the electronic component in Equation (15.65).

The simplest reaction field model is a spherical cavity, where only the lowest-order electric moment of the molecule is taken into account. For a net charge q in a cavity of radius a, the difference in energy between a vacuum and a medium with a dielectric constant of ε is given by the *Born* model:[106]

$$\Delta G_{\text{elec}}(q) = -\left(1 - \frac{1}{\varepsilon}\right)\frac{q^2}{2a} \tag{15.76}$$

It can be noted that the Born model predicts equal solvation energies for positive and negative ions of the same size, which is not the observed behavior in solvents such as water. The reciprocal dependence on the dielectric constant furthermore means that the calculated solvent effect is sensitive to the variation of ε in the low dielectric limit but is virtually unaffected by large differences in the high dielectric limit. Changing ε from 1 to 2 gives a factor of $^1/_2$ in Equation (15.76) but there is virtually no difference between a solvent with a dielectric constant of 30 (e.g. acetonitrile) and one with a dielectric constant of 78 (e.g. water), although in actual experiments there may be a significant difference.

Using partial atomic charges in Equation (15.76) is often called the *Generalized Born* (GB) model, which has been used especially in connection with force field methods in the *Generalized Born/Surface Area* (GB/SA) model.[107,108] In this case, the Coulomb interaction between the partial charges (Equation (2.21)) is combined with the Born formula by means of a function f_{ij} depending on the internuclear distance and Born radii for each of the two atoms, a_i and a_j:

$$
\begin{aligned}
&G_{\text{elec}}(Q_i, Q_j) = -\left(1 - \frac{1}{\varepsilon}\right) \frac{Q_i Q_j}{f_{ij}} \\
&f_{ij} = \sqrt{r_{ij}^2 + a_{ij}^2 \mathrm{e}^{-D}} \quad ; \quad a_{ij}^2 = a_i a_j \quad ; \quad D = \frac{r_{ij}^2}{4a_{ij}^2}
\end{aligned}
\tag{15.77}
$$

The effective Born radius for a given atom depends on the nature and position of all the atoms. The dependence on the other atoms is in practice relatively weak, and updates of the a_i parameters can be done at suitable intervals, for example when updating the non-bonded list in an optimization or simulation. The boundary between the solute and solvent is usually taken as a modified van der Waals surface generated from the unification of atomic van der Waals radii scaled by a suitable factor. The cavity/dispersion terms are parameterized according to the SAS, as in Equation (15.67). The GB/SA model provides a very fast method of incorporating solvent effects, and it is furthermore relatively easy to formulate gradients of the energy function, making it possible to perform optimizations and simulations. It has been shown to reproduce the results from Poisson–Boltzmann calculations rather accurately, but it should be noted that the results are somewhat sensitive to the magnitude of the partial charges.

The dipole in a spherical cavity is known as the *Onsager* model,[109] which for a dipole moment of μ leads to an energy stabilization given by

$$
\Delta G_{\text{elec}}(\mu) = -\frac{\varepsilon - 1}{2\varepsilon + 1} \frac{\mu^2}{a^3}
\tag{15.78}
$$

The *Kirkwood* model[110] refers to a general multipole expansion in a spherical cavity, while the *Kirkwood–Westheimer* model arises for an ellipsoidal cavity.[111]

The charge distribution of the molecule can be represented either as atom-centered partial charges or as a multipole expansion. The lowest-order approximation for a neutral molecule considers only the dipole moment. This may be a quite poor approximation and fails completely for symmetric molecules that do not have a dipole moment. It is often necessary to extend the expansion up to order six or more in order to obtain converged results, that is including dipole, quadrupole, octupole, etc., moments. Furthermore, only for small and symmetric molecules can the approximation of a spherical or ellipsoidal cavity be considered realistic. The use of the Born/Onsager/Kirkwood models should therefore only be considered as a rough estimate of the solvent effects, and quantitative results can rarely be obtained.

15.6.4 Self-Consistent Reaction Field Models

A classical description of the molecule **M** in Figure 15.9 can be a force field with (partial) atomic charges, while a quantum description involves calculation of the electronic wave function. The latter may be either a semi-empirical model, such as PM6, or more sophisticated electronic structure methods, that is HF, DFT, MCSCF, MP2, CCSD, etc. When a quantum description of **M** is employed, the calculated electric moments induce charges in the dielectric medium, which in turn acts back on the molecule, causing the wave function to respond and thereby changing the electric moments, etc. The interaction with the solvent model must thus be calculated by an iterative procedure, leading to various *Self-Consistent Reaction Field* (SCRF) models.

The interaction of a fixed dipole moment with a polarizable medium is given by Equation (15.78). This, however, is not an SCRF model, as the dipole moment and stabilization are not calculated in a self-consistent way. When the backpolarization of the medium is taken into account, the dipole moment changes, depending on how polarizable the molecule is. Taking only the first-order effect into account, the stabilization is given by

$$\Delta G_{\text{elec}}(\mu) = -\frac{\varepsilon - 1}{2\varepsilon + 1}\frac{\mu^2}{a^3}\left[1 - \frac{\varepsilon - 1}{2\varepsilon + 1}\frac{2\alpha}{a^3}\right]^{-1} \tag{15.79}$$

Here α is the molecular polarizability, that is the first-order change in the dipole moment with respect to an electric field. In the SCRF model the full polarization is taken into account, that is the initial dipole moment generates a polarization of the medium, which changes the dipole moment, which in turn generates a slightly different polarization, etc.

For spherical or ellipsoidal cavities the Poisson equation can be solved analytically, but for molecular-shaped surfaces it must be done numerically. This is typically done by reformulating it in terms of a surface integral over surface charges and solving this numerically by dividing the surface into smaller fractions called tesserae, each having an associated charge $\sigma(\mathbf{r}_s)$. The surface charges are related to the electric field **F** (the derivative of the potential ϕ) perpendicular to the surface by

$$4\pi\varepsilon\sigma(\mathbf{r}_s) = (\varepsilon - 1)\mathbf{F}(\mathbf{r}_s) \tag{15.80}$$

Once $\sigma(\mathbf{r}_s)$ is determined, the associated potential is added as an extra term to the Hamiltonian operator:

$$\phi_\sigma(\mathbf{r}) = \int \frac{\sigma(\mathbf{r}_s)}{|\mathbf{r} - \mathbf{r}_s|} d\mathbf{r}_s \tag{15.81}$$

$$\mathbf{H} = \mathbf{H}_0 + \phi_\sigma \tag{15.82}$$

The potential ϕ_σ from the surface charge is given by the molecular charge distribution (Equation (15.81)), but also enters the Hamiltonian and thus influences the molecular wave function. The procedure is therefore iterative.

For the case of the Onsager model (spherical cavity, dipole moment only) the term added to the molecular Hamiltonian operator is given by

$$\phi_\sigma = -\mathbf{r} \cdot \mathbf{R} \tag{15.83}$$

Here $\mathbf{r}$ is the dipole moment *operator* (i.e. the position vector) and $\mathbf{R}$ is proportional to the molecular dipole moment, with the proportionality constant depending on the radius of the cavity and the dielectric constant:

$$\mathbf{R} = g\boldsymbol{\mu}$$
$$g = \frac{2(\varepsilon - 1)}{(2\varepsilon + 1)a^3} \tag{15.84}$$

The ϕ_σ operator at the HF level of theory corresponds to the addition of an extra term to the Fock matrix elements (Section 3.5):

$$F_{\alpha\beta} = \langle \chi_\alpha | \mathbf{F} | \chi_\beta \rangle - g\boldsymbol{\mu}\langle \chi_\alpha | \mathbf{r} | \chi_\beta \rangle \tag{15.85}$$

The additional integrals are just expectation values of x, y and z coordinates, and their inclusion requires very little additional computational effort. Generalization to higher-order multipoles is straightforward.

In connection with electronic structure methods (i.e. a quantum description of **M**), the term SCRF is quite generic and it does not by itself indicate a specific model. Typically, however, the term is used for models where the cavity is either spherical or ellipsoidal, the charge distribution is represented as a multipole expansion, often terminated at quite low orders (e.g. only including the charge and dipole terms), and the cavity/dispersion contributions are neglected. Such a treatment can only be used for a qualitative estimate of the solvent effect, although relative values may be reasonably accurate if the molecules are polar (dominance of the dipole electrostatic term) and sufficiently similar in size and shape (cancellation of the cavity/dispersion terms).

The cavity size in the Born/Onsager/Kirkwood models strongly influences the calculated stabilization, but there is no consensus on how to choose the cavity radius. In some cases, the molecular volume is calculated from the experimental density of the solvent and the cavity radius is defined by equating the cavity volume to the molecular volume. The cavity size may alternatively be derived from the (experimental) dielectric constant and the calculated dipole moment and polarizability.[112] The underlying assumption in all these models is that the molecule is roughly spherical or ellipsoidal, which is only generally true for small compact molecules.

More sophisticated models employ molecular-shaped cavities, but there is again no consensus on the exact procedure. The cavity is often defined based on van der Waals radii of the atoms in the molecule multiplied by an empirical scale factor ~1.2. The molecular volume may alternatively be calculated directly from the electronic wave function, for example by using an isodensity surface corresponding to a value of 10^{-3}–10^{-4}.

The *Polarizable Continuum Model* (PCM) employs a van der Waals cavity formed by interlocking atomic van der Waals radii scaled by an empirical factor, a detailed description of the electrostatic potential, and parameterizes the cavity/dispersion contributions based on the surface area.[113] Several slightly different implementations have been published, of which the *Integral Equation Formalism PCM* (IEFPCM) is the most general.[114]

The *COnductor-like Screening MOdel* (COSMO) also employs molecular-shaped cavities and represents the electrostatic potential by partial atomic charges. COSMO was originally implemented for semi-empirical methods but has also been used in connection with *ab initio* methods. It may be considered as a limiting case of the PCM model, where the dielectric constant is set to infinity. The COSMO-RS (RS for Real Solvents) includes additional terms in order to model, for example, hydrogen bonding in terms of the surface charges.[115]

The *Solvation Models* (SMx, x being a version number) developed by Cramer and Truhlar are generalized Born-type models,[116] where the partial atomic charges are calculated from a wave function and the dispersion/cavity terms in Equation (15.65) are parameterized based on the solvent exposed surface area (Equation (15.67)). The version number of these models reflects increasingly sophisticated parameterizations.

The *Composite Method for Implicit Representation of Solvent* (CMIRS) model employs a cavity defined by an isodensity surface and describes the electrostatic component as in the PCM model, but parameterize the dispersion, exchange–repulsion and hydrogen bonding in terms of descriptors derived from the wave function.[117] The dispersion is modeled after the van Vorhis–Vydrov DFT formula (Section 6.5.7):

$$\Delta G_{\text{dispersion}} = A\int_{\text{solute}} \frac{\rho(\mathbf{r})I(\mathbf{r},\delta)}{g(\mathbf{r})\left(g(\mathbf{r})+\sqrt{\rho_{\text{solvent}}(\mathbf{r})}\right)}\mathbf{dr} \tag{15.86}$$

$$g(\mathbf{r}) = \sqrt{\rho(\mathbf{r}) + \frac{3K}{4\pi}\left(\frac{\nabla\rho(\mathbf{r})}{\rho(\mathbf{r})}\right)^4} \quad ; \quad I(\mathbf{r},\delta) = \int_{\text{solvent}} [(\mathbf{r}-\mathbf{r}')^6 + \delta^6]^{-1}\mathbf{dr}' \tag{15.87}$$

Here K is a fitting parameter taken from the van Vorhis–Vydrov work, $\rho_{solvent}$ is a solvent specific constant, while the damping parameter δ and A are fitted to experimental reference data. The integral over the solute in Equation (15.86) is over the interior of the cavity as well as the exterior of the cavity, where the electron density is non-negligible. The integral over the solvent in Equation (15.87) is only over the space outside the cavity. The exchange–repulsion is taken to be proportional to the integral of the gradient of the electron density outside the cavity, that is it represents a measure of the amount of solute electron density that is not contained in the cavity and thus interacts repulsively with the surrounding solvent:

$$\Delta G_{\text{exchange}} = B\int_{\text{solvent}} |\nabla\rho(\mathbf{r})|\mathbf{dr} \tag{15.88}$$

The hydrogen bonding is parameterized in term of the largest negative and positive electric fields perpendicular to the cavity surface:

$$\Delta G_{\text{H-bond}} = C|F_{\text{min}}|^{\gamma} + DF_{\text{max}}^{\gamma} \tag{15.89}$$

The B, C, D and γ are fitting parameters.

It should be noted that the parameterization of continuums solvent models, such as, for example, the cavity size defined by the atomic radii, is against experimental free energies, which implicitly include entropy and finite temperature effects. These effects should therefore not be added from calculated frequencies (Equations (14.45) and (14.49)), as this effectively would be a double counting. Furthermore, the use of the ideal-gas rigid-rotor harmonic-oscillator approximation for calculating finite temperature effects is unlikely to be a good approximation for condensed phases. The parameterization against experimental values also has as a consequence that the results are only strictly valid at the temperature where the experimental data have been obtained (normally 298 K) and implicitly include conformational averaging.

The "mixed" solvent models, where the first solvation shell is accounted for by including a number of solvent molecules, implicitly include the solute–solvent cavity/dispersion terms, although the corresponding terms between the solvent molecules and the continuum are usually neglected. Once discrete solvent molecules are included, however, the problem of configuration sampling arises. Furthermore, a parameterization of the continuum model against experimental data must be done by explicitly taking the first solvation shell into account. Nevertheless, the first solvation shell is in many cases by far the most important, and mixed models may yield substantially better results than pure continuum models, at the price of an increased computational cost.

Given the diversity of the various SCRF models, and the fact that solvation energies in water may range from a few kJ/mol for, say, ethane to perhaps several hundred kJ/mol for an ion, it is difficult to evaluate just how accurately continuum methods may in principle be able to represent solvation. It seems clear, however, that molecular-shaped cavities must be employed, the electrostatic polarization needs a description either in terms of atomic charges or quite high order multipoles and cavity, dispersion and exchange terms must be included. Properly parameterized, such models appear to be able to give absolute values with an accuracy of a few kJ/mol.[118] Comparison with results obtained by explicit solvent modeling, however, suggests that the electrostatic component is underestimated for continuum models by roughly a factor of two, while the non-bonded part is essentially uncorrelated with the surface area.[119]

Inclusion of solvent effects may change the geometry, charge distribution and conformational preferences. Employing a PCM-type solvation water model in connection with the B3LYP/aug-cc-pVTZ method, for example, leads to an increase of the C=O bond length in acetamide by 0.015 Å, while the C—N bond is reduced by a similar amount. The calculated dipole moment correspondingly changes from 3.9 to 5.2 Debye. Since solvation preferentially stabilizes the more polar systems, it may also change the conformational preference of molecules. Using the above computational model, for example, changes the energy difference between the *anti* (no dipole moment by symmetry) and *gauche* (gas phase dipole moment of 2.8 Debye) conformations of 1,2-dichloroethane from 6.7 kJ/mol in the gas phase to 1.1 kJ/mol in solution. Molecular properties are in many cases also sensitive to the environment, but a detailed discussion of this is outside the scope of this book.[39,40]

References

1 M. P. Allan and D. J. Tildesley, *Computer Simulations of Liquids* (Clarendon Press, Oxford, 1987).
2 J. M. Haile, *Molecular Dynamics Simulation* (John Wiley & Sons, 1991).
3 A. R. Leach, *Molecular Modelling. Principles and Applications* (Longman, 2001).
4 D. Frenkel and B. Smith, *Understanding Molecular Simulations* (Academic Press, 2002).
5 W. F. Van Gunsteren and H. J. C. Berendsen, *Angewandte Chemie International Edition in English* **29** (9), 992–1023 (1990).
6 W. L. Jorgensen, *Advances in Chemical Physics* **70**, 469 (1988).
7 N. Metropolis, A. W. Rosenbluth, M. N. Rosenbluth, A. H. Teller and E. Teller, *Journal of Chemical Physics* **21** (6), 1087–1092 (1953).
8 W. F. van Gunsteren and A. E. Mark, *Journal of Chemical Physics* **108** (15), 6109–6116 (1998).
9 S. Duane, A. D. Kennedy, B. J. Pendleton and D. Roweth, *Physics Letters B* **195** (2), 216–222 (1987).
10 Y. Sugita and Y. Okamoto, *Chemical Physics Letters* **314** (1–2), 141–151 (1999).
11 D. J. Earl and M. W. Deem, *Physical Chemistry Chemical Physics* **7** (23), 3910–3916 (2005).
12 T. H. Click, A. Liu and G. A. Kaminski, *Journal of Computational Chemistry* **32** (3), 513–524 (2011).
13 V. I. Manousiouthakis and M. W. Deem, *Journal of Chemical Physics* **110** (6), 2753–2756 (1999).
14 S. Bottaro, W. Boomsma, K. E. Johansson, C. Andreetta, T. Hamelryck and J. Ferkinghoff-Borg, *Journal of Chemical Theory and Computation* **8** (2), 695–702 (2011).
15 I. R. McDonald, *Molecular Physics* **23** (1), 41 (1972).
16 L. Verlet, *Physical Review* **159** (1), 98 (1967).
17 H. C. Andersen, *Journal of Chemical Physics* **72** (4), 2384–2393 (1980).

18 S. Toxvaerd, *Physical Review E* **47** (1), 343–350 (1993).

19 J. P. Ryckaert, G. Ciccotti and H. J. C. Berendsen, *Journal of Computational Physics* **23** (3), 327–341 (1977).

20 B. Hess, H. Bekker, H. J. C. Berendsen and J. Fraaije, *Journal of Computational Chemistry* **18** (12), 1463–1472 (1997).

21 H. C. Andersen, *Journal of Computational Physics* **52** (1), 24–34 (1983).

22 H. J. C. Berendsen, J. P. M. Postma, W. F. Vangunsteren, A. Dinola and J. R. Haak, *Journal of Chemical Physics* **81** (8), 3684–3690 (1984).

23 S. Nose, *Molecular Physics* **52** (2), 255–268 (1984).

24 W. G. Hoover, *Physical Review A* **31** (3), 1695–1697 (1985).

25 D. G. Truhlar, R. Steckler and M. S. Gordon, *Chemical Reviews* **87** (1), 217–236 (1987).

26 L. Sun and W. L. Hase, in *Reviews in Computational Chemistry*, edited by K. B. Lipkowitz, R. Larter, T. R. Cundari and D. B. Boyd (John Wiley & Sons, 2003), Vol. 19, pp. 79–146.

27 A. Gonzalezlafont, T. N. Truong and D. G. Truhlar, *Journal of Chemical Physics* **95** (12), 8875–8894 (1991).

28 G. H. Peslherbe and W. L. Hase, *Journal of Chemical Physics* **104** (20), 7882–7894 (1996).

29 Y. Y. Chuang, M. L. Radhakrishnan, P. L. Fast, C. J. Cramer and D. G. Truhlar, *Journal of Physical Chemistry A* **103** (25), 4893–4909 (1999).

30 T. D. Kuehne, *Wiley Interdisciplinary Reviews – Computational Molecular Science* **4** (4), 391–406 (2014).

31 X. S. Li, J. M. Millam and H. B. Schlegel, *Journal of Chemical Physics* **113** (22), 10062–10067 (2000).

32 R. Car and M. Parrinello, *Physical Review Letters* **55** (22), 2471–2474 (1985).

33 J. Hutter, *Wiley Interdisciplinary Reviews – Computational Molecular Science* **2** (4), 604–612 (2012).

34 P. Tangney, *Journal of Chemical Physics* **124** (4), 044111 (2006).

35 I. F. W. Kuo, C. J. Mundy, M. J. McGrath and J. I. Siepmann, *Journal of Chemical Theory and Computation* **2** (5), 1274–1281 (2006).

36 T. D. Kuehne, M. Krack, F. R. Mohamed and M. Parrinello, *Physical Review Letters* **98** (6), 066401 (2007).

37 H. B. Schlegel, S. S. Iyengar, X. S. Li, J. M. Millam, G. A. Voth, G. E. Scuseria and M. J. Frisch, *Journal of Chemical Physics* **117** (19), 8694–8704 (2002).

38 J. M. Herbert and M. Head-Gordon, *Physical Chemistry Chemical Physics* **7** (18), 3269–3275 (2005).

39 G. D. Billing and K. V. Mikkelsen, *Introduction to Molecular Dynamics and Chemical Kinetics* (John Wiley & Sons, 1996).

40 G. D. Billing and K. V. Mikkelsen, *Advanced Molecular Dynamics and Chemical Kinetics* (John Wiley & Sons, 1997).

41 G. D. Billing, *Physical Chemistry Chemical Physics* **4** (13), 2865–2877 (2002).

42 R. P. A. Bettens and M. A. Collins, *Journal of Chemical Physics* **111** (3), 816–826 (1999).

43 Y. P. Liu, D. H. Lu, A. Gonzalezlafont, D. G. Truhlar and B. C. Garrett, *Journal of the American Chemical Society* **115** (17), 7806–7817 (1993).

44 G. E. Moyano and M. A. Collins, *Journal of Chemical Physics* **121** (20), 9769–9775 (2004).

45 G. D. Billing, *Molecular Physics* **89** (2), 355–372 (1996).

46 R. P. Bell, *Transactions of the Faraday Society* **55** (1), 1–4 (1959).

47 R. P. Bell, *The Tunnel Effect in Chemistry* (Chapman and Hall, 1980).

48 E. Wigner, *Zeitschrift Fur Physikalische Chemie-Abteilung B – Chemie Der Elementarprozesse Aufbau Der Materie* **19** (2/3), 203–216 (1932).

49 J. C. Corchado, J. Espinosagarcia, W. P. Hu, I. Rossi and D. G. Truhlar, *Journal of Physical Chemistry* **99** (2), 687–694 (1995).

50 A. Fernandez-Ramos, J. A. Miller, S. J. Klippenstein and D. G. Truhlar, *Chemical Reviews* **106** (11), 4518–4584 (2006).
51 R. Meana-Pañeda, D. G. Truhlar and A. Fernández-Ramos, *Journal of Chemical Theory and Computation* **6** (1), 6–17 (2010).
52 J. Kaestner, *Wiley Interdisciplinary Reviews – Computational Molecular Science* **4** (2), 158–168 (2014).
53 A. D. Bochevarov, E. F. Valeev and C. D. Sherrill, *Molecular Physics* **102** (1), 111–123 (2004).
54 S. Bubin, M. Pavanelo, W.-C. Tung, K. L. Sharkey and L. Adamowicz, *Chemical Reviews* **113** (1), 36–79 (2013).
55 P. Cassam-Chenai, B. Suo and W. Liu, *Physical Review A* **92** (1), 012502 (2015).
56 E. Deumens and Y. Öhrn, *Journal of Physical Chemistry A* **105** (12), 2660–2667 (2001).
57 M. Nest, *Chemical Physics Letters* **472** (4–6), 171–174 (2009).
58 G. Fiorin, M. L. Klein and J. Henin, *Molecular Physics* **111** (22–23), 3345–3362 (2013).
59 G. M. Torrie and J. P. Valleau, *Journal of Computational Physics* **23** (2), 187–199 (1977).
60 E. A. Carter, G. Ciccotti, J. T. Hynes and R. Kapral, *Chemical Physics Letters* **156** (5), 472–477 (1989).
61 G. Ciccotti and M. Ferrario, *Molecular Simulation* **30** (11–12), 787–793 (2004).
62 R. Rajamani, K. J. Naidoo and J. L. Gao, *Journal of Computational Chemistry* **24** (14), 1775–1781 (2003).
63 B. Isralewitz, M. Gao and K. Schulten, *Current Opinion in Structural Biology* **11** (2), 224–230 (2001).
64 S. K. Ludemann, V. Lounnas and R. C. Wade, *Journal of Molecular Biology* **303** (5), 797–811 (2000).
65 A. Laio and M. Parrinello, *Proceedings of the National Academy of Sciences of the United States of America* **99** (20), 12562–12566 (2002).
66 E. Darve and A. Pohorille, *Journal of Chemical Physics* **115** (20), 9169–9183 (2001).
67 D. Hamelberg, J. Mongan and J. A. McCammon, *Journal of Chemical Physics* **120** (24), 11919–11929 (2004).
68 L. Maragliano and E. Vanden-Eijnden, *Chemical Physics Letters* **426** (1–3), 168–175 (2006).
69 H. Bekker, *Journal of Computational Chemistry* **18** (15), 1930–1942 (1997).
70 C. L. Brooks, B. M. Pettitt and M. Karplus, *Journal of Chemical Physics* **83** (11), 5897–5908 (1985).
71 M. Bergdorf, C. Peter and P. H. Hünenberger, *Journal of Chemical Physics* **119** (17), 9129–9144 (2003).
72 P. J. Steinbach and B. R. Brooks, *Journal of Computational Chemistry* **15** (7), 667–683 (1994).
73 P. P. Ewald, *Annalen Der Physik* **64** (3), 253–287 (1921).
74 U. Essmann, L. Perera, M. L. Berkowitz, T. Darden, H. Lee and L. G. Pedersen, *Journal of Chemical Physics* **103** (19), 8577–8593 (1995).
75 N. M. Fischer, P. J. van Maaren, J. C. Ditz, A. Yildirim and D. van der Spoel, *Journal of Chemical Theory and Computation* **11** (7), 2938–2944 (2015).
76 L. Greengard and V. Rokhlin, *Journal of Computational Physics* **73** (2), 325–348 (1987).
77 H. G. Petersen, D. Soelvason, J. W. Perram and E. R. Smith, *Journal of Chemical Physics* **101** (10), 8870–8876 (1994).
78 W. Weber, P. H. Hünenberger and J. A. McCammon, *Journal of Physical Chemistry B* **104** (15), 3668–3675 (2000).
79 M. A. Kastenholz and P. H. Hünenberger, *Journal of Physical Chemistry B* **108** (2), 774–788 (2004).
80 P. E. Smith and B. M. Pettitt, *Journal of Chemical Physics* **95** (11), 8430–8441 (1991).
81 J. D. Chodera, *Journal of Chemical Theory and Computation* **12** (4), 1799–1805 (2016).
82 H. Flyvbjerg and H. G. Petersen, *Journal of Chemical Physics* **91** (1), 461–466 (1989).
83 K. V. Mikkelsen, P. Linse, P. O. Astrand and G. Karlstrom, *Journal of Physical Chemistry* **98** (33), 8209–8215 (1994).
84 B. Guillot, *Journal of Chemical Physics* **95** (3), 1543–1551 (1991).
85 M. Adler and P. Beroza, *Journal of Chemical Information and Modeling* **53** (8), 2065–2072 (2013).

86 T. P. Straatsma, *Reviews in Computational Chemistry* **9**, 81 (1996).
87 N. Hansen and W. F. van Gunsteren, *Journal of Chemical Theory and Computation* **10** (7), 2632–2647 (2014).
88 C. Chipot and D. A. Pearlman, *Molecular Simulation* **28** (1–2), 1–12 (2002).
89 M. R. Shirts, J. W. Pitera, W. C. Swope and V. S. Pande, *Journal of Chemical Physics* **119** (11), 5740–5761 (2003).
90 R. W. Zwanzig, *Journal of Chemical Physics* **22** (8), 1420–1426 (1954).
91 D. A. Pearlman and P. A. Kollman, *Journal of Chemical Physics* **91** (12), 7831–7839 (1989).
92 H. Hu, R. H. Yun and J. Hermans, *Molecular Simulation* **28** (1–2), 67–80 (2002).
93 J. G. Kirkwood, *Journal of Chemical Physics* **3** (5), 300–313 (1935).
94 P. A. Kollman and K. M. Merz, *Accounts of Chemical Research* **23** (8), 246–252 (1990).
95 P. Kollman, *Chemical Reviews* **93** (7), 2395–2417 (1993).
96 J. Aqvist, C. Medina and J. E. Samuelsson, *Protein Engineering* **7** (3), 385–391 (1994).
97 M. Almlof, B. O. Brandsdal and J. Aqvist, *Journal of Computational Chemistry* **25** (10), 1242–1254 (2004).
98 W. E. Miranda, S. Y. Noskov and P. A. Valiente, *Journal of Chemical Information and Modeling* **55** (9), 1867–1877 (2015).
99 C. J. Cramer and D. G. Truhlar, *Chemical Reviews* **99** (8), 2161–2200 (1999).
100 J. Tomasi, B. Mennucci and R. Cammi, *Chemical Reviews* **105** (8), 2999–3093 (2005).
101 G. Scalmani, M. J. Frisch, B. Mennucci, J. Tomasi, R. Cammi and V. Barone, *Journal of Chemical Physics* **124** (9) (2006).
102 J. B. Foresman, T. A. Keith, K. B. Wiberg, J. Snoonian and M. J. Frisch, *Journal of Physical Chemistry* **100** (40), 16098–16104 (1996).
103 R. C. Harris and B. M. Pettitt, *Journal of Chemical Theory and Computation* **11** (10), 4593–4600 (2015).
104 G. Lamm, in *Reviews in Computational Chemistry*, edited by K. B. Lipkowitz, R. Larter, T. R. Cundari and D. B. Boyd (John Wiley & Sons, 2003), Vol. 19, pp. 147–365.
105 N. A. Baker, in *Reviews in Computational Chemistry*, edited by K. B. Lipkowitz, R. Larter and T. R. Cundari (John Wiley & Sons, 2005), Vol. 21, pp. 349–379.
106 M. Born, *Zeitschrift Fur Physik* **1**, 45–48 (1920).
107 W. C. Still, A. Tempczyk, R. C. Hawley and T. Hendrickson, *Journal of the American Chemical Society* **112** (16), 6127–6129 (1990).
108 D. Bashford and D. A. Case, *Annual Review of Physical Chemistry* **51**, 129–152 (2000).
109 L. Onsager, *Journal of the American Chemical Society* **58**, 1486–1493 (1936).
110 J. G. Kirkwood, *Journal of Chemical Physics* **2** (7), 351 (1934).
111 J. G. Kirkwood and F. H. Westheimer, *Journal of Chemical Physics* **6** (9), 506–512 (1938).
112 Y. Luo, H. Agren and K. V. Mikkelsen, *Chemical Physics Letters* **275** (3–4), 145–150 (1997).
113 M. Cossi, V. Barone, R. Cammi and J. Tomasi, *Chemical Physics Letters* **255** (4–6), 327–335 (1996).
114 E. Cances, B. Mennucci and J. Tomasi, *Journal of Chemical Physics* **107** (8), 3032–3041 (1997).
115 A. Klamt, V. Jonas, T. Burger and J. C. W. Lohrenz, *Journal of Physical Chemistry A* **102** (26), 5074–5085 (1998).
116 A. V. Marenich, C. J. Cramer and D. G. Truhlar, *Journal of Chemical Theory and Computation* **9** (1), 609–620 (2013).
117 A. Pomogaeva and D. M. Chipman, *Journal of Chemical Theory and Computation* **10** (1), 211–219 (2014).
118 V. Barone, M. Cossi and J. Tomasi, *Journal of Chemical Physics* **107** (8), 3210–3221 (1997).
119 J. Wagoner and N. A. Baker, *Journal of Computational Chemistry* **25** (13), 1623–1629 (2004).

16

Qualitative Theories

Although sophisticated electronic structure methods may be able to accurately predict a molecular structure or the outcome of a chemical reaction, the results are often hard to rationalize. Generalizing the results to other similar systems therefore becomes difficult. Qualitative theories, on the other hand, are unable to provide accurate results but they may be useful for gaining *insight*, for example why a certain reaction is favored over another. They also provide a link to many concepts used by experimentalists. Frontier molecular orbital theory considers the interaction of the orbitals of the reactants and attempts to predict relative reactivities by second-order perturbation theory. It may also be considered as a simplified version of the Fukui function, which considered how easily the total electron density can be distorted. The Woodward–Hoffmann rules allow a rationalization of the stereochemistry of certain types of reactions, while the more general qualitative orbital interaction model can often rationalize the preference for certain molecular structures over other possible arrangements.

16.1 Frontier Molecular Orbital Theory

Frontier Molecular Orbital (FMO) theory attempts to predict relative reactivity based on properties of the reactants. It is commonly formulated in terms of perturbation theory, where the energy change in the initial stage of a reaction is estimated and "extrapolated" to the transition state.[1] For a reaction where two different modes of reaction are possible, this may be illustrated as shown in Figure 16.1.

The reaction mode that involves the *least* energy change in the initial stage is assumed also to have the lowest activation energy. FMO theory uses a low-order perturbation expansion with the reactants as the unperturbed reference, and it is clear that such a treatment can only be used to follow the reaction a short part of the whole reaction pathway.

Introduction to Computational Chemistry, Third Edition. Frank Jensen.

Companion Website: http://www.wiley.com/go/jensen/computationalchemistry3

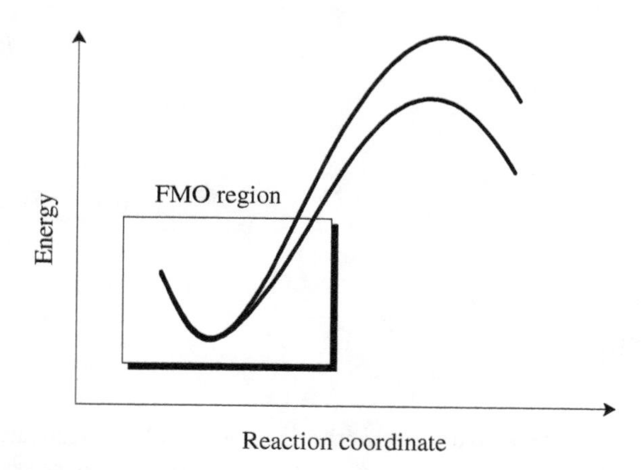

Figure 16.1 FMO region of a reaction profile.

The change in the energy can be derived from second-order perturbation theory (Section 4.8) and is given by[2]

$$\Delta E = -\sum_{\mathrm{A,B}}^{\mathrm{atoms}} (\rho_\mathrm{A} + \rho_\mathrm{B})\langle \chi_\mathrm{A}|\mathbf{V}|\chi_\mathrm{B}\rangle\langle \chi_\mathrm{A}|\chi_\mathrm{B}\rangle + \sum_{\mathrm{A,B}}^{\mathrm{atoms}} \frac{Q_\mathrm{A} Q_\mathrm{B}}{R_\mathrm{AB}} + \left(\sum_{i\in\mathrm{A}}^{\substack{\mathrm{occ.}\\ \mathrm{MO}}} \sum_{a\in\mathrm{B}}^{\substack{\mathrm{vir.}\\ \mathrm{MO}}} + \sum_{i\in\mathrm{B}}^{\substack{\mathrm{occ.}\\ \mathrm{MO}}} \sum_{a\in\mathrm{A}}^{\substack{\mathrm{vir.}\\ \mathrm{MO}}} \right) \frac{2\left(\sum_{\alpha}^{\mathrm{AO}} c_{\alpha i} c_{\alpha a} \langle \chi_{\alpha i}|\mathbf{V}|\chi_{\alpha a}\rangle\right)^2}{\varepsilon_i - \varepsilon_a} \tag{16.1}$$

Here A and B denote atoms in each of the two interacting molecules. The **V** operator contains all the potential energy operators from both molecules and the $\langle \chi_\mathrm{A}|\mathbf{V}|\chi_\mathrm{B}\rangle$ integral is a "resonance"-type integral between two atomic orbitals, one from each molecule. The ρ_A is the electron density on atom A and the first term in (16.1) represents a repulsion ($\langle \chi_\mathrm{A}|\mathbf{V}|\chi_\mathrm{B}\rangle$ is a negative quantity) between occupied MOs (steric repulsion). This will usually lead to a net energy barrier for a reaction. The second term represents an attraction or repulsion between charged parts of the molecules, Q_A being the (net) charge on atom A. The last term is a stabilizing interaction ($\varepsilon_i - \varepsilon_a < 0$) due to mixing of occupied MOs on one molecule with unoccupied MOs on the other, $c_{\alpha i}/c_{\alpha a}$ being MO coefficients and $\varepsilon_i/\varepsilon_a$ MO energies. The summation is over all pairs of occupied/unoccupied MOs.

If we are comparing reactions that have approximately the same steric requirements, the first term is roughly constant. If the species are very polar, the second term will dominate and the reaction is *charge controlled*. This means, for example, that an electrophilic attack is likely to occur at the most negative atom or, in a more general sense, along a path where the electrostatic potential is most negative. If the molecules are non-polar, the third term in Equation (16.1) will dominate and the reaction is said to be *orbital controlled*. This means that the reaction will occur where the molecular orbital coefficients are largest.

All other things being equal, the largest contribution to the double summation over orbital pairs in the third term will arise when the denominator is smallest. This corresponds to the *Highest Occupied Molecular Orbital* (HOMO) and the *Lowest Unoccupied Molecular Orbital* (LUMO) pair of orbitals. FMO theory considers only this one contribution in the whole summation. From a purely numerical consideration this is certainly not a good approximation: the contributions from all the other pairs are much larger than the single HOMO–LUMO term. Nevertheless, it is possible to rationalize many

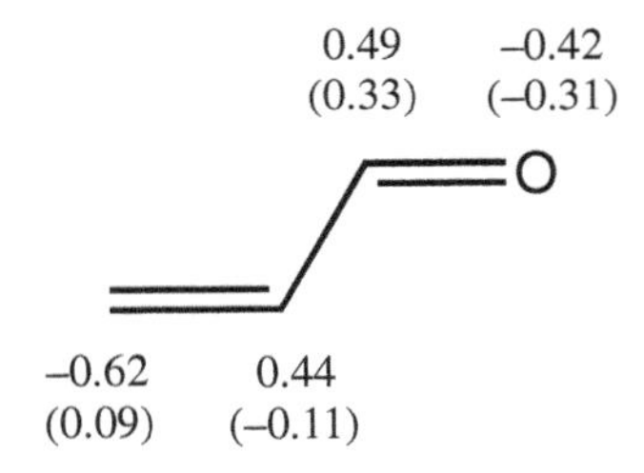

Figure 16.2 AM1 LUMO coefficients for acrolein with net charges in parenthesis.

trends in terms of FMO theory and thus the result justifies the means. If we furthermore consider a matrix element $\langle \chi_{\alpha i}|\mathbf{V}|\chi_{\alpha a.}\rangle$ to be non-zero only between atoms, where new bonds are being formed (where it is furthermore assumed to be roughly constant), the deciding factor becomes a sum over products of MO coefficients from the HOMO on one fragment with LUMO coefficients on the other. A few examples should help clarify this.

The reaction of a nucleophile involves the addition of electrons to the reactant, that is interaction of the HOMO of the nucleophile with the LUMO of the reactant. If there is more than one possible center of attack, the preferred reaction mode is predicted to occur on the atom having the largest LUMO coefficient. Figure 16.2 shows that the orbital component shows preference for addition to the 4-position of acrolein (as a model for unsaturated carbonyl compounds in general), with the second most reactive position being C_2. The net charges, however, prefer position 2, as it is the most positive carbon. Experimentally, it is found that attack at the 4-position is usually favoured (especially with "soft" nucleophiles such as organocuprates), but addition at the 2-position is also observed (and may dominate with "hard" nucleophiles such as organolithium compounds).[3] This is consistent with the reaction switching from being orbital controlled to charge controlled as the nucleophile becomes more ionic.

Similarly, the reaction of an electrophile will involve the HOMO of the reactant, that is the reaction should occur preferentially on the atom having the largest HOMO coefficient. The coefficients for furan shown in Figure 16.3 indicate that electrophilic substitution should preferentially occur at the 2-position, again in agreement with experimental results.[4]

Consider now the reaction between butadiene and ethylene, where both 2 + 2 and 4 + 2 reaction modes are possible. The qualitative appearances of the butadiene HOMO and ethylene LUMO are given in Figure 16.4. The MO coefficients are given as a, b and c, where $a > b > c$.

For the 2 + 2 pathway the FMO sum becomes $(ab - ac)^2 = a^2(b - c)^2$ while for the 4 + 2 reaction it is $(ab + ab)^2 = a^2(2b)^2$. As $(2b)^2 > (b - c)^2$, it is clear that the 4 + 2 reaction has the largest stabilization, and therefore increases *least* in energy in the initial stages of the reaction (Equation (16.1)), remembering that the steric repulsion will cause a net increase in energy). The 4 + 2 reaction should consequently have the lowest activation energy and therefore occur more easily than the 2 + 2. This

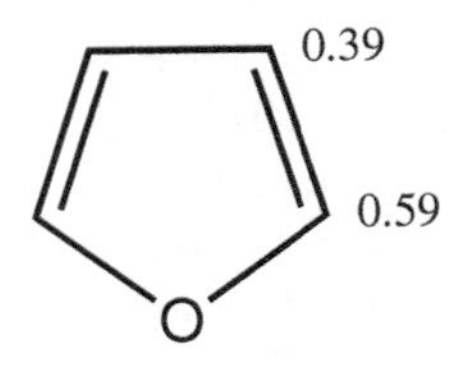

Figure 16.3 AM1 HOMO coefficients for furan.

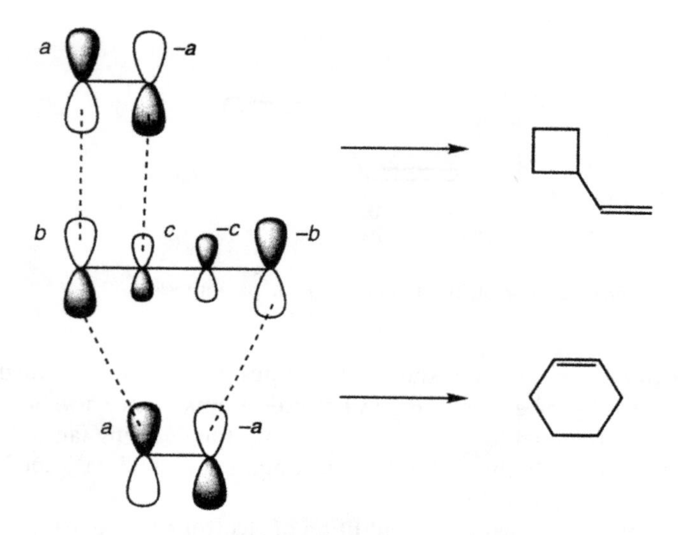

Figure 16.4 FMO theory favors the 4 + 2 over the 2 + 2 reaction.

is indeed what is observed: the Diels–Alder reaction occurs readily but cyclobutane formation is not observed between non-polar dienes and dieneophiles.

The appearance of the difference in MO energies in the denominator in Equation (16.1) suggests that a smaller gap between the diene HOMO and dieneophile LUMO in a Diels–Alder reaction should lower the activation energy. If the diene is made more electron-rich (electron-donating substituents) or the dieneophile more electron-deficient (electron-withdrawing substituents), the reaction should proceed faster. This is indeed the observed trend. For the reaction between cyclopentadiene and cyanoethylenes (mono-, di-, tri- and tetra-substituted), the correlation is reasonably quantitative, as shown in Figure 16.5.[5]

This is of course a rather extreme example, as the reaction rates differ by a factor of $\sim 10^7$, and rate differences by over a factor of 100 are observed for quite similar HOMO–LUMO differences. For a more varied set of compounds where the reaction rates are more similar, the correlation is often quite poor.

FMO theory can also be used for explaining the stereochemistry of the Diels–Alder reaction, as can be illustrated by the reaction between 2-methylbutadiene and cyanoethylene. These may react to give two different products, the "*para*" and/or "*meta*" isomer.

The MO coefficients for the p-orbitals on the butadiene HOMO and ethylene LUMO (taken from AM1 calculations) are given in Figure 16.6. The FMO sum for the "*para*" isomer is $(0.594 \times 0.682 + 0.517 \times 0.552)^2 = 0.690$, while the sum for the "*meta*" isomer is $(0.594 \times 0.552 + 0.517 \times 0.682)^2 = 0.680$. FMO theory thus predicts that the "*para*" isomer should dominate, as is indeed observed (experimental ratio 70 : 30). If cyanoethylene is replaced by 1,1-dicyanoethylene, the LUMO coefficients change to 0.708 and −0.511. The corresponding "*para*" and "*meta*" FMO sums change to 0.685 and 0.670, respectively, that is a larger difference between the two isomers. This is again reflected in the experimental data, where the ratio is 91 : 9. The regiochemistry is thus determined by matching the two largest sets of coefficients and the two smaller sets, rather than making two sets of large/small.

FMO theory was developed at a time when detailed calculations of reaction paths were infeasible. As many sophisticated computational models, and methods for actually locating the transition state, have become widespread, the use of FMO arguments for predicting reactivity has declined. The primary goal of computational chemistry, however, is not to provide *numbers*, but to provide

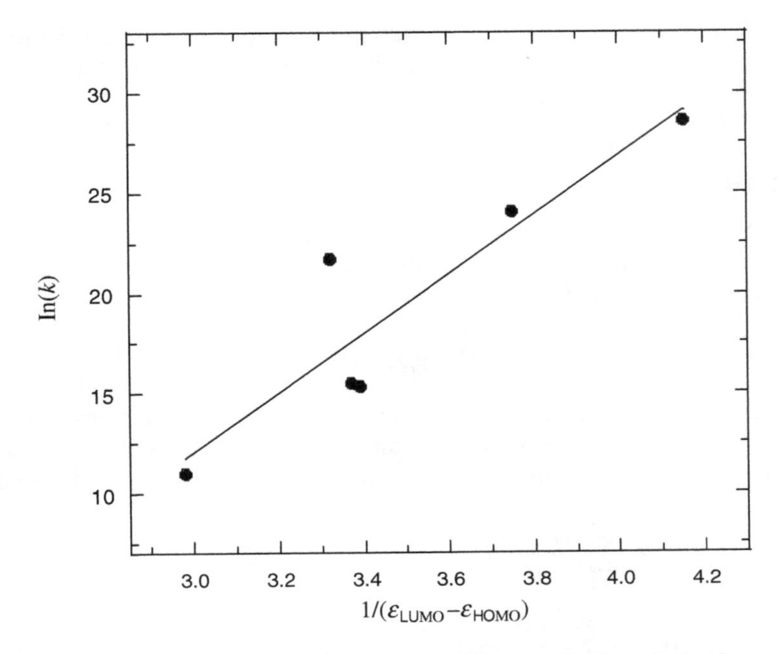

Figure 16.5 Correlation between reaction rates and FMO energy differences (AM1 calculations).

understanding. As such, FMO theory still forms a conceptual model that can be used for rationalizing trends without having to perform time-consuming calculations.

16.2 Concepts from Density Functional Theory

The success of FMO theory is not because the neglected terms in the second-order perturbation expansion (Equation (16.1)) are especially small; an actual calculation will reveal that they completely swamp the HOMO–LUMO contribution. The deeper reason is that the shapes of the HOMO and LUMO resemble features in the total electron density, which determines the reactivity. There are also other quantities derived from density functional theory that directly relate to the properties and reactivity of molecules, and these are discussed in this section.[6]

A reaction will in general involve a change in the electron density, which may be quantified in terms of the *Fukui function*:[7]

$$f(\mathbf{r}) = \frac{\partial \rho(\mathbf{r})}{\partial N_{\text{elec}}} \tag{16.2}$$

0.594 0.682 −0.517 −0.552 CN CN CN

Major Minor

Figure 16.6 FMO rationalizes the stereochemistry of substituted Diels–Alder reactions.

The Fukui function indicates the change in the electron density at a given position when the number of electrons is changed. We may define two finite difference versions of the function, corresponding to addition or removal of an electron:

$$f_+(\mathbf{r}) = \rho_{N+1}(\mathbf{r}) - \rho_N(\mathbf{r}) \quad (16.3)$$

$$f_-(\mathbf{r}) = \rho_N(\mathbf{r}) - \rho_{N-1}(\mathbf{r}) \quad (16.4)$$

The f_+ function is expected to reflect the initial part of a nucleophilic reaction and the f_- function an electrophilic reaction, that is the reaction will typically occur where the $f_\pm$ function is large.[8] For radical reactions the appropriate function is an average of f_+ and f_-:

$$f_0(\mathbf{r}) = \frac{1}{2}(f_+(\mathbf{r}) - f_-(\mathbf{r})) = \frac{1}{2}(\rho_{N+1}(\mathbf{r}) - \rho_{N-1}(\mathbf{r})) \quad (16.5)$$

The change in the electron density for each atomic site can be quantified by using the change in the atomic charges, although this of course suffers from the usual problems of defining atomic charges, as discussed in Chapter 10. The $f_\pm$ functions may also be written in terms of orbital contributions:

$$f_+(\mathbf{r}) = \phi^2_{\text{LUMO}}(\mathbf{r}) + \sum_{i=1}^{N_{\text{elec}}} \frac{\partial\phi_i^2(\mathbf{r})}{\partial n_i} \quad (16.6)$$

$$f_-(\mathbf{r}) = \phi^2_{\text{HOMO}}(\mathbf{r}) + \sum_{i=1}^{N_{\text{elec}}-1} \frac{\partial\phi_i^2(\mathbf{r})}{\partial n_i} \quad (16.7)$$

In the frozen MO approximation the last terms are zero and the Fukui functions are given directly by the contributions from the HOMO and LUMO. The preferred site of attack is therefore at the atom(s) with the largest MO coefficients in the HOMO/LUMO, in exact agreement with FMO theory. The Fukui function(s) may be considered as the equivalent (or generalization) of FMO methods within density functional theory (Chapter 6).

In the *Quantum Theory of Atoms In Molecules* approach (Section 10.3), the Laplacian ∇^2 (trace of the second-derivative matrix with respect to the coordinates) of the electron density measures the local increase or decrease of electrons. Specifically, if $\nabla^2\rho$ is negative, it marks an area where the electron density is locally concentrated, and therefore susceptible to attack by an electrophile. Similarly, if $\nabla^2\rho$ is positive, it marks an area where the electron density is locally depleted, and therefore susceptible to attack by a nucleophile. It has in general been found that a map of negative values of $\nabla^2\rho$ resembles the shape of the HOMO and a map of positive values of $\nabla^2\rho$ resembles the shape of the LUMO.

The fact that features in the total electron density are closely related to the shapes of the HOMO and LUMO provides a much better rationale than the perturbation derivation as to why FMO theory works as well as it does. It should be noted, however, that improvements in the wave function do not necessarily lead to better performance of the FMO method. Indeed, the use of MOs from semi-empirical methods or low-level *ab initio* methods often works better than data from more sophisticated methods. Part of the reason for this is how the HOMO and LUMO are calculated.

The HOMO is the highest occupied molecular orbital, usually from an HF calculation, but could also arise from a DFT calculation. The energy and shape of the HOMO will converge to a specific value and shape if the basis set is improved, and as long as the system is well described by a single determinant wave function, this will provide a good representation of the change in electron density corresponding to removal of an electron with the least amount of energy (Koopmans' theorem). The LUMO, on the other hand, corresponds to an orbital that makes no contribution to the total energy of the molecular system, and its energy and shape depend on the employed basis set.[9] It represents the function with the lowest eigenvalue of the employed Hamilton operator in the function space spanned by the basis set, once the occupied orbitals have been projected out. In a complete basis set there will always be an orbital with zero energy corresponding to describing an unbound electron, and for many molecular systems this will have the lowest eigenvalue of the virtual orbitals, but clearly not carrying any molecular information. If a low- or medium-quality basis set composed of atom centered functions resembling atomic orbitals are employed, such as, for example, a DZP quality Gaussian basis set (Section 5.2), the occupied molecular orbitals will be combinations of the atomic orbitals and the orthogonal function space describing the virtual orbitals will be restrained to antibonding combinations of low-energy atomic orbitals and combinations of higher energy atomic orbitals. In this constrained function space, the basis function combination with the lowest eigenvalue often resembles the HOMO that would arise if an additional electron was added to the system, that is a good description of the change in electron density arising from attachment of an additional electron. If an extended basis set is employed, however, it is often quite arbitrary which of the virtual orbitals acquires the lowest eigenvalue. This is especially problematic if the basis set contains diffuse (small-exponent) functions, as these can be combined to resemble free-electron solutions, or if the basis set consists of functions not resembling atomic orbitals (e.g. plane waves, Section 5.5). A well-defined LUMO can be calculated as the HOMO of the corresponding system with one additional electron, but this of course requires two separate calculations and that the system actually is capable of binding an extra electron. Schmidt and coworker have suggested a procedure for transforming the canonical virtual orbitals into *Valence Virtual Orbitals* (VVOs) that are insensitive to the basis set quality and can be used in place of the actual canonical LUMO.[9] HOMO and LUMO are inherently concepts of one-determinantal methods (HF or DFT), while Fukui functions can be calculated for any type of wave function since they are defined in terms of density differences.

Besides the already mentioned Fukui function, there are a couple of other commonly used concepts that can be connected with density functional theory (Chapter 6).[10] The *electronic chemical potential* μ is given as the first derivative of the energy with respect to the number of electrons, which in a finite difference version is given as the average of the ionization potential (IP) and electron affinity (EA). Except for a difference in sign, this is also the Mulliken definition of *electronegativity* χ:[11]

$$-\mu = \chi = \frac{\partial E}{\partial N_{\text{elec}}} \approx \frac{1}{2}(\text{IP} + \text{EA}) \tag{16.8}$$

It should be noted that there are several other definitions of electronegativity, which do not necessarily agree on the ordering of the elements.[12]

The second derivative of the energy with respect to the number of electrons is the *hardness* η (the inverse quantity η^{-1} is called the *softness*), which again may be approximated in terms of the ionization potential and electron affinity:

$$\eta = \frac{1}{2}\frac{\partial^2 E}{\partial N_{\text{elec}}^2} \approx \frac{1}{2}(\text{IP} - \text{EA}) \tag{16.9}$$

The electrophilicity, which measures the total ability to attract electrons, is defined as[13]

$$\omega = \frac{\mu^2}{2\eta} \approx \frac{(\text{IP} + \text{EA})^2}{4(\text{IP} - \text{EA})} \tag{16.10}$$

A local version of the electrophilicity can be obtained by multiplying ω with the relevant Fukui function. These concepts play an important role in the *Hard and Soft Acid and Base* (HSAB) principle, which states that hard acids prefer to react with hard bases, and vice versa.[14] By means of Koopmans' theorem (Section 3.4) the hardness is related to the HOMO–LUMO energy difference. A "hard" molecule thus has a large HOMO–LUMO gap and is expected to be chemically unreactive, that is hardness is related to chemical stability. A small HOMO–LUMO gap, on the other hand, indicates a "soft" molecule, and from second-order perturbation theory it also follows that a small gap between occupied and unoccupied orbitals will give a large contribution to the polarizability (Section 11.7.1), that is softness is a measure of how easily the electron density can be distorted by external fields, for example generated by another molecule. In terms of the perturbation Equation (16.1), a hard–hard interaction is primarily charge controlled, while a soft–soft interaction is orbital controlled. Both FMO and HSAB theories may be considered as being limiting cases of chemical reactivity described by the Fukui function.[8]

16.3 Qualitative Molecular Orbital Theory

Frontier molecular orbital theory is closely related to various schemes of qualitative orbital theory where interactions between fragment MOs are considered.[15,16] Ligand field theory, as commonly used in systems involving coordination to metal atoms, can be considered as a special case where only the d-orbitals on the metal and selected orbitals of the ligands are considered.

Two interacting orbitals will in general produce two new orbitals having lower and higher energies than the non-interacting orbitals. The magnitude of the changes is determined by the orbital energy difference $\varepsilon_a - \varepsilon_b$ and the overlap S_{ab}. The overlap depends on the symmetries of the orbitals (orbitals of different symmetry have zero overlap) and the distance between them (the shorter the distance, the larger the overlap). The energies of the new orbitals can be calculated from the variational principle, and the qualitative result is shown in Figure 16.7:

$$\Delta \propto \frac{|S_{ab}|}{|\varepsilon_a - \varepsilon_b|} \tag{16.11}$$

There are two important features. The change in orbital energies is dependent on the magnitude of the overlap, $|S_{ab}|$, and is inversely proportional to the energy difference of the original orbitals,

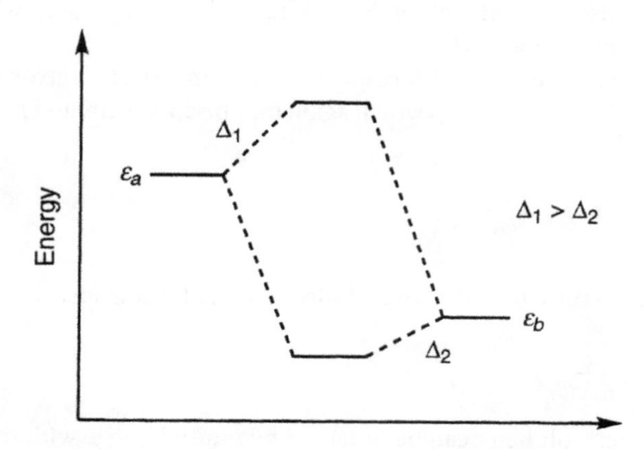

Figure 16.7 Linear combination of two orbitals leads to two new orbitals with different energies.

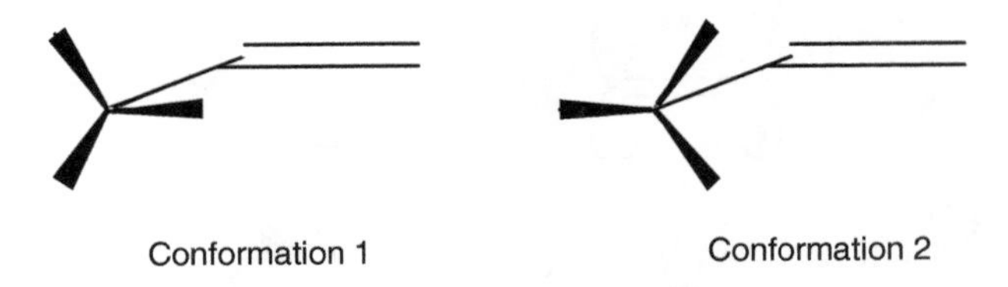

Figure 16.8 Possible propene conformations.

$|\varepsilon_a - \varepsilon_b|$. Furthermore, the effect is largest for the highest energy orbital (antibonding combination), that is $\Delta_1 > \Delta_2$.

If the two initial orbitals contain one, two or three electrons, the interaction will lead to a lower energy, with the stabilization being largest for the case of two electrons (e.g. a filled orbital interacting with an empty orbital). If both initial orbitals are fully occupied, the interaction will be destabilizing, that is a steric-type repulsion. By adapting a set of HOMO and LUMO orbitals for atomic or molecular fragments, the favorable interactions may be identified based on overlap and energy considerations. Qualitative MO theory may thus be considered as an intramolecular form of FMO theory, with suitably chosen fragments.

Consider, for example, the two conformations for propene shown in Figure 16.8: which should be the more stable?

By "chemical intuition", the most important interaction is likely to be between the (filled) hydrogen s-orbitals and the (empty) π-orbital. The CH_3 group as a fragment has C_{3v} symmetry and the three proper (symmetry adapted) linear combinations of the s-orbitals, together with the antibonding π-orbital, are given in Figure 16.9. The ϕ_1 orbital is lowest in energy, while the ϕ_2 and ϕ_3 orbitals are degenerate in perfect C_{3v} symmetry.

The ϕ_1 and ϕ_2 orbitals have a different symmetry than the π^*-orbital and can consequently not interact ($S = 0$). The interaction of the ϕ_3 orbital with the π^* system in the two conformations is shown in Figure 16.10.

The overlap between the nearest carbon p-orbital and ϕ_3 is the largest contribution, but it is the same in the two conformations. The overlap with the distant carbon p-orbital is of opposite sign and largest in conformation 2, since the distance is shorter. The total overlap between the ϕ_3 and

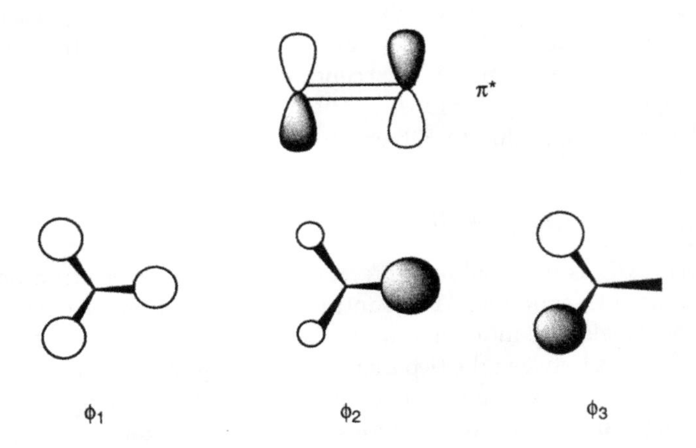

Figure 16.9 Fragment orbitals for propene.

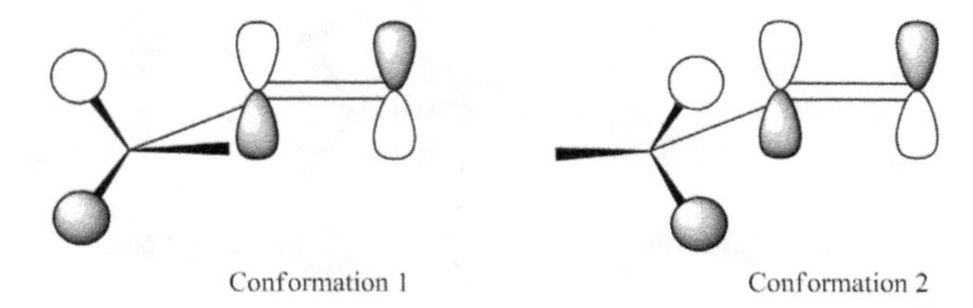

Figure 16.10 Fragment orbital interaction.

π^*-orbitals is thus largest for conformation 1, which implies a larger stabilizing interaction, and it should consequently be lowest in energy. Indeed, conformation 2 is a transition structure for interconverting equivalent conformations corresponding to 1.

It is important to realize that whenever qualitative or frontier molecular orbital theory is invoked, the description is within the orbital (Hartree–Fock or density functional) model for the electronic wave function. In other words, rationalizing a trend in computational results by qualitative MO theory is only valid if the effect is present at the HF or DFT level. If the majority of the variation is due to electron correlation, an explanation in terms of interacting orbitals is not appropriate.

16.4 Energy Decomposition Analyses

The interacting fragment orbital analysis can be put on more quantitative terms by performing explicit energy decomposition analysis of HF or DFT wave functions.[17] The *Extended Transition State* (ETS) approach decomposes the energy change into four terms:[18–20]

$$\Delta E^{\text{ETS}} = \Delta E_{\text{prep}} + \Delta E_{\text{el}} + \Delta E_{\text{Pauli}} + \Delta E_{\text{orb}} \tag{16.12}$$

The energy change can, for example, be formation of a bond and the analysis can be performed at various points along the reaction path. The preparation energy ΔE_{prep} describes the cost for perturbing the nuclear geometry from the optimum for the fragment to that of the species of interest. The electrostatic term ΔE_{el} describes the Coulomb interaction between the two fragment charge densities, while ΔE_{Pauli} describes the repulsion due to antisymmetrization and re-normalization of the two fragment wave functions when they are combined into one. Finally, ΔE_{orb} describes the stabilizing interaction due to mixing of occupied and unoccupied orbitals of the two fragments, which includes polarization and charge-transfer terms. The two central terms, ΔE_{el} and ΔE_{Pauli}, may loosely be associated with the concept of steric repulsion.

An alternative decomposition, due to Kitaura and Morokuma (KM), partitions the interaction energy into five terms:[21]

$$\Delta E^{\text{KM}} = \Delta E_{\text{el}} + \Delta E_{\text{pol}} + \Delta E_{\text{CT}} + \Delta E_{\text{ex}} + \Delta E_{\text{mix}} \tag{16.13}$$

The electrostatic term ΔE_{el} is the Coulomb interaction between the electron densities of the fragments, analogous to the corresponding ETS quantity. The polarization term ΔE_{pol} describes the stabilization due to induced electric moments, while the charge transfer term ΔE_{CT} is a stabilization due to the transfer of charge between the two fragments. The exchange term ΔE_{ex} is analogous to the Pauli term in Equation (16.12), describing the repulsion due to the exchange energy arising from the antisymmetrization of the fragment wave functions. Finally, the mix term ΔE_{mix} contains the residual interaction not accounted for by the first four terms. The KM energy decomposition is most useful for analyzing weak interactions; for strong interactions the mixing term often accounts for a significant part of the total interaction, thus obscuring the decomposition. The polarization and charge-transfer terms furthermore become numerically unstable as the size of the basis set is increased. The KM energy decomposition has been combined with the NBO analysis (Section 10.5.1) in the *Natural Energy Decomposition Analysis* (NEDA), which partly alleviates some of the instabilities in the original method.[22]

The *Reduced Variable Space* (RVS) method can be considered as a modified version of the KM decomposition, where the proper antisymmetry of the total wave function is enforced.[23] This necessitates combining the electrostatic and exchange energies into a single $\Delta E_{\text{el-ex}}$ term:

$$\Delta E^{\text{RVS}} = \Delta E_{\text{el-ex}} + \Delta E_{\text{pol}} + \Delta E_{\text{CT}} + \Delta E_{\text{mix}} \tag{16.14}$$

The polarization and charge-transfer terms are as a consequence also altered and the mixing term is substantially reduced in magnitude.

The *Absolute Localized Molecular Orbital* (ALMO),[24] *Block-Localized Wave* (BLW)[25] and *Density-based Energy Decomposition Analysis* (DEDA)[26] methods decomposed the interaction energy into an interaction between the frozen densitites of the fragments ΔE_{FRZ}, and polarization and charge-transfer terms:

$$\Delta E^{\text{ALMO/BLW/DEDA}} = \Delta E_{\text{FRZ}} + \Delta E_{\text{pol}} + \Delta E_{\text{CT}} \tag{16.15}$$

The ΔE_{FRZ} thus includes the electrostatic and exchange terms. If the interaction energy is calculated with both HF and correlated wave functions, the effect of dispersion/correlation may be extracted as a simple difference. The main difference between ALMO/BLW and DEDA is that the former perform the decomposition within the non-orthogonal space of the combined fragment orbitals, while DEDA constructs a variationally optimized single Slater determinant with orthogonal orbitals that exactly reproduce the electron densities of the isolated fragments. The electrostatic interaction of these frozen densities defines the ΔE_{FRZ} component, ΔE_{pol} is the energy lowering by allowing the fragment densities to relax, while ΔE_{CT} is the additional energy lowering by allowing electron density to move between the fragments. The polarization and charge-transfer terms both correspond to mixing of the occupied and virtual orbitals, where polarization indicates orbital mixing within the same fragment while charge-transfer indicates orbital mixing between fragments. The assignment of orbitals or electron density as belonging to a specific fragment is somewhat arbitrary, as discussed in Section 5.13 and Chapter 10, and the partitioning between these two energy terms consequently depends on the exact scheme and often has a strong basis set dependence. A procedure where these terms are evaluated by a limited expansion into a set of electric field perturbation orbitals has been proposed, and this is much less basis set dependent.[27]

The above approaches *decompose* the total interaction energy into terms, but the interaction can alternatively be *estimated* from the fragment wave functions. Within the NBO framework (Section 10.5.1) the interaction energies between occupied and virtual orbitals can be estimated from second-order perturbation theory (compared to Equation (16.11)):

$$\Delta E_{ia}^{(2)} = \frac{n_i |F_{ia}|^2}{\varepsilon_i - \varepsilon_a} \tag{16.16}$$

Here F_{ia} is a Fock matrix element between occupied orbital i and virtual orbital a, n_i is the natural occupation number and $\varepsilon_i / \varepsilon_a$ are orbital energies.

The *Symmetry-Adapted Perturbation Theory* (SAPT)[28] is a more rigorous approach where the interaction energy is calculated by partitioning the total Hamiltonian into a term for each fragment and an interfragment interaction potential $\mathbf{V}_{AB}$:

$$\mathbf{H} = \mathbf{H}_A + \mathbf{H}_B + \lambda \mathbf{V}_{AB} \tag{16.17}$$

The $\mathbf{V}_{AB}$ operator contains interactions between electrons/nuclei on one fragment with electrons/nuclei on the other fragment and can be written as follows, with $\mathbf{r}$ and $\mathbf{R}$ being electron and nuclear coordinates, respectively:

$$\mathbf{V}_{AB} = \sum_a^A \sum_b^B \left(\frac{1}{|\mathbf{r}_a - \mathbf{r}_b|} - \frac{Z_a}{|\mathbf{r}_b - \mathbf{R}_a|} - \frac{Z_b}{|\mathbf{r}_a - \mathbf{R}_b|} + \frac{Z_a Z_b}{|\mathbf{R}_a - \mathbf{R}_b|} \right) \tag{16.18}$$

The total wave function is written as a product of the two fragment wave functions, and the interaction energy is evaluated in terms of the perturbation parameter λ, which eventually is set to 1, and taking the proper antisymmetry of the total wave function into account by a projection operator. The effect of the perturbation operator is denoted as induction and includes polarization and charge transfer in the above notations, while the effect of antisymmetrization is denoted as exchange. Using a superscript to denote the perturbation order, the interaction energy can be written as a sum of terms:

$$\Delta E^{\mathrm{SAPT}} = \Delta E^{(1)}_{\mathrm{el}} + \Delta E^{(1)}_{\mathrm{ex}} + \Delta E^{(2)}_{\mathrm{ind}} + \Delta E^{(2)}_{\mathrm{disp}} + \Delta E^{(2)}_{\mathrm{ex\text{-}ind}} + \Delta E^{(2)}_{\mathrm{ex\text{-}disp}} + \cdots \tag{16.19}$$

The electrostatic and exchange terms arise at first order in the intermolecular perturbation parameter λ while induction and dispersion (interfragment correlation) effects appear at second order along with their exchange components. When the A/B fragments are described at the single determinant HF or DFT level, the method is denoted SAPT0 while SAPT2 denotes that electron correlation is taking into account to second order for each fragment. The latter method necessitates a double perturbation expansion in both the λ parameter and corresponding parameters for the fluctuation potential for each fragment. When intrafragment electron correlation is included, each of the terms in Equation (16.19) may contain several contributions that can be classified acording to the perturbation order of the fluctuation potential in the fragments. A number of SAPT variations exist where selected higher-order terms and/or corrections are included. When using the SAPT formalism in connection with a DFT description of the fragments, it is important to employ a range-separated version of the exchange functional to avoid spurious errors due to the incorrect limiting form of the exchange potential. The SAPT approach is capable of calculating intermolecular potentials with an accuracy that often rivals brute force CCSD(T) calculations.

Energy analyses and decompositions are often used as reference data for parameterization of advanced force field methods, as discussed in Section 2.2.8.

16.5 Orbital Correlation Diagrams: The Woodward–Hoffmann Rules

The *Woodward–Hoffmann* (W–H) rules are qualitative statements regarding relative activation energies for two possible modes of reaction, which may have different stereochemical outcomes.[29,30] For simple systems, the rules may be derived from a conservation of orbital symmetry, but they may also be generalized by an FMO treatment with conservation of bonding. Let us illustrate the Woodward–Hoffmann rules with a couple of examples, the preference of the 4 + 2 over the 2 + 2 product for the reaction of butadiene with ethylene and the ring-closure of butadiene to cyclobutene.

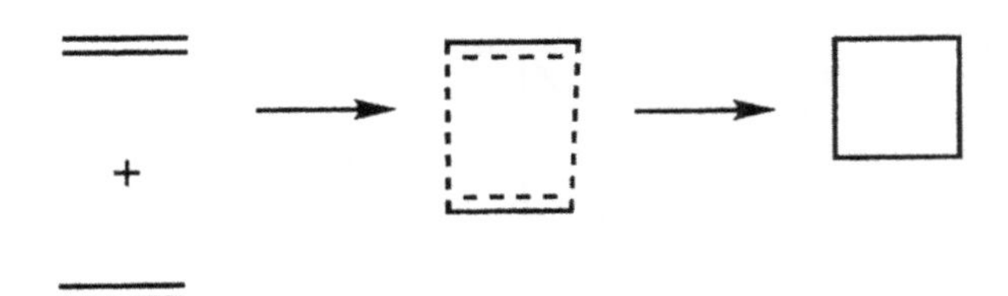

Figure 16.11 Reaction of two ethylenes to form cyclobutane under C_{2v} symmetry.

A face-to-face reaction of two π-orbitals to form a cyclobutane involves the formation of two new C—C σ-bonds. The reaction may be imagined to occur under the preservation of symmetry, in this case C_{2v}, that is *concerted* (one-step, no intermediates) and *synchronous* (both bonds are formed at the same rate) (see Figure 16.11).

Both the reactant and product orbitals may be classified according to their behaviour with respect to the two mirror planes present, being either *Symmetric* (no change of sign) or *Antisymmetric* (change of sign). The energetic ordering of the orbitals follows from a straightforward consideration of the bonding/antibonding properties. Since orbitals of different symmetries cannot mix, conservation of orbital symmetry establishes the correlation between the reactant and product sides.

The *orbital correlation diagram* shown in Figure 16.12 indicates that an initial electron configuration of $(\pi_1 + \pi_2)^2(\pi_1 - \pi_2)^2$ (ground state for the reactant) will end up as a doubly excited configuration $(\sigma_1 + \sigma_2)^2(\sigma^*_1 + \sigma^*_2)^2$ for the cyclobutane product.[31] This by itself indicates that the reaction should be substantially uphill in terms of energy. It may be put on a more sound theoretical footing by looking at the *state correlation diagram* in Figure 16.13.

The ground state wave function for the whole system (all four active and the remaining core and valence electrons) is symmetric with respect to both mirror planes, while the first excited state is antisymmetric. The intended correlation is indicated with dashed lines, the lowest energy configuration for the reactant correlates with a doubly excited configuration of the product, and vice versa. Since these configurations have the same symmetry (SS), an avoided crossing is introduced, leading to a significant barrier for the reaction. The presence of a reaction barrier due to symmetry conservation for the orbitals makes this a Woodward–Hoffmann *forbidden* reaction. The reaction for the excited state, however, does not encounter a barrier and is therefore denoted an *allowed* reaction.

The same conclusion may be reached directly from a consideration of the frontier orbitals. Formation of two new σ-bonds requires interaction of the HOMO of one fragment with the LUMO on the other. When the interaction is between orbital lobes on the same side (*suprafacial*) of each fragment (2s + 2s), this leads to the picture shown in Figure 16.14.

It is clearly seen that the HOMO–LUMO interaction leads to the formation of one bonding and one antibonding orbital, that is this is not a favorable interaction. The FMO approach also suggests that the 2 + 2 reaction may be possible if it could occur with bond formation on opposite sides (*Antarafacial*) for one of the fragments. Although the 2s + 2a reaction is Woodward–Hoffmann allowed, it is sterically so hindered that thermal 2 + 2 reactions in general are not observed (see Figure 16.15). Photochemical 2 + 2 reactions, however, are well known.[32]

The 4s + 2s reaction of a diene with a double bond can in a concerted and synchronous reaction be envisioned to occur with the preservation of C_s symmetry (Figure 16.16) and the corresponding orbital correlation diagram is shown in Figure 16.17.

In this case the orbital correlation diagram shows that the lowest energy electron configuration in the reactant, $(\pi_1)^2(\pi_2)^2(\pi_3)^2$, correlates directly with the lowest energy electron configuration in

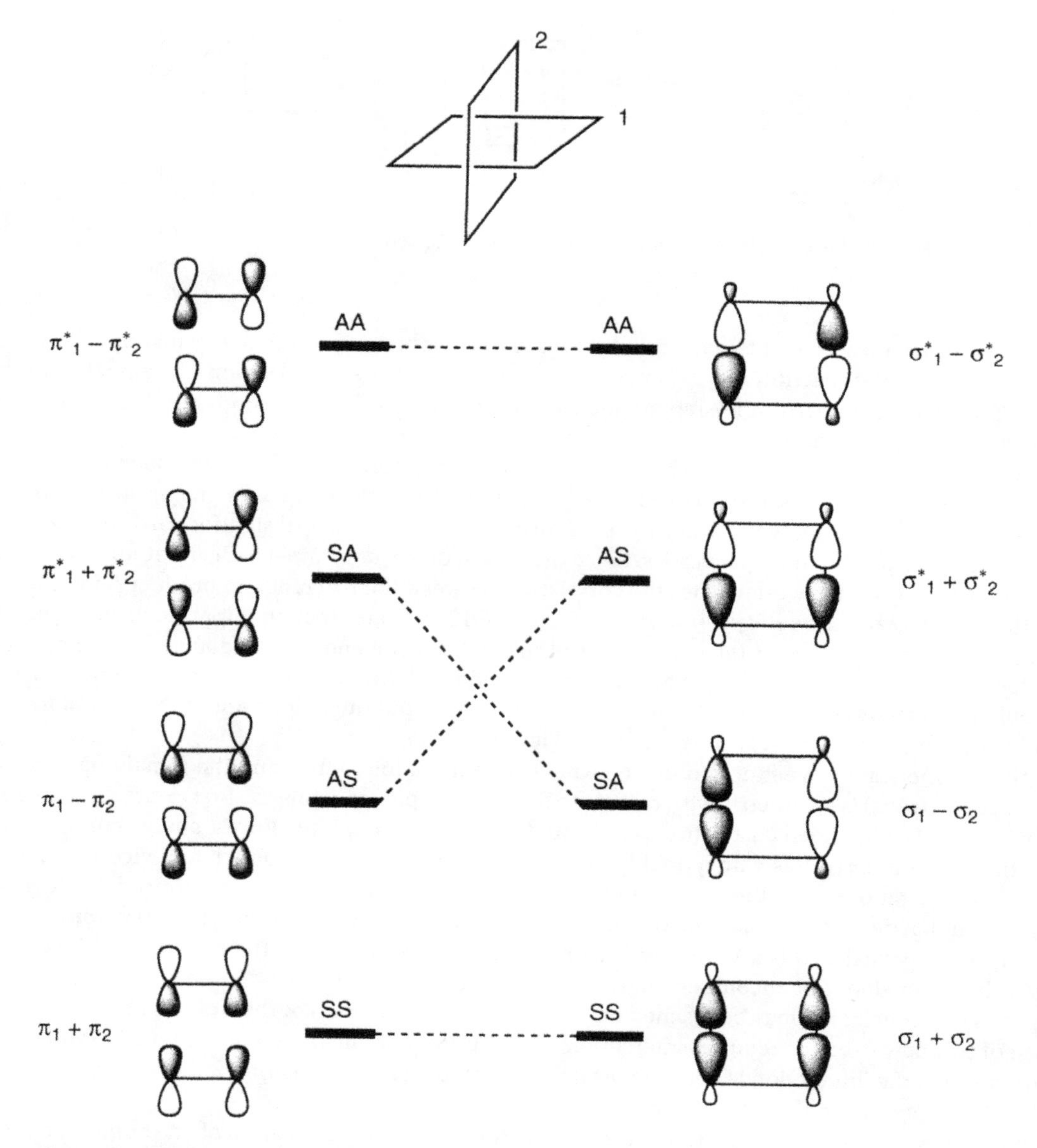

Figure 16.12 Orbital correlation diagram for cyclobutane formation.

the product, $(\sigma_1)^2(\sigma_2)^2(\pi_1)^2$. This is also shown by the corresponding state correlation diagram, Figure 16.18.

In this case, there is no energetic barrier due to unfavourable orbital correlation, although other factors lead to an activation energy larger than zero. The direct correlation of ground state configurations for the reactant and product indicates a (relatively) easy reaction, and is therefore an allowed reaction. The lowest excited state for the reactant, however, does not correlate with the lowest excited product state and the photochemical reaction is consequently forbidden. The FMO approach again

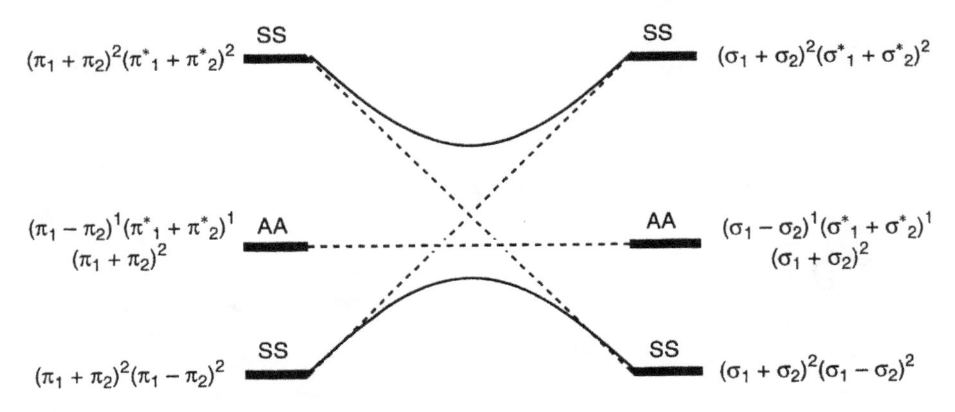

Figure 16.13 State correlation diagram for cyclobutane formation.

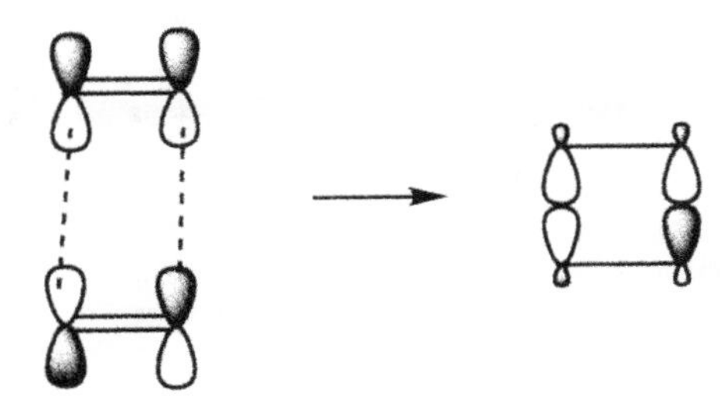

Figure 16.14 2s + 2s HOMO–LUMO interaction leading to two new σ-bonds.

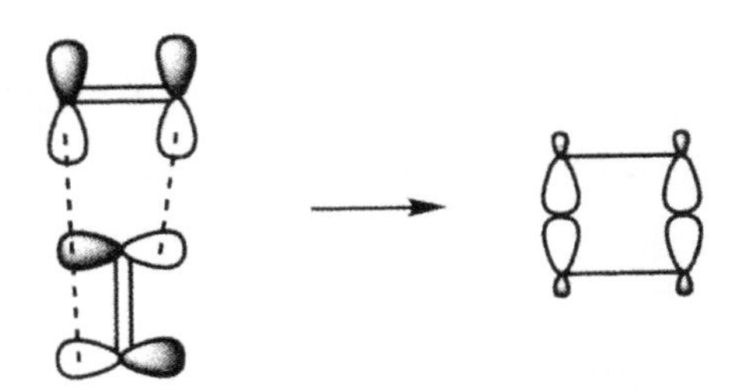

Figure 16.15 2s + 2a HOMO–LUMO interaction leading to two new σ-bonds.

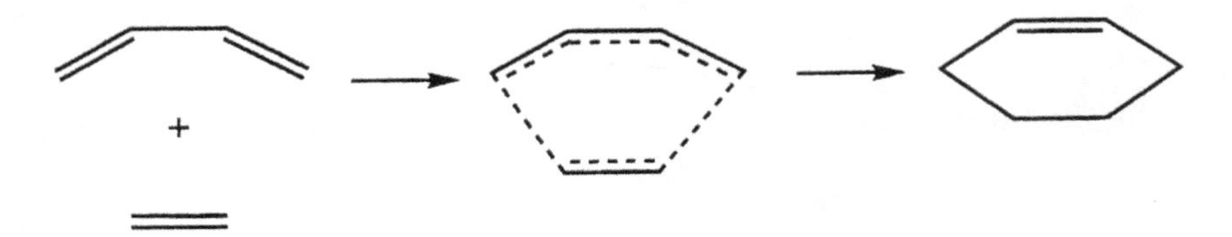

Figure 16.16 Reaction of butadiene and ethylene to form cyclohexene under C_s symmetry.

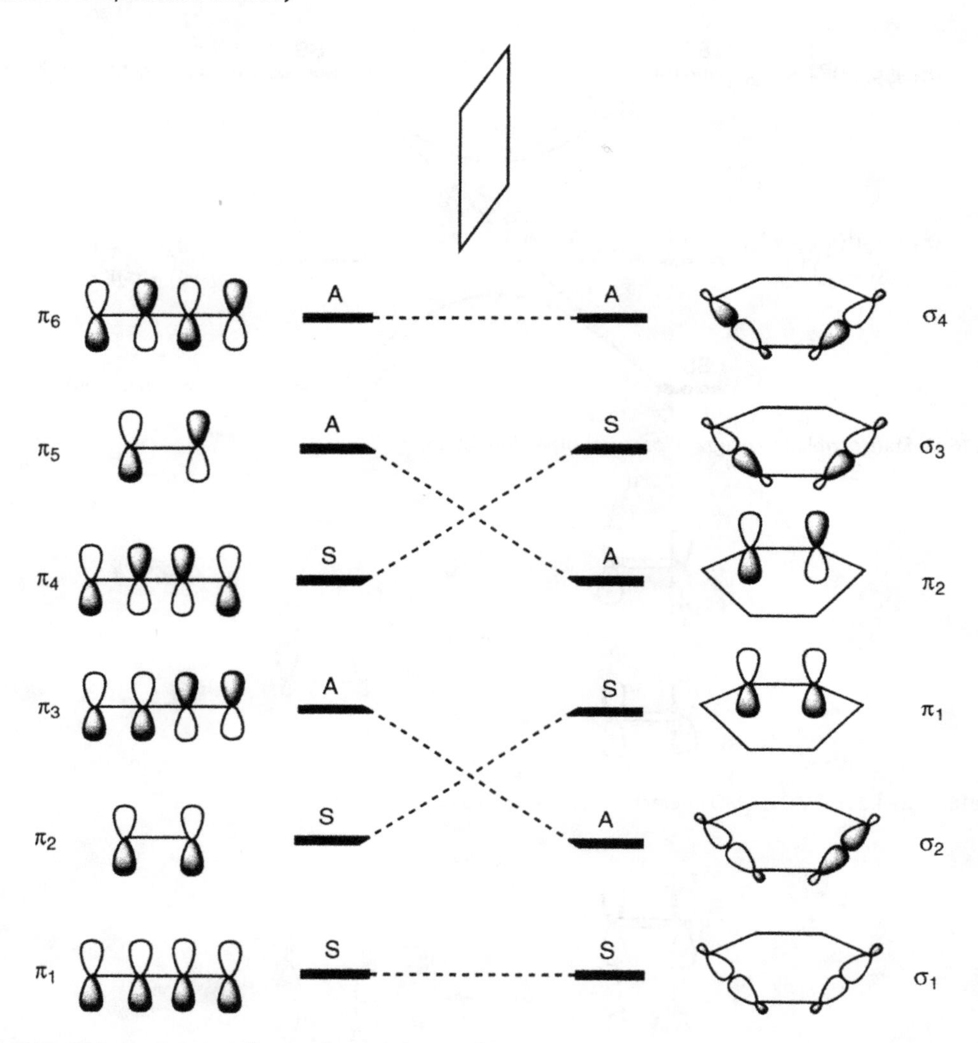

Figure 16.17 Orbital correlation diagram for cyclohexene formation.

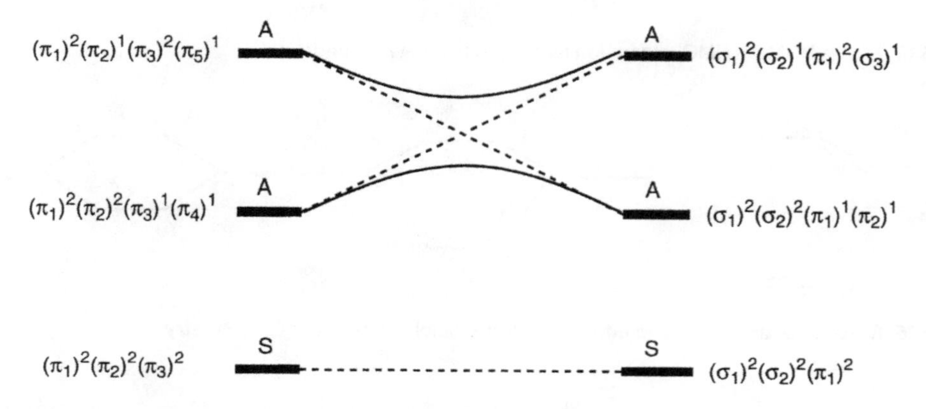

Figure 16.18 State correlation diagram for cyclohexene formation.

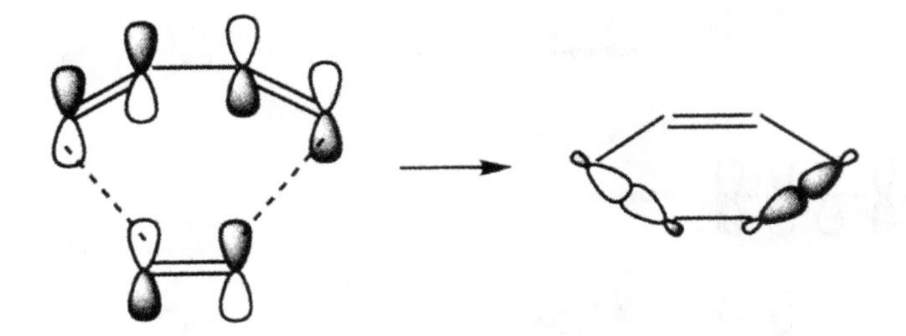

Figure 16.19 4s + 2s HOMO–LUMO interaction leading to two new σ-bonds.

indicates that the 4s + 2s interaction should lead directly to formation of two new bonding σ-bonds, that is this is an allowed reaction (Figure 16.19).

The preference for a concerted 4s + 2s reaction is experimentally supported by observations that show that the stereochemistry of the diene and dieneophile is carried over to the product, for example a *trans,trans*-1,4-disubstituted diene results in the two substituents ending up in a *cis* configuration in the cyclohexene product.[33]

The ring-closure of a diene to a cyclobutene can occur with rotation of the two termini in the same (*Conrotatory*) or opposite (*Disrotatory*) directions. For suitably substituted compounds, these two reaction modes lead to products with different stereochemistry (see Figure 16.20).

The disrotatory path has C_s symmetry during the whole reaction, while the conrotatory mode preserve C_2 symmetry. The orbital correlation diagrams for the two possible paths are shown as Figures 16.21 and 16.22.

It is seen that only the conrotatory path directly connects the reactant and product ground state configurations. Taking into account also the excited states leads to the state correlation diagram in Figure 16.23.

The conrotatory path is Woodward–Hoffmann allowed for a thermal reaction, while the corresponding photochemical reaction is predicted to occur in a disrotatory fashion.

The same conclusion may again be reached by considering only the HOMO (Figure 16.24). For the conrotatory path the orbital interaction leads directly to a bonding orbital, while the orbital phases for the disrotatory motion lead to an antibonding orbital.

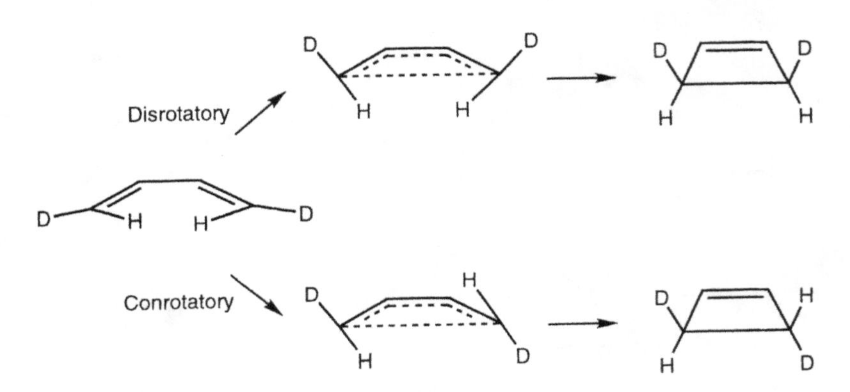

Figure 16.20 Two possible modes of closing a diene to cyclobutene.

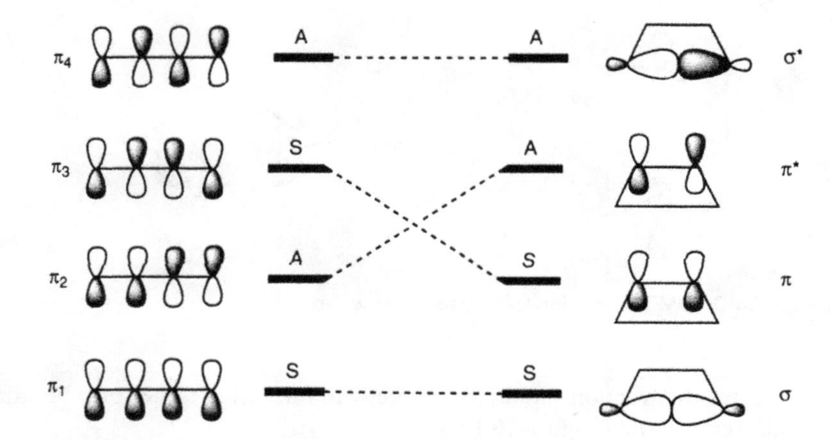

Figure 16.21 Orbital correlation diagram for the disrotatory ring-closure of butadiene.

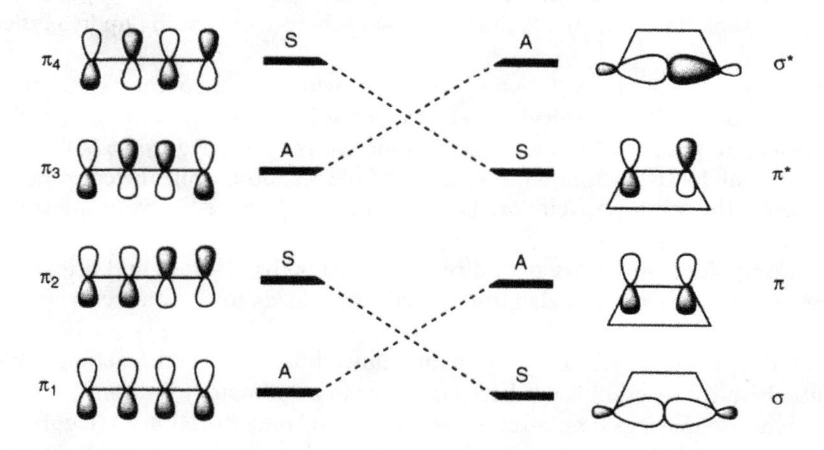

Figure 16.22 Orbital correlation diagram for the conrotatory ring-closure of butadiene.

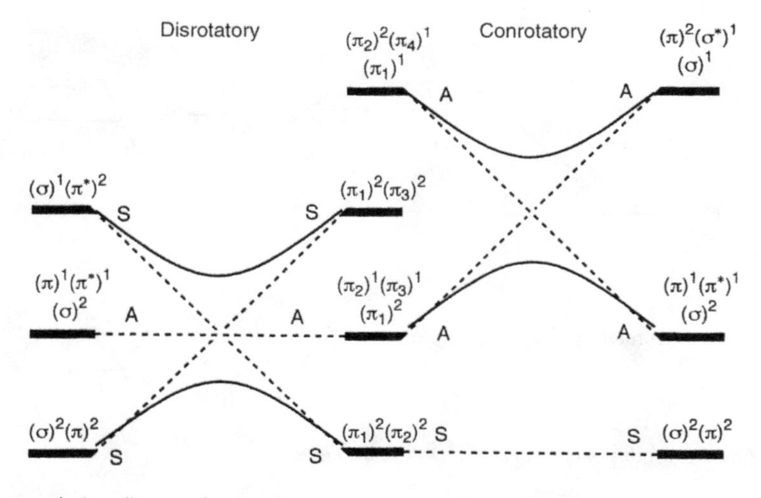

Figure 16.23 State correlation diagram for the dis- and conrotatory ring-closure of butadiene.

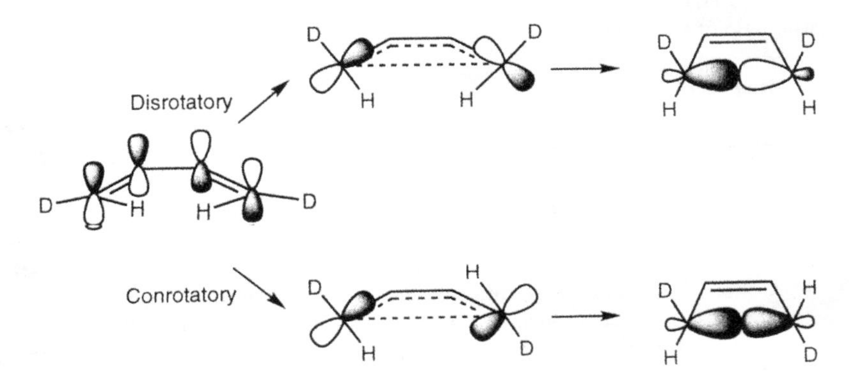

Figure 16.24 HOMO orbital for the ring-closure of butadiene.

While the orbital and state diagrams can only be rigorously justified in the simple parent system where symmetry is present, the addition of substituents normally only alters the shape of the relevant orbitals slightly. The nodal structure of the orbitals is preserved for a large range of substituted systems, and the "preservation of bonding" displayed by the FMOtype diagrams consequently have a substantially wider predictive range. It may be used for analyzing reactions where there is no symmetry element present under the whole reaction, as in, for example, the [1,5]-hydrogen shift in 1,3-pentadiene (Figure 16.25).

In the suprafacial migration the interaction of the pentadienyl radical singly occupied orbital with the hydrogen s-orbital is seen to involve breaking and making bonds where the orbital phases match. For the antarafacial path, however, the orbital in the product ends up being antibonding, that is a [1,5]-hydrogen migration is predicted to occur suprafacially, in agreement with experiments.[34]

In the general case, the transferring group may migrate with either *retention* or *inversion* of its stereochemistry. A [1,5]-CH_3 migration, for example, is thermally allowed if it occurs suprafacially with retention of the CH_3 configuration, or if it occurs antarafacially with inversion of the methyl group (see Figure 16.26).

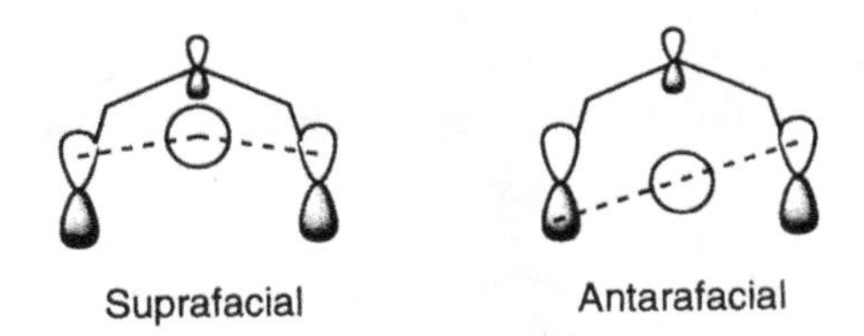

Figure 16.25 FMO interactions for the [1,5]-hydrogen shift in 1,3-pentadiene.

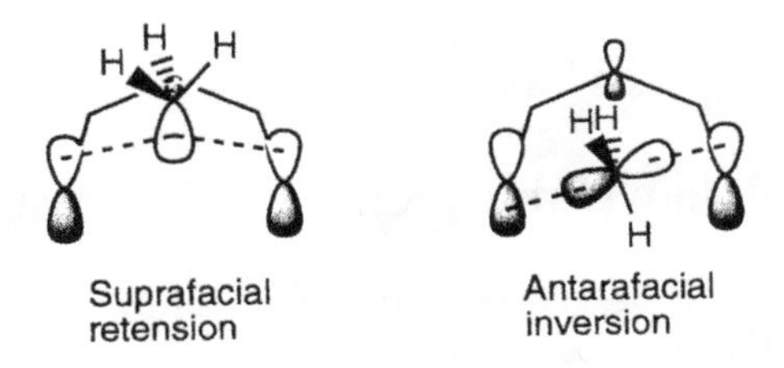

Figure 16.26 FMO interactions for allowed modes of the [1,5]-methyl shift in 1,3-hexadiene.

Table 16.1 Woodward–Hoffmann allowed reactions.

Reaction type	Number of electrons	Thermally allowed	Photochemically allowed
Ring-closure	$4n$	Conrotatory	Disrotatory
	$4n + 2$	Disrotatory	Conrotatory
Cycloadditions	$4n$	Supra–antara *or* antara–supra	Supra–supra *or* antara–antara
	$4n + 2$	Supra–supra *or* antara–antara	Supra–antara *or* antara–supra
Migrations	$4n$	Antara–retention *or* supra–inversion	Supra–retention *or* antara–inversion
	$4n + 2$	Supra–retention *or* antara–inversion	Antara–retention *or* supra–inversion

The Woodward–Hoffmann allowed reactions can be classified according to how many electrons are involved and whether the reaction occurs thermally or photochemically, as shown in Table 16.1.

The state correlation diagrams give an indication of the minimum theoretical level necessary for describing a reaction. For allowed reactions, the reactant configuration smoothly transforms into the product configuration by a continuous change of the orbitals, and they are consequently reasonably described by a single-determinant wave function along the whole reaction path. Forbidden reactions, on the other hand, necessarily involve at least two configurations since there is no continuous orbital transformation that connects the reactant and product ground states. Such reactions therefore require MCSCF-type wave functions for a qualitative correct description.

While the state correlation diagram for the 2s + 2s reaction (Figure 16.13) indicates that the photochemical reaction should be allowed (and cyclobutanes are indeed observed as one of the products from such reactions), the implication that the product ends up in an excited state is not correct. Although the reaction starts out on the excited surface, it will at some point along the reaction path return to the lowest energy surface, and the product is formed in its ground state. The transition from the upper to the lower energy surface will normally occur at a geometry where the two surfaces "touch" each other, that is they have the same energy, and this is known as a *conical intersection*.[35] Achieving the proper geometry for a transition between the two surfaces is often the dynamical bottleneck, and a conical intersection may be considered the equivalent of a TS for a photochemical reaction. As conical intersections involve two energy surfaces, MCSCF-based methods are required and non-adiabatic coupling elements (Section 3.1) are important. Locating a geometry corresponding to a conical intersection for a multidimensional system may be done using constrained optimization techniques (Section 13.5).

The product of a pericyclic reaction will in some cases itself be subject to a further pericyclic rearrangement. Such cascade reactions are often synthetically useful, as they may form complicated products in a single step. Depending on the exact system, two pericyclic reactions may occur with an intermediate or in a single kinetic step, but with a very asynchronous bond breaking/formations. M. T. Reetz has coined the name *dyotropic* reaction for two sigmatropic shifts occurring in tandem,[36] while the term *bispericyclic* has been used in the more general case.[37,38]

16.6 The Bell–Evans–Polanyi Principle/Hammond Postulate/Marcus Theory

The simpler the idea, the more names, could be the theme of this section.[39] The overriding idea is simple: for similar reactions, the more exothermic (endothermic) reaction will have the lower (higher)

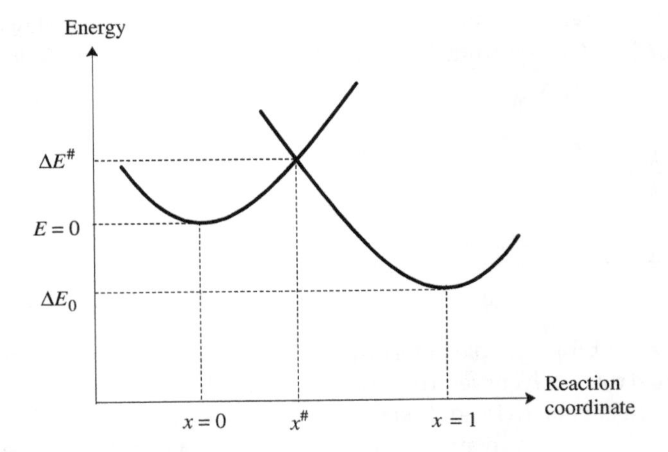

Figure 16.27 Transition structure as the intersection of two parabolas.

activation energy. This was formulated independently by *Bell*, *Evans* and *Polanyi* (BEP) in the 1930s, and is commonly known as the *BEP principle*.[40,41] The *Hammond postulate* relates the position of the transition structure to the exothermicity: for similar reactions, the more exothermic (endothermic) reaction will have the earlier (later) TS.[42] Compared with FMO theory, which tries to estimate relative activation energies from the *reactant* properties, the BEP principle tries to estimate relative activation energies from *product* properties (reaction energies). The above qualitative statements have been put on a more quantitative footing by the *Marcus equation*. This equation was originally derived for electron transfer reactions,[43] but it has since been shown that the same equation can be derived from a number of different assumptions, three of which will be illustrated below.

Let us assume a reaction coordinate x running from 0 (reactant) to 1 (product). The energy of the reactant as a function of x is taken as a simple parabola with a "force constant" of a. The energy of the product is also taken as a parabola with the same force constant, but offset by the reaction energy ΔE_0. The position of the TS ($x^{\neq}$) is taken as the point where the two parabolas intersect, as shown in Figure 16.27.

The TS position is calculated by equating the two energy expressions:

$$
\begin{aligned}
E_{\text{reactant}} &= a\,(x)^2\,; \quad E_{\text{product}} = a\,(x-1)^2 + \Delta E_0 \\
a(x^{\neq})^2 &= a(x^{\neq}-1)^2 + \Delta E_0 \\
x^{\neq} &= \frac{1}{2} + \frac{\Delta E_0}{2a}
\end{aligned}
\tag{16.20}
$$

For a thermoneutral reaction ($\Delta E_0 = 0$) the TS is exactly halfway between the reactant and product (as expected), while it becomes earlier and earlier as the reaction becomes more and more exothermic (ΔE_0 negative). The activation energy is given by

$$
\begin{aligned}
\Delta E^{\neq} &= E(x^{\neq}) = a\left(\frac{1}{2} + \frac{\Delta E_0}{2a}\right)^2 \\
\Delta E^{\neq} &= \frac{a}{4} + \frac{\Delta E_0}{2} + \frac{\Delta E_0^2}{4a}
\end{aligned}
\tag{16.21}
$$

Let us define the activation energy for a (possible hypothetical) thermoneutral reaction as the *intrinsic activation energy*, $\Delta E_0^{\neq}$.[44] As seen from Equation (16.21), $a = 4\Delta E_0^{\neq}$. The TS position and activation energy expressed in terms of $\Delta E_0^{\neq}$ are given in

$$x^{\neq} = \frac{1}{2} + \frac{\Delta E_0}{8\Delta E_0^{\neq}} \tag{16.22}$$

$$\Delta E^{\neq} = \Delta E_0^{\neq} + \frac{\Delta E_0}{2} + \frac{\Delta E_0^2}{16\Delta E_0^{\neq}} \tag{16.23}$$

Equation (16.23) is, except for a couple of terms, related to solvent reorganization, the Marcus equation. It should be noted that such curve-crossing models have been used in connection with VB methods to rationalize chemical reactivity and selectivity in a more general sense.[45,46]

The central idea in the Marcus treatment is that the activation energy can be decomposed into a component characteristic of the reaction type, the intrinsic activation energy and a correction due to the reaction energy, being different from zero. Similar reactions should have similar intrinsic activation energies, and the Marcus equation obeys both the BEP principle and the Hammond postulate. Except for very exo- or endothermic reactions (or a very small $\Delta E_0^{\neq}$), the last term in the Marcus equation is small, and it is seen that roughly half the reaction energy enters the activation energy. Note, however, that the activation energy is a parabolic function of the reaction energy. Thus for sufficiently exothermic reactions the equation predicts that the activation energy should *increase* as the reaction becomes *more* exothermic. The turnover occurs when $\Delta E_0 = -4\Delta E_0^{\neq}$. Much research has gone into proving such an "inverted" region, but experiments with very exothermic reactions are difficult to perform.[47]

An alternative way of deriving the Marcus equation is again to assume a reaction coordinate running from 0 to 1. The intrinsic activation energy is taken as a parabola centered at $x = {}^1\!/_2$. The reaction energy is taken as progressing linearly along the reaction coordinate. Adding these two contributions and evaluating the position of the TS and the activation energy in terms of ΔE_0 and $\Delta E_0^{\neq}$ again leads to the Marcus equation.

Actually, the assumptions can be made even more general. The energy as a function of the reaction coordinate can always be decomposed into an "intrinsic" term, which is symmetric with respect to $x = {}^1\!/_2$, and a "thermodynamic" contribution, which is antisymmetric. Denoting these two energy functions h_2 and h_1, it can be shown that the Marcus equation can be derived from the "square" condition, $h_2 = h_1{}^2$.[48] The intrinsic and thermodynamic parts do not have to be parabolas and linear functions, as shown in Figure 16.28, they can be *any* type of function. As long as the intrinsic part is the square of the thermodynamic part, the Marcus equation is recovered.

The idea can be taken one step further. The h_2 function can always be expanded in a power series of even powers of h_1, that is $h_2 = c_2 h_1^2 + c_4 h_1^4 + \ldots$. The exact values of the c-coefficients only influence the appearance of the last term in the resulting Marcus-like equation (Equation (16.23)). As already mentioned, this is usually a small correction anyway. For reactions where the reaction energy is less than or similar to the activation energy, there is thus a quite general theoretical background for the following statement: *For similar reactions, the difference in activation energy is roughly half the difference in reaction energy.* The trouble here is the word "similar". How similar should reactions be in order for the intrinsic activation energy to be constant? And how do we calculate or estimate the intrinsic activation energy? We will return to the latter question shortly.

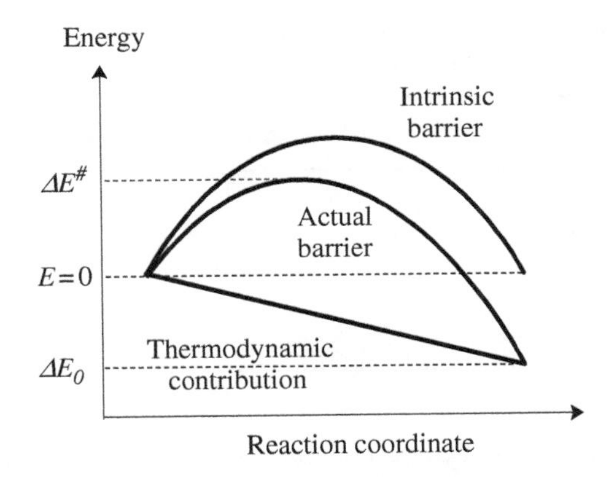

Figure 16.28 Decomposition of a reaction barrier into a parabola and a linear term.

The Marcus equation provides a nice conceptual tool for understanding trends in reactivity.[49] Consider, for example, the degenerate Cope rearrangement of 1,5-hexadiene and the ring-opening of Dewar benzene (bicyclo-[2,2,0]hexa-2,5-diene) to benzene (see Figure 16.29).

The experimentally observed activation energies are 142 and 96 kJ/mol, respectively.[50,51] The Cope reaction is an example of a Woodward–Hoffmann allowed reaction ([3,3]-sigmatropic shift) while the ring-opening of Dewar benzene is a Woodward–Hoffmann forbidden reaction (the cyclobutene ring-opening must necessarily be disrotatory, otherwise the benzene product ends up with a *trans* double bond). Why does a forbidden reaction have a lower activation energy than an allowed reaction? This is readily explained by the Marcus equation. The Cope reaction is thermoneutral (reactant and product are identical) and the activation energy is purely intrinsic, while the ring-opening is exothermic by 297 kJ/mol, and therefore has an intrinsic barrier of 218 kJ/mol. The "forbidden" reaction occurs only because it has a huge driving force in terms of a much more stable product, while the allowed reaction occurs even without a net energy gain.

The goal of understanding chemical reactivity is to be able to predict how the activation energy depends on properties of the reactant and product. Decomposing the activation energy into two terms, an intrinsic and a thermodynamic contribution, does not solve the problem. The reaction energy is relatively easy to obtain, from experiments, various theoretical methods or estimates based on additivity. However, how does one estimate the intrinsic activation energy? It is purely a theoretical concept – the activation energy for a thermoneutral reaction. But most reactions are *not* thermoneutral and there is no way of measuring such an intrinsic activation energy. For a series of "closely

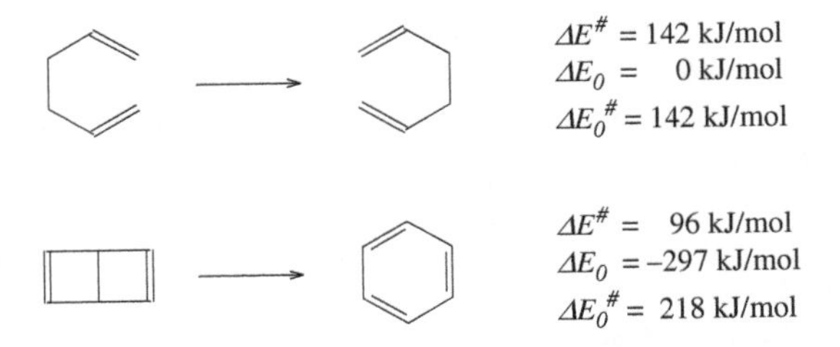

Figure 16.29 The Cope rearrangement and Dewar benzene ring-opening reaction.

Table 16.2 Comparing experimental activation barriers (kJ/mol) to those calculated by the Marcus equation for the reaction $OH^- + CH_3X$.

	$\Delta G^\ddagger$ (identity)	ΔG_0 (exp.)	$\Delta G^\ddagger$ (exp.)	$\Delta G^\ddagger$ (Marcus)
OH^-	170			
F^-	133	−94	109	108
Cl^-	111	−92	103	98
Br^-	99	−98	95	90
I^-	97	−89	97	93

related" reactions it may be assumed to be constant, but the question then becomes: how closely related should reactions be?

Alternatively, it may be assumed that the intrinsic component can be taken as an average of the two corresponding *identity* reactions. Consider, for example, the S_N2 reaction of OH^- with CH_3Cl. The two identity reactions are $OH^- + CH_3OH$ and $Cl^- + CH_3Cl$. These two reactions are thermoneutral and their activation energies, which are purely intrinsic, can in principle be measured by isotopic substitution (e.g. $^{35}Cl^- + CH_3{}^{37}Cl \rightarrow CH_3{}^{35}Cl + {}^{37}Cl^-$). From the reaction energy for the $OH^- + CH_3Cl$ reaction and the assumption that the intrinsic barrier is the average of the two identity reactions, the activation energy can be calculated. An example of the accuracy of this procedure for the series of S_N2 reactions $OH^- + CH_3X$ is given in Table 16.2.

Again this averaging procedure can only be expected to work when the reactions are sufficiently "similar", which is difficult to quantify *a priori*. The Marcus equation is therefore more a conceptual tool for explaining trends than for deriving quantitative result.

16.7 More O'Ferrall–Jencks Diagrams

The BEP/Hammond/Marcus treatment only considers changes due to energy differences between the reactant and product, that is changes in the TS position *along* the reaction coordinate. It is often useful also to include changes that may occur in a direction *perpendicular* to the reaction coordinate. Such two-dimensional diagrams are associated with the names of *More O'Ferrall* and *Jencks* (MOJ diagrams).[44,52–54]

Consider, for example, the Cope rearrangement of 1,5-hexadiene. Since the reaction is degenerate the TS will have D_{2h} symmetry (the lowest energy TS has a conformation resembling a chair-like cyclohexane). It is, however, not clear how strong the forming and breaking C—C bonds are at the TS. If they both are essentially full C—C bonds, the reaction may be described as bond formation followed by bond breaking. The TS therefore has a 1,4-biradical character, as illustrated by path **B** in Figure 16.30. Alternatively, the C—C bonds may be very weak at the TS, corresponding to a situation where bond breaking occurs before bond formation, and the TS can be described as two weakly interacting allyl radicals (path **C**). The intermediate situation, where both bonds are roughly half formed/broken can be described as having a delocalized structure similar to benzene, that is an "aromatic"-type TS (path **A**).

In such MOJ diagrams the x- and y-coordinates are normally taken to be bond orders ρ (Section 10.1) or $(1 - \rho)$ for the breaking and forming bonds, such that the coordinates run from 0 to 1. A third axis corresponding to the energy is implied, but rarely drawn.

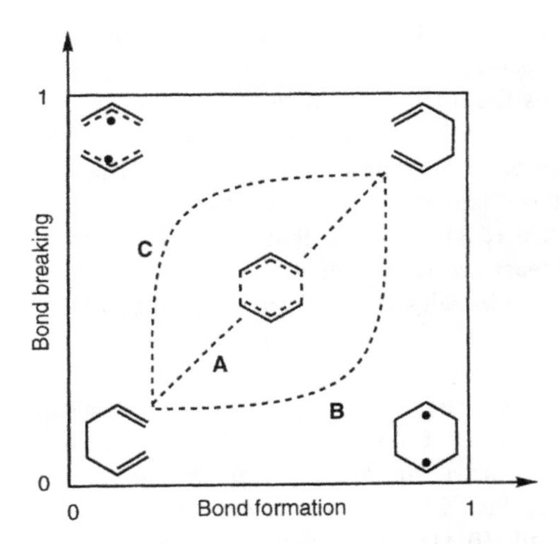

Figure 16.30 MOJ diagram for the Cope rearrangement of 1,5-hexadiene.

At the TS, the energy along the reaction path is a maximum, while it is a minimum in the perpendicular direction(s). A one-dimensional cut through the (0, 0) and (1, 1) corners for path **A** in Figure 16.30 thus corresponds to Figure 16.28. A similar cut through the (0, 1) and (1, 0) corners will display a normal (as opposed to inverted) parabolic behavior, with the TS being at the minimum on the curve. The whole energy surface corresponding to Figure 16.29 will have the qualitative appearance as shown in Figure 16.31.

There is good evidence that the Cope reaction in the parent 1,5-hexadiene has an "aromatic"-type TS, corresponding to path **A** in Figure 16.30, that is a "central" or "diagonal" reaction path. The importance of MOJ diagrams is that they allow a qualitative prediction of changes in the TS structure for a series of similar reactions. The addition of substituents that stabilize the product relative to the

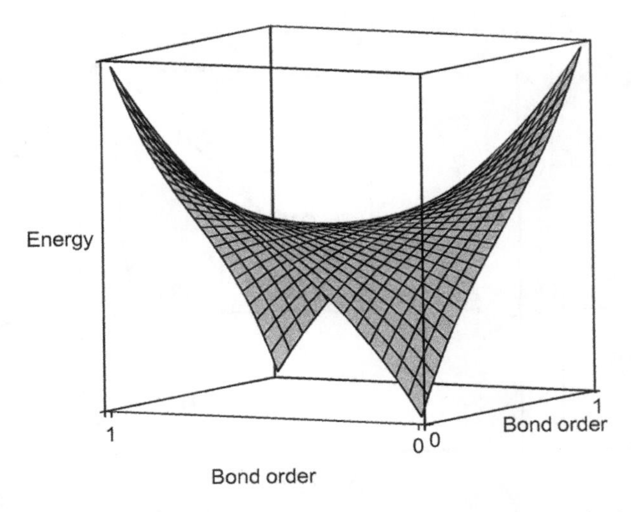

Figure 16.31 MOJ diagram corresponding to Figure 16.30 with the energy as the vertical axis.

reactant corresponds to a lowering of the (1, 1) corner, thereby moving the TS closer to the (0, 0) corner, that is towards the reactant. The one-dimensional BEP/Hammond/Marcus treatment thus corresponds to changes along the (0, 0)–(1, 1) diagonal.

Substituents that do not change the overall reaction energy may still have an influence on the TS geometry. Consider, for example, 2,5-diphenyl-1,5-hexadiene. The reaction is still thermoneutral but the phenyl groups will preferentially stabilize the 1,4-biradical structure, that is lower the energy of the (1, 0) corner. From Figure 16.31 it is clear that this will lead to a TS that is shifted toward this corner, that is moving the reaction from path **A** towards **B** in Figure 16.30. Similarly, substituents that preferentially stabilize the bis-allyl radical structure (such as 1,4-diphenyl-1,5-hexadiene) will perturb the reaction toward path **C**, since the (0, 1) corner is lowered in energy relative to the other corners.

From such MOJ diagrams it can be inferred that changes in the system that alter the relative energy *along* the reaction diagonal (lower-left to upper-right) imply changes in the TS in the *opposite* direction. Changes that alter the relative energy *perpendicular* to the reaction diagonal (upper-left to lower-right) imply changes in the TS in the *same* direction as the perturbation.

The structures in the (1, 0) and (0, 1) corners are not necessarily stable species; they may correspond to hypothetical structures. In the Cope rearrangement, it appears that the reaction only involves a single TS, independent of the number and nature of substituents. The reaction path may change from **B** → **A** → **C** depending on the system, but there are no intermediates along the reaction coordinate.

In other cases, one or both of the perpendicular corners may correspond to a minimum on the potential energy surface, and the reaction mechanism can change from being a one-step reaction to a two-step one. An example of this would be elimination reactions (see Figure 16.32). The x-axis in this case corresponds to the breaking bond between carbon and hydrogen, while the y-axis is the breaking bond between the other carbon and the leaving group.

An E2-type reaction has simultaneous breaking of the C—H and C—L bonds while forming the B—H bond, and corresponds to the diagonal path **A** in Figure 16.32. Path **C** involves initial loss of the leaving group to form a carbocation (upper-left corner), followed by loss of H^+ (which is picked up by the base), that is this corresponds to an E1-type mechanism involving two TSs and an intermediate.

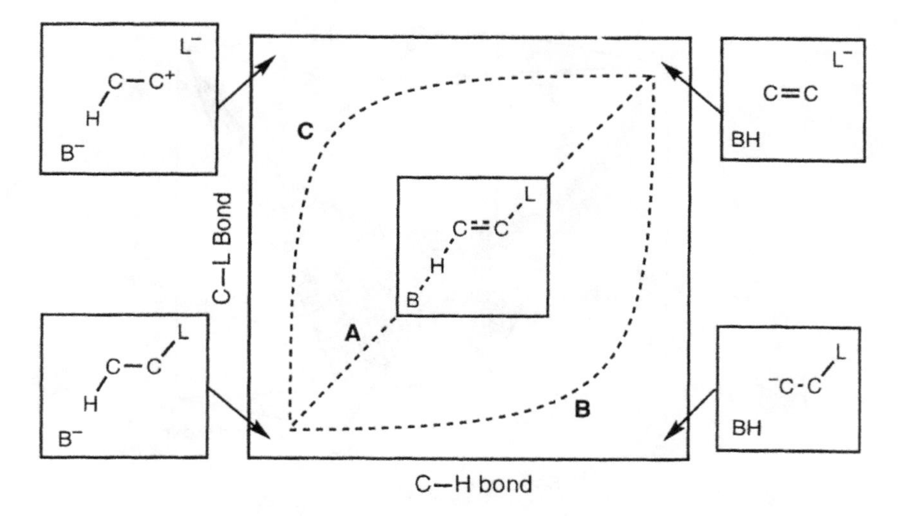

Figure 16.32 MOJ diagram for elimination reactions.

Path **B**, on the other hand, involves formation of a carbanion, followed by elimination of the leaving group in a second step, that is an $E1_{cb}$ mechanism. Substituents that stabilize the carbocation thus shift the reaction from an E2- to an E1-type mechanism, while anionic stabilizing substituents will shift the reaction towards an $E1_{cb}$ path.

In principle MOJ diagrams can be extended to more dimensions, for example by also including the B–H bond order in the above elimination reaction, but this is rarely done, not least because of the problems of illustrating more than two dimensions.

References

1 I. Fleming, *Frontier Orbitals and Organic Chemical Reactions* (John Wiley & Sons, 1976).
2 A. Devaquet, *Molecular Physics* **18** (2), 233 (1970).
3 A. Alexakis, C. Chuit, M. Commerconbourgain, J. P. Foulon, N. Jabri, P. Mangeney and J. F. Normant, *Pure and Applied Chemistry* **56** (1), 91–98 (1984).
4 G. Marino, in *Advances in Heterocyclic Chemistry*, edited by A. R. Katritzky and A. J. Boulton (Academic Press, 1971), Vol. 13, pp. 235–314.
5 J. Sauer, *Angewandte Chemie – International Edition* **6** (1), 16 (1967).
6 P. Geerlings, F. De Proft and W. Langenaeker, *Chemical Reviews* **103** (5), 1793–1873 (2003).
7 R. G. Parr and W. T. Yang, *Journal of the American Chemical Society* **106** (14), 4049–4050 (1984).
8 Y. Li and J. N. S. Evans, *Journal of the American Chemical Society* **117** (29), 7756–7759 (1995).
9 M. W. Schmidt, E. A. Hull and T. L. Windus, *Journal of Physical Chemistry A* **119** (41), 10408–10427 (2015).
10 F. De Proft, S. B. Liu and R. G. Parr, *Journal of Chemical Physics* **107** (8), 3000–3006 (1997).
11 R. S. Mulliken, *Journal of Chemical Physics* **2** (11), 782 (1934).
12 K. T. Giju, F. De Proft and P. Geerlings, *Journal of Physical Chemistry A* **109** (12), 2925–2936 (2005).
13 R. G. Parr, L. Von Szentpaly and S. B. Liu, *Journal of the American Chemical Society* **121** (9), 1922–1924 (1999).
14 R. G. Pearson, *Journal of the American Chemical Society* **85** (22), 3533 (1963).
15 T. A. Albright, J. K. Burdett and M.-H. Whangbo, *Orbital Interactions in Chemistry* (John Wiley & Sons, 1985).
16 A. Rauk, *Orbital Interaction Theory of Organic Chemistry* (John Wiley & Sons, 1994).
17 M. J. S. Phipps, T. Fox, C. S. Tautermann and C.-K. Skylaris, *Chemical Society Reviews* **44** (10), 3177–3211 (2015).
18 T. Ziegler and A. Rauk, *Theoretica Chimica Acta* **46** (1), 1–10 (1977).
19 F. M. Bickelhaupt and E. J. Baerends, in *Reviews in Computational Chemistry*, edited by K. B. Lipkowitz and D. B. Boyd (John Wiley & Sons, 2000), Vol. 15, pp. 1–86.
20 F. M. Bickelhaupt, *Journal of Computational Chemistry* **20** (1), 114–128 (1999).
21 K. Morokuma, *Accounts of Chemical Research* **10** (8), 294–300 (1977).
22 E. D. Glendening and A. Streitwieser, *Journal of Chemical Physics* **100** (4), 2900–2909 (1994).
23 W. Chen and M. S. Gordon, *Journal of Physical Chemistry* **100** (34), 14316–14328 (1996).
24 R. Z. Khaliullin, E. A. Cobar, R. C. Lochan, A. T. Bell and M. Head-Gordon, *Journal of Physical Chemistry A* **111** (36), 8753–8765 (2007).
25 Y. Mo, P. Bao and J. Gao, *Physical Chemistry Chemical Physics* **13** (15), 6760–6775 (2011).
26 Q. Wu, *Journal of Chemical Physics* **140** (24), 244109 (2014).
27 P. R. Horn and M. Head-Gordon, *Journal of Chemical Physics* **143** (11), 114111 (2015).
28 B. Jeziorski, R. Moszynski and K. Szalewicz, *Chemical Reviews* **94** (7), 1887–1930 (1994).

29 R. B. Woodward and R. Hoffmann, *The Conservation of Orbital Symmetry* (Academic Press, 1970).
30 R. B. Woodward and R. Hoffmann, *Angewandte Chemie – International Edition* **8** (11), 781 (1969).
31 H. Hirao, *Wiley Interdisciplinary Reviews – Computational Molecular Science* **1** (3), 337–349 (2011).
32 N. Turro, *Modern Molecular Photochemistry* (The Benjamin/Cummings Publishing Co., 1978).
33 J. G. Martin and R. K. Hill, *Chemical Reviews* **61** (6), 537–562 (1961).
34 W. R. Roth, J. Konig and K. Stein, *Chemische Berichte – Recueil* **103** (2), 426 (1970).
35 F. Bernardi, M. Olivucci and M. A. Robb, *Chemical Society Reviews* **25** (5), 321 (1996).
36 M. T. Reetz, *Angewandte Chemie – International Edition* **11** (2), 129 (1972).
37 J. Limanto, K. S. Khuong, K. N. Houk and M. L. Snapper, *Journal of the American Chemical Society* **125** (52), 16310–16321 (2003).
38 D. H. Nouri and D. J. Tantillo, *Journal of Organic Chemistry* **71** (10), 3686–3695 (2006).
39 W. P. Jencks, *Chemical Reviews* **85** (6), 511–527 (1985).
40 R. P. Bell, *Proceedings of the Royal Society of London A: Mathematical, Physical and Engineering Sciences* **154** (882), 414–429 (1936).
41 M. G. Evans and M. Polanyi, *Transactions of the Faraday Society* **32** (2), 1333–1359 (1936).
42 G. S. Hammond, *Journal of the American Chemical Society* **77** (2), 334–338 (1955).
43 R. A. Marcus, *Journal of Physical Chemistry* **72** (3), 891 (1968).
44 C. F. Bernasconi, *Accounts of Chemical Research* **20** (8), 301–308 (1987).
45 S. Shaik and P. C. Hiberty, in *Reviews in Computational Chemistry*, edited by K. B. Lipkowitz, R. Larter and T. R. Cundari (John Wiley & Sons, 2004), Vol. 20, pp. 1–100.
46 A. Pross, *Theoretical and Physical Principles of Organic Reactivity* (John Wiley & Sons, 1995).
47 D. M. Guldi and K. D. Asmus, *Journal of the American Chemical Society* **119** (24), 5744–5745 (1997).
48 J. Donnella and J. R. Murdoch, *Journal of the American Chemical Society* **106** (17), 4724–4735 (1984).
49 V. Aviyente, H. Y. Yoo and K. N. Houk, *Journal of Organic Chemistry* **62** (18), 6121–6128 (1997).
50 W. V. E. Doering, V. G. Toscano and G. H. Beasley, *Tetrahedron* **27** (22), 5299 (1971).
51 M. J. Goldstein and R. S. Leight, *Journal of the American Chemical Society* **99** (24), 8112–8114 (1977).
52 R. A. More OFerrall, *Journal of the Chemical Society B – Physical Organic* (2), 274 (1970).
53 W. P. Jencks, *Chemical Reviews* **72** (6), 705 (1972).
54 S. S. Shaik, H. B. Schlegel and S. Wolfe, *Theoretical Aspects of Physical Organic Chemistry. The SN2 Mechanism* (John Wiley & Sons, 1992).

17

Mathematical Methods

Computational chemistry relies on computers to solve the complicated mathematical equations describing the physics behind the models. The language for deriving and describing these models is mathematics, and this chapter summarizes some of the commonly used mathematical concepts and techniques used in computational chemistry.

17.1 Numbers, Vectors, Matrices and Tensors

Some physical quantities, such as the total molecular mass or charge, can be specified by a single number, referring to the magnitude of the quantity in a given set of units. The mathematical term for such a number is a *scalar*. Other quantities require a *set* of numbers, such as, for example, three scalars for specifying the position of a particle in a coordinate system. A coordinate system is defined by the *origin* ("zero point"), the *directions* of the coordinate axes and the *units* along the axes. Two common examples are Cartesian $\{x, y, z\}$ and spherical polar $\{r, \theta, \varphi\}$ systems (Figure 17.1). The same point in space can be specified either by the Cartesian x, y, z coordinates (0.500, 0.866, 1.000) or by the spherical polar r, θ, φ coordinates (2, 30, 60) with angles measured in degrees.

The direction from the origin to the point specified by the three coordinates represents a *vector*, having a length and a direction. Another example of a 3-vector is the velocity of a particle (v_x, v_y, v_z), or alternatively $(\partial x/\partial t, \partial y/\partial t, \partial z/\partial t)$. For a system with N particles, the positions or velocities of all

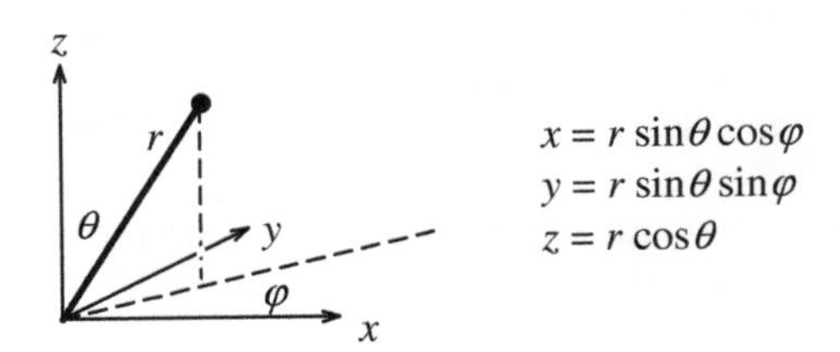

Figure 17.1 Cartesian and spherical polar coordinate systems.

Introduction to Computational Chemistry, Third Edition. Frank Jensen.

Companion Website: http://www.wiley.com/go/jensen/computationalchemistry3

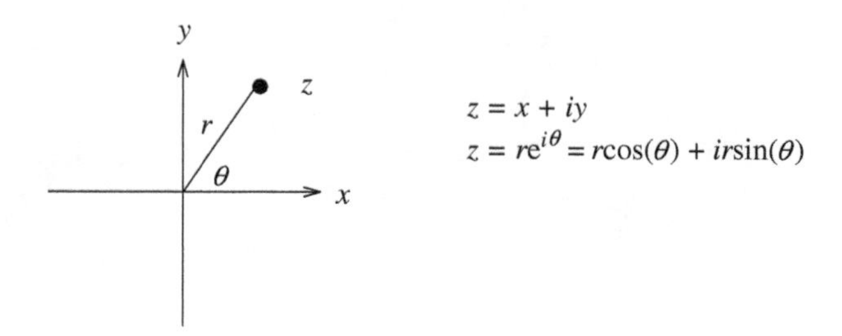

Figure 17.2 An imaginary number interpreted as a point in a two-dimensional coordinate system.

particles can be specified by a vector of length 3*N*, that is $(x_1, y_1, z_1, x_2, y_2, \ldots, y_N, z_N)$ or $(v_{x1}, v_{y1}, v_{z1}, v_{x2}, v_{y2}, \ldots, v_{yN}, v_{zN})$:

$$\{v_{x1} \quad v_{y1} \quad v_{z1} \quad v_{x2} \quad \cdots\} = \left\{\frac{\partial x_1}{\partial t} \quad \frac{\partial y_1}{\partial t} \quad \frac{\partial z_1}{\partial t} \quad \frac{\partial x_2}{\partial t} \quad \cdots\right\} \tag{17.1}$$

The notation for such vectors is often generalized to simply $\mathbf{x} = (x_1, x_2, x_3, \ldots, x_{N-1}, x_N)$, where *N* now refers to the total number of elements, that is equal to 3*N* in the above notation.

A *complex number z* can be interpreted as a 2-vector in an *xy*-coordinate system, $z = x + iy$, where i is the symbol for $\sqrt{-1}$ and x and y are real numbers. Here x and y are referred to as the real and imaginary parts of z. Alternatively, the complex number can be associated with polar coordinates, that is the distance r from the origin and the angle θ relative to the real axis, as shown in Figure 17.2.

The *complex conjugate* of a complex number z is denoted by z^* and is obtained by changing the sign of the imaginary part, that is $z^* = x - iy$ or equivalently $z^* = re^{-i\theta}$.

The concept of complex numbers can be generalized to hypercomplex numbers, with the next level being a 4-vector, called a *quarternion*, that is $q = q_0 + iq_1 + jq_2 + kq_3$, with q_0, q_1, q_2, q_3 being real numbers. A quarternion has a real part, q_0, and the three imaginary components q_1, q_2, q_3. The latter can be considered as a vector in a three-dimensional space, where each of the unit vectors has the property $i^2 = j^2 = k^2 = -1$. As one moves up in dimensions in this generalization, common mathematical laws gradually get lost. Quarternions, for example, do not obey the commutative law ($q_a q_b \neq q_b q_a$), while octonions (8-vectors) in addition do not obey the associative law ($(q_a q_b)q_c \neq q_a(q_b q_c)$). Quarternions are encountered, for example, in relativistic (4-component) quantum mechanics, and they also form a more natural basis for parameterizing the rotation of a three-dimensional structure, rather than the traditional three Euler angles.[1] The latter involves trigonometric functions that are both computationally expensive to evaluate and display singularities. Furthermore, the quarternion formulation treats all the coordinate axes as equivalent, while the Euler parameterization makes the z-axis a special direction.

Vectors can also arise from mathematical operations on functions of coordinates, such as, for example, the gradient being the first derivative of an energy function:

$$\{g_1 \quad g_2 \quad g_3 \quad g_4 \quad \cdots\} = \left\{\frac{\partial E}{\partial x_1} \quad \frac{\partial E}{\partial x_2} \quad \frac{\partial E}{\partial x_3} \quad \frac{\partial E}{\partial x_4} \quad \cdots\right\} \tag{17.2}$$

In such cases, it is implicit that the first gradient element is the derivative with respect to the first variable, etc.

The second derivative of the energy is an ordered two-dimensional set of numbers, called a *matrix*:

$$\begin{pmatrix} H_{11} & H_{12} & \cdots \\ H_{21} & H_{22} & \cdots \\ \vdots & \vdots & \ddots \end{pmatrix} = \begin{pmatrix} \dfrac{\partial^2 E}{\partial x_1^2} & \dfrac{\partial^2 E}{\partial x_1 \partial x_2} & \cdots \\ \dfrac{\partial^2 E}{\partial x_2 \partial x_1} & \dfrac{\partial^2 E}{\partial x_2^2} & \cdots \\ \vdots & \vdots & \ddots \end{pmatrix} \tag{17.3}$$

The third derivative of the energy is an ordered three-dimensional set of numbers, called a *tensor*, which can be arranged in a cube. Corresponding higher-order derivatives can be thought of as ordered sets of numbers in "hypercubes", called d-order tensors for a hypercube of dimension d.

Ordered sets of numbers may collectively be called tensors, with a matrix being a second-order tensor and a vector a first-order tensor. Since tensors of order higher than two are relatively rare, the terms vector and matrix are more commonly used. Sometimes it is also convenient to consider a vector as a $1 \times N$ or $N \times 1$ matrix and a scalar as a 1×1 matrix.

The conversion between a $1 \times N$ and an $N \times 1$ vector, or from an $M \times N$ matrix to an $N \times M$ matrix, is done by *transposition*, indicated by a superscript ^{t}. Transposition simply interchanges the ijth element with the jith element:

$$\begin{aligned} \begin{pmatrix} a & b \\ c & d \end{pmatrix}^t &= \begin{pmatrix} a & c \\ b & d \end{pmatrix} \\ \begin{pmatrix} a & b & c \\ d & e & f \end{pmatrix}^t &= \begin{pmatrix} a & d \\ b & e \\ c & f \end{pmatrix} \end{aligned} \tag{17.4}$$

If the matrix elements are complex, the *adjoint* matrix is defined by complex conjugation of the elements followed by transposition, and is denoted by a superscript †. *Hermitian* matrices are very common in quantum chemistry and are defined as being self-adjoint, that is $\mathbf{A} = \mathbf{A}^\dagger$. If all the matrix elements are real, the matrix is called *symmetric*, that is $\mathbf{A} = \mathbf{A}^t$.

The addition and subtraction of matrices, which now encompass vectors as well, is directly the addition and subtraction of the elements, analogous to the rules for scalars:

$$\begin{pmatrix} a_1 \\ a_2 \\ \vdots \end{pmatrix} + \begin{pmatrix} b_1 \\ b_2 \\ \vdots \end{pmatrix} = \begin{pmatrix} a_1 + b_1 \\ a_2 + b_2 \\ \vdots \end{pmatrix} \tag{17.5}$$

$$\begin{pmatrix} a_{11} & a_{12} & \cdots \\ a_{21} & a_{22} & \cdots \\ \vdots & \vdots & \ddots \end{pmatrix} + \begin{pmatrix} b_{11} & b_{12} & \cdots \\ b_{21} & b_{22} & \cdots \\ \vdots & \vdots & \ddots \end{pmatrix} = \begin{pmatrix} a_{11} + b_{11} & a_{12} + b_{12} & \cdots \\ a_{21} + b_{21} & a_{22} + b_{22} & \cdots \\ \vdots & \vdots & \ddots \end{pmatrix} \tag{17.6}$$

The multiplication of matrices, however, is somewhat different. In standard matritx multiplications, the ijth element in the product is formed by multiplying the elements of the ith row with the elements of the jth coloumn and adding all the terms. For the multiplication of two 2×2 matrices the result is given in

$$\begin{pmatrix} a_{11} & a_{12} \\ a_{21} & a_{22} \end{pmatrix} \begin{pmatrix} b_{11} & b_{12} \\ b_{21} & b_{22} \end{pmatrix} = \begin{pmatrix} a_{11}b_{11} + a_{12}b_{21} & a_{11}b_{12} + a_{12}b_{22} \\ a_{21}b_{11} + a_{22}b_{21} & a_{21}b_{12} + a_{22}b_{22} \end{pmatrix} \tag{17.7}$$

Note that this means that matrix multiplication is not necessarily commutative:

$$\begin{pmatrix} a_{11} & a_{12} \\ a_{21} & a_{22} \end{pmatrix} \begin{pmatrix} b_{11} & b_{12} \\ b_{21} & b_{22} \end{pmatrix} \neq \begin{pmatrix} b_{11} & b_{12} \\ b_{21} & b_{22} \end{pmatrix} \begin{pmatrix} a_{11} & a_{12} \\ a_{21} & a_{22} \end{pmatrix} \tag{17.8}$$

This is perhaps most easily seen by multiplying two rectangular matrices. The result of multiplying a 2×3 matrix with a 3×2 matrix is a 2×2 matrix:

$$\begin{pmatrix} a_{11} & a_{12} & a_{13} \\ a_{21} & a_{22} & a_{23} \end{pmatrix} \begin{pmatrix} b_{11} & b_{12} \\ b_{21} & b_{22} \\ b_{31} & b_{32} \end{pmatrix} = \begin{pmatrix} c_{11} & c_{12} \\ c_{21} & c_{22} \end{pmatrix} \tag{17.9}$$

while the result of multiplying a 3×2 matrix with 2×3 matrix is a 3×3 matrix:

$$\begin{pmatrix} b_{11} & b_{12} \\ b_{21} & b_{22} \\ b_{31} & b_{32} \end{pmatrix} \begin{pmatrix} a_{11} & a_{12} & a_{13} \\ a_{21} & a_{22} & a_{23} \end{pmatrix} = \begin{pmatrix} c_{11} & c_{12} & c_{13} \\ c_{21} & c_{22} & c_{23} \\ c_{31} & c_{32} & c_{33} \end{pmatrix} \tag{17.10}$$

Even for square matrices, however, the matrix product **AB** is not necessarily equal to **BA**.

In some cases the matrix elements are multiplied together element by element, which is called an *entry-wise* product, and is denoted with a "dot" between the two matrices:

$$\begin{pmatrix} a_{11} & a_{12} & \cdots \\ a_{21} & a_{22} & \cdots \\ \vdots & \vdots & \ddots \end{pmatrix} \cdot \begin{pmatrix} b_{11} & b_{12} & \cdots \\ b_{21} & b_{22} & \cdots \\ \vdots & \vdots & \ddots \end{pmatrix} = \begin{pmatrix} a_{11}b_{11} & a_{12}b_{12} & \cdots \\ a_{21}b_{21} & a_{22}b_{22} & \cdots \\ \vdots & \vdots & \ddots \end{pmatrix} \tag{17.11}$$

For vectors, which can be considered $1 \times N$ or $N \times 1$ matrices, the result of multiplying a $1 \times N$ matrix with an $N \times 1$ matrix is a 1×1 matrix, or a scalar. This is called an *inner* or *dot* product:

$$\mathbf{a}^t\mathbf{b} = \begin{pmatrix} a_1 & a_2 \end{pmatrix} \begin{pmatrix} b_1 \\ b_2 \end{pmatrix} = (a_1 b_1 + a_2 b_2) \tag{17.12}$$

The length (or *norm*) of a vector follows directly from the interpretation of a vector as a directional line from the origin to a point in space, and is defined as the square root of the dot product of the vector with itself. If the vector components are complex numbers the transposition is replaced by the adjoint instead:

$$|\mathbf{a}| = \sqrt{\mathbf{a}^t\mathbf{a}} = \sqrt{a_1^2 + a_2^2 + a_3^2 + \cdots} \tag{17.13}$$

The above norm is more specifically known as the Euclidian or 2-norm. The generalization to the p-norm is shown in

$$|\mathbf{a}|_p = \sqrt[p]{\sum_i |a_i|^p} \tag{17.14}$$

The 2-norm is by far the most common, but the 1-norm, which is the sum of absolute value of the a_i-elements, and the infinity-norm, which is simply the largest a_i-element, are also used frequently.

The "physical" interpretation of a dot product of two vectors is related to the angle between them; specifically for two vectors of unit length, the dot product is the cosine of the angle:

$$\mathbf{a}^{t}\mathbf{b} = |\mathbf{a}||\mathbf{b}| \cos \alpha \tag{17.15}$$

A dot product of +1 means that the two (unit) vectors are aligned, a value of −1 means that they are aligned, but pointing in opposite directions, while a dot product of 0 means that the two vectors are orthogonal.

The opposite of a dot product is multiplication of an $N \times 1$ matrix with a $1 \times N$ matrix to give an $N \times N$ matrix, and is called an *outer* product:

$$\mathbf{a}\mathbf{b}^{t} = \begin{pmatrix} a_1 \\ a_2 \end{pmatrix} \begin{pmatrix} b_1 & b_2 \end{pmatrix} = \begin{pmatrix} a_1 b_1 & a_1 b_2 \\ a_2 b_1 & a_2 b_2 \end{pmatrix} \tag{17.16}$$

The inner and outer products of two vectors produce a scalar and a matrix, respectively. Two 3-vectors may also be multiplied together to generate a new 3-vector, a procedure called a *vector* or *cross* product, with the result given by

$$\begin{Bmatrix} x_1 \\ y_1 \\ z_1 \end{Bmatrix} \times \begin{Bmatrix} x_2 \\ y_2 \\ z_2 \end{Bmatrix} = \begin{Bmatrix} y_1 z_2 - z_1 y_2 \\ z_1 x_2 - x_1 z_2 \\ x_1 y_2 - y_1 x_2 \end{Bmatrix} \tag{17.17}$$

The vector product gives a vector perpendicular to both of the original vectors with a length of $|\mathbf{a}|\mathbf{b}|\sin \alpha$ (compare with Equation (17.15)), and is therefore zero if the two original vectors are aligned. It follows trivially that $\mathbf{a} \times \mathbf{a} = 0$ for any vector $\mathbf{a}$ and that $\mathbf{a} \times \mathbf{b} = -\mathbf{b} \times \mathbf{a}$.

The generalization of the inner (dot) product from vectors to tensors is usually called *tensor contraction*, where the contraction decreases the tensor dimension by the number of contraction indices. Calculation of the Fock matrix elements in Equation (3.56) is an example of a contraction of the four-dimension tensor containing the two-electron integrals with the two-dimensional density matrix, giving a two-dimension Fock matrix. The generalization of the outer product from vectors to tensors is called a *tensor* or *dyadic product* (often denoted by ⊗), where the tensor product increases the dimension of the resulting tensor to the sum of the dimensions of the two tensors.

A matrix *determinant* is denoted $|\mathbf{A}|$ and is given explicitly for the 2×2 and 3×3 cases in the following equations:

$$|\mathbf{A}| = \begin{vmatrix} a_{11} & a_{12} \\ a_{21} & a_{22} \end{vmatrix} = a_{11}a_{22} - a_{12}a_{21} \tag{17.18}$$

$$|\mathbf{A}| = \begin{vmatrix} a_{11} & a_{12} & a_{13} \\ a_{21} & a_{22} & a_{23} \\ a_{31} & a_{32} & a_{33} \end{vmatrix} = \begin{matrix} a_{11}a_{22}a_{33} - a_{11}a_{23}a_{32} - a_{12}a_{21}a_{33} + \\ a_{12}a_{23}a_{31} + a_{13}a_{21}a_{32} - a_{13}a_{22}a_{31} \end{matrix} \tag{17.19}$$

The determinant of larger matrices is similarly given as a sum of $N!$ terms, each being the product of N elements. A convenient procedure for the evaluation consists of decomposing the determinant according to a row (or column), with each element being multiplied by a subdeterminant, formed by removing the elements of the corresponding rows and columns, and a factor $(-1)^{(i+j)}$:

$$\begin{vmatrix} a_{11} & a_{12} & a_{13} \\ a_{21} & a_{22} & a_{23} \\ a_{31} & a_{32} & a_{33} \end{vmatrix} = a_{11}\begin{vmatrix} a_{22} & a_{23} \\ a_{32} & a_{33} \end{vmatrix} - a_{12}\begin{vmatrix} a_{21} & a_{23} \\ a_{31} & a_{33} \end{vmatrix} + a_{13}\begin{vmatrix} a_{21} & a_{22} \\ a_{31} & a_{32} \end{vmatrix} \tag{17.20}$$

This procedure can be applied recursively until only 1×1 determinants remain. Only square matrices have determinants, and determinants have a number of important properties:

1. Interchanging two rows or columns in a matrix changes the sign of the determinant. This property is used for parameterizing wave functions in terms of Slater determinants, as the wave function antisymmetry is thereby automatically fulfilled (Section 3.2).
2. Adding a row (or a fraction thereof) to another row leaves the determinant unchanged, and similarly for columns. This allows, for example, representation of a wave function either in terms of canonical or localized molecular orbitals (Section 10.4).
3. If two rows or columns are identical except for a multiplicative constant, the determinant is zero. This is easily seen, since one of these rows/columns can be made into a zero vector by subtraction of the two, and expansion according to this zero row/column by Equation (17.20) will give zero. Such matrices may arise owing to *linear dependencies* of the rows or columns.

Division by matrices is done formally by multiplying with the *inverse* of a matrix, where the inverse is defined such that multiplication of a matrix with its inverse produces a unit matrix:

$$\mathbf{A}^{-1}\mathbf{A} = \mathbf{A}\mathbf{A}^{-1} = \mathbf{I} = \begin{pmatrix} 1 & 0 & 0 & \cdots \\ 0 & 1 & 0 & \cdots \\ 0 & 0 & 1 & \cdots \\ \vdots & \vdots & \vdots & \ddots \end{pmatrix} \tag{17.21}$$

The elements of a matrix inverse are given by the elements of the matrix itself and the inverse of the matrix determinant. Specifically, the ijth element in the $\mathbf{A}^{-1}$ matrix is given as the inverse of $|\mathbf{A}|$ times the determinant of the submatrix corresponding to removing the jth row and ith column, and a factor $(-1)^{(i+j)}$. Note that the transposition of the ijth element in the inverse matrix is formed from the submatrix corresponding to the jith element in the original matrix. For a 3×3 matrix, this is exemplified by the b_{1j} elements in

$$\mathbf{A} = \begin{pmatrix} a_{11} & a_{12} & a_{13} \\ a_{21} & a_{22} & a_{23} \\ a_{31} & a_{32} & a_{33} \end{pmatrix} \; ; \; \mathbf{A}^{-1} = |\mathbf{A}|^{-1}\begin{pmatrix} b_{11} & b_{12} & b_{13} \\ b_{21} & b_{22} & b_{23} \\ b_{31} & b_{32} & b_{33} \end{pmatrix}$$
$$b_{11} = \begin{vmatrix} a_{22} & a_{23} \\ a_{32} & a_{33} \end{vmatrix}, \quad b_{12} = -\begin{vmatrix} a_{12} & a_{13} \\ a_{32} & a_{33} \end{vmatrix}, \quad b_{13} = \begin{vmatrix} a_{12} & a_{13} \\ a_{22} & a_{23} \end{vmatrix} \tag{17.22}$$

It thus follows that only square matrices with determinants different from zero have an inverse matrix. Rectangular matrices can be defined to have a *generalized* inverse matrix. Such generalized inverse matrices and more complicated matrix algebra, such as calculating functions of matrices, are considered in the next section.

A square matrix can be considered as composed of row/column vectors. Matrices where each row/column vector has a unit length (Equation (17.13)) and is orthogonal to all other row/column vectors (Equation (17.15)) are called *unitary*, and have determinants equal to $e^{i\theta}$ (Figure 17.2 with

Table 17.1 Some special matrices, names and properties.

Name	Properties
Unit	$\mathbf{I}, \mathrm{I}_{ij} = \delta_{ij}$
Complex conjugate	$\mathbf{A}^*$, complex conjugate all elements
Transposed	$\mathbf{A}^t$, interchange elements ij and ji
Adjoint	$\mathbf{A}^\dagger$, interchange and complex conjugate elements ij and ji
Symmetric	$\mathbf{A}^t = \mathbf{A}$
Antisymmetric	$\mathbf{A}^t = -\mathbf{A}$
Hermitian	$\mathbf{A}^\dagger = \mathbf{A}$
Anti-Hermitian	$\mathbf{A}^\dagger = -\mathbf{A}$
Inverse	$\mathbf{A}^{-1}\mathbf{A} = \mathbf{A}\mathbf{A}^{-1} = \mathbf{I}$
Orthogonal	$\lvert\mathbf{A}\rvert = \pm 1$, real elements, $\mathbf{A}^{-1} = \mathbf{A}^t$
Unitary	$\lvert\mathbf{A}\rvert = e^{i\theta}$, complex elements, $\mathbf{A}^{-1} = \mathbf{A}^\dagger$

$r = 1$). A unitary matrix, where all the elements are real (i.e. not complex), is called *orthogonal* and has a determinant equal to +1 or −1. We will in general use the unitary notation, although in most cases the matrices are actually orthogonal. Orthogonal matrices have the appealing property that the inverse is simply the transposed matrix, $\mathbf{A}^{-1} = \mathbf{A}^t$; for unitary matrices the transposition must be accompanied by a complex conjugation also, that is $\mathbf{A}^{-1} = \mathbf{A}^\dagger$. The names and properties of some special matrices are shown in Table 17.1.

Matrices arise, for example, in solving a system of linear equations, where the formal solution can be obtained by multiplication with $\mathbf{A}^{-1}$ on both sides:

$$
\begin{aligned}
a_{11}x_1 + a_{12}x_2 + \cdots + a_{1n}x_n &= b_1 \\
a_{21}x_1 + a_{22}x_2 + \cdots + a_{2n}x_n &= b_2 \\
\vdots \quad \vdots \quad \vdots \quad \vdots \quad \vdots \quad \vdots\vdots \quad &\vdots = \vdots \\
a_{n1}x_1 + a_{n2}x_2 + \cdots + a_{nn}x_n &= b_n \\
\mathbf{A}\mathbf{x} = \mathbf{b} \quad \Leftrightarrow \quad \mathbf{x} &= \mathbf{A}^{-1}\mathbf{b}
\end{aligned}
\tag{17.23}
$$

It thus follows that a solution only exists if $\mathbf{A}^{-1}$ exists, that is if $\lvert\mathbf{A}\rvert$ is non-zero. In actual calculations, it is rare that matrix determinants are exactly zero. If $\lvert\mathbf{A}\rvert$ is very small, the solution vector $\mathbf{x}$ becomes sensitive to small details in the original $\mathbf{A}$ matrix. Such systems are called *ill-conditioned* and should be treated by *singular value decomposition*, as described in the next section.

For the special case of the right-hand side ($\mathbf{b}$-vector) in Equation (17.23) being zero, only the trivial $\mathbf{x} = \mathbf{0}$ solution exists *if* $\mathbf{A}^{-1}$ exists. A non-trivial solution is therefore only possible if $\mathbf{A}^{-1}$ does *not* exist, which is equivalent to the condition that $\lvert\mathbf{A}\rvert$ is zero. Linear dependence in the $\mathbf{A}$ matrix is thus a *condition* for a non-trivial solution and the resulting $\mathbf{x}$-vector is obtained as a parametric solution of one or more variables. These parameters can be fixed, for example, by requiring that the $\mathbf{x}$-vector(s) are normalized and mutually orthogonal.

17.2 Change of Coordinate System

In many cases it is possible to simplify a problem by choosing a particular coordinate system. It is therefore important to be able to describe how vectors and matrices change when switching from one coordinate system to another.

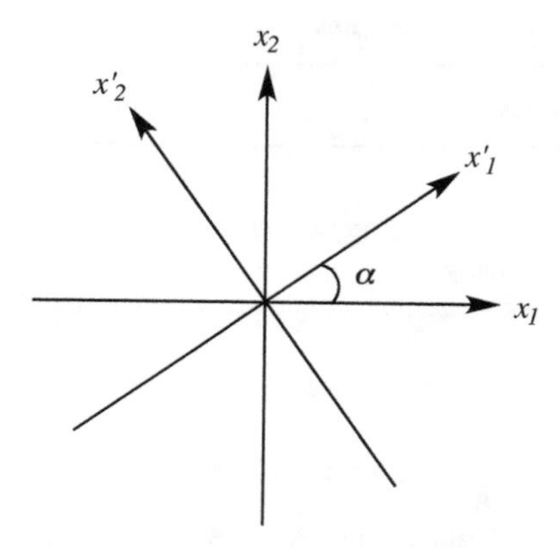

Figure 17.3 Rotation of a coordinate system.

Some coordinate transformations are *non-linear*, such as converting from a Cartesian to a spherical polar system. Here the r, θ, φ coordinates are related to the x, y, z coordinates by square root and trigonometric functions, as shown in Figure 17.1. Other coordinate transformations are *linear*, with the new coordinates given as linear combinations of the old ones. A linear transformation can be described as a *rotation* of the coordinate system (Figure 17.3).

For the 2 × 2 case, the new coordinates x'_1 and x'_2 are related to the original x_1 and x_2 coordinates by means of a 2 × 2 matrix containing cosines and sines of the rotational angle α:

$$\begin{pmatrix} x'_1 \\ x'_2 \end{pmatrix} = \begin{pmatrix} \cos\alpha & \sin\alpha \\ -\sin\alpha & \cos\alpha \end{pmatrix} \begin{pmatrix} x_1 \\ x_2 \end{pmatrix} \tag{17.24}$$

The rotation matrix is an orthogonal (unitary) matrix **U**, since the rows/columns are orthonormal. The significance of a unitary matrix is that it describes a rotation of the coordinate system without changing the length of the coordinate axes. An orthogonal matrix with a determinant of −1 describes a rotation of the coordinate system, followed by inverting the directions of the coordinate axis, that is an improper rotation in the language of point group symmetry.

The connection between the primed and unprimed coordinate systems is given by the unitary matrix in Equation (17.24), and can be written as

$$\mathbf{x}' = \mathbf{U}\mathbf{x} \tag{17.25}$$

The inverse operation $\mathbf{U}^{-1}$ corresponds to rotation with the angle $-\alpha$ and backtransforms the primed coordinates to the unprimed ones:

$$\begin{pmatrix} x_1 \\ x_2 \end{pmatrix} = \begin{pmatrix} \cos\alpha & -\sin\alpha \\ \sin\alpha & \cos\alpha \end{pmatrix} \begin{pmatrix} x'_1 \\ x'_2 \end{pmatrix}$$
$$\mathbf{x} = \mathbf{U}^{-1}\mathbf{x}' \tag{17.26}$$

It is easily verified that the matrix product $\mathbf{U}^{-1}\mathbf{U}$ gives a 2 × 2 unit matrix. The coordinate system can alternatively be considered as spanned by basis vectors arranged as columns in an **X** matrix, in which case the unitary transformations can be written as $\mathbf{X}'=\mathbf{X}\mathbf{U}$.

Consider now a (multidimensional) linear function $\mathbf{f}$ defined by the action of a matrix $\mathbf{A}$ on a vector $\mathbf{x}$:

$$\mathbf{f} = \mathbf{Ax} \tag{17.27}$$

In the rotated coordinate system the corresponding connection is given by

$$\mathbf{f}' = \mathbf{A}'\mathbf{x}' \tag{17.28}$$

By using the transformations (17.25) between the two coordinate systems and the fact that a unit matrix of the form $\mathbf{U}^{-1}\mathbf{U}$ can be freely inserted, we get

$$\begin{aligned}
\mathbf{f} &= \mathbf{Ax} \\
\mathbf{f} &= \mathbf{A}(\mathbf{U}^{-1}\mathbf{U})\mathbf{x} \\
\mathbf{Uf} &= \mathbf{UA}(\mathbf{U}^{-1}\mathbf{U})\mathbf{x} \\
(\mathbf{Uf}) &= (\mathbf{UAU}^{-1})(\mathbf{Ux}) \\
\mathbf{f}' &= (\mathbf{UAU}^{-1})\mathbf{x}'
\end{aligned} \tag{17.29}$$

Changing the coordinate system thus changes a matrix by pre- and postmultiplication of a unitary matrix and its inverse, a procedure called a *similarity transformation*. Since the $\mathbf{U}$ matrix describes a rotation of the coordinate system in an arbitrary direction, one person's $\mathbf{U}$ may be another person's $\mathbf{U}^{-1}$. There is thus no significance whether the transformation is written as $\mathbf{U}^{-1}\mathbf{AU}$ or $\mathbf{UAU}^{-1}$, and for an orthogonal transformation matrix ($\mathbf{U}^{-1} = \mathbf{U}^{t}$), the transformation may also be written as $\mathbf{U}^{t}\mathbf{AU}$ or $\mathbf{UAU}^{t}$.

For the case of a symmetric ($A_{12} = A_{21}$) 2×2 matrix, the similarly transformed matrix elements are given as

$$\begin{gathered}
\mathbf{A}' = \mathbf{UAU}^{-1} \\
\begin{pmatrix} A'_{11} & A'_{12} \\ A'_{12} & A'_{22} \end{pmatrix} = \begin{pmatrix} \cos\alpha & \sin\alpha \\ -\sin\alpha & \cos\alpha \end{pmatrix} \begin{pmatrix} A_{11} & A_{12} \\ A_{12} & A_{22} \end{pmatrix} \begin{pmatrix} \cos\alpha & -\sin\alpha \\ \sin\alpha & \cos\alpha \end{pmatrix} \\
A'_{11} = A_{11}\cos^2\alpha + A_{22}\sin^2\alpha + 2A_{12}\cos\alpha\sin\alpha \\
A'_{22} = A_{22}\cos^2\alpha + A_{11}\sin^2\alpha - 2A_{12}\cos\alpha\sin\alpha \\
A'_{12} = A_{12}\left(\cos^2\alpha - \sin^2\alpha\right) + (A_{22} - A_{11})\cos\alpha\sin\alpha
\end{gathered} \tag{17.30}$$

The off-diagonal element A'_{12} can be made to vanish by choosing a specific rotational angle, as shown in

$$\begin{aligned}
A'_{12} = A_{12}(\cos^2\alpha - \sin^2\alpha) + (A_{22} - A_{11})\cos\alpha\sin\alpha &= 0 \\
A_{12}(\cos^2\alpha - \sin^2\alpha) &= (A_{11} - A_{22})\cos\alpha\sin\alpha \\
A_{12}(\cos 2\alpha) &= (A_{11} - A_{22})\left(\tfrac{1}{2}\sin 2\alpha\right) \\
\tan(2\alpha) &= \frac{2A_{12}}{(A_{11} - A_{22})}
\end{aligned} \tag{17.31}$$

In the new coordinate system, the $\mathbf{A}'$ matrix is simplified, as it only contains diagonal elements:

$$\mathbf{A}' = \begin{pmatrix} A'_{11} & A'_{12} \\ A'_{12} & A'_{22} \end{pmatrix} = \begin{pmatrix} \lambda_1 & 0 \\ 0 & \lambda_2 \end{pmatrix} = \mathbf{\Lambda} \tag{17.32}$$

An $N \times N$ Hermitian (or real symmetric) matrix can always be brought to a diagonal form by a multidimensional rotation of the coordinate system, and there are efficient standard computational procedures for diagonalizing matrices. The simplest method consists of an iterative series of 2×2 rotations as in Equation (17.30), which reduces the off-diagonal elements to zero.The rotational matrix $\mathbf{U}$ thus contains elements corresponding to products of cosines and sines of rotational angles.

The elements of the $\mathbf{A}$ matrix in the diagonal form (λ) are called *eigenvalues* and the columns of the unitary rotation matrix are called *eigenvectors*. In matrix notation the diagonalization can be written as

$$\mathbf{\Lambda} = \mathbf{UAU}^{-1} \tag{17.33}$$

A Hermitian matrix will always have real eigenvalues and orthogonal eigenvectors. Matrix diagonalizations play an important role in many areas of computational chemistry, and scientific computations in general, since they correspond to selecting a coordinate system where the variables are (approximately) independent of each other. Furthermore, the magnitude of the eigenvalues indicates the variation along that particular direction. For applications with many variables, it may be possible to describe a significant fraction of the whole variation by taking only a few selected eigenvector directions into account, and this forms the basis for *principal component analysis*, as discussed in Section 18.4.3, and can be generalized to *tensor decompositions*, as discussed in Section 17.6.3.

It can be shown that a matrix determinant is independent of a change in the coordinate system, and in the diagonal representation the determinant is simply the product of the eigenvalues. A non-zero determinant is thus equivalent to all the eigenvalues being different from zero. Furthermore, the *trace* of a matrix, defined as the sum of the diagonal elements, is also invariant to a change in the coordinate system, as can be verified for the 2×2 case from Equation (17.30). In the diagonal representation the trace is given by the sum of the eigenvalues.

An alternative way of introducing matrix eigenvalues and eigenvectors is to require non-zero $\mathbf{x}$-solutions to

$$\begin{aligned} \mathbf{Ax} &= \lambda\mathbf{x} \\ (\mathbf{A} - \lambda\mathbf{I})\mathbf{x} &= \mathbf{0} \end{aligned} \tag{17.34}$$

This is a set of linear equations in the form of Equation (17.23), with the right-hand side being zero, and a non-trivial solution therefore only exists when the determinant is zero:

$$|\mathbf{A} - \lambda\mathbf{I}| = 0 \tag{17.35}$$

Expansion of the determinant (17.35) according to Equation (17.20) produces an Nth-order polynomial in λ, which can be solved to give N roots (eigenvalues). If some of these are identical, they are called *degenerate* eigenvalues. For each eigenvalue, Equation (17.34) can be solved to produce the corresponding eigenvector. In the non-degenerate case (all λ_i being different) the one free parameter can be fixed by normalization. For degenerate eigenvectors, the normalization condition must be augmented with a mutual orthogonality condition (Section 17.4) in order to fix all the free parameters.

In the coordinate system ($\mathbf{x}'$) where the $\mathbf{A}'$ matrix is diagonal, it is easy to see that Equation (17.35) holds since the ($\mathbf{A}' - \lambda\mathbf{I}$) matrix has at least one column consisting of only zeros. In this diagonal representation, it is furthermore clear that the eigenvectors are simply unit vectors along the primed coordinate axes, and the eigenvectors in the unprimed coordinate system are therefore given by the

$$\mathbf{A} = \begin{pmatrix} a_{11} & a_{12} & \cdots \\ a_{21} & a_{22} & \cdots \\ \vdots & \vdots & \ddots \end{pmatrix} \xrightarrow{\mathbf{U}} \Lambda = \begin{pmatrix} \lambda_1 & 0 & \cdots \\ 0 & \lambda_2 & \cdots \\ \vdots & \vdots & \ddots \end{pmatrix} \xrightarrow{f}$$

$$f(\Lambda) = \begin{pmatrix} f(\lambda_1) & 0 & \cdots \\ 0 & f(\lambda_2) & \cdots \\ \vdots & \vdots & \ddots \end{pmatrix} \xrightarrow{\mathbf{U}^{-1}} f(\mathbf{A})$$

Figure 17.4 Construction of functions of matrices.

elements of the $\mathbf{U}^t$ transformation matrix:

$$\mathbf{A}'\mathbf{x}' = \lambda\mathbf{x}'$$

$$\begin{pmatrix} \lambda_1 & 0 \\ 0 & \lambda_2 \end{pmatrix}\begin{pmatrix} 1 \\ 0 \end{pmatrix} = \lambda_1\begin{pmatrix} 1 \\ 0 \end{pmatrix} \quad ; \quad \begin{pmatrix} \lambda_1 & 0 \\ 0 & \lambda_2 \end{pmatrix}\begin{pmatrix} 0 \\ 1 \end{pmatrix} = \lambda_2\begin{pmatrix} 0 \\ 1 \end{pmatrix} \tag{17.36}$$

$$\mathbf{x} = \mathbf{U}^t\mathbf{x}'$$

While the polynomial method can be used for solving small eigenvalue problems by hand, all computational implementations rely on iterative similarity transform methods for bringing the matrix to a diagonal form. The simplest of these is the *Jacobi* method, where a sequence of 2×2 rotations analogous to Equations (17.30) to (17.32) can be used to bring all the off-diagonal elements below a suitable threshold value.

In the diagonal form, the matrix $\mathbf{\Lambda}$ contains only elements along the diagonal. The diagonal elements can be treated like regular numbers, allowing calculation of functions of matrices (see Figure 17.4). Calculating, for example, $\mathbf{A}^{1/2}$ proceeds by first transforming it to a diagonal form, taking the square root of the diagonal elements and backtransforming to the original coordinate system. This procedure in general allows calculation of functions of matrices, such as $e^{\mathbf{A}}$, $\ln(\mathbf{A})$ or $\cos(\mathbf{A})$.

This also provides an alternative way of calculating the inverse of a matrix by simply taking the inverse of the eigenvalues in the diagonal representation and backtransforming the matrix to the original representation.

A unitary matrix can always be parameterized as the exponential of an anti-Hermitian matrix $\mathbf{X}$:

$$\mathbf{U} = e^{\mathbf{X}} \quad ; \quad \mathbf{X}^\dagger = -\mathbf{X} \tag{17.37}$$

The unitarity follows from

$$\mathbf{U}^\dagger\mathbf{U} = (e^{\mathbf{x}})^\dagger(e^{\mathbf{x}}) = (e^{\mathbf{x}^\dagger})(e^{\mathbf{x}}) = (e^{-\mathbf{x}})(e^{\mathbf{x}}) = 1 \tag{17.38}$$

The $\mathbf{X}$ matrix in the general case contains complex elements, but these can be separated into an antisymmetric real part (${}^{\mathrm{R}}x_{ij} = -\,{}^{\mathrm{R}}x_{ji}$) and a symmetric imaginary part (${}^{\mathrm{I}}x_{ij} = {}^{\mathrm{I}}x_{ji}$). For the two-dimensional example in Equation (17.24), the $\mathbf{X}$ matrix contains the rotation angle α:

$$\mathbf{X} = \begin{pmatrix} 0 & \alpha \\ -\alpha & 0 \end{pmatrix} \tag{17.39}$$

The absolute sign is again arbitrary, and there is no difference between defining the unitary matrix as $e^{\mathbf{X}}$ or $e^{-\mathbf{X}}$, as this simply implies changing the sign for α in the $\mathbf{X}$ matrix. The connection between $\mathbf{X}$ and $\mathbf{U}$ is illustrated in Figure 17.4 and construction of $\mathbf{U}$ involves diagonalization of $\mathbf{X}$

(to give eigenvalues of $\pm\, i\alpha$), exponentiation (to give complex exponentials that may be written as $\cos\alpha \pm i\sin\alpha$), followed by backtransformation:

$$\mathbf{U} = e^{\mathbf{X}} = \begin{pmatrix} \cos\alpha & \sin\alpha \\ -\sin\alpha & \cos\alpha \end{pmatrix} \tag{17.40}$$

In the general case, the **X** matrix contains off-diagonal elements corresponding to angles for rotating all pairs of variables.

The **A** matrix may in some cases have eigenvalues that are zero or nearly so. The number of non-zero eigenvalues is called the *rank* of the matrix **A** and corresponds to the number of independent rows/columns in the matrix. In actual applications, it is rare that an eigenvalue is exactly zero, but a very small value will clearly give numerical problems for constructing matrices such as $\mathbf{A}^{-1}$ or $\ln(\mathbf{A})$. The ratio between the largest and smallest eigenvalue is called the *condition number*, and large values ($>10^6$) indicate that the **A** matrix is close to having linear dependencies. *Singular value decomposition* (Section 17.6.3) constructs $\mathbf{A}^{-1}$ by inverting only those eigenvalues larger than a suitable threshold and setting the rest to zero, before backtransformation to the original coordinate system.

For an $N \times M$ ($N > M$) rectangular matrix **A**, a *generalized inverse* can be defined by the matrix $(\mathbf{A}^t\mathbf{A})^{-1}\mathbf{A}^t$. Such generalized inverse matrices correspond to obtaining the best solution in a least squares sense for an overdetermined system of linear equations, for example, those that arise in statistical applications (Section 18.4.2). Consider, for example, a system of equations analogous to Equation (17.23), but with more b solution elements than x variables ($n > m$):

$$\begin{aligned} A_{11}x_1 + A_{12}x_2 + \cdots + A_{1m}x_m &= b_1 \\ A_{21}x_1 + A_{22}x_2 + \cdots + A_{2m}x_m &= b_2 \\ \vdots \quad \vdots \quad \vdots \quad \vdots \quad \vdots \quad \vdots \quad \vdots \quad \vdots &= \vdots \\ A_{n1}x_1 + A_{n2}x_2 + \cdots + A_{nm}x_m &= b_n \end{aligned} \tag{17.41}$$

$$\begin{aligned} \mathbf{Ax} &= \mathbf{b} \\ \mathbf{A}^t\mathbf{Ax} &= \mathbf{A}^t\mathbf{b} \\ \mathbf{x} &= (\mathbf{A}^t\mathbf{A})^{-1}\mathbf{A}^t\mathbf{b} \end{aligned}$$

Multiplication from the left by $\mathbf{A}^t$ and by the inverse of $\mathbf{A}^t\mathbf{A}$ leads to the formal solution, that is $(\mathbf{A}^t\mathbf{A})^{-1}\mathbf{A}^t$ acts as the inverse to the rectangular **A** matrix.

17.2.1 Examples of Changing the Coordinate System

From the "separability" theorem (Section 1.6.3) it follows that if an operator (e.g. the Hamiltonian) depending on N coordinates can be written as a sum of operators that only depend on one coordinate, the corresponding N coordinate wave function can be written as a product of one-coordinate functions and the total energy as a sum of energies:

$$\begin{aligned} \mathbf{H}(x_1, x_2, x_3, \ldots)\Psi(x_1, x_2, x_3, \ldots) &= E_{\text{tot}}\Psi(x_1, x_2, x_3, \ldots) \\ \mathbf{H}(x_1, x_2, x_3, \ldots) &= \sum_i \mathbf{h}_i(x_i) \\ \mathbf{h}_i(x_i)\phi_i(x_i) &= \varepsilon_i\phi_i(x_i) \\ E_{\text{tot}} &= \sum_i \varepsilon_i \\ \Psi(x_1, x_2, x_3, \ldots) &= \prod_i \phi_i(x_i) \end{aligned} \tag{17.42}$$

Instead of solving one equation with N variables, the problem is transformed into solving N equations with only one variable. When the operator is transformed into a matrix representation, the separation is equivalent to finding a coordinate system where the representation is diagonal.

Consider a matrix $\mathbf{A}$ expressed in a coordinate system $(x_1, x_2, x_3, \ldots, x_N)$. The coordinate axes are the x_i vectors, and these may be simple Cartesian axes, or one-variable functions, or many-variable functions. The matrix $\mathbf{A}$ is typically defined by an operator working on the coordinates. Some examples are:

1. The force constant matrix in Cartesian coordinates (Section 14.5.3)
2. The Fock matrix in basis functions (atomic orbitals, Section 3.5)
3. The CI matrix in Slater determinants (Section 4.2).

Finding the coordinates where these matrices are diagonal corresponds to finding:

1. The vibrational normal coordinates
2. The molecular orbitals
3. The state coefficients, that is CI wave function(s).

The coordinate axes are usually orthonormal, but this is not a requirement, since they can be orthogonalized by the methods in Section 17.4.

17.2.2 Vibrational Normal Coordinates

The potential energy is approximated by a second-order Taylor expansion around the stationary geometry $\mathbf{x}_0$:

$$V(\mathbf{x}) \approx V(\mathbf{x}_0) + \left(\frac{\mathrm{d}V}{\mathrm{d}\mathbf{x}}\right)^{\mathrm{t}} (\mathbf{x}-\mathbf{x}_0) + \tfrac{1}{2}(\mathbf{x}-\mathbf{x}_0)^{\mathrm{t}} \left(\frac{\mathrm{d}^2 V}{\mathrm{d}\mathbf{x}^2}\right)(\mathbf{x}-\mathbf{x}_0) \tag{17.43}$$

The energy for the expansion point, $\mathbf{V}(\mathbf{x}_0)$, may be chosen as zero, and the first derivative is zero since $\mathbf{x}_0$ is a stationary point:

$$V(\Delta\mathbf{x}) = \tfrac{1}{2}\Delta\mathbf{x}^{\mathrm{t}}\mathbf{F}\Delta\mathbf{x} \tag{17.44}$$

Here $\mathbf{F}$ is a $3N_{\mathrm{atom}} \times 3N_{\mathrm{atom}}$ (force constant) matrix containing the second derivatives of the energy with respect to the coordinates.

The nuclear Schrödinger equation for an N_{atom} system is given by

$$\left[-\sum_{i=1}^{3N_{\mathrm{atom}}}\left(\frac{1}{2m_i}\frac{\partial^2}{\partial x_i^2}\right) + \tfrac{1}{2}\Delta\mathbf{x}^{\mathrm{t}}\mathbf{F}\Delta\mathbf{x}\right]\Psi_{\mathrm{nuc}} = E_{\mathrm{nuc}}\Psi_{\mathrm{nuc}} \tag{17.45}$$

Equation (17.45) is first transformed to mass-dependent y-coordinates by a $\mathbf{G}$ matrix containing the inverse square root of atomic masses:

$$\begin{gathered} y_i = \sqrt{m_i}\Delta x_i \quad ; \quad \frac{\partial^2}{\partial y_i^2} = \frac{1}{m_i}\frac{\partial^2}{\partial x_i^2} \quad ; \quad G_{ij} = \frac{1}{\sqrt{m_i m_j}} \\ \left[-\sum_{i=1}^{3N_{\mathrm{atom}}}\left(\frac{1}{2}\frac{\partial^2}{\partial y_i^2}\right) + \tfrac{1}{2}\mathbf{y}^{\mathrm{t}}(\mathbf{F}\cdot\mathbf{G})\,\mathbf{y}\right]\Psi_{\mathrm{nuc}} = E_{\mathrm{nuc}}\Psi_{\mathrm{nuc}} \end{gathered} \tag{17.46}$$

A unitary transformation is then introduced that diagonalizes the **F**·**G** (entrywise product, Equation (17.11)) matrix, yielding eigenvalues ε_i and eigenvectors $\mathbf{q}_i$. The kinetic energy operator is still diagonal in these coordinates:

$$\begin{aligned}
\mathbf{q} &= \mathbf{U}\mathbf{y} \\
\left[-\sum_{i=1}^{3N_{\text{atom}}}\left(\frac{1}{2}\frac{\partial^2}{\partial q_i^2}\right)+\tfrac{1}{2}\mathbf{q}^{\text{t}}\left(\mathbf{U}\left(\mathbf{F}\cdot\mathbf{G}\right)\mathbf{U}^{\text{t}}\right)\mathbf{q}\right]\Psi_{\text{nuc}} &= E_{\text{nuc}}\Psi_{\text{nuc}} \\
\left[-\sum_{i=1}^{3N_{\text{atom}}}\left(\frac{1}{2}\frac{\partial^2}{\partial q_i^2}+\frac{1}{2}\varepsilon_i q_i^2\right)\right]\Psi_{\text{nuc}} &= E_{\text{nuc}}\Psi_{\text{nuc}} \\
\left[\sum_{i=1}^{3N_{\text{atom}}}\mathbf{h}_i\left(q_i\right)\right]\Psi_{\text{nuc}} &= E_{\text{nuc}}\Psi_{\text{nuc}}
\end{aligned} \tag{17.47}$$

In the **q**-coordinate system, the *vibrational normal coordinates*, the $3N_{\text{atom}}$-dimensional Schrödinger equation can be separated into $3N_{\text{atom}}$ one-dimensional Schrödinger equations, which are just in the form of a standard harmonic oscillator, with the solutions being Hermite polynomials in the **q**-coordinates. The eigenvectors of the **F**·**G** matrix are the (mass-weighted) vibrational normal coordinates and the eigenvalues ε_i are related to the vibrational frequencies as follows (analogous to Equation (14.36)):

$$\nu_i = \frac{1}{2\pi}\sqrt{\varepsilon_i} \tag{17.48}$$

When this procedure is carried out in Cartesian coordinates, there should be six (five for a linear molecule) eigenvalues of the **F**·**G** matrix being exactly zero, corresponding to the translational and rotational modes. In real calculations, however, these values are not exactly zero. The three translational modes usually have "frequencies" very close to zero, typically less than 0.01 cm^{-1}. The deviation from zero is due to the fact that numerical operations are only carried out with a finite precision and the accumulations of errors will typically give inaccuracies in ν of this magnitude. The residual "frequencies" for the rotational modes, however, may often be as large as 10–50 cm^{-1}. This is due to the fact that the geometry cannot be optimized to a gradient of exactly zero, again due to numerical considerations. Typically, the geometry optimization is considered converged if the root mean square (RMS) gradient is less than $\sim 10^{-4}$–10^{-5} au, corresponding to the energy being converged to $\sim 10^{-5}$–10^{-6} au. The residual gradient shows up as vibrational frequencies for the rotations of the above magnitude. If there are real frequencies of the same magnitude as the "rotational frequencies", mixing may occur and result in inaccurate values for the "true" vibrations. For this reason, the translational and rotational degrees of freedom are normally removed by projection (Section 17.4) from the force constant matrix before diagonalization.

The calculation of vibrational frequencies by the above procedure rests on the Born–Oppenheimer decoupling of the nuclear and electronic motions, and the masses in Equation (17.46) are therefore expected to be *nuclear* masses. All programs, however, use *atomic* masses, which differ from nuclear masses by the addition of the electron masses. This is consistent with the Born–Oppenheimer approximation that the electrons follow the nuclear motion "instantaneously". The core electrons are closely associated with the nuclei and are expected to obey the Born–Oppenheimer approximation, but the valence electrons in a molecule are delocalized, and it therefore becomes problematic to assign (exactly) Z electrons to a nucleus with atomic number Z in order to arrive at the atomic mass.[2] The Born–Oppenheimer approximation thus leads to ambiguities in the atomic masses in Equation (17.40) with a magnitude of $\sim 10^{-4}$, corresponding to ambiguities in vibrational frequencies of a few tenths of a cm^{-1}. The diagonal Born–Oppenheimer correction (Section 3.1) in comparison leads to vibrational frequency corrections of a few hundreds of a cm^{-1}.[3] Vibrational frequencies therefore have an accuracy limit of ~0.1 cm^{-1} within the Born–Oppenheimer approximation and attempts to refine the electronic potential energy surface or solve the nuclear vibrational problem beyond this accuracy have little physical relevance.

If the stationary point is a minimum on the energy surface, the eigenvalues of the **F** and **F·G** matrices are all positive. If, however, the stationary point is a transition state (TS), one (and only one) of the eigenvalues is negative. This corresponds to the energy being a maximum in one direction and a minimum in all other directions. The "frequency" for the "vibration" along the eigenvector with a negative eigenvalue will formally be imaginary, as it is the square root of a negative number (Equation (17.48)). The corresponding eigenvector is the direction leading downhill from the TS towards the reactant and product. *At the TS, the eigenvector for the imaginary frequency is the reaction coordinate.* The whole reaction path may be calculated by sliding downhill to each side from the TS. This can be performed by taking a small step along the TS eigenvector, calculating the gradient and taking a small step in the negative gradient direction. The negative of the gradient always points downhill and by taking a sufficiently large number of such steps an energy minimum is eventually reached. This is equivalent to a steepest descent minimization, but more efficient methods are available (see Section 13.8 for details). The reaction path in mass-weighted coordinates is called the *Intrinsic Reaction Coordinate* (IRC).

The vibrational Hamiltonian is completely separable within the harmonic approximation, with the vibrational energy being a sum of individual energy terms and the nuclear wave function being a product of harmonic oscillator functions (a Hermite polynomial in the normal coordinates). When anharmonic terms are included in the potential, the Hamiltonian is no longer separable and the resulting nuclear Schrödinger equation can be solved by techniques completely analogous to those used for solving the electronic problem. The vibrational SCF method is analogous to the electronic Hartree–Fock method, with the nuclear harmonic oscillator functions playing the same role as the orbitals in electronic structure theory. Corrections beyond the mean-field approximation can be added by configuration interaction, perturbation theory or coupled cluster methods.[4]

It should be noted that the force constant matrix can be calculated at any geometry, but the transformation to normal coordinates is only valid at a stationary point, that is where the first derivative is zero. At a non-stationary geometry, a set $3N_{atom} - 7$ *generalized frequencies* may be defined by removing the gradient direction from the force constant matrix (e.g. by projection techniques, Section 17.4) before transformation to normal coordinates.

17.2.3 Energy of a Slater Determinant

The variational problem is to minimize the energy of a single Slater determinant by choosing suitable values for the molecular orbital (MO) coefficients, under the constraint that the MOs remain orthonormal. With ϕ being an MO written as a linear combination of the basis functions (atomic orbitals) χ, this leads to a set of secular equations, **F** being the Fock matrix, **S** the overlap matrix and **C** containing the coefficients (Section 3.5):

$$\begin{gathered}\phi_i = \sum_{\alpha=1}^{M_{basis}} c_{\alpha i}\chi_\alpha \\ F_{\alpha\beta} = \langle \chi_\alpha |\mathbf{F}| \chi_\beta \rangle \quad ; \quad S_{\alpha\beta} = \langle \chi_\alpha / \chi_\beta \rangle \\ \mathbf{FC} = \mathbf{SC}\boldsymbol{\varepsilon}\end{gathered} \tag{17.49}$$

The basis functions (coordinate system) in this case are non-orthogonal, with the overlap elements contained in the **S** matrix. By multiplying from the left by $\mathbf{S}^{-1/2}$ and inserting a unit matrix written in the form $\mathbf{S}^{-1/2}\,\mathbf{S}^{1/2}$, Equation (17.49) may be reformulated as

$$\begin{aligned}\{\mathbf{S}^{-1/2}\mathbf{F}\mathbf{S}^{-1/2}\}\{\mathbf{S}^{1/2}\mathbf{C}\} &= \{\mathbf{S}^{-1/2}\mathbf{S}^{1/2}\}\{\mathbf{S}^{1/2}\mathbf{C}\}\,\varepsilon \\ \mathbf{F}'\mathbf{C}' &= \mathbf{C}'\varepsilon\end{aligned} \tag{17.50}$$

This equation is now in a standard form for determining the eigenvalues of the $\mathbf{F}'$ matrix. The eigenvectors contained in $\mathbf{C}'$ can then be backtransformed to the original coordinate system ($\mathbf{C} = \mathbf{S}^{-1/2}\mathbf{C}'$). This is an example of a symmetrical orthogonalization (Section 17.4) of the initial coordinate system, the basis functions χ. Solving Equation (17.49) corresponds to rotating the original space of basis functions into one of molecular orbitals where the Fock matrix is diagonal.

17.2.4 Energy of a CI Wave Function

The variational problem may again be formulated as a secular equation, where the coordinate axes are many-electron functions (Slater determinants) Φ_i, which are orthogonal (Section 4.2):

$$\begin{aligned}\Psi_{\mathrm{CI}} &= \sum_i a_i\Phi_i \\ H_{ij} &= \langle\Phi_i|\mathbf{H}|\Phi_j\rangle \\ \mathbf{Ha} &= \mathbf{Ea}\end{aligned} \tag{17.51}$$

The **a** matrix contains the coefficients of the CI wave function. This problem may again be considered as selecting a basis where the Hamiltonian operator is diagonal (Equation (4.7) and Figure 4.5). In the initial coordinate system, the Hamiltonian matrix will have many off-diagonal elements, but it can be diagonalized by a suitable unitary transformation. The diagonal elements are energies of many-electron CI wave functions, being approximations to the ground and exited states. The corresponding eigenvectors contain the expansion coefficients a_i.

17.2.5 Computational Considerations

The time required for diagonalizing a matrix grows as the *cube* of the size of the matrix and the amount of computer memory necessary for storing the matrix grows as the *square* of the size. Diagonalizing matrices with dimensions up to ~100 takes insignificant amounts of time, unless there are extraordinarily many such matrices. Matrices with dimensions up to ~1000 pose no particular problems, although some consideration should be made as to whether the time required for diagonalization is significant relative to other operations. Matrices larger than ~1000 require consideration, as the storage and computational time for determining all eigenvalues and eigenvectors rapidly becomes a computational bottleneck. In many cases, however, only one or a few eigenvalues and eigenvectors are desired. A prime example is solving the CI matrix Equation (17.51), where the dimension may be 10^6–10^9, but only the lowest eigenvalue and eigenvector are required, since this is the ground state energy and wave function. Such problems may be solved by iterative techniques, of which variations of the *Davidson* method[5] are commonly used.

The basic idea in iterative methods for extracting eigenvalues and eigenvectors of a matrix **A** is to project the problem on to a smaller subspace and solve the subspace diagonalization problem by standard methods. The error in the resulting solution vector is quantified by a residual vector in the full space and terminated if this is small enough, or otherwise used for increasing the dimension of the subspace. The procedure for determining the lowest eigenvalue and eigenvector can be formulated as follows:

1. The desired equation to be solved can be written as

$$\mathbf{Ax} = \lambda\mathbf{x} \tag{17.52}$$

2. Select a (small) initial set K of orthonormal trial vectors $\mathbf{b}_i$. These can be simply unit vectors or obtained by diagonalization of a small subsection of the $\mathbf{A}$ matrix, and in the simplest case, the space is just a single unit vector. These vectors are used to expand the current approximation to the eigenvector $\mathbf{x}$:

$$\mathbf{x} = \sum_{i=1}^{K} \alpha_i \mathbf{b}_i \tag{17.53}$$

3. Project the large $\mathbf{A}$ matrix on to the subspace to form a smaller $\mathbf{B}$ matrix of dimension $K \times K$ with elements given by

$$b_{ij} = \mathbf{b}_i^{\mathrm{t}} \mathbf{A} \mathbf{b}_j \tag{17.54}$$

4. Diagonalize the $\mathbf{B}$ matrix. The lowest eigenvalue is the current estimate of the eigenvalue λ and the corresponding eigenvector contains the expansion coefficients α_i in Equation (17.53).
5. Form a residual vector $\mathbf{r}$ to measure how close the current $\mathbf{x}$ vector is to being an eigenvector in the full space:

$$\mathbf{r} = (\mathbf{A} - \lambda\mathbf{I})\mathbf{x} \tag{17.55}$$

6. Terminate if $\mathbf{r}$ is sufficiently small, for example measured by its norm.
7. If $\mathbf{r}$ is not sufficiently small, the eigenvalue Equation (17.52) is written in terms of a perturbed eigenvalue and eigenvector:

$$\mathbf{A}(\mathbf{x} + \boldsymbol{\delta}) = (\lambda + \lambda')(\mathbf{x} + \boldsymbol{\delta}) \tag{17.56}$$

Expanding Equation (17.56) and ignoring the quadratic $\lambda'\boldsymbol{\delta}$ term gives

$$\begin{aligned}
(\mathbf{A} - \lambda\mathbf{I})\boldsymbol{\delta} &= -(\mathbf{A} - \lambda\mathbf{I})\mathbf{x} + \lambda'\mathbf{x} \\
(\mathbf{A} - \lambda\mathbf{I})\boldsymbol{\delta} &= -\mathbf{r} + \lambda'\mathbf{x} \\
\boldsymbol{\delta} &= -(\mathbf{A} - \lambda\mathbf{I})^{-1}(\mathbf{r} - \lambda'\mathbf{x})
\end{aligned} \tag{17.57}$$

8. The solution in Equation (17.57) formally requires the inverse of the $(\mathbf{A} - \lambda\mathbf{I})$ matrix, but this is computationally more expensive than solving Equation (17.52) itself. The key point is therefore to obtain a useful approximate solution to Equation (17.57) by a computationally inexpensive method. The Davidson method corresponds to ignoring the $\lambda'\mathbf{x}$ term and use a diagonal approximation for the $\mathbf{A}$ matrix. The latter allows a straightforward inversion of the diagonal $(\mathbf{A}_0 - \lambda\mathbf{I})$ matrix:

$$\boldsymbol{\delta} = -(\mathbf{A}_0 - \lambda\mathbf{I})^{-1}\mathbf{r} \tag{17.58}$$

9. The correction vector from Equation (17.58) is orthonormalized to the current subspace vectors $\mathbf{b}_i$ and included in the subspace in the next iteration.

As the iterative sequence approaches convergence, the correction vector from Equation (17.58) can become near-linear dependent with $\mathbf{x}$ and the subsequent orthonormalization is prone to numerical instability. An improved correction vector can be obtained by keeping the $\lambda'\mathbf{x}$ term in Equation (17.57) and requiring that $\boldsymbol{\delta}$ is orthogonal to $\mathbf{x}$ such that it is guaranteed that it will improve the subspace.[6] The correction vector from Equation (17.58), and thus the overall convergence rate, is furthermore sensitive to the $\mathbf{A}$ matrix being sufficiently diagonal dominant, as only the diagonal elements are used. Several modifications of the original Davidson algorithm have been proposed where either selected subsections of $\mathbf{A}$ are included to improve the estimate of $(\mathbf{A} - \lambda\mathbf{I})^{-1}$ or a suitable precondition matrix is used to make the problem more diagonal dominant.

An attractive feature of the Davidson algorithm is that it can easily be modified to extract a selected eigenvalue other than the lowest by choosing a different eigenvalue and eigenvector from the subspace diagonalization in step 4. It can also be used to simultaneously extract more than one eigenvector by calculating several residual and new expansion vectors in each iteration. In the CI example, this would correspond to determining both the ground and a few of the lowest excited states in the same iterative sequence. This of course requires that the initial selection of subspace vectors is at least the desired number of eigenvectors, and the subspace should also be capable of providing a good zeroth-order representation of the desired eigenvectors, as the iterative sequence otherwise may converge on undesired solutions.

The main computational effort in the Davidson algorithm is the formation of the reduced space $\mathbf{B}$ matrix, as this involves matrix–vector multiplications with the dimension of the large $\mathbf{A}$ matrix. The $\mathbf{A}$ matrix, however, does not need to be stored at any point; only the result of multiplying elements in $\mathbf{A}$ with elements in $\mathbf{b}$ is required. For the CI example, the matrix elements are given in terms of two-electron integrals and there are often significantly fewer of these than CI matrix elements. The $\mathbf{Ab}$ matrix–vector product can thus be calculated by a single pass through the list of two-electron integrals, a procedure often called *integral-driven* or *direct CI*. A potential limitation is the requirement of storing K $\mathbf{b}$ vectors, each having the dimension of the $\mathbf{A}$ matrix. If this is problematic, the iterative sequence can be restarted at suitable intervals, with the current $\mathbf{x}$ vector becoming the starting $\mathbf{b}$ vector in the new iterative sequence.

17.3 Coordinates, Functions, Functionals, Operators and Superoperators

A *function* is a recipe for producing a scalar from another set of scalars, for example calculating the energy E from a set of (nuclear) coordinates $\mathbf{x}$:

$$E(\mathbf{x}) \therefore \mathbf{x} \rightarrow E \tag{17.59}$$

A *functional* is a recipe for producing a scalar from a function, for example calculating the exchange–correlation energy E_{xc} from an electron density ρ depending on a set of (electronic) coordinates $\mathbf{x}$:

$$E_{xc}[\rho(\mathbf{x})] \therefore \mathbf{x} \rightarrow \rho \rightarrow E_{xc} \tag{17.60}$$

An *operator* is a recipe for producing a function from another function, for example the kinetic energy operator acting on a wave function. The operator in this case consists of differentiating the function twice with respect to the coordinates, adding the results and dividing by twice the particle mass:

$$\begin{aligned} \mathbf{T}\Psi(\mathbf{x}) &= \mathbf{T}\Psi(x, y, z) \\ \mathbf{T} &= \frac{1}{2m}\left(\frac{\partial^2}{\partial x^2} + \frac{\partial^2}{\partial y^2} + \frac{\partial^2}{\partial z^2}\right) \end{aligned} \tag{17.61}$$

A *superoperator* is a recipe for producing an operator from another operator. This level of abstraction is rarely used, but is, for example, employed in some formulations of propagator theory:

$$\hat{\mathbf{O}}(\mathbf{h}(f(\mathbf{x}))) \tag{17.62}$$

In the abstract function (or operator) space, often called a *Hilbert* space, it is possible to consider the functions (or operators) as vectors. The *bra* –*ket* notation is defined as

$$\begin{aligned} bra &: \langle \mathbf{f}| = \mathbf{f}^*(\mathbf{x}) \\ ket &: |\mathbf{f}\rangle = \mathbf{f}(\mathbf{x}) \end{aligned} \tag{17.63}$$

The equivalent of a vector dot product is defined as the integral of the product of the two functions:

$$\langle \mathbf{f}|\mathbf{g}\rangle = \int \mathbf{f}^*(\mathbf{x})\mathbf{g}(\mathbf{x})\mathrm{d}\mathbf{x} \tag{17.64}$$

The combination of a bra and a ket is called a *bracket*, and the bracket notation is often also used for the dot product of regular coordinate vectors (Equation (17.12)). By analogy with coordinate vectors, the bracket of two functions measures the "angle" or overlap between the functions, with a value of zero indicating that the two functions are orthogonal.

The norm of a function is defined as the square root of the bracket of the function with itself:

$$|\mathbf{f}(\mathbf{x})| = \sqrt{\langle \mathbf{f}|\mathbf{f}\rangle} = \sqrt{\int \mathbf{f}^*(\mathbf{x})\mathbf{f}(\mathbf{x})\mathrm{d}\mathbf{x}} \tag{17.65}$$

The bracket notation can also be used in connection with functions and operators, as in the example below:

$$\langle \mathbf{f}|\mathbf{O}|\mathbf{g}\rangle = \int \mathbf{f}^*(\mathbf{x})\mathbf{O}\mathbf{g}(\mathbf{x})\mathrm{d}\mathbf{x} \tag{17.66}$$

Operator algebra shares the characteristic with matrix algebra; indeed, matrices can be considered as representations of operators in a given set of functions (coordinate system).

Most operators in computational chemistry are *linear*:

$$\mathbf{O}(c\mathbf{f}) = c\mathbf{O}\mathbf{f} \tag{17.67}$$

$$\mathbf{O}(\mathbf{f}+\mathbf{g}) = \mathbf{O}\mathbf{f} + \mathbf{O}\mathbf{g} \tag{17.68}$$

Operators are *associative*, but not necessarily *commutative*:

$$\mathbf{O}(\mathbf{P}\mathbf{Q}) = (\mathbf{O}\mathbf{P})\mathbf{Q} \tag{17.69}$$

$$\mathbf{O}\mathbf{P} \neq \mathbf{P}\mathbf{O} \tag{17.70}$$

The *commutator* of two operators is denoted with square brackets and defined as

$$[\mathbf{O},\mathbf{P}] = \mathbf{O}\mathbf{P} - \mathbf{P}\mathbf{O} \tag{17.71}$$

Two operators are said to commute when Equation (17.71) is zero.

Operator eigenvalues and eigenfunctions are defined analogously to Equation (17.34):

$$\mathbf{O}\mathbf{f} = \lambda\mathbf{f} \tag{17.72}$$

Operators analogous to matrices (Equation (17.29)) can be subjected to a similarity transformation. A commonly encountered example is shown below, where an operator **O** is similarity transformed by the exponential **P** operator:

$$\mathbf{O}' = \mathrm{e}^{-\mathbf{P}}\mathbf{O}\mathrm{e}^{\mathbf{P}} \tag{17.73}$$

The $e^{\mathbf{P}}$ operator is defined in terms of its Taylor expansion:

$$e^{\mathbf{P}} = \sum_{n=0}^{\infty} \frac{1}{n!}\mathbf{P}^n \tag{17.74}$$

The similarity transformed operator $\mathbf{O}'$ can with the definition in Equation (17.74) be expressed as an inifinite series of nested commutators, known as the *Baker–Campbell–Hausdorff expansion*:

$$\mathbf{O}' = \mathbf{O} + [\mathbf{O},\mathbf{P}] + \frac{1}{2!}[[\mathbf{O},\mathbf{P}],\mathbf{P}] + \frac{1}{3!}[[[\mathbf{O},\mathbf{P}],\mathbf{P}],\mathbf{P}] + \cdots \tag{17.75}$$

If $\mathbf{P}$ is anti-Hermitian, the $e^{\mathbf{P}}$ corresponds to a unitary transformation (Equation (17.37)).

17.3.1 Differential Operators

Differential operators, which describe the first- and higher-order variations of functions, represent an important class of operators in computational chemistry. For a simple one-dimensional function, the first derivative is given by the normal rules for differentiation:

$$f(x) = x^2 \Rightarrow \frac{\mathrm{d}f}{\mathrm{d}x} = 2x \tag{17.76}$$

For a multidimensional *scalar function* (i.e. each point in space is associated with a number), the first derivative is a vector containing all the partial derivatives. The corresponding operator is called the *gradient* and is denoted by ∇:

$$\begin{aligned} f(x,y,z) &= x + y^2 + z^3 + xy + xz \\ \nabla f(x,y,z) &= \left\{ \frac{\partial f}{\partial x} \quad \frac{\partial f}{\partial y} \quad \frac{\partial f}{\partial z} \right\} = \{1+y+z \quad 2y+x \quad 3z^2+x\} \end{aligned} \tag{17.77}$$

The gradient vector points in the direction where the function increases most.

Two choices are possible for defining the first derivative of a *vector function* (i.e. each point in space is associated with a vector). The *divergence* is denoted by $\nabla\cdot$ and produces a *scalar*:

$$\nabla \cdot f(x,y,z) = \frac{\partial f_x}{\partial x} + \frac{\partial f_y}{\partial y} + \frac{\partial f_z}{\partial z} \tag{17.78}$$

The divergence measures how much the vector field "dilutes" or "contracts" at a given point. Alternatively, the first derivative may be defined as the *curl*, denoted with $\nabla\times$, which produces a *vector*:

$$\nabla \times f(x,y,z) = \left\{ \left(\frac{\partial f_z}{\partial y} - \frac{\partial f_y}{\partial z} \right) \quad \left(\frac{\partial f_x}{\partial z} - \frac{\partial f_z}{\partial x} \right) \quad \left(\frac{\partial f_y}{\partial x} - \frac{\partial f_x}{\partial y} \right) \right\} \tag{17.79}$$

The curl describes how fast the vector field rotates, that is how rapidly and in which direction the field changes.

Given the above three definitions of first derivatives, there are nine possible combinations for defining second derivatives. Four of these are invalid since the gradient only works on scalar fields and the divergence and curl only work on vector fields. Two of the remaining five combinations can be shown to be zero, leaving only three interesting combinations. Figure 17.5 indicates the action of the three first derivatives (conversion from/to scalar and vector) and their combinations.

		First		
		$\underset{s\to v}{\nabla}$	$\underset{v\to s}{\nabla\cdot}$	$\underset{v\to v}{\nabla\times}$
	$\underset{s\to v}{\nabla}$	I	$\nabla\nabla\cdot$	I
Second	$\underset{v\to s}{\nabla\cdot}$	∇^2	I	0
	$\underset{v\to v}{\nabla\times}$	0	I	$\nabla\times\nabla\times$

Figure 17.5 Possibilities for second derivatives of multidimensional functions, with *I* indicating an invalid combination.

The divergence of the gradient is commonly denoted the *Laplacian* and is, for example, involved in the (non-relativistic) quantum mechanical kinetic energy operator. It operates on a scalar function and produces a scalar function:

$$\nabla \cdot \nabla f(x,y,z) = \nabla^2 f(x,y,z) = \frac{\partial^2 f}{\partial x^2} + \frac{\partial^2 f}{\partial y^2} + \frac{\partial^2 f}{\partial z^2} \tag{17.80}$$

The Laplacian measures the local depletion or concentration of the function. The two other combinations produce a vector from a vector function and are used less commonly.

17.4 Normalization, Orthogonalization and Projection

The vectors (functions) of a coordinate system may in some cases be given naturally by the problem, and these are not always normalized or orthogonal. For computational purposes, however, it is often advantageous to work in an orthonormal coordinate system. We first note that normalization is trivially obtained by simply scaling each vector by the inverse of its length:

$$\begin{aligned} \langle \mathbf{x}|\mathbf{x}\rangle &= N^2 \\ \mathbf{x}' &= N^{-1}\mathbf{x} \\ \langle \mathbf{x}'|\mathbf{x}'\rangle &= 1 \end{aligned} \tag{17.81}$$

The orthogonalization of a set of non-orthogonal vectors can be done in many ways, but the two most commonly used are *Gram–Schmidt* and *symmetrical* orthogonalization.

The *Gram–Schmidt* procedure selects an initial vector and then sequentially constructs additional orthogonal vectors by removing (projecting out) the component(s) along all previous vectors, and re-normalizing the remaining component, as shown by

$$\begin{aligned} \mathbf{x}'_1 &= N_1^{-1}\mathbf{x}_1 \\ \mathbf{x}'_2 &= N_2^{-1}\left(\mathbf{x}_2 - \langle \mathbf{x}_2|\mathbf{x}'_1\rangle \mathbf{x}'_1\right) \\ \mathbf{x}'_3 &= N_3^{-1}\left(\mathbf{x}_3 - \langle \mathbf{x}_3|\mathbf{x}'_1\rangle \mathbf{x}'_1 - \langle \mathbf{x}_3|\mathbf{x}'_2\rangle \mathbf{x}'_2\right) \\ \mathbf{x}'_k &= N_k^{-1}\left(\mathbf{x}_k - \sum_{i=1}^{k-1} \langle \mathbf{x}_k|\mathbf{x}'_i\rangle \mathbf{x}'_i\right) \end{aligned} \tag{17.82}$$

It should be noted that the final set of orthogonal vectors will depend on the selection of the first vector and the order in which the remaining vectors are orthogonalized, although the total space spanned will of course be the same.

A *symmetrical* orthogonalization corresponds to a transformation that has the property $\mathbf{O}^\dagger\mathbf{S}\mathbf{O} = \mathbf{I}$, where $\mathbf{S}$ contains the overlap elements of all the $\mathbf{x}_i$ coordinate vectors:

$$S_{ij} = \langle \mathbf{x}_i | \mathbf{x}_j \rangle \tag{17.83}$$

One such transformation is given by the inverse square root of the overlap matrix ($\mathbf{O} = \mathbf{S}^{-1/2}$) and is often called a *Löwdin orthogonalization*.[7] It has the property that the orthogonalized set of vectors is the least changed relative to the original set of vectors, and this is used, for example, in solving the self-consistent field equations in Hartree–Fock and Kohn–Sham theories, and for performing the Löwdin population analysis (Section 10.1):

$$\mathbf{x}' = \mathbf{S}^{-1/2}\mathbf{x} \tag{17.84}$$

A variation of the Löwdin procedure allows a weighting of the original vectors, such that the similarity of the original and orthogonalized set of vectors is not uniformly distributed over all vectors, but selected vectors are chosen to change less than others. This corresponds to the transformation shown below, where $\mathbf{W}$ is a diagonal matrix containing weighting factors, and this is used, for example, in the *natural bond analysis* (Section 10.5.1):[8]

$$\mathbf{x}' = \mathbf{W}(\mathbf{W}\mathbf{S}\mathbf{W})^{-1/2}\mathbf{x} \tag{17.85}$$

A second variation is a *canonical orthogonalization*, where the transformation is done by a unitary matrix $\mathbf{U}$ containing the eigenvectors obtained by diagonalizing the overlap matrix and weighting by the inverse square root of the eigenvalues:

$$\mathbf{x}' = (\mathbf{U}\boldsymbol{\Lambda}^{-1/2})\mathbf{x} \tag{17.86}$$

The advantage of a canonical orthogonalization is that it allows for handling (near-) linear dependencies in the basis by truncating the transformation matrix by removal of columns with eigenvalues smaller than a suitable cutoff value.

In multidimensional coordinate systems it may often be advantageous to work in a subset of the full coordinate system. The component of a vector function $\mathbf{f}$ along a specific (unit) coordinate vector $\mathbf{x}_k$ is given by the projection

$$f_k = \langle \mathbf{f} | \mathbf{x}_k \rangle \tag{17.87}$$

For a matrix representation of an operator, $\mathbf{A}$, the projection onto the $\mathbf{x}_k$ subspace is given by pre- and post-multiplying with a $\mathbf{Q}_k$ matrix defined as the outer product of $\mathbf{x}_k$, or the function equivalent in a ket-bra notation:

$$\mathbf{Q}_k = \mathbf{x}_k\mathbf{x}_k^{\mathrm{t}} = |\mathbf{x}_k\rangle\langle\mathbf{x}_k| \tag{17.88}$$

$$\mathbf{A}_k = \mathbf{Q}_k^{\mathrm{t}}\mathbf{A}\mathbf{Q}_k \tag{17.89}$$

The reverse process, removing the $\mathbf{x}_k$ subspace, is done by projecting the $\mathbf{x}_k$ subspace direction out by the complementary matrix $\mathbf{P}_k$, with $\mathbf{I}$ being a unit matrix:

$$\mathbf{P}_k = \mathbf{I} - \mathbf{Q}_k \tag{17.90}$$

$$\mathbf{f}_{\neg k} = \mathbf{P}_k\mathbf{f} = \mathbf{f} - \langle \mathbf{f}|\mathbf{x}_k\rangle\mathbf{x}_k \tag{17.91}$$

$$\mathbf{A}_{\neg k} = \mathbf{P}_k^{\mathrm{t}}\mathbf{A}\mathbf{P}_k \tag{17.92}$$

Projection of larger subspaces can be done similarly by adding more vectors to the projection matrix. For the case of removing the translational and rotational degrees of freedom from the vibrational normal coordinates, for example, the (normalized) vector describes a translation in the x-direction:

$$\mathbf{t}_x^{\mathrm{t}} = \frac{1}{\sqrt{N_{\mathrm{atom}}}}(1, 0, 0, 1, 0, 0, 1, 0, 0, \ldots) \tag{17.93}$$

The superscript ^{t} indicates that $\mathbf{t}_x^{\mathrm{t}}$ is a row vector. The $\mathbf{T}_x$ matrix removes the direction corresponding to translation in the x-direction:

$$\mathbf{T}_x = \mathbf{I} - \mathbf{t}_x\mathbf{t}_x^{\mathrm{t}} \tag{17.94}$$

Extending this to include vectors for all three translational and rotational modes gives a projection matrix for removing the six (five) translational and rotational degrees of freedom:

$$\mathbf{P} = \mathbf{I} - \mathbf{t}_x\mathbf{t}_x^{\mathrm{t}} - \mathbf{t}_y\mathbf{t}_y^{\mathrm{t}} - \mathbf{t}_z\mathbf{t}_z^{\mathrm{t}} - \mathbf{r}_a\mathbf{r}_a^{\mathrm{t}} - \mathbf{r}_b\mathbf{r}_b^{\mathrm{t}} - \mathbf{r}_c\mathbf{r}_c^{\mathrm{t}} \tag{17.95}$$

The $\mathbf{r}$ vectors are derived from the atomic coordinates and principal axes of inertia determined by diagonalization of the matrix of inertia (Equation (14.32)).[9] By forming the matrix product $\mathbf{P}^{\mathrm{t}}\mathbf{F}\mathbf{P}$, the translation and rotational directions are removed from the force constant matrix, and consequently the six (five) trivial vibrations become exactly zero (within the numerical accuracy of the machine).

The *resolution of the identity* (Section 4.11) method can be considered as an internal double projection on to an auxiliary set of functions $\mathbf{k}_i$. This may, for example, allow separation of a composite operator:

$$\begin{aligned}
\text{Value} &= \langle \mathbf{f}|\mathbf{O}_1\mathbf{O}_2|\mathbf{g}\rangle \\
\mathbf{I} &= \sum_{ij} |\mathbf{k}_i\rangle M_{ij}\langle \mathbf{k}_j| \quad ; \quad M_{ij} = \langle \mathbf{k}_i|\mathbf{k}_j\rangle^{-1} \\
\text{Value} &= \sum_{ij}\langle \mathbf{f}|\mathbf{O}_1|\mathbf{k}_i\rangle M_{ij}\langle \mathbf{k}_j|\mathbf{O}_2|\mathbf{g}\rangle
\end{aligned} \tag{17.96}$$

When the auxiliary set of functions is complete, the procedure is an exact identity, but the use of a finite number of functions in practice makes this an approximation, which of course can be controlled by the size of the auxiliary basis set. Resolution of identity methods are very closely connected to tensor decomposition methods, described in Section 17.6.3.

17.5 Differential Equations

Many of the fundamental equations in physics (and science in general) are formulated as differential equations. The desired mathematical function is typically known to obey some relationship in terms of its first and/or second derivatives. The task is to solve this differential equation to find the function itself. A complete treatment of the solution of differential equations is beyond the scope of this book and only a simplified introduction is given here. Furthermore, we will only discuss solutions of differential equations with one variable. In most cases the physical problem gives rise to a differential equation involving many variables, but prior to solution these can often be (approximately) decoupled by separation of the variables, as discussed in Section 1.6.

17.5.1 Simple First-Order Differential Equations

Differential equations where the right-hand side only depends on the variable are normally straightforward to solve. A simple example is shown below, where the first derivative of the unknown function

f is equal to the value of the variable x times a constant c:

$$\frac{\mathrm{d}f}{\mathrm{d}x} = cx \tag{17.97}$$

The equation can be solved formally by moving dx to the right-hand side and integrating:

$$\begin{aligned}\int \mathrm{d}f &= c\int x\,\mathrm{d}x \\ f &= c\left(\tfrac{1}{2}x^2 + a\right)\end{aligned} \tag{17.98}$$

The integral of df is f itself, while the integral of x dx is $\frac{1}{2}x^2$, except that any constant a can be added. This is completely general: a first-order differential equation will give one additional integration constant that must be determined by some other means, for example from knowing the functional value at some point. That the function in Equation (17.98) indeed is a solution to the differential equation in Equation (17.97) can be verified by differentiation. The same technique can be applied if the right-hand side is a more complicated function of x, although in some cases the resulting integral cannot be found analytically, in which case numerical methods can be employed.

17.5.2 Less Simple First-Order Differential Equations

Differential functions where the right-hand side depends only on the variable are relatively simple. A slightly more difficult problem arises when the right-hand side depends on the function itself:

$$\frac{\mathrm{d}f}{\mathrm{d}x} = cx \tag{17.99}$$

The task is now to find a function that upon differentiation gives the same function, except for a multiplicative constant. Formally it can be solved as by separating the variables and integrating:

$$\begin{aligned}\int f^{-1}\mathrm{d}f &= c\int \mathrm{d}x \\ \ln(f) &= cx + a \\ f &= \mathrm{e}^{cx+a} = \mathrm{e}^{a}\mathrm{e}^{cx} \\ f &= A\mathrm{e}^{cx}\end{aligned} \tag{17.100}$$

The solution is an exponential function where the integration constant can be written as a multiplicative factor A. It may again be verified that the solution indeed satisfies the original differential equation by differentiation.

17.5.3 Simple Second-Order Differential Equations

A second-order differential equation involves the second derivative of the function

$$\frac{\mathrm{d}^2 f}{\mathrm{d}x^2} = cx \tag{17.101}$$

Since the second derivative may be written as two consecutive differentiations, it can formally be solved by applying the above technique twice. The first integration gives the same solution as in

Equation (17.98), with an integration constant a_1:

$$\begin{aligned}\frac{d^2f}{dx^2} &= \frac{d}{dx}\left(\frac{df}{dx}\right) = cx \\ \left(\frac{df}{dx}\right) &= c\int x\,dx \\ \frac{df}{dx} &= \frac{1}{2}cx^2 + a_1\end{aligned} \tag{17.102}$$

The second integration is now analogous to a simple first-order differential equation:

$$\begin{aligned}\frac{df}{dx} &= \frac{1}{2}cx^2 + a_1 \\ \int df &= \frac{1}{2}c\int x^2 dx + a_1\int dx \\ f &= \frac{1}{6}cx^3 + a_1x + a_2\end{aligned} \tag{17.103}$$

Solving the second-order differential equation produces two integration constants, which must be assigned based on knowledge of the function at two points.

17.5.4 Less Simple Second-Order Differential Equations

Analogously to first-order differential equations, second-order differential equations may have the function itself on the right-hand side, as, for example, in

$$\frac{d^2f}{dx^2} = cf \tag{17.104}$$

Another example is when the right-hand side involves both the function and its first derivative:

$$\frac{d^2f}{dx^2} = c_1f + c_2\frac{df}{dx} \tag{17.105}$$

The right-hand side may also involve both the unknown function f and another known function of the variable, such as x^2:

$$\frac{d^2f}{dx^2} = c_1f + c_2x^2 \tag{17.106}$$

Equations of this type are representative of the Schrödinger equation, although in this case it is often written in a slightly different form with the kinetic energy operator plus the potential energy on the left-hand side, and with the c_1 constant written as an energy ε:

$$\frac{d^2f}{dx^2} + cx^2 = \varepsilon f \tag{17.107}$$

Equation (17.107), for example, arises for a harmonic oscillator, where the potential energy depends on the square of the variable.

The task in these cases is to find a function that upon differentiation twice gives some combination of the same function, its derivative and variable. Such differential equations cannot be solved by the above "separation of variables" technique. A detailed discussion of how to solve second-order differential equations is beyond the scope of this book, but we will consider two special cases that often arise in computational chemistry.

17.5.5 Second-Order Differential Equations Depending on the Function Itself

A second-order differential equation with the function itself on the right-hand side times a positive constant is, for example, involved in solving the radial part of the Schrödinger equation for the hydrogen atom:

$$\frac{d^2f}{dx^2} = c^2 f \tag{17.108}$$

For reasons that will become clear shortly, we have written the constant as c^2, rather than c. By reference to the corresponding first-order equation (Equation (17.100)), we may guess that a possible solution is an exponential function:

$$f = Ae^{cx} \tag{17.109}$$

That this is indeed a solution can be verified by explicit differentiation twice. Recognizing that the corresponding exponential function with a negative argument is also a solution, we can write a more general solution as a linear combination of the two:

$$f = A_1 e^{cx} + A_2 e^{-cx} \tag{17.110}$$

This contains two constants, A_1 and A_2, as required for a solution to a second-order differential equation, and it is indeed the complete solution. The two integration constants A_1 and A_2 must be assigned based on physical arguments. For the radial part of the hydrogen atom, for example, the A_1 constant is zero, since the wave function must be finite for all values of x, and A_2 becomes a normalization constant.

A slightly different situation arises when the differential equation contains a negative constant on the right-hand side, such as that involved in solving the angular part of the Schrödinger equation for the hydrogen atom:

$$\frac{d^2f}{dx^2} = -c^2 f \tag{17.111}$$

The solutions are analogous to those above, except for the presence of a factor i in the exponentials:

$$f = A_1 e^{icx} + A_2 e^{-icx} \tag{17.112}$$

However, since complex exponentials can be combined to give sine and cosine functions, the complete solution can also be written as a linear combination of real functions:

$$f = B_1 \sin(cx) + B_2 \cos(cx) \tag{17.113}$$

The constants A_1/A_2 or B_1/B_2 must again be assigned based on physical arguments.

17.6 Approximating Functions

Although the fundamental mathematical equations describing a physical phenomenon are often very compact, such as, for example, the Schrodinger equation written in operator form, $\mathbf{H}\Psi = E\Psi$, their application to all but the simplest model systems usually leads to equations that cannot be solved in analytical form. Even if the equations could be solved, one may only be interested in the solution for a certain limited range of variables. It is therefore in many cases of interest to obtain an approximate solution, and preferably in a form where the accuracy of the solution can be improved in a systematic fashion. We will here consider four approaches for obtaining such approximate solutions:

1. *Taylor expansion.* The real function is approximated by a polynomial that is constructed such that it becomes more and more accurate the closer the variable is to the expansion point. Away from the expansion point, the accuracy can be improved by including more terms in the polynomial.
2. *Basis set expansion.* The unknown function is written as a (linear) combination of known functions. The accuracy is determined by the number and mathematical form of the expansion functions. In contrast to a Taylor expansion, which has an error that increases as the variable is removed from the expansion point, a basis set expansion tends to distribute the error over the whole variable range. The error can be reduced by adding more functions in the expansion.
3. *Grid representation.* This is similar to expansion in a basis set, except that the known functions are points (δ-functions) rather than continuous functions. The accuracy is determined by the number of grid points and their location.
4. *Tensor decomposition.* The function representation in a given (large) basis set is compressed into a smaller basis set, which is derived in systematic fashion from the large basis set.

17.6.1 Taylor Expansion

The idea in a *Taylor expansion* is to approximate the unknown function by a polynomial centered at an expansion point x_0, typically at or near the "center" of the variable of interest. The coefficients of an Nth-order polynomial are determined by requiring that the first N derivatives match those of the unknown function at the expansion point. For a one-dimensional case this can be written as

$$f(x) = f(x_0) + \left.\frac{\partial f}{\partial x}\right|_{x_0} (x - x_0) + \frac{1}{2!}\left.\frac{\partial^2 f}{\partial x^2}\right|_{x_0} (x - x_0)^2 + \frac{1}{3!}\left.\frac{\partial^3 f}{\partial x^3}\right|_{x_0} (x - x_0)^3 + \cdots \tag{17.114}$$

For a many-dimensional function, the corresponding second-order expansion can be written as

$$f(\mathbf{x}) = f(\mathbf{x}_0) + \mathbf{g}^{\mathrm{t}}(\mathbf{x} - \mathbf{x}_0) + \tfrac{1}{2}(\mathbf{x} - \mathbf{x}_0)^{\mathrm{t}}\mathbf{H}(\mathbf{x} - \mathbf{x}_0) + \cdots \tag{17.115}$$

Here $\mathbf{g}^{\mathrm{t}}$ is a transposed vector (gradient) containing all the partial first derivatives and $\mathbf{H}$ is the (Hessian) matrix containing the partial second derivatives. In many cases, the expansion point x_0 is a stationary point for the real function, making the first derivative disappear, and the zeroth-order term can be removed by a shift of the origin:

$$f(x) \approx \tfrac{1}{2}k_2(x - x_0)^2 + \tfrac{1}{6}k_3(x - x_0)^3 + \tfrac{1}{24}k_4(x - x_0)^4 + \cdots \tag{17.116}$$

A Taylor expansion is an approximation to the real function by a polynomial terminated at order N. For a given (fixed) N, the Taylor expansion becomes a better approximation as the variable x approaches x_0. For a fixed point x at a given distance from x_0, the approximation can be improved by including more terms in the polynomial. Except for the case where the real function is a polynomial, however, the Taylor expansion will always be an approximation. Furthermore, as one moves away from the expansion point, the rate of convergence slows down, that is more and more terms are required to reach a given accuracy. At some point the expansion may become divergent, that is even inclusion of all terms up to infinite order does not lead to a well-defined value; this point is called the *radius of convergence*. It is determined by the distance from the expansion point to the nearest point (which may be in the complex plane) where the function has a singularity. A Taylor expansion of the function ln $(1 + x)$ around $x = 0$, for example, has a convergence radius of 1, as the logarithm function is not defined for $x = -1$. Attempting to approximate ln $(1 + x)$ by a Taylor expansion for x-values near -1 or 1 will thus require inclusion of a very large number of terms and will not converge if $x \geq 1$.

A specific example of a Taylor expansion is the molecular energy as a function of the nuclear coordinates. The real energy function is quite complex, but for describing a stable molecule at sufficiently low temperatures, only the functional form near the equilibrium geometry is required. Terminating the expansion at second order corresponds to modeling the nuclear motion by harmonic vibrations, while higher-order terms introduce anharmonic corrections.

For illustration, we will consider the Morse potential in reduced units.

$$y_{\text{Morse}} = (1 - e^{-(x-1)})^2 \tag{17.117}$$

Figure 17.6 shows the second-, third- and fourth-order Taylor approximations to the Morse potential.

Approximating the real function by a second-order polynomial forms the basis for the Newton–Raphson optimization techniques described in Section 13.2.

17.6.2 Basis Set Expansion

An alternative way of modeling an unknown function is to write it as a linear combination of a set of known functions, often called a *basis set*. The basis functions may or may not be orthogonal:

$$f(x) = \sum_{i=1}^{M} c_i \chi_i(x) \tag{17.118}$$

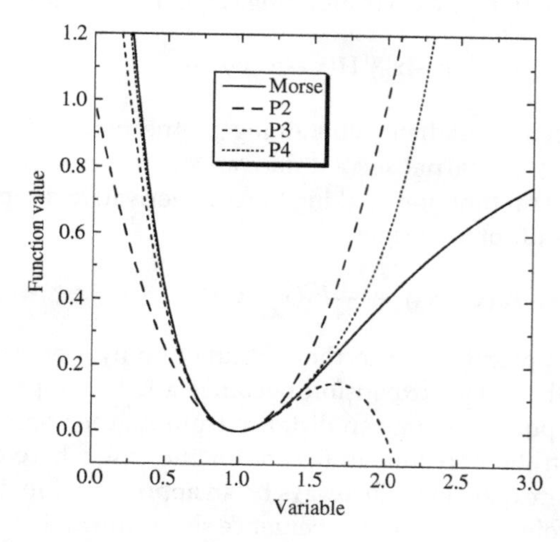

Figure 17.6 Taylor approximations to the Morse potential.

This corresponds to describing the function f in an M-dimensional space of the basis functions χ. For a fixed basis set size M, only the components of f that lie within this space can be described, and f is therefore approximated. As the size of the basis set M is increased, the approximation becomes better since more and more components of f can be described. If the basis set has the property of being *complete*, the function f can be described to any desired accuracy, provided that a sufficient number of functions are included. The expansion coefficients c_i are often determined either by variational or perturbational methods. For the expansion of the molecular orbitals in a Hartree–Fock wave function, for example, the coefficients are determined by minimizing the total energy.

The basis set expansion can be illustrated by using polynomials as basis functions for reproducing the Morse potential in Equation (17.117), that is the approximating function is given by

$$f(x) = \sum_{i=0}^{M} a_i(x-1)^i \tag{17.119}$$

The fitting coefficients a_i can be determined by requiring that the integrated difference in a certain range $[a,b]$ is a minimum:

$$ErrF = \int_a^b (y_{\text{Morse}} - f(x))^2 dx \tag{17.120}$$

Taking the range to be either [0.5, 2.0] or [0.2, 2.5] produces the fits for a second-, third- and fourth-order polynomials shown in Figure 17.7.

Note that the polynomial no longer has the same minimum as the Morse potential but provides a smaller *average* error over the fitting range than the corresponding Taylor polynomial. The Taylor expansion provides an exact fit at the expansion point, which rapidly deteriorates as the variable moves away from the reference point, while the basis set expansion provides a rather uniform accuracy over the fitting range, at the price of sacrificing local accuracy.

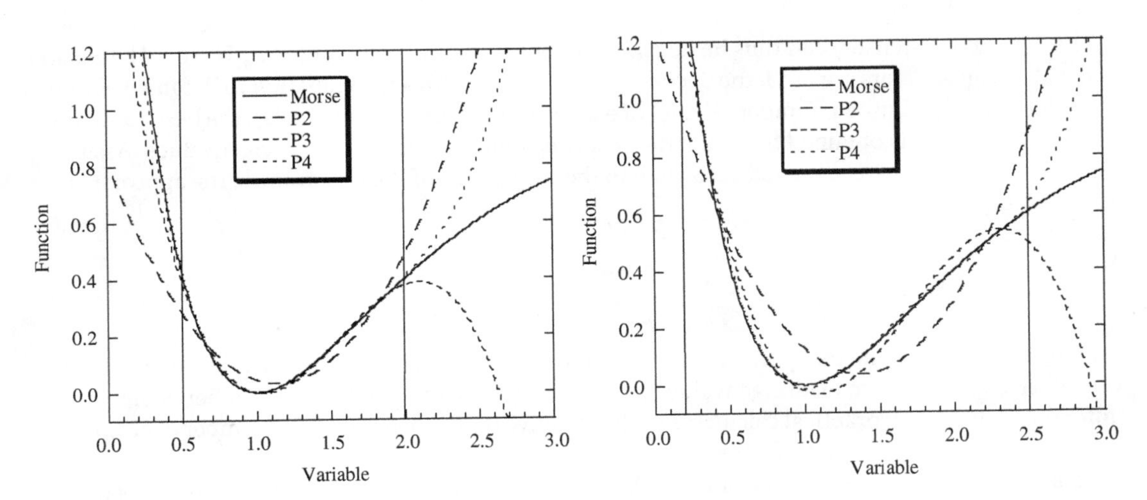

Figure 17.7 Approximations to the Morse potential by expansion in a polynomial basis set; vertical lines indicate the range used for the fitting.

17.6.3 Tensor Decomposition Methods

The goal of many statistical and computational methods can be viewed as an information compression: how can the number of variables required to describe a given property to a given accuracy be reduced by forming new descriptors from the original set of variables? This has connections to optimization theory and can often be formulated as minimization or maximization of a suitable target function. It can alternatively be formulated in terms of *tensor decomposition* methods.[10] For a 2-tensor, a matrix, the *Principal Component Analysis* (PCA, Section 18.4.3)) is a prototypical example of a tensor decomposition method and corresponds operationally to selecting the new variables as the matrix eigenvectors. PCA thus derives the new variables from properties of the variables themselves. The *Partial Least Squares* (PLS, Section 18.4.4)) method attempts to derive a new set of variables by also taking into account information about the property that the variables should describe. Tensor decomposition can be considered as a generalization of PCA to arrays of dimension larger than 2 and *Higher-Order PLS* (HOPLS) is a corresponding generalization of PLS.[11] PCA and PLS methods are heavily used in correlation analyses, as described in Sections 18.4 and 18.5, which also contain an illustrative example. Only the PCA generalization will be discussed in the following, starting with the simplest case of a square symmetric matrix.

An $N \times N$ symmetric (real) matrix **A** can be diagonalized by a unitary transformation **U**, as shown below (same as Equation (17.33)):

$$\mathbf{\Lambda} = \mathbf{U}\mathbf{A}\mathbf{U}^{-1} \quad ; \quad \mathbf{A} = \mathbf{U}^{-1}\mathbf{\Lambda}\mathbf{U} \tag{17.121}$$

The **A** matrix can be written as a sum of outer products of the (normalized) eigenvectors $\mathbf{u}_\alpha$ contained as column vectors in the **U** matrix:

$$\begin{aligned} \mathbf{A} &= \sum_{\alpha=1}^{N} \lambda_\alpha \mathbf{u}_\alpha \mathbf{u}_\alpha^{\mathrm{t}} \\ A_{ij} &= \sum_{\alpha=1}^{N} \lambda_\alpha u_{i\alpha} u_{j\alpha} \end{aligned} \tag{17.122}$$

The **A** matrix (2-tensor) can thus be decomposed into a sum of products of $\mathbf{u}_\alpha$ vectors (1-tensors), and if all N eigenvectors are used, the decomposition is exact. An approximation to **A** can be obtained by including only a limited number ($K < N$) of eigenvectors in Equation (17.122); the best approximation with K vectors is obtained by taking those corresponding to the K largest eigenvalues. Assuming that the eigenvectors are arranged according to the magnitude of the eigenvalues, the approximation at level K is given by

$$\mathbf{A}^K = \sum_{\alpha=1}^{K} \lambda_\alpha \mathbf{u}_\alpha \mathbf{u}_\alpha^{\mathrm{t}} \tag{17.123}$$

One can loosely define the (percent-wise) error at level K by the sum of the neglected $N - K$ eigenvalues relative to the sum of all eigenvalues (= trace of **A**). The rank of **A** is the number of eigenvalues different from zero. In practical calculations, the effective rank is defined as the number of eigenvalues larger than a predefined cutoff factor.

Equation (17.123) is the essence of PCA. A matrix representing the property of interest is decomposed into vector components ranked in terms of their importance by their eigenvalues, and only a limited number of components are used to capture the information to a given accuracy.

The generalization of (17.121) to arbitrary matrices is called *Singular Value Decomposition* (SVD):

$$\mathbf{A} = \mathbf{U}\boldsymbol{\Lambda}V^{\mathrm{t}} \tag{17.124}$$

A (real) **A** matrix can be written as a product of two unitary matrices **U** and **V**, and a diagonal matrix $\boldsymbol{\Lambda}$ containing non-negative numbers. If **A** is an $m \times n$ rectangular matrix, then **U** and **V** are square matrices of dimension $m \times m$ and $n \times n$, respectively, while $\boldsymbol{\Lambda}$ is a rectangular matrix of dimension $m \times n$ (**U** may also be defined to be of dimension $m \times n$, in which case $\boldsymbol{\Lambda}$ has dimension $n \times n$). Equations (17.125) and (17.126) show that **U** and **V** contain eigenvectors of the $\mathbf{AA}^{\mathrm{t}}$ and $\mathbf{A}^{\mathrm{t}}\mathbf{A}$ matrices, and $\boldsymbol{\Lambda}$ contains the square root of the eigenvalues of these eigenvalue equations, which therefore necessarily are non-negative:

$$\mathbf{AA}^{\mathrm{t}} = \mathbf{U}\boldsymbol{\Lambda}V^{\mathrm{t}}\mathbf{V}\boldsymbol{\Lambda}^{\mathrm{t}}\mathbf{U}^{\mathrm{t}} = \mathbf{U}\boldsymbol{\Lambda}^2\mathbf{U}^{\mathrm{t}} \tag{17.125}$$

$$\mathbf{A}^{\mathrm{t}}\mathbf{A} = \mathbf{V}\boldsymbol{\Lambda}^{\mathrm{t}}\mathbf{U}^{\mathrm{t}}\mathbf{U}\boldsymbol{\Lambda}V^{\mathrm{t}} = \mathbf{V}\boldsymbol{\Lambda}^2\mathbf{V}^{\mathrm{t}} \tag{17.126}$$

Equation (17.122) can be generalized to

$$\begin{aligned} \mathbf{A} &= \sum_{\alpha=1} \lambda_\alpha \mathbf{u}_\alpha \mathbf{v}_\alpha^{\mathrm{t}} \\ A_{ij} &= \sum_{\alpha=1}^{R} \lambda_\alpha u_{i\alpha} v_{j\alpha} \end{aligned} \tag{17.127}$$

The effective rank of **A** is the number of λ values larger than a predefined cutoff factor and the best rank R approximation to **A** is obtained by taking the first R SVD components.

Tensor decomposition methods can be viewed as a generalization of SVD to dimensions larger than 2, but in contrast to SVD there is no unique method for the decomposition. The rank of a tensor is furthermore not a unique and well-defined quantity, as the rank in different dimensions may be different and may depend on the method of decomposition. For illustration purposes we will in the following assume a tensor **M** of dimension 3 with N elements and (effective) rank R ($R \leq N$) in all dimensions. This can be generalized to higher dimensions and different numbers of elements and ranks in different dimensions, at the price of introducing further indices and summations.

The equivalent of Equation (17.127) is Equation (17.128), where a 3-tensor **M** is decomposed into a sum of outer products of three vectors. The scalar s determines the weight of each product component and corresponds to the singular value λ, but is often included in the definition of the **u**, **v**, **w** vectors:

$$\begin{aligned} \mathbf{M}^R &= \sum_{\alpha=1}^{R} s_\alpha \mathbf{u}_\alpha \otimes \mathbf{v}_\alpha \otimes \mathbf{w}_\alpha \\ M_{ijk}^R &= \sum_{\alpha=1}^{R} s_\alpha u_{i\alpha} v_{j\alpha} w_{k\alpha} \end{aligned} \tag{17.128}$$

This decomposition is called *canonical decomposition* (CANDECOMP), *parallel factorization* (PARAFAC) or *canonical polyadic* (CP). In contrast to the SVD procedure for the twodimensional case, there is no deterministic method for performing the decomposition. The decomposition can be determined iteratively by minimizing the error function defined by

$$ErrF(R) = \sum_{ijk}^{N} \left(M_{ijk} - \sum_{\alpha}^{R} s_\alpha u_{i\alpha} v_{j\alpha} w_{k\alpha} \right)^2 \tag{17.129}$$

The effective rank R must be determined by sequentially including an increasing number of **u**, **v**, **w** vectors, until the value of the error function is below a chosen cutoff. The minimization can be performed by the *alternating least squares* procedure, where $ErrF$ is minimized by a sequence of iterations each consisting of: (1) keeping j and k fixed and minimizing with respect to i, (2) keeping i and k fixed and minimizing with respect to j, (3) keeping i and j fixed and minimizing with respect to k. Alternatively, the minimization can be performed by gradient methods described in Section 13.2. The minimization of $ErrF$ is unfortunately an ill-posed problem and numerically not stable, and the objective function may contain multiple minima.

An alternative decomposition method is the *Tucker decomposition* shown below, where **m** is a tensor of the same dimension as **M**, but only of size R (i.e. **m** contains only R^{d} elements instead of N^{d} elements):

$$\begin{aligned} \mathbf{M}^R &= \sum_{\alpha\beta\gamma=1}^{R} \mathbf{m} \cdot (\mathbf{u}_\alpha \otimes \mathbf{v}_\beta \otimes \mathbf{w}_\gamma) \\ M_{ijk}^R &= \sum_{\alpha\beta\gamma=1}^{R} m_{\alpha\beta\gamma} u_{i\alpha} v_{j\beta} w_{k\gamma} \end{aligned} \tag{17.130}$$

The advantage of the Tucker decomposition is that it can be calculated by a series of SVD procedures, and thus does not suffer from the numerical problems of the CP decomposition. The disadvantage is that the Tucker method retains a tensor **m** of dimension d, although only the size of the (effective) rank R, and is thus primarily useful for tensors of low (3 or 4) dimensions. The CP decomposition can be considered as a Tucker decomposition where the **m** tensor is super-diagonal and has the same rank in all dimensions, and thus reduces to the scalar s in Equation (17.128).

The Tucker decomposition retains a coupling between (all) the **u**, **v**, **w** component vectors (the **m** tensor), while the CP procedure completely decouples all the **u**, **v**, **w** component vectors. An intermediate method is the *Tensor-Train* (TT) or *Matrix Product State* (MPS) decomposition defined below, which can be viewed as a Tucker decomposition where only a sequential coupling of two-dimensional tensors (matrices) is retained:

$$\begin{aligned} \mathbf{M}^R &= \sum_{\alpha\beta=1}^{R} \mathbf{u}_\alpha \otimes \mathbf{v}_{\alpha\beta} \otimes \mathbf{w}_\beta \\ M_{ijk}^R &= \sum_{\alpha\beta=1}^{R} u_{i\alpha} v_{j\alpha\beta} w_{k\beta} \end{aligned} \tag{17.131}$$

The first and last components are one-dimensional (vectors) while all intermediate components are two-dimensional (matrices). The TT/MPS decomposition can be calculated by a series of numerically stable SVD procedures, where $d - 1$ tensor indices are initially grouped together and subsequently ungrouped one-by-one after each SVD. Note that this decomposition depends on the order in which the indices are selected for ungrouping.

17.6.4 Examples of Tensor Decompositions

Tensor decomposition methods, especially in the form of PCA, are widespread in computational chemistry, but often carry other labels. PLS methods are widely used in correlation analyses

(Section 18.4.4), where the response to be described is known in advance, but are less used in computational chemistry, since the response often is the (unknown) result of the decomposition. The PCA/PLS notation, however, may provide a conceptual framework for classifying many procedures. Some examples are:

1. The generation of non-redundant natural internal coordinates from a large set of primitive internal coordinates for geometry optimization (Section 13.3). A set of primitive internal coordinates $\mathbf{p}$ can be defined from a set of Cartesian atomic coordinates ($\mathbf{x}_{3N}$) by including all distances, angles and torsional angles (or other types of internal coordinates) between atoms within predefined distance criteria, that is $\mathbf{p} = f(\mathbf{x}_{3N})$. This set of internal coordinates will normally be (much) larger than the $3N_{atom} - 6$ non-redundant internal coordinates. The $\mathbf{B}$ matrix contains the first derivative of the internal coordinates with respect to the Cartesian coordinates, and performing a PCA on the $\mathbf{B}^t\mathbf{B}$ matrix will provide $3N_{atom} - 6$ eigenvalues different from zero; the corresponding eigenvectors are the $3N_{atom} - 6$ non-redundant linear combinations of the primitive internal coordinates:

$$\mathbf{p} = f(\mathbf{x}_{3N}) \quad ; \quad B_{i\alpha} = \frac{\partial p_i}{\partial x_\alpha} \tag{17.132}$$

2. Vibrational normal modes can be considered as the principle components of the harmonic vibrational Hamilton operator or as the PLS components of mass-weighted coordinates for describing the vibrational energy.
3. Hartree–Fock (or Kohn–Sham) Self-Consistent Field calculations. The original descriptors are the non-orthogonal Gaussian basis functions and the principle components are the eigenvectors of the overlap matrix $\mathbf{S}$. PCA is thus used to transform the HF matrix equation into regular form ($\mathbf{FC} = \mathbf{SCE}$ into $\mathbf{F}'\mathbf{C}' = \mathbf{C}'\mathbf{E}$) and for detecting/eliminating linear dependence in the basis set. The SCF procedure can be considered as a non-linear PLS (linear and quadratic dependence) procedure. In the first SCF iteration the Fock matrix is non-diagonal and the density matrix have non-zero eigenvalues for all MOs. When converged, the Fock matrix is diagonal and the density matrix has $^1/_2N_{\text{elec}}$ eigenvalues = 2, while the rest (unoccupied MOs) have eigenvalues = 0. The final PLS components are the MOs (occupied and virtual), where only $^1/_2N_{\text{elec}}$ (occupied) MOs are required to describe the electronic Hamilton operator, instead of M_{basis} MOs.
4. Natural orbitals from a (correlated) wave function (Section 10.5). Löwdin showed that the orbitals obtained by diagonalizing the density matrix from a full-CI wave function provide a systematic approach for obtaining the best approximation using a fixed number of (natural) orbitals by including those with largest eigenvalues (occupation numbers). Natural orbitals can thus be considered as a PCA of the density matrix.
5. Configuration interaction. The CI wave functions can be considered as the principle components of the electronic Hamilton operator or as the PLS components in the space of Slater determinants for describing the electronic energy.
6. Natural transition orbitals correspond to an SVD decomposition of the transition density matrix arising in excited state calculations and produce a compact representation of the "hole" and "particle" nature of the excitation (Section 4.14.1).[12] Alternatively, a PCA can be performed on the density difference matrix, producing attachment/detachment natural orbitals, which again provide an easy visualization of the nature of an electronic excitation.[12]
7. The generation of better descriptors in QSAR models (Section 18.4). A large number of descriptors for a given property collected in an $\mathbf{X}$ matrix may be internally correlated and unsuitable for generating linear correlation models. Performing a PCA on the $\mathbf{X}^t\mathbf{X}$ matrix allows extraction

of linear combinations of descriptors that are mutually orthogonal and ordered according to the largest variation in the descriptors. A PLS decomposition instead extracts new variables from the $(\mathbf{y}^t\mathbf{X})^t(\mathbf{y}^t\mathbf{X})$ matrix, where $\mathbf{y}$ contains the response to be described.

8. Description of molecular dynamics trajectories by a PCA analysis of the time-dependent correlation matrix of atomic coordinates. The complicated overall motion of a large structure (e.g. protein) from an MD trajectory can often effectively be described by a few (3–5) PCA components displaying collective motion of many/all atoms. The selection of collective variables for steering/guiding an MD simulation can be considered as trying to guess a few important PLS components.
9. The collection of all two-electron integrals in a basis set χ can be viewed as a four-dimensional tensor $\mathbf{G}$ with elements G_{ijkl} as defined below (same as Equation (3.62) except for a change in index labels):

$$\int \chi_i(1)\chi_j(1)\left(\frac{1}{|\mathbf{r}_1 - \mathbf{r}_2|}\right)\chi_k(2)\chi_l(2)\mathrm{d}\mathbf{r}_1\mathrm{d}\mathbf{r}_2 = (\chi_i\chi_j\,|\,\chi_k\chi_l) = G_{ijkl} \tag{17.133}$$

The Cholesky decomposition approximates the 4-tensor as a sum of products of 3-tensors using a single external variable:

$$\mathbf{G}^R = \sum_{\alpha=1}^{R} (\chi_i\chi_j\,|\,\chi'_\alpha)(\chi'_\alpha\,|\,\chi_k\chi_l) \tag{17.134}$$

The resolution of identity approach approximates the 4-tensor as a sum of products of two 3-tensors with a 2-tensor using two external variables. The 2-tensor can be split into two components and included in the 3-tensors:

$$\begin{aligned}\mathbf{G}^R &= \sum_{\alpha\beta=1}^{R} (\chi_i\chi_j\,|\,\chi'_\alpha)\left(\chi'_\alpha\,|\,\chi'_\beta\right)^{-1}\left(\chi'_\beta\,|\,\chi_k\chi_l\right)\\ &= \sum_{\alpha\beta=1}^{R} \left[(\chi_i\chi_j\,|\,\chi'_\alpha)\left(\chi'_\alpha\,|\,\chi'_\beta\right)^{-1/2}\right]\left[\left(\chi'_\alpha\,|\,\chi'_\beta\right)^{-1/2}\left(\chi'_\beta\,|\,\chi_k\chi_l\right)\right]\end{aligned} \tag{17.135}$$

The resolution of identity for simplifying three-electron integrals (Section 4.11) can similarly be considered as a tensor decomposition, where a six-dimensional tensor is decomposed into products of four-dimensional quantities.

10. The coupled cluster amplitudes for the doubly excited state is similarly a four-dimensional tensor; an attempt has been made to decompose these into simpler quantities.

The attractive aspect of tensor decomposition methods is the reduced complexity, but the algebra involved in performing calculations with reduced rank tensors need careful analysis. Multiplications of two tensors of ranks R_1 and R_2 formally produce a new tensor of rank $R_1 \cdot R_2$, but it is quite likely that this can be decomposed into a new tensor of lower rank. The iterative solution of the coupled cluster equations, for example, consists of a series of tensor products, and each such product must be subjected to a tensor decomposition before progressing to the next step in order to keep the complexity at a minimum. This clearly requires a numerically stable decomposition procedure and careful control of the numerical accuracy in each step in order to keep the whole algorithm numerically stable.

17.7 Fourier and Laplace Transformations

Transforming functions between different coordinate systems can often simplify the description. It may also be advantageous in some cases to switch between different *representations* of a function. A function in real space, for example, can be transformed into a reciprocal space, where the coordinate axes have units of inverse length. Similarly, a function defined in time may be transformed into a representation of inverse time or frequency. Such interconversions can be done by *Fourier transformations*. The Fourier transform g of a function f is defined as

$$g(k) = \int_{-\infty}^{\infty} f(r)\mathrm{e}^{-ikr}\mathrm{d}r \tag{17.136}$$

The inverse transformation is given as

$$f(r) = \frac{1}{2\pi}\int_{-\infty}^{\infty} g(k)\mathrm{e}^{ikr}\mathrm{d}k \tag{17.137}$$

The factor of 2π can also be included as the square root in both the forward and reverse transformation or included in the complex exponent.

While the integral form of the Fourier transform is useful in analytical work, the computational implementation is often done by a discrete representation of the function(s) on a grid, in which case the integrals are replaced by a sum over grid points:

$$g(k_n) = \sum_{n=0}^{N-1} f(r_n)\mathrm{e}^{-ikr_n/N} \tag{17.138}$$

In a straightforward implementation of the discrete Fourier transform, the computational time increases as the square of the number of grid points. If, however, the number of grid points is an integer power of two, that is $N_{\text{grid}} = 2^m$, the Fourier transform can be done recursively, and is called the *Fast Fourier Transform* (FFT). The FFT involves (only) a computational effort proportional to $N_{\text{grid}} \ln N_{\text{grid}}$, which is a substantial reduction relative to the general case of N_{grid}^2 for large values of N_{grid}.

Fourier transforms are often used in connection with periodic functions, for example for evaluating the kinetic energy operator in a density functional calculation where the orbitals are expanded in a plane wave basis.

The *Laplace transform* is defined as below, where the integral can again be approximated as a finite sum in practical applications:

$$g(k) = \int_{0}^{\infty} f(r)\mathrm{e}^{-kr}\mathrm{d}r \tag{17.139}$$

The inverse orbital energy differences in the MP2 method (Section 4.12), for example, can be rewritten as the sum over the auxiliary variable r, and a sufficient accuracy can often be obtained by including only a few points in the sum.[13]

17.8 Surfaces

A one-dimensional function $f(x)$ can be visualized by plotting the functional value against the variable x. A two-dimensional function $f(x, y)$ can similarly be visualized by plotting the functional value against the variables x and y. However, since plotting devices (paper or an electronic screen) are inherently two-dimensional, the functional value must be plotted in a pseudo-three-dimensional fashion, with the three-dimensional object being imagined by the viewer's brain. Functions with more than two variables cannot readily be visualized.

Functions in computational chemistry typically depend on many variables, often several hundreds, thousands or millions. For analysis purposes, it is possible to visualize the behavior of such functions in a reduced variable space, that is keeping some (most) of the variables constant. Figure 17.8 shows the value of the acrolein LUMO (lowest unoccupied molecular orbital) in a two-dimensional cut 1 Å above the molecular plane.[14] The magnitude and sign of the orbital is plotted along the third perpendicular dimension.

An alternative way of visualizing multivariable functions is to condense or contract some of the variables. An electronic wave function, for example, is a multivariable function, depending on $3N$ electron coordinates. For an independent-particle model, such as Hartree–Fock or density functional theory, the total (determinantal) wave function is built from N orbitals, each depending on three coordinates:

$$\Phi_{\text{HF/DFT}} = \frac{1}{\sqrt{N!}} \begin{vmatrix} \phi_1(1) & \phi_2(1) & \cdots & \phi_N(1) \\ \phi_1(2) & \phi_2(2) & \cdots & \phi_N(2) \\ \vdots & \vdots & \ddots & \vdots \\ \phi_1(N) & \phi_2(N) & \cdots & \phi_N(N) \end{vmatrix} \tag{17.140}$$

The electron density can be obtained by integrating the coordinates for $N - 1$ electrons, giving a function depending on only three coordinates:

$$\rho(x, y, z) = \int \Phi^2(x_1, y_1, z_1, x_2, y_2, \ldots, z_N) \mathrm{d}x_2 \mathrm{d}y_2 \cdots \mathrm{d}z_N \tag{17.141}$$

Functions of three variables can be visualized by generating a surface in the three-dimensional space corresponding to a constant value of the function, for example $\rho(x, y, z) =$ constant. Such surfaces can be plotted, again using the viewer's brain for generating the illusion of a three-dimensional object. The full three-dimensional figure can be visualized by plotting surfaces for different values. Figure 17.9

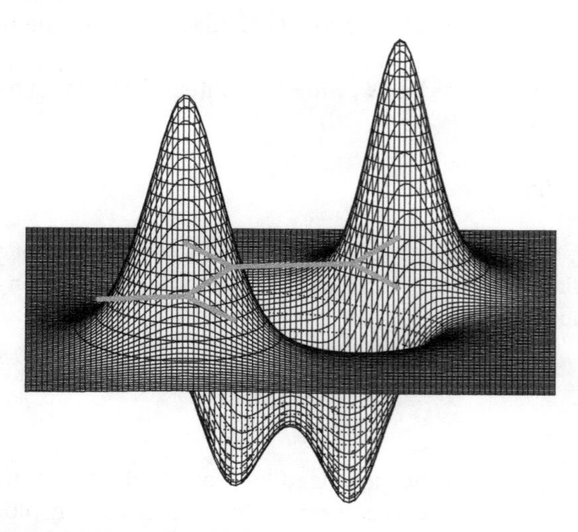

Figure 17.8 Representation of the acrolein LUMO orbital.

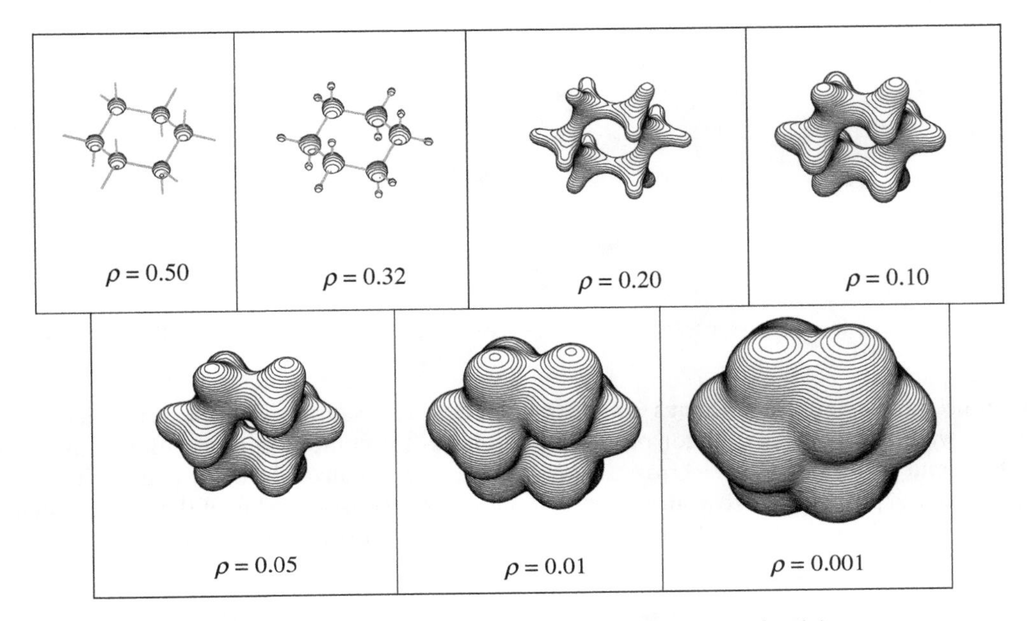

Figure 17.9 Surfaces corresponding to the indicated value for the electron density of cyclohexane.

shows the total electron density of cyclohexane, plotted for decreasing density values. The scales of the seven plots are the same, that is the sizes of the surfaces are directly comparable.

The first box corresponds to $\rho = 0.50$ and only the core electrons for the carbon atoms are visible. For $\rho = 0.32$ the hydrogens also appear and bonds can be recognized for $\rho = 0.20$. Further reduction in the electron density level used for generating the surface obscures the bonding information and for $\rho = 0.001$ there is little information about the underlying molecular structure left. A surface generated by a constant value of the electron density defines the size and shape of a molecule, but the exact size and shape clearly depend on the value chosen. It can be noted that an isocontour value around 0.001 has often been taken to represent a van der Waals-type surface.

More information can be added to surface plots by color-coding. Orbitals, for example, have a sign associated with the overall shape, which can be visualized by adding two different colors or grey shading to the surface corresponding to a constant (numerical) value of the orbital. Figure 17.10 shows the acrolein LUMO in a grey-coded surface representation, which can be compared with the two-dimensional plot shown in Figure 17.8.

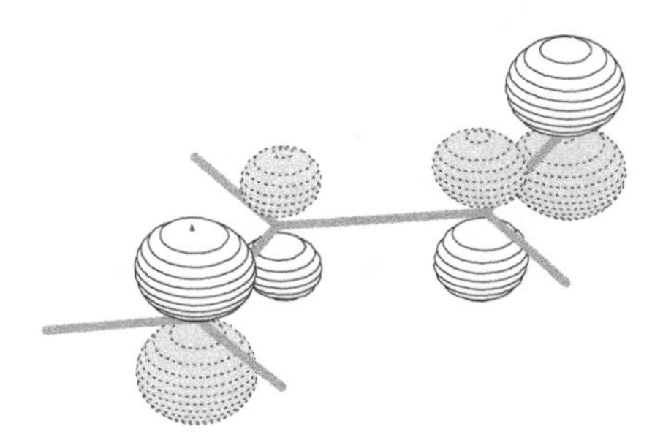

Figure 17.10 Grey-coded surface representation of the acrolein LUMO.

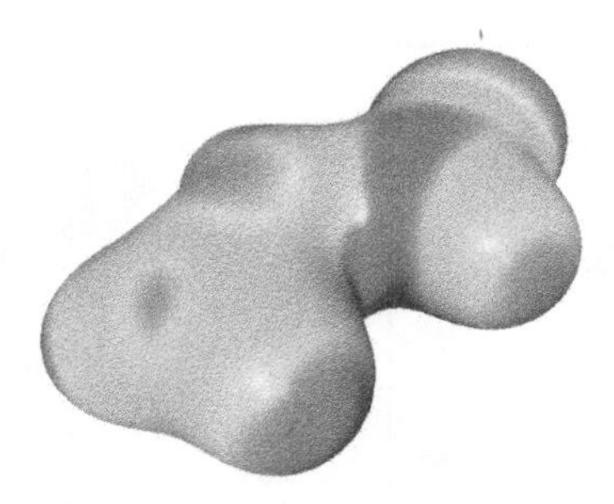

Figure 17.11 Electrostatic potential superimposed on a surface corresponding to a fixed value of the electron density for acrolein.

Other properties have a continuous range of values, not just a sign. An example is the electrostatic potential, which by itself is a function of three coordinates. The combined information of the molecular shape and the value of the electrostatic potential can be visualized by color-coding the value of the electrostatic potential on to a surface corresponding to a constant value of the electron density, as illustrated in Figure 17.11 for the acrolein molecule (color figure available on web site).

References

1 G. R. Kneller, *Molecular Simulation* 7 (1–2), 113–119 (1991).
2 W. Kutzelnigg, *Molecular Physics* **105** (19–22), 2627–2647 (2007).
3 N. C. Handy and A. M. Lee, *Chemical Physics Letters* **252** (5–6), 425–430 (1996).
4 O. Christiansen, *Journal of Chemical Physics* **120** (5), 2149–2159 (2004).
5 E. R. Davidson, *Journal of Computational Physics* **17** (1), 87–94 (1975).
6 J. Olsen, P. Jorgensen and J. Simons, *Chemical Physics Letters* **169** (6), 463–472 (1990).
7 P. O. Lowdin, *Journal of Chemical Physics* **18** (3), 365–375 (1950).
8 A. E. Reed, R. B. Weinstock and F. Weinhold, *Journal of Chemical Physics* **83** (2), 735–746 (1985).
9 W. H. Miller, N. C. Handy and J. E. Adams, *Journal of Chemical Physics* **72** (1), 99–112 (1980).
10 U. Benedikt, H. Auer, M. Espig, W. Hackbusch and A. A. Auer, *Molecular Physics* **111** (16–17), 2398–2413 (2013).
11 Q. Zhao, L. Zhang and A. Cichocki, *Wiley Interdisciplinary Reviews – Data Mining and Knowledge Discovery* **4** (2), 104–115 (2014).
12 A. Dreuw and M. Head-Gordon, *Chemical Reviews* **105** (11), 4009–4037 (2005).
13 M. Haser, *Theoretica Chimica Acta* **87** (1–2), 147–173 (1993).
14 G. Schaftenaar and J. H. Noordik, *Journal of Computer-Aided Molecular Design* **14** (2), 123–134 (2000).

18

Statistics and QSAR

18.1 Introduction

An essential component of calculations is to calibrate new methods and to use the results of calculations to predict or rationalize the outcome of experiments. Both of these types of investigation compare two types of data and the interest is in characterizing how well one set of data can represent or predict the other. Unfortunately, one or both sets of data usually contain "noise", and it may be difficult to decide whether a poor correlation is due to noisy data or to a fundamental lack of connection. Statistics is a tool for quantifying such relationships. We will start with some philosophical considerations and move into elementary statistical measures before embarking on more advanced tools.

The connection between reality and the outcome of a calculation can be illustrated as shown in Figure 18.1.

A specific example for "*Reality*" could be the (experimental) atomization energy of a molecule, defined as the energy required to separate a molecule into atoms, which is equivalent to the total binding energy. The atomization energy is closely related to the heat of formation, differing only by the zero-point reference state and neglect of vibrational effects. The zero point for the atomization energy scale is the isolated atoms, while it is the elements in their most stable form (e.g. H_2 and N_2) for the heat of formation. Since the dissociation energies for the reference molecules can also be measured, the atomization energy is an experimental observable quantity.

It is important to realize that each element in Figure 18.1 contains errors, and these can be either *systematic* or *random*. A systematic error is one due either to an inherent bias or to a user-introduced error. A random error is, as the name implies, a non-biased deviation from the "true" result. A systematic error can be removed or reduced, once the source of the error is identified. A random error, also sometimes called a *statistical* error, can be reduced by averaging the results of many measurements.

Model → Parameters → Computational implementation → Results ↔ Reality

Hartree–Fock → Basis set → Various cutoffs → Total energies ↔ Atomization energy

Figure 18.1 Relationship between *Model* and *Reality*.

Introduction to Computational Chemistry, Third Edition. Frank Jensen.

Companion Website: http://www.wiley.com/go/jensen/computationalchemistry3

Note that random errors can be truly random, for example due to thermal fluctuations or a cosmic ray affecting a detector, but may also be due to many small unrecognized systematic errors adding up to an apparent random noise.

Experimental measurements may contain both systematic and random errors. The latter can be quantified by repeating the experiment a number of times and taking the deviation between these results as a measure for the uncertainty of the (average) result. Systematic errors, however, are difficult to identify. One possibility for detecting these is to measure the same quantity by different methods or to use the same method in different laboratories. The literature is littered with independent investigations reporting conflicting results for a given quantity, each with error bars smaller than the deviation between the results. Such cases clearly indicate that at least one of the experiments contains unrecognized systematic errors.

Theory almost always contains "errors", but these are called "approximations" in the community. The Hartree–Fock method, for example, systematically underestimates atomization energies since it neglects electron correlation and the correlation energy is larger for molecules than for atoms. For other properties, the Hartree–Fock method has the same fundamental flaw, neglect of electron correlation, but this may not necessarily lead to systematic errors. For energy barriers for rotation around single bonds, which are differences between two energies for the same molecule with (slightly) different geometries, the contribution from the correlation energy often leads to near cancellation, and Hartree–Fock calculations do not systematically over- or underestimate rotational barriers.

The use of a basis set also introduces a systematic error but the direction depends on the specific basis set and the molecule to hand. For a system composed of second row elements (such as C, N, O), the isolated atoms can be completely described with *s*- and *p*-functions at the Hartree–Fock level, but molecules require the addition of higher angular momentum (polarization) functions. Using a basis set containing only *s*- and *p*-functions will systematically underestimate the atomization energy, while a basis set containing few *s*- and *p*-functions but many polarization functions may overestimate the atomization energy. In principle one should chose a balanced basis set, defined as one where the error for the molecule is almost the same as for the atoms, but since the number of basis functions of each kind necessarily is quantized (one cannot have a fractional number of basis functions), this is not rigorously possible, and will depend on the computational level in any case. A very large (complete) basis set will fulfill the balance criterion but this is usually impossible in practice. An example of a (systematic) user error is the use of one basis set for the molecule and another for the atoms, as is sometimes done by inexperienced users of electronic structure methods.

The computational implementation of the Hartree–Fock method involves a choice of a specific algorithm for calculating the integrals and solving the HF equations. Furthermore, various cutoff parameters are usually implemented for deciding whether to neglect certain integral contributions and a tolerance is set for deciding when the iterative HF equations are considered to be converged. Since computers perform arithmetic with a finite precision, given by the number of bits chosen to represent a number, this introduces truncation errors, which are of a random nature (roughly as many negative as positive deviations for a large sample). These random errors can be reduced by increasing the number of bits per number, by using smaller cutoffs and convergence criteria, but can never be completely eliminated. Usually these errors can be controlled and reduced to a level where they are insignificant compared with the other approximations, such as neglect of electron correlation and the use of a finite basis set.

Janes and Rendell have reported interesting work using *interval arithmetic* to rigorously quantify the numerical errors in a basis set Hartree–Fock calculation.[1] Interval arithmetic corresponds to replacing the value of a given quantity by two numbers that represent upper and lower bounds of the finite precision representation of the value. Such calibration studies are of interest as calculations move to ever larger systems. Linear scaling methods allow electronic structure calculations to be carried out with tens of thousands of basis functions, and these calculations formally require handling $\sim 10^{20}$ two-electron integrals. Although most of these are numerically small and can be neglected without problems, the potential accumulative numerical noise in such calculations is daunting. Based on extrapolations from smaller systems, the above work suggests that the relative energy error for a system with 10^5 basis functions may be $\sim 10^{-8}$, and since the total energy for such a system will be large, the numerical uncertainty in the absolute energy may be $\sim 10^{-3}$ au (~3 kJ/mol), which is comparable to the often quoted "chemical accuracy" level.

An alternative approach for quantifying numerical errors is to add small amounts of random numerical noise at various stages in a calculation and observe how sensitive the final results are towards the perturbations.[2] While interval arithmetic focuses on the worst case errors, the random noise approach allows for error cancellation and may be useful in locating potentially numerical unstable algorithms and for testing whether, for example, single precision algorithms may be sufficient for selected components of a given calculation.

18.2 Elementary Statistical Measures

Statistics is a tool for characterizing a large amount of data by a few key quantities and it may therefore also be considered as an information compression. Consider a data set containing N data points with values x_i ($i = 1, 2, 3, \ldots, N$). One important quantity is the *mean* or *average* value, denoted with either a bar or an angle bracket:

$$\bar{x} = \langle x \rangle = \frac{1}{N} \sum_{i=1}^{N} x_i \tag{18.1}$$

The average is the "middle" point, or the "center of gravity", of the data set but it does not tell how wide the data point distribution is. The data sets {3.0, 4.0, 5.0, 6.0, 7.0} and {4.8, 4.9, 5.0, 5.1, 5.2} have the same average value of 5.0.

The mean may depend on an external parameter, such as time. In a molecular dynamics simulation, for example, the average energy (NVT ensemble) or temperature (NVE ensemble) will depend on the simulation time. Indeed, a plot of the average energy or temperature against time can be used as a measure of whether the system is sufficiently (thermal) equilibrated to provide a realistic model.

The *width* or the *spread* of the data set can be characterized by the second moment, the *variance*:

$$\sigma^2 = \frac{1}{N-1} \sum_{i=1}^{N} (x_i - \bar{x})^2 \tag{18.2}$$

The "normalization" factor is $N - 1$ when the average is calculated from Equation (18.1); if the exact average is known from another source, the normalization is just N. For large samples the difference between the two is minute and the normalization is often taken as N. The square root of the variance (i.e. σ) is called the *standard deviation*. The above two data sets have standard deviations of 1.6 and 0.2, clearly showing that the first set contains elements further from the mean than the second.

If the distribution of the data is given by a Gaussian function ($\exp(-ax^2)$), then σ determines how large a percentage of the data is within a given range of the mean. Specifically, 68% of the data is within one σ of the mean, 95% is within 2σ and 99.7% is within 3σ. The measured result for experimental quantities is often given by the notation $\langle x \rangle \pm \sigma$. The σ is loosely called the "error bar", reflecting the common procedure of drawing a line stretching from $\langle x \rangle - \sigma$ to $\langle x \rangle + \sigma$ in a plot diagram. Note, however, that for a Gaussian data distribution there is a 32% chance that the actual value is outside this interval. Furthermore, for actual data, the distribution is rarely exactly Gaussian. Note also that the standard deviation depends inversely on the square root of the number of data points, that is for truly random errors, the standard deviation can be reduced by increasing the number of points. Error bars from experiments, however, are often based on only a few (2–10) data points.

The third and fourth moments are called the *skew* and *kurtosis*:

$$\text{Skew} = \frac{1}{N}\sum_{i=1}^{N}\left(\frac{x_i - \bar{x}}{\sigma}\right)^3 \tag{18.3}$$

$$\text{Kurtosis} = \frac{1}{N}\sum_{i=1}^{N}\left(\frac{x_i - \bar{x}}{\sigma}\right)^4 - 3 \tag{18.4}$$

These quantities are dimensionless, in contrast to the first and second moments (mean and variance). The skew, kurtosis and corresponding higher moments are rarely used.

The mean and variance are closely connected with the qualifiers *accurate* and *precise*. An accurate measure is one where the mean is close to the real value. A precise measure is one that has a small variance. The goal is an accurate and precise measurement (many data points close to the "real" value). An accurate but imprecise measurement (good mean, large variance) indicates large random and small systematic errors, while a precise but inaccurate measurement (poor mean, small variance) indicates small random but large systematic errors. This is illustrated in Figure 18.2.

Some data sets are (almost) symmetric, such as the digits in the phone book of a large city, containing almost the same number of elements below and above the mean value. Others may be asymmetric, for example containing many points slightly below the mean, but relatively few with much larger values than the mean (e.g. the income profile for the US population or the Boltzmann energy distribution). Higher-order moments such as the skew can be used to characterize such cases. Two alternative

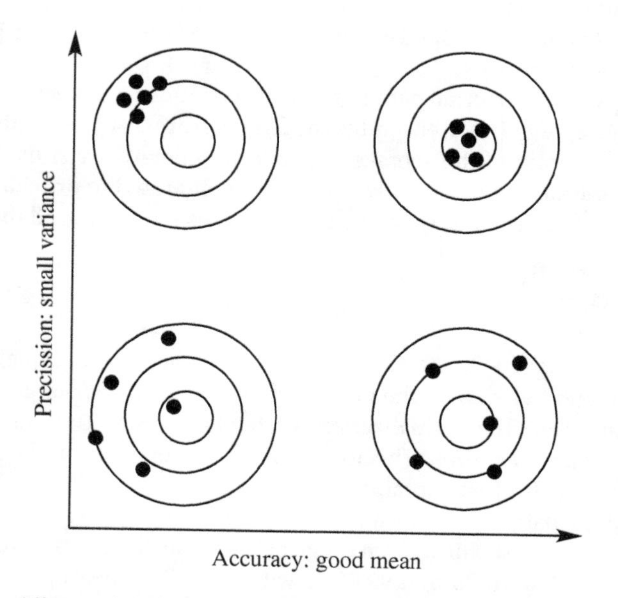

Figure 18.2 Illustrating the difference between the qualifiers: accurate and precise.

quantities can also be used, the *median* and the *mode*. The *median* is the *value in the middle* of the data points, that is 50% of the data are below the median and 50% are above. The *mode* is the *most probable element*, that is the one that occurs with the highest frequency. In some cases, there may be more than one maximum in the probability distribution, for example a *bimodal* distribution for a probability function with two maxima. The median and mean are identical for a symmetric distribution, and a large difference between these two quantities thus indicates an asymmetric distribution.

One should be aware that essential information can easily be lost in the data compression of a statistical analysis. European women, for example, have on average 1.5 children, but none have 1.5 children (but 0, 1, 2, 3, . . . children). Such "paradoxes" are at the root of characterizing statistics as "a precise and concise method of communicating half-truths in an inaccurate way".

18.3 Correlation between Two Sets of Data

It is often of interest to quantify whether one type of data is connected with another type, that is whether the data points from one set can be used to predict the other. We will denote two such data sets **x** and **y**, and ask whether there is a function *f*(**x**) that can model the **y** data. When the function *f*(**x**) is defined or known *a priori*, the question is how well the function can reproduce the **y** data.

Two quantities are commonly used for qualifying the "goodness of fit", the *Root Mean Square* (RMS) deviation and the *Mean Absolute Deviation* (MAD), which for a set of *N* data points are defined by

$$\mathrm{RMS} = \sqrt{\frac{1}{N}\sum_{i=1}^{N} (y_i - f(x_i))^2} \tag{18.5}$$

$$\mathrm{MAD} = \frac{1}{N}\sum_{i=1}^{N} |y_i - f(x_i)| \tag{18.6}$$

The MAD represents a uniform weighting of the errors for each data point, while the RMS quantity has a tendency to be dominated by the (few) points with the largest deviations. The maximum absolute deviation (MaxAD) is also a useful quantity, as it measures the worst-case performance of the model. A more detailed description of the deviations can be provided by a *box plot*, shown in Figure 18.3, where the median is indicated by the central line, the upper and lower quartiles are indicated by the box and the minimum and maximum correspond to the lower and upper lines.

Note that the function *f*(**x**) can be very complicated, for example calculating the atomization energy by the Hartree–Fock method from a set of nuclear coordinates (Figure 18.1).

When the functional form *f*(**x**) is unknown, correlation analysis may be used to seek an approximate function connecting the two sets of data. The simplest case corresponds to a linear correlation (Figure 18.4):

$$y_i = f(x_i) = ax_i + b \tag{18.7}$$

We want to determine the a (slope) and b (intersection) parameters to give the best possible fit, that is in a plot of y against x, we seek the best straight line.

The familiar least-squares linear fit arises by defining the "best" line as the one that has the smallest deviation between the actual y_i-points and those derived from Equation (18.7), and taking the error

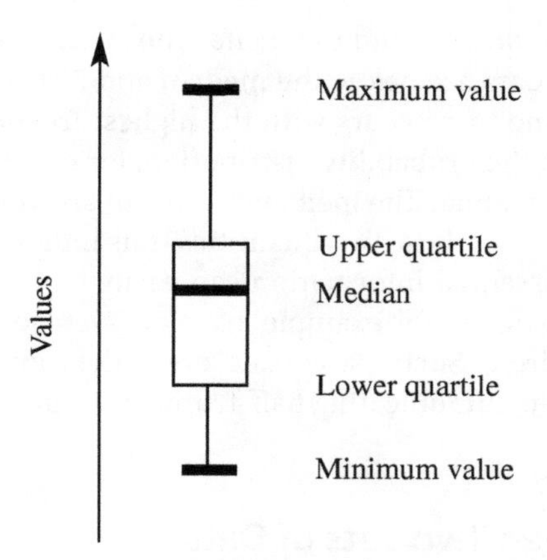

Figure 18.3 Illustrating how a box plot can visualize the spread of a data set.

to be the deviation squared. The equations defining a and b can be derived by minimizing (setting the first derivations to zero) an error function:

$$ErrF = \sum_{i=1}^{N} w_i(y_i - ax_i - b)^2$$
$$\left.\begin{matrix} \dfrac{\partial ErrF}{\partial a} = 0 \\ \dfrac{\partial ErrF}{\partial b} = 0 \end{matrix}\right\} \Rightarrow a, b \tag{18.8}$$

We note in passing that the minimum number of data points is two, since there are two fitting parameters a and b, that is the correlation between any two points can be modeled by a straight line. The data points can be weighted by w_i factors, for example related to the uncertainty with which the y_i

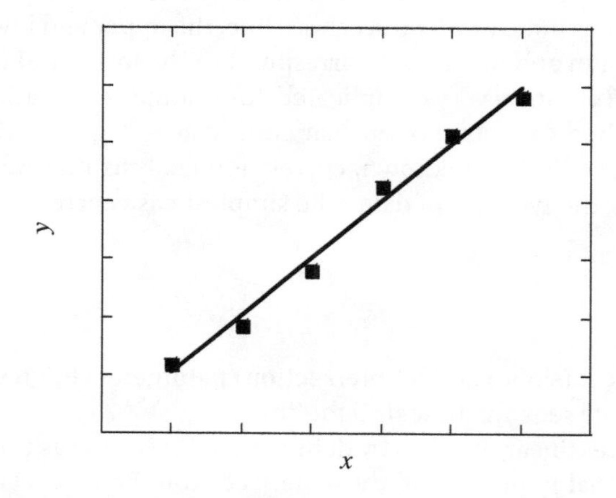

Figure 18.4 Illustrating an approximate linear correlation between x and y.

data points are measured. The actual equations for a and b can be written in several different ways. One convenient form is by introducing some auxiliary sum-functions:

$$S = \sum_{i=1}^{N} w_i \quad ; \quad S_x = \sum_{i=1}^{N} w_i x_i \quad ; \quad S_y = \sum_{i=1}^{N} w_i y_i$$
$$S_{xx} = \sum_{i=1}^{N} w_i x_i^2 \quad ; \quad S_{yy} = \sum_{i=1}^{N} w_i y_i^2 \quad ; \quad S_{xy} = \sum_{i=1}^{N} w_i x_i y_i \tag{18.9}$$
$$K = \left(SS_{xx} - S_x^2\right)^{-1}$$

The optimum a and b parameters are given in terms of these quantities by

$$a = K(SS_{xy} - S_x S_y) \quad ; \quad b = K(S_{xx}S_y - S_x S_{xy}) \tag{18.10}$$

The associated variances are given by

$$\sigma_a^2 = KS \quad ; \quad \sigma_b^2 = KS_{xx} \tag{18.11}$$

The "goodness of fit" for such xy-plots is commonly measured by the *correlation coefficient*, R, which is defined by

$$R = \frac{\sum_{i=1}^{N} (x_i - \bar{x})(y_i - \bar{y})}{\sqrt{\sum_{i=1}^{N} (x_i - \bar{x})^2}\sqrt{\sum_{i=1}^{N} (y_i - \bar{y})^2}} \tag{18.12}$$

The correlation coefficient is by construction constrained to the interval $[-1, 1]$, where $R = 1$ indicates that all points lie exactly on a line with a positive slope ($a > 0$), $R = -1$ indicates that all points lie exactly on a line with a negative slope ($a < 0$), while $R = 0$ indicates two sets of uncorrelated data. Note that the "correlation coefficient" is often given as R^2 instead, which of course is constrained to the interval $[0, 1]$ and measures the fraction of the variation described by the linear model.

The error function in Equation (18.8) is defined by the *vertical* distance, that is assuming that the error is located mainly in the **y** data set. If both data sets have similar errors, the *perpendicular* distance from the data points to the line can be used instead in the error function. This, however, leads to somewhat complicated non-linear equations for the fitting parameters[3] and is rarely used.

Non-linear correlations (e.g. a quadratic correlation, $f(x) = ax^2 + bx + c$) can be treated completely analogously to the linear case above, by defining an error function and setting the first derivatives with respect to the fitting parameters to zero. Non-linear correlations, however, are used much less than linear ones, for five reasons:

1. Many non-linear connections can be linearized by a suitable variable transformation. An exponential dependence, for example, can be made linear by taking the logarithm ($y = a\,e^{bx}$ becomes $\ln(y) = \ln(a) + bx$).
2. Increasing the number of fitting parameters will always produce a better fit, since the fitting function becomes more flexible (a quadratic fitting function has three parameters, while a linear one has only two). The data set, however, usually contains noise (random errors), and a more flexible fitting function may simply try to fit the noise rather than improving the fit of the "true" data. For polynomial fitting functions of increasing degree, oscillations of the fitting function between the data points are often seen, which is a clear indication of "overfitting".

3. Any function connecting two sets of data can be Taylor expanded and, to a first approximation, the connection will be linear. All correlations will therefore be linear within a sufficiently "small" interval.
4. For functions where the fitting parameters enter in a linear fashion, the equations defining the parameters can be solved analytically. For a function with non-linear parameters, however, the resulting equations must be solved by iterative techniques, with the risk of divergence or convergence to a non-optimal solution (multiple minima problem).
5. If a specific non-linear behavior is anticipated, a new non-linear variable can be defined, which can be used as a linear correlation parameter. The one-variable non-linear function $f(x) = ax^2 + bx + c$, for example, can be made into a two-variable linear function by defining $z = x^2$, and thus $f(x,z) = az + bx + c$.

Points 2 to 4 suggest that a non-linear fitting function should only be attempted if there are sound theoretical reasons for expecting a non-linear correlation between the data. One such example is the often observed parabolic dependence of the biological activity on the lipophilicity for a series of compounds. Compounds with a low lipophilicity will have difficulty entering the cells and therefore often have a low activity. Compounds with a high lipophilicity, however, will have a tendency to accumulate in the fat cells and therefore also have a low activity. A quadratic dependence with a negative second-order term is therefore expected based on physical arguments.

18.4 Correlation between Many Sets of Data

In the previous section there were only two sets of data, the $\mathbf{y}$ data we wanted to model and the variable $\mathbf{x}$, each being a vector of dimension $N \times 1$. In many cases, however, there may be several sets of $\mathbf{x}$ variables $(\mathbf{x}_1, \mathbf{x}_2, \mathbf{x}_3, \ldots, \mathbf{x}_M)$, each of which can potentially describe some of the variation in the $\mathbf{y}$ data set, and we thus seek a function $f(\mathbf{x}_1, \mathbf{x}_2, \mathbf{x}_3, \ldots, \mathbf{x}_M)$ capable of modeling $\mathbf{y}$.[4] There may also be several different sets of $\mathbf{y}$ data that we want to model with the same $\mathbf{x}$ descriptors, but for simplicity we will only consider a single set of $\mathbf{y}$ data.

Using the concepts from Chapter 17, the $\mathbf{x}$ vectors can be thought of as being the (non-orthogonal and unnormalized) basis vectors of an M-dimensional coordinate system, in which we want to model $\mathbf{y}$ by a linear vector. The coefficients derived by the fitting process can be considered as coordinates for the vector modeling the $\mathbf{y}$ data along each of the $\mathbf{x}$ axes. In *Multiple Linear Regression*, the basis vectors $\mathbf{x}$ are usually non-orthogonal, while *Principal Component Analysis* performs an orthogonalization of the vector space. *Partial Least Squares* in addition attempts to align the coordinate axes with the variation in the $\mathbf{y}$ data.

The $\mathbf{x}$ variables can be arranged into an $\mathbf{X}$ matrix with dimension $N \times M$:

$$\mathbf{X} = \{\mathbf{x}_1 \quad \mathbf{x}_2 \quad \mathbf{x}_3 \quad \cdots \quad \mathbf{x}_M\} = \begin{Bmatrix} x_{1,1} & x_{2,1} & x_{3,1} & \cdots & x_{M,1} \\ x_{1,2} & x_{2,2} & x_{3,2} & \cdots & x_{M,2} \\ x_{1,3} & x_{2,3} & x_{3,3} & \cdots & x_{M,3} \\ \vdots & \vdots & \vdots & \ddots & \vdots \\ x_{1,N} & x_{2,N} & x_{3,N} & \cdots & x_{M,N} \end{Bmatrix} \tag{18.13}$$

The $\mathbf{x}$ descriptors are often derived from many different sources and may have different units, means and variances. Prior to any correlation analysis, each $\mathbf{x}$ (and $\mathbf{y}$) vector is usually *centered* to have a mean value of zero (i.e. subtracting the mean value from each vector element), as this eliminates any constant terms in the correlation analysis and focuses on describing the variation in the $\mathbf{y}$ data.

Each **x** vector may also be *scaled* with a suitable factor to take into account, for example, different units for the variables. This, however, is non-trivial and requires careful consideration. A common procedure, which avoids a user decision, is to normalize each **x** vector to have a variance of 1, a procedure called *autoscaling*. Autoscaling equalizes the variance of each descriptor and can thus amplify random noise in the sample data and reduce the importance of a variable having a large response and a good correlation with the **y** data.

18.4.1 Quality Measures

Analogous to the correlation coefficient in Equation (18.12), we want a measure of the quality of fit produced by a given correlation model. Two commonly used quantities are the *Predicted REsidual Sum of Squares* (*PRESS*) and the correlation coefficient R^2 defined by the normalized *PRESS* value and the variance of the **y** data (σ_y^2):

$$PRESS = \sum_{i=1}^{N} \left(y_i^{\text{actual}} - y_i^{\text{predicted}} \right)^2 \tag{18.14}$$

$$\sigma_y^2 = \frac{1}{N} \sum_{i=1}^{N} \left(y_i^{\text{actual}} - \bar{y} \right)^2 \tag{18.15}$$

$$R^2 = 1 - \frac{PRESS}{N\sigma_y^2} = 1 - \frac{\sum_{i=1}^{N} \left(y_i^{\text{actual}} - y_i^{\text{predicted}} \right)^2}{\sum_{i=1}^{N} \left(y_i^{\text{actual}} - \bar{y} \right)^2} \tag{18.16}$$

R^2 thus measures how well the model reproduces the variation in **y**, compared with a model that just predicts the average **y** value for all data points.

A straightforward inclusion of more and more **x** variables having some relationship with **y** in a correlation analysis will necessarily monotonically increase the amount of **y** variation described and thus produce an R^2 converging towards 1. Inclusion of variables that primarily serve to describe the noise in the **y** data, however, will lead to a model with *less* predictive value for the real variation. This is clearly something that should be avoided but in many cases it is not obvious when the additional components included primarily serve to model the noise in the **y** data. To make an unbiased judgment, it is of interest to introduce a quantity that does not measure how well the variables can *fit* the **y** data, but one that measures how well the variables can *predict* the **y** data. Since we are ultimately interested in predicting **y** data from the independent variables, such *cross-validations* are more useful quantities.

One possible measure is to make a correlation analysis using only $N - 1$ data points and ask how well this model predicts the point left out. This can be performed for each of the data points and thus requires a total of N correlation analyses to be performed. Such "*leave-one-out*" or "*jackknife*" models give an independent measure of the predictive power of a correlation model using a given number of variables. Either the *PRESS* calculated from summing over all the N correlation models or the *predictive* correlation coefficient Q^2, defined analogously to R^2 in Equation (18.16), can be used to measure the predictive capabilities of the model. Q^2 has in analogy with R^2 a maximum value of 1 but can achieve negative values for models with poor predictive capabilities (i.e. if the prediction is worse than predicting the average **y** value for all data points). A small *PRESS* or a large Q^2 value thus indicates a model with good predictive powers and models with Q^2 of 0.5–0.6 are often considered acceptable.

The leave-one-out cross-correlation model tends to overestimate the predictive capabilities, as the predicted fraction of the sample is only $1/N$, which rapidly approaches zero for a large sample. Alternative models can be generated by randomly leaving out, for example, k data points, rather than just one, or by forming subgroups of the data set, and either leaving one point out in each group or leaving all points in one group out.

Intuitively one expects the model with the largest R^2 (or Q^2) value to be the best, but this overlooks the fact that the model attempts to fit (experimental) data that contains uncertainties (experimental errors), often of poorly known magnitude. The maximum correlation coefficient is related to the uncertainty and the spread in the data:[5]

$$R^2_{\text{max}} = 1 - \left(\frac{\sigma_{\text{uncertainty}}}{\sigma_{\text{data}}}\right)^2 \tag{18.17}$$

Although $\sigma_{\text{uncertanity}}$ usually is unknown, Equation (18.17) shows that one should be cautious when trying to fit data with a small spread and potentially large (relative) uncertainties. Furthermore, constructing models with correlation coefficients close to (or exceeding) $R^2{}_{\text{max}}$ is clearly a sign of overfitting. While $R^2{}_{\text{max}}$ is independent on the sample size, the confidence interval of the correlation coefficient depends inversely on the number of data points.[6] Two models with correlation coefficients of 0.80 and 0.70 may thus be equally "accurate" in a statistical sense. Whether a model with correlation coefficient R_1 is statistically better than a model with correlation coefficient R_2 depends on the sample size, the difference between R_1 and R_2 and the magnitude of R_1. The required sample size increases as the magnitude of R_1 decreases and as the R_1, R_2 difference becomes smaller. Differentiating between two models each with low predictability is thus more difficult than between two models of high predictability. Even fairly high correlation coefficients, like 0.80 and 0.70, usually require several hundred data points for selecting the model with the higher correlation coefficient with confidence. These limitations should be kept in mind when, for example, selecting one QSAR model over another, where both models may have correlation coefficients in the 0.60–0.70 range and constructed from perhaps 20–30 data points, and the selected model is subsequently used for predictions.

18.4.2 Multiple Linear Regression

For multiple-descriptor data sets, one could use the methods in Section 18.3 to derive a correlation between $\mathbf{y}$ and $\mathbf{x}_1$, between $\mathbf{y}$ and $\mathbf{x}_2$, between $\mathbf{y}$ and $\mathbf{x}_3$, etc., to find the $\mathbf{x}_k$ data set that gives the best correlation with $\mathbf{y}$. It is very likely, however, that one of the other $\mathbf{x}$ variables can describe part of the $\mathbf{y}$ variation, which is not described by $\mathbf{x}_k$, and a third $\mathbf{x}$ variable describing some of the remaining variation, etc. Since the $\mathbf{x}$ vectors may be internally correlated, however, the second most important $\mathbf{x}$ vector found in a one-to-one correlation is not necessarily the most important once the most important $\mathbf{x}_k$ vector has been included. In the vector space language, the $\mathbf{x}$ vectors form a non-orthogonal coordinate system.

In order to use all the information in the $\mathbf{x}$ variables, a *Multiple Linear Regression* (MLR) of the type indicated in the following equations can be attempted:

$$\mathbf{y} = \sum_{k=1}^{M} a_k \mathbf{x}_k + \mathbf{e} \tag{18.18}$$

$$\mathbf{y} = \mathbf{X}\mathbf{a} + \mathbf{e} \tag{18.19}$$

Note that each data set ($\mathbf{y}$ and $\mathbf{x}_k$) is a vector containing N data points, and the constant corresponding to b in Equation (18.7) is eliminated if the data are centered with a mean value of zero. The $\mathbf{e}$ vector contains the residual variation in $\mathbf{y}$ that cannot be described by the $\mathbf{x}$ variables, that is this is a vector outside the space spanned by the $\mathbf{x}$ vectors. Since the expansion coefficients are multiplied directly on to the $\mathbf{x}_k$ variables, MLR is independent of a possible scaling of the $\mathbf{x}_k$ data (a scaling just affects the magnitude of the a_k coefficients but does not change the correlation).

The number of fitting parameters is M, and N must therefore be larger than or equal to M; in practice one should not attempt fitting unless $N > 5M$, as overfitting is otherwise a strong possibility. The optimum fitting coefficients contained in the $\mathbf{a}$ vector can be determined by requiring that the norm of the error vector is minimized:

$$\frac{\partial \|\mathbf{e}\|^2}{\partial \mathbf{a}} = \frac{\partial \|\mathbf{y} - \mathbf{X}\mathbf{a}\|^2}{\partial \mathbf{a}} = 0 \tag{18.20}$$

This leads to a solution in terms of the generalized inverse (Section 17.2) of the $\mathbf{X}$ matrix:

$$\mathbf{a} = (\mathbf{X}^t\mathbf{X})^{-1}\mathbf{X}^t\mathbf{y} \tag{18.21}$$

This procedure works well as long as there are relatively few $\mathbf{x}$ variables that are not internally correlated. In reality, however, it is very likely that some of the $\mathbf{x}$ vectors describe almost the same variation, and in such cases there is a large risk of overfitting the data. This can also be seen from the solution vector in Equation (18.21); the $(\mathbf{X}^t\mathbf{X})^{-1}$ matrix has dimension $M \times M$ and will be poorly conditioned (Section 17.2) if the $\mathbf{x}$ vectors are (almost) linearly dependent. Note that the presence of (experimental) noise in the $\mathbf{x}$ data can often mask the linear dependence and MLR methods are therefore sensitive to noisy $\mathbf{x}$ data.

MLR works best if $N \gg M$ and when the $\mathbf{x}$ data are not internally correlated. If either of these criteria is not fulfilled, one can try to form MLR models by selecting subsets of the descriptors. Searching all possible combinations of descriptors from a total of M data sets rapidly leads to a large number of possibilities, which may be impossible to search in a systematic fashion. Global optimization methods such as genetic algorithms or simulated annealing (Sections 13.6.3 and 13.6.4) can be used to hunt for the best combination of number and types of descriptors. Such optimizations clearly should focus on maximizing the Q^2 value and not the R^2 value. One may also consider weighting Q^2 with a factor depending (inversely) on the number of components, as a slight increase in Q^2 by including one or more components may not be statistical significant. Alternatively, and somewhat more systematically, one of the methods in the next sections can be used.

18.4.3 Principal Component Analysis

Multiple linear regression cannot easily handle situations where $M > N$ or when the $\mathbf{x}$ variables are (almost) linearly correlated. These problems can be removed by introducing a modified set of descriptors, often called *principal components* or *latent variables*. The idea is to extract linear combinations of the $\mathbf{x}$ variables that are orthogonal and ranked according to their variation and only use a limited set of these variables for performing the correlation with the $\mathbf{y}$ variables. In the vector space language, the $\mathbf{x}$ coordinate system is transformed into an orthogonal set of coordinate axes in order to simplify the description of the best linear vector modeling the $\mathbf{y}$ data. Simplify in this context means that there are fewer coefficients that are significantly different from zero. Performing an orthogonalization produces a vector space describing the same fundamental coordinate system but with basis vectors that are rotated with respect to the original $\mathbf{x}$ vectors.

The eigenvalues of the $\mathbf{X}^t\mathbf{X}$ matrix (Equation (18.13)) contain information on the correlation between the **x** variables: an eigenvalue of zero indicates that one column can be written as a linear combination of the other columns and a small non-zero value indicates that one column contains almost redundant information. The eigenvector corresponding to the largest eigenvalue describes the linear combination of **x** descriptors having the largest variation in the **x** data, the eigenvector corresponding to the second largest eigenvalue has the second largest variations, etc. Furthermore, since the eigenvectors are orthogonal, different eigenvectors describe different parts of the variation.

In *Principal Component Analysis* (PCA) methods the orthogonalization of the **x** vectors is done using the eigenvectors of the $\mathbf{X}^t\mathbf{X}$ matrix (a canonical orthogonalization, Equation (17.86), without the eigenvalue weighting). If we denote the orthogonalized basis vectors by **z**, we can write the connection as

$$\mathbf{Z} = \mathbf{X}\mathbf{U} \tag{18.22}$$

The **U** matrix contains the eigenvectors of the $\mathbf{X}^t\mathbf{X}$ matrix and it is convenient to let the ordering of the **z** vectors reflect the magnitude of the $\mathbf{X}^t\mathbf{X}$ eigenvalues, such that $\mathbf{z}_1$ is associated with the largest eigenvalue, $\mathbf{z}_2$ with the second largest eigenvalue, etc. We note that if there are $\mathbf{X}^t\mathbf{X}$ eigenvalues close to zero then the effective dimension of the coordinate system is less than M. The **z** vectors are referred to as the *principal components* (of the **x** variables). The **U** matrix depends on the relative magnitudes of the individual **x** vectors and the **z** vectors therefore depend on a possible scaling of the original **x** descriptors. Figure 18.5 shows a two-dimensional example where the points for the two non-orthogonal **x** descriptors have an internal correlation and display a similar variation along the two directions. The new **z** variables are orthogonal and most of the variation is now located in the $\mathbf{z}_1$ variable, while $\mathbf{z}_2$ describes a much smaller variation.

The idea in PCA is to use the **z** components as the descriptive variables:

$$\mathbf{y} = \mathbf{Z}\mathbf{b} + \mathbf{e} \tag{18.23}$$

Here **b** is a vector containing the fitting coefficients (analogous to **a** in Equation (18.19)) and **e** is again the vector containing the residual error. The b_k coefficients can, in contrast to Equation (18.21), be obtained by a simple projection of the **y** vector on to each **z** component since the latter are orthogonal. If the **x** variables are normalized, this corresponds to (Equation (18.21) where $\mathbf{Z}^t\mathbf{Z}$ is a unit matrix)

$$\mathbf{b} = \mathbf{Z}^t\mathbf{y} \tag{18.24}$$

If all M **z** vectors are used the result is identical to using the original **x** variables (i.e. MLR). However, the premise of the PCA method is to include only a few **z** variables and to select them according to

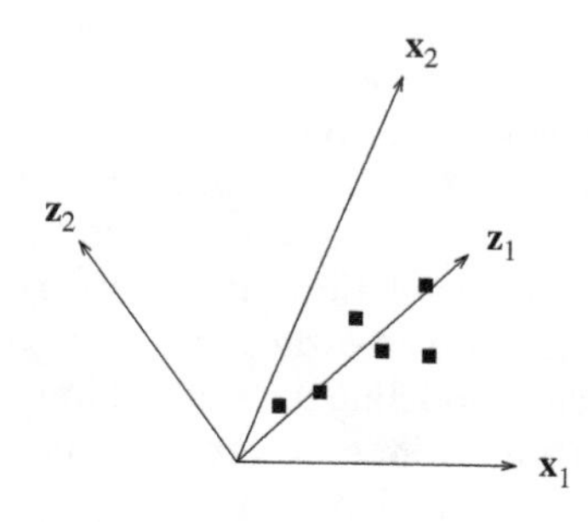

Figure 18.5 Illustrating the relationship between the original (**x**) and latent (**z**) variables.

the $\mathbf{X}^t\mathbf{X}$ eigenvalues. A sequence of linear correlation models are constructed by including more and more $\mathbf{z}$ vectors, first (only) $\mathbf{z}_1$, then including also $\mathbf{z}_2$, then $\mathbf{z}_3$, etc. At each stage the predictive capabilities of the model are calculated, for example quantified by Q^2. If the original $\mathbf{x}$ data have a reasonable correlation with the $\mathbf{y}$ data, then a plot of Q^2 against the number of variables included will typically display an initial steep increase, but then level off or even start to decrease slightly as the number of latent variables is increased. The point where Q^2 levels off indicates that the optimal number of components has been reached, that is at this point the predictive power of the model cannot be increased further by including more components.

The main problem with the PCA method is that some of the $\mathbf{x}$ variables may not be particularly good at describing the $\mathbf{y}$ variables, that is the first few PCA vectors describing the largest variation among the $\mathbf{x}$ variables may correlate poorly with the variation in the $\mathbf{y}$ data. In such cases, a global optimization search can be made for a model based on a relatively small number of components selected from all the PCA vectors with eigenvalues above a suitable cutoff.

Since the PCA vectors depend on the relative magnitude of the original $\mathbf{x}$ variable, and thus, for example, on the units employed for each variable, the variables should have comparable variances or be normalized to unit variance. Another problem is that the PCA vectors constructed from diagonalization of the $\mathbf{X}^t\mathbf{X}$ matrix involve all of the original $\mathbf{x}$ descriptors, that is all the original variables are required even for a correlation models having only a few $\mathbf{z}$ vectors as descriptors. In *sparse PCA* methods, the $\mathbf{z}$ variables are restricted to have a user-defined upper limit on the number of non-zero elements, while still describing as much of the variation in the $\mathbf{X}^t\mathbf{X}$ matrix as possible. This, however, is computationally a much harder problem (NP-hard) as it cannot be obtained simply by diagonalizing a matrix.

18.4.4 Partial Least Squares

The *Partial Least Squares* (PLS, also sometimes called *Projection to Latent Structures*) method in analogy with PCA performs an orthogonalization of the $\mathbf{x}$ vectors to form new latent variables. In contrast to PCA, however, PLS attempts to align the new coordinate axes with the variation in $\mathbf{y}$, rather than simply describing the variation in the $\mathbf{x}$ variables. This is obtained by weighting the $\mathbf{X}$ matrix with the $\mathbf{y}$ vector prior to diagonalization, that is the $\mathbf{U}$ matrix in Equation (18.22) is obtained by diagonalization of the $\mathbf{X}^t\mathbf{y}\mathbf{y}^t\mathbf{X}$ (equivalent to $(\mathbf{y}^t\mathbf{X})^t(\mathbf{y}^t\mathbf{X})$) matrix instead of $\mathbf{X}^t\mathbf{X}$.[7] This ensures that the $\mathbf{z}$-vectors will be selected in an order that is biased towards describing the variation in $\mathbf{y}$. The only difference between PCA and PLS is thus in how the latent $\mathbf{z}$ variables are generated, either from diagonalization of the $\mathbf{X}^t\mathbf{X}$ matrix or from the corresponding $\mathbf{y}$-weighted matrix. Expressed in the statistical language, the PLS components describe the largest covariance between $\mathbf{X}$ and $\mathbf{y}$.

The PCA components can computationally be obtained by a single diagonalization of the $\mathbf{X}^t\mathbf{X}$ matrix while the PLS components must be extracted iteratively. The latter can be done by sequentially generating a $\mathbf{z}$ vector associated with the largest (only one non-zero) eigenvalue of the $\mathbf{X}^t\mathbf{y}\mathbf{y}^t\mathbf{X}$ matrix, project this vector out of the $\mathbf{X}$ matrix (projection of the $\mathbf{y}$ matrix can also be done but is unnecessary) and calculate the new $\mathbf{X}^t\mathbf{y}\mathbf{y}^t\mathbf{X}$ matrix. Since only the $\mathbf{z}$ vector corresponding to the largest eigenvalue is required, it is computationally inefficient to diagonalize the full $\mathbf{X}^t\mathbf{y}\mathbf{y}^t\mathbf{X}$ matrix multiple times, especially if the $\mathbf{X}$ matrix is large, and the required $\mathbf{z}$ vector can instead be extracted by other algorithms, of which the NIPALS (Non-linear Iterative PArtial Least-Squares) algorithm is commonly used. Taking the correlation model as in Equation (18.23) ($\mathbf{y} = \mathbf{Zb} + \mathbf{e}$), the NIPALS algorithm sequentially extracts PLS components by a series of matrix–matrix multiplications. If we denote vectors in the current PLS iteration as $\mathbf{u}$, $\mathbf{z}$, $\mathbf{p}$ ($\mathbf{u}$ and $\mathbf{z}$ being column vectors in the matrices $\mathbf{U}$ and $\mathbf{Z}$) and b as the coefficient (entry in the $\mathbf{b}$ vector), the steps in the NIPALS algorithm for a single $\mathbf{y}$ vector can be written as follows:

$$\begin{aligned}
&1.\ \mathbf{u} = \mathbf{X}^t\mathbf{y} \\
&2.\ \mathbf{u} \leftarrow \mathbf{u}/|\mathbf{u}| \\
&3.\ \mathbf{z} = \mathbf{X}\mathbf{u} \\
&4.\ \mathbf{z} \leftarrow \mathbf{z}/|\mathbf{z}| \\
&5.\ b = \mathbf{y}^t\mathbf{z} \\
&6.\ \mathbf{p} = \mathbf{X}^t\mathbf{z} \\
&7.\ \mathbf{X} \leftarrow \mathbf{X} - \mathbf{z}\mathbf{p}^t
\end{aligned} \tag{18.25}$$

The PLS components will naturally be ordered according to their ability to describe the $\mathbf{y}$ variation, alleviating the necessity to perform a combinatorial search for which components to use in the correlation model. For optimal cases, a plot of Q^2 against the number of PLS components will rapidly reach a maximum and provide a compact model with good predictive capabilities.

A disadvantage of the PLS method is the inherent bias toward selecting latent variables describing noise in the $\mathbf{y}$ data, that is $\mathbf{x}$ variables that only have a small internal variation but correlate with the noise in the $\mathbf{y}$ data are selected as important. For this reason, $\mathbf{x}$ variables with small internal variance over the $\mathbf{y}$ data points are often removed from the descriptor data set prior to performing the PLS analysis. This pre-selection procedure, however, requires user involvement and it is not always easy to decide which variables to remove. Unfortunately, the predictive capabilities of a PLS model are often sensitive to elimination of one or more $\mathbf{x}$ variables. A global optimization scheme may again be employed in such cases, that is performing a search for which $\mathbf{x}$ components to *remove* from the PLS analysis in order to provide a model with a high Q^2 value. Other variations such as sparse PLS methods are also being developed, but require more sophisticated optimization techniques for calculating the latent components.

18.4.5 Illustrative Example

Consider the following $\mathbf{y}$ and $\mathbf{X}$ matrices where the variables have been centered to give a mean of zero:

$$\mathbf{y} = \begin{pmatrix} 0.1 \\ 0.1 \\ 0.9 \\ -1.1 \end{pmatrix} \quad ; \quad \mathbf{X} = \begin{pmatrix} 1.0 & 1.0 \\ -1.0 & -1.0 \\ 0.1 & -0.1 \\ -0.1 & 0.1 \end{pmatrix} \tag{18.26}$$

Clearly the first two rows show a large variation in x with no change in y, that is these variables are not related to the response. The last two rows display correlation and anticorrelation, respectively, with the y data in equal amounts. Solving with MLR Equation (18.21) gives the solution vector $\mathbf{a}$ in Equation (18.27), which shows that both $\mathbf{x}$ columns are equally important in describing the $\mathbf{y}$ variation. The difference between the actual and predicted $\mathbf{y}$ indicates that there is a residual variation $\mathbf{e}$ that cannot be modeled by the $\mathbf{x}$ variables:

$$\mathbf{a} = \begin{pmatrix} 5 \\ -5 \end{pmatrix} \quad ; \quad \mathbf{y}_{\text{pred}} = \mathbf{X}\mathbf{a} = \begin{pmatrix} 0.0 \\ 0.0 \\ 1.0 \\ -1.0 \end{pmatrix} \quad ; \quad \mathbf{e} = \begin{pmatrix} 0.1 \\ 0.1 \\ -0.1 \\ -0.1 \end{pmatrix} \tag{18.27}$$

Diagonalizing the $\mathbf{X}^t\mathbf{X}$ matrix to construct the principal components gives eigenvalues of 4.00 and 0.04, with corresponding (unnormalized) eigenvectors (1, 1) and (1, −1). In the transformed coordinate system, the (unnormalized) principal components $\mathbf{z}$ (Equation (18.22)) can be written as

$$\mathbf{Z} = \begin{pmatrix} 1.0 & 0.0 \\ -1.0 & 0.0 \\ 0.0 & 0.1 \\ 0.0 & -0.1 \end{pmatrix} \tag{18.28}$$

Clearly the eigenvector corresponding to the largest eigenvalue (the first principal component, i.e. the first column in $\mathbf{Z}$) has a zero overlap with the $\mathbf{y}$ data, while the second eigenvector accounts for all the important variation in the original $\mathbf{x}$ variables. The PLS components arising from diagonalization of the $\mathbf{X}^t\mathbf{y}\mathbf{y}^t\mathbf{X}$, on the other hand, have eigenvalues of 0.0 and 0.08 with corresponding (unnormalized) eigenvectors (1, 1) and (1, −1). The transformed $\mathbf{X}$ matrix is identical to Equation (18.28), except that it is now the *second* column that corresponds to the *largest* eigenvalue. The direction associated with the largest eigenvalue in the PCA case has an eigenvalue of zero in the PLS case, showing that it contains no information of the $\mathbf{y}$ variation. In both the PCA and PLS cases, the $\mathbf{y}$ variation that can be described is contained in only one component, but the PLS procedure identifies the most important component as the first extracted component, while PCA has the information in the second principal component. The predicted $\mathbf{y}$ is identical to the MLR results (Equation (18.27)) for both PCA and PLS; the only difference is that only one component is required, rather than two.

18.5 Quantitative Structure–Activity Relationships (QSAR)

One important application of PCA and PLS techniques is in the lead development and optimization of new drugs in the pharmaceutical industry. From the basic chemical and physical theories, it is clear that there must be a connection between the structure of a molecule and its biological activity. It is also clear, however, that the connection is very complex and it is very unlikely that the biological activity can ever be predicted completely *a priori*. A drug taken orally, for example, must first survive the digestive enzymes and acidic conditions in the stomach, cross over into the blood stream, possibly also cross over into the brain, diffuse to the target protein without binding to other proteins, bind the target protein at a specific site and with a large binding constant, and finally have a suitable half-life in the organism before being degraded into non-toxic components. Each of these quantities depends on different parts of the molecular structure and the combined effect is therefore very difficult to predict. Each quantity, however, may possibly be correlated with (different) molecular properties, but adequate data for each effect is rarely available.

In the initial stages of developing a new drug, the focus is usually on the binding to the target protein and having a suitable lipophilicity, the latter to ensure a reasonable transfer rate for crossing between the blood and cells. A lead compound is somehow determined, more often than not by serendipity. From this lead, a small initial pool of compounds is synthesized and their biological activity, of which the binding constant to the target protein is an important quantity, is measured by a suitable biological assay. At this point, statistical methods are often used to correlate the molecular structure and properties to the observed activity, a *Quantitative Structure–Activity Relationship* (QSAR),[8] thereby providing a tool for "guessing" which new modified compounds should be tried next.

In a traditional QSAR study, a variety of molecular descriptors are included. Typically, these include a measure of the lipophilicity (often taken as the partition coefficient between water and 1-octanol), electronic and steric substituent parameters (Hammett and Taft constants[9]) and $pK_{a/b}$ values for acidic and/or basic groups. These are rather obvious molecular descriptors, but many other less obvious descriptors have also been used, such as the molecular weight, IR frequencies, dipole moments, NMR chemical shifts, etc. The philosophy is to include, rather indiscriminately, as many descriptors as can easily be generated and then let the statistical method sort out which of these are actually important. With many of the descriptors having only a remote connection with the measured activity, classical or MLR correlations are clearly not suitable methods. PCA or PLS methods are better at detecting poor descriptors and dealing with the fact that the measured biological activities often have rather large uncertainties.

Classical QSAR methods focus on correlating experimental activities with experimental descriptors. This allows an identification of important structural features, such as having a pK_a value close to 5 or having an electron-withdrawing group in a *para*-position of a phenyl ring, thereby limiting the field of possible new compounds to prepare and test. The focus has increasingly been on "virtual" (*in silico*) screening, that is correlating experimental activities with theoretical descriptors. If a good QSAR model can be constructed from such data, this allows prediction of the biological activities of molecules that exist only as a model in the computer. The activity of many thousands of (possibly computer-generated) structures can thus be predicted and only those that are predicted to be reasonably active need to be synthesized. The theoretical descriptors can be similar to those in traditional QSAR methods, for example replacing the experimental water–octanol partition coefficient with a corresponding theoretical estimate, etc. The so-called 3D-QSAR models, however, are more representative of these QSAR methods, and the *COmparative Molecular Field Analysis* (*COMFA*) method was one of the first examples of these.[10]

The molecular descriptors are in the COMFA method taken as steric and electric fields calculated at a large number of points surrounding each molecule. The molecule is placed in a "box" and a few thousand points are selected between the surface of the molecule and a few van der Waals radii outwards. The steric repulsion or attraction from a probe atom (typically a carbon atom) is calculated at each of these points by a force field van der Waals term. The electric attraction or repulsion is similarly calculated by placing a point charge with magnitude +1 in each point. Specific interaction descriptors can be generated by selecting specific force field atom types to probe, for example, hydrogen bonding acceptor/donor capabilities. The complete set of molecular descriptors thus consists of a few thousand data points, representing steric, electric and property specific interactions of the molecule with other (possible) atoms in the near vicinity. These data are clearly strongly correlated; the value at a given point will be very close to the average of the neighboring points. Deriving QSAR models with such large sets of data descriptors is only possible using PCA and PLS methods. Such 3D-QSAR methods are primarily used when the structure or identity of the receptor protein is unknown. If the protein and binding site is known from an X-ray structure, the testing of possible drug candidates can be done by docking methods, as described in Section 13.7.

The main problem with 3D-QSAR methods such as COMFA is the alignment of the molecules in the test set.[11] First, it must be assumed that each of the molecules binds to the protein at the same site. Second, it is not always clear that all the molecules bind to the active site in the same overall orientation. Third, if the molecule has several conformations available, one has to guess or estimate which conformation actually binds to the protein. Furthermore, even if the overall orientation of all the molecules is assumed to be the same, the specific alignment is not unambiguous. Even for molecules having the same major structural features, one can choose either to align on the best RMS fit of all atoms, on only the non-hydrogen atoms or on only the atoms in a substructure of the molecule (e.g. a phenyl ring). Figure 18.6 illustrates the ambiguity for aligning the compound on the right with that on the left and whether the imine or nitro group should be used for alignment.

When the molecular structure of the compounds in the training set has few or no common elements, the alignment may instead be done based on, for example, the electric moments (dipole,

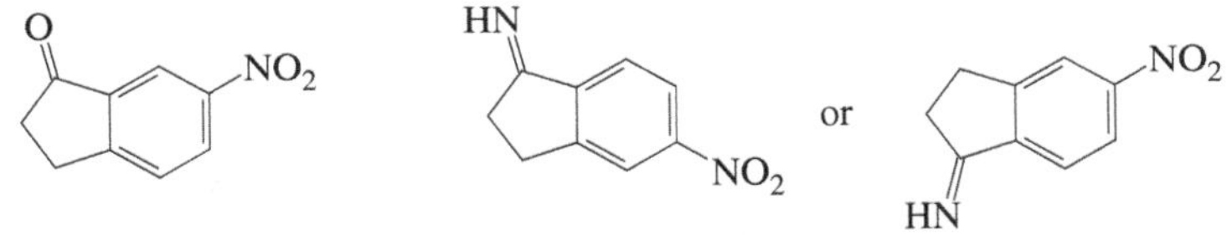

Figure 18.6 Illustrating the alignment problem in COMFA methods.

quadrupole, etc.) or on the electrostatic potential on a suitable molecular surface.[12] If certain common interaction elements, like hydrogen bonding or hydrophobic interactions, can be identified for a large class of active compounds, such pharmacophor elements may also be used for alignment.

Since the alignment of the molecules influences the values calculated at the steric and electric grid points, this is a feature that influences the statistical correlation. If a successful QSAR model can be obtained from such data, however, the model will provide information on which of the grid points are important in a steric and electric sense. Analysis of such data provides a virtual mapping of the receptor, that is identifying regions of the active site where the drug candidates should have large/small groups and where they should have positively/negatively charged groups.

Pharmacophore modeling can be considered as a coarse-grained version of 3D-QSAR methods, where so-called *pharmacophore elements* such as hydrogen donors/acceptors, hydrophobic interaction sites, aromatic moieties, anion/cation sites, etc., are used as molecular descriptors, rather than steric and electrostatic interaction points used in the COMFA method. The methodology and challenges (selection of conformations and molecular alignment), however, are very similar, and both types of methods aim at building a virtual active site based only on information from a series of compounds displaying activity toward a (unknown) receptor.[13] A given ligand in a particular conformation can be characterized by its pharmacophor elements and a vector containing distances between each pair of elements. From a list of ligands, which are known to be active, pharmacophor modeling tried to identify a set of pharmacophor elements with the same pair-wise distance (within a given threshold). If such a set can be found that includes all ligands, it is called a *common pharmacophor*. Inactive or weakly active ligands with the same pharmacophor can be included in the model to map out exclusion volume regions, where steric interactions lead to weak binding or non-binding. The end result of a pharmacophore model is a collection of pharmacophore elements, and possibly steric exclusion regions, in a three-dimensional arrangement that can be used for virtual screening of new drug candidates.

18.6 Non-linear Correlation Methods

MLR, PCA and PLS methods construct linear correlation models between the multidimensional **x** and **y** data. There are no technical problems in also constructing non-linear correlation models, but, as mentioned in Section 18.3, this should only be done if there are good arguments based on the underlying problem for expecting a specific non-linear dependence. This may be possible for multivariable regression methods analogous to MLR, but is more difficult for PCA/PLS-based methods, as these form linear combinations of the original descriptors. Non-linear correlation models may instead be constructed by *Artificial Neural Network* or *Symbolic Regression* methods.

An *Artificial Neural Network* is modeled after the way the brain processes information by forming multiple connections between neurons, with the strength of each connection being a variable that controls the flow of information.[14] A large variety of artificial neural networks are in use, but the basic principle is illustrated in Figure 18.7.

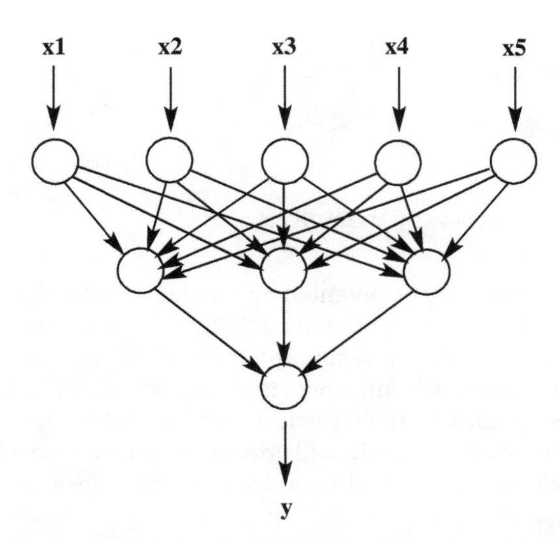

Figure 18.7 Illustrating the information flow in an artificial neural network containing a single hidden layer between the input and output neurons.

The $\mathbf{x}_k$ descriptors provide data to the input layer of neurons, and this information is passed to a hidden layer of neurons, with a separate weighting factor for each connection. The hidden layer neurons in turn pass the information to the output neuron, which produces the response **y**, again with separate weighting factors for each connection. The connections are usually non-linear functions containing parameters that are adjusted iteratively by training the network to produce a correlation between the input descriptors and (known) output variables. Figure 18.7 illustrates that the number of parameters in even quite simple networks rapidly becomes large, and this leads to the risk of overfitting, that is the network simply learns to "remember" the connections between input and desired output, rather than extracting a (non-linear) correlation model that can be used in a predictive sense. Cross-validation is therefore an essential component in ensuring that overfitting does not occur. Artificial neural networks have proven to be very powerful in establishing non-linear correlation models, but they have the disadvantage that it is difficult to obtain insight into why or how the models work, that is extract information of which descriptors are the most important for the predicted property.

Symbolic Regression is an interesting method for constructing non-linear correlation models when there is little or no *a priori* information regarding the specific non-linear functional behavior. Symbolic regression methods employ a genetic type of global optimization (Section 13.6.4) using a (large) set of mathematical operations and associated parameters as the genes and attempt to find a combination of functions and parameters that can fit the data.[15] The disadvantage is that many such combinations may exist and many (all) may correspond to pure data fitting without any physical significance. The advantage is that a particular solution can be written analytically in terms of standard mathematical functions, which may provide insight into the problem.

18.7 Clustering Methods

A large set of data often needs to be reduced or compressed before further analysis or refinement. This could, for example, be a number of poses from docking a ligand into a protein target or a (large) number of structures arising from a force field conformational search that are targeted for further refinement by a sequence of electronic structure methods to identify a number of low-energy conformations. A system composed of the order of hundreds of atoms often has thousands or millions of possible conformations, which can be searched near-exhaustively by force field methods, but subsequent refinements by more accurate methods are limited to a few hundred conformations. Since relative energies by standard force fields often have errors in the 20–30 kJ/mol range, all structures within a ~30 kJ/mol window of the lowest energy structure should be refined by a sequence of increasingly accurate methods and pruning of high-energy structures at each level. In many cases, however, it is only the overall structure of the whole systems that is of interest, and conformations that only differ in terms of, for example, an OH group orientation are irrelevant. It is thus of interest to group structures together into clusters, which by some measure are sufficiently similar that a single representative structure can be extracted and used in the further refinement. Structures from different clusters should furthermore be sufficiently dissimilar that a collection of a single representative structure from each cluster represents the full diversity of the original set of data.

There are many methods for cluster analyses, but we will only give a brief overview of the general principles and provide some examples of algorithms commonly used in computational chemistry. The three key elements in a cluster analysis are:

1. The *similarity* between two elements.
2. The *linkage* between clusters.
3. The *clustering* algorithm.

The similarity is for continuous variables commonly defined in terms of a distance in variable space, and similarity and distance are often used interchangeably, but it should be kept in mind that they are inversely related (a large similarity implies a short distance and vice versa). Many types of similarity measures can be defined, but we will restrict the exemplification to some that have a clear geometrical interpretation.

Consider a set of N elements each containing M variables. In the above example, N is the number of conformations and M is the number of atomic coordinates. The similarity between each pair of elements could be taken as the root mean square difference between all atomic coordinates when optimally aligned. This is known as the Euclidian distance (2-norm), but could alternatively also be defined in terms of the corresponding squared distance, the absolute distance (1-norm) or the maximum element distance (infinity-norm) (Section 17.1). The similarity can be calculated over all atoms or a selected subset of atoms (e.g. only non-hydrogen atoms).

The linkage between clusters containing only a single element is unique, being just the similarity, but the linkage between clusters containing more than one element can be defined in several different ways, and four commonly used definitions are:

1. *Single link*: the shortest distance (largest similarity) between any pair of elements belonging to two clusters (clusters are linked by a single element).
2. *Complete link*: the longest distance (smallest similarity) between any pair of elements belonging to two clusters (clusters are linked by all elements).
3. *Average link*: the average over distances between all pairs of elements belonging to each of the two clusters (similarity between elements in different clusters, then average).
4. *Centroid link*: the distance between the centroid of the two clusters (average over elements within the same clusters, then similarity).

These differences are illustrated in Figure 18.8.

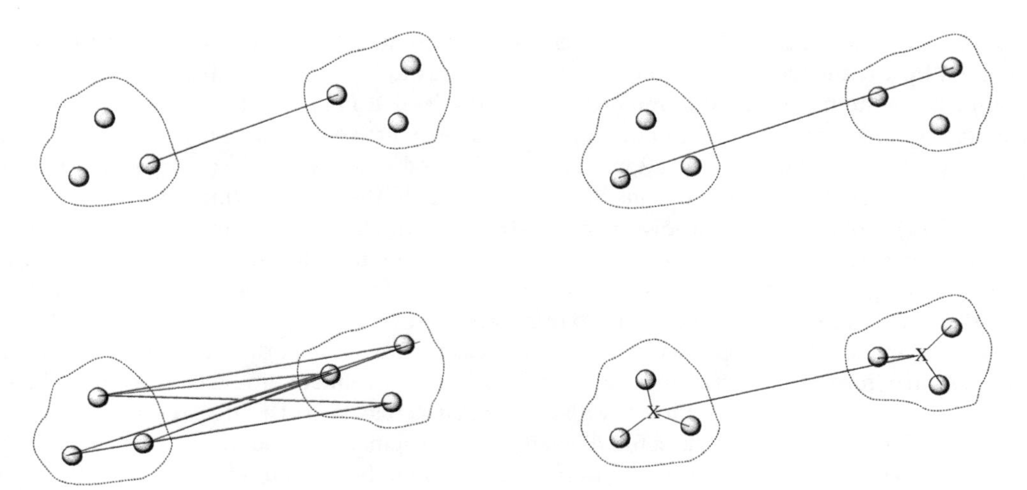

Figure 18.8 Illustrating the differences between the Single link (upper left), Complete link (upper right), Average link (lower left) and Centroid link (lower right) clustering methods.

The Single link method tends to be sensitive to outliers and noise in the data, which may link an otherwise distinct cluster, while the Complete link method tends to favor globular clusters and can break large clusters. The Average link method has intermediate characteristics between these two methods. The Centroid link method has the disadvantage that the similarity measure for sequentially joining clusters may be non-monotonic (the distance for joining clusters at level $C+1$ may be smaller than for joining at level C), since the centroid position changes when an additional element is included.

The clustering algorithm tries to answer two related questions: what is the "optimum" partitioning of the N data into a predefined C set of clusters and what is the "optimum" number of clusters that the data set can be partitioned into? The quotation marks around the words optimum indicate that there is not a unique definition and that the answer depends on how optimum is defined, that is the clustering algorithm. The clustering can furthermore be *exclusive*, where a given element belongs to only one cluster, or *overlapping*, where an element can belong to more than one cluster, perhaps with a weighting factor (then usually called fuzzy clustering).

We will in the following only consider exclusive clustering and Figure 18.9 can be used to illustrate the ambiguities in cluster analyses. The 12 data points can visually be partitioned into four, three or two clusters, but other choices are clearly also possible.

The optimum C clustering of N elements is one where each element has a smaller distance to the centroid that it belongs to than to other centroids, usually called *K-means* clustering. Using the squared 2-norm distance as the similarity measure, this can be formulated as a minimization of an object function:

$$ObjF = \sum_{c=1}^{C} \sum_{\mathbf{x}_k \in c}^{N_c} (\mathbf{x}_c - \mathbf{x}_k)^2 \tag{18.29}$$

$$\mathbf{x}_c = \frac{1}{N_c} \sum_{\mathbf{x}_k \in c}^{N_c} \mathbf{x}_k \tag{18.30}$$

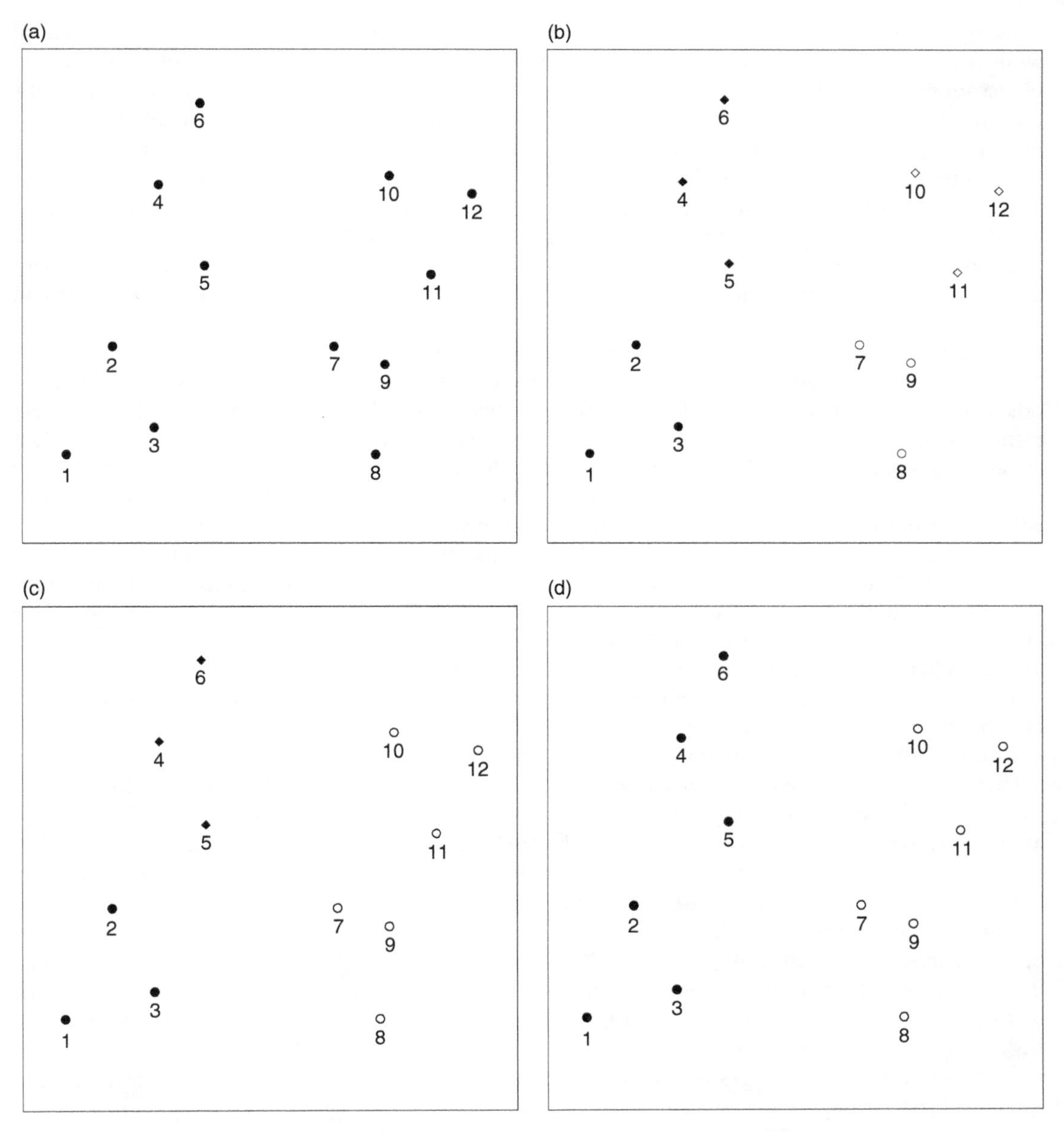

Figure 18.9 Three possible clusterings of the data in (a): (b) into four clusters, (c) into three clusters and (d) into two clusters.

This definition, however, leads to an NP-hard optimization problem, as there are $K_{N,C}$ possible ways of distributing N elements into C clusters, which even for small N and C values is infeasible to search exhaustively. Practical algorithms employ an initial guess of C centroids ($\mathbf{x}_c$) that iteratively are refined until convergence, where each iteration consists of assigning all elements to a specific centroid based on the similarity and recalculating the centroid position from Equation (18.30). This usually converges in a few iterations and corresponds to finding a (local) minimum of the object function, while a search in the full combinatorial space would locate the global minimum. Unfortunately the object function usually has many local minima and different starting guesses often lead to different local minima. Global search procedures discussed in Section 13.6 can be used to locate a best estimate of the global minimum. Variations of the algorithm consist of starting with a small number of clusters and sequentially split and recombine clusters in order to provide better estimates of the initial centroid positions at a given cluster level. The centroid position ($\mathbf{x}_c$) corresponds to the average of all elements in the cluster, but this may be a poor representation of the physical objects in the cluster. Averaging the (Cartesian) coordinates of several molecular conformations, for example, may yield an average structure that has non-physical internal coordinates. Using the median (Section 18.2) instead of the average as the cluster centroid may in such cases work better, especially if the centroid structure subsequently is used as a representative element for the whole cluster.

The global optimization problem in K-means clustering is bypassed in *hierarchical* clustering methods, where the number of clusters is sequentially reduced from N to 1 based on the similarity matrix elements. Initially the two elements with the smallest distance (largest similarity) are grouped into a cluster, the similarity matrix is recalculated based on the chosen linkage, the element with the smallest distance to another element/cluster is assigned to a cluster, etc., and this leads to a deterministic algorithm for grouping N elements into C clusters. Alternatively, but used less frequently, the algorithm can be run in reverse by initially grouping all elements into one cluster and sequentially increasing the number of clusters to N. The result of a hierarchical clustering can be visualized as a dendrogram showing which elements are grouped together as a function of the similarity for joining a cluster at a given level. Using a squared Euclidian norm as the similarity and single linkage on the data in Figure 18.9(a) gives the dendrogram shown in Figure 18.10.

The first cluster is formed by elements 7 and 9, the second cluster consists of elements 10 and 12, etc. The dendrogram suggests that four clusters (Figure 18.9(b)) could be a reasonable partitioning, but three clusters (Figure 18.9(c)) is clearly also a partitioning that should be considered.

Hierarchical clustering has the disadvantage that the assignment of an element to a given cluster cannot be undone at a later stage, even though the element at a given cluster level may be closer to another cluster centroid than the one to which it belongs. Another disadvantage is that calculation of the full similarity matrix and its sequential update at each cluster level leads to an algorithmic $N^2 \log N$ complexity, which makes it unsuitable for large data sets.

The optimum clustering of N elements for a given cluster size can be quantified in terms of an object function, as shown in Equation (18.29). Assuming that the global optimization problem has been solved for all cluster sizes from N to 1, the object function will increase monotonically from zero to some maximum value as a function of C. Such a plot may display a "kink" for a given cluster size, suggesting that this is the "natural" or "optimum" number of clusters, but this often yields inconclusive results. Kelley *et al.* have suggested[16] that the optimum partitioning can be found as the minimum of a penalty function depending on the number of clusters and the average spread of the elements within each cluster containing more than one element. The spread of the elements in a specific cluster c is defined in Equation (18.31) and the corresponding average spread over all clusters containing more than one element (C') is given in Equation (18.32).

$$S_c = \frac{1}{N_c(N_c-1)/2} \sum_{j=1}^{N_c} \sum_{k<j}^{N_c} (\mathbf{x}_j - \mathbf{x}_k)^2 \tag{18.31}$$

$$\overline{S_C} = \frac{1}{C'} \sum_{c=1}^{C'} S_c \tag{18.32}$$

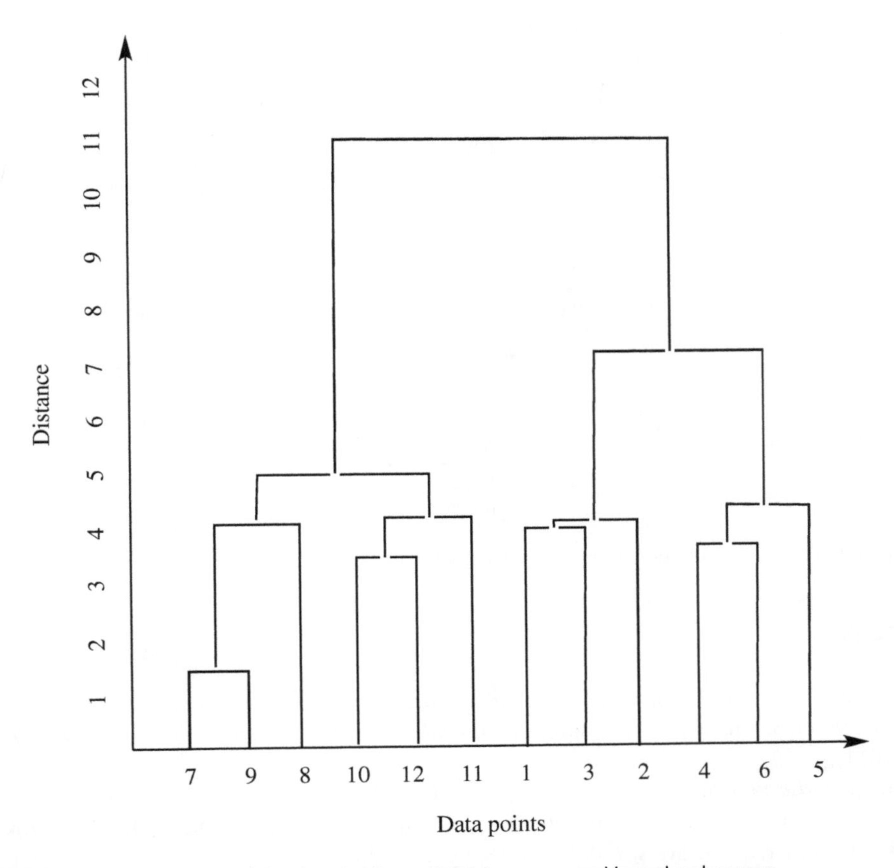

Figure 18.10 Hierachical clustering of the data in Figure 18.9 (a) represented by a dendrogram.

The average spread at each cluster level C is collected during the clustering process and normalized at the end to be in the range of 1 to $N - 1$:

$$\overline{S_C^N} = \left(\frac{N-2}{Max(\overline{S_C}) - Min(\overline{S_C})} \right) (\overline{S_C} - Min(\overline{S_C})) + 1 \tag{18.33}$$

A "good" clustering has all elements assigned to a few clusters, each having very similar elements, that is few clusters each with a small spread. Kelley *et al.* proposed that the optimum number of clusters can be obtained by finding the minimum of a penalty function P_C written as a sum of the normalized average spread and the total number of clusters (including clusters having only one element):

$$P_C = \overline{S_C^N} + C \tag{18.34}$$

P_C has a value of N for $N - 1$ and 1 clusters and a plot for the data in Figure 18.9(a) is shown in Figure 18.11.

The minimum P_C value occurs for four clusters, but the value for three clusters is only marginally larger, which agrees with the conclusion drawn from the dendrogram in Figure 18.9 that four or three clusters are a reasonable partitioning of the 12 elements.

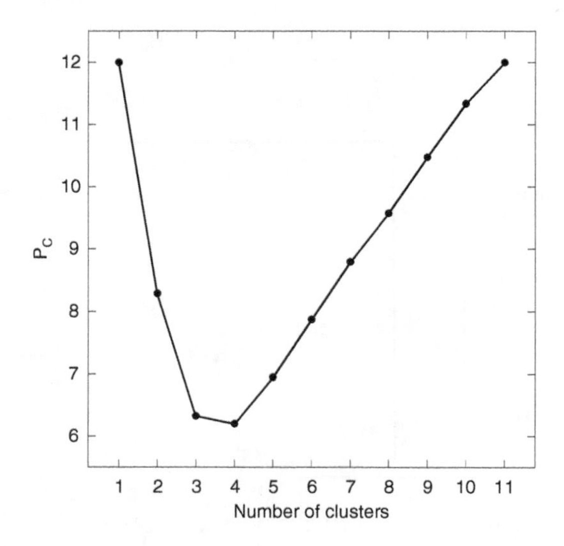

Figure 18.11 Plot of the Kelley function for the data in Figure 18.9(a).

References

1 P. P. Janes and A. P. Rendell, *Journal of Chemical Theory and Computation* 7 (6), 1631–1639 (2011).
2 G. Knizia, W. Li, S. Simon and H.-J. Werner, *Journal of Chemical Theory and Computation* 7 (8), 2387–2398 (2011).
3 D. York, *Canadian Journal of Physics* **44** (5), 1079 (1966).
4 G. R. Famini and L. Y. Wilson, in *Reviews in Computational Chemistry*, edited by K. B. Lipkowitz and D. B. Boyd (John Wiley & Sons, 2002), Vol. 18, pp. 211–255.
5 C. Kramer, T. Kalliokoski, P. Gedeck and A. Vulpetti, *Journal of Medicinal Chemistry* **55** (11), 5165–5173 (2012).
6 H. A. Carlson, *Journal of Chemical Information and Modeling* **53** (8), 1837–1841 (2013).
7 S. Wold, M. Sjostrom and L. Eriksson, *Chemometrics and Intelligent Laboratory Systems* **58** (2), 109–130 (2001).
8 A. Cherkasov, E. N. Muratov, D. Fourches, A. Varnek, I. I. Baskin, M. Cronin, J. Dearden, P. Gramatica, Y. C. Martin, R. Todeschini, V. Consonni, V. E. Kuz'min, R. Cramer, R. Benigni, C. Yang, J. Rathman, L. Terfloth, J. Gasteiger, A. Richard and A. Tropsha, *Journal of Medicinal Chemistry* **57**, 4977–5010 (2013).
9 C. Hansch, A. Leo and R. W. Taft, *Chemical Reviews* **91** (2), 165–195 (1991).
10 R. D. Cramer, D. E. Patterson and J. D. Bunce, *Journal of the American Chemical Society* **110** (18), 5959–5967 (1988).
11 F. Melani, P. Gratteri, M. Adamo and C. Bonaccini, *Journal of Medicinal Chemistry* **46** (8), 1359–1371 (2003).
12 L. Moretti and W. G. Richards, *Bioorganic and Medicinal Chemistry Letters* **20** (19), 5887–5890 (2010).
13 J. H. Van Drie, *Wiley Interdisciplinary Reviews: Computational Molecular Science* **3** (5), 449–464 (2013).
14 S. Agatonovic-Kustrin and R. Beresford, *Journal of Pharmaceutical and Biomedical Analysis* **22** (5), 717–727 (2000).
15 C. Ferreira, *Complex Systems* **13** (2), 87–129 (2001).
16 L. A. Kelley, S. P. Gardner and M. J. Sutcliffe, *Protein Engineering* **9** (11), 1063–1065 (1996).

19

Concluding Remarks

The real world is very complex and a complete description is therefore also very complicated. Only by imposing a series of often quite stringent limitations and approximations can a problem be reduced in complexity such that is may be analyzed in some detail, such as, for example, by calculations. A chemical reaction in the laboratory may involve perhaps 10^{20} molecules surrounded by 10^{24} solvent molecules, in contact with a glass surface and interacting with gases (N_2, O_2, CO_2, H_2O, etc.) in the atmosphere. The whole system will be exposed to a flux of photons of different frequency (light) and a magnetic field (from the earth), and possibly also a temperature gradient from external heating. The dynamics of all the particles (nuclei and electrons) is determined by relativistic quantum mechanics and the interaction between particles is governed by quantum electrodynamics. In principle the gravitational and strong (nuclear) forces should also be considered. For chemical reactions in biological systems, the number of different chemical components will be large, involving various ions and assemblies of molecules behaving intermediately between solution and solid state (e.g. lipids in cell walls).

Except for a couple of rather extreme areas (such as the combination of general relativity and quantum mechanics or the unification of gravitation with the strong and electroweak interactions), we believe that all the fundamental physics is known. The "only" problem is that the real world contains so many (different) components interacting by complicated potentials that a detailed description is impossible.

As this book hopefully has given some insight into, the key is to know what to neglect or approximate when trying to obtain answers to specific questions in predefined systems. For chemical problems, only the electrostatic force needs to be considered; the gravitational interaction is a factor of 10^{39} weaker and can be completely neglected. Similarly, the strong nuclear force is so short-ranged that is has no effect on chemical phenomena (although the brief claims regarding "cold" fusion for a period seemed to contradict this). The weak interaction, which is responsible for radioactive decay by the $n \rightarrow p + e$ process, is also much smaller than the electrostatic, although there have been various estimates of whether it might give rise to symmetry breaking (i.e. preference for one enantiomer over its mirror image), which possibly could be detected experimentally. Similarly, the earth's magnetic field is so tiny that only under very special circumstances can it have any detectable influence on the outcome of a chemical reaction.

For electronic structure calculations within the wave function approach, the starting point is usually an independent-particle model, which for electrons is the Hartree–Fock model. The results from

Introduction to Computational Chemistry, Third Edition. Frank Jensen.

Companion Website: http://www.wiley.com/go/jensen/computationalchemistry3

this model can be improved by adding electron correlation corrections and increasing the basis set. The resulting two-dimensional diagram shown in Figure 4.3 indicates that the "exact" result can be obtained by systematically increasing the level of sophistication along both axes until convergence is reached. Usually the desired level of accuracy is such that the convergence cannot be reached owing to limitations in computational resources, and the results thus suffer from approximations in the one-particle (basis set) and many-particle (configurations) space. Even if the residual errors could be reduced below the target accuracy, the "exact" solution is still subject to a number of limitations that must also be considered in order to compare with the experimental results. These may, for example, be:

1. Neglect of relativistic effects by using the Schrödinger instead of the Dirac equation. This is reasonably justified in the upper part of the periodic table but not in the lower half. For some phenomena, such as spin–orbit coupling, there is no classical counterpart and only a relativistic treatment can provide an understanding. The relativistic effects may be incorporated by a one-component (mass–velocity and Darwin terms), two-component (spin–orbit) or full four-component methods (Figure 9.2).
2. The effects of the environment, such as solvent effects. These may be modeled, for example, by a continuum model, by treating the solvent as an ensemble of classical particles (QM/MM methods) or by including them in the quantum description (e.g. Car–Parrinello methods).
3. Vibrational corrections. For energies, this would typically be inclusion of zero-point energies while for properties this may correspond to vibrational averaging. The corrections may again be done at several levels of accuracy, for example using a harmonic approximation or also including anharmonic effects.
4. Finite temperature effects. This would correspond, for example, to a molecular structure not being represented as a fixed geometry but rather as an ensemble of structures corresponding to an average over accessible geometries at a given temperature.
5. Non-Born–Oppenheimer effects. The assumption of a rigorous separation of nuclear and electronic motions is in most cases a quite good approximation and there is a good understanding of when it will fail. Methods for going beyond the Born–Oppenheimer approximation are still somewhat limited in terms of generality and applicability.
6. Quantum effects for the nuclei. One may argue that the vibrational effects are the most important of these, but in some cases other effects such as tunneling may also be important.
7. Quantum mechanics being replaced (wholly or partly) by classical mechanics. For electrons such an approximation would lead to disastrous results, but for nuclei (atoms) the quantum effects are sufficiently small that in most cases they can be neglected.
8. Approximating the intermolecular interactions to only include two-body effects, for example electrostatic forces are only calculated between pairs of fixed atomic charges in force field techniques. Alternatively, the discrete interactions between molecules may be treated only in an average fashion, by using Langevin dynamics instead of molecular dynamics.
9. Calculating ensemble or time averages over a relatively small number of points (perhaps a few million) and a limited number of particles (perhaps a few hundred), instead of something that approaches the macroscopic sample of perhaps 10^{20} molecules/configurations.
10. Finite temperature being reduced to zero kelvin, that is the use of static structures to represent molecules, rather than treating them as an ensemble of molecules in a distribution of states (translational, rotational and vibrational) corresponding to a (macroscopic) temperature.
11. Making approximations in the Hamiltonian describing the system, for example semi-empirical electronic structure methods or density functional theory.

12. Approximating external fields (electric or magnetic) by only considering their linear components. For normal conditions, this will be a quite good approximation, but this is not the case in, for example, intense laser fields or for very large molecules.
13. Treating the nuclei as point particles. In reality a nucleus has a finite size ($\sim 10^{-15}$ m) and since the electrons can penetrate the nucleus the potential felt inside the nucleus will differ from that of a point particle and consequently will lead to changes in the electronic energy.
14. QED corrections. The interaction between charged particles is normally described by the Coulomb interaction, but when the quantization of the field is considered, there are additional higher-order correction terms.

Most of these approximations are mainly of a computational nature, as there are well-defined methods available for going beyond the approximations, but they are computationally too demanding. The key is therefore to be able to evaluate what level of theory (i.e. which approximations are appropriate) is required for obtaining results that are sufficiently accurate to provide useful information about the question at hand. Hopefully this book has given a few clues as how to select a suitable method for a given problem.

Appendix A

Notation

The following contains a list of the symbols used in the text. It is not possible, nor desirable, to enforce a unique symbol notation, and the same symbols can therefore represent different quantities in different contexts. Attempts have been made to have consistent notation within each chapter, but tradition and common usage dictate differences in notation between chapters.

Bold quantities are operators, vectors, matrices or tensors. Plain symbols are scalars.

α	Morse, Hill or screening parameter
$\boldsymbol{\alpha}$	Polarizability
$\alpha\beta$	Spin functions
$\boldsymbol{\alpha\beta}$	Dirac 4×4 spin matrices
$\alpha\beta\gamma\delta$	Summation indices for basis functions
$\alpha, \beta, \gamma, \delta, \zeta$	Basis function exponents
α_A, β_{AB}	Hückel parameters for atom A, and between atoms A and B
a	Born radius for solvation cavities
$abcd$	Summation indices for virtual MOs
$a_n, a_i, b_i, c_i,\ldots$	Expansion coefficients
$\mathbf{a}$	Acceleration
A, B, C,…	General parameters or coefficients
A	Helmholtz free energy
$\mathbf{A}$	Dipole–quadrupole polarizability
$\mathbf{A}$	Antisymmetrizing operator
$\mathbf{A}$	Vector potential
$\mathbf{A}$	Hyperfine coupling constant
$\boldsymbol{\beta}$	First hyperpolarizability
β_μ	Resonance parameter in semi-empirical theory
$\mathbf{B}$	Dipole–quadrupole hyperpolarizability
$\mathbf{B}$	Magnetic field (magnetic induction)
χ	Basis functions (atomic orbitals) in *ab initio* methods
χ	Electronegativity

Introduction to Computational Chemistry, Third Edition. Frank Jensen.

Companion Website: http://www.wiley.com/go/jensen/computationalchemistry3

χ^{B}	Out-of-plane angle for atom B
$\boldsymbol{\chi}$	Magnetic susceptibility
X	Gauge including basis function
c	Speed of light
$c_{\alpha i}$	MO expansion coefficients
C_{xy}	Correlation function
$\mathbf{C}$	Quadrupole polarizability
$\mathbf{C}$	Matrix of coefficients
δ	A small variation or quantity
δ_{ij}	Kronecker delta ($\delta_{ij} = 1$ for $i = j$, $\delta_{ij} = 0$ for $i \neq j$)
$\delta(\mathbf{r})$	Dirac delta function ($\delta(\mathbf{r}) = 0$ for $\mathbf{r} \neq 0$)
Δ	A finite difference or quantity
d	Distance
D	Dissociation energy
$D_{\alpha\beta}$	Density matrix element in AO basis
$\mathbf{D}$	Density matrix
$\mathbf{D}$	Zero field splitting tensor
ε	Matrix eigenvalue (energy related)
ε	van der Waals parameter
ε	Dielectric constant
ε	Small perturbation
ε	Energy, for one electron or as an individual term in a sum
$\varepsilon[\rho]$	Energy functional per electron
$\boldsymbol{\varepsilon}$	Energy matrix
$\mathbf{e}$	Error vector
E	Energy, many particles or terms
E_{e}	Electronic energy
$E[\rho]$	Energy functional
EA	Electron affinity
φ	Block orbital
ϕ	Molecular orbital
ϕ	Electrostatic potential
Φ	Slater determinant or similar approximate wave function
Φ_i^a, Φ_{ij}^{ab}	Excited Slater determinants
f, g	General functions
$F_{ij}, F_{\alpha\beta}$	Fock matrix element in MO and AO basis
$\mathbf{F}$	Electric field
$\mathbf{F}$	Force
$\mathbf{F}$	Force constant matrix
$\mathbf{F}$	Fock operator or Fock matrix
γ_{AB}	Two-electron matrix element in semi-empirical methods
$\boldsymbol{\gamma}$	Second hyperpolarizability
γ_k	Reduced density matrix of order k
g_{i}	Degeneracy factor
g_{e}	Electronic g-factor
g_{A}	Nuclear g-factor

$\mathbf{g}$	Two-electron repulsion operator
$\mathbf{g}$	Gradient (first derivative)
$\mathbf{g}$	ESR g-tensor
G	Gibbs free energy
G_{xy}	Coulomb-type matrix elements in semi-empirical theory ($x, y = s, p, d$)
$\mathbf{G}$	Matrix containing square root of inverse atomic masses
$\mathbf{G}$	Tensor containing two-electron integrals
η	Absolute hardness
h	Planck's constant
$h_{ij}, h_{\alpha\beta}$	Matrix element of a one-electron operator in MO and AO basis
$h_{\mu\nu}$	Matrix element of a one-electron operator in semi-empirical theory
$\mathbf{h}$	Core or other effective one-electron operator
H	Enthalpy
H_{ij}	Matrix element of a Hamiltonian operator between Slater determinants
H_{xy}	Exchange-type matrix elements in semi-empirical theory ($x, y = s, p, d$)
$\mathbf{H}$	Hessian (second derivative matrix of a function)
$\mathbf{H}, \mathbf{H}_{\mathrm{e}}, \mathbf{H}_{\mathrm{n}}$	Hamiltonian operator or Hamiltonian matrix (general, electronic, nuclear)
$ijkl$	Summation indices for occupied MOs
I	Moment of inertia
$\mathbf{I}$	Unit matrix
$\mathbf{I}$	Nuclear magnetic moment or spin
IP, I_{μ}	Ionization potential
J_{ij}	Coulomb integral
$\mathbf{J}$	Spin–spin coupling matrix
$\mathbf{J}$	Coulomb operator
$J[\rho]$	Coulomb functional
κ	Lagrange multiplier
κ	Transmission coefficient
κ	Compressibility constant
k	Boltzmann's constant
k	Rate constant
$k, k^{\mathrm{AB}\ldots}$	Force constant (for atoms A, B,…)
$\mathbf{k}$	Wave vector
K_{ij}	Exchange integral
$\mathbf{K}$	Anharmonic constants (third derivative)
$\mathbf{K}$	Exchange operator
$K[\rho]$	Exchange functional
λ	Lagrange multiplier
λ	Matrix eigenvalue (general)
λ	General perturbation or scaling factor
λ	Hessian shift parameter
$\mathbf{\Lambda}$	Diagonal matrix
l, L	Angular momentum quantum number
L	Lagrange function
L	Path length
$\mathbf{l}, \mathbf{L}$	Angular momentum operator

μ	Mulliken electronegativity, chemical potential
μ	Reduced or fictive mass
μ_B	Bohr magneton
μ_N	Nuclear magneton
$\mu, \nu, \lambda, \sigma$	Basis functions (atomic orbitals) in semi-empirical methods
$\boldsymbol{\mu}$	Dipole moment
m	Mass, general or electron mass
$\mathbf{m}$	Magnetic dipole moment
M	Number of basis functions or configurations
M_A	Nuclear mass
ν	Vibrational frequency
n_i	Orbital occupation number
N	Number of particles
N_A	Avogadro's number
$\mathbf{O}, \mathbf{P}, \mathbf{Q}$	General operators
$\boldsymbol{\pi}$	Generalized momentum operator
Π	Product of diagonal elements in a Slater determinant
$\mathbf{p}$	Momentum operator or vector
P	Probability
P	Pressure
P_i	Legendre polynomial
$\mathbf{P}, \mathbf{P}_{ij}$	Permutation operators (permuting indices i and j)
$\mathbf{P}, \mathbf{P}_1, \mathbf{P}_2$	Perturbation operators (one- and two-electron)
q	Charge on a particle (integer)
q	Partition function (one particle)
$\mathbf{q}$	Normal or generalized coordinate
Q	Atomic charge (can be fractional), fitted or from population analysis
Q	Partition function (many particles)
Q	Predictive correlation coefficient
$\mathbf{Q}$	Quadrupole moment in traced form
ρ	Electron density, density matrix
ρ	Bond order
r, θ, ϕ	Polar coordinates
r_{ij}	Distance between electrons i and j
$\mathbf{r}$	Position vector(s), general or electronic
R	Trust radius
R	Gas constant
R	Correlation coefficient
$R, R_{ij}, R_{AB}, R^{AB}$	Distance between atoms or nuclei, i and j or A and B
$\mathbf{R}$	Position vector, nuclear
σ	van der Waals collision parameter
σ	Order of rotational subgroup
σ	Charge density
σ^2	Variance
$\boldsymbol{\sigma}$	NMR shielding tensor
$\boldsymbol{\sigma}_{x,y,z}$	Pauli 2×2 spin matrices

$\mathbf{s}$	Electron spin operator
S	Entropy
S	Switching function
S	Action
$S_{\alpha\beta}$	Overlap matrix element in AO basis
$\mathbf{S}, \mathbf{S}^2$	Total spin and spin squared operators
$\theta(t)$	Heaviside step function ($\theta(t) = 0$ for $t < 0$, $\theta(t) = 1$ for $t > 0$)
θ^{ABC}	Angle between atoms A, B and C
$\mathbf{\Theta}$	Quadrupole moment in traceless form
$\Theta_{S,i}^{N}$	Spin coupling function
τ	Heat or pressure bath coupling parameter
τ	Phase factor
τ	Imaginary time variable
τ	Orbital kinetic energy density
t	Time
Δt	Small (finite) time step
t_i^a, t_{ij}^{ab}	Cluster amplitudes
$\mathbf{t}$	Translational vector
$\mathbf{t}$	Normalized gradient or tangent vector
T	Temperature
$\mathbf{T}$	Transition matrix
$\mathbf{T}, \mathbf{T}_1, \mathbf{T}_2, \ldots$	Cluster operator (general, single, double,…, excitations)
$\mathbf{T}, \mathbf{T}_\mathbf{e}, \mathbf{T}_\mathbf{n}$	Kinetic energy operator (general, electronic, nuclear)
$T[\rho]$	Kinetic energy functional, exact
$T_s[\rho]$	Kinetic energy functional, calculated from a Slater determinant
U	Internal energy
U	Bias potential
U_i	Matrix element of a semi-empirical one-electron operator, usually parameterized
$\mathbf{U}$	Unitary matrix
$\mathbf{v}$	Velocity
V	Volume
V, V_{AB}, V^{AB}	Potential energy (between atoms A and B)
V_{ij}	Coulomb potential between particles i and j
V_n	Potential energy parameter
$\mathbf{v}_{eff}, \mathbf{V}_{eff}$	Effective potential operators (one- or multielectron)
$\mathbf{V}, \mathbf{V}_{ee}, \mathbf{V}_{ne}, \mathbf{V}_{nn}$	Potential (Coulomb) energy operator (general, electron–electron, nuclear–electron, nuclear–nuclear)
ω	Electrophilicity
ω	Range separator parameter
ω	Excitation energy
ω	Frequency associated with an electric or magnetic field
ω^{ABCD}	Torsional angle between atoms A, B, C and D
ϖ	Phase factor
$\mathbf{\Omega}$	Two-electron operator
w_A	Weighting factor for atom A

W	Energy or quasi-energy of an approximate wave function
W_i	Perturbation energy correction at order i
W_k	Wigner intracule of order k
$W_{\alpha\beta}$	Energy-weighted density matrix element in AO basis
$\mathbf{W}$	Matrix containing weighting factors
$\mathbf{W}$	Energy-weighted density matrix
ξ	Collective variable
ξ	Molecular surface parameter for calculating solvation energies
$\boldsymbol{\xi}$	Magnetizability
x_i, y_i, z_i	Cartesian coordinates for particle i
Δx_i	Component in a vector
$\boldsymbol{\Xi}$	Hexadecapole moment
X, Y	Cardinal numbers
Ψ, Ψ_e, Ψ_n	Exact or multideterminant wave function (general, electronic, nuclear)
ζ	Spin polarization
ζ	Friction coefficient
Z	Nuclear charge, exact
Z'	Nuclear charge, reduced by the number of core electrons
$\langle n\|$	Bra, referring to a function characterized by quantum number n
$\|n\rangle$	Ket, referring to a function characterized by quantum number n
$\langle n\|\mathbf{O}\|m\rangle$	Bracket (matrix element) of operator **O** between functions n and m
$\langle\mathbf{O}\rangle$	Average value of **O**
$\|\mathbf{O}\|$	Norm or determinant of **O**
$\langle\langle\mathbf{O}; \mathbf{P}\rangle\rangle$	Propagator of **O** and **P**
$[\mathbf{O}, \mathbf{P}]$	Commutator of **O** and **P** ($[\mathbf{O}, \mathbf{P}] = \mathbf{OP} - \mathbf{PO}$)
∇	Gradient operator
∇^2	Laplace operator
$\cdot$	Entry-wise matrix product or tensor contraction
$\times$	Cross-product
$\otimes$	Dyadic (outer) product
$\nabla\cdot$	Divergence operator
$\nabla\times$	Curl operator
^{t}	Vector transposition
$^\dagger, ^*$	Complex conjugate

Appendix B

The Variational Principle

The *Variational Principle* states that an approximate wave function has an energy that is above or equal to the exact energy. The equality holds only if the wave function is exact. The proof is as follows.

Assume that we know the exact solutions to the Schrödinger equation:

$$\mathbf{H}\Psi_i = E_i\Psi_i, \qquad i = 0, 1, 2, \ldots, \infty \tag{B.1}$$

There are infinitely many solutions and we assume that they are labeled according to their energies, E_0 being the lowest. Since the $\mathbf{H}$ operator is Hermitian, the solutions form a *complete* basis. We may furthermore choose the solutions to be orthogonal and normalized:

$$\langle\Psi_i|\Psi_j\rangle = \delta_{ij} \tag{B.2}$$

An approximate wave function can be expanded in the exact solutions, since they form a complete set:

$$\Phi = \sum_{i=0}^{\infty} a_i\Psi_i \tag{B.3}$$

The energy of an approximate wave function is calculated as

$$W = \frac{\langle\Phi|\mathbf{H}|\Phi\rangle}{\langle\Phi|\Phi\rangle} \tag{B.4}$$

Inserting the expansion (B.3) we obtain

$$W = \frac{\sum_{i=0}^{\infty}\sum_{j=0}^{\infty} a_i a_j \langle\Psi_i|\mathbf{H}|\Psi_j\rangle}{\sum_{i=0}^{\infty}\sum_{j=0}^{\infty} a_i a_j \langle\Psi_i|\Psi_j\rangle} \tag{B.5}$$

Using the fact that $\mathbf{H}\Psi_i = E_i\Psi_i$ and the orthonormality of the Ψ_i (Equations (B.1) and (B.2)), we obtain

$$W = \frac{\sum_{i=0}^{\infty} a_i^2 E_i}{\sum_{i=0}^{\infty} a_i^2} \tag{B.6}$$

Introduction to Computational Chemistry, Third Edition. Frank Jensen.

Companion Website: http://www.wiley.com/go/jensen/computationalchemistry3

The variational principle states that $W \geq E_0$ or, equivalently, $W - E_0 \geq 0$:

$$W - E_0 = \frac{\sum_{i=0}^{\infty} a_i^2 E_i}{\sum_{i=0}^{\infty} a_i^2} - E_0 = \frac{\sum_{i=0}^{\infty} a_i^2 (E_i - E_0)}{\sum_{i=0}^{\infty} a_i^2} \geq 0 \tag{B.7}$$

Since a_i^2 is always positive or zero and $E_i - E_0$ is always positive or zero (E_0 is by definition the lowest energy), this completes the proof. Furthermore, in order for the equal sign to hold, all $a_{i\neq0} = 0$ since $E_{i\neq0} - E_0$ is non-zero (neglecting degenerate ground states). This in turns means that $a_0 = 1$, owing to the normalization of Φ, and consequently the wave function is the exact solution.

This proof shows that any approximate wave function will have an energy above or equal to the exact ground state energy. There is a related theorem, known as *MacDonald's Theorem*, which states that the nth root of a set of secular equations (e.g. a CI matrix) is an upper limit to the $(n - 1)$th excited exact state, within the given symmetry subclass.[1] In other words, the lowest root obtained by diagonalizing a CI matrix is an upper limit to the lowest exact wave functions, the second root is an upper limit to the exact energy of the first excited state, the third root is an upper limit to the exact second excited state, and so on.

The Hohenberg–Kohn Theorems

The electron density is in wave mechanics given by the square of the wave function integrated over $N - 1$ electron coordinates and the wave function is determined by solving the Schrödinger equation. For a system of N_{nuclei} nuclei and N_{elec} electrons, the electronic Hamiltonian operator contains the terms shown below:

$$\mathbf{H}_e = -\sum_{i=1}^{N_{\text{elec}}} \tfrac{1}{2}\nabla_i^2 - \sum_{i=1}^{N_{\text{elec}}}\sum_{A=1}^{N_{\text{nuclei}}} \frac{Z_A}{|\mathbf{R}_A - r_i|} + \sum_{i=1}^{N_{\text{elec}}}\sum_{j>i}^{N_{\text{elec}}} \frac{1}{|\mathbf{r}_i - \mathbf{r}_j|} + \sum_{A=1}^{N_{\text{nuclei}}}\sum_{B=1}^{N_{\text{nuclei}}} \frac{Z_A Z_B}{|\mathbf{R}_A - \mathbf{R}_B|} \tag{B.8}$$

The last term is a constant within the Born–Oppenheimer approximation. It is seen that the Hamiltonian operator is uniquely determined by the number of electrons and the potential created by the nuclei, $\mathbf{V}_{\text{ne}}$, that is the nuclear charges and positions. This means that the ground state wave function (and thereby the electron density) and ground state energy are also given uniquely by these quantities.

Assume now that two different external potentials (which may be from nuclei), $\mathbf{V}_{\text{ext}}$ and $\mathbf{V}'_{\text{ext}}$, result in the same electron density, ρ. Two different potentials imply that the two Hamiltonian operators are different, $\mathbf{H}$ and $\mathbf{H}'$, and the corresponding lowest energy wave functions are different, Ψ and Ψ'. Taking Ψ' as an approximate wave function for $\mathbf{H}$ and using the variational principle yields

$$\begin{aligned}
&\langle \Psi' | \mathbf{H} | \Psi' \rangle > E_0 \\
&\langle \Psi' | \mathbf{H}' | \Psi' \rangle + \langle \Psi' | \mathbf{H} - \mathbf{H}' | \Psi' \rangle > E_0 \\
&E_0' + \langle \Psi' | \mathbf{V}_{\text{ext}} - \mathbf{V}'_{\text{ext}} | \Psi' \rangle > E_0 \\
&E_0' + \int \rho(\mathbf{r}) \left(\mathbf{V}_{\text{ext}} - \mathbf{V}'_{\text{ext}}\right) \mathrm{d}\mathbf{r} > E_0
\end{aligned} \tag{B.9}$$

Similarly, taking Ψ as an approximate wave function for $\mathbf{H}'$ yields

$$E_0 - \int \rho(\mathbf{r}) \left(\mathbf{V}_{\text{ext}} - \mathbf{V}'_{\text{ext}}\right) \mathrm{d}\mathbf{r} > E_0' \tag{B.10}$$

Addition of these two inequalities gives $E'_0 + E_0 > E'_0 + E_0$, showing that the assumption was wrong. In other words, for the ground state there is a one-to-one correspondence between the electron density and the nuclear potential, and thereby also with the Hamiltonian operator and the energy. In the language of density functional theory, the energy is a unique functional of the electron density, $E[\rho]$.

Using the electron density as a parameter, there is a variational principle analogous to that in wave mechanics. Given an approximate electron density ρ' (assumed to be positive definite everywhere) that integrates to the number of electrons and originates from a proper antisymmetric wave function, the energy given by this density is an upper bound to the exact ground state energy, provided that the exact functional is used:

$$\begin{aligned} &\int \rho'(\mathbf{r})\mathrm{d}\mathbf{r} = N_{\mathrm{elec}} \\ &E_0[\rho'] \geq E_0[\rho] \end{aligned} \tag{B.11}$$

The Adiabatic Connection Formula

The Hellmann–Feynman theorem (Equation (11.40)) is given by

$$\frac{\partial}{\partial \lambda}\langle \Psi_\lambda | \mathbf{H}_\lambda | \Psi_\lambda \rangle = \left\langle \Psi_\lambda \left| \frac{\partial \mathbf{H}_\lambda}{\partial \lambda} \right| \Psi_\lambda \right\rangle \tag{B.12}$$

With the Hamiltonian in Equation (6.10), this gives

$$\begin{aligned} \mathbf{H}_\lambda &= \mathbf{T} + \mathbf{V}_{\mathrm{ext}}(\lambda) + \lambda \mathbf{V}_{\mathrm{ee}} \\ \frac{\partial}{\partial \lambda}\langle \Psi_\lambda | \mathbf{H}_\lambda | \Psi_\lambda \rangle &= \left\langle \Psi_\lambda \left| \frac{\partial \mathbf{V}_{\mathrm{ext}}(\lambda)}{\partial \lambda} + \mathbf{V}_{\mathrm{ee}} \right| \Psi_\lambda \right\rangle \end{aligned} \tag{B.13}$$

Integrating over λ between the limits 0 and 1 corresponds to smoothly transforming the non-interacting reference to the real system:

$$\begin{aligned} \int_0^1 \frac{\partial}{\partial \lambda}\langle \Psi_\lambda | \mathbf{H}_\lambda | \Psi_\lambda \rangle \mathrm{d}\lambda &= \int_0^1 \left\langle \Psi_\lambda \left| \frac{\partial \mathbf{V}_{\mathrm{ext}}(\lambda)}{\partial \lambda} + \mathbf{V}_{\mathrm{ee}} \right| \Psi_\lambda \right\rangle \mathrm{d}\lambda \\ \langle \Psi_1 | \mathbf{H}_1 | \Psi_1 \rangle - \langle \Psi_0 | \mathbf{H}_0 | \Psi_0 \rangle = E_1 - E_0 &= \int_0^1 \left\langle \Psi_\lambda \left| \frac{\partial \mathbf{V}_{\mathrm{ext}}(\lambda)}{\partial \lambda} + \mathbf{V}_{\mathrm{ee}} \right| \Psi_\lambda \right\rangle \mathrm{d}\lambda \end{aligned} \tag{B.14}$$

The integration is done under the assumption that the density remains constant, that is Ψ_0 and Ψ_1 yield the *same* density. For the term involving the external potential, this allows the integration to be written in terms of the two limits:

$$\begin{aligned} \int_0^1 \left\langle \Psi_\lambda \left| \frac{\partial \mathbf{V}_{\mathrm{ext}}(\lambda)}{\partial \lambda} \right| \Psi_\lambda \right\rangle \mathrm{d}\lambda &= \int \rho(\mathbf{r}) \left(\int_0^1 \frac{\partial \mathbf{V}_{\mathrm{ext}}(\mathbf{r}, \lambda)}{\partial \lambda} \mathrm{d}\lambda \right) d\mathbf{r} \\ &= \int \rho(\mathbf{r})(\mathbf{V}_{\mathrm{ext}}(1) - \mathbf{V}_{\mathrm{ext}}(0))\mathrm{d}\mathbf{r} \\ &= \int \rho(\mathbf{r})\mathbf{V}_{\mathrm{ext}}(1)\mathrm{d}\mathbf{r} - \int \rho(\mathbf{r})\mathbf{V}_{\mathrm{ext}}(0)\mathrm{d}\mathbf{r} \end{aligned} \tag{B.15}$$

The energy of the non-interacting (E_0) system is given as follows since $\mathbf{V}_{\mathrm{ee}}$ makes no contribution:

$$E_0 = \langle \Psi_0 | \mathbf{T} | \Psi_0 \rangle + \int \rho(\mathbf{r})\mathbf{V}_{\mathrm{ext}}(0)\mathrm{d}\mathbf{r} \tag{B.16}$$

Combining Equations (B.14), (B.15) and (B.16) yields

$$E_1 = \langle \Psi_0 | \mathbf{T} | \Psi_0 \rangle + \int \rho(\mathbf{r}) \mathbf{V}_{\mathrm{ext}}(1) \mathrm{d}\mathbf{r} + \int_0^1 \langle \Psi_\lambda | \mathbf{V}_{\mathrm{ee}} | \Psi_\lambda \rangle \mathrm{d}\lambda \tag{B.17}$$

Using the fact that $\mathbf{V}_{\mathrm{ext}}(1) = \mathbf{V}_{\mathrm{ne}}$ and the definition of E_1 (Equation (6.9)) we obtain

$$J[\rho] + E_{\mathrm{xc}}[\rho] = \int_0^1 \langle \Psi_\lambda | \mathbf{V}_{\mathrm{ee}} | \Psi_\lambda \rangle \mathrm{d}\lambda \tag{B.18}$$

The exchange–correlation energy can thus be obtained by integrating the electron–electron interaction over the λ variable and subtracting the Coulomb part. The right-hand side of Equation (B.18) can be written in terms of the second-order reduced density matrix, Equation (6.18), and the definition of the exchange–correlation hole in Equation (6.37) allows the Coulomb energy to be separated out:

$$\begin{aligned}
\langle \Psi_\lambda | \mathbf{V}_{\mathrm{ee}} | \Psi_\lambda \rangle &= \frac{1}{2} \int \frac{\rho_2(\lambda, \mathbf{r}_1, \mathbf{r}_2)}{|\mathbf{r}_1 - \mathbf{r}_2|} \mathrm{d}\mathbf{r}_1 \mathrm{d}\mathbf{r}_2 \\
&= \frac{1}{2} \int \frac{\rho_1(\mathbf{r}_1)\rho_1(\mathbf{r}_2)}{|\mathbf{r}_1 - \mathbf{r}_2|} \mathrm{d}\mathbf{r}_1 \mathrm{d}\mathbf{r}_2 + \frac{1}{2} \int \frac{\rho_1(\mathbf{r}_1) h_{\mathrm{xc}}(\lambda, \mathbf{r}_1, \mathbf{r}_2)}{|\mathbf{r}_1 - \mathbf{r}_2|} \mathrm{d}\mathbf{r}_1 \mathrm{d}\mathbf{r}_2 \\
&= J[\rho] + \frac{1}{2} \int \frac{\rho_1(\mathbf{r}_1) h_{\mathrm{xc}}(\lambda, \mathbf{r}_1, \mathbf{r}_2)}{|\mathbf{r}_1 - \mathbf{r}_2|} \mathrm{d}\mathbf{r}_1 \mathrm{d}\mathbf{r}_2
\end{aligned} \tag{B.19}$$

Defining $\mathbf{V}_{\mathrm{xc}}^{\mathrm{hole}}$ as below gives the *adiabatic connection formula* (Equation (6.67)):

$$\begin{aligned}
\mathbf{V}_{\mathrm{xc}}^{\mathrm{hole}}(\lambda, \mathbf{r}_1) &= \frac{1}{2} \int \frac{h_{\mathrm{xc}}(\lambda, \mathbf{r}_1, \mathbf{r}_2)}{|\mathbf{r}_1 - \mathbf{r}_2|} \mathrm{d}\mathbf{r}_2 \\
E_{\mathrm{xc}} &= \int_0^1 \left\langle \Psi_\lambda \left| \mathbf{V}_{\mathrm{xc}}^{\mathrm{hole}}(\lambda) \right| \Psi_\lambda \right\rangle \mathrm{d}\lambda \\
E_{\mathrm{xc}} &= \int \rho(\mathbf{r}) \left(\int_0^1 \mathbf{V}_{\mathrm{xc}}^{\mathrm{hole}}(\lambda, \mathbf{r}) d\lambda \right) \mathrm{d}\mathbf{r}
\end{aligned} \tag{B.20}$$

Reference

1 J. K. L. MacDonald, *Physical Review* **43** (10), 0830–0833 (1933).

Appendix C

Atomic Units

It is convenient in electronic structure calculations to work in the *atomic unit* (au) system, which is defined by setting $m_e = e = \hbar = 1$. From these values follow related quantities, as shown in Table C.1

Table C.1 The atomic unit system.

Symbol	Quantity	Value in au	Value in SI units
m_e	Electron mass	1	9.110×10^{-31} kg
e	Electron charge	1	1.602×10^{-19} C
t	Time	1	2.419×10^{-17} s
$\hbar$	$h/2\pi$ (atomic momentum unit)	1	1.055×10^{-34} J s
h	Planck's constant	2π	6.626×10^{-34} J s
a_0	Bohr radius (atomic distance unit)	1	5.292×10^{-11} m
E_H	Hartree (atomic energy unit)	1	4.360×10^{-18} J
c	Speed of light	137.036	2.998×10^{8} m/s
α	Fine structure constant ($= e^2/\hbar c\ 4\pi\varepsilon_0 = 1/c$)	0.00729735	0.00729735
μ_B	Bohr magneton ($= e\hbar/2m_e$)	1/2	9.274×10^{-24} J/T
μ_N	Nuclear magneton	2.723×10^{-4}	5.051×10^{-27} J/T
$4\pi\varepsilon_0$	Vacuum permittivity	1	1.113×10^{-10} C^2/J m
μ_0	Vacuum permeability ($4\pi/c^2$)	6.692×10^{-4}	1.257×10^{-6} N s^2/C^2

Introduction to Computational Chemistry, Third Edition. Frank Jensen.

Companion Website: http://www.wiley.com/go/jensen/computationalchemistry3

Appendix D

Z-Matrix Construction

All calculations need a molecular geometry as input. This is commonly given by one of the following three methods:

1. Cartesian coordinates
2. Internal coordinates
3. Via a graphical interface.

Generating Cartesian coordinates by hand is only realistic for small molecules. If, however, the geometry is taken from outside sources, such as an X-ray structure, Cartesian coordinates are often the natural choice. Similarly, a graphical interface produces a set of Cartesian coordinates for the underlying program, which carries out the actual calculation.

Generating internal coordinates such as bond lengths and angles by hand is relatively simple, even for quite large molecules. One widely used method is the *Z-matrix*, where each atom is specified in terms of a distance, angle and torsional angle to other atoms. It should be noted that internal coordinates are *not* necessarily related to the actual bonding, they are only a convenient method for specifying the geometry. The internal coordinates are usually converted to Cartesian coordinates before any calculations are carried out. Geometry optimizations, however, are often done in internal coordinates in order to remove the six (five) translational and rotational degrees of freedom.

Construction of a *Z*-matrix begins with a drawing of the molecule and a suitable numbering of the atoms. *Any* numbering will result in a valid *Z*-matrix, although assignment of numerical values to the parameters is greatly facilitated if the bonding and symmetry of the molecule is considered when the numbering is performed (see the examples below). The *Z*-matrix specifies the position of each atom in terms of a distance, an angle and a torsional angle relative to other atoms. The first three atoms, however, are slightly different. The first atom is always positioned at the origin of the coordinate system. The second atom is specified as having a distance to the first atom, and is placed along one of the Cartesian axes (usually x or z). The third atom is specified by a distance to either atom 1 or 2, and an angle to the other atom. All subsequent atoms need a distance, an angle and a torsional angle to uniquely specify the position. The atoms are normally identified either by the chemical symbol or by their atomic number.

Introduction to Computational Chemistry, Third Edition. Frank Jensen.

Companion Website: http://www.wiley.com/go/jensen/computationalchemistry3

If the molecular geometry is optimized by the program then only rough estimates of the parameters are necessary. In terms of internal coordinates, this is fairly easy. Some typical bond lengths (Å) and angles are given below:

A—H: A=C: 1.10; A=O, N: 1.00; A=S, P: 1.40
A—B: A, B=C, O, N: 1.40–1.50
A=B: A, B=C, O, N: 1.20–1.30
A≡B: A, B=C, N: 1.20
A—B: A=C, B=S, P: 1.80

Angles around sp^3-hybridized atoms: 110°
Angles around sp^2-hybridized atoms: 120°
Angles around sp-hybridized atoms: 180°

Torsional angles around sp^3-hybridized atoms: separated by 120°
Torsional angles around sp^2-hybridized atoms: separated by 180°

Such estimates allow specification of molecules with up to 50–100 atoms fairly easily. For larger molecules, however, it becomes cumbersome, and the molecule may instead be built from pre-optimized fragments. This is typically done by means of a graphical interface, that is the molecule is pieced together by selecting fragments (such as amino acids) and assigning the bonding between the fragments.

Below are some examples of how to construct *Z*-matrices. Figure D.1 shows acetaldehyde.

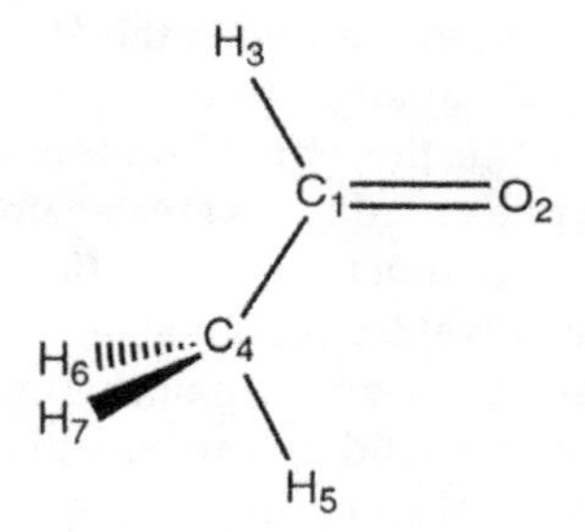

Figure D.1 Atom numbering for acetaldehyde.

C1	0	0.00	0	0.0	0	0.0
O2	1	1.20	0	0.0	0	0.0
H3	1	1.10	2	120.0	0	0.0
C4	1	1.50	2	120.0	3	180.0
H5	4	1.10	1	110.0	2	0.0
H6	4	1.10	1	110.0	2	120.0
H7	4	1.10	1	110.0	2	−120.0

The definition of the torsional angles is illustrated in Figure 2.7. To emphasize the symmetry (C_s) of the above conformation, the Z-matrix may also be given in terms of *symbolic* variables, where variables that are equivalent by symmetry have identical names.

```
C1
O2   1   R1
H3   1   R2   2   A1
C4   1   R3   2   A2   3    D1
H5   4   R4   1   A3   2    D2
H6   4   R5   1   A4   2    D3
H7   4   R5   1   A4   2   -D3

R1 = 1.20
R2 = 1.10
R3 = 1.50
R4 = 1.10
R5 = 1.10
A1 = 120.0
A2 = 120.0
A3 = 110.0
A4 = 110.0
D1 = 180.0
D2 = 0.0
D3 = 120.0
```

Some important things to notice:

1. Each atom must be specified in terms of atoms already defined, that is relative to atoms *above*.
2. Each specification atom can only be used *once* in each line.
3. The specification in terms of distance, angle and torsional angle has *nothing* to do with the bonding in the molecule; for example the torsional angle for C_4 in acetaldehyde is given to H_3, but there is no bond between O_2 and H_3. A Z-matrix, however, is usually constructed such that the distances, angles and torsional angles follow the bonding. This makes it much easier to estimate reasonable values for the parameters.
4. *Distances* should always be *positive* and *angles* always in the range 0°–180°. Torsional angles may be taken in the range −180°–180° or 0°–360°.
5. The symbolic variables show explicitly which parameters are constrained to have the same values, that is H_6 and H_7 are symmetry equivalent and must therefore have the same distances and angles, and a sign difference in the torsional angle. Although the R4 and R5 (and A3 and A4) parameters have the same values initially, they will be different in the final optimized structure.

The limitation that the angles must be between 0° and 180° introduces a slight complication for linear arrays of atoms, such as the cyano group in acetonitrile. Specification of the nitrogen in terms of a distance to C_2 and an angle to C_1 does not allow a unique assignment of a torsional angle since the C_1—C_2—N_6 angle is linear, which makes the torsional angle undefined. There are two methods for solving this problem, either by specifying N_6 relative to C_1 with a long distance:

Figure D.2 Atom numbering for acetaldehyde.

```
C1
C2   1   R1
H3   1   R2   2   A1
H4   1   R2   2   A1   3    D1
H5   1   R2   2   A1   3   -D1
N6   1   R3   3   A2   4    D2
```

R1 = 1.50
R2 = 1.10
R3 = 2.70
A1 = 110.0
A2 = 110.0
D1 = 120.0
D2 = 120.0

Note that the variables imply that the molecule has C_{3v} symmetry. Alternatively, a *Dummy* Atom (X) may be introduced.

Figure D.3 Atom numbering for acetonitrile including a dummy atom.

```
C1
C2   1   R1
H3   1   R2   2   A1
H4   1   R2   2   A1   3    D1
H5   1   R2   2   A1   3   -D1
X6   2   R3   1   A2   3    D2
N7   2   R4   6   A3   1    D3
```

R1 = 1.50
R2 = 1.10
R3 = 1.00
R4 = 1.20
A1 = 110.0
A2 = 90.0
A3 = 90.0
D1 = 120.0
D2 = 0.0
D3 = 180.0

A dummy atom is just a point in space and has no significance in the actual calculation. The above two *Z*-matrices give identical Cartesian coordinates. The R3 variable has arbitrarily been given a distance of 1.00 and the D2 torsional angle of 0.0° is also arbitrary – any other values may be substituted without affecting the coordinates of the real atoms. Similarly, the A2 and A3 angles should just add up to 180°; their individual values are not significant. The function of a dummy atom in this case is to break up the problematic 180° angle into two 90° angles. It should be noted that the introduction of dummy atoms does not increase the number of (non-redundant) parameters, although there are formally three more variables for each dummy atom. The dummy variables may be identified by excluding them from the symbolic variable list or by explicitly forcing them to be non-optimizable parameters.

When a molecule is symmetric, it is often convenient to start the numbering with atoms lying on a rotation axis or in a symmetry plane. If there are no real atoms on a rotation axis or in a mirror plane, dummy atoms can be useful for defining the symmetry element. Consider, for example, the cyclopropenyl system, which has D_{3h} symmetry. Without dummy atoms, one of the C—C bond lengths will be given in terms of the two other C—C distances and the C—C—C angle, and it will be complicated to force the three C—C bonds to be identical. By introducing two dummy atoms to define the C_3 axis, this becomes easy.

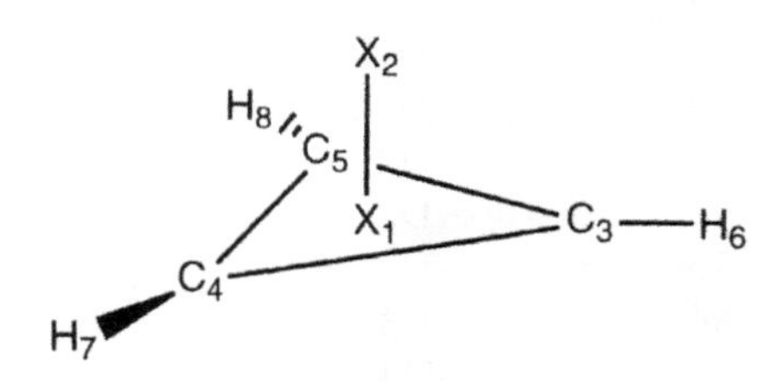

Figure D.4 Atom numbering for the cyclopropyl system.

```
X1
X2   1   1.00
C3   1   R1    2   90.0
C4   1   R1    2   90.0   3    120.0
C5   1   R1    2   90.0   3   -120.0
H6   1   R2    2   90.0   3      0.0
H7   1   R2    2   90.0   3    120.0
H8   1   R2    2   90.0   3   -120.0
```

```
R1 = 0.80
R2 = 1.90
```

In this case there are only two genuine variables, the others are fixed by symmetry.

Let us finally consider two Z-matrices for optimization to transition structures, the Diels–Alder reaction of butadiene and ethylene, and the [1,5]-hydrogen shift in (Z)-1,3-pentadiene. To enforce the symmetries of the TSs (C_s in both cases) it is again advantageous to use dummy atoms.

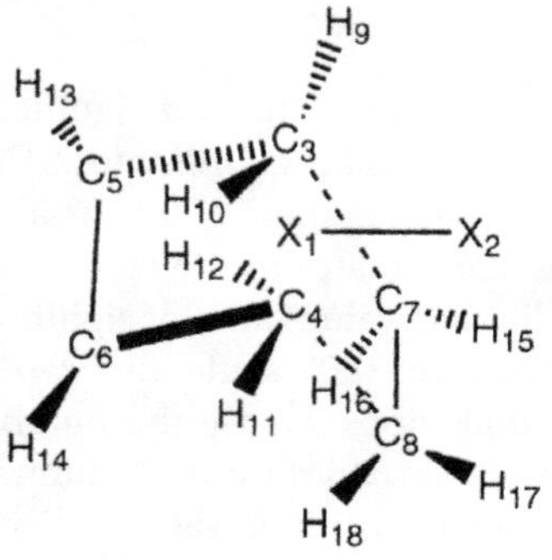

Figure D.5 Atom numbering for the transition structure of the Diels–Alder reaction of butadiene and ethylene.

```
X1
X2    1   1.00
C3    1   R1     2   90.0
C4    1   R1     2   90.0   3   180.0
C5    3   R2     1   A1     2   180.0
C6    4   R2     1   A1     2   180.0
C7    3   R3     1   A2     2   D1
C8    4   R3     1   A2     2   −D1
H9    3   R4     5   A3     6   D2
H10   3   R5     5   A4     6   −D3
H11   4   R4     6   A3     5   −D2
H12   4   R5     6   A4     5   D3
H13   5   R6     3   A5     1   −D4
H14   6   R6     4   A5     1   D4
H15   7   R7     8   A6     4   D5
H16   7   R8     8   A7     4   −D6
H17   8   R7     7   A6     3   −D5
H18   8   R8     7   A7     3   D6

R1 = 1.40
R2 = 1.40
R3 = 2.20
```

R4 = 1.10
R5 = 1.10
R6 = 1.10
R7 = 1.10
R8 = 1.10
A1 = 60.0
A2 = 70.0
A3 = 120.0
A4 = 120.0
A5 = 120.0
A6 = 120.0
A7 = 120.0
D1 = 60.0
D2 = 170.0
D3 = 30.0
D4 = 170.0
D5 = 100.0
D6 = 100.0

The mirror plane is defined by the two dummy atoms and the fixing of the angles and torsional angle of the first two carbons. The torsional angles for atoms C5 and C6 are dummy variables as they only define the orientation of the plane of the first four carbon atoms relative to the dummy atoms, and may consequently be fixed at 180°. Note that the C_5—C_6 and C_7—C_8 bond distances are given implicitly in terms of the R2/A1 and R3/A2 variables. The presence of such "indirect" variables means that some experimentation is necessary for assigning proper values to the "direct" variables. The forming C—C bond is given directly as one of the *Z*-matrix variables, R3, which facilitates a search for a suitable start geometry for the TS optimization, for example by running a series of constrained optimizations with fixed R3 distances.

The [1,5]-hydrogen shift in (*Z*)-1,3-pentadiene is an example of a "narcissistic" reaction, with the reactant and product being identical. The TS is therefore located exactly at the halfway point and has a symmetry different from either the reactant or product. By suitable constraints on the geometry the TS may therefore be located by a *minimization* within a symmetry-constrained geometry.

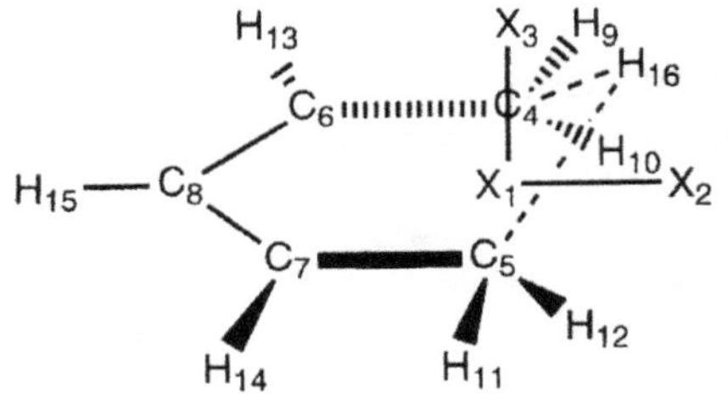

Figure D.6 Atom numbering for the transition structure for the [1,5]-hydrogen shift in (Z)-1,3-pentadiene.

```
X1
X2    1   1.00
X3    1   1.00   2   90.0
C4    1   R1     2   90.0   3    90.0
C5    1   R1     2   90.0   3   -90.0
C6    4   R2     1   A1     2   180.0
C7    5   R2     1   A1     2   180.0
C8    1   R3     3   A2     2   180.0
H9    4   R4     6   A3     8   -D1
H10   4   R5     6   A4     8    D2
H11   5   R4     7   A3     8    D1
H12   5   R5     7   A4     8   -D2
H13   6   R6     4   A5     1    D3
H14   7   R6     5   A5     1   -D3
H15   1   R7     3   A6     2   180.0
H16   1   R8     2   A7     3    0.0

R1 = 1.30
R2 = 1.40
R3 = 2.10
R4 = 1.10
R5 = 1.10
R6 = 1.10
R7 = 3.20
R8 = 0.70
A1 = 80.0
A2 = 90.0
A3 = 120.0
A4 = 120.0
A5 = 120.0
A6 = 90.0
A7 = 60.0
D1 = 160.0
D2 = 60.0
D3 = 160.0
```

The mirror plane is defined by the dummy atoms. The migrating hydrogen H_{16} is not allowed to move out of the plane of symmetry and must consequently have the same distance to C_4 and C_5. A minimization will locate the lowest energy structure within the given C_s symmetry, and a subsequent frequency calculation will reveal that the optimized structure is a TS, with the imaginary frequency belonging to the a'' representation (breaking the symmetry).

Appendix E

First and Second Quantization

The notation used in this book is in terms of *first quantization*. The electronic Hamilton operator, for example, is written as

$$\mathbf{H}_e = -\sum_{i}^{N_{\text{elec}}} \tfrac{1}{2}\nabla_i^2 - \sum_{a}^{N_{\text{nuclei}}}\sum_{i}^{N_{\text{elec}}} \frac{Z_a}{|\mathbf{R}_a - \mathbf{r}_i|} + \sum_{i}^{N_{\text{elec}}}\sum_{j>i}^{N_{\text{elec}}} \frac{1}{|\mathbf{r}_i - \mathbf{r}_j|} + \sum_{a}^{N_{\text{nuclei}}}\sum_{b>a}^{N_{\text{nuclei}}} \frac{Z_a Z_b}{|\mathbf{R}_a - \mathbf{R}_b|} \tag{E.1}$$

A single determinant wave function is given as an antisymmetized product of orbitals (Equation (3.21)):

$$\Phi = \mathbf{A}[\phi_1(1)\phi_2(2)\cdots\phi_N(N)] \tag{E.2}$$

A matrix element of the Hamilton operator over the wave function in Equation (E.2) can be written as a sum over one- and two-electron integrals (Equation (3.32)):

$$\langle\Phi|\mathbf{H}_e|\Phi\rangle = \sum_{i}^{N_{\text{elec}}} \langle\phi_i|\mathbf{h}_i|\phi_i\rangle + \frac{1}{2}\sum_{ij}^{N_{\text{elec}}} (\langle\phi_i\phi_j|\mathbf{g}|\phi_i\phi_j\rangle - \langle\phi_i\phi_j|\mathbf{g}|\phi_j\phi_i\rangle) + V_{\text{nn}} \tag{E.3}$$

There is another commonly used notation known as *second quantization*.[1,2] The wave function in this language is written as a series of *creation operators* acting on the *vacuum state*. A creation operator $\mathbf{a}_i^\dagger$ working on the vacuum state generates an (occupied) molecular orbital i:

$$\phi_i = \mathbf{a}_i^\dagger|0\rangle \tag{E.4}$$

The determinantal wave function in Equation (E.2) can be generated as

$$\Phi = \mathbf{a}_1^\dagger\mathbf{a}_2^\dagger\cdots\mathbf{a}_N^\dagger|0\rangle \tag{E.5}$$

The opposite of a creation operator is an *annihilation operator* $\mathbf{a}_i$, which removes orbital i from the wave function it is acting on. The product of operators $\mathbf{a}_j^\dagger\mathbf{a}_i$ removes orbital i and creates orbital j, that

Introduction to Computational Chemistry, Third Edition. Frank Jensen.

Companion Website: http://www.wiley.com/go/jensen/computationalchemistry3

is it replaces the occupied orbital i with unoccupied orbital j. The antisymmetry of the wave function is built into the operators by the following anticommutator relationships:

$$[\mathbf{a}_i^\dagger, \mathbf{a}_j^\dagger]_+ = \mathbf{a}_i^\dagger \mathbf{a}_j^\dagger + \mathbf{a}_j^\dagger \mathbf{a}_i^\dagger = 0 \tag{E.6}$$

$$[\mathbf{a}_i, \mathbf{a}_j]_+ = \mathbf{a}_i \mathbf{a}_j + \mathbf{a}_j \mathbf{a}_i = 0 \tag{E.7}$$

$$[\mathbf{a}_i^\dagger, \mathbf{a}_j]_+ = \mathbf{a}_i^\dagger \mathbf{a}_j + \mathbf{a}_j \mathbf{a}_i^\dagger = \delta_{ij} \tag{E.8}$$

Equations (E.6) and (E.7) show that $\mathbf{a}_i^\dagger \mathbf{a}_i^\dagger = \mathbf{a}_i \mathbf{a}_i = 0$.

The Hamilton operator in Equation (E.1) is in a second quantization given as (note the summation now is over the full set of orbitals)

$$\mathbf{H}_e = \sum_{pq}^{N_{orb}} h_{pq} \mathbf{a}_p^\dagger \mathbf{a}_q + \frac{1}{2} \sum_{pqrs}^{N_{orb}} g_{pqrs} \mathbf{a}_p^\dagger \mathbf{a}_r^\dagger \mathbf{a}_q \mathbf{a}_s + V_{nn}$$
$$h_{pq} = \langle \phi_p | \mathbf{h}_i | \phi_q \rangle \quad ; \quad g_{pqrs} = \langle \phi_p \phi_r | \mathbf{g} | \phi_q \phi_s \rangle \tag{E.9}$$

The exponential orbital transformation in Equation (3.65) can be written as

$$\boldsymbol{\phi}' = \boldsymbol{\phi} \mathrm{e}^{\mathbf{X}} \tag{E.10}$$

$$\mathbf{X} = \sum_{ai} x_{ai} \mathbf{a}_a^\dagger \mathbf{a}_i \quad ; \quad x_{ai} = -x_{ia} \tag{E.11}$$

A CCSD wave function (Equation (4.65)) can be written as

$$|\Psi_{CCSD}\rangle = \mathrm{e}^{\mathbf{T}_1 + \mathbf{T}_2} |\Phi_{HF}\rangle \tag{E.12}$$

$$\mathbf{T}_1 = \sum_{ai} t_i^a \mathbf{a}_a^\dagger \mathbf{a}_i \tag{E.13}$$

$$\mathbf{T}_2 = \frac{1}{4} \sum_{aibj} t_{ij}^{ab} \mathbf{a}_a^\dagger \mathbf{a}_i \mathbf{a}_b^\dagger \mathbf{a}_j \tag{E.14}$$

References

1 P. Jørgensen and J. Simons, *Second Quantization-Based Methods in Quantum Chemistry* (Academic Press, 1981).
2 T. Helgaker, P. Jørgensen and J. Olsen, *Molecular Electronic Structure Theory* (John Wiley & Sons, 2000).

Index

Introduction to Computational Chemistry, Third Edition. Frank Jensen.

Companion Website: http://www.wiley.com/go/jensen/computationalchemistry3

c

d

e

f

g

h

o

p

t

www.ingramcontent.com/pod-product-compliance
Lightning Source LLC
Chambersburg PA
CBHW080549270125
20834CB00018B/219
* 9 7 8 1 1 1 8 8 2 5 9 9 0 *